Cellular and Molecular Biology of Metals

Edited by
Rudolfs K. Zalups
James Koropatnick

CRC Press
Taylor & Francis Group
Boca Raton London New York

CRC Press is an imprint of the
Taylor & Francis Group, an **informa** business

The cover displays a space-filled rendering of rat Metallothionein II as generated from the Brookhaven Protein Databank file 4MT2 (submitted by A.H. Robbins and C.D. Stout). The image was rendered by Rudolfs K. Zalups, Ph.D. using the molecular drawing program Chimera, which is supplied by University of California at San Francisco.

CRC Press
Taylor & Francis Group
6000 Broken Sound Parkway NW, Suite 300
Boca Raton, FL 33487-2742

First issued in paperback 2019

ISBN-13: 978-1-4200-5997-7 (hbk)
ISBN-13: 978-1-138-37265-8 (pbk)

Library of Congress Cataloging-in-Publication Data

Cellular and molecular biology of metals / editors, Rudolfs K. Zalups and D. James Koropatnick.
p. ; cm.
Includes bibliographical references and index.
ISBN 978-1-4200-5997-7 (hardcover : alk. paper)
1. Metals--Physiological effect. 2. Metals--Metabolism. 3. Metals--Toxicology. 4. Cytochemistry. I. Zalups, Rudolfs K. II. Koropatnick, Donald James, 1952- III. Title.
[DNLM: 1. Metals--metabolism. 2. Metals--toxicity. 3. Biological Transport--physiology. 4. Cell Physiological Phenomena--drug effects. 5. Molecular Biology--methods. QV 275 C393 2010]

QP532.C45 2010
615.9'253--dc22 2009053540

Visit the Taylor & Francis Web site at
http://www.taylorandfrancis.com

and the CRC Press Web site at
http://www.crcpress.com

Contents

Preface v
Editors vii
Contributors ix

Chapter 1 Cellular Inorganic Chemistry Concepts and Examples 1
David H. Petering, Rajendra Kothinti, Jeffrey Meeusen, and Ujala Rana

Chapter 2 Molecular and Cellular Biology of Mercury in the Kidneys 35
Rudolfs K. Zalups and Christy C. Bridges

Chapter 3 Essential and Toxic Metal Transport in the Liver 79
Nazzareno Ballatori, David A. Aremu, and Michael S. Madejczyk

Chapter 4 Molecular and Cell Biology of Lead 113
Bruce A. Fowler

Chapter 5 Transport and Biological Impact of Manganese 127
Keith M. Erikson, Khristy J. Thompson, and Michael Aschner

Chapter 6 Metallothionein and Metal Homeostasis 143
James Koropatnick

Chapter 7 Cellular and Molecular Biology of Iron-Binding Proteins 167
Danuta S. Kalinowski and Des R. Richardson

Chapter 8 The Regulatory and Signaling Functions of Zinc Ions in Human Cellular Physiology 181
Hajo Haase and Wolfgang Maret

Chapter 9 Glutathione, Protein Thiols, and Metal Homeostasis 213
Lawrence H. Lash

Chapter 10 Ionic and Molecular Mimicry and the Transport of Metals 241
Christy C. Bridges and Rudolfs K. Zalups

Chapter 11 Heavy Metal Transport and Detoxification in Crustacean Gastrointestinal and Renal Epithelial Cells 295

Gregory A. Ahearn

Chapter 12 Metals and Cell Adhesion Molecules 327

Alan R. Parrish and Walter C. Prozialeck

Chapter 13 Iron Metabolism and Disease 351

Giada Sebastiani and Kostas Pantopoulos

Chapter 14 Metal Influences on Immune Function 379

Michael A. Lynes, Kathryn Pietrosimone, Gregory Marusov, Douglas V. Donaldson, Clare Melchiorre, Xiuyun Yin, David A. Lawrence, and Michael J. McCabe

Index 415

Preface

Despite all the progress being made in the fields of molecular and cellular biology, the role and effects of metal ions on cellular homeostasis in the various organs of mammals are only beginning to be truly defined. Working with divalent and trivalent metals and metals with higher valency in biological systems can be particularly challenging because of the complex and, under certain conditions, transient bonding interactions that metal ions can undergo. It is particularly challenging to follow metal ions in their complex biological journey from the environment in tissues and cells. That journey commonly involves association of metals with extracellular ligands that are either specific to particular metal species or promiscuous in their associations with metals, then entry into the cytosolic compartment of target cells. Cell entry requires metals to traverse the cellular plasma membrane, often through the interaction of metals, their ligands, or both, with transporter molecules or by mechanisms independent of transporters. Intracellular metal ions then associate with intracellular molecules in specific compartments to signal their presence and trigger cellular responses to that presence, and to carry out physiological functions as essential components of cellular enzymes and structural molecules.

Moreover, without the availability of radioactive forms of certain metal ions, accurate measurement of metal content within target cells and their subcellular compartments and organelles exceeds the sensitivity, accuracy, and reproducibility of current quantitative and qualitative analytical methods to measure these metallic species. With the continued decrease in commercially available isotopes of various metals, new challenges are being imposed on the next generation of molecular and cellular biologists. We rely on them for new methods and experimental strategies to discover how mammalian cells detect, take up, use, and excrete metals to maximize their extraordinarily valuable reductive and oxidative capacity for cellular function while minimizing their capacity for harm—and to exploit that knowledge for therapeutic benefit and to avoid metal-induced damage.

Our rationale for this volume stems from the ever-shifting sands of opportunity to compile a written summary of the state of knowledge in metal metabolism and homeostasis in target cells. We have compiled the current perspectives of experts in the areas of transport and handling, metabolism, and transcriptional regulatory activity of a number of metal ions of high current interest in the scientific literature.

Unlike our previous volume (*Molecular Biology and Toxicology of Metals*, published in 2000 by Taylor & Francis), which focused on the toxicology effects of a number of metals, the present volume concentrates primarily on physiological mechanisms underlying metal ion handling with respect to homeostasis, enzyme activity, transcriptional regulation, and other events designed to avoid toxicity and enhance cellular function. In view of the long life (indeed, the immortality) of metal ions, their capacity to both nurture and damage living systems, and their exceptional value as molecular redox tools in the hands of cellular molecules, the subject continues to both fascinate and generate new knowledge with the potential to reframe our understanding of cellular function.

James Koropatnick
Rudolfs K. Zalups

Editors

Rudolfs K. Zalups attended the State University of New York (SUNY) College at Brockport as an undergraduate where he received a B.S. in mathematics and electronic music. He later received a M.S. in zoology from SUNY Brockport and a doctorate in human anatomy and cell biology at the University of Western Ontario in London, Ontario, Canada. He continued his training as a fellow and instructor at the Mayo Clinic, Yale University School of Medicine, University of Maryland School of Medicine and the University of Rochester School of Medicine and Dentistry. Later, he joined the faculty of a newly formed medical school, Mercer University School of Medicine (MUSM), in Macon, Georgia, where he is currently a full professor.

James Koropatnick obtained his Ph.D. in genetics at the University of British Columbia (UBC) in 1980 followed by a postdoctoral Alberta Heritage Medical Research Foundation Fellowship at the University of Calgary until 1984. He came to the University of Western Ontario (UWO) and the Lawson Health Research Institute in London, Ontario, Canada in 1985 and is now a Distinguished Oncology Scientist, director of the LRCP Cancer Research Laboratory Program, director of the UWO/Canadian Institutes of Health Research (CIHR) Strategic Training Program in Cancer Research and Technology Transfer, assistant director of the Lawson Health Research Institute, and a full professor in the UWO Department of Oncology with cross-appointments in the UWO Departments of Microbiology and Immunology, Physiology and Pharmacology, and Pathology. He is also the chief scientific officer of Sarissa, Inc. (a biotech discovery company) and sits on the Scientific Advisory Boards of Critical Outcomes Therapeutics, Inc., ID Laboratories, Inc., and MedVax Pharma, Inc.

Contributors

Gregory A. Ahearn
Department of Biology
University of North Florida
Jacksonville, Florida

David A. Aremu
Department of Environmental Medicine
University of Rochester School of Medicine
Rochester, New York

Michael Aschner
Departments of Pediatrics and Pharmacology
Vanderbilt Kennedy Center
Vanderbilt University Medical Center
Nashville, Tennessee

Nazzareno Ballatori
Department of Environmental Medicine
University of Rochester School of Medicine
Rochester, New York

Christy C. Bridges
Division of Basic Medical Sciences
Mercer University School of Medicine
Macon, Georgia

Douglas V. Donaldson
Department of Molecular and Cell Biology
University of Connecticut
Storrs, Connecticut

Keith M. Erikson
Department of Nutrition
University of North Carolina–Greensboro
Greensboro, North Carolina

Bruce A. Fowler
Agency for Toxic Substances and Disease Registry
Atlanta, Georgia

Hajo Haase
Institute of Immunology
RWTH Aachen University
Aachen, Germany

Danuta S. Kalinowski
Department of Pathology and Bosch Institute
University of Sydney
Sydney, New South Wales, Australia

James Koropatnick
Departments of Oncology, Microbiology and Immunology, Physiology and Pharmacology, and Pathology
The University of Western Ontario
London, Ontario, Canada

Rajendra Kothinti
Department of Chemistry and Biochemistry
University of Wisconsin–Milwaukee
Milwaukee, Wisconsin

Lawrence H. Lash
Department of Pharmacology
Wayne State University School of Medicine
Detroit, Michigan

David A. Lawrence
Wadsworth Center
New York Department of Health
Albany, New York

Michael A. Lynes
Department of Molecular and Cell Biology
University of Connecticut
Storrs, Connecticut

Michael S. Madejczyk
Department of Environmental Medicine
University of Rochester School of Medicine
Rochester, New York

Wolfgang Maret
Department of Preventive Medicine & Community Health and Anaesthesiology
The University of Texas Medical Branch
Galveston, Texas

Gregory Marusov
Department of Molecular and Cell Biology
University of Connecticut
Storrs, Connecticut

Michael J. McCabe
Department of Environmental Medicine
University of Rochester
Rochester, New York

Jeffrey Meeusen
Department of Chemistry and Biochemistry
University of Wisconsin–Milwaukee
Milwaukee, Wisconsin

Clare Melchiorre
Department of Molecular and Cell Biology
University of Connecticut
Storrs, Connecticut

Kostas Pantopoulos
Lady Davis Institute for Medical Research
Sir Mortimer B. Davis Jewish General Hospital
Montreal, Quebec, Canada

and

Department of Medicine
McGill University
Montreal, Quebec, Canada

Alan R. Parrish
College of Medicine
Texas A&M Health Science Center
College Station, Texas

David H. Petering
Department of Chemistry and Biochemistry
University of Wisconsin–Milwaukee
Milwaukee, Wisconsin

Kathryn Pietrosimone
Department of Molecular and Cell Biology
University of Connecticut
Storrs, Connecticut

Walter C. Prozialeck
Department of Pharmacology
Midwestern University
Downers Grove, Illinois

Ujala Rana
Department of Chemistry and Biochemistry
University of Wisconsin–Milwaukee
Milwaukee, Wisconsin

Des R. Richardson
Department of Pathology
Bosch Institute
University of Sydney
Sydney, New South Wales, Australia

Giada Sebastiani
Department of Digestive Diseases, Hepatology and Clinical Nutrition
Dell'Angelo Hospital
Venice, Italy

and

Venetian Institute of Molecular Medicine (VIMM)
Padova, Italy

Khristy J. Thompson
Department of Biology
Elizabethtown College
Elizabethtown, Pennsylvania

Xiuyun Yin
Department of Molecular and Cell Biology
University of Connecticut
Storrs, Connecticut

Rudolfs K. Zalups
Division of Basic Medical Sciences
Mercer University School of Medicine
Macon, Georgia

1 Cellular Inorganic Chemistry Concepts and Examples

David H. Petering, Rajendra Kothinti, Jeffrey Meeusen, and Ujala Rana

CONTENTS

1.1 Introduction 2
1.2 Introduction to Inorganic Biochemistry Related to Metal Ion Trafficking 2
1.2.1 Metal-Ligand Binding 3
1.2.2 Ligand Substitution 4
1.2.3 Metal Ion Exchange 5
1.2.4 Adduct Formation 6
1.2.5 Redox Reaction 6
1.3 Metal-Ligand Binding and Ligand Substitution Chemistry 7
1.3.1 Cellular Zn^{2+} Trafficking in Relation to Apo-Metallothionein 7
1.3.2 Apo-MT Reactivity with the Zn-Proteome 10
1.3.3 *In Vitro* Reactions of Apo-MT and GSH with Zn-Proteins: A Model for Ligand Substitution Deactivation 10
1.3.4 Nutrient Zn^{2+} Deficiency and Apo-MT: Implications for Zn^{2+} Trafficking 11
1.3.5 Utilization of Metal Binding Ligands to Induce Intracellular Metal Ion Deficiency 13
1.3.6 Pharmacologic Ligands that Compete for Toxic Metal Ions 15
1.3.7 Metallothionein and Intracellular Competition for Toxic Metal Ions 15
1.4 Metal Ion Exchange Chemistry 16
1.4.1 Zn-Proteins that Undergo Metal Ion Exchange with Cd^{2+}: Metallothionein 16
1.4.2 Zn-Proteins that Undergo Metal Ion Exchange with Cd^{2+}: Zn-Finger Proteins 18
1.5 Adduct Formation 20
1.5.1 Fluorescent Zn^{2+} Sensors in Relation to Zn^{2+} Trafficking 20
1.5.2 Fluorescent Zn^{2+} Sensor TSQ 21
1.5.3 What is TSQ Imaging in Cells? 21
1.6 Redox Reactions 22
1.6.1 Iron Trafficking and Pharmaceutical Ligands for Fe Overload Disorder 22
1.6.2 Zn-Metallothionein and Sulfhydryl Group Redox Chemistry 24
1.6.2.1 Reactions with Disulfides 24
1.6.2.2 Reactions with Nitric Oxide Species 24
1.7 Conclusions 25
Acknowledgments 26
References 26

1.1 INTRODUCTION

Cells and organisms require many different metal ions. As many as 3000 proteins in the human proteome utilize Zn^{2+} for structural or catalytic purposes [1,2]. Hundreds of proteins employ iron and copper [3,4]. When these and other metal ions are unavailable nutritionally or their metabolism is deranged, the consequences can be severe.*

Several nonessential, toxic metals consistently appear in the list of environmental pollutants of most concern for human health [5–7]. Still, some of the best anticancer therapeutic agents are metallodrugs or otherwise interact with metals as part of their mechanism of action [8–11]. Nevertheless, despite the challenges and opportunities, surprisingly few scientists study metal ions in biological systems. Fewer still focus on questions in metallobiology from a chemical perspective, striving to link chemistry with biology. Thus, physiological or pathological studies may conclude that a metal ion or complex *causes* a particular cellular outcome and delineate changes that ensue upon perturbation of the metallic species but never define the actual site where the inorganic chemistry takes place. For example, zinc deficiency *causes* defects in immune response, *causes* apoptosis, and *inhibits* cell proliferation, but the molecular sites that undergo depopulation of Zn^{2+} and start complex cascades of reactions leading to these outcomes are largely unknown [12–14]. Or, Pb^{2+} and CH_3Hg^+ *induce* neurotoxicity that exhibits well-established phenotypes [15,16]. However, relatively little is known about the specific binding sites occupied by these ions and how such interactions initiate and perpetuate toxicity.

This chapter offers an excursion into metallobiochemical research aimed at revealing the importance of the chemical perspective for studying and understanding metallobiological processes. The topics reflect the authors' interests in relation to subjects addressed in this monograph. The discussion begins with a general introduction to inorganic reaction classes. Then several topics are used to illustrate a combined chemical-cellular approach to investigating metallobiological problems related to metal ion metabolism.

1.2 INTRODUCTION TO INORGANIC BIOCHEMISTRY RELATED TO METAL ION TRAFFICKING

Cells present themselves to researchers as remarkably complex, endlessly integrated entities. Until recently, biochemists gained information and understanding about cellular chemistry by studying individual metabolic reactions and cellular structures. As new technologies emerged, scientists began studying collectives such as the genome and the proteome, with the aim of comprehending how cell structures interact and work together to generate the basic living system, the cell.

The "omics" perspective now extends to virtually any grouping of molecules within the metabolome (all of the metabolites in the cell), including the glycosylome, the lipidome, and the *metalome* [1–4]. At first sight, one wonders what rationale might justify grouping diverse metal ions into the *metalome*. In a sentence: All are small, positively charged ions that are *metabolized* by a small set of general inorganic reaction mechanisms.

Metabolism means the collection of reactions that govern the organized cellular uptake, distribution, and efflux of metal ions (M) that link their presence in cells to their localization in specific sites, where they participate in a huge array of structures and reactions. Used in this way, the *metabolism* of metal ions is called *trafficking*. As charged entities, metal ions exist in aqueous

* Abbreviations: CA, carbonic anhydrase C; DEA/NO, diethylamine nonoate; DTNB, 5,5*N*-dithio-bis(2-nitrobenzoate); EGTA, (2,2′-oxypropylene-dinitrilo)tetracetic acid; FRET, fluorescent resonance energy transfer; green fluorescent protein; ICPMS, inductively coupled plasma mass spectrometry; MT, metallothionein; PAGE, polyacrylamide gel electrophoresis; PYR, pyrithione, 2-mercaptopyridine-*N*-oxide; SNAP, *S*-nitrosyl-acetylpenicillamine; TPEN [*N,N,NNNN*-tetrakis(2-pyridylmethyl)-ethylenediamine]; TSQ, *N*-(6-methoxy-8-quinolyl)-*p*-toluensulfonamide.

solution either in aquated form neutralized by an equivalent number of negatively charged ions or as complexes with charged or polar ligand molecules that bind metal ions through electron-rich metal ion binding sites involving N, O, and/or S atoms. Trafficking of biologically essential metal ions from outside the cell to the final sites of functional activity such as metalloproteins consists conceptually of a series of directed reactions that involve metal-ligand species at every step along each pathway.

The grand conceptual problem in metal ion trafficking may be posed as follows:

A substantial number of metal ions or metallic species play key roles in cellular processes. Their properties range from those of alkali metal ions to the left of the periodic table to transition metal ions such as $Fe^{2+,3+}$, Zn^{2+}, and $Cu^{1+,2+}$. In cells, they confront a multitude of metal ion binding ligands, both their natural binding sites and many other potential sites that compete for binding. The latter exist simply because proteins (amine, imidazole, carboxyl, and thiol groups) and nucleic acids (phosphate and base nitrogen and oxygen substituents) are replete with groups that display significant affinity for metal ions. In this heterogeneous environment, how are specific pathways that deliver metal ions from outside the cell to their ultimate binding sites favored?

The entrance into the cell and the pathological activity of toxic or therapeutic metal ions or metal complexes, such as Cd^{2+}, Pb^{2+}, and *cis*-diamminedichloro-Pt(II), must also be based on similar principles of metal ion trafficking, involving intracellular binding sites and the formation of metal-ligand complexes that are not normally part of the cellular milieu.

1.2.1 Metal-Ligand Binding

The generalized trafficking reactions consist of the following processes, in which M and L are metal ion and metal binding ligand, respectively [17]. Each reaction is characterized by an equilibrium (stability) constant (K) and rate constants (k_1, k_{-1}) for the forward and reverse reactions:

$$M + L \rightleftharpoons M\text{-}L, \quad K, k_1, k_{-1}, \tag{1.1a}$$

$$iM + \Sigma L_i \rightleftharpoons \Sigma M\text{-}L_i. \tag{1.1b}$$

Reaction 1.1a describes the association of the metal ion with a ligand. This reaction and its equilibrium constant, K, and rate constants for formation and dissociation, k_1 and k_{-1}, comprise fundamental information about biological M-L complexes that can be used to assess the comparative energetic favorability of binding sites for particular metal ions and the kinetic stability of the product complexes. As one moves from left to right in the periodic table, metal ions progressively prefer to bind to oxygen, then nitrogen, and finally sulfhydryl ligands [18]. The same trend operates on moving down the table within elemental families.

Alkali (Na^+, K^+) and substantial concentrations of alkaline earth (Mg^{2+}, Ca^{2+}) metal ions exist in cells as free metal ions because the equilibrium constants with cellular ligands are relatively small to modest and the rates of formation (k_1) and dissociation (k_{-1}) are rapid [19]. As such, the succession of formation and dissociation reactions, conceived for a variety of ligands (ΣL_i, sum of many intracellular ligands, Reaction 1.1b), constitutes a primary means of distributing M among binding sites ($\Sigma M\text{-}L_i$) according to equilibrium stability. In contrast, for transition metal ions such as $Fe^{2+,3+}$, Zn^{2+}, or $Cu^{1+,2+}$ and toxic, heavy metal ions including Cd^{2+}, Hg^{2+}, or Pb^{2+}, the concentration of free metal ion may be vanishingly small because cells contain many natural metal ion binding sites with large equilibrium constants for M as well as an abundance of lower affinity sites that, nevertheless, represent a very large combined affinity for M [20]. In this situation, it becomes paramount to understand the mechanisms by which native metal ions (metal-ligand complexes) reach specific sites and toxic metal ions either localize selectively or distribute nonspecifically within the cell.

1.2.2 Ligand Substitution

$$\text{M-L} + \text{L} \rightleftharpoons \text{M-L} + \text{L}, \quad K, k_1, k_{-1}, \tag{1.2a}$$

$$\text{M-L} + \Sigma\text{L}_i \rightleftharpoons \text{L} + \Sigma\text{M-L}_i, \quad K, k_1, k_{-1}. \tag{1.2b}$$

Reaction 1.2a represents the most general means by which transition and toxic metal ions move from one site to another. For transition metal ions, M-L complexes that have large equilibrium constants (Reaction 1.1a) may still be kinetically reactive in Reaction 1.2a (large k_1). As such, their Rate of distribution among ligands would not be rate limited by small dissociation rate constants in Reaction 1.2a, implied by the large thermodynamic stability of M-L or the inherent inorganic properties of the metal ion. For example, the documented trafficking of Cu from cell membrane to metalloprotein binding site is characterized by a series of ligand substitution reactions that successively transfer Cu(I) from one thermodynamically stable binding site to another (Figure 1.1) [21]:

$$\text{Cu(I)-L}_1 + \text{L}_2 \rightleftharpoons \text{Cu(I)-L}_2 + \text{L}_1, \tag{1.3}$$

$$\text{Cu(I)-L}_2 + \text{L}_3 \rightleftharpoons \text{Cu(I)-L}_3 + \text{L}_2, \tag{1.4}$$

$$\text{Cu(I)-L}_3 + \text{apo-protein} \rightarrow \text{Cu(I)-protein} + \text{L}_3. \tag{1.5}$$

Each of these reactions must be thermodynamically favorable and kinetically feasible. Moreover, since each Cu-protein terminates a specific pathway of copper trafficking, there would seem to be kinetic barriers to interpathway Cu(I) transfer as in Reaction 1.6:

$$\text{Cu(I)-L}_{1,\text{pathway 1}} + \text{L}_{2,\text{pathway 2}} \rightleftharpoons \text{Cu(I)-L}_{2,\text{pathway 2}} + \text{L}_{1,\text{pathway 1}}. \tag{1.6}$$

A particularly stringent test of the forbidden nature of such reactions occurs when the metal binding protein metallothionein is present in cells as a metal-unsaturated protein (apo-MT) [22]. The very large affinity of apo-MT for Cu(I) suggests that Reaction 1.6 is thermodynamically favorable when $\text{L}_{2,\text{pathway 2}}$ represents apo-MT. Yet, the metal-unsaturated pool of MT contains little, if any, Cu(I). Nor does its presence seem to perturb Cu metabolism.

Ligand substitution reactions also provide a general route by which nonessential metal ions and metal ion complexes, either toxic contaminants or pharmacological agents, gain access to target molecules and, on binding to them, modify their biological activity. On examining the Hg^{2+} and CH_3Hg^+ stability constants with molecules containing N, O, and S ligating groups, each species displays enormous preference for sulfhydryl group-containing ligands [23]. Nevertheless, the

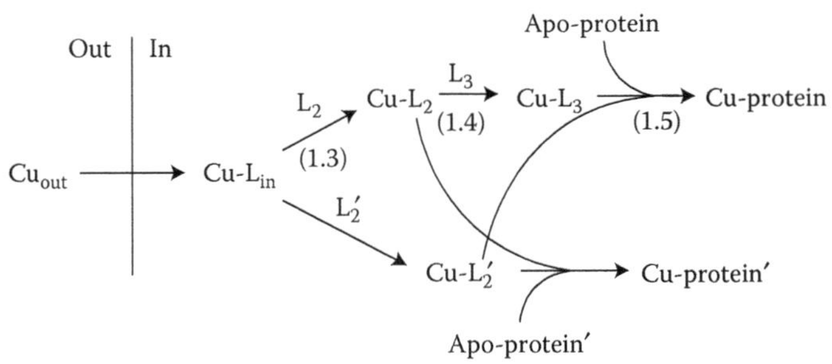

FIGURE 1.1 Generalized copper trafficking by ligand substitution with chaperones (L). Numbers in parentheses refer to reactions in text.

affinities of an array of sulfhydryl ligands for mercury are similar and the ligand substitution rates are rapid [24]. Thus, CH_3Hg^+ readily distributes among competing sulfhydryl-containing sites:

$$CH_3Hg\text{-}SR + R'SH \rightleftharpoons CH_3Hg\text{-}SR' + RSH. \quad (1.7)$$

In this case, mercurial localization must depend on other factors such as the contribution of the methyl group to the equilibrium or kinetic stability of $CH_3Hg\text{-}SR'$ species.

Similarly, the antitumor drug *cis*-dichlorodiammine Pt(II) reacts with DNA guanine bases through ligand substitution reactions, leading to cytotoxic DNA adduct species [25,26]:

$$(NH_3)_2PtCl_2 \rightleftharpoons (NH_3)_2PtCl(H_2O) \rightleftharpoons (NH_3)_2Pt(H_2O)_2, \quad (1.8)$$

$$(NH_3)_2Pt(H_2O)_2 + N\text{-}N \text{ (adjacent } N^7\text{-nitrogens of guanine)} \rightarrow (NH_3)_2Pt(N\text{-}N). \quad (1.9)$$

In this case, the rate limiting reactions are the dissociation of Cl^- ions, followed by the rapid substitution of guanine nitrogens for bound water molecules. Studies of the reactivity of the platinum drug with alternative ligand binding sites in the cell, for example, demonstrate that *cis*-dichlorodiammine Pt(II) reacts faster with metallothionein than with DNA because the thiolate compound can directly attack the dichloro species [27–29]. Thus, mechanisms of drug resistance may involve sulfhydryl-containing molecules such as metallothionein or glutathione that react with the drug and inactivate it toward further reaction with DNA or other sites [30,31]:

$$(NH_3)_2PtCl_2 + 2RS^- \rightleftharpoons (NH_3)_2Pt(SR)_2 + 2Cl^-. \quad (1.10)$$

1.2.3 Metal Ion Exchange

$$M_1 + M_2\text{-}L_2 \rightleftharpoons M_1\text{-}L_2 + M_2, \quad K, k_1, k_{-1}, \quad (1.11a)$$

$$M_1\text{-}L_1 + M_2\text{-}L_2 \rightleftharpoons M_2\text{-}L_1 + M_1\text{-}L_2. \quad (1.11b)$$

Metal ion exchange Reactions 1.11a and 1.11b represent a class of reactions that essential metal ions must and do avoid during trafficking so that the selective binding of specific metals to particular sites is achieved. But in the face of exposure to toxic metal ions, this type of reaction becomes a primary consideration. Competition between essential and toxic metal ions for physiologically important metal ion binding sites is thought to comprise a major category of reaction leading to cell injury. Thus, acute Pb^{2+} exposure in humans *causes* anemia due to the lack of protoporphyrin IX for heme synthesis and hemoglobin formation [32]. Pb^{2+} or a Pb-ligand complex inhibits δ-amino-levulenic acid dehydratase (ALAD) by displacing active site Zn^{2+} from the enzyme, resulting in a Pb-enzyme that is inactive and unable to participate in porphyrin synthesis [33]:

$$Pb^{2+} + Zn\text{-}ALAD \rightleftharpoons Pb\text{-}ALAD + Zn^{2+}. \quad (1.12)$$

In the case of Cd^{2+}, its metal ion exchange reaction with Zn-metallothionein (Zn-MT) serves as the primary means to protect cells from Cd^{2+} toxicity [34–36]:

$$Cd^{2+} + Zn\text{-}MT \rightleftharpoons Cd\text{-}MT + Zn^{2+}. \quad (1.13)$$

Indeed, in some cases, Cd-substituted Zn-proteins can undergo direct metal ion exchange with Zn-MT, resulting in reactivation of the Cd-impaired protein, as has been seen in the case

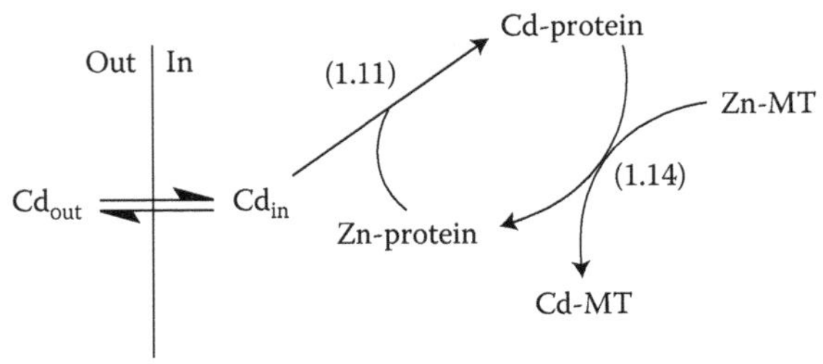

FIGURE 1.2 Metal ion exchange between metalloprotein and metallothionein. Numbers in parentheses refer to reactions in text.

of a Cd-modified Zn-finger protein, tramtrack, and Cd-carbonic anhydrase (CA) (Figure 1.2) [37,38]:

$$\text{Cd-tramtrack} + \text{Zn-MT} \rightleftharpoons \text{Zn-tramtrack} + \text{Cd-MT}. \qquad (1.14)$$

1.2.4 Adduct Formation

$$\text{M-L} + \text{L}' \rightleftharpoons \text{L}'\text{-M-L}, \quad K, k_1, k_{-1}. \qquad (1.15)$$

Reaction 1.15 symbolizes the association of metal ion binding ligands with metal ion centers of M-L complexes to form ternary complexes. The importance of this type of reaction for cellular chemistry remains to be seen. Nevertheless, the presence of millimolar concentrations of glutathione with its prominent sulfhydryl group begs the question of whether it interacts with metalloprotein metal binding sites that are ligand unsaturated (e.g., Zn-CA).

Considering the reactions of some xenobiotic metal complexes or metal ion binding ligands with cells, the formation of adduct species is an attractive means of bringing these species into association with particular sites and molecules in the cell. For instance, in the reaction of pyridoxal-thiosemicarbazonato-Cu(II) (Cu(II)-PTSC) with cells, electron spin resonance (ESR) spectroscopy provides clear evidence that the metal complex initially forms an adduct species and then undergoes redox chemistry that may account for its strong cytotoxic behavior [39]:

$$\text{Cu(II)-PTSC} + \text{L}' \rightarrow \text{L}'\text{-Cu(II)-PTSC}. \qquad (1.16)$$

The cellular adduct can be modeled by GS-Cu(II)-PTSC, in which GS is glutathione. Once formed, it may undergo internal oxidation reduction, resulting in the formation of GSSG and Cu(I)-PTSC that reacts with O_2 to initiate the production of reactive oxygen species and regenerate Cu(II)-PTSC for further reaction with the reduction equivalents of the glutathione pool.

1.2.5 Redox Reaction

$$\text{M-L}_{\text{oxidized}} + \text{C}_{\text{reduced}} \rightarrow \text{M-L}_{\text{reduced}} + \text{C}_{\text{oxidized}}, \quad K, k_1, k_{-1}. \qquad (1.17)$$

The redox reactions generalized in Reaction 1.17 play key roles in the chemistry of metal ions with multiple, accessible oxidation states such as $Fe^{2+,3+}$ and $Cu^{1+,2+}$ as well as metal complexes that involve redox-active thiolate ligands. In this context, unregulated oxidation–reduction reactions commonly lead to a cell injury center, such as metal-catalyzed O_2 activation in the Fenton reaction,

$$H_2O_2 + Fe(II) \rightarrow OH^- + OH^\bullet + Fe(III), \quad (1.18)$$

$$Fe(III) + e^- \rightarrow Fe(II), \quad (1.19)$$

and chromate, $Cr(VI)O_4^{2-}$, reduction to redox-active Cr(V) and DNA binding Cr(III) [39–41]:

$$Cr(VI) + n\text{RSH} \rightarrow Cr(VI)(SR)_n, \quad (1.20)$$

$$Cr(VI)(SR)_n + RS^- \rightarrow Cr(V)(SR)_n + RS^\bullet\ (1/2RSSR), \quad (1.21)$$

$$Cr(V)(SR)_n + RS^- \rightarrow Cr(IV)(SR)_n + RS^\bullet, \quad (1.22)$$

$$Cr(IV)(SR)_n + RS^- \rightarrow Cr(III)(SR)_n + RS^\bullet. \quad (1.23)$$

With this general introduction to cellular inorganic chemistry in hand, several topics will be considered in more depth, with the aim of illustrating how cellular and chemical perspectives can work effectively together to gain enhanced understanding.

1.3 METAL-LIGAND BINDING AND LIGAND SUBSTITUTION CHEMISTRY

1.3.1 Cellular Zn^{2+} Trafficking in Relation to Apo-Metallothionein

Zn^{2+} is utilized as structural or catalytic elements of many proteins [1,2]. It is increasingly recognized that changes in Zn^{2+} concentration or distribution contribute to dynamic organismic processes. Nutrient Zn^{2+} depletion inhibits cell proliferation, growth, and normal development [14,42]. Zn^{2+} redistributes during the generalized stress response [43]. Glutamatergic neuron firing is accompanied by the release of Zn^{2+} [44]. Apoptosis is associated with intracellular Zn^{2+} reorganization [45]. All these phenomena involve facets of Zn^{2+} trafficking, broadly understood.

The conceptual problem of Zn^{2+} trafficking is imposing. Bertini et al. estimate that the human genome includes sequence information for about 150 Cu-proteins, 250 nonheme Fe-proteins, and 2800 Zn-proteins [1–4]. How organized, controlled movement of Zn^{2+} from the cell's exterior to nearly 3000 nascent apo-Zn-proteins takes place is unknown (Figure 1.3). Nor is there much insight into how organismal Zn^{2+} is reorganized in response to physiological or pathological stimuli.

A primary Zn^{2+} trafficking requirement is to make available sufficient Zn^{2+} to populate apo-Zn^{2+} protein binding sites, so that the reaction

$$i\text{Zn} + \Sigma\text{apo-Zn-proteins}_i \rightarrow \Sigma\text{Zn-proteins}_i \quad (1.24)$$

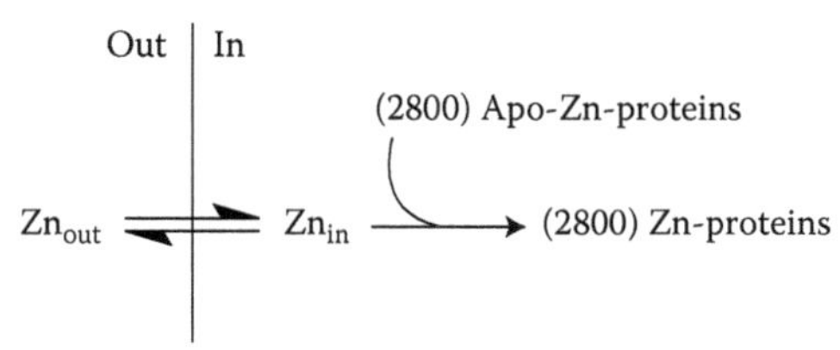

FIGURE 1.3 Intracellular formation of the Zn-proteome.

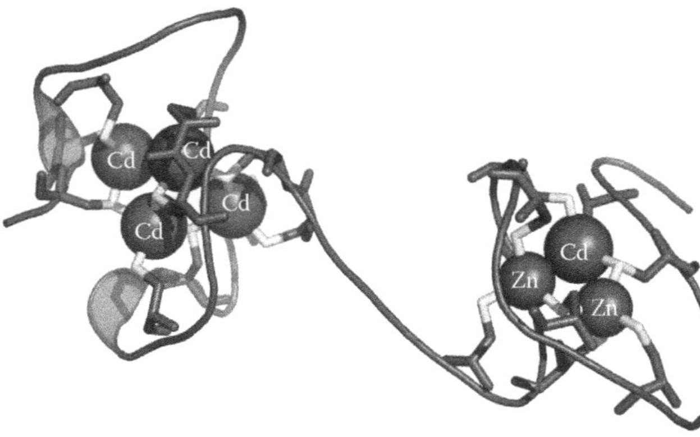

FIGURE 1.4 X-ray crystallographic structure of Cd_5, Zn_2-MT (Data from Robbins, A.H. et al., *J. Mol. Biol.*, 221, 1269, 1991). Metal ions (Cd^{2+}, larger; Zn^{2+}, smaller) and sulfhydryl ligands (light gray).

takes place under demand conditions (i.e., synthesis of apo-Zn-proteins), because cells may contain virtually no free Zn^{2+} [20]. Coupling this reaction to an extracellular source through Zn transporters would allow the biosynthesis of new apo-Zn-proteins to energetically "pull" the appropriate amount of Zn^{2+} into cells and cellular compartments (Reactions 1.25 and 1.26) and into Zn-proteins (Reaction 1.24):

$$Zn^{2+}_{out} \rightarrow Zn^{2+}_{in,cytoplasm}, \tag{1.25}$$

$$Zn^{2+}_{in,cytoplasm} \rightarrow Zn^{2+}_{in,compartment\ a}. \tag{1.26}$$

Even this is a complicated chemical process, because extracellular Zn^{2+}_{out} exists at least in part in the form of Zn-ligand complexes. Nevertheless, when dividing cells are bathed in a medium containing a basal concentration of Zn^{2+}, they accumulate Zn^{2+} commensurate with the need to constitute the Zn-proteome in the new cells.

Mammalian metallothionein is a small protein that is distinguished by the presence of 20 cysteinyl residues within its 60–70 amino acid composition [46]. They are positioned along the sequence such that seven Zn^{2+} or Cd^{2+} or combinations of these metal ions bind to the 20 sulfhydryl groups to form two metal-thiolate clusters that each comprise the interior of one of the protein's two domains (Figures 1.4 and 1.5) [47]. Because the protein was viewed as a potent binding site for these and other metal ions, for many years little thought was given to the possibility that metal-unsaturated MT (apo-MT) might exist in cells. Nevertheless, several studies show that a

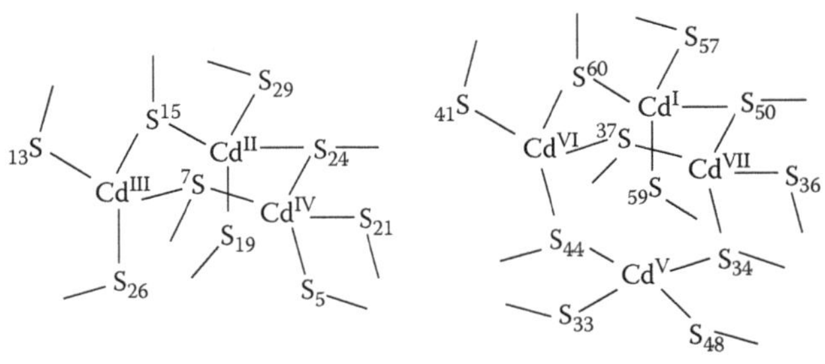

FIGURE 1.5 Metal-thiolate cluster structure of a metallothionein binding site.

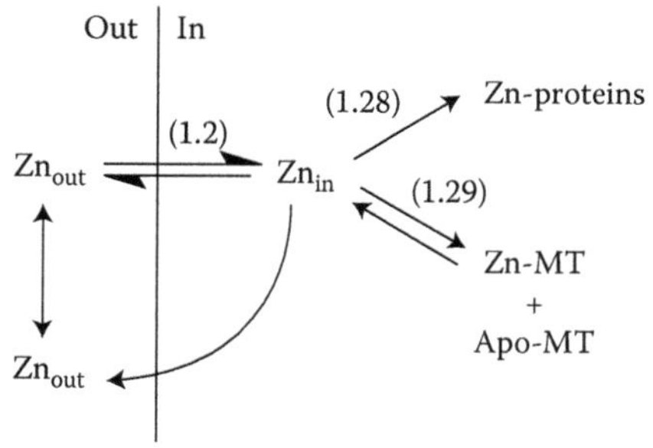

FIGURE 1.6 Zn^{2+} proteomic trafficking and metallothionein: MT as a unique Zn-protein.

substantial pool of apo-MT can exist in the steady state in a variety of mammalian cells under a variety of conditions (normal tissues, basal cell culture, tissues, and cells after induction of MT protein synthesis with Zn^{2+}, Cd^{2+}, Bi^{2+}, or dexamethasone, and in the hearts of MT-transgenic mice) [22,48,49].

Apo-MT's mere existence indicates that the above reactions and mechanism to draw Zn^{2+} into the cell based on newly synthesized apo-Zn-proteins acting as the energetic sink are insufficient to explain Zn^{2+} trafficking [50]. Its presence also signals that MT does not behave like a typical member of the Zn-proteome (Figure 1.6). Otherwise, apo-MT with its equilibrium constant in the range of $10^{11.2}$ per Zn^{2+} (Reaction 1.1a) would surely act as a thermodynamically favorable Zn^{2+} acceptor from the extracellular medium or in competition with other apo-Zn-protein sites as they are synthesized (Reaction 1.11a) [51–54]. Indeed, even induction of metallothionein synthesis by Zn^{2+} itself, which generates large increases in cellular Zn-MT, results in an MT pool that is about one-third unsaturated in Zn^{2+} [22]. Thus, an intracellular pool of Zn^{2+}-unsaturated MT can exist even in the presence of an elevated reservoir of extracellular Zn^{2+}. Evidently, the external and internal Zn^{2+} pools are not in steady state based on equilibrium properties of Zn^{2+} transport, Zn-proteins, and Zn^{2+} binding ligands in the two compartments:

$$Zn^{2+}_{out} \rightarrow Zn^{2+}_{in}, \quad k_1, k_{-1}, \tag{1.27}$$

$$Zn^{2+}_{in} + \text{apo-protein}_{in} \rightarrow \text{Zn-protein}_{in}, \tag{1.28}$$

$$Zn^{2+}_{in} + \text{apo-MT}_{in} \rightarrow \text{Zn-MT}_{in}, \quad K = 10^{11.2}. \tag{1.29}$$

Otherwise, Reaction 1.29, which is highly favorable, would readily take place.

If an equilibrium argument does not apply, perhaps there is a kinetic explanation, involving k_1 or k_{-1}, where these rate constants probably refer to independent influx and efflux transport processes, not to the binding of Zn^{2+} to apo-MT, which occurs in milliseconds (Figure 1.6) [55]. Here, one would need to posit that there is an effective efflux process that competes with apo-MT for Zn^{2+}_{in} and prevents the accumulation of Zn^{2+} in the MT pool. Although plausible, these explanations make it more difficult to understand Reaction 1.24 and how cellular Zn^{2+} uptake can be coupled with the continual constitution of Zn-proteins from biosynthesized apo-protein as cells divide and require a new Zn-proteome complement. If the Zn^{2+} transport systems, themselves, limit Zn^{2+} availability to apo-MT, would they not also do so to other apo-Zn-proteins? Furthermore, would not apo-MT be a formidable competitor of Zn^{2+} trafficking mechanisms for Zn^{2+}_{in} and Zn^{2+} bound to proteins, ensuring in the steady state that MT gained its complete complement of Zn^{2+}? Alternatively, to avoid this problem, it might be argued that MT is not a terminal repository of Zn^{2+} but instead participates in the trafficking of Zn^{2+}_{in}, resulting in Zn-protein constitution [22]. In that case, MT's Zn^{2+} saturation state would reflect the relative rates of accumulation of Zn^{2+} from the external medium and transfer to other Zn^{2+} binding sites. This option is considered in Section 1.3.4.

1.3.2 Apo-MT Reactivity with the Zn-Proteome

The demonstration of apo-MT in a variety of cells poses the fundamental question: how is the intracellular array of Zn-proteins maintained in the midst of a large pool of high-affinity, unsaturated Zn^{2+} binding sites in MT [50]? Why does not the following aggregate ligand substitution reaction occur?

$$\text{apo-MT} + \Sigma\text{Zn-proteins}_\text{I} \rightarrow \text{Zn-MT} + \Sigma\text{apo-proteins}_\text{I}, \quad K, k_1, k_{-1}. \tag{1.30}$$

The reactivity of the isolated Zn-proteome (100 μM Zn^{2+}) from U87 cells with apo-MT (stability constant $10^{11.2}$) was compared with a series of multidentate ligands for Zn^{2+} that vary in stability constant at pH 7.4 from $10^{15.6}$ for TPEN [*N,N,N′,N′*-(2-pyridylethyl)ethylenediammine] to $10^{8.8}$ for EGTA [(2,2′-oxypropylene-dinitrilo)tetraacetic acid], a relative of EDTA, all present at 200 μM Zn^{2+} binding sites [50]. Strikingly, 30-min incubation of the Zn-proteome with apo-MT resulted in little or no reaction (0–5%), whereas TPEN and EGTA extracted about 30% of proteomic Zn^{2+}. According to the listed stability constants, apo-MT is thermodynamically competent to react with at least 30% of the proteome Zn^{2+}. Since it did not undergo much, if any, reaction, the explanation must be that for kinetic reasons (i.e., there is no feasible pathway of reaction) apo-MT is unable to undergo ligand substitution with the Zn-proteome, a remarkable conclusion.

Also surprising was the large fraction of proteomic Zn^{2+} that is accessible to a ligand like EGTA that displays intermediate affinity for Zn^{2+}. This result suggests that on a thermodynamic basis, substantial proteomic Zn^{2+} might be available to glutathione (GSH), a Zn^{2+} binding ligand present in millimolar concentrations in cells. However, in the actual experiment as above with GSH concentration set at 4 mM, only 0–10% of the proteome's Zn^{2+} underwent reaction. Thus, it seems that the stability of the observed Zn-proteome array is dependent on its pervasive lack of reactivity with apo-MT, GSH, and, perhaps, other possible Zn^{2+} binding sites. Conversely, the maintenance of the apo-MT pool is a result of apo-MT's lack of kinetic reactivity in Reaction 1.30. That being the case, if there are regulated ligand substitution pathways that deliver Zn^{2+}_{out} to newly synthesized apo-Zn-proteins and do not interact with apo-MT, one could rationalize the dual existence of apo-MT and a growing pool of intracellular Zn-proteome as cells divide. These experiments provide a chemical hypothesis to rationalize the coexistence of apo-MT and Zn-proteome. It will be necessary to devise discerning cellular experiments to test this idea.

1.3.3 *In Vitro* Reactions of Apo-MT and GSH with Zn-Proteins: A Model for Ligand Substitution Deactivation

Apo-MT (10–30 μM) and GSH (2 mM), like TPEN (5 μM), readily react with the Zn-finger transcription factor, Zn-Spl (50 nM Zn^{2+}) [50]. Zn-Spl is a representative of the most common type of eukaryotic DNA binding protein that binds to Zn^{2+} through two cysteinyl sulfhydryl groups and two histidinyl imidazole nitrogens and interacts with DNA primarily through its helical element (Figure 1.7) [56,57]. Yet, cells that contain apo-MT and mM GSH also possess active Zn-Spl. One explanation supported by *in vitro* studies is that Zn-Spl, bound to its cognate DNA binding sites (GC rich), is unreactive with either apo-MT or GSH at these same concentrations (e.g., Reaction 1.31):

$$\text{apo-MT} + \text{Zn-Spl-DNA}_{\text{GC rich}} \rightarrow \text{Zn-MT} + \text{apo-Spl} + \text{DNA}_{\text{GC rich}}. \tag{1.31}$$

Zn-Spl-$DNA_{\text{GC rich}}$ displays increasing reactivity with TPEN in the reaction mixture between 10 and 1000 μM that still does not match that of free Zn-Spl with 5 μM TPEN [50].

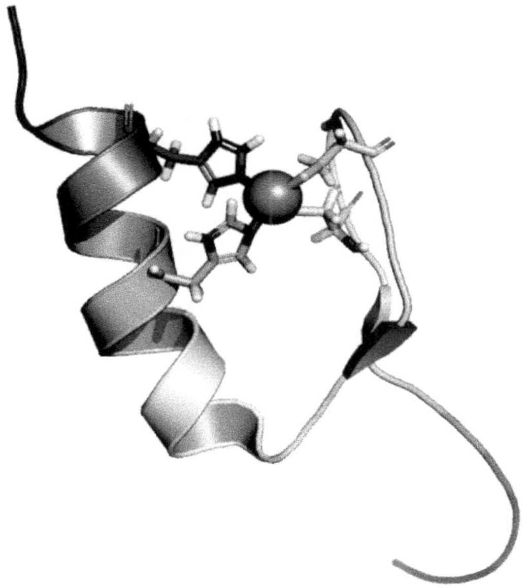

FIGURE 1.7 Generic Zn-finger motif in the DNA binding region of Sp1 and other transcription factors. Two cysteine sulfhydryl groups in the β turn and two imidazole nitrogens of histidine within the helix bind Zn^{2+}. Based on Zn-Sp1 structure (Data from Narayan, V.A., Kriwacki, R.W., and Caradonna, J.P., *J. Biol. Chem.*, 272, 7801, 1997).

Structures of Zn-finger-DNA complexes reveal that bound Zn^{2+} is accessible to the solvent, and thus might be expected to be reactive with competing ligands such as apo-MT. This suggests that the lack of reactivity of Zn-Sp1-$DNA_{GC\ rich}$ with apo-MT and other ligands has a thermodynamic, not kinetic, origin. The energetic problem can be seen by decomposing the overall reaction into two reactions:

$$\text{Zn-Sp1-DNA}_{\text{GC rich}} \rightarrow \text{Zn-Sp1} + \text{DNA}_{\text{GC rich}}, \tag{1.32}$$

$$\text{Zn-Sp1} + \text{apo-MT} \rightarrow \text{Zn-MT} + \text{apo-Sp1}. \tag{1.33}$$

The favorability of the second reaction is overcome by the unfavorability of the first, the dissociation of Zn-Sp1 from its DNA binding site. Thus, part of the ligand substitution unreactivity of the Zn-proteome with apo-MT may be due to Zn-protein binding interactions with other molecules.

1.3.4 Nutrient Zn^{2+} Deficiency and Apo-MT: Implications for Zn^{2+} Trafficking

Cells encountering a Zn^{2+}-deficient extracellular medium might release Zn^{2+} to the extracellular medium as proteins degrade and Reactions 1.27 through 1.29 shift to the left. This does not occur in a variety of mammalian tissues or cells. Instead, at the level of overall proteomic Zn^{2+}, the only major pool of Zn^{2+} that declines in the liver or kidney of juvenile male rats during the course of a 30-day feeding of a Zn^{2+}-deficient diet is Zn-containing metallothionein (Table 1.1) [58,59]. On reintroduction of Zn^{2+} to the diet, the full complement of Zn^{2+} is rapidly restored to the kidney protein [58]. Based on such results, metallothionein is a unique, labile depot of Zn^{2+} that is rapidly responsive to extracellular Zn^{2+} status (Figure 1.6).

Similar behavior has been observed in mouse Ehrlich ascites carcinoma [60,61]. The MT pool contains about 15% of proteomic Zn^{2+} [62]. When animals were maintained on a Zn^{2+}-deficient diet for 35 days, tumor growth halted, in agreement with earlier studies showing the stringent

TABLE 1.1
Male Rat Kidney Cytosolic Distribution of Zn[a]

	Zn-Proteome (I)	Superoxide Dismutase (II)	Metallothionein (III)
Zn^{+b}	4.5 ± 0.1[c]	2.1 ± 0.2	2.9 ± 0.3
Zn^{-}	4.2 ± 0.8	2.2 ± 0.5	0.7 ± 0.2

[a] Sephadex G-75 separation of cytosol. Fraction I contains most of Zn-proteome, fraction II includes superoxide dismutase, and fraction III represents metallothionein.
[b] Zn^{+}, 30-day treatment with semipurified diet containing 20 µg Zn/g diet; Zn^{-}, the same treatment with diet containing less than 1 µg Zn/g diet.
[c] µg Zn/g wet weight tissue.

requirement that cancerous tumors have for dietary Zn^{2+} [63]. Nevertheless, quiescent Ehrlich cells remained viable. Their gross proteomic Zn^{2+} was unchanged except for the complete loss of MT-bound Zn^{2+}. Unexpectedly, the cells contained a concentration of apo-MT protein that was equivalent to the Zn-MT concentration of untreated control cells.

Related results have been seen in cultured cells. When Ehrlich cells were incubated in a Zn^{2+}-depleted medium, Zn^{2+} left MT with a halftime of 1 h, leaving a pool of apo-MT that degraded with a halftime of 6 h [64]. The much faster halftime for Zn^{2+} loss indicates that active metal chemistry, not simply protein biodegradation, governs the loss of Zn^{2+} from the protein. We hypothesize that MT-bound Zn^{2+} becomes available for other apo-Zn-proteins through ligand substitution Reaction 1.34 (Figure 1.8) or as a participant in a chain of ligand substitution reactions as portrayed in Figure 1.1:

$$\text{Zn-MT} + \text{apo-Zn-protein} \rightarrow \text{apo-MT} + \text{Zn-protein}. \quad (1.34)$$

An inquiry into the steady-state flux of Zn^{2+} through the MT pool under Zn^{2+} normal conditions demonstrates a similar halftime for the residence of Zn^{2+} in Zn-MT in LLC-PK1 cells as observed in Ehrlich cells under Zn-deficient conditions [22]. In this experiment, the movements of stable Zn^{2+} isotopes into and out of Zn-MT were analyzed:

$$^{66}\text{Zn}^{2+}_{\text{out}} + {}^{67}\text{Zn-MT} \rightarrow {}^{66}\text{Zn-MT} + {}^{67}\text{Zn}^{2+}_{\text{in}}. \quad (1.35)$$

Occurring under steady-state conditions of Zn^{2+} trafficking, this experiment also supported the remarkable kinetic lability of Zn^{2+} with respect to MT, consistent with MT serving as an active intermediate in Zn^{2+} trafficking.

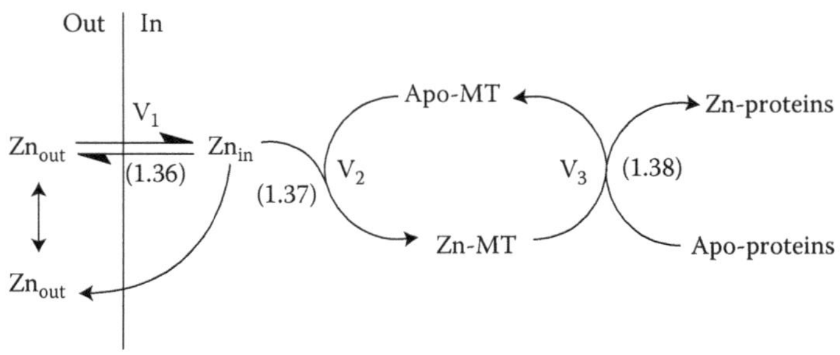

FIGURE 1.8 Zn^{2+} trafficking and metallothionein: metallothionein as a participant in trafficking.

Another reaction suggests the close connection between MT-bound Zn^{2+} and cell proliferation. U373 cells normally contain a measurable amount of apo-MT [65]. Its Zn^{2+} status was readily modulated by first blocking cell division with the cell cycle inhibitor thymidine, leading to the appearance of fully saturated Zn-MT. Once the thymidine block was lifted and cells begin to divide, MT returned to its Zn^{2+} free state. These events were interpreted in terms of the differential requirement of dividing and stationary cells for Zn^{2+}. Considering Reactions 1.36 through 1.38, if the rate of uptake of Zn^{2+} into MT is slower than the rate of Zn^{2+} transfer to other Zn-proteins ($V_1,V_2 > V_3$), apo-MT will be observed. If these relative rates reverse, Zn-MT will be detected (Figure 1.8):

$$Zn^{2+}_{out} \rightarrow Zn^{2+}_{in} \qquad V_1, \qquad (1.36)$$

$$Zn^{2+}_{in} + \text{apo-MT} \rightarrow \text{Zn-MT} \qquad V_2, \qquad (1.37)$$

$$\text{Zn-MT} + \text{apo-protein} \rightarrow \text{apo-MT} + \text{Zn-protein} \qquad V_3. \qquad (1.38)$$

The relative rates of these reactions would determine whether Zn- or apo-MT is present in particular cells, which vary in their concentration of apo-MT [22,48,49]. If $V_1,V_2 > V_3$, Zn-MT would predominate; if $V_3 > V_1,V_2$, apo-MT would be observed.

These results point to a conspicuous role for MT in Zn^{2+} trafficking in proliferative cells. It remains to reconcile the emerging experimental information about Zn-MT as a major, kinetically reactive pool of Zn^{2+} with the MT gene knock-out results that indicate that MT is not necessary for cell proliferation during development [66,67]. That conclusion itself needs to be tempered by the observation that modest as well as severe zinc deficiency in MT-null mice conspicuously raises the level of birth defects, suggesting that MT is important for Zn^{2+} trafficking in proliferative cells and tissues [68–70].

1.3.5 Utilization of Metal Binding Ligands to Induce Intracellular Metal Ion Deficiency

This discussion of Zn^{2+} trafficking has included considerations of Zn^{2+} deficiency as well as the reactivity of the Zn-proteome with competing metal binding ligands. At an organismic level, metal deficiency results from the lack of availability of extracellular nutrient metal ion for uptake and incorporation into cells. Models for this condition should similarly withhold the metal ion from cells. Ideally, establishment of metal ion deficiency would utilize a defined growth medium that is constituted of reagent grade chemicals that can be further purified as needed, so that the target metal ion can be selectively removed from the medium. Unfortunately, many cells do not grow in the absence of serum and are not amenable to induction of metal ion deficiency by this method.

An alternative approach is to introduce a powerful metal chelating agent into the cell culture medium as a substitute for extracellular metal ion deficiency in order to selectively deplete media of a particular metal ion (Figure 1.9). For this approach to be successful, one of two conditions need to be met. First, the competing ligand must be confined to the extracellular medium where it binds the target metal ion and sequesters it in that compartment, preventing its transport into cells. Second, the ligand may enter the cell where it selectively competes only for the transit pool of the metal ion that is moving from the plasma cell membrane to its specific sites of binding in metalloproteins.

$$\text{M-L}_{\text{extracellular nutrient source}} + \text{L}' \rightarrow \text{M-L}'_{\text{extracellular}} + \text{L}_{\text{extracellular nutrient source}}, \qquad (1.39)$$

$$\text{M-L}_{\text{intracellular trafficking}} + \text{L}' \rightarrow \text{M-L}'_{\text{intracellular}} + \text{L}_{\text{intracellular trafficking}}. \qquad (1.40)$$

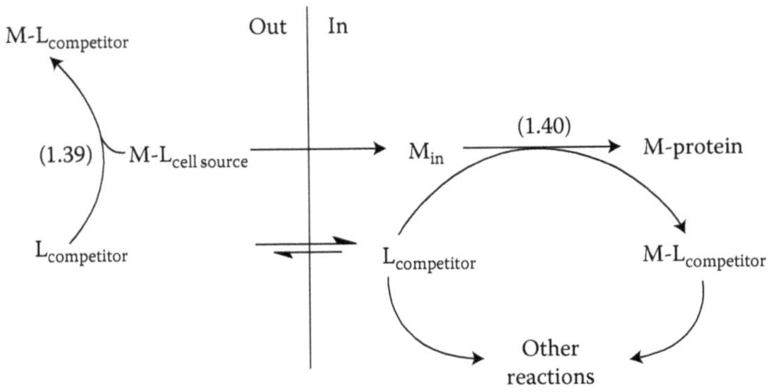

FIGURE 1.9 Use of metal binding ligands to induce metal ion deficiency: possible sites of reaction that mimic nutrient metal ion deficiency.

Ligands used for this purpose include TPEN for Zn^{2+} deficiency and desferrioximine for iron deficiency [71,72]. In the case of TPEN, which is recognized and used as a cell-permeant ligand, the implicit assumption is that TPEN does not disturb intracellular Zn^{2+} distribution but only intercepts Zn^{2+} that is undergoing trafficking. The results summarized above clearly show that TPEN competes successfully for a substantial fraction of the proteome's Zn^{2+} (Section 1.3.2). It does not innocently confine its ligand substitution reactions to Zn^{2+} in the trafficking pathway to intracellular Zn-proteins. Thus, it not only deprives cells of Zn^{2+}, but also disrupts the functional activities of Zn-proteins. As a result, its cellular effects such as the induction of apoptosis cannot be equated readily to nutrient Zn^{2+} deficiency [73].

Incubation of cells with desferrioxamine has been employed as an agent that induces cellular Fe deficiency because it has been assumed to be localized in the extracellular medium, where it competes with transferrin (Tr) for Fe(III) [74,75]. However, desferrioxamine gains entrance to cells as well [76]. Once inside, it may behave like TPEN and attack functional Fe-proteins. To the extent that this happens, the mechanistic linkage between cells deprived of a nutrient metal ion and cells treated with a potent metal ion chelator is broken.

An approach that appears to meet the objective of inducing nutrient Fe deficiency is based on an observation from many years ago that a facile way of liberating Fe(III) from transferrin is to reduce it with a mild reductant in the presence of a ligand that stabilizes Fe(II), such as bathophenanthroline-disulfonate (4,7-ρ-sulfonyl-phenyl-1,10-phenanthroline, BPS) [77]. A key in the process is the much lower affinity of Fe(II) for the transferrin ligand set than Fe(III) [78,79]. Another is the use of a high-affinity Fe(II) binding ligand, BPS, in the reaction to overcome the highly unfavorable reduction potentials of the two Fe(III) ions bound to Tr that fall below –0.5 V [80]. Using ascorbate as the reducing agent, Fe(III)-Tr in serum-containing culture media releases transferrin-bound Fe to BPS, which is rigorously localized in the extracellular medium because of its overall negative charge [81]:

$$\text{Fe(III)}_2\text{-Tr} + \text{ascorbate}_{\text{reduced}} + 6\text{BPS}^{2-} \rightarrow 2\text{Fe(III)(BPS)}_3^{3-} + \text{apo-Tr} + \text{ascorbate}_{\text{oxidized}}. \quad (1.41)$$

Cells treated with ascorbate and BPS remain viable, rapidly lose ferritin-Fe, see their Tr-receptor mRNA increase, and maintain their ferritin mRNA concentration as expected for the Fe-deficient state [81]. Iron deficiency established in this way inhibits the cytotoxicity of doxorubicin, thereby demonstrating the requirement of intracellular Fe for the activity of this anticancer agent in cell culture [81]. The ambiguities associated with using chelating agents in this fashion extend beyond the authenticity of the metal-deficient states that they induce (Figure 1.9). The distinct possibility exists that ligands interact with other metal ions or participate in other reactions that are unrelated to metal ions and their trafficking. TPEN binds other transition metal ions as well, such as Fe^{2+} and

Cu^{2+}. Does TPEN's presence in cells impinge on the trafficking and function of metal ions other than Zn^{2+}? In the case of desferrioxamine, its potential to participate in oxidation–reduction reactions provides a caution that it might serve as a site for redox chemistry that contributes to oxidative stress [82,83].

The problem of nonspecificity of metal chelation properties is emphasized by the multiple ways in which 1,10-phenanthroline (PHEN) has been used, each one presented as if PHEN were a specific ligand for a single metal ion [84]. PHEN is said to bind Zn^{2+} and alter Zn-dependent enzyme activity, protein kinase signaling, and apoptosis [85–88]; it binds Fe, perturbing energy metabolism, associating with mitochondrial "chelatable" Fe^{2+}, modulating oxidative stress, and downregulating Fe-dependent macrophage activity [89–92].

Another serious question arises: do the resultant metal complexes of competing ligands such as TPEN and desferrioximine, themselves, display biological activity [83]? A variety of antitumor ligands such as thiosemicarbazone and bleomycin ligands acquire metal ions and their cytotoxic capability on interacting with organisms [93,94]. Might that happen with ligands designed to extract metal ions from cells?

1.3.6 Pharmacologic Ligands that Compete for Toxic Metal Ions

Substantial effort has been devoted to developing ligands that remove toxic metal ions from tissues. For example, dithiol ligands such as BAL (2,3,-dimercaptopropanol) and 2,3-dimercaptopropane sulfonate have been used to chelate Hg^{2+}; 2,3-dimercaptosuccinate was developed for Pb^{2+}; and dithiocarbamates (DTCs) have been synthesized for Cd^{2+} [95,96].

DTC ligands effectively remove Cd^{2+} previously deposited in rodent kidney, presumably by extracting Cd^{2+} from MT. Forming a neutral $Cd(DTC)_2$ complex, the product could readily escape from cells and be excreted. However, experiments also revealed that DTC ligands can modulate organ concentrations of essential trace metals such as zinc and iron and enhance the excretion of Cu [97]. These effects might follow the intermediate formation of the respective DTC metal complexes. Indeed, an early study had shown that diethyldithiocarbamate is toxic to cells to the extent that it can complex Cu(II) and probably undergo intracellular redox reaction, producing oxidative stress [98]. Later, a related compound, pyrrolidine dithiocarbamate, was also described as a reagent that delivers Cu into cells and causes related cytotoxic responses, including apoptosis [99,100]. Such pleotropic effects of DTCs represent an alert that various metal binding ligands may not display specificity of reaction with cells and tissues and that ligands can be activated by reaction with essential transition metal ions.

1.3.7 Metallothionein and Intracellular Competition for Toxic Metal Ions

The biosynthesis of the powerful ligand, apo-MT, in response to Cd^{2+} exposure represents the cell's effort to compete for Cd^{2+} that otherwise becomes bound to sensitive sites and causes injury [101]. For example, Cd^{2+} causes kidney failure, characterized by the inability to resorb essential nutrients from the glomerular filtrate [102]. Among these is glucose, transported by sodium–glucose cotransporters (SGLT) I and II. Cd^{2+} downregulates transporter activity by inhibiting the synthesis of SGLT I and II mRNA [103–105]. Underlying this effect is the loss of the DNA binding capacity of a key transcription factor for the SGLT genes, the zinc-finger protein, Zn-Spl [106]. The chemical site of reaction of Cd^{2+} might be Zn-Spl or some other protein that controls the activity of Zn-Spl. In either case, in kidney cortical cells exposed to Cd^{2+}, downregulation of SGLT mRNAs progresses in parallel with the synthesis of metallothionein and the sequestration of most of the cell's Cd^{2+} load in the protein [36]. In contrast, preinduction of MT with Zn^{2+} completely prevents inhibition of SGLT-mediated uptake of glucose [36]. Evidently, MT cannot reverse the inhibitory process, but, if it is present to intercept incoming Cd^{2+}, toxicity can be prevented. Section 1.4.2 considers the reaction of Cd^{2+} with Zn-finger proteins and why the product might be unreactive with metallothionein.

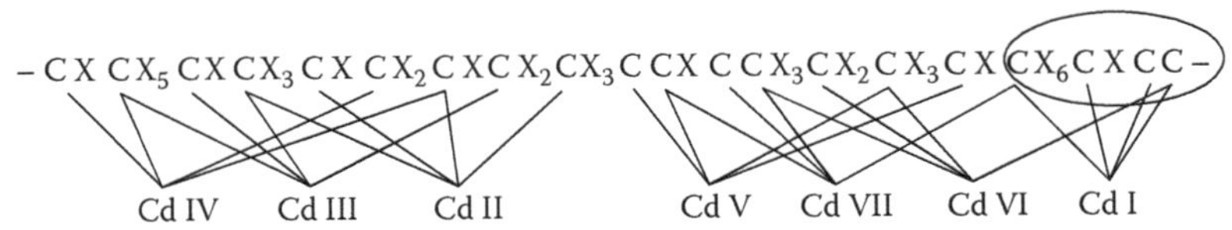

FIGURE 1.10 Metal-thiolate connectivities in metallothionein. Cd numbering refers to NMR resonance numbering as in Figure 1.5. Encircled are the only four contiguous cysteine residues that bind to the same Cd^{2+}.

In another study, the ligand (L) substitution reactions of apo-MT and EDTA with Cd-CA were compared [38]:

$$\text{Cd-CA} + \text{L} \rightarrow \text{Cd-L} + \text{apo-MT}. \tag{1.42}$$

These reactions are bimolecular and involve the direct attack of the competing ligand on the Cd^{2+} binding site. Thus, it was surprising that apo-MT, which is much bulkier than EDTA, nevertheless reacts considerably more rapidly with Cd-CA than EDTA. Indeed, considering the lack of reactivity of apo-MT with the Zn-proteome (Section 1.3.2), it is striking that apo-MT reacts with Cd-CA at all.

It was speculated that the C-terminal region of the apo-MT peptide chain might be particularly active in ligand substitution chemistry. Residues 49–61 contain the only four cysteines that bind to a single Cd^{2+} in the fully saturated protein (Figure 1.10). According to computational studies, peptide$_{49-61}$ can bind and fold around Cd^{2+} in the same conformation that exists in Cd_7-MT [107]. Furthermore, being located at one end of the molecule, the 49–61 sequence should experience less steric hindrance than other parts of the molecule as apo-MT interacts with Cd-CA. In fact, synthetic peptide$_{49-61}$ extracts Cd^{2+} from Cd-CA at the time of mixing, confirming its efficacy as a competing ligand and indicating that the rest of the peptide chain in the holoprotein offers a significant steric barrier to reaction.

1.4 METAL ION EXCHANGE CHEMISTRY

Cells contain myriad metal ion binding sites that are occupied by essential metals as well as by many other adventitious sites that might accommodate metal ions should they present themselves. Foreign metal ions entering cells either compete with essential metal ions for binding to metalloproteins or bind to vacant sites. Common metalloenzymes are relatively inert to metal ion exchange [51]. Thus, the halftime for exchange for the reaction of Zn-carboxypeptidase with Cd^{2+} is 23 h even though the product, Cd-carboxypeptidase, is more stable than the Zn-protein. The same kinetic reluctance to react applies to the reaction of Cd-CA with Zn^{2+}. Such results suggest that Cd^{2+} metal ion exchange reactivity may not be widespread within the Zn-proteome.

1.4.1 Zn-Proteins that Undergo Metal Ion Exchange with Cd^{2+}: Metallothionein

Mammalian metallothionein was discovered as a Cd^{2+} binding protein [108]. Interestingly, the Cd-containing protein is generally isolated as a mixed Cd,Zn-MT with variable ratios of the two metal ions. Why this occurs when the formation of homogeneous Cd-MT would suffice to protect cells from exposure to Cd^{2+} has not been experimentally resolved [101]. How it occurs is quite remarkable.

Experimental animals injected with Cd^{2+} rapidly accumulate the metal ion in liver and promptly form Cd-MT [109]. Over time, Cd-MT is replaced with Cd,Zn-MT, which remains as the product for an indefinite period of steady-state biosynthesis and degradation of the MT protein. The first process may involve the reaction of newly synthesized apo-MT with Cd^{2+} or the metal ion exchange reaction of apo-MT that has acquired Zn^{2+} with Cd^{2+}. The latter reaction *in vitro* is a prime example of facile metal ion exchange that results in the sequestration of Cd^{2+} within the two metal-thiolate clusters that define the MT structure (Figure 1.5):

$$Zn_7\text{-MT} + 7Cd^{2+} \rightarrow Cd_7\text{-MT} + 7Zn^{2+}. \quad (1.43)$$

The reaction is a multistep process: the Zn_3S_9 cluster reacts more rapidly and undergoes exchange as a unit [110]. In contrast, the Zn_4S_{11} cluster experiences complete exchange in three separable kinetic steps:

$$Zn_4S_{11} + 2Cd^{2+} \rightarrow Zn_2,Cd_2S_{11} + 2Zn^{2+}, \quad (1.44)$$

$$Zn_2,Cd_2S_{11} + Cd^{2+} \rightarrow Zn,Cd_3S_{11} + Zn^{2+}, \quad (1.45)$$

$$Zn,Cd_3S_{11} + Cd^{2+} \rightarrow Cd_4S_{11} + Zn^{2+}. \quad (1.46)$$

According to the kinetic analysis, each reaction involves the formation of a Cd^{2+}-bound intermediate that breaks down to yield the product. For example,

$$Zn_2,Cd_2S_{11} + Cd^{2+} \rightarrow Cd\text{-}Zn_2,Cd_2\text{-MT}, \quad (1.47)$$

$$Cd\text{-}Zn_2,Cd_2\text{-MT} \rightarrow Zn,Cd_3S_{11} + Zn^{2+}. \quad (1.48)$$

As described above, the structures of Cd_7-MT and Cd_5,Zn_2-MT reveal that the two metal-thiolate clusters comprise the interior of the two independent metallothionein domains and that peptide folding depends on the positions of clustered sulfhydryl groups with respect to the peptide backbone (Figures 1.4 and 1.10). This is a highly unusual situation. The clusters are partially buried by the backbone and side chains but also retain some sulfur atom solvent accessibility (Figure 1.11). The extent of solvent accessibility is thought to contribute to the capacity of each cluster to undergo reactions such as metal ion exchange [111,112]. Thus, it is hypothesized that Cd^{2+} competes for solvent-available

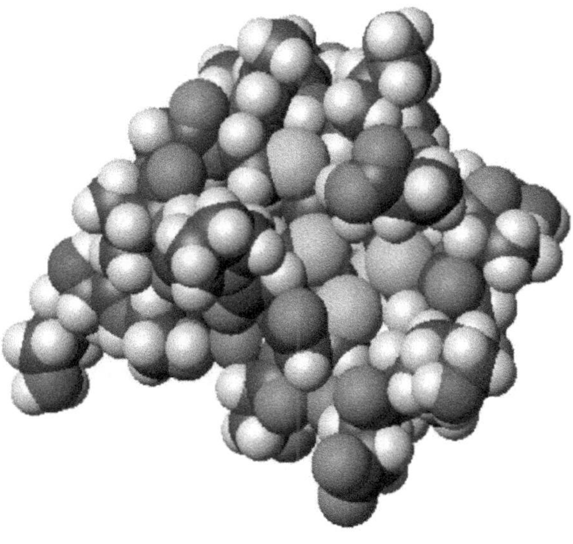

FIGURE 1.11 Solvent accessibility of metal-thiolate cluster sulfhydryl groups in Cd_4,S_{11} domain. The large, central, light gray atoms are sulfurs.

sulfhydryl groups that are bound to Zn^{2+}, binding weakly to some of them and then proceeding to fully exchange thiolate ligands with Zn^{2+}. Although Zn^{2+} is very strongly bound to MT ($K = 10^{11.2}$), a rapid reaction with Cd^{2+} ensues [51,113]. Possibly, the presence of thiolate ligands that bridge two Zn^{2+} ions in the cluster is the key to its metal exchange reactivity. Such bridging ligands may be particularly reactive with Cd^{2+} and potentially able to undergo bond breaking that makes them accessible to Cd^{2+}:

$$\text{-Zn}_a\text{-S}_{\text{bridging}}\text{-Zn}_b\text{-} + Cd^{2+} \rightarrow \text{Zn}_a\text{-S-Cd} + \text{Zn}_b\text{-}, \tag{1.49}$$

$$\text{Zn}_a\text{-S-Cd} + \text{Zn}_b\text{-} \rightarrow \text{-Zn}_a\text{-S}_{\text{bridging}}\text{-Cd}_b\text{-} + Zn^{2+}. \tag{1.50}$$

These reactions generate homogeneous Cd_7-MT. To form the mixed metal species, the obvious hypothesis is that the Cd^{2+}-Zn^{2+} exchange illustrated in Reactions 1.44 through 1.46 is truncated at some intermediate point, leaving the mixed metal protein. But that reaction does not give rise to native distribution of the metal ions among the MT binding sites [114]. Instead, the reaction of Cd_7-MT with Zn_7-MT results in the steady-state form of the protein that is observed in tissues [114]. The reaction involves interprotein metal ion exchange on an unprecedented scale:

$$n\text{Cd}_7\text{-MT} + m\text{Zn}_7\text{-MT} \rightarrow (n+m)\text{Zn}_{7m/(m+n)},\text{Cd}_{7n/(m+n)}\text{-MT}. \tag{1.51}$$

Metal ion exchange results in mixed metal clusters with Cd^{2+} favored in the four-metal cluster and Zn^{2+} in the three-metal cluster. In this reaction, it is thought that clusters within the two proteins interact through thiolate exchange modeled after Reactions 1.52 through 1.54:

$$\text{-Zn-S-}_{\alpha\text{-domain, protein 1}} + \text{-Cd-S-}_{\beta\text{-domain, protein 2}} \rightarrow \text{-Zn-S}_{\alpha,1}\text{-Cd-S}_{\beta,2}, \tag{1.52}$$

$$\text{-Zn-S}_{\alpha,1}\text{-Cd-S}_{\beta,2} \rightarrow \text{-Cd-S}_{\alpha,1}\text{-Zn-S}_{\beta,2}, \tag{1.53}$$

$$\text{-Cd-S}_{\alpha,1}\text{-Zn-S}_{\beta,2} \rightarrow \text{-Cd-S-}_{\alpha\text{-domain, protein 1}} + \text{-Zn-S-}_{\beta\text{-domain, protein 2}}. \tag{1.54}$$

Supporting this model is the fact that in the crystal, metallothionein protein monomers align head (α) to tail (β) and appear to establish dimer interactions. Indeed, the protein does form a dimer in solution that is detected physically and perturbs the protein's reactivity by altering accessibility to the clusters [115,116].

1.4.2 Zn-Proteins that Undergo Metal Ion Exchange with Cd^{2+}: Zn-Finger Proteins

Zn-finger proteins also display substantial reactivity with Cd^{2+}. In at least one instance, described in Section 1.3.7, direct reaction of Cd^{2+} with a Zn-finger transcription factor may lie at the heart of the mechanism of kidney toxicity of Cd^{2+}. The most common type of Zn-finger transcription factor, in which Zn^{2+} binds to two cysteinyl sulfhydryl groups and two histidinyl imidazole nitrogens, readily exchanges its Zn^{2+} for Cd^{2+}. Zn-tramtrack mentioned above (Reaction 1.14), Zn-MTF-1, the metal response element binding transcription factor that regulates MT synthesis, transcription factor IIIA (TFIIIA), the original Zn-finger transcription factor prototype, and Zn-Sp1, a protein widely involved in eukaryotic promoter activity, all readily react with Cd^{2+}, resulting in diminution in DNA binding activity [37,50,117,118]. In each case, the reaction is presumed to be simple metal ion exchange, for example,

$$Cd^{2+} + \text{Zn-TFIIIA} \rightarrow \text{Cd-TFIIIA} + Zn^{2+}. \tag{1.55}$$

In terms of possible physiological significance, it must at least be shown that Zn^{2+} displacement occurs under approximately stoichiometric concentrations of Cd^{2+} and the Zn-finger protein. The need for a large excess of Cd^{2+} would bring into question whether such concentrations could be attained in viable cells. Even if the reaction occurs at comparable concentrations of reactants, the problem of site selectivity for cellular metal ion exchange is a difficult one to address. Why this reaction when Cd^{2+} might react with a huge number of Zn-finger proteins of similar structure (ca. 10^3) as well as multiple adventitious metal binding sites [1]?

Cd^{2+} lies just below Zn^{2+} in the periodic table and, as a result, displays similar chemical properties to Zn^{2+}. Indeed, comparing the affinity of the Zn^{2+} and Cd^{2+} complexes of finger 3 (F3) of TFIIIA with its DNA binding site, Zn-F3 binds only 10 times stronger than Cd-F3 [119]. This result suggests that the substitution of Cd^{2+} for Zn^{2+} induces only minor structural perturbations in the finger conformation.

Two-dimensional NMR spectroscopy was used to determine the structures of Zn- and Cd-mF3, in which mF3 is a peptide closely related to F3 that forms a sufficiently stable complex with Cd^{2+} so that its NMR structure can be defined [119]. Figure 1.12 compares the two structures. Each metal ion is coordinated to two thiolate groups and two imidazole nitrogens. However, the conformation of the imidazole planes is markedly different in the two structures. Both structures exhibit the $\beta\beta\alpha$ peptide structure that is typical for C_2H_2 Zn-finger peptides with the cysteinyl ligands located in the β strand segments and the histidine imidazole side chains in the α-helix[120]. The helix serves as the primary DNA binding element. Nevertheless, because the imidazole ligands interact with Zn^{2+} and Cd^{2+} in somewhat different orientations, the two helices, themselves, are differentially folded. By projecting their DNA binding side chains in different directions (e.g., the arginine side chain in Figure 1.12), the two peptides display differential DNA binding affinities.

Interestingly, when native Zn_9-TFIIIA reacts with Cd^{2+}, the product contains less than stoichiometric concentrations of Cd^{2+} that are refractory to reaction with EDTA [117]. When more than one tandem Zn-finger (Zn-F_N) is present in a peptide, it is hypothesized that the following pair of reactions occurs:

$$Cd^{2+} + (Zn\text{-}F_N)\text{-}(Zn\text{-}F_{N+1}) \rightarrow (Cd\text{-}F_N)\text{-}(Cd\text{-}F_{N+1}) + 2Zn^{2+}, \tag{1.56}$$

$$(Cd\text{-}F_N)\text{-}(Cd\text{-}F_{N+1}) \rightarrow F_N\text{-}Cd\text{-}F_{N+1} + Cd^{2+}. \tag{1.57}$$

In the first, Cd^{2+} displaces Zn^{2+} and maintains the C_2H_2 coordination environment. In the second, Cd^{2+}, which in comparison with Zn^{2+} favors sulfhydryl over nitrogen ligand coordination, establishes a more stable CdS_4 site by binding to the thiolate groups of two adjacent fingers, F_N and F_{N+1}.

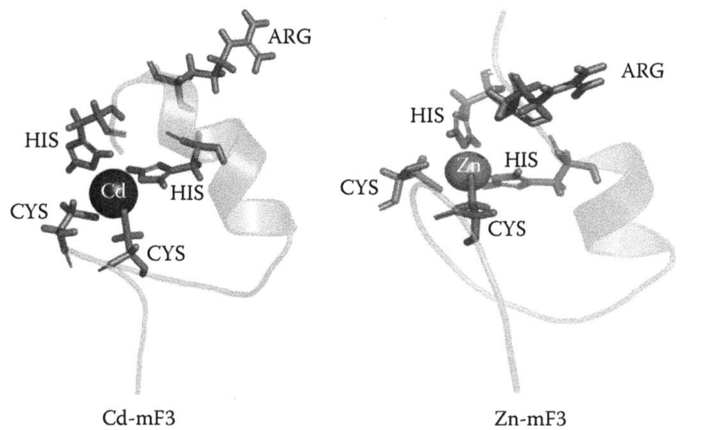

FIGURE 1.12 NMR two-dimensional spectral structures of Zn-fingers containing Zn^{2+} and Cd^{2+}. Structures based on modification of finger 3 of transcription factor IIIA (Data from Krepkiy, D., Försterling, F.H., and Petering, D.H., *Chem. Res. Toxicol.*, 17, 863, 2004). Cysteine and histidine ligands shown along with key DNA binding residue (arginine) in a helix side chain.

The possibility that Cd^{2+} reorganizes the Zn^{2+} metal binding site as in Reaction 1.57 suggests that another metal ion, Pb^{2+}, which also inactivates Zn-finger proteins and has high affinity for sulfhydryl groups, might do so in a similar way [117,121].

Zn-MT participates in metal ion exchange reactions with other metalloproteins besides its self-exchange reactions (Reaction 1.51). As indicated in Section 1.2.3, Zn-MT undergoes reaction with Cd-tramtrack to form Cd-MT and Zn-tramtrack and, in the process, restores the specific DNA binding capacity of tramtrack [37]. Zn-MT metal exchange does not occur with Cd-Sp1 and at best very slowly with Cd-transcription factor IIIA, possibly because hypothetical interfinger CdS_4 sites are sufficiently stable to render the reaction energetically unfavorable [50,122].

1.5 ADDUCT FORMATION

The role of adduct formation in cellular inorganic chemistry has received less attention than other categories of inorganic reactions. Nevertheless, in many ligand substitution reactions, the formation of adduct species comprises part of the mechanistic pathway:

$$\text{M-L}_1 + \text{L}_2 \rightarrow \text{L}_2\text{-M-L}_1 \qquad \text{(adduct)}, \tag{1.58}$$

$$\text{L}_2\text{-M-L}_1 \rightarrow \text{M-L}_2 + \text{L}_1 \qquad \text{(substitution)}. \tag{1.59}$$

For example, in the reaction of Cd_7- or Zn_7-MT with EDTA, each cluster initially forms an adduct species with EDTA, in which one or more EDTA ligand atoms replace sulfhydryl groups in the metal-thiolate clusters [123]. Conceivably, in some reactions, ligand substitution stops at the stage of adduct formation. In this context, whenever a metal binding ligand is used, the possibility should be considered that adduct formation plays a key role in the observed cellular effects of certain metal binding ligands, as well as or perhaps instead of complete substitution.

1.5.1 Fluorescent Zn^{2+} Sensors in Relation to Zn^{2+} Trafficking

Recently, Zn^{2+} trafficking has been invoked in a variety of physiological processes such as apoptosis and neuronal synapse activity, suggesting that it plays dynamic roles in cells beyond its responsibilities when bound to proteins [124–128]. In order to provide chemical tools to observe such time-dependent activities, increasing effort has been invested in synthesizing and using fluorescent Zn^{2+} sensors. In principle, these sensors are accumulated by cells and, once inside, innocuously report on "chelatable" Zn^{2+} distribution and changes in its distribution during cellular activities and their perturbations [124–133]. This approach to observing cellular Zn^{2+} localization and trafficking follows the highly successful development and application of fluorescent Ca^{2+} sensors [134,135].

The ideal sensor is a passive observer of Zn^{2+} status that does not significantly perturb Zn^{2+} trafficking or distribution [136]. In the parlance of the field, L_{sensor} reacts with "free," "labile," "chelatable," or "accessible" Zn^{2+} ($Zn\text{-}L_{reactive}$).

$$\text{Zn-L}_{\text{reactive}} + \text{L}_{\text{sensor}} \rightarrow \text{Zn-L}_{\text{sensor}} + \text{L}_{\text{reactive}}. \tag{1.60}$$

These adjectives suggest readily available Zn^{2+} that is weakly or modesty associated with binding sites. Nevertheless, it is evident that the extent of formation of $Zn\text{-}L_{sensor}$ depends on the properties of particular $L_{sensors}$ and $Zn\text{-}L_{reactive}$, their equilibrium affinity for Zn^{2+}, kinetic reactivity with one another, as well as intracellular concentration. Thus, in principle, each chemically distinct type of sensor might access different pools of intracellular Zn^{2+} and, in so doing, alter facets of steady-state or transient Zn^{2+} distribution.

Typically, reports of experiments that utilize such sensors describe cells initially stained with a Zn^{2+} fluorophore that undergo a change in fluorescence intensity after some stimulus. An increase

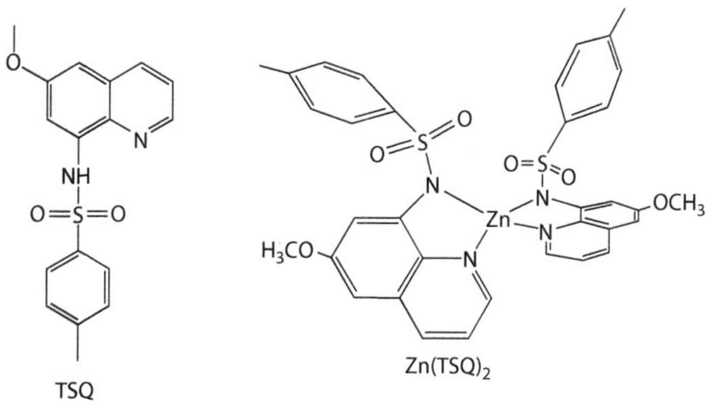

FIGURE 1.13 Structure of Zn-fluorophore, TSQ, and its Zn^{2+} complex.

in fluorescent emission intensity is ascribed to the liberation of "free" or "accessible" Zn^{2+} within the cells. The repertoire of experimental methods employed with such sensors may be quite sophisticated and include, for example, kinetic measurements [137]. And that is about as much as fluorescent measurements can reveal when employed alone to monitor Zn^{2+} trafficking. Left unresolved is the basic chemical question: what is being imaged?

1.5.2 Fluorescent Zn^{2+} Sensor TSQ

The prototypical Zn^{2+} sensor is *N*-(6-methoxy-8-quinolyl)-*p*-toluensulfonamide (TSQ), a molecule linking a metal binding domain with a fluorescent moiety (Figure 1.13) [138,139]. Together with its close relative Zinquin, the two sensors comprise the most commonly used type of Zn^{2+} fluorescent probe [129,140]. TSQ reacts with Zn^{2+}, resulting in a drastic increase in the intensity of its fluorescence emission. On incubation with cells, TSQ produces an asymmetric pattern of fluorescence that is characterized by the appearance of punctate bodies, thought to contain high concentrations of Zn^{2+} (Figure 1.14). In the absence of TSQ, the entire microscopic field would be black.

To understand what TSQ, Zinquin, or other types of Zn^{2+} sensors image in cells, it is necessary to understand their relevant chemistry. Two studies of the widely used Zn^{2+} sensor Zinquin examined what the authors considered to be the plausible or potential cellular chemistry of this sensor that needs to be considered in order to interpret its cellular behavior [131,132]. Zn $(Zinquin)_2$'s log apparent stability constant is 13.5 at pH 7.2 and 25°C, and its log adduct formation constant with a number of $Zn\text{-}L_{reactive}$ species ranges from 1 to 8 [130,141]. On a competitive equilibrium basis therefore, Zinquin cannot compete for Zn^{2+} bound to metalloproteins such as CA (log $K \sim 12$) [142]. It was concluded that this sensor could image nM–pM concentrations of cellular free Zn^{2+}. Alternatively, adduct species at the upper end of the stability range might be stable in cells.

1.5.3 What is TSQ Imaging in Cells?

Pig kidney, LLC-PK_1 cells exposed to TSQ display an intense punctate staining surrounding the nucleus, extending out into the cytoplasm (Figure 1.14). Determination of cell TSQ-Zn^{2+} concentration by quantitative fluorescence spectrophotometry demonstrated that approximately 8% of the cellular complement of Zn^{2+} was detected as a TSQ complex (total: 200 nmol/10^8 cells or ca. 500 μM, based on literature values for volumes of cultured mammalian cells) [143]. This is a remarkably large percentage and incommensurate with the view that TSQ detects nM–pM concentrations of Zn^{2+} [130].

The common assumption has been that TSQ gains access to $Zn^{2+}_{reactive}$ (Reaction 1.61),

$$Zn\text{-}L_{reactive} + 2TSQ \rightarrow Zn(TSQ)_2, \qquad (1.61)$$

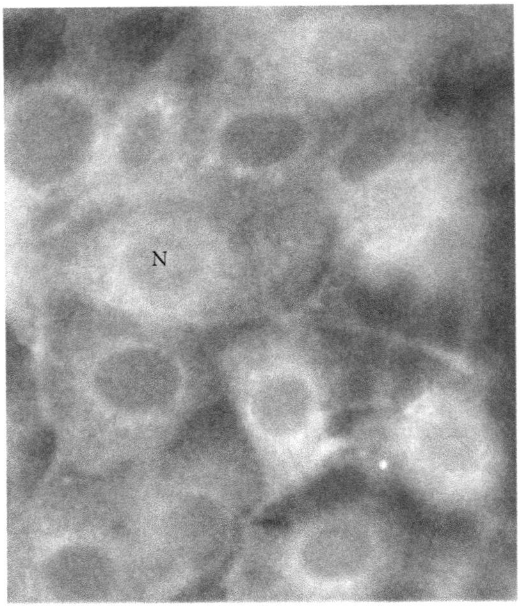

FIGURE 1.14 Fluorescent image of LLC-PK$_1$ cells stained with TSQ. Nucleus (N) and punctate staining about the nucleus.

and forms the low-molecular-weight product $Zn(TSQ)_2$. The speciation of cytosolic Zn-TSQ species was probed using gel filtration chromatography to separate Zn^{2+} high molecular weight (Zn-proteome pool), metallothionein (MT), or low molecular weight (LMW) fractions [143]. Although $Zn(TSQ)_2$ would be anticipated to migrate with the LMW fractions, all of the Zn-TSQ fluorescence localized with the Zn-proteome, consistent with the reaction of TSQ with Zn-proteins to form ternary adducts [141]:

$$\text{Zn-protein}_{\text{reactive}} + \text{TSQ} \rightarrow \text{TSQ-Zn-protein}_{\text{reactive}}. \quad (1.62)$$

Confirmation of this hypothesis came from the demonstration that cellular and isolated Zn-proteome fluorescence was not due to free $Zn(TSQ)_2$, which has its fluorescence emission maximum at 492 nm, but stemmed from adducts that exhibit blue-shifted emission maxima near 470 nm, as previously described [141]. The conclusion that TSQ images Zn-proteins and not "accessible" Zn^{2+}, at least in some cells, completely changes the interpretation of TSQ cellular fluorescence and brings into question what other sensors are imaging.

1.6 REDOX REACTIONS

The final category of reactions focuses on oxidation–reduction processes. As with other facets of cellular inorganic chemistry, there is still much to be learned about these reactions.

1.6.1 Iron Trafficking and Pharmaceutical Ligands for Fe Overload Disorder

Fe-trafficking mechanisms, like those for Zn^{2+}, remain largely unresolved. The single exception is the transferrin-dependent uptake of Fe from the extracellular medium into mammalian cells. That set of reactions is summarized as

$$\text{Fe(III)}_2\text{-Tr}\cdot\text{Tr-receptor}_{\text{extracellular}} \rightarrow \text{Fe(III)}_2\text{-Tr}\cdot\text{Tr-receptor}_{\text{extracellular}}, \quad (1.63)$$

$$\text{Fe(III)}_2\text{-Tr}\cdot\text{Tr-receptor}_{\text{extracellular}} \rightarrow \text{Fe(III)}_2\text{-Tr}\cdot\text{Tr-receptor}_{\text{vesicle}}, \quad (1.64)$$

$$\text{Fe(III)}_2\text{-Tr} - \text{Tr-receptor}_{\text{vesicle}} + n\text{H}^+ \rightarrow \text{apo-Tr(H}^+)_n - \text{Tr-receptor}_{\text{vesicle}} + 2\text{Fe}^{3+}_{\text{vesicle}}, \quad (1.65)$$

$$2\text{Fe}^{3+}_{\text{vesicle}} + 2e^- \rightarrow 2\text{Fe}^{2+}_{\text{vesicle}}, \quad (1.66)$$

$$2\text{Fe}^{2+}_{\text{vesicle}} \rightarrow 2\text{Fe}^{2+}_{\text{cytosol}}. \quad (1.67)$$

Fe(III) in the extracellular medium is presented to the cell primarily as $\text{Fe(III)}_{1\text{-}2}$-Tr [144]. Fe-loaded Tr binds to the exterior of the cell by associating with the transferrin receptor [145,146]. The adduct is internalized as the plasma membrane invaginates and forms an endosomal vesicle such that the complex is located inside the vesicle [147,148]. How Fe(III) dissociates from Tr has been the subject of much study since each Fe(III) bound to Tr is characterized by a very large stability constant that is greater than 10^{22} [149]. Initially, it was thought that the lower internal pH of the vesicle was sufficient to reduce the strength of association as H^+ competed for binding to Fe(III) ligands [147,148]. Indeed, given enough time, the pH of the vesicle, on the order of 5.5, is sufficient to shift the binding equilibrium toward free Fe^{3+}:

$$\text{Fe(III)}_2\text{-Tr} + n\text{H}^+ \rightarrow \text{apo-Tr(H}^+)_n + 2\text{Fe}^{3+}. \quad (1.68)$$

But the reaction occurs too slowly to explain the cellular turnover of Fe(III)_2-Tr, which can be on the order of 10^2 s [150,151]. That is, the reaction as written was kinetically too sluggish to account for the observed cellular dynamics.

Studies of this reaction carried out with Fe(III)_2-Tr bound to the transferrin receptor either in endosomes or in synthetic vesicles reveal that Reaction 1.65 occurs one to two orders of magnitude faster than Reaction 1.68 at acidic pH, which would qualify it as a kinetically competent step in the cellular turnover of transferrin-bound iron [152,153]. Once released, Fe^{3+} would then be reduced in the endosome by some unidentified mechanism and transported to the cytoplasm by the divalent metal ion transporter one [152,154,155]. The apparent rates of reduction and transport of Fe^{3+} experimentally dissociated from the transferrin•transferrin receptor complex with different reductants are considerably slower than the H^+-dependent release of Fe(III) from the complex and might not qualify as a kinetically feasible step in the transferrin-iron cycle [155].

An appealing alternative hypothesis is based on the findings that under some circumstances, transferrin-bound Fe(III) can be reduced and dissociated [77,150]. Might Fe^{3+} reduction that is posited to occur after dissociation of Fe^{3+} from Tr actually take place with $\text{Fe(III)}_2\text{Tr}\cdot\text{Tr-receptor}_{\text{vesicle}}$ bound Fe(III)? Perhaps, Reactions 1.63 through 1.67 are combined into a single process:

$$\text{Fe(III)}_2\text{-Tr}\cdot\text{Tr-receptor}_{\text{vesicle}} + n\text{H}^+ + 2e^- \rightarrow \text{apo-Tr(H}^+)_n\cdot\text{Tr-receptor}_{\text{vesicle}} + 2\text{Fe}^{2+}_{\text{vesicle}}. \quad (1.69)$$

Reaction 1.69 involving the N-terminal lobe of transferrin occurs five times faster at pH 6.0 than at pH 7.4, indicating that acidification of the endosome speeds up Fe(III) reduction as well as Fe(III) dissociation from transferrin [156]. However, the need for a strong Fe(II) acceptor such as BPS to drive the reaction in the face of the very unfavorable reduction potentials of Fe(III)Tr makes this a problematic cellular mechanism [80]. Recently, Crumbliss and colleagues determined that Fe(III)_2Tr binding to its receptor raises its reduction potential at pH 5.8 of about 200 mV to greater than –300 mV, placing it within reach of reduced nicotine adenine dinucleotide (NADH) [150]. Clearly, the interaction between Tr and Tr-receptor alters the

$$\text{Fe(III)}_2\text{-Tr}\cdot\text{Tr-receptor}_{\text{vesicle}} + n\text{H}^+ + 2e^- \rightarrow \text{apo-Tr(H}^+)_n\cdot\text{Tr-receptor}_{\text{vesicle}} + 2\text{Fe}^{3+}_{\text{vesicle}} \quad (1.70)$$

properties of the Fe(III) binding sites under acidic conditions. Whether this extends to enhancing the kinetics of Reaction 1.69 sufficiently to qualify it as a dual mechanism for iron dissociation and reduction remains to be established.

Considerations of Fe trafficking mechanisms suggest approaches to addressing the problem of iron overload, which results largely from the treatment of the genetic disease of thalassemia [157]. Extensive effort has been devoted to encouraging the discovery of drugs that chelate Fe and facilitate its excretion [157]. A measure of success has been achieved with the siderophore desferrioxamine in humans and other agents in experimental model systems [157]. Still, strategies to address this issue have focused almost exclusively on Fe(III) chelating agents because the primary molecules involved in Fe transport and storage, transferrin and ferritin, are Fe(III) binders. Given the discussion above, it seems that molecules that can intercept and bind Fe(II) deserve equal attention. For example, 2-formylpyridine thiosemicarbazone forms a relatively redox-stable Fe(II) complex and its derivative, 5-hydroxy-2-formylpyridine thiosemicarbzone (L), mobilizes substantial amounts of Fe from cancer patients as what appears to be $Fe(II)L_2$ [158,159].

1.6.2 Zn-Metallothionein and Sulfhydryl Group Redox Chemistry

1.6.2.1 Reactions with Disulfides

A common ligand in biometal complexes is the sulfhydryl group (RSH) of cysteine. Besides its large affinity for transition and other heavy metal ions, it is redox active and its sulfhydryl groups are readily oxidized to the level of disulfide or beyond. Should an oxidizing agent target metal-bound thiol, for example, its conversion to an oxidized form would be accompanied by substantially reduced metal binding affinity and the increased likelihood that the metal ion will dissociate.

Because of its unique complement of Zn^{2+} and cysteinyl thiolate groups as well as its significant concentration in many cells, metallothionein has been singled out as a probable, prominent site of sulfhydryl group reaction with oxidizing agents. For example, reaction of the thiol groups of Zn-MT with DTNB [(5,5*N*-dithiobis(2-nitrobenzoate), RS-SR], a thiol-disulfide interchange reaction that is formally a redox process, results in the release of Zn^{2+} from the protein [160]:

$$\text{Zn-SMT} + \text{RS-SR} \rightarrow \text{MTS-SR} + \text{RS}^- + \text{Zn}^{2+}. \quad (1.71)$$

It has been proposed that this reaction has physiological significance as a means to mobilize MT-bound Zn^{2+} when glutathione disulfide (GS-SG) replaces DTNB [161]. However, an early study showed that this redox process occurs with very slow kinetics, presumably because GSH's bulky structure presents a formidable steric barrier to efficient reaction with Zn-MT [162]. Even at millimolar concentrations of GS-SG, the reaction halftime is on the order of 24 h. Thus, it is unclear how this reaction might contribute to ongoing Zn^{2+} trafficking.

1.6.2.2 Reactions with Nitric Oxide Species

A second example offers a more complicated cellular picture. Zn^{2+} fluorescent sensors such as TSQ have been employed to investigate the impact of NO donors on cells [163,164]. Typically, exposure to NO results in fluorescence enhancement that has been interpreted as reaction of the sensor with Zn^{2+} lost from MT upon reaction of its thiol groups with an NO metabolite. The thiol groups (RH^-) might have been modified to form *S*-nitrosothiolate (RS-NO) or disulfide species. In one experiment, the kinetics of Zn^{2+} release detected by the sensor (Zinquin) were very fast, with the reaction going to completion in about 50–100 s [163].

The fast reaction observed in this cellular study seems inconsistent with *in vitro* experiments that reveal much slower kinetics of reaction of Zn_7-MT (isoform 2) with NO donors [165]. In summary, under anaerobic conditions, Zn-MT (Zn_7-MT 2) is unreactive with NO supplied by the rapid decomposition of DEA-NO (diethylaminenonoate) or the photochemical decay of SNAP (*S*-nitrosylacetylpenicillamine, RS-NO) (Reaction 1.72). Moreover, Zn-MT fails to react with similar concentrations of SNAP under aerobic conditions in the absence of light (Reaction 1.73):

$$\text{RS-NO} + \text{light} \rightarrow \text{RS}^\bullet + \text{NO}^\bullet, \tag{1.72}$$

$$\text{RS}^-_{\text{MT}} + \text{R'S-NO} \rightarrow \text{RS}_{\text{MT}}\text{-NO} + \text{R'S}^-. \tag{1.73}$$

Zn-MT slowly reacts and releases its Zn^{2+} in the presence of aerobic DEA-NO or aerobic, light-activated SNAP or *S*-nitrosylglutathione (Reactions 1.74 through 1.76). Each of these reactions involves the liberation of NO followed by its reaction with O_2 to produce NO_2 and N_2O_3. Spectral analysis of the product did not detect the presence of *S*-nitrosyl groups (340 nm absorbance maximum), suggesting that the reaction, instead,

$$2\text{NO}^\bullet + \text{O}_2 \rightarrow 2\text{NO}_2^\bullet, \tag{1.74}$$

$$\text{RS}^-_{\text{MT}} + \text{NO}_2^\bullet \rightarrow \text{RS}^\bullet_{\text{MT}}\ (\tfrac{1}{2}\text{RS-SR}) + \text{NO}_2^-, \tag{1.75}$$

$$\text{RS}^\bullet_{\text{MT}} + \text{NO}^\bullet \rightarrow \text{RS}_{\text{MT}}\text{-NO} \tag{1.76}$$

produced disulfides (Reactions 1.74 and 1.75). Thus, at best, Zn-MT sluggishly reacts with large concentrations of NO under aerobic conditions. In contrast, apo-MT reacts very rapidly with both anaerobic NO and SNAP, in agreement with the expected higher reactivity of free than Zn^{2+}-bound sulfhydryl groups.

The argument that Zn-MT is a target of NO is bolstered by clever experiments in which cells were transfected with a plasmid that expressed metallothionein linked at N and C termini to two mutant green fluorescent proteins (GFPs) [166,167]. When present in the Zn^{2+}-bound form, the GFPs underwent fluorescent resonance energy transfer (FRET) that could be observed in cells. On exposing these transfected cells to NO, the FRET was lost, indicative of the reaction of the Zn-MT chimeric protein to unfold the protein, lengthening the average distance between the fluorophores and abrogating the FRET.

Other experiments exposed cells in cell culture to DEA-NO and then determined its impact on the Zn^{2+} and thiol content of the proteome, MT, and glutathione pools after gel filtration chromatography [165]. No evidence of Zn-MT reaction was obtained but apo-MT underwent extensive modification. Strikingly, both the proteomic and glutathione pools of sulfhydryl groups were reactive with the NO donor. Indeed, the extent of reaction in each pool paralleled its respective concentration in the cell. These analytical measurements emphasize that even if cellular Zn-MT reacts with NO metabolites, it does so in the midst of a bewildering, global reaction of the cell's thiol complement with NO donors. It is difficult to assign cellular target specificity to Zn-MT when much of the entire thiol metabolome reacts with DEA-NO.

1.7 CONCLUSIONS

Initial inquiries into the content of the metal-dependent proteome confirm its great size and suggest comparable complexity of its metal trafficking mechanisms [1–4]. Defining the roles and actions of cellular essential metals is especially difficult because generally they operate through their association with macromolecules and not as independent entities. Investigators are presented with added challenges when attempting to understand how xenobiotic metallic species, toxic metals and metallodrugs, react in the cellular milieu because one cannot presume the existence of established pathways of reaction. In this context, complementary chemical and cellular studies are useful to test proposed hypotheses: model chemical experiments probe whether cellular observations can be attributed to particular chemical reactions. More difficult but equally important, experiments with cells offer the opportunity of determining whether chemical models resemble actual cellular chemistry. Together, they offer an effective strategy in order to understand cellular metal trafficking and function.

ACKNOWLEDGMENTS

The authors acknowledge the support of National Institute of Environmental Health Sciences grants ES-04026 and ES-04184 as well as the support of the University of Wisconsin-Milwaukee through its Research Growth Initiative.

REFERENCES

1. Andreini, C., Banci, L., Bertini, I., and Rosato, A., Counting the zinc-proteins encoded in the human genome, *J. Proteome Res.*, 5, 196, 2006.
2. Andreini, C., Banci, L., Bertini, I., and Rosato, A., Zinc through the three domains of life, *J. Proteome Res.*, 5, 3173, 2006.
3. Andreini, C., Banci, L., Bertini, I., Elmi, S., and Rosato, A., Non-heme iron through the three domains of life, *Proteins*, 67, 317, 2007.
4. Andreini, C., Banci, L., Bertini, I., and Rosato, A., Occurrence of copper proteins through the three domains of life: A bioinformatic approach, *J. Proteome Res.*, 7, 209, 2008.
5. Satarug, S. and Moore, M.R., Adverse health effects of chronic exposure to low-level cadmium in foodstuffs and cigarette smoke, *Environ. Health Perspect.*, 112, 1099, 2004.
6. Mergler, D., Anderson, H.A., Chan, L.H., Mahaffey, K.R., Murray, M., Sakamoto, M., and Stern, A.H., The panel on health risks and toxicological effects of methylmercury. Methylmercury exposure and health effects in humans: A worldwide concern, *Ambio*, 36, 3, 2007.
7. White, L.D., Cory-Slechta, D.A., Gilbert, M.E., Tiffany-Castiglioni, E., Zawia, N.H., Virgolini, M., Rossi-George, A., Lasley, S.M., Qian, Y.C., and Basha, M.R., New and evolving concepts in the neurotoxicology of lead, *Toxicol. Appl. Pharmacol.*, 225, 1, 2007.
8. Kelland, L., The resurgence of platinum-based cancer chemotherapy, *Nat. Rev. Cancer*, 7, 573, 2007.
9. Petering, D.H., Xia, C., and Antholine, W.E., Metal ion dependent antibiotics in chemotherapy, In: A. Sigel and H. Sigel (eds), *Metal Ions in Biological Systems*, Vol. 42, p. 463, Marcel Dekker, NY, 2004.
10. Xu, X., Sutak, R., and Richardson, D.R., Iron chelation by clinically relevant anthracyclines: Alteration in expression of iron-regulated genes and atypical changes in intracellular iron distribution and trafficking, *Mol. Pharmacol.*, 73, 833, 2008.
11. Nyayapati, S., Afshan, G., Lornitzo, F., Byrnes, R.W., and Petering, D.H., Depletion of cellular iron by BPS and ascorbate: Effect on toxicity of adriamycin, *Free Rad. Biol. Med.*, 20, 319, 1996.
12. Fraker, P.J. and King, L.E., Reprogramming of the immune system during zinc deficiency, *Ann. Rev. Nutr.*, 24, 277, 2004.
13. Clegg, M.S., Hanna, L.A., Niles, B.J., Momma, T.Y., and Keen, C.L., Zinc deficiency-induced cell death, *IUBMB Life*, 57, 661, 2005.
14. MacDonald, R.S., The role of zinc in growth and cell proliferation, *J Nutr.*, 130(5S Suppl), 1500S, 2000.
15. White, L.D., Cory-Slechta, D.A., Gilbert, M.E., Tiffany-Castiglioni, E., Zawia, N.H., Virgolini, M., Rossi-George, A., Lasley, S.M., Qian, Y.C., and Basha, M.R., New and evolving concepts in the neurotoxicology of lead, *Toxicol. Appl. Pharmacol.*, 225, 1, 2007.
16. LoPachin, R.M. and Barber, D.S., Synaptic cysteine sulfhydryl groups as targets of electrophilic neurotoxicants, *Toxicol Sci.*, 94, 240, 2006.
17. Petering, D.H. and Antholine, W.E., Copper toxicity: Speciation and reactions of copper in biological systems, *Rev. Biochem. Toxicol.*, 9, 225, 1988.
18. Lippard, S.J. and Berg, J.M., *Principles of Bioinorganic Chemistry*, Chap. 2, University Sciences Books, Mill Valley, CA, 1994.
19. Fraústo da Silva, F. and Williams, R.J.P., The principles of the uptake and chemical speciation of the elements in biology, In: *The Biological Chemistry of the Elements*, Chap. 2, Oxford University Press, Oxford, 2001.

20. Outten, C.E. and O'Halloran, T.V., Femtomolar sensitivity of metalloregulatory proteins controlling zinc homeostasis, *Science*, 292, 2488, 2001.
21. Huffman, D.L. and O'Halloran, T.V., Function, structure, and mechanism of intracellular copper trafficking proteins, *Ann. Rev. Biochem.*, 70, 677, 2001.
22. Petering, D.H., Zhu, J., Krezoski, S., Meeusen, J., Kiekenbush, C., Krull, S., Specher, T., and Dughish, M., Apo-metallothionein emerging as a major player in the cellular activities of metallothionein, *Exp. Biol. Med. (Maywood)*, 231, 1528, 2006.
23. Petering, D.H. and Petering, H.G., A molecular basis for metal toxicity, In: R.S. Bhatnager (ed.), *Molecular Basis of Environmental Toxicity*, Chap. 20, Ann Arbor Science Publications, Ann Arbor, MI, 1980.
24. Rabinstein, D.L. and Reid, R.S., Nuclear magnetic resonance studies of the solution chemistry of metal complexes. 20. Ligand-exchange kinetics of methylmercury(II)–thiol complexes, *Inorg. Chem.*, 23, 1246, 1984.
25. Howe-Grant, M.E. and Lippard, S.J., *Metal Ions in Biological Systems*, H. Sigel (ed.), Chap 2, Vol. 11, Marcel Dekker, Basel, 1980.
26. Petering, D.H., Antholine, W.E., and Saryan, L.A., Metal complexes as antitumor agents, In: R.M. Ottenbrite and G.B. Butler (eds), *Anticancer and Interferon Agents, Synthesis and Properties*, Chap. 7, Marcel Dekker, New York, 1984.
27. Kraker, A., Schmidt, J., Krezoski, S., and Petering, D.H., Binding of *cis*-dichlorodiammine platinum(II) to metallothionein in Ehrlich cells, *Biochem. Biophys. Res. Commun.*, 130, 786, 1985.
28. Pattanaik, A., Bachowski, G., Laib, J., Lemkuil, D., Shaw, C.F., 3rd, Petering, D.H., Hitchcock, A., and Saryan, L., Properties of the reaction of *cis*-dichlorodiammineplatinum(II) with metallothionein, *J. Biol. Chem.*, 267, 16121, 1992.
29. Lemkuil, D.C., Nettesheim, D., Shaw, C.F., 3rd, and Petering, D.H., Reaction of Cd_7-metallothionein with *cis*-dichlorodiammine platinum (II), *J. Biol. Chem.*, 269, 24792, 1994.
30. Kelley, S.L., Basu, A., Teicher, B.A., Hacker, M.P., Hamer, D.H., and Lazo, J.S., Overexpression of metallothionein confers resistance to anticancer drugs, *Science*, 241, 1813, 1988.
31. Rabik, C.A. and Dolan, M.E., Molecular mechanisms of resistance and toxicity associated with platinating agents, *Cancer Treat. Rev.*, 33, 9, 2007.
32. Lubran, M.M., Lead toxicity and heme biosynthesis, *Ann. Clin. Lab. Sci.*, 10, 402, 1980.
33. Finelli, V.N., Klauder, D.S., Karaffa, M.H., and Petering, H.G., Interaction of zinc and lead on δ-aminolevulinate dehydratase, *Biochem. Biophys. Res. Commun.*, 65, 303, 1975.
34. Liu, J., Squibb, K.S., Akkerman, M., Nordberg, G.F., Lipsky, M., and Fowler, B.A., Cytotoxicity, zinc protection, and stress protein induction in rat proximal tubule cells exposed to cadmium chloride in primary cell culture, *Ren. Fail.*, 18, 867, 1996.
35. Day, F.A., Funk, A.E., and Brady, F.O., *In vivo* and *ex vivo* displacement of zinc from metallothionein by cadmium and by mercury, *Chem. Biol. Interact.*, 50, 159, 1984.
36. Blumenthal, S., Lewand, D., Sochanik, A., Krezoski, S., and Petering, D.H., Inhibition of Na(+)-glucose cotransport in kidney cortical cells by cadmium and copper: Protection by zinc, *Toxicol. Appl. Pharmacol.*, 129, 177, 1994.
37. Roesijadi, G., Bogumil, R., Vasák, M., and Kägi, J.H., Modulation of DNA binding of a tramtrack zinc finger peptide by the metallothionein–thionein conjugate pair, *J. Biol. Chem.*, 273, 17425, 1998.
38. Ejnik, J., Muñoz, A., Gan, T., Shaw, C.F. 3rd, and Petering, D.H., Interprotein metal ion exchange between cadmium-carbonic anhydrase and apo- or zinc-metallothionein, *J. Biol. Inorg. Chem.*, 4, 784, 1999.
39. Lay, P.A. and Levina, A., Kinetics and mechanism of chromium(VI) reduction to chromium(III) by l-cysteine in neutral aqueous solutions, *Inorg. Chem.*, 35, 7709, 1996.
40. Krepkiy, D., Antholine, W.E., and Petering, D.H., Properties of the reaction of chromate with metallothionein, *Chem. Res. Toxicol.*, 16, 750, 2003.
41. Zhitkovich, A., Voitkun, V., and Costa, M., Glutathione and free amino acids form stable complexes with DNA following exposure of intact mammalian cells to chromate, *Carcinogenesis*, 16, 907, 1995.
42. Keen, C.L., Hanna, L.A., Lanoue, L., Uriu-Adams, J.Y., Rucker, R.B., and Clegg, M.S., Developmental consequences of trace mineral deficiencies in rodents: Acute and long-term effects, *J. Nutr.*, 133(5 Suppl 1), 1477S, 2003.
43. Oh, S.H., Deagen, J.T., Whanger, P.D., and Weswig, P.H., Biological function of metallothionein. V. Its induction in rats by various stresses, *Am. J. Physiol.*, 234, E282, 1978.
44. Kay, A.R., Neyton, J., and Paoletti, P., A startling role for synaptic zinc, *Neuron*, 52, 572, 2006.

45. Formigari, A., Irato, P., and Santon, A., Zinc, antioxidant systems and metallothionein in metal-mediated apoptosis: Biochemical and cytochemical aspects, *Comp. Biochem. Physiol. C Toxicol. Pharmacol.*, 146, 443, 2007.
46. Petering, D.H., Krezoski, S., and Tabatabai, N., Metallothionein toxicology: Metal ion trafficking and cellular protection in "metallothioneins and related chelators," In: A. Sigel, H. Sigel, and R.K.O. Sigel (eds), *Metal Ions in Life Sciences*, Vol. 5, The Royal Society of Chemistry, Cambridge, UK, 2009.
47. Robbins, A.H., McRee, D.E., Williamson, M., Collett, S.A., Xuong, N.H., Furey, W.F., Wang, B.C., and Stout, C.D., Refined crystal structure of Cd, Zn metallothionein at 2.0 A resolution, *J. Mol. Biol.*, 221, 1269, 1991.
48. Pattanaik, A., Shaw, C.F., 3rd, Petering, D.H., Garvey, J., and Kraker, A.J., Basal metallothionein in tumors: Widespread presence of apoprotein, *J. Inorg. Biochem.*, 54, 91, 1994.
49. Yang, Y., Maret, W., and Vallee, B.L., Differential fluorescence labeling of cysteinyl clusters uncovers high tissue levels of thionein, *Proc. Natl. Acad. Sci. USA*, 98, 5556, 2001.
50. Rana, U., Kothinti, R., Meeusen, J., Tabatabai, N.M., Krezoski, S., and Petering, D.H., Zinc binding ligands and cellular zinc trafficking: Apo-metallothionein, glutathione, TPEN, proteomic zinc, and Zn-Sp1, *J. Inorg. Biochem.*, 102, 489, 2008.
51. Otvos, J.D., Petering, D.H., and Shaw, C.F., Structure–reactivity relationships of metallothionein, a unique metal binding protein, *Comm. Inorg. Chem.*, 9, 1, 1989.
52. Petering, D.H. and Shaw, C.F., 3rd, Stability constants and related equilibrium properties of metallothioneins, *Methods Enzymol.*, 205, 475, 1991.
53. Roschitzki, B. and Vašák, M., Redox labile site in a Zn_4 cluster of Cu_4,Zn_4-metallothionein-3, *Biochemistry*, 42, 9822, 2003.
54. Krezel, A. and Maret, W., Dual nanomolar and picomolar Zn(II) binding properties of metallothionein, *J. Am. Chem. Soc.*, 129, 10911, 2007.
55. Ejnik, J., Robinson, J., Zhu, J., Försterling, H., Shaw, C.F., and Petering, D.H., Folding pathway of apo-metallothionein induced by Zn^{2+}, Cd^{2+} and Co^{2+}, *J. Inorg. Biochem.*, 88, 144, 2002.
56. Brayer, K.J. and Segal, D.J., Keep your fingers off my DNA: Protein–protein interactions mediated by C_2H_2 zinc finger domains, *Cell. Biochem. Biophys.*, 50, 111, 2008.
57. Narayan, V.A., Kriwacki, R.W., and Caradonna, J.P., Structures of zinc finger domains from transcription factor Sp1. Insights into sequence-specific protein–DNA recognition, *J. Biol. Chem.*, 272, 7801, 1997.
58. Petering, D.H., Loftsgaarden, J., Schneider, J., and Fowler, B., Metabolism of cadmium, zinc and copper in the rat kidney: The role of metallothionein and other binding sites, *Environ. Health Perspect.*, 54, 73, 1984.
59. Petering, D.H. and Fowler, B.A., unpublished information.
60. Minkel, D.T., Dolhun, P.J., Calhoun, B.L., Saryan, L.A., and Petering, D.H., Zinc deficiency and growth of Ehrlich ascites tumor, *Cancer Res.*, 39, 2451, 1979.
61. Saryan, L.A., Minkel, D.T., Dolhun, P.J., Calhoun, B.L., Wielgus, S., Schaller, M., and Petering, D.H., Effects of zinc deficiency on cellular processes and morphology in Ehrlich ascites tumor cells, *Cancer Res.*, 39, 2457, 1979.
62. Koch, J., Wielgus, S., Shankara, B., Saryan, L.A., Shaw, C.F., and Petering, D.H., Zinc-, copper- and cadmium-binding protein in Ehrlich ascites tumour cells, *Biochem. J.*, 189, 95, 1989.
63. Kraker, A.J., Krakower, G., Shaw, C.F., 3rd, Petering, D.H., and Garvey, J.S., Zinc metabolism in Ehrlich cells: Properties of a metallothionein-like zinc-binding protein, *Cancer Res.*, 48, 3381, 1988.
64. Krezoski, S.K., Villalobos, J., Shaw, C.F., 3rd, and Petering, D.H., Kinetic lability of zinc bound to metallothionein in Ehrlich cells, *Biochem. J.*, 255, 483, 1988.
65. Petering, D.H., Quesada, A., Dughish, M., Eisch, S., Gan, T., Lemkuil, D., Pattanaik, A., et al., *Metallothionein in Tumor and Host: Intersections of Zinc Metabolism, the Stress Response, and Tumor Therapy in Metallothionein-III*, K. Suzuki (ed.), p. 329, Birkhäuser Verlag, Basel, 1993.
66. Michalska, A.E. and Choo, K.H., Targeting and germ-line transmission of a null mutation at the metallothionein I and II loci in mouse, *Proc. Natl. Acad. Sci. USA*, 90, 8088, 1993.
67. Masters, B.A., Kelly, E.J., Quaife, C.J., Brinster, R.L., and Palmiter, R.D., Targeted disruption of metallothionein I and II genes increases sensitivity to cadmium, *Proc. Natl. Acad. Sci. USA*, 91, 584, 1994.
68. Andrews, G.K. and Geiser, J., Expression of the mouse metallothionein-I and -II genes provides a reproductive advantage during maternal dietary zinc deficiency, *J. Nutr.*, 129, 1643, 1999.
69. Kelly, E.J., Quaife, C.J., Froelick, G.J., and Palmiter, R.D., Metallothionein I and II protect against zinc deficiency and zinc toxicity in mice, *J. Nutr.*, 126, 1782, 1996.

70. Lee, D.K., Geiser, J., Dufner-Beattie, J., and Andrews, G.K., Pancreatic metallothionein-I may play a role in zinc homeostasis during maternal dietary zinc deficiency in mice, *J. Nutr.*, 133, 45, 2003.
71. Haase, H., Mazzatti, D.J., White, A., Ibs, K.H., Engelhardt, G., Hebel, S., Powell, J.R., and Rink, L., Differential gene expression after zinc supplementation and deprivation in human leukocyte subsets, *Mol. Med.*, 13, 362, 2007.
72. Ho, E., Courtemanche, C., and Ames, B.N., Zinc deficiency induces oxidative DNA damage and increases p53 expression in human lung fibroblasts, *J. Nutr.*, 133, 2543, 2003.
73. Chai, F., Truong-Tran, A.Q., Evdokiou, A., Young, G.P., and Zalewski, P.D., Intracellular zinc depletion induces caspase activation and p21 Waf1/Cip1 cleavage in human epithelial cell lines, *J. Infect. Dis.*, 182(Suppl 1), S85, 2000.
74. Fu, D. and Richardson, D.R., Iron chelation and regulation of the cell cycle: 2 mechanisms of posttranscriptional regulation of the universal cyclin-dependent kinase inhibitor p21CIP1/WAF1 by iron depletion, *Blood*, 110, 752, 2007.
75. Nurtjahja-Tjendraputra, E., Fu, D., Phang, .J.M., and Richardson, D/R., Iron chelation regulates cyclin D1 expression via the proteasome: A link to iron deficiency-mediated growth suppression, *Blood*, 109, 4045, 2007.
76. Richardson, D., Ponka, P., and Baker, E., The effect of the iron(III) chelator, desferrioxamine, on iron and transferrin uptake by the human malignant melanoma cell, *Cancer Res.*, 54, 685, 1994.
77. Ankel, E. and Petering, D.H., Iron-chelating agents and the reductive removal of iron from transferrin, *Biochem. Pharmacol.*, 29, 1833, 1980.
78. Aisen, P. and Listowski, I., Iron transport and storage proteins, *Ann. Rev. Biochem.*, 49, 357, 1980.
79. Harris, W.R., Estimation of the ferrous-transferrin binding constants based on thermodynamic studies of nickel(II)-transferrin, *J. Inorg. Biochem.*, 27, 41, 1986.
80. Kretchmar, S.A., Reyes, Z.E., and Raymond, K.N., The spectroelectrochemical determination of the reduction potential of diferric serum transferrin, *Biochim. Biophys. Acta*, 956, 85, 1988.
81. Nyayapati, S., Afshan, G., Lornitzo, F., Byrnes, R.W., and Petering, D.H., Depletion of cellular iron by BPS and ascorbate: Effect on toxicity of adriamycin, *Free Radic. Biol. Med.*, 20, 319, 1996.
82. Borg, D.C. and Schaich, K.M., Prooxidant action of desferrioxamine: Fenton-like production of hydroxyl radicals by reduced ferrioxamine, *J. Free Radic. Biol. Med.*, 2, 237, 1986.
83. Bergamini, S., Rota, C., Staffieri, M., Tomasi, A., and Iannone, A., Prooxidant activity of ferrioxamine in isolated rat hepatocytes and linoleic acid micelles, *Chem. Res. Toxicol.*, 12, 365, 1999.
84. Krishnamurti, C., Saryan, L.A., and Petering, D.H., Effects of ethylenediaminetetraacetic acid and 1,10-phenanthroline on cell proliferation and DNA synthesis of Ehrlich ascites cells, *Cancer Res.*, 40, 4092, 1980.
85. Shomron, N., Malca, H., Vig, I., and Ast, G., Reversible inhibition of the second step of splicing suggests a possible role of zinc in the second step of splicing, *Nucleic Acids Res.*, 30, 4127, 2002.
86. Borchardt, T., Schmidt, C., Camarkis, J., Cappai, R., Masters, C.L., Beyreuther, K., and Multhaup, G., Differential effects of zinc on amyloid precursor protein (APP) processing in copper-resistant variants of cultured Chinese hamster ovary cells, *Cell. Mol. Biol. (Noisy-le-grand)*, 46, 785, 2000.
87. Kiss, Z., The zinc chelator 1,10-phenanthroline enhances the stimulatory effects of protein kinase C activators and staurosporine, but not sphingosine and H_2O_2, on phospholipase D activity in NIH 3T3 fibroblasts, *Biochem. J.*, 298, 93, 1994.
88. Hughes, F.M., Jr. and Cidlowski, J.A., Regulation of apoptosis in S49 cells, *J. Steroid Biochem. Mol. Biol.*, 49, 303, 1994.
89. Gerber, E., Bredy, A., and Kahl, R., Ortho-phenanthroline modulates enzymes of cellular energy metabolism, *Toxicology*, 110, 85, 1996.
90. Petrat, F., Weisheit, D., Lensen, M., de Groot, H., Sustmann, R., and Rauen, U., Selective determination of mitochondrial chelatable iron in viable cells with a new fluorescent sensor, *Biochem. J.*, 362, 137, 2002.
91. Barbouti, A., Doulias, P.T., Zhu, B.Z., Frei, B., and Galaris, D., Intracellular iron, but not copper, plays a critical role in hydrogen peroxide-induced DNA damage, *Free Radic. Biol. Med.*, 31, 490, 2001.
92. Harhaji, L., Vuckovic, O., Miljkovic, D., Stosic-Grujicic, S., and Trajkovic, V., Iron down-regulates macrophage anti-tumour activity by blocking nitric oxide production, *Clin. Exp. Immunol.*, 137, 109, 2004.
93. Petering, H.G., Buskirk, H.H., and Crim, J.A., The effect of dietary mineral supplements of the rat on the antitumor activity of 3-ethoxy-2-oxobutyraldehyde bis(thiosemicarbazone), *Cancer Res.*, 27, 1115, 1967.
94. Radtke, K., Lornitzo, F.A., Byrnes, R.W., Antholine, W.E., and Petering, D.H., Iron requirement for cellular DNA damage and growth inhibition by hydrogen peroxide and bleomycin, *Biochem. J.*, 302, 655, 1994.

95. Blanusa, M., Varnai, V.M., Piasek, M., and Kostial, K., Chelators as antidotes of metal toxicity: Therapeutic and experimental aspects, *Curr. Med. Chem.*, 12, 2771, 2005.
96. Jones, M.M., Cherian, M.G., Singh, P.K., Basinger, M.A., and Jones, S.G., A comparative study of the influence of vicinal dithiols and a dithiocarbamate on the biliary excretion of cadmium in rat, *Toxicol. Appl. Pharmacol.*, 110, 241, 1991.
97. Jones, M.M., Basinger, M.A., Holscher, M.A., and Singh, P.K., Effect of dithiocarbamates and dithiocarbamate-induced cadmium mobilization on essential trace metal metabolism in the female rat, *Fundam. Appl. Toxicol.*, 19, 432, 1992.
98. Ujjani, B., Lyman, S., Winkelmann, D., Antholine, W., Petering, D.H., and Jones, M.M., Enhancement of cytotoxicity of bleomycin by dithiocarbamates: Formation of bis(dithiocarbamato) Cu(II), *J. Inorg. Biochem.*, 38, 81, 1990.
99. Nobel, C.I., Kimland, M., Lind, B., Orrenius, S., and Slater, A.F., Dithiocarbamates induce apoptosis in thymocytes by raising the intracellular level of redox-active copper, *J. Biol. Chem.*, 270, 26202, 1995.
100. Chen, J., Du, C., Kang, J., and Wang, J., Cu^{2+} is required for pyrrolidine dithiocarbamate to inhibit histone acetylation and induce human leukemia cell apoptosis, *Chem. Biol. Interact.*, 171, 26, 2008.
101. Klaassen, C.D., Liu, J., and Choudhuri, S., Metallothionein: An intracellular protein to protect against cadmium toxicity, *Ann. Rev. Pharmacol. Toxicol.*, 39, 267, 1999.
102. Thévenod, F., Nephrotoxicity and the proximal tubule. Insights from cadmium, *Nephron Physiol.*, 93, 87, 2003.
103. Blumenthal, S.S., Lewand, D.L., Buday, M.A., Kleinman, J.G., Krezoski, S.K., and Petering, D.H., Cadmium inhibits glucose uptake in primary cultures of mouse cortical tubule cells, *Am. J. Physiol.*, 258, F1625, 1990.
104. Tabatabai, N.M., Blumenthal, S.S., Lewand, D.L., and Petering, D.H., Differential regulation of mouse kidney sodium-dependent transporters mRNA by cadmium, *Toxicol. Appl. Pharmacol.*, 177, 163, 2001.
105. Tabatabai, N.M., Blumenthal, S.S., Lewand, D.L., and Petering, D.H., Mouse kidney expresses mRNA of four highly related sodium–glucose cotransporters: Regulation by cadmium, *Kidney Int.*, 64, 1320, 2003.
106. Tabatabai, N.M., Blumenthal, S.S., and Petering, D.H., Adverse effect of cadmium on binding of transcription factor Sp1 to the GC-rich regions of the mouse sodium–glucose cotransporter 1, SGLT1, promote, *Toxicology*, 207, 369, 2005.
107. Muñoz, A., Laib, F., Petering, D.H., and Shaw, C.F., 3rd, Characterization of the cadmium complex of peptide 49–61: A putative nucleation center for cadmium-induced folding in rabbit liver metallothionein IIA, *J. Biol. Inorg. Chem.*, 4(4), 495–507, 1999.
108. Margoshes, M. and Vallee, B.L., A cadmium protein from equine kidney cortex, *J. Am. Chem. Soc.*, 79, 4813–4814, 1957.
109. Winge, D.R., Premakumar, R., and Rajagopalan, K.V., Studies on the zinc content of Cd-induced thionein, *Arch. Biochem. Biophys.*, 188, 466, 1978.
110. Ejnik, J., Shaw, C.F., 3rd, and Petering, D.H., Kinetics and mechanism of Zn^{2+} displacement from metallothionein by Cd^{2+}, submitted for publication.
111. Muñoz, A., Försterling, F.H., Shaw, C.F., 3rd, and Petering, D.H., Structure of the (113)Cd(3)beta domains from *Homarus americanus* metallothionein-1: Hydrogen bonding and solvent accessibility of sulfur atoms, *J. Biol. Inorg. Chem.*, 7, 713, 2002.
112. Ejnik, J.W., Muñoz, A., DeRose, E., Shaw, C.F., 3rd, and Petering, D.H., Structural consequences of metallothionein dimerization: Solution structure of the isolated Cd4-alpha-domain and comparison with the holoprotein dimer, *Biochemistry*, 42, 8403, 2003.
113. Petering, D.H. and Shaw, C.F., 3rd., Stability constants and related equilibrium properties of metallothioneins, *Methods Enzymol.*, 205, 475, 1991.
114. Nettesheim, D.G., Engeseth, H.R., and Otvos, J.D., Products of metal exchange reactions of metallothionein, *Biochemistry*, 24, 6744, 1985.
115. Otvos, J.D., Liu, X., Li, H., Shen, G., and Basti, M., Dynamic aspects of metallothionein structure, In: K. Suzuk, T. Imura, and M. Kimura (eds), *Metallothionein III: Biological Roles and Medical Implications*, p. 57, Birkhäuser Verlag, Basal, 1993.
116. Ejnik, J.W., Muñoz, A., DeRose, E., Shaw, C.F., 3rd, and Petering, D.H., Structural consequences of metallothionein dimerization: Solution structure of the isolated Cd_4-alpha-domain and comparison with the holoprotein dimer, *Biochemistry*, 42, 8403, 2003.
117. Huang, M., Krepkiy, D., Hu, W., and Petering, D.H., Zn-, Cd-, and Pb-transcription factor IIIA: Properties, DNA binding, and comparison with TFIIIA-finger 3 metal complexes, *J. Inorg. Biochem.*, 98, 775, 2004.

118. Bittel, D., Dalton, T., Samson, S.L., Gedamu, L., and Andrews, G.K., The DNA binding activity of metal response element-binding transcription factor-1 is activated *in vivo* and *in vitro* by zinc, but not by other transition metals, *J. Biol. Chem.*, 273, 7127, 1998.
119. Krepkiy, D., Försterling, F.H., and Petering, D.H., Interaction of Cd^{2+} with Zn finger 3 of transcription factor IIIA: Structures and binding to cognate DNA, *Chem. Res. Toxicol.*, 17, 863, 2004.
120. Pavletich, N.P. and Pabo, C.O., Zinc finger-DNA recognition: Crystal structure of a Zif268-DNA complex at 2.1 A, *Science*, 252, 809, 1991.
121. Zawia, N.H., Crumpton, T., Brydie, M., Reddy, G.R., and Razmiafshari, M., Disruption of the zinc finger domain: A common target that underlies many of the effects of lead, *Neurotoxicology*, 21, 1069, 2000.
122. Huang, M., Shaw, C.F., 3rd, and Petering, D.H., Interprotein metal exchange between transcription factor IIIa and apo-metallothionein, *J. Inorg. Biochem.*, 98, 639, 2004.
123. Gan, T., Munoz, A., Shaw, C.F., 3rd, and Petering, D.H., Reaction of $^{111}Cd_7$-metallothionein with EDTA. A reappraisal, *J. Biol. Chem.*, 270, 5339, 1995.
124. Truong-Tran, A.Q., Ruffin, R.E., and Zalewski, P.D., Visualization of labile zinc and its role in apoptosis of primary airway epithelial cells and cell lines, *Am. J. Physiol. Lung Cell. Mol. Physiol.*, 279, L1172, 2000.
125. Smith, P.J., Wiltshire, M., Davies, S., Chin, S.F., Campbell, A.K., and Errington, R.J., DNA damage-induced [Zn(2+)](i) transients: Correlation with cell cycle arrest and apoptosis in lymphoma cells, *Am. J. Physiol. Cell Physiol.*, 283, C609, 2002.
126. Budde, T., Minta, A., White, J.A., and Kay, A.R., Imaging free zinc in synaptic terminals in live hippocampal slices, *Neuroscience*, 79, 347, 1997.
127. Frederickson, C.J., Suh, S.W., Silva, D., Frederickson, C.J., and Thompson, R.B., Importance of zinc in the central nervous system: The zinc-containing neuron, *J. Nutr.*, 130, 1471S, 2000.
128. St. Croix, C.M., Wasserloos, K.J., Dineley, K.E., Reynolds, I.J., Levitan, E.S., and Pitt, B.R., Nitric oxide-induced changes in intracellular zinc homeostasis are mediated by metallothionein/thionein, *Am. J. Physiol. Lung Cell. Mol. Physiol.*, 282, L185, 2002.
129. Zalewski, P.D., Forbes, I.J., and Betts, W.H., Correlation of apoptosis with change in intracellular labile Zn(II) using Zinquin [(2-methyl-8-*p*-toluenesulphonamido-6-quinolyloxy)acetic acid], a new specific fluorescent probe for Zn(II), *Biochem. J.*, 296, 403, 1993.
130. Fahrni, C.J. and O'Halloran, T.V., Aqueous coordination chemistry of quinoline-based fluorescence probes for the biological chemistry of zinc, *J. Am. Chem. Soc.*, 121, 11448, 1999.
131. Chang, C.J., Jaworski, J., Nolan, E.M., Sheng, M., and Lippard, S.J., A tautomeric zinc sensor for ratiometric fluorescence imaging: Application to nitric oxide-induced release of intracellular zinc, *Proc. Natl. Acad. Sci. USA*, 101, 1129, 2004.
132. Reynolds, I.J., Fluorescence detection of redox-sensitive metals in neuronal culture: Focus on iron and zinc, *Ann. N. Y. Acad. Sci.*, 1012, 27, 2004.
133. Nolan, E.M., Ryu, J.W., Jaworski, J., Feazell, R.P, Sheng, M., and Lippard, S.J., Zinspy sensors with enhanced dynamic range for imaging neuronal cell zinc uptake and mobilization, *J. Am. Chem. Soc.*, 128, 15517, 2006.
134. Tsien, R.Y., Pozzan, T., and Rink, T.J., Homeostasis in intact lymphocytes: Cytoplasmic free calcium monitored with a new, intracellularly trapped fluorescent indicator, *J. Cell. Biol.*, 94, 325, 1982.
135. O'Connor, N. and Silver, R.B., Ratio imaging: Practical considerations for measuring intracellular Ca^{2+} and pH in living cells, *Methods Cell. Biol.*, 81, 415, 2007.
136. Petering, D., Zinc fluorescent probes, *Chemtracts B Inorg. Chem.*, 17, 569, 2004.
137. Kay, A.R. and Tóth, K., Influence of location of a fluorescent zinc probe in brain slices on its response to synaptic activation, *J. Neurophysiol.*, 95, 1949, 2006.
138. Frederickson, C.J., Perez-Clausell, J., and Danscher, G., Zinc-containing 7S-NGF complex. Evidence from zinc histochemistry for localization in salivary secretory granules, *J. Histochem. Cytochem.*, 35, 579, 1987.
139. Nitzan, Y.B., Sekler, I., and Silverman, W.F., Histochemical and histofluorescence tracing of chelatable zinc in the developing mouse, *J. Histochem. Cytochem.*, 52, 529, 2004.
140. Colvin, R.A., Laskowski, M., and Fontaine, C.P., Zinquin identifies subcellular compartmentalization of zinc in cortical neurons. Relation to the trafficking of zinc and the mitochondrial compartment, *Brain Res.*, 1085, 1, 2006.
141. Hendrickson, K.M., Geue, J.P., Wyness, O., Lincoln, S.F., and Ward, A.D., Coordination and fluorescence of the intracellular Zn^{2+} probe [2-methyl-8-(4-toluenesulfonamido)-6-quinolyloxy]acetic acid (Zinquin A) in ternary Zn^{2+} complexes, *J. Am. Chem. Soc.*, 125, 3889, 2003.

142. Lindskog, S. and Malmstrom, B.G., Metal binding and catalytic activity in bovine carbonic anhydrase, *J. Biol. Chem.*, 237, 1129, 1962.
143. Meeusen, J., Tomasiewicz, H., and Petering, D.H., What are zinc fluorophores imaging? Basal reaction of TSQ with cells, submitted for publication.
144. Hall, D.R., Hadden, J.M., Leonard, G.A., Bailey, S., Neu, M., Winn, M., and Lindley, P.F., The crystal and molecular structures of diferric porcine and rabbit serum transferrins at resolutions of 2.15 and 2.60 A, respectively, *Acta Crystallogr. D Biol. Crystallogr.*, 58, 70, 2002.
145. Aisen, P., Transferrin receptor 1, *Int. J. Biochem. Cell Biol.*, 36, 2137, 2004.
146. Cheng, Y., Zak, O., Aisen, P., Harrison, S.C., and Walz, T., Structure of the human transferrin receptor–transferrin complex, *Cell*, 116, 565, 2004.
147. Dautry-Varsat, A., Ciechanover, A., and Lodish, H.F., pH and the recycling of transferrin during receptor-mediated endocytosis, *Proc. Natl. Acad. Sci. USA*, 80, 2258, 1983.
148. Klausner, R.D., Ashwell, G., van Renswoude, J., Harford, J.B., and Bridges, K.R., Binding of apo-transferrin to K562 cells: Explanation of the transferrin cycle, *Proc. Natl. Acad. Sci. USA*, 80, 2263, 1983.
149. Martin, R.B., Savory, J., Brown, S., Bertholf, R.L., and Wills, M.R., Transferrin binding of Al^{3+} and Fe^{3+}, *Clin. Chem.*, 33, 405, 1987.
150. Dhungana, S., Taboy, C.H., Zak, O., Larvie, M., Crumbliss, A.L., Aisen, P., Redox properties of human transferrin bound to its receptor, *Biochemistry*, 43, 205, 2004.
151. Ciechanover, A., Schwartz, A.L., Dautry-Varsat, A., and Lodish, H.F., Kinetics of internalization and recycling of transferrin and the transferrin receptor in a human hepatoma cell line. Effect of lysosomotropic agents, *J. Biol. Chem.*, 258, 9681, 1983.
152. Núñez, M.T., Gaete, V., Watkins, J.A., and Glass, J., Mobilization of iron from endocytic vesicles. The effects of acidification and reduction, *J. Biol. Chem.*, 265, 6688, 1990.
153. Hémadi, M., Ha-Duong, N.T., and El Hage Chahine, J.M., The mechanism of iron release from the transferrin–receptor 1 adduct, *J. Mol. Biol.*, 358, 1125, 2006.
154. Touret, N., Furuya, W., Forbes, J., Gros, P., and Grinstein, S., Dynamic traffic through the recycling compartment couples the metal transporter Nramp2 (DMT1) with the transferrin receptor, *J. Biol. Chem.*, 278, 25548, 2003.
155. Watkins, J.A., Altazan, J.D., Elder, P., Li, C.Y., Nunez, M.T., Cui, X.X., and Glass, J., Kinetic characterization of reductant dependent processes of iron mobilization from endocytic vesicles, *Biochemistry*, 31, 5820, 1992.
156. Nyayapati, S. and Petering, D.H., Reductive removal of iron from transferrin II., submitted for publication.
157. Chaston, T.B. and Richardson, D.R., Iron chelators for the treatment of iron overload disease: Relationship between structure, redox activity, and toxicity, *Am. J. Hematol.*, 73, 200, 2003.
158. Antholine, W., Knight, J., Whelan, H., and Petering, D.H., Studies of the reaction of 2-formylpyridine thiosemicarbazone and its iron and copper complexes with biological systems, *Mol. Pharmacol.*, 13, 89, 1977.
159. DeConti, R.C., Toftness, B.R., Agrawal, K.C., Tomchick, R., Mead, J.A., Bertino, J.R., Sartorelli, A.C., and Creasey, W.A., Clinical and pharmacological studies with 5-hydroxy-2-formylpyridine thiosemicarbazone, *Cancer Res.*, 32, 1455, 1972.
160. Li, T.Y., Minkel, D.T., Shaw, C.F., 3rd, and Petering, D.H., On the reactivity of metallothioneins with 5,5′-dithiobis-(2-nitrobenzoic acid), *Biochem. J.*, 193, 441, 1981.
161. Maret, W., Oxidative metal release from metallothionein via zinc-thiol/disulfide interchange, *Proc. Natl. Acad. Sci. USA*, 91, 237, 1991.
162. Savas, M.M., Shaw, C.F., 3rd, and Petering, D.H., The oxidation of rabbit liver metallothionein-II by 5,5′-dithiobis(2-nitrobenzoic acid) and glutathione disulfide, *J. Inorg. Biochem.*, 52, 235, 1993.
163. St Croix, C.M., Wasserloos, K.J., Dineley, K.E., Reynolds, I.J., Levitan, E.S., and Pitt, B.R., Nitric oxide-induced changes in intracellular zinc homeostasis are mediated by metallothionein/thionein, *Am. J. Physiol. Lung Cell. Mol. Physiol.*, 282, L185, 2002.
164. Chang, C.J., Jaworski, J., Nolan, E.M., Sheng, M., and Lippard, S.J., A tautomeric zinc sensor for ratiometric fluorescence imaging: Application to nitric oxide-induced release of intracellular zinc, *Proc. Natl. Acad. Sci. USA*, 101, 1129, 2004.
165. Zhu, J. , Meeusen, J., Krezoski, S., and Petering, D.H., Reactivity of Zn-, Cd- and apo-metallothionein with nitric oxide compounds: In vitro and cellular comparison, *Chemical Research in Toxicology*, accepted, 2009.

166. Pearce, L.L., Gandley, R.E., Han, W., Wasserloos, K., Stitt, M., Kanai, A.J., McLaughlin, M.K., Pitt, B.R., and Levitan, E.S., Role of metallothionein in nitric oxide signaling as revealed by a green fluorescent fusion protein, *Proc. Natl. Acad. Sci. USA*, 97, 477, 2000.
167. St Croix, C.M., Stitt, M.S., Leelavanichkul, K., Wasserloos, K.J., Pitt, B.R., and Watkins, S.C., Nitric oxide-induced modification of protein thiolate clusters as determined by spectral fluorescence resonance energy transfer in live endothelial cells, *Free Radic. Biol. Med.*, 37, 785, 2004.

2 Molecular and Cellular Biology of Mercury in the Kidneys

Rudolfs K. Zalups and Christy C. Bridges

CONTENTS

2.1 Introduction 36
2.2 Molecular Transport of Mercury in the Kidneys 38
2.2.1 Renal Disposition of Mercury 38
2.2.2 Proximal Tubular Uptake and Transport of Mercury 39
2.2.2.1 Role of Endocytosis in the Proximal Tubular Absorption of Mercury 39
2.2.2.2 Role of Low-Molecular-Weight Thiols in the Proximal Tubular Absorption of Mercury 39
2.2.3 Mechanisms Participating in the Luminal Uptake of Mercury 40
2.2.3.1 Role of γ-Glutamyltransferase 40
2.2.3.2 GSH *S*-Conjugates of Mercury in the Lumen of the Proximal Tubule 40
2.2.3.3 Role of Cysteinylglycinase 42
2.2.3.4 Cleavage Products of GSH *S*-Conjugates of Mercury as Transportable Forms at the Luminal Plasma Membrane 43
2.2.3.5 Mercuric Conjugates of Cys as a Primary Transportable Form of Mercury at the Luminal Plasma Membrane 43
2.2.3.6 Mercuric Conjugates of Hcy as a Transportable Form of Mercury at the Luminal Plasma Membrane 44
2.2.3.7 Molecular "Mimicry" and Homology 46
2.2.4 Mechanisms of Basolateral Uptake of Mercury 47
2.2.4.1 Role of Organic Anion Transport Systems 47
2.2.4.2 Role of the Dicarboxylate Transporter 48
2.2.4.3 Mercuric Conjugates of GSH as Possible Transportable Forms of Mercury at the Basolateral Membrane 49
2.2.4.4 Mercuric Conjugates of Cys as Transportable Forms of Mercury at the Basolateral Membrane 49
2.2.4.5 Mercuric Conjugates of Hcy and NAC as Transportable Forms of Mercury at the Basolateral Membrane 50
2.2.4.6 Membrane Transporters Implicated in the Basolateral Uptake of Thiol *S*-Conjugates of Mercury in Proximal Tubular Cells 50
2.2.5 Role of MRP Proteins in the Luminal Export of Mercuric Ions 51
2.2.6 Role of the Liver in the Renal Tubular Uptake of Mercury 52
2.2.7 Intracellular Distribution of Mercury in Renal Epithelial Cells 53
2.3 Urinary Excretion of Mercury 53
2.4 Subcellular Molecular Interactions in Renal Epithelial Cells 55
2.4.1 Mercury and Intracellular MT Metabolism 55
2.4.2 Mercury and Intracellular GSH Metabolism 56
2.4.3 Lipid Peroxidation, Oxidative Stress, and Mercury 57
2.4.4 Renal Mitochondrial Function and Mercury 58

2.4.5 Intracellular Distribution of Calcium Ions and Mercury....59
2.4.6 Alterations in ($Na^+ + K^+$)-Stimulated ATPase Induced by Mercury....59
2.4.7 Molecular Interactions between Mercuric Ions and Water Channels....60
2.4.8 Influence of Mercury on Heme Metabolism....60
2.4.9 Expression of Stress Proteins Induced by Mercury....61
2.4.10 Interactions between Mercury and the Cytoskeleton....61
2.5 Mercury and Autoimmunity....62
2.6 Factors Modifying the Renal Handling of Mercury....62
2.6.1 Intracellular Thiols and the Renal Accumulation and Toxicity of Mercury....62
2.6.2 Extracellular Thiols and the Renal Accumulation and Toxicity of Mercury....64
2.6.3 Reductions in Functioning Nephrons and Compensatory Tubular Hypertrophy and the Renal Disposition and Toxicity of Mercury....67
Acknowledgment....68
References....68

2.1 INTRODUCTION

Of the toxic heavy metals, mercury is unique in that it can exist in several physical states and chemical forms. For instance, elemental (metallic) mercury is a liquid at room temperature. Additionally, at room temperature or higher temperatures, elemental mercury is readily converted to mercury vapor at normal atmospheric pressure (760 torr). Mercury can also exist in cationic form, with an oxidation state of 1+ (mercurous) or 2+ (mercuric). However, the most common cationic form of mercury encountered in occupational and environmental settings is the mercuric form, which can exist with a valence of 0, 1+, or 2+. In general, the valence of a mercuric ion depends on whether the ion is covalently bonded to a carbon atom of an organic group, such as a methyl, ethyl, or phenyl group. Monomethylmercury (CH_3Hg^+) is, without question, the most prevalent organic form of mercury encountered in the environment. Methylation of inorganic mercuric ions by microorganisms in soil and water accounts for the preponderance of methylmercury in the environment [1].

Global expansion of industrialization along with corresponding increases in environmental contamination during the twentieth century have led to the increased risk of humans and animals being exposed to the various forms of mercury. Even in present-day society, where environmental issues are of great concern to many, mercury continues to be an environmental toxicant that is disseminated ubiquitously throughout the world, largely due to the burning of fossil fuels and changing atmospheric conditions.

Although mercuric ions have adverse effects in a number of organs and tissues, the kidneys are one of the primary targets where mercuric species are taken up and accumulated, especially following exposure to inorganic forms of mercury (Hg^{2+}). More importantly, the kidneys are particularly at risk of becoming intoxicated following exposure to mercury due to both the burden of mercury taken up by the kidneys and the level of oxygen consumption that occurs along the nephron. Both inorganic and organic forms of mercury are potentially nephrotoxic, with the organic forms of mercury being somewhat less nephrotoxic (on a per mole basis) than inorganic mercurous or mercuric compounds. The corporal distribution of organic mercury is more diffuse than that of inorganic forms. In addition to its adverse effects in the kidneys, organic forms of mercury have significant adverse effects in other tissues, organs, and systems, such as in hematopoietic tissues and the central and peripheral nervous systems [1,2]. It should be stressed that there are significant differences in systemic distribution, and patterns of biological and toxicological effects, between inorganic and organic forms of mercury. These differences are most likely related to mechanisms participating in the transport and metabolism of these species of mercury in target cells [3,4].

Regarding the biological impact of mercuric ions in target tissues and organs, one must take into account the bonding characteristics of these ions. After exposure to mercury vapor or an organic mercuric compound, oxidation of these forms of mercury is promoted in systemic circulation and/or in target cells. Once these forms undergo oxidation, mercuric ions are able to bond to nucleophilic groups on a number of molecules. In fact, mercuric ions have the greatest predilection to bond to reduced sulfur atom(s) on endogenous thiol-containing molecules, such as glutathione (GSH), cysteine (Cys), homocysteine (Hcy), *N*-acetylcysteine (NAC), metallothionein (MT), and albumin. The affinity constant for mercuric ions bonding to reduced sulfur atoms and thiolate anions is on the order of 10^{15}–10^{20}. By contrast, the affinity constants for mercuric ions bonding to oxygen- or nitrogen-containing ligands (such as carbonyl or amino groups) are about 10 orders of magnitude lower. Accordingly, it is reasonable to consider much, if not all, of the biological effects of inorganic and organic forms of mercury in terms of their interactions with molecules possessing one or more sulfhydryl (SH) group(s).

Experimental evidence indicates that in the presence of an excess of a low-molecular-weight thiol-containing molecule in aqueous solution, mercuric ions have a tremendously high propensity toward linear II coordination with two of these molecules. For example, when GSH inorganic and mercuric ions are in a ratio as little as 2:1 in aqueous solution (at room temperature), each mercuric ion forms a linear II coordinate covalent complex with two molecules of GSH by bonding to the sulfur atom on the cysteinyl residue of each of the two GSH molecules. Organic mercurials, such as monomethylmercury, have the propensity to form a linear I coordinate covalent complex with thiol-containing molecules. Despite the thermodynamic stability of the (linear I or II) coordinate covalent bonds formed between mercuric ions and various thiol-containing molecules in aqueous solution, the bonding characteristics between mercuric ions and these thiol-containing molecules appear to be more labile within biological systems [5]. A host of complex factors, such as thiol and/or other nucleophilic competition and exchange, likely serve as cogent arguments for the apparent labile nature of bonding that occurs between mercuric ions and certain thiol-containing molecules in certain tissues and cellular compartments. For example, once mercuric ions enter systemic circulation after exposure, most of the ions in plasma bond to albumin, which is the most abundant protein in plasma and contains a single SH group that can bond readily to a mercuric ion [6–9]. However, these mercuric ions do not remain bonded to these proteins for very long, as evidenced by 40% of the body burden of inorganic mercury being taken up within the kidneys in as little as 1 h after exposure. Current evidence indicates that this uptake is due to transport mechanisms involving low-molecular-weight thiol *S*-conjugates or mercury [4].

As mentioned above, the rapid decrease in plasma mercury occurs concurrently with a rapid rate of uptake of mercuric ions in the kidneys (and liver). Because current evidence indicates that small endogenous thiol *S*-conjugates form with mercuric species and that these complexes are involved in the transport of mercury in the kidneys, one can only conclude that mercuric ions must be transferred from the plasma proteins (mainly albumin) to these low-molecular-weight thiols by one or more ligand-exchange mechanism(s) that have yet to be defined. Additional support for this notion comes from both experimental and clinical studies demonstrating the effectiveness of certain thiol-containing pharmacological agents, such as penicillamine, *N*-acetylpenicillamine, *meso*-2,3-dimercaptosuccinic acid (DMSA), 2,3-dimercapto-1-propanesulfonic acid (DMPS), dithioerythritol, and dithiothreitol, in reversing or protecting against toxic effects of mercury-containing compounds. The therapeutic action of these thiol-containing mercuric chelators is fundamentally premised on, and best explained by, the ability of these agents to compete for inorganic or organic mercuric ions from endogenous ligands via nucleophilic competition and exchange, thereby forming new thiol–mercury complexes.

The dose-dependent toxic effects of inorganic mercury are extremely high in various renal systems, including rats and rabbits treated with mercuric chloride [10–12], renal cortical slices [13], isolated segments of rabbit proximal tubules [14,15], freshly isolated proximal tubular cells from rats [16], and primary cultures of renal cortical cells from rats [17,18]. A threshold effect is generally observed in these systems, in that cellular death occurs rapidly and extensively over a narrow dose- or concentration-range. Above the dose or concentration where renal cellular death begins to be induced, the level of cellular death progresses rapidly. In some systems, an all-or-none response is observed. This does not mean that subtoxic doses of mercury do not have biochemical or physiological effects. A possible explanation for the threshold effect and the subsequent steep dose–response curve is that endogenous protective ligands, such as GSH, which likely behave as intracellular buffers preventing functional changes, are depleted as a result of binding to mercuric ions. Above the dose or concentration of mercuric ions where renal cellular injury is induced, it is likely that the pools of intracellular thiol buffers become depleted, thus permitting mercuric or mercurous ions to bind to critical nucleophilic groups in target cells. Bonding of mercuric ions to critical nucleophilic groups on vital cellular proteins ultimately leads to oxidative events resulting in functional impairment and death in the target renal epithelial cells. Intracellular SH-containing proteins such as MT (especially newly synthesized apothionein) likely also play protective roles by potentially bonding to mercuric ions entering into target renal epithelial cells. Accordingly, designing therapeutic regimens to treat the nephropathy induced by mercuric ions is of great importance. In order to design these therapeutic regimens, it is essential to understand the mechanisms involved in the uptake, intracellular binding, and cellular elimination of mercuric ions in target renal epithelial cells, namely those lining the proximal tubule. In addition to seeking a better understanding of the chemical properties of mercury-containing compounds and the intracellular buffering capacity in cells of target and nontarget organs, other factors must be considered to better define the biochemical and molecular mechanisms of action of mercury-containing compounds in the kidney. Particular attention needs to be paid to the potential role of "molecular mimicry" and the species of mercury involved in the renal (proximal) tubular uptake and transport of mercuric ions.

Susceptibility to the injurious effects of mercury can be modified by a number of intracellular and extracellular factors. These very factors serve as the theoretical basis for most of the currently employed therapeutic strategies. Physiological or pathological alterations in cellular function, particularly in the kidney(s) and liver, also play important roles in modifying susceptibility to mercury-induced renal injury.

2.2 MOLECULAR TRANSPORT OF MERCURY IN THE KIDNEYS

In most (if not all) mammals, the kidneys are principal organs where mercuric ions accumulate, especially following exposure to elemental or inorganic forms of mercury [10,11,19–31]. Experimental evidence indicates that mercuric ions are taken up in the kidneys very rapidly. As much as 50% of a non-nephrotoxic, intravenous dose (0.5 μmol kg^{-1}) of inorganic mercury localizes in the kidneys of rats within a few hours after exposure. Mercuric ions also accumulate in the kidneys after exposure to organic forms of mercury [26,32–39] However, the level of accumulation is much less than that after exposure to inorganic or elemental forms of mercury. For example, only about 10% of an intravenous dose of mercury has been shown to be present in the combined renal mass of rats 24 h after the administration of a non-nephrotoxic (5 mg kg^{-1}) dose of methylmercury [40].

2.2.1 Renal Disposition of Mercury

Previous findings from numerous studies indicate that both inorganic and organic forms of mercury accumulate primarily in the cortex and outer stripe of the outer medulla of the kidneys [10,12,19,22,29,30,41–51]. Histochemical and autoradiographic data from mice and rats [30,51–57] and tubular microdissection data from rats and rabbits [19,29] provide evidence that the majority of the renal burden of mercury can be accounted for by mercuric ions being taken up by, and accumulated in, the epithelial cells lining the three segments (S1, S2, and S3) of the proximal tubule. Deposits of mercury have also been localized along the renal proximal tubule of monkeys exposed to elemental mercury from dental amalgams [58]. Although the segments of the proximal tubule appear to be the primary sites where mercuric ions are taken up and accumulated, one cannot exclude, at present, the possibility that other segments of the nephron and/or collecting duct may take up, accumulate, and transport inorganic and/or organic forms of mercury.

Presumed mercuric deposits have also been demonstrated in the renal proximal tubules of rats and mice treated with organic forms of mercury, such as methylmercury [52,56,57]. Additionally, there are findings demonstrating that a significant fraction of the mercury in the kidneys of animals exposed to methylmercury is in the inorganic form [32,33,40,59,60]. This indicates that organic mercuric ions are oxidized to inorganic mercuric ions prior to and/or after they localize in the proximal tubular epithelial cells. Furthermore, there is evidence indicating that intracellular oxidation of methylmercury occurs [61], although the precise nature of this intracellular conversion of methylmercury remains uncertain.

2.2.2 Proximal Tubular Uptake and Transport of Mercury

2.2.2.1 Role of Endocytosis in the Proximal Tubular Absorption of Mercury

A number of theories about the mechanism(s) by which inorganic and organic forms of mercury gain entry into epithelial cells of the three segments of the renal proximal tubule have been examined over the past several decades. Nearly three decades ago, Madsen [62], and then later Zalups and Barfuss [63], put forth the hypothesis that some mercuric ions gain entry into proximal tubular cells by a process involving endocytosis of filtered mercury–albumin complexes. As stated above, albumin is the most abundant protein in plasma and it has an SH group on a terminal cysteinyl residue. This SH group can serve as a binding site for both inorganic and organic mercuric ions. However, other nucleophilic sites on albumin may also serve as binding sites for mercuric ions. Previous findings show that the largest percentage of mercury in plasma, shortly after intravenous exposure to inorganic mercury, is bound to acid-precipitable proteins, such as albumin [6–8,64]. Since endocytotic processes occur primarily at the luminal plasma membrane of proximal tubular cells, both mercuric ions and albumin would have to be filtered at the glomerulus before they could enter into the lumen of the proximal tubule. In most mammals, there is normally a very low sieving coefficient for albumin at renal glomeruli. Interestingly, though, significant amounts of protein, mainly albumin, are indeed filtered during each day. Thus, the notion of albumin–mercury complexes being filtered at the glomerulus is reasonable. In fact, Madsen [62] showed that when rats were made proteinuric by the proximal tubular toxicant gentamicin, subsequent exposure to inorganic mercury resulted in significant urinary excretion of mercury, where mercuric ions were bonded mainly to excreted molecules of albumin. Assuming that the proteinuria (induced by gentamicin) was not due to increased glomerular permeability, these data would suggest that a significant fraction of inorganic mercury filtered into the proximal tubular lumen is bound to albumin.

Zalups and Barfuss [63] later attempted to implicate an albumin conjugate of inorganic mercury in the luminal uptake of inorganic mercury. In their study, the simultaneous renal disposition of inorganic mercury and albumin was studied after administering mercuric conjugates of albumin containing both [^{125}I]-albumin and [$^{203}Hg^{2+}$]. Unfortunately, the data provided insufficient evidence to implicate endocytosis of a mercuric conjugate of albumin as a primary mechanism in the luminal uptake of mercury. Conversely, there remain insufficient data to exclude endocytosis of a mercuric conjugate of albumin as a minor mechanism in the proximal tubular uptake of mercuric ions.

2.2.2.2 Role of Low-Molecular-Weight Thiols in the Proximal Tubular Absorption of Mercury

Recent studies have provided substantive evidence implicating several membrane transporters and low-molecular-weight thiol *S*-conjugates of mercury in the proximal tubular uptake of mercuric ions. Data from these studies indicate that the proximal tubular uptake of mercuric ions involves two distinct pathways: at the luminal membrane [65–73] and at the basolateral membrane [67,70–72,74–85].

2.2.3 Mechanisms Participating in the Luminal Uptake of Mercury

2.2.3.1 Role of γ-Glutamyltransferase

There is convincing evidence linking the activity of the brush-border enzyme γ-glutamyltransferase in the luminal uptake of inorganic mercury, and to a lesser extent organic forms of mercury. Within the kidneys, γ-glutamyltransferase is localized predominantly, if not exclusively, in the luminal (brush-border) membrane of the epithelial cells lining all three segments of the proximal tubule. The principal, if not sole, role of this extracellular-facing luminal plasma enzyme is to cleave hydrolytically the γ-glutamylcysteine bond on molecules of GSH in the proximal tubular lumen.

An important consideration in the degradation of GSH in the lumen along the length of the proximal tubule is that there is significant axial heterogeneity in the amount of enzyme present in the luminal plasma membrane among the three segments. This heterogeneity appears to correlate with the surface area and surface density of the luminal plasma membrane in the epithelial cells lining each of the three portions of the proximal tubule. The implication of this heterogeneity is that differential levels of degradation of GSH along the proximal tubule have a profound impact on thiol metabolism along the three segments of the proximal tubule, which most likely impacts the manner in which mercuric ions are handled in both the intracellular and extracellular compartments in these proximal tubular segments.

Much of the evidence implicating the activity of γ-glutamyltransferase in the renal tubular uptake of mercury comes from *in vivo* experiments in which γ-glutamyltransferase was inhibited by alkylation with *L*-(α*S*,5*S*)-α-amino-3-chloro-4,5-dihydro-5-isoxazoleacetic acid (acivicin). Data from these experiments show clearly that when the activity of γ-glutamyltransferase is inhibited in the kidneys by pretreatment with acivicin, profound effects occur in the renal disposition of administered mercury. More specifically, pretreatment with acivicin significantly decreases renal tubular uptake and/or accumulation of mercury, which correspondingly results in enhanced urinary excretion of mercury in mice [86,87] and rats [75,88,89] treated with inorganic mercury, in mice given methylmercury [87], or in mice exposed to mercury vapor [90,91]. Enhanced urinary excretion of GSH has also been documented in acivicin-pretreated rats that were subsequently injected with inorganic mercury [88].

Cannon et al. [65,66] provided the most direct evidence to date implicating the role of γ-glutamyltransferase in the luminal uptake of mercury in the proximal tubule. They demonstrated, in isolated perfused S2 segments of the rabbit proximal tubule, that inhibition of γ-glutamyltransferase (by direct application of acivicin to the luminal plasma membrane) causes significant reductions in the luminal uptake (disappearance flux, J_D) and cellular accumulation of mercuric ions when they are in the form of linear II, coordinate covalent, GSH *S*-conjugates.

Collectively, both the *in vivo* and *in vitro* data described above provide strong evidentiary support for the hypothesis that a significant fraction of the mercuric ions taken up by proximal tubular epithelial cells is accomplished by a luminal absorptive mechanism dependent on the actions of γ-glutamyltransferase. Figure 2.1 summarizes mechanisms in the luminal uptake of mercuric ions that involve the activity of γ-glutamyltransferase.

2.2.3.2 GSH S-Conjugates of Mercury in the Lumen of the Proximal Tubule

One implication of the data obtained during *in vivo* inhibition of γ-glutamyltransferase is that some pool of mercuric ions present in the lumen of the proximal tubule exists in the form of a GSH *S*-conjugate of inorganic mercury (GSH-Hg-GSH) just prior to being taken up. Although it remains unclear at present where these mercuric conjugates of GSH are formed prior to arriving in the lumen of the proximal tubule, one must consider the possibility that some of these conjugates are formed outside the kidneys and enter into the lumen of the proximal tubule via glomerular filtration. There are a few reasons to suspect that this may occur. First, the formation of mercuric conjugates of GSH in the plasma (after exposure to mercuric compounds) is theoretically possible because the concentration of this thiol-containing molecule in the plasma of rats has been estimated to be

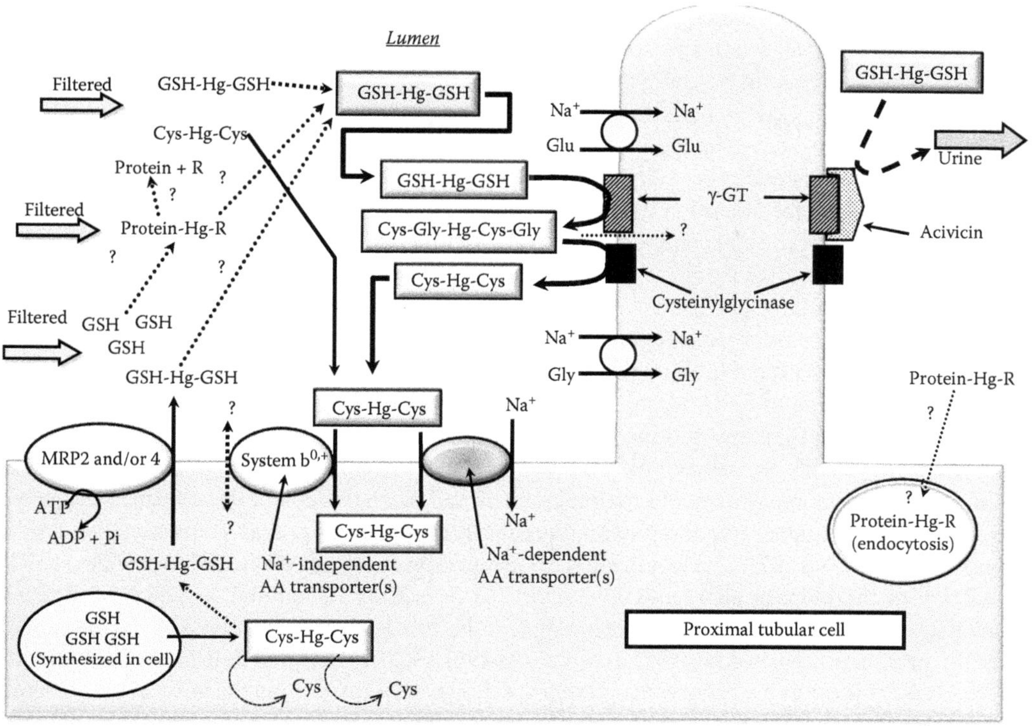

FIGURE 2.1 Mechanisms involved in the luminal uptake of inorganic mercury in renal proximal tubular epithelial cells. The fundamental premise underlying this scheme is the presence of mercuric GSH *S*-conjugates of mercury (GSH-Hg-GSH) in the proximal tubular lumen. What remains unclear at present is how these conjugates gain entry into the luminal compartment. At present, one cannot exclude the possibility that these mercuric complexes enter the tubular lumen by glomerular filtration, tubular secretion, and/or formation in the luminal compartment. The primary mechanism involved in the luminal uptake of mercuric ions along proximal tubular segments involves the participation of the brush-border enzyme γ-glutamyltranspeptidase (also known as glutamyltransferase). When there is near-complete inhibition of the enzyme (with the alkylating agent acivicin), luminal uptake of mercury decreases significantly, while urinary excretion of mercury and GSH increases significantly. γ-Glutamyltranspeptidase is believed to hydrolytically cleave the γ-glutamylcysteine bond on molecules of GSH bonded to mercuric ions (via the sulfur atom of the cysteinyl residue). There is also the possibility that mixed mercuric conjugates may be present in the tubular lumen, but that these have not been included in this figure. According to evidence gathered to date, the resulting cysteinylglycine *S*-conjugate of mercury (CysGly-Hg-CysGly) can potentially enter proximal tubular epithelial cells by two pathways. The most likely pathway involves the additional catalytic cleavage of the cysteinylglycinyl bond on molecules of cysteinylglycine bonded to a mercuric ion by the dehydropeptidase-1, cysteinylglycinase, which is also present in the luminal membrane. The second pathway may involve the transport of a mercuric *S*-conjugate of cysteinylglycine into the proximal tubular cell by one of the sodium-dependent peptide transport systems. However, due to the abundance of dehydropeptidase activity on the luminal membrane of proximal tubular epithelial cells, it seems unlikely that any appreciable amount of transport of a mercuric conjugate of cysteinylglycine would occur under normal circumstances. Subsequent to the actions of cysteinylglycinase, a Cys *S*-conjugate (Cys-Hg-Cys) remains in the lumen. Strong evidence indicates that this resulting mercuric *S*-conjugate of Cys is taken up by proximal tubular epithelial cells via amino transporters in the luminal plasma membrane. Both sodium-dependent and sodium-independent amino transport systems appear to be involved. One cannot exclude, however, the possibility that some type of a mercuric conjugate of GSH is also taken up intact by one of the luminal transport systems. This scheme also shows the potential for endocytosis of mercuric conjugates of albumin, although if this occurs, this mechanism does not likely account for an appreciable fraction of the uptake of mercuric ions. GSH is synthesized in proximal tubular epithelial cells and may be secreted into the tubular lumen of proximal tubular segments, where GSH may compete for the mercuric ions (by a mechanism of thiol competition) carried into the tubular lumen by albumin or other plasma proteins. Mercuric conjugates of Cys formed in blood may also be filtered into the tubular lumen.

approximately 10 μM [92], which provides a pool of GSH to form conjugates with mercuric ions in plasma. Second, the liver is a major source for GSH in the body, and mercuric conjugates of GSH have been shown to form in hepatocytes. Once formed, these conjugates may enter into systemic circulation along with GSH, where they can then be delivered to the kidneys. Third, the size and shape of GSH *S*-conjugates of inorganic mercury are such that they can pass through the glomerular filtration barrier unimpeded.

Another important consideration is that a significant fraction of the pool of GSH *S*-conjugates of mercury present in the luminal compartment of proximal tubules is actually formed in the lumen of the proximal tubule via mechanisms involving some form of thiol competition. Support for this notion comes from isolated perfused tubule data demonstrating that approximately 75% of the GSH synthesized *de novo* in *pars recta* segments of proximal tubules is secreted into the tubular lumen [93,94]. This pool of secreted GSH could theoretically provide a high enough concentration of GSH to permit the formation of GSH *S*-conjugates of mercury in the lumen of the proximal tubule by a mechanism involving thiol competition.

An additional possibility deserving consideration is that mercuric conjugates of GSH are formed within the cytosolic compartment of proximal tubules and that a fraction of these formed conjugates are actually secreted intact into the proximal tubular lumen. There are data from mice that tend to support this hypothesis [87]. The localization of the multidrug resistance associated proteins, MRP2 and MRP4, in the kidneys also tends to support the possibility of luminal secretion of mercuric *S*-conjugates of GSH. These proteins are present in the brush-border membrane of epithelial cells lining the proximal tubule of rats [95] and humans [96]. Both MRP2 and MRP4 are ATP-binding cassette (ABC) transport proteins, which have been shown to be involved in the intracellular to extracellular transport of GSH *S*-conjugates of various metabolites and xenobiotics at the canalicular membrane of hepatocytes [97]. Based on what is currently known about MRP2 and MRP4, it seems reasonable to hypothesize that intracellular mercuric *S*-conjugates of GSH are also transported (in a secretory manner) by these proteins in both hepatocytes and proximal tubular epithelial cells.

In a recent series of studies, we provide strong evidence supporting a role for MRP2 in the proximal tubular secretion of species of both inorganic mercury and methylmercury. In particular, we demonstrated, in MRP2-deficient (TR^-) rats and in inside-out membrane vesicles from sf9 cells transfected with hMRP2, that DMPS *S*-conjugates of inorganic mercury or methylmercury and *S*-conjugates of DMSA are substrates that can be transported from the intracellular environment of proximal tubular epithelial cells into the tubular lumen by MRP2. As referred to above, the luminal plasma membrane of proximal tubular epithelial cells also contains MRP4.

2.2.3.3 Role of Cysteinylglycinase

Inasmuch as cysteinylglycine is a product of GSH degradation by γ-glutamyltransferase, one must consider the possibility that a cysteinylglycine *S*-conjugate may serve as a transportable substrate at the luminal plasma membrane of proximal tubular epithelial cells. In a recent study, Cannon et al. [65,66] examined this possibility in isolated perfused S2 segments of the rabbit proximal tubule. Their findings show that near-complete inhibition of cysteinylglycinase with the dehydropeptidase-1 inhibitor cilastatin significantly reduced the luminal uptake of inorganic mercuric ions when cysteinylglycine *S*-conjugates of inorganic mercury were delivered into the luminal compartment. These findings are apparently the first to support the hypothesis that when inorganic mercuric ions are conjugated to cysteinylglycine, much of the luminal absorption of mercury is also linked to the actions of the dehydropeptidase-1 (cysteinylglycinase), which cleaves the peptide bond between the cysteinyl and glycinyl residues.

Cannon et al. [65,66] discovered, however, that inhibition of luminal dehydropeptidases did not completely prevent the luminal uptake of mercury when it was perfused through the lumen in the form of a mercuric conjugate of cysteinylglycine. These findings indicate that some level of transport of mercuric conjugates of cysteinylglycine may actually occur at the luminal membrane when luminal dehydropeptidases are inhibited. However, before one can make any definitive conclusions about potential transport of mercuric conjugates of cysteinylglycine in the proximal tubule *in vivo*, additional factors need to be considered, such as axial heterogeneity in the handling of GSH, cysteinylglycine and mercuric conjugates of GSH, and cysteinylglycine along the entire length of the proximal tubule. In fact, as referred to above, recent findings from isolated perfused tubule studies demonstrate that there is significant heterogeneity in the synthesis, secretion, and/or transport of GSH along the length of the rabbit proximal tubule [93,94].

2.2.3.4 Cleavage Products of GSH *S*-Conjugates of Mercury as Transportable Forms at the Luminal Plasma Membrane

Since a considerable fraction of the luminal uptake of mercuric ions by proximal tubular cells is linked to the activity of γ-glutamyltransferase and the presence of GSH *S*-conjugates of mercury in the tubular lumen, the actual luminal uptake of mercuric ions likely involves the transport of some cleavage product formed by the action of γ-glutamyltransferase. As mentioned above, one such product might be a cysteinylglycine *S*-conjugate, which could be transported potentially by one of the small-peptide transport systems in the luminal plasma membrane. However, because of the high level of activity of luminal membrane dehydropeptidases (such as cysteinylglycinase), one would predict that little, if any, of such a mercuric species could serve as a transportable substrate *in vivo*. Accordingly, based on the abundance of both γ-glutamyltransferase and cysteinylglycinase in the luminal plasma membrane of proximal tubular epithelial cells, a Cys *S*-conjugate of mercury is the most likely species of mercury taken up at the luminal membrane.

It should be stressed that sequential enzymatic degradation of GSH to cysteinylglycine, and then to Cys, has been demonstrated *in vitro*, while mercuric ions were bonded to the sulfur atom of the cysteinyl residue of GSH molecules that were being degraded enzymatically [98].

2.2.3.5 Mercuric Conjugates of Cys as a Primary Transportable Form of Mercury at the Luminal Plasma Membrane

Numerous sets of recent findings indicate that mercuric conjugates of Cys, such as Cys-Hg-Cys, are likely one of the primary species of mercury transported across the luminal plasma membrane into the cytosolic compartment of proximal tubular cells. For example, there are *in vivo* data showing that the renal uptake and accumulation of inorganic mercury [69,70] and the level of renal tubular injury induced by inorganic mercury [99] were increased in animals when the inorganic mercury was administered as a mercuric conjugate of Cys. In addition, there are *in vitro* data showing that mercuric ions gained entry into renal brush-border membrane vesicles far more readily when they were in the form of mercuric conjugates of Cys than when they were in the form of mercuric conjugates of GSH or even mercuric chloride [71]. Subsequently, Cannon et al. [65,66] provided isolated perfused tubular data showing that Cys *S*-conjugates of inorganic mercury are transportable species of mercury at the luminal plasma membrane of proximal tubular epithelial cells. More specifically, these investigators demonstrated that the rates of luminal uptake (disappearance flux) of mercuric ions in isolated perfused proximal tubular segments were approximately twofold or more greater when Cys *S*-conjugates of inorganic mercury were present in the luminal compartment than when either GSH or cysteinylglycine *S*-conjugates of inorganic mercury were present in the lumen. They also showed that mercuric conjugates of Cys are taken up by at least two separate amino acid transport systems, with at least one of them being sodium dependent and the other being sodium independent [66].

Additional data by Cannon and colleagues [65,66] indicate that one or more of the same transport systems involved in the luminal uptake of the disulfide amino acid cystine may be involved in the luminal uptake of Cys *S*-conjugates of inorganic mercury. Additionally, these investigators demonstrated that addition of an excess of L-lysine to a perfusate containing a low micromolar concentration of newly formed Cys *S*-conjugates of inorganic mercury caused an approximate 50% decrease in the net rate of luminal uptake of inorganic mercury in the S2 segments of the rabbit proximal tubule. To put these findings into context, Schafer and Watkins [100] established previously in

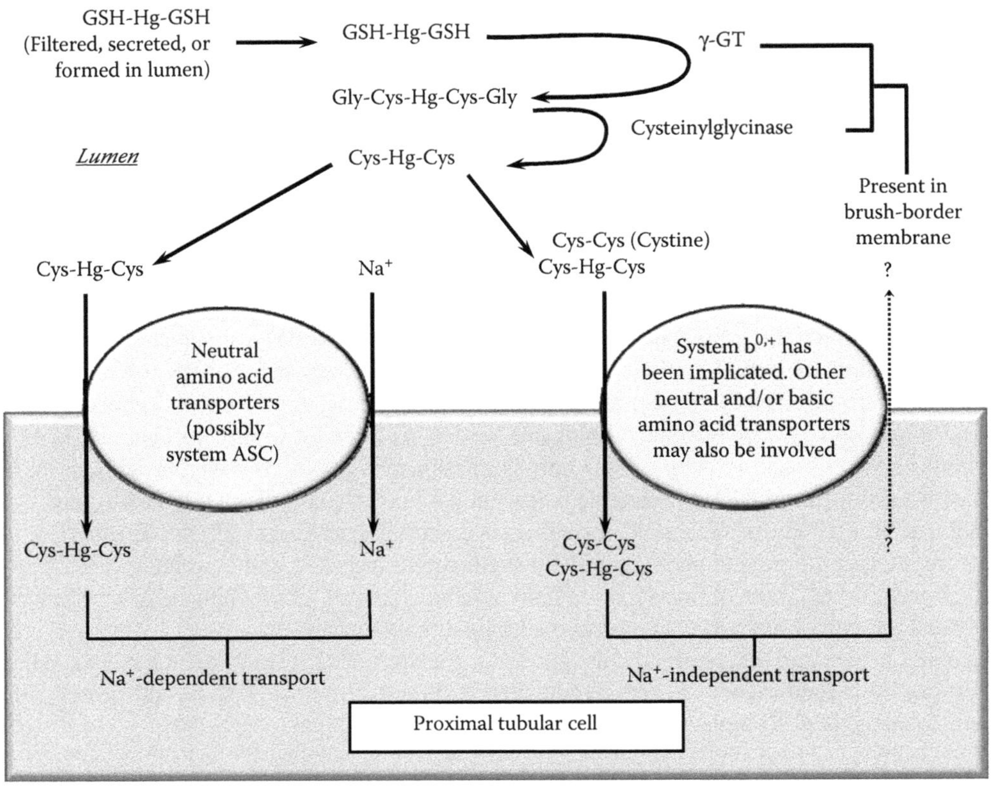

FIGURE 2.2 Additional details on the mechanisms involved in the luminal uptake of the Cys *S*-conjugates of mercury (Cys-Hg-Cys) in proximal tubular segments that are outlined in Figure 2.1. At least two amino acid transporters have been implicated by isolated perfused tubule studies. One of these appears to be a sodium-dependent transporter, and at least one of them is a sodium-independent transporter. A potential candidate for the sodium-dependent transporter involved in the uptake of Cys-Hg-Cys is system ASC. Among the sodium-independent transporters, system $b^{0,+}$ has been implicated in the absorptive transport of Cys-Hg-Cys. A current hypothesis regarding the mechanism by which Cys-Hg-Cys can gain entry into the proximal tubular epithelial cells is by a mechanism involving molecular homology (also referred to as "molecular mimicry"). According to this hypothesis, Cys-Hg-Cys is homologous to the amino acid cystine and enters proximal tubular epithelial cells through one of the primary sodium-independent transporters involved in the absorption of the disulfide amino acid, cystine. Recent evidence from *Xenopus laevis* oocytes microinjected with the cRNA for, and from renal epithelial cells transfected stably with, system $b^{0,+}$ indicates that system $b^{0,+}$ is capable of transporting Cys-Hg-Cys.

isolated perfused S2 segments that L-lysine (3 mM) inhibits the luminal uptake of cystine (300 μM) by approximately 50%. Their findings suggest that some component of the luminal absorption of cystine occurs through a transporter shared by the dibasic amino acid lysine. Overall, the aforementioned findings indicated that a significant fraction of the luminal uptake of Cys *S*-conjugates of inorganic mercury is transported into proximal tubular epithelial cells by one of the same transport systems involved in the luminal absorptive transport of cystine and lysine. Current knowledge of the mechanisms involved in the luminal uptake of mercuric conjugates of Cys by proximal tubules is outlined in Figure 2.2.

2.2.3.6 Mercuric Conjugates of Hcy as a Transportable Form of Mercury at the Luminal Plasma Membrane

The SH-containing amino acid Hcy is an important nonprotein thiol that is present in plasma. In fact, elevations in the plasma concentration of Hcy (known as hyperhomocysteinemia) can occur from abnormalities in the metabolism of methionine (Met) or Hcy, which can have serious deleterious effects.

Normally, Hcy forms from the intracellular metabolism of Met and is broken down by one of two pathways: remethylation or transsulfuration. In the remethylation pathway, a methyl group, from N^5-methyltetrahydrofolate or betaine, is transferred to Hcy to re-form methionine. Alternatively, Hcy can enter the transsulfuration pathway, where it is broken down into Cys and α-ketobutyrate via the sequential actions of cystathionine-β-synthase and γ-cystathionase [101]. Alterations in the metabolism of Hcy or Met leading to hyperhomocysteinemia are important clinically in that chronic elevation in the plasma levels of Hcy can play a contributory role in the induction of cardiovascular disease in some humans [102].

A number of studies have implicated the kidney as a major site of Hcy metabolism [101,103–107]. The enzymes involved in both the remethylation and transsulfuration pathways of Hcy metabolism have been identified in the kidney [108]. However, *in vivo* and *in vitro* studies have demonstrated that the primary route by which Hcy is broken down is via transsulfuration and that this metabolism occurs primarily in proximal tubular epithelial cells [104,105].

Interestingly, proximal tubular cells are also the primary sites where inorganic mercury (Hg^{2+}) accumulates and exerts its toxic effects. Owing to the strong electrophilic properties of Hg^{2+}, mercuric ions are carried around in the plasma as conjugates of thiol-containing biomolecules. Some of the extracellular thiols that have been implicated in binding Hg^{2+} include Cys and GSH. *In vitro* studies from our laboratory have demonstrated that mercuric conjugates of Cys, 2-amino-3-(2-amino-2-carboxy-ethylsulfanylmercuricsulfanyl)-propionic acid (Cys-Hg-Cys), which are structurally similar to the amino acid cystine (Cys-*S*-*S*-Cys), are taken up at the luminal plasma membrane of proximal tubular cells by the amino acid transporter, system $b^{0,+}$ [109]. This heterodimeric amino acid transporter is comprised of a light chain, $b^{0,+}$AT, and a heavy chain, rBAT. In the renal proximal tubular epithelium, it is found on the apical plasma membrane [110] and has been shown to transport cystine as well as a variety of neutral and cationic amino acids [111,112]. Figure 2.3 illustrates how system $b^{0,+}$ is believed to be involved in the luminal uptake of Cys *S*-conjugates of mercury in proximal tubular epithelial cells.

Indirect evidence from *in vivo* studies indicates that when the mercuric conjugate of Hcy, 2-amino-4-(3-amino-3-carboxy-propylsulfanylmercuricsulfanyl)-butyric acid (Hcy-Hg-Hcy), is administered intravenously to rats, some of the administered Hg^{2+} is taken up at both the luminal and basolateral plasma membranes of renal (proximal) tubular epithelial cells as Hcy-Hg-Hcy [70]. Although the mechanism for this uptake has not yet been defined, the participation of luminal amino acid transporters appears to be a logical possibility. Some support for this notion comes in part from recent *in vitro* data showing that the transport of Hcy and homocystine (Hcy-*S*–*S*-Hcy) in the kidneys is mediated by a high-affinity transporter with a substrate specificity similar to that characterized for the cystine transporter, system $b^{0,+}$ [113]. Moreover, recent *in vitro* studies in retinal pigment epithelial (RPE) cells have demonstrated that Hcy is a substrate of system $b^{0,+}$ [114]. As homocystine and cystine are structural homologs, it is logical to postulate that homocystine may be transported by the same transporter (i.e., system $b^{0,+}$) that facilitates the uptake of cystine and the mercuric conjugate Cys-Hg-Cys. Given this and the structural similarity between Cys-Hg-Cys and Hcy-Hg-Hcy, we hypothesize that system $b^{0,+}$ can mediate the uptake of Hcy-Hg-Hcy in renal epithelial cells expressing this transport system.

Bridges and Zalups [115] tested the hypothesis that mercuric conjugates of Hcy, primarily as 2-amino-4-(3-amino-3-carboxy-propylsulfanylmercuricsulfanyl)-butyric acid (Hcy-Hg-Hcy), are taken up by the same system that mediates the uptake of Cys-Hg-Cys, that is, the sodium-independent transporter, system $b^{0,+}$. To test this hypothesis, we analyzed the saturation kinetics, time dependence, temperature dependence, and substrate specificity of the transport of Hcy-Hg-Hcy in Madin-Darby canine kidney (MDCK) II cells stably transfected with both subunits of system $b^{0,+}$ ($b^{0,+}$AT and rBAT). These data show that Hcy-*S*-Hg-*S*-Hcy is taken up at the luminal plasma membrane of renal epithelial cells by the amino acid transporter, system $b^{0,+}$. Furthermore, our data indicate that

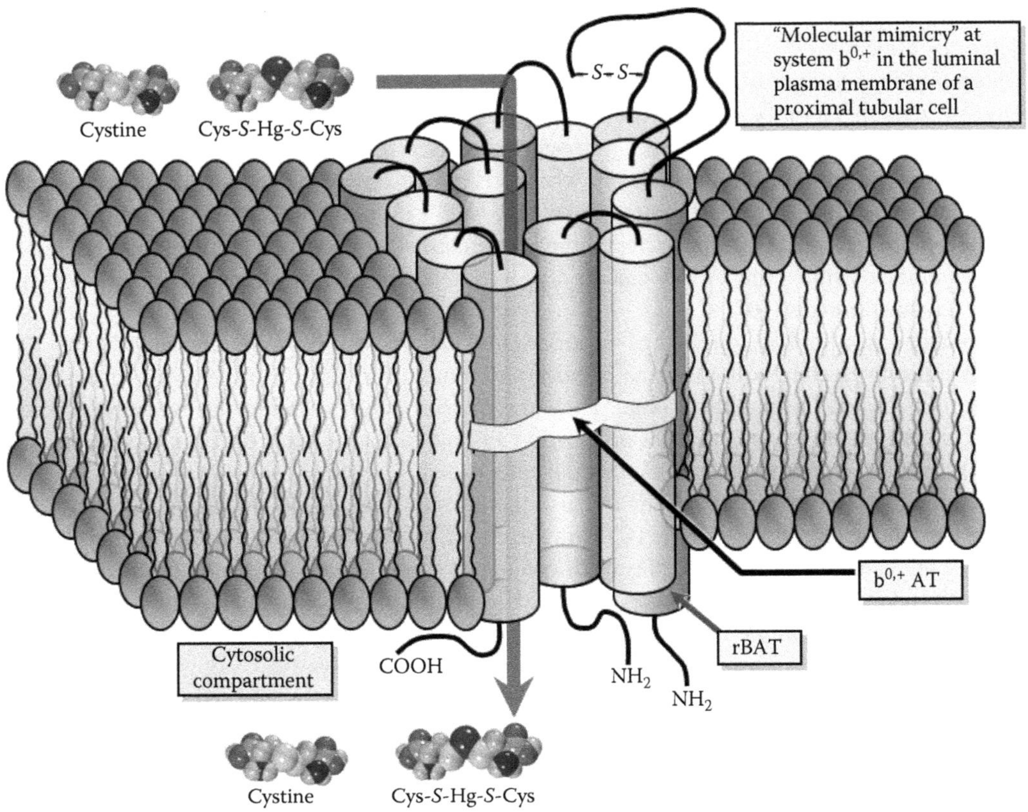

FIGURE 2.3 Model of a portion of the luminal plasma membrane containing the sodium-independent, heterodimeric (consisting of the subunits rBAT and $b^{0,+}$AT), amino acid transporter, system $b^{0,+}$ (consisting of the subunits rBAT and $b^{0,+}$AT). Three-dimensional space-fill renderings of cystine and the complex formed by two molecules of Cys bonded to an inorganic mercuric ion are provided. This figure illustrates the homology between these two molecular species and provides a putative pathway by which Cys-Hg-Cys gains entry into the cytosolic compartment of proximal tubular epithelial cells.

this transporter likely plays a role in the nephropathy induced following exposure to inorganic forms of mercury. Our data are the first to implicate a specific membrane transporter (present in the luminal plasma membrane of proximal tubular cells) in the transport and toxicity of mercuric conjugates of Hcy.

2.2.3.7 Molecular "Mimicry" and Homology

Based on the structural similarities between cystine and the linear II coordinate covalent complex formed by the bonding of two molecules of Cys to a single mercuric ion, the laboratories of Zalups and Barfuss hypothesized that the absorptive luminal transport of this mercuric complex occurs by a mechanism involving molecular homology (or "mimicry"). They postulated that Cys *S*-conjugates of inorganic mercury act as a molecular homolog or "mimic" of the amino acid cystine at the site of one or more transporter(s) responsible for the luminal uptake of cystine [65,109]. Moreover, Bridges and Zalups [115] recently provided evidence that the Hcy *S*-conjugate of Hg^{2+} (Hcy-Hg-Hcy) displays some form of molecular homology at the binding site of one or more luminal amino acid transporters in proximal tubular epithelial cells.

One of the more likely participating amino acid transporters believed to participate in the absorptive transport of the Cys and Hcy *S*-conjugates of Hg^{2+} is system $b^{0,+}$, which is a sodium-independent transporter involved in the proximal tubular absorption of cystine and lysine. Bridges et al. [115] provided the most definitive evidence implicating system $b^{0,+}$ in the luminal absorption of Cys-Hg-Cys, (*R*)-2-amino-3-((*R*)-2-amino-2carboxy-ethylsulfanylmercuricsulfanyl)-propionic acid. In addition, Bridges and Zalups [115] provided similar evidence implicating system $b^{0,+}$ in the luminal absorption of Hcy-Hg-Hcy, 2-amino-4-(3-amino-3-carboxy-propylsulfanylmercuricsulfanyl)-butyric acid.

System $b^{0,+}$ is a heterodimeric amino acid transporter comprised of a light chain, $b^{0,+}$AT, and a heavy chain, rBAT [111,112]. More importantly, this amino acid transporter has a high affinity for cystine and, in the kidneys, is localized exclusively in the luminal plasma membrane of the target epithelial cells, that is, proximal tubular cells [110,116,117]. Based on these characteristics, Bridges et al. [109] and Bridges and Zalups [115] provided evidence supporting the hypothesis that system $b^{0,+}$ is capable of transporting mercuric ions, in the form of Cys-*S*-Hg-*S*-Cys or Hcy-*S*-Hg-*S*-Hcy, into the intracellular compartment of proximal tubular epithelial cells.

The data of Bridges et al. [109] and Bridges and Zalups [115] show that in type II, MDCK cells transfected stably with the human isoform of system $b^{0,+}$, Cys-Hg-Cys and Hcy-Hg-Hcy are transported readily across the luminal plasma membrane. Moreover, substrate specificity data suggest strongly that Cys-Hg-Cys acts as a molecular mimic of cystine. The substrate specificity data of Bridges and Zalups [115] suggest that Hcy-Hg-Hcy acts as a molecular mimic of cystine and/or homocystine. Overall, these data provide the first direct line of molecular evidence implicating the participation of a specific membrane luminal transporter of proximal tubular cells in the transport of thiol *S*-conjugates of Hg^{2+}.

Molecular homology, or what some have referred to as "molecular mimicry," is not a novel concept. In a previous review, Clarkson [118] discusses the concept of mercury, and other divalent metals, forming complexes with biological molecules that can mimic endogenous molecules structurally. For example, the complex formed between methylmercury and Cys has been hypothesized to "mimic" the amino acid methionine, as a means to gain entry into the central nervous system via specific amino acid transporter(s) handling methionine. Evidence supporting this specific hypothesis comes largely from studies on the uptake and/or transport of methylmercury by astrocytes [119–121] and the endothelial cells lining the blood–brain barrier [119,122].

Another potential transportable molecular homolog may occur when inorganic mercury or methylmercury binds to GSH. The complex formed when two molecules of GSH bind to a single mercuric ion may also prove to be a functional molecular homolog of GSH disulfide (GSSG).

2.2.4 Mechanisms of Basolateral Uptake of Mercury

2.2.4.1 Role of Organic Anion Transport Systems

In addition to the large body of evidence indicating that mercuric ions are taken up at the luminal membrane of proximal tubular cells, there is substantial evidence indicating that mercuric ions are also taken up at the basolateral membrane of these cells. Approximately 40% of the dose of inorganic mercury is normally taken up by the total renal mass of rats during the initial hour after the intravenous injection of a nontoxic dose of mercuric chloride [11,31,63,68–70,72,82,83,123,124]. Current evidence indicates that approximately 40–60% of this renal burden of mercury can be attributed to a basolateral mechanism [70–72,75–78,81,82,84,99]. It should be stressed that this applies only to doses of inorganic mercury that are non-nephrotoxic. Under conditions where the dose is increased to levels that induce renal tubular injury, the percentage of the dose found in the kidneys (at various times after exposure) decreases. This is due in part to the necrosis of tubular epithelial cells and the subsequent release and excretion of cytosolic mercury [11,99].

One of the first lines of substantial evidence indicating a basolateral mechanism in the renal tubular uptake of inorganic mercury comes from a study by Zalups and Minor [72]. In this study, the uptake and disposition of administered inorganic mercury were evaluated in rats in which glomerular filtration had been reduced to negligible levels in one or both kidneys, by using pretreatment with mannitol in combination with ureteral ligation [72]. It was demonstrated that induction of "stop-flow" conditions by these pretreatments caused an approximate 40% decrease in the net uptake and accumulation of inorganic mercury during the first hour after the administration of a 0.5 μmol kg^{-1} intravenous dose of mercuric chloride. These findings indicate that a major fraction of the renal tubular uptake of inorganic mercury occurred via a basolateral mechanism. They also demonstrated that pretreatment with *para*-aminohippurate (PAH), which is a specific competitive substrate for the renal organic anion transporter (OAT) [125–130], caused significant reductions in the acute renal tubular uptake and accumulation of inorganic mercury in normal animals and in animals that had one or both ureters ligated. In fact, the combination of ureteral ligation and pretreatment with PAH caused an approximate 85% reduction in the net uptake and accumulation of inorganic mercury during the first hour after injection of mercuric chloride. These findings suggest that the majority of the basolateral uptake of inorganic mercury was being inhibited by PAH, which implicates the OAT as the primary mechanism in the basolateral uptake of inorganic mercury. Data from other recent studies have confirmed that the basolateral uptake of inorganic mercury occurs in the kidneys and that the primary mechanism involved is linked to the activity of the organic anion transport system [67,70–72,75–84].

There are also data implicating the activity of the OAT in the basolateral uptake of organic mercuric compounds. These data show that the renal uptake and/or accumulation [131] and toxicity [132] of methylmercury are reduced significantly in mice pretreated with probenecid, which is another competitive substrate and inhibitor of the OAT in renal proximal tubules. The most convincing data implicating the role of one or more OATs in the basolateral uptake of certain thiol *S*-conjugates of Hg^{2+} and CH_3Hg^+ come from a series of recent studies utilizing MDCK cells or oocytes from *Xenopus laevis* expressing the human or rat isoform(s) of the organic anion transporter 1 (hOAT1) or 3 (hOAT3) [67,74,78–81].

2.2.4.2 Role of the Dicarboxylate Transporter

In an early study, Clarkson and Magos [24] demonstrated that pretreatment with the dicarboxylate maleate caused dose-dependent reductions in the net renal accumulation of inorganic mercury when it was given as a Cys–mercury complex (100 μg Hg kg^{-1}). Unfortunately, it is not clear from this study whether the changes in the renal disposition of mercury were due to the inhibitory effects of maleate on renal cellular metabolism [133–135] or whether they were due to direct effects at the site of a transporter of mercury. Interestingly, they found that fumarate (an isomer of maleate) did not have the same effects as maleate, which suggests isomer specificity.

More recently, Zalups and Barfuss [83] demonstrated that pretreatment with small (4–6 carbon) aliphatic dicarboxylates, such as succinate, glutarate, or adipate (but not malonate), inhibited the renal (basolateral) uptake of intravenously administered inorganic mercury in a dose-dependent manner both in normal rats and in rats that had their ureters ligated. Zalups and Barfuss put forth hypotheses by which dicarboxylates influence the renal tubular uptake, transport, and accumulation of inorganic mercury. Some of the details of these hypotheses are provided below.

Current evidence indicates that the OAT is driven by an organic anion/dicarboxylic acid (dicarboxylate) exchange (reviewed by Pritchard and Miller [136] and Dantzler et al. [137–139]). It appears that intracellular generation of α-ketoglutarate (from normal metabolic processes) contributes to the creation of an intracellular chemical gradient favoring the movement of this dicarboxylate out of the cell. α-Ketoglutarate is transported out of proximal tubular cells at the basolateral membrane by exchanging with organic anions at the site of the organic anion exchanger when the gradient becomes sufficiently high enough. There is evidence indicating that a significant fraction of the α-ketoglutarate (and other dicarboxylic acids) that exits proximal tubular cells at the organic exchanger reenters the cells across the basolateral membrane via a sodium-dicarboxylic acid cotransporter. This cotransport system is driven by the sodium gradient generated by the ($Na^+ + K^+$)-stimulated ATPase. Although it is not exactly clear by which mechanism(s) succinate, glutarate, or adipate inhibits the renal tubular uptake of inorganic mercury, it seems likely that an excess of any of these dicarboxylates in the extracellular compartment creates competition for the sodium-dependent entry of α-ketoglutarate at the site of the dicarboxylic acid cotransporter. Reduction in the basolateral uptake of α-ketoglutarate would likely cause a decrease in the intracellular concentration of this dicarboxylate. This, in turn, would decrease the chemical gradient favoring the movement of α-ketoglutarate out of the proximal tubular epithelial cell in exchange for the uptake of an organic anion from the plasma. The net result would be a decreased rate of uptake of organic anions and presumably mercuric conjugates of Cys and/or GSH that are transported at this site. Since dicarboxylates are themselves organic anions, an excess of these molecules in the extracellular fluid likely also creates direct competition with whatever form of mercury that is putatively transported by the OAT, and thus contributes to a decreased rate of uptake of mercury at the basolateral membrane. There is evidence that both adipate and glutarate, but not succinate or malonate, can compete with α-ketoglutarate at the site of the organic anion transport system [127–130,136].

2.2.4.3 Mercuric Conjugates of GSH as Possible Transportable Forms of Mercury at the Basolateral Membrane

Molecules of GSH have a net negative charge at physiological pH. Because of this charge and its size, GSH has been postulated to be a substrate at the site of the OAT. Support for this comes in part from the studies of Lash and Jones, who demonstrated transport of GSH (as an intact tripeptide) in basolateral membrane vesicles (isolated from the renal cortex of rats) by a mechanism that was sodium dependent and that could be blocked by probenecid [140,141]. They have also demonstrated the basolateral transport of certain organic *S*-conjugates of GSH, such as *S*-(1,2-dichlorovinyl)glutathione, into proximal tubular epithelial cells by a probenecid-sensitive mechanism [142].

Since both GSH and certain *S*-conjugates of GSH appear to be transported across the basolateral membrane of proximal tubular cells by the organic anion transport system, it seems plausible that mercuric *S*-conjugates of GSH may also be transported across the basolateral membrane by this same transport system. There are some findings from a recent study, in which mercuric conjugates of GSH were administered to rats that had undergone bilateral ureteral ligation, that support this contention [77]. The data show that the basolateral uptake of inorganic mercury was greater when it was administered in the form of a mercuric conjugate of GSH than when it was administered as mercuric chloride.

2.2.4.4 Mercuric Conjugates of Cys as Transportable Forms of Mercury at the Basolateral Membrane

Despite the fact that Cys has a net neutral charge at physiological pH, it has become highly relevant to consider that inorganic or organic mercuric conjugates of Cys are transportable species at the site of the OAT. The relevance for this consideration comes from studies in which organic *S*-conjugates of Cys have been shown to be taken up at the basolateral membrane of proximal tubular cells by a mechanism consistent with the activity of the OAT. For example, Lash and Anders [143] demonstrated that organic *S*-conjugates of Cys [such as *S*-(1,2-dichlorovinyl)-L-cysteine] were taken up by isolated proximal tubular epithelial cells from rats by a sodium-dependent, probenecid- and PAH-sensitive transport system. More recently, Dantzler et al. [144], using isolated proximal tubules from rabbits, also demonstrated that certain organic *S*-conjugates of Cys were taken up at the basolateral membrane by a probenecid- and PAH-sensitive transport mechanism.

Based on these findings, it seems logical to believe that mercuric conjugates of Cys are transported into proximal tubular epithelial cells at the basolateral membrane by the organic anion transport system. Several sets of data tend to support this hypothesis. For example, one set of data shows that the rates of association and transport of inorganic mercury in basolateral membrane vesicles (isolated from the kidneys of rats) tend to be greater when the vesicles are exposed to mercuric conjugates of Cys than when they are exposed to mercuric chloride [71]. Additional support for this hypothesis comes indirectly from an *in vivo* study [76]. First of all, data from this study show that bilateral ureteral ligation caused an approximate one half reduction in the net renal accumulation of mercury in control rats treated with a low 0.5 μmol kg^{-1} dose of mercuric chloride and in rats coadministered this dose of inorganic mercury with a fourfold greater (2.0 μmol kg^{-1}) amount of L-Cys. More importantly, the findings also show that the net renal accumulation of mercury was greater in animals treated with inorganic mercury plus Cys than in animals treated with mercuric chloride, while the relative intrarenal distribution of mercury was similar in both groups of rats. Furthermore, pretreatment with PAH was shown to cause a significant decrease in the renal uptake of mercury in rats that had their ureters ligated and that were administered inorganic mercury plus Cys [77]. The most reasonable explanation for these findings is that by injecting mercuric conjugates of Cys in animals that have had their ureters ligated, more of these conjugates, than are formed normally when inorganic mercury is administered as mercuric chloride, are made available at the site of the OAT (and possibly other basolateral transporters, such as basolateral amino acid transporters) to promote the uptake of mercury.

2.2.4.5 Mercuric Conjugates of Hcy and NAC as Transportable Forms of Mercury at the Basolateral Membrane

Although a considerable body of experimental evidence points to mercuric conjugates of Cys and GSH being highly relevant species of mercury involved in the luminal and basolateral uptake of inorganic mercury along the proximal tubule (after exposure to mercuric chloride), previous *in vivo* studies have provided evidence that other thiols, especially homologs of Cys, such as Hcy and NAC, can significantly influence the manner in which mercuric ions are handled in the kidneys [70,77]. For example, when inorganic mercury was administered with Hcy, a much lower level of uptake of mercury occurred at the luminal membrane relative to that which occurred at the basolateral membrane. Even greater differences in the levels of luminal uptake versus basolateral uptake of mercury were detected when rats were treated with inorganic mercury and NAC. When inorganic mercury was administered with this negatively charged molecule (NAC), virtually the entire renal tubular uptake of mercury occurred at the basolateral membrane, and the majority of this uptake could be inhibited by pretreatment with PAH. In fact, irrespective of how inorganic mercury was administered, the majority of the basolateral uptake of mercury was inhibited by pretreatment with PAH, which implicates the activity of the organic anion transport system in the basolateral uptake of inorganic mercury under all of the experimental conditions studied.

In addition to the high level of basolateral uptake of mercury in the kidneys of the animals treated with inorganic mercury and NAC, the amount of mercury excreted in 24 h was shown to be at least 45–50% greater than that in groups of rats treated with any other nonprotein thiol or mercuric chloride. The overall findings from these rats indicate that the negative charge on NAC likely promotes the rapid transport of mercuric conjugates of NAC into proximal tubular cells at the site of the OAT, while it prevents or impedes the uptake of these mercuric conjugates at the luminal plasma membrane, which promotes the urinary excretion of mercury.

2.2.4.6 Membrane Transporters Implicated in the Basolateral Uptake of Thiol *S*-Conjugates of Mercury in Proximal Tubular Cells

At present, it appears that inorganic mercuric and methylmercuric conjugates of low-molecular-weight thiol ligands are the most likely species of mercury taken up at the basolateral membrane by the organic anion transporters 1 (OAT1) and 3 (OAT3). Current molecular evidence, from MDCK cells transfected with hOAT1 and oocytes from *Xenopus laevis* microinjected with cRNA for hOAT1, rOAT, or rOAT3, indicates that Cys, Hcy, and NAC *S*-conjugates of Hg^{2+} and CH_3Hg^+ are the most likely mercuric substrates transported by OAT1 and/or OAT3 [67,74,78–81]. More specifically, when MDCK cells or *Xenopus laevis* oocytes were manipulated genetically to express OAT1 or OAT3 protein, the cells gained the ability to transport the aforementioned mercuric species. Figure 2.4 summarizes the mechanisms involved in the basolateral uptake of mercuric species.

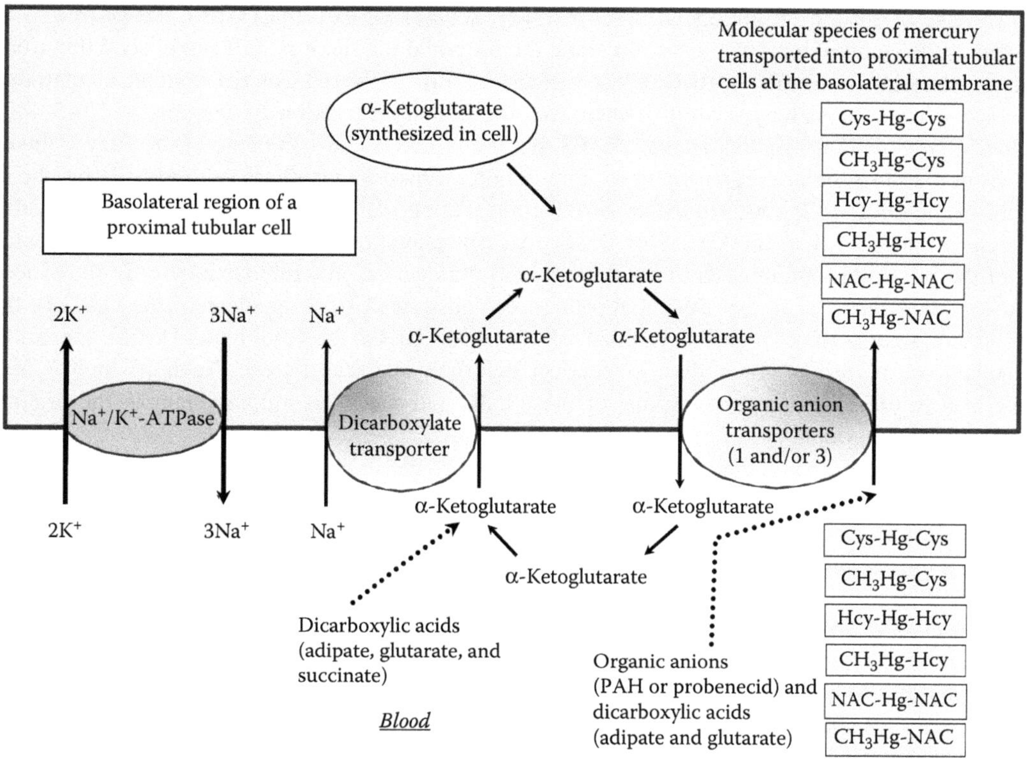

FIGURE 2.4 Diagram outlining the putative roles of both the organic anion and the dicarboxylic acid transport systems in the basolateral uptake of both methylmercury and inorganic mercury along the proximal tubule. Based on the current state of knowledge regarding the organic anion transport system, intracellular generation of α-ketoglutarate (from normal metabolic processes) creates a chemical gradient facilitating the movement of this dicarboxylate out of the cell. When intracellular concentrations of α-ketoglutarate become high enough, α-ketoglutarate exits proximal tubular cells at the basolateral membrane by exchanging with organic anions at one of the organic anion transporters (OATs). After being transported into the extracellular compartment, a significant fraction of α-ketoglutarate molecules is taken back up into proximal tubular cells via a sym-port carrier protein driven by the extracellular to intracellular sodium gradient generated by the Na^+–K^+–ATPase. According to the scheme presented, methylmercury and inorganic mercury enter proximal tubular epithelial cells as Cys, Hcy, and NAC *S*-conjugates via the organic anion transport system in exchange for intracellular α-ketoglutarate. Support for this notion comes from the fact that the basolateral uptake of mercury can be inhibited by PAH, probenecid, and dicarboxylic acids. The scheme presented also shows that succinate, glutarate, and adipate compete with α-ketoglutarate at the site of the dicarboxylic acid transporter. It appears that both glutarate and adipate, but not succinate, can influence the basolateral uptake of mercury by theoretically acting at two sites. Evidence indicates that these two dicarboxylic acids can compete for the uptake of substrates transported by the OAT and other dicarboxylic acids (such as α-ketoglutarate) transported by the sodium-dicarboxylic acid sym-port system. Succinate, on the other hand, appears to influence the basolateral uptake of mercury by competing only at the site of the dicarboxylic acid transporter. Preferential uptake of succinate over α-ketoglutarate would result in a decrease in the intracellular concentration of α-ketoglutarate, which would decrease the driving force behind the activity of the OAT.

2.2.5 Role of MRP Proteins in the Luminal Export of Mercuric Ions

Only recently have investigators been able to examine the question of whether mercuric species are transported out of proximal tubular epithelial cells into the tubular luminal membrane. With the development and progress of molecular technologies, investigators are better equipped to evaluate which chemical forms of mercury serve as potential substrates for export transporters on the luminal membrane.

The use of certain dithiol chelators of mercury, such as DMSA and DMPS, has proved to be useful in the study of mercury secretion along the proximal tubule. It is well established that treatment of humans (and rats) with DMPS or DMSA significantly reduces the content of mercuric species within the kidneys, while it increases the urinary excretion of mercury [85,145–149]. Experimental evidence indicates that DMPS, but not a mercuric *S*-conjugate of mercury, is taken up from the peritubular compartment (or circulation *in vivo*) by proximal tubular cells by one or more of the OATs [85]. Once in the cytosolic compartment, DMPS or DMSA binds to accessible mercuric ions; and then the DMPS or DMSA *S*-conjugate(s) of mercury formed is/are extracted from the proximal tubular cells in a unidirectional manner into the tubular lumen. Both reduced and oxidized forms of native DMPS appear to be substrates of one or more of the OATs in the basolateral membrane of proximal tubular cells [85,150,151]. On the other hand, DMSA appears to be taken up mainly by the sodium-dependent dicarboxylate transporter (NaC2), which is also localized in the basolateral membrane of these cells and serves as a component of the organic anion transport system [152].

Owing to experimental data implicating the multidrug resistance protein 2 (MRP2) in the hepatobiliary transport of mercuric species [153], we have suggested that this carrier protein participates in the secretory transport of DMPS and DMSA *S*-conjugates of Hg^{2+} and CH_3Hg^+ from the intracellular compartment of proximal tubular cells into the tubular lumen [154–156]. MRP2 is a member of the ABC transporter superfamily and is localized in the luminal plasma membrane of proximal tubular cells [95,96,157]. This carrier protein has been hypothesized to play a role in the excretion of metabolic wastes and xenobiotics that have been extracted from blood [4,158]. The activity of MRP2 appears to be dependent on the cotransport of GSH in the secretion of certain compounds. Interestingly, Tanaka-Kagawa et al. [87] suggest that the secretion of Hg^{2+} from proximal tubular cells also appears to require the cotransport of GSH. These data suggest that MRP2 may be responsible for at least part of the proximal tubular elimination of Hg^{2+}.

We have demonstrated recently that the export protein MRP2 participates in the process of secreting DMPS and DMSA *S*-conjugates of Hg^{2+} from the intracellular compartment of proximal tubular cells into the lumen [154,155]. MRP2 has also been implicated in the transport of NAC *S*-conjugates of CH_3Hg^+ [159]. Moreover, recent findings from both MRP2-deficient (TR-) rats and membrane vesicles isolated from Sf9 cells transfected with hMRP2 indicate that MRP2 is also capable of transferring DMPS or DMSA *S*-conjugates of CH_3Hg^+ from the intracellular compartment of proximal tubular segments into the luminal compartment [154].

Although recent data provide evidence for MRP2 being capable of transporting mercuric species into the luminal compartment of proximal tubules, other transporters such as MRP4 may also play an important role in the secretion of mercuric compounds. Further research is clearly needed to examine this possibility.

2.2.6 Role of the Liver in the Renal Tubular Uptake of Mercury

Interestingly, certain aspects of yet unknown hepatic function appear to play roles in the renal uptake and transport of mercuric compounds. For example, Tanaka et al. [86] showed that when the hepatic content of GSH was depleted with 1,2-dichloro-4-nitrobenzene, and mercuric chloride was administered subsequently, there was a significant diminution in the renal uptake and/or accumulation of mercuric species in mice. In addition, we have shown in a series of studies that biliary ligation or cannulation, prior to the administration of a nontoxic dose of mercuric chloride, resulted in a decrease in the renal tubular uptake and accumulation of inorganic mercury in rats [160–163]. Overall, these findings indicate that some aspects of hepatic function are linked to the renal tubular uptake and/or accumulation of inorganic forms of mercury. The hepatic synthesis and secretion of GSH seems to be linked strongly to the renal tubular uptake of mercuric species. Additional studies are clearly needed to better understand the role of the liver in the renal tubular uptake of mercuric species.

2.2.7 Intracellular Distribution of Mercury in Renal Epithelial Cells

Once inorganic forms of mercury gain entry into proximal tubular cells, they distribute apparently throughout all intracellular pools [60,62,88,164,165]. Subcellular fractionation studies using the renal cortical tissue from rats treated acutely or chronically with mercuric chloride indicate that mercury distributes in nuclear, lysosomal, mitochondrial, brush-border and supernatant fractions, with the nuclear fraction containing the greatest amount of mercury among the organelle fractions [62,166]. Similar findings have also been obtained using homogenates of renal cortical tissue from normal and uninephrectomized rats treated with mercuric chloride [62,164–166]. In these studies, however, the cytosolic fraction was found to contain the greatest content of mercury.

Interestingly, the relative specific content of mercury was shown to increase more in the renal lysosomal fraction than in other cellular fractions of the renal tissue of rats when the animals were made proteinuric with an aminoglycoside [62] or when they were treated chronically with mercuric chloride [166]. Increases in the lysosomal content of mercury may reflect the fusion of primary lysosomes with endocytotic or cytosolic vesicles containing complexes of mercuric ions bound to proteins.

2.3 URINARY EXCRETION OF MERCURY

Urinary excretion and fecal excretion of mercury are the principal means by which humans and other mammals eliminate the different forms of mercury from the body. Under most circumstances, a greater fraction of a dose of mercury is excreted in the feces than in the urine early after exposure [3,4,149,167,168]. In rats, more than twice as much inorganic mercury is excreted in the feces than in the urine during the initial days after exposure to a non-nephrotoxic dose of mercuric chloride [4]. Less than 10% of the administered dose is excreted in the urine during this time. In one study, rats injected intravenously with a non-nephrotoxic dose of mercuric chloride had excreted about 20% of the dose in the urine and 30% of the dose in the feces during the initial 54 days after injection [4]. The low level in the urinary excretion of mercury is due to two principal factors: the avid uptake of mercuric ions and the retention of accumulated mercuric ions in proximal tubular segments.

Following exposure to organic forms of mercury, even less mercury is excreted in the urine than after exposure to inorganic forms of mercury. For example, both normal and uninephrectomized rats were shown to excrete only about 3% of the dose of mercury in the urine after the initial 7 days following the intravenous injection of a low dose (5 mg × kg^{-1}) of methylmercury [40]. By contrast, greater than 15% of the administered dose was excreted in the feces during the same period of time. In a study by Smith et al. [169], the cumulative fecal excretion of mercury over 70 days was much greater than the cumulative urinary excretion of mercury in seven adult, male, human subjects who received a tracer amount of ^{203}Hg-labeled methylmercury intravenously. More specifically, about 30% of the dose was excreted in the feces, while only about 4% of the dose was excreted in the urine.

In a couple of early reports [170,171], mercury was stated to have appeared in the urine prior to administered inulin (a plant sugar that is freely filtered, but is neither absorbed nor secreted along the nephron). Clarkson and Magos [24] interpreted these findings to indicate that aspects of the urinary excretion of mercury represented mercury that had been secreted from the blood into the tubular lumen by a transepithelial mechanism. This was a reasonable view, considering that there had been a published report claiming that approximately 99% of the mercury in plasma was not filterable [22]. Based on recent data, however, it appears that much more than 1% of the mercury in plasma is filtered into the proximal tubule lumen [4] and that the mechanisms involved in the urinary excretion of mercury are less clear than once thought. However, a significant body of recent experimental evidence indicates that Hg^{2+} and CH_3Hg^+ can indeed be transported from the peritubular blood to the luminal compartment of proximal tubular segments by mechanisms involving the organic anion transport system on the basolateral plasma membrane and one or more MRP proteins in the luminal plasma membrane [4,67,70,75–84,155–157]. What remains unclear at present is the nature of bonding interactions that occur as inorganic and organic forms of mercury are secreted from the blood into the proximal tubular lumen.

It should be emphasized that although 95–99% (depending on animal species and experimental conditions) of the mercury in plasma is bound to albumin (and other plasma proteins), a significant fraction of albumin is filtered at the glomerulus. Thus, substantial amounts of mercury could theoretically gain access to the luminal compartment of proximal tubules by filtration of a mercury–albumin complex. There is some indirect *in vivo* evidence supporting this notion. Madsen [62] demonstrated, in rats made proteinuric by gentamicin (presumably by decreasing the absorptive capacity of the proximal tubular epithelium by cellular necrosis), that a preponderance of mercury excreted in the urine was associated with albumin. A fundamental assumption in these findings, however, is that the preponderance of albumin associated with mercury in the urine came from glomerular filtration rather than intercellular leak. In contrast to the findings of Madsen [62], Clarkson and Magos [24] found that about 70% of the mercury excreted in urine by rats treated with sodium maleate, subsequent to the exposure of inorganic mercury, was not bound to protein. This finding is actually not that surprising, because much of the mercury excreted in the urine probably originated from cellular stores, and thus was likely bound to low-molecular-weight thiols (such as GSH).

Some insight into the mechanisms involved in the urinary excretion of mercury has been gained by experimental maneuvers that cause the urinary excretion of mercury to increase. In most cases, the increased urinary excretion of mercury is associated with decreased luminal absorption of mercury and/or luminal elimination or extraction of accumulated mercury along the proximal tubule (and/or other segments of the nephron). Some examples of these maneuvers are listed below.

In an early study by Clarkson and Magos [24], pretreatment of female rats with sodium maleate, prior to the injection of a low 100 $\mu g\ kg^{-1}$ dose of mercury in the form of mercuric chloride or a mercury–Cys complex, was shown to cause the urinary excretion of mercury to increase and the renal accumulation of mercury to decrease. Sodium maleate was used because it caused "profound metabolic disturbances in renal cells." They also found that administration of sodium maleate after treatment with mercury caused the renal content of mercury to decrease and the urinary excretion of mercury to increase.

As mentioned above, the urinary excretion of mercury also increases dramatically when renal γ-glutamyltransferase is inhibited prior to the administration of inorganic mercury [75,88,162,163]. Much of the mercury excreted in urine following the inhibition of γ-glutamyltransferase appears to be associated with GSH, which suggests that mercuric conjugates of GSH are present in the proximal tubular lumen [172]. Current evidence indicates that the increased urinary excretion of mercury associated with the inhibition of γ-glutamyltransferase is due mainly to decreased luminal absorption and transport of mercury along the proximal tubule [65,75,86,88,89,91].

When inorganic mercury is applied to the luminal membrane of proximal tubular epithelial cells as a mercuric conjugate of NAC [70,85], DMPS [85], 2,3-dimercaptosuccinic acid (DMSA) [149] or MT [173], urinary excretion of mercury increases greatly due to the lack of luminal uptake of these mercuric conjugates. In general, it appears that when mercuric ions are bound to organic ligands possessing a net negative charge, the mercuric conjugates of these molecules are not taken up readily at the luminal membrane and, in turn, are excreted in the urine. When DMPS is administered after exposure to mercury, the urinary excretion of mercury also increases greatly. Recent evidence (obtained from isolated perfused proximal tubular segments) indicates that the increased urinary excretion of mercury that occurs under these conditions results from unidirectional extraction of mercury from within or on proximal tubular epithelial cells into the tubular lumen. It is likely that the increased urinary excretion of mercury induced by treatment with DMSA occurs by a similar mechanism.

Urinary excretion of mercury has been shown to increase after proximal tubular necrosis is induced by mercury or other agents [24,174,175], This is due largely to mercury being released from, or not being absorbed by, necrotic or degenerating proximal tubular epithelial cells [31,62]. More studies are needed to better define the factors and mechanisms involved in the urinary excretion of mercury and mercury-containing compounds.

Despite all the studies that have been carried out to date, very little is really known about the mechanisms involved in the urinary excretion of inorganic and organic forms of mercury. The major questions that still need to be addressed include the following: (1) What are the magnitudes and rates at which mercury is filtered at the glomerulus? (2) To what extent is filtered mercury taken up by proximal tubular epithelial cells? Alternatively, to what extent is filtered mercury excreted in the urine? (3) What is/are the chemical form(s) of mercury excreted in the urine? (4) Is some of the mercury that is excreted in the urine added to the luminal fluid by a *trans*-epithelial secretory mechanism (as has been suggested previously) [4,24,63,68,70,75,77,176,177]?

2.4 SUBCELLULAR MOLECULAR INTERACTIONS IN RENAL EPITHELIAL CELLS

2.4.1 Mercury and Intracellular MT Metabolism

After gaining entry into the intracellular compartment of target cells, mercuric ions are effective inducers of MT 1 and 2 [178,179]. MTs are a group of small intracellular metal-binding proteins with an approximate molecular weight of 6000–7000 Da. Each MT 1 or 2 molecule contains numerous cysteinyl residues and can bind up to seven atoms of various divalent cations, including ionic forms of mercury, cadmium, zinc, copper, silver, and platinum. Administration of a single, daily, nontoxic dose of mercuric chloride over several days has been shown to cause a near doubling in the concentration of MT in the renal cortex or outer stripe of the outer medulla in rats [41]. Induction of the synthesis of MT 1 and 2 proteins in renal epithelial cells *in vivo* has also been demonstrated in rats exposed to elemental mercury vapor over the course of several days [23]. The nature of induction of MTs by mercury appears to be tissue selective [179].

After exposure to mercury vapor, this form of mercury is eventually converted into inorganic mercury in the body, which accumulates predominantly in the kidneys, suggesting that the induction of MT in the kidneys following exposure to elemental mercury may actually be mediated by inorganic mercury. The induction of MTs in the kidney by inorganic mercury likely involves increased transcription of *MT-1* and *MT-2* genes via interaction of zinc-dependent metal transcription factor(s) [MTF(s)] and *cis*-acting DNA elements termed metal-responsive elements (MREs) resident in the promoter region of the MT genes. Additionally, there is post-transcriptional control of translation of new messenger RNA into MT 1 and MT 2 proteins, especially in renal epithelial cells [4].

New insights into the relationships between the cellular content of mercury and the expression of MT have been recently discovered in both the kidneys and the liver [179]. Evidence indicates that the retention of inorganic mercury by renal tubular epithelial cells is associated with the continual induction of MT. More specifically, the rate of transcription of *MT-1* and *MT-2* genes was as great at 2 weeks after treatment with mercury as it was 1 day after treatment (a single 0.5 μmol kg^{-1} dose of mercuric chloride given to rats). In addition, renal levels of MT 1 and MT 2 proteins remained elevated throughout 2 weeks post-treatment, during which time the renal burden of mercury decreased by only about 26% and the cumulative urinary excretion of mercury was equal to about 24% of the dose of mercury. By contrast, hepatic levels of mercury, MT proteins, and rates of transcription of *MT-1* and *MT-2* genes were shown to decrease continually over the initial 2 weeks after treatment. The rates of transcription of MT genes in the liver correlated highly with the amount of MT proteins in the liver. In the kidneys, however, no correlation between the rates of transcription of MT genes and MT proteins was detected, suggesting that some form of post-transcriptional events is involved in the expression of MT protein in the kidneys subsequent to exposure to inorganic mercury.

2.4.2 Mercury and Intracellular GSH Metabolism

GSH is one of the more important antioxidants found in the intracellular compartment of cells. The molecule is very unique inasmuch as it is the only peptide containing a peptide bond involving the γ-carbon of a glutamyl residue. Another importance of GSH metabolism is that the γ-glutamyl bond can be broken or cleaved only by enzymatic hydrolysis mediated by the membrane enzyme γ-glutamyltransferase. GSH is generally present in millimolar concentrations in both hepatocytes and proximal tubular epithelial cells, but is also present in the extracellular environment at much lower concentrations [92].

In addition to a host of other toxicants, both inorganic and organic forms of mercury can also significantly affect intracellular GSH metabolism in the kidneys. Effects on GSH homeostasis can be observed acutely after single treatments with mercuric compounds, and are concentration dependent. Several sets of *in vivo* and *in vitro* data demonstrate increases in the intracellular contents of GSH in renal tubular epithelial cells after the administration of relatively low toxic or nontoxic doses of either methylmercury [180] or inorganic mercury [12,16,181–183]. At higher doses of inorganic mercury, decreases in the renal content of GSH (which are often substantial) are observed, likely because of induction of apoptotic and/or necrotic events in proximal tubular epithelial cells.

Dose-dependent effects of inorganic mercury on renal GSH metabolism have been demonstrated in male Sprague-Dawley rats that received one of several intravenous nontoxic or nephrotoxic doses of mercuric chloride [12]. At the level of the whole kidney, or in the renal cortex or the outer stripe of the outer medulla, nontoxic (0.5 μmol kg^{-1}) or moderately nephrotoxic (2 μmol kg^{-1}) doses of mercuric chloride apparently induce significant increases in the renal cellular concentration of GSH. This effect is most marked in the outer stripe of the outer medulla, where the concentration of GSH increases by as much as 85%. The toxicological significance of this finding relates to the fact that the outer stripe of the outer medulla is an important zone in which mercuric species are taken up and induce deleterious effects in the terminal portions of the *pars recta* of proximal tubules. At a 3 μmol kg^{-1} nephrotoxic dose of mercuric chloride, the concentrations of GSH in the renal cortex and outer stripe of the outer medulla have been shown to be similar to those in saline-treated control rats [12].

Since the cellular content of GSH is under feedback control, the large increases in renal GSH observed after treatment with inorganic mercury suggest that subtoxic or moderately toxic doses of inorganic mercury induce the synthesis of GSH via the activity of γ-glutamylcysteine synthetase (GCS), which is the rate-limiting enzyme involved in the intracellular synthesis of GSH. The data of Lash and Zalups [184] support this hypothesis. They found that the activity of GCS was increased in renal proximal tubular epithelial cells isolated from rats treated with inorganic mercury, relative to that in proximal tubular cells isolated from control rats. Further support for the hypothesis comes from a study by Woods et al. [180]. They showed that the mRNA for GCS increased (by 4.4-fold) in kidneys of male Fischer 344 rats treated with methylmercury hydroxide for 3 weeks. Thus, both inorganic and organic mercury appear to be capable of inducing the synthesis of GSH via the activity of GCS.

In addition to causing upregulation of GCS, inorganic mercury alters, in a dose-dependent manner, the activity of other GSH-dependent enzymes. The effects of inorganic mercury on these enzymes differ, depending on whether a nontoxic, a moderately toxic, or a highly toxic dose is administered. Nontoxic doses of inorganic mercury apparently cause increases in the activities of GSH disulfide reductase and GSH peroxidase in isolated epithelial cells from both proximal tubular and distal tubular regions of the rat nephron [184]. In contrast, one group of investigators [185] observed marked decreases in the activities of renal GSH disulfide reductase and GSH peroxidase in male rats treated chronically (15 days) with a relatively high dose of mercuric chloride (5 mg $HgCl_2$ day^{-1}, per os). They also found apparent adaptive increases in catalase activity. Similarly, others have found significant decreases in the activity of GSH disulfide reductase after the administration of highly nephrotoxic doses of mercuric chloride (10 μmol $HgCl_2$ kg^{-1}, subcutaneously (sc)

[181]; 15 μmol $HgCl_2$ kg^{-1}, sc [182]; and 4 mg $HgCl_2$ kg^{-1}, sc [186]). While two groups of investigators found small (20–35%), but statistically significant, decreases in the activity of GSH peroxidase [181,186], another group did not detect any change in the activity of this enzyme [182]. It is important to keep in mind, however, that it becomes nearly impossible to interpret *in vivo* data obtained from renal tissue in which there has been extensive cellular injury and death. When there is extensive renal tubular necrosis, decreases in the content of an enzyme or molecule of interest (in samples of renal tissue) can be accounted for simply by the release and excretion of the cytoplasmic contents from dead epithelial cells.

2.4.3 Lipid Peroxidation, Oxidative Stress, and Mercury

Findings from several studies suggest that an important mechanism involved in renal cellular injury induced by either *in vivo* or *in vitro* exposure to inorganic or organic forms of mercury involves oxidative stress. The high affinity of mercuric ions binding to thiols naturally suggests that the ensuing depletion of intracellular thiols (especially GSH) either directly causes or predisposes proximal tubular cells to oxidative stress. Furthermore, other cellular antioxidants, including ascorbic acid and vitamin E, have been reported to be depleted in the kidneys of rats treated with mercuric chloride [182]. The activities of several antioxidant enzymes also appear to be markedly diminished after *in vivo* exposure of rats to nephrotoxic doses of mercuric chloride. For example, it has been reported that administration of mercuric chloride to male Sprague-Dawley rats caused marked decreases in the activities of superoxide dismutase, catalase, GSH peroxidase, and GSH disulfide reductase in the renal cortex [186].

Decreases in the activities of these protective enzymes would be expected to enhance the susceptibility of renal epithelial cells to oxidative injury. There has been some disagreement as to whether mercuric ions themselves cause oxidative injury or whether they merely predispose renal epithelial to other factors that produce oxidative stress. Fukino et al. [182] found that thiobarbiturate reactants, which indicate the occurrence of lipid peroxidation, were markedly increased in renal cortical homogenates obtained from rats 12 h after a subcutaneous injection of a nephrotoxic (15 μmol kg^{-1}) dose of mercuric chloride. Gstraunthaler et al. [186] observed increases in the formation of malondialdehyde in renal cortical homogenates obtained from mercuric chloride-treated rats (relative to homogenates generated from control rats treated only with cumene hydroperoxide). Because the two groups of rats were administered similar doses of mercuric chloride, concentration dependence cannot be invoked to explain the difference in observed responses. Based on these findings, it appears that inorganic mercury can enhance the ability of other agents to induce lipid peroxidation.

There are close relationships between the maintenance of normal renal function, renal cellular content of GSH, and cellular redox status and the generation of ATP in mitochondria. These relationships served as an impetus for Lund et al. [187] to investigate the role of mercury-induced oxidative stress in the mitochondria of renal epithelial cells as a mechanism for mercury-induced renal cellular injury. More specifically, they investigated the effects of inorganic mercury on the production of hydrogen peroxide by renal cortical mitochondria isolated from rats. Depending on the supply and coupling site specificity of respiratory substrates, variable increases in the formation of hydrogen peroxide were observed; incubation of isolated mitochondria with 30 nmol mercuric chloride mg $protein^{-1}$ increased the formation of hydrogen peroxide fourfold at the ubiquinone-cytochrome *b* region and twofold at the NADH dehydrogenase region. Additionally, iron-dependent lipid peroxidation was increased 3.5-fold at the NADH dehydrogenase region and by 25% at the ubiquinone-cytochrome *b* region. Intramitochondrial GSH was decreased in a time- and concentration-dependent manner by mercuric chloride. In fact, at a concentration of 12 nmol mercury mg $protein^{-1}$, the content of GSH in mitochondria was depleted completely within 30 min, suggesting that targeting of mitochondrial GSH by mercury may be responsible for the intramitochondrial oxidative stress. Lund et al. [188] also demonstrated that production of hydrogen peroxide, depletion of GSH, and lipid peroxidation increased in mitochondria (isolated from renal cortical homogenates of rats treated *in vivo* with mercuric chloride) after the addition of an appropriate respiratory substrate. These findings support *in vitro* data and lead one to suggest that mercury-induced oxidative stress within mitochondria is an important mechanism involved in renal tubular injury induced by mercury.

2.4.4 Renal Mitochondrial Function and Mercury

As described above, Lund et al. [187,188] demonstrated that inorganic mercury interferes with mitochondrial respiratory function, causing increased production of hydrogen peroxide in the mitochondria, particularly at coupling site II of the electron transport chain. Their findings indicate that an oxidative stress localized in the mitochondria may be responsible for mercury-induced inhibition of various energy-dependent processes in renal epithelial cells.

In an earlier series of studies, Weinberg et al. [189,190] compared the effects of mercuric chloride on mitochondrial function *in vitro* after either *in vivo* or *in vitro* treatment with mercuric chloride. When mitochondria were isolated from male Sprague-Dawley rats and then treated *in vitro* with inorganic mercury [189,190], a marked uncoupling of respiration (i.e., an increase in state 4 rate of oxygen consumption) and a significant decrease in the rate of substrate-stimulated respiration (i.e., state 3 respiration) were observed. Additionally, uptake of atractyloside-insensitive ADP and the activities of both basal- and Mg^{2+}-activated oligomycin-sensitive ATPase were markedly increased by inorganic mercury. These *in vitro* effects occurred with a threshold concentration of mercuric chloride of 2 nmol mg^{-1} protein. Similarly, when renal cortical mitochondria were isolated from rats treated *in vivo* with mercuric chloride (5 mg kg^{-1}, sc), the most prominent effects detected were inhibition of ADP uptake and decreases in the rates of state 3 and uncoupler-stimulated respiration [189,190]. These effects were not attributed to interaction of mercury with mitochondria during the isolation procedure. However, with both *in vivo* and *in vitro* treatment, inorganic mercury was not readily washed out of mitochondria, suggesting binding between mercuric ions and thiol-containing molecules in the mitochondria.

Chavez and Holguin [191] and Chavez et al. [192] also reported uncoupling of mitochondrial respiration after either *in vivo* or *in vitro* treatment of male Wistar rats with mercuric chloride. Consistent with this finding, they found that inorganic mercury induced calcium efflux from mitochondria, oxidation of pyridine nucleotides, and a collapse of the membrane potential. Chavez and Holguin [191] found that inorganic mercury bonded to mitochondrial protein in a concentration-dependent manner, with saturation at approximately 9 nmol Hg^{2+} mg $protein^{-1}$. The finding supports the notion that a mechanism by which mercury induces mitochondrial injury is by the formation of complexes between mercuric ions and mitochondrial SH groups.

Jung et al. [193] employed ATP depletion by different chemical agents in microdissected nephron segments to localize the nephron site specificity of injury. They found that 1 μM mercuric chloride produced a significant depletion of intracellular ATP exclusively in S2 segments; nephron segments derived from the other regions of the proximal tubule (i.e., S1 or S3) or distal nephron (such as the distal convoluted tubule or the medullary thick ascending limb of the loop of Henle) were not as sensitive to ATP depletion after incubation with inorganic mercury. This pattern agrees with histopathological data, which demonstrate that the *pars recta* of the proximal tubule is the primary target of inorganic mercury, although the S3 segment is also part of the *pars recta* and becomes intoxicated by mercury *in vivo*. These data tend to support the conclusion that renal mitochondria are early intracellular targets of inorganic mercury. This is logical considering the extremely high content of SH-containing proteins in both the mitochondrial matrix and the inner mitochondrial membrane.

Zalups et al. [15] studied the effects of inorganic mercury on mitochondrial function in suspensions of isolated segments of renal proximal tubules from the rabbit. When proximal tubular segments were exposed to mercuric chloride, in the absence of extracellular thiols, marked time- and concentration-dependent inhibition of nystatin-stimulated oxygen consumption was demonstrated in association with mitochondrial toxicity in an intact *in vitro* renal cellular model. Additionally, inhibition of oxygen consumption by mercuric chloride preceded the development of irreversible cellular injury, as assessed by release of LDH from the tubular segments, suggesting that inhibition of cellular energetics is a critical component of the nephrotoxic response to inorganic mercury.

2.4.5 Intracellular Distribution of Calcium Ions and Mercury

Efflux of calcium ions from the renal mitochondria of rats has been demonstrated following *in vivo* and *in vitro* exposure to inorganic mercury [191,192]. The importance of maintaining appropriate intracellular concentrations of calcium for proper cellular function is well documented, suggesting that the prominent effects of mercury on mitochondrial calcium status may play an important role in the acute nephropathy induced by mercury.

Smith et al. [18] used primary cultures of renal tubular cells from rabbits (mostly of proximal tubular origin) as an *in vitro* model system to study the effects of inorganic mercury on the intracellular distribution of ionic calcium. They used the fluorescent dye FURA 2 to quantitate the cytosolic content of free ionized calcium. Treatment of cells with low concentrations (2.5–10 μM) of inorganic mercury produced a 2–10-fold increase in the intracellular content of calcium. In contrast, exposure of cells to higher concentrations (25–100 μM) of inorganic mercury produced an initial, rapid, 10–12-fold increase in intracellular calcium, and then the levels of calcium returned quickly to about twice those in control cells. This was followed subsequently by a second, more gradual increase in the intracellular content of calcium that was dependent on the presence of extracellular calcium. Cytotoxicity was also associated with this phase of increase in intracellular calcium and was similarly dependent on the presence of extracellular calcium. Increases in the cytosolic content of calcium that were independent of extracellular concentrations of calcium were primarily due to the release of intracellular calcium ions from nonmitochondrial intracellular stores, presumably derived from the endoplasmic reticulum. The subsequent decrease in intracellular calcium may be due to buffering processes, such as uptake through microsomal (Ca^{2+} + Mg^{2+})-ATPase or through the mitochondrial uniporter. The dependence of the slow, late phase increase in cytosolic calcium on extracellular calcium associated with higher concentrations of inorganic mercury suggests that nonlethal effects of inorganic mercury in renal cells are associated with redistribution of intracellular stores of calcium. However, the toxic effects of inorganic mercury are associated with changes in the permeability of the plasma membrane.

2.4.6 Alterations in (Na^+ + K^+)-Stimulated ATPase Induced by Mercury

Cellular plasma membranes contain a large number of proteins possessing SH groups that are critical for enzymatic activity and membrane structure [194]. Among these is the well-studied (Na^+ + K^+)-stimulated ATPase, which is situated in the basolateral membrane of epithelial cells in both the proximal and distal regions of the nephron. This membrane enzyme has been shown to be inhibited markedly by alkylation or oxidation of its SH group. Anner and colleagues [195–197] conducted a detailed series of studies on the interaction between mercury-containing compounds and purified and reconstituted ATPase protein from the renal outer medulla of the rat, rabbit, and sheep. To determine molecular details of the interaction between mercury-containing compounds and (Na^+ + K^+)-stimulated ATPase, studies with purified and reconstituted enzymes rather than intact renal epithelial cells or renal tubules had to be conducted.

Anner et al. [197] showed that a number of mercury-containing compounds, including mercuric chloride, mersalyl, and *p*-mercuribenzene-sulfonic acid, potently inhibited the activity of ATPase by binding to a site distinct from that at which the cardiac glycosides (e.g., digoxin and ouabain) bind. The binding of inorganic mercury was concentration dependent and was modulated by the addition of chelators of heavy metal ions, such as EDTA or DMPS, indicating that the binding of inorganic mercury to the enzyme is reversible.

Imesch et al. [196] showed that the inactivation of ($Na^+ + K^+$)-stimulated ATPase by mercuric chloride (0.1–100 μM) apparently loosens the interaction between subunits of the ATPase molecule, thereby altering the sensitivity of the enzyme to extracellular drugs, hormones, and antibodies.

Moreover, Anner and Moosmayer [195] showed that the binding of inorganic mercury to the ($Na^+ + K^+$)-stimulated ATPase molecule occurs primarily at the cytosolic surface. Binding of mercury was closely correlated with inhibition of uptake of ^{86}Rb, indicating that the metal-binding site is critical to the active transport function of the ATPase.

An important extension of these studies will be to design experiments that investigate the effects of mercury on ($Na^+ + K^+$)-stimulated ATPase function in more intact renal systems, such as isolated perfused tubular segments or isolated cells. Considering the potency of the interaction and the fact that the plasma membrane is a very early target site for mercury, it is likely that this interaction will be important in the mechanism of mercury-induced renal cellular injury. It is also likely that SH groups on other membrane proteins, particularly those in the epithelial cells lining the proximal tubule, interact with mercury and may play a role in the nephropathy induced by mercury.

2.4.7 Molecular Interactions between Mercuric Ions and Water Channels

Based on the current body of knowledge, virtually all of the aquaporins (water channels) found in mammals, except for AQP4, have been shown to be sensitive to the action of mercury [198]. AQP1 is present in the proximal tubule, the thin descending limb of the loop of Henle and vasa recta; AQP2, which is the vasopressin-regulated water channel, AQP3, and AQP4 are found in the collecting duct; AQP6 is found in the papilla; and AQP7 is found in the proximal tubule [199]. The binding of mercuric ions to these aquaporins results in the blockade of their function. This blockade is likely one of the mechanisms by which mercuric compounds, including mercurial diuretics, induce polyurea or diuresis. Levy et al. [200] hypothesized that mercurial diuretics functioned by having the mercuric ion in the diuretic molecule bond to critical SH and other nucleophilic groups on the tubular epithelial cells following the cleavage of the mercury–carbon bond in the diuretic molecule. A likely target is one of the cysteinyl residues in one or more of the different types of aquaporin molecules.

It is currently thought that the blockade of water channel function is indeed due to a critical change in the conformation of the protein, which results after the binding of mercury to the SH group of one or more cysteinyl residue(s) in that protein. Recent data collected using site-directed mutagenesis in *Xenopus* oocytes provide evidence that Cys-11 is the mercury-sensitive residue in AQP3 [201] (which is found in the basolateral membrane of the collecting duct) and is involved in the transport of water and small molecules like urea. The effects of mercury on water channel function have been shown to be reversible using chelators such as 2-mercaptoethanol [198]. It is currently unclear, however, whether the molecular interactions that occur between mercuric ions and aquaporins play a mechanistic role in the nephropathy induced by mercury-containing compounds.

2.4.8 Influence of Mercury on Heme Metabolism

Porphyrinuria has been shown to occur following *in vivo* exposure to inorganic mercury [202]. The porphyrinogenic properties of mercury-containing compounds were initially attributed to metal-induced alterations in the regulation of enzymes involved in heme biosynthesis or degradation in target cells. However, since the magnitude of porphyrin excretion during prolonged exposure to either methylmercury or inorganic mercury is greater than can be accounted for by changes in heme metabolism alone, Woods et al. [202,203] invoked alternative biochemical mechanisms to explain their findings. They showed that mercuric ions promoted free radical-mediated oxidation of reduced porphyrins. The mechanism involved depletion or interference of normal antioxidants in renal epithelial cells, such as endogenous thiols like GSH. Furthermore, the ability of inorganic mercury and

GSH to react with endogenously produced reactive oxygen metabolites, from both hepatic and renal mitochondria of rats, was correlated with porphyrinogen oxidation.

An important clinical application of this effect of mercury is illustrated in a study by Bowers et al. [204], who evaluated patterns of urinary excretion of porphyrin in male Fischer 344 rats as a diagnostic tool to assess exposure to inorganic mercury or methylmercury. Evaluation of the urinary excretion of porphyrins is a noninvasive method that can be applied to human populations suspected of being exposed to mercury-containing compounds [205].

2.4.9 Expression of Stress Proteins Induced by Mercury

Various environmental stimuli, including toxic chemicals, increase the synthesis of a class of proteins known as stress proteins. Goering et al. [206] evaluated the effect of a nephrotoxic dose of mercuric chloride (1 mg kg^{-1}) on patterns of protein synthesis in the kidneys of male Sprague-Dawley rats. Enhanced *de novo* synthesis of 70 and 90 kDa molecular mass proteins were detected as early as 2 h after exposure to inorganic mercury, and maximal increases in protein levels were observed at 4–8 h post-treatment. By 16 h postinjection, the rates of synthesis of the stress proteins decreased back toward basal levels. Changes in protein expression also occurred in the liver, but were of smaller magnitudes and were not observed until 16–24 h postinjection.

Goering et al. [206] concluded that alterations in the expression of stress proteins precede overt renal injury and are target organ specific, suggesting that they may serve as biomarkers of renal injury. Furthermore, once the biological functions of these proteins are identified, a more complete understanding of the early effects of mercury can be obtained.

The 65 and 70 kDa heat shock proteins were shown to be associated with renal tubular necrosis in rats treated with mercuric chloride [207]. In control rats, the 65 kDa heat shock proteins were present in the cytoplasm of podocytes and proximal convoluted tubules, while the 70 kDa heat shock proteins were found in the cytoplasm and nuclei of podocytes, cortical convoluted tubules, and collecting ducts. Increased expression of the 65 kDa protein was seen in mitochondria, nuclear chromatin, and nucleoli, and overexpression of the 70 kDa heat shock protein was seen in the cytoplasm, mitochondria, lysosomes, cytoskeleton, nuclear chromatin, and nucleoli in cortical tubular epithelial cells. During the postregenerative phase after treatment with mercuric chloride, the level of expression of the 65 and 70 kDa heat shock proteins was similar to that found in control animals. These findings appear to indicate that induction of 65 and 70 kDa heat shock proteins is a significant component of the nephropathy induced by inorganic mercury.

2.4.10 Interactions between Mercury and the Cytoskeleton

Very little is known about the interactions between inorganic or organic mercuric ions and the cytoskeleton in renal epithelial cells. However, there are data from nonrenal cells indicating that mercury can have a significant effect on the cytoskeleton. Miura et al. [208] demonstrated that inorganic mercury and methylmercury inhibit the *in vitro* polymerization of tubulin. They also demonstrated, in mouse glioma cells, that methylmercury disrupts the microtubular network at an early stage of growth inhibition. Sager and Syversen [209] also demonstrated that disruption of microtubules occurs in the neuroblastoma, glioma, and fibroblast cell-lines when they are exposed to methylmercury. Neuroblastoma cells appear to be particularly sensitive to the microtubular disruption induced by methylmercury. Microtubular damage has also been reported in lymphocytes exposed to methylmercury [210]. It was suggested by Vogel et al. [211] that methylmercury inhibits microtubular assembly by binding to free SH groups on the ends and surface of the microtubules. Addition of the chelator DMSA appears to promote the reassembly of microtubules in cells exposed to methylmercury [209], presumably by removing mercuric ions from critical SH groups.

The potential for the various forms of mercury mediating some form of toxic effect(s) in renal epithelial cells via interactions with cytoskeletal elements remains a possibility. Inasmuch as there are numerous homeostatic functions, in addition to providing structural integrity to cells, which are carried out by various cytoskeletal components, one must consider the potential effects of mercury on the cytoskeleton when evaluating the mechanisms involved in the nephropathy induced by mercury.

2.5 MERCURY AND AUTOIMMUNITY

Evidence from experiments utilizing rabbits [212], inbred Brown-Norway rats [213], and a genetic cross between Brown-Norway [214,215] and Lewis rats indicates that multiple exposures to inorganic forms of mercury can lead to the production of antibodies against the glomerular basement membrane and results in an immunologically mediated membranous glomerular nephritis [215,216]. This glomerular nephropathy is characterized by the binding of antibodies to the glomerular basement membrane, followed by the deposition of immune complexes in glomeruli [212,213,217]. Additional evidence from several strains of both mice and rats exposed repeatedly to inorganic mercury indicates that deposition of immune complexes can occur in the mesangium and glomerular basal lamina, which leads to an immune-complex glomerulonephritis [54,55,215,218,219]. Whether mercuric compounds can induce an autoimmune glomerulonephritis in humans is not clear at present.

It should be pointed out that a majority of the cases of glomerulonephritis (of immunological origin) in humans are classified as idiopathic. Thus, until research proves otherwise, it remains possible that some forms of glomerulonephritis may be linked to exposure to mercury or other environmental and occupational toxicants.

The autoimmunity induced by mercury likely reflects some complex effects of mercuric ions on cell signaling and gene expression events in immune cells, such as in monocytes and lymphocytes. For example, Koropatnick and Zalups [220] recently demonstrated that exposure of human monocytes to low, nontoxic doses of the inorganic mercury causes a rapid suppression of activation signaling events that are normally induced in these cells by lipopolysaccharide or phorbol ester.

2.6 FACTORS MODIFYING THE RENAL HANDLING OF MERCURY

2.6.1 Intracellular Thiols and the Renal Accumulation and Toxicity of Mercury

Two major intracellular thiols, GSH and MT, appear to be important in regulating the renal accumulation of mercury and, ultimately, the susceptibility to mercury-induced renal cellular injury. It is likely that other molecules within cells, including the large supply of non-MT, protein-SHs, play some role in the renal cellular accumulation and toxicity of mercury.

Intracellular concentrations of GSH can be manipulated readily within a relatively brief period of time. Several investigators have employed diethyl maleate to conjugate GSH, thereby lowering the amount of intracellular GSH available to interact with mercuric ions. Johnson [221], Baggett and Berndt [222], Berndt et al. [88], Zalups and Lash [223], and Lash and Zalups et al. [17,162,163] demonstrated that depletion of intracellular GSH or nonprotein thiols is accompanied by decreases in the renal accumulation of inorganic mercury in animals treated with mercuric chloride. In studies by Berndt and colleagues [88,172], depletion of intracellular GSH appeared to increase the severity of renal injury induced by treatment with mercuric chloride. Zalups and Lash [12] also found a close correlation between intrarenal concentrations of GSH and the accumulation of inorganic mercury. There are some conflicting findings on the effects of diethyl maleate from the laboratory of Girardi and Elias [224], who reported increases in the renal accumulation of inorganic mercury in mice treated with this compound. Recently, Zalups and Lash [223] and Zalups et al. [17,162,163] demonstrated, in rats, that acute depletion of GSH in the kidneys and liver by treatment with diethyl maleate caused significant decreases in the renal uptake and accumulation of mercury during the initial hour after the administration of a low nontoxic dose of mercuric chloride. Interestingly, while the renal accumulation of mercury decreased following treatment with diethyl maleate, the net hepatic accumulation of mercury increased. Thus, the depletion of renal and hepatic GSH has mixed effects on the disposition of mercury.

In other experiments, Tanaka-Kagawa et al. [87] lowered the intracellular content of GSH in the kidneys of mice by administering buthionine sulfoximine (BSO) (a potent inhibitor of GCS, which is the rate-limiting enzyme in the intracellular synthesis of GSH) and then inhibited the extracellular degradation of GSH by γ-glutamyltransferase using acivicin. They observed no changes in accumulation of either inorganic mercury or methylmercury when compared with control animals. Zalups and Lash [223] and Zalups et al. [163] also showed, in rats, that acute depletion of renal GSH with BSO does not affect the early aspects of the accumulation of inorganic mercury in the kidneys. By contrast, Zalups et al. [162,163] demonstrated that pretreatment with BSO did cause significant decreases in the net renal content of mercury 24 h after treatment with inorganic mercury [162,163]. These findings indicate that there are significant temporal factors with respect to the effects of BSO on the renal disposition of mercury.

In studies where acivicin was used to inhibit γ-glutamyltransferase, Berndt et al. [88] and Zalups [75] showed in rats, and Tanaka et al. [86] showed in mice, that the urinary excretion of GSH and inorganic mercury increased after inhibition of GSH degradation. Tanaka-Kagawa et al. [87] also found that the urinary excretion of inorganic mercury increased while the renal accumulation of either inorganic mercury or methylmercury decreased.

Tanaka et al. [86] found that when mice were pretreated with 1,2-dichloro-4-nitrobenzene to deplete the hepatic content of GSH (prior to injection of mercuric chloride), there was a marked reduction in the renal accumulation of mercury and a significant decrease in the level of renal cellular injury induced by inorganic mercury. These findings suggest that hepatically synthesized GSH and the activity of γ-glutamyltransferase are involved in the renal uptake of mercury. Additional findings from a set of recent studies, in which bile flow was either diverted or prevented from entering the small intestine of rats, demonstrate that some aspect of hepatic function is linked to a component of the renal uptake and accumulation of mercury [160,161].

Increases in the intracellular contents of GSH and other nonprotein thiols can be achieved by several means. Girardi and Elias [224] reported that treatment of mice with NAC caused decreased intracellular accumulation of inorganic mercury in both the kidneys and liver. Inasmuch as hepatic transport of inorganic mercury with GSH has been established in the liver, higher intracellular contents of GSH would be expected to provide increased numbers of ligands for binding to inorganic mercury. The seemingly paradoxical results of Girardi and Elias [224] and the discrepancies described above suggest that the intrarenal disposition of mercury-containing compounds must be regulated by a more complex array of factors than the availability of reduced GSH.

Acute biliary ligation has also been shown to cause significant increases in the renal and hepatic content of GSH in rats [162,163]. Zalups et al. [162] suggested that the observed increased renal concentration of GSH induced by biliary ligation was due to a hepatic mechanism. They believed that as the concentration of GSH in the biliary canaliculi increased (after biliary ligation), the transport of GSH out of the hepatocytes was redirected down a concentration gradient into sinusoidal blood. They also believed that as GSH was continually added to the blood, plasma concentrations of this thiol increased, which provided more GSH to be taken up at the luminal and basolateral membranes of proximal tubular epithelial cells in the kidneys. Interestingly, biliary ligation was shown to cause the net accumulation of mercury in the liver to increase and the net accumulation of mercury in the kidneys to decrease during the initial 24 h after intravenous injection of 0.5 μmol $HgCl_2$ kg^{-1}. What makes these findings interesting is that the renal accumulation of mercury was decreased despite an increased renal cellular content of GSH, which is contrary to what one might expect. It was postulated that the decreased renal accumulation of mercury in animals that had undergone biliary ligation was not due to the content of GSH in the kidney, but rather to the content of GSH in the liver, where the accumulation of mercury had increased. These findings also confirm that some aspects of hepatic function play a role in the renal disposition of mercury.

Additional experiments by Tanaka-Kagawa [87], in which intracellular levels of MT were modulated, may provide some clarification of the contradictory reports on the effects of GSH depletion on the renal accumulation of mercury. These investigators found that induction of renal MT with $Bi(NO_3)_3$ diminished the ability of acivicin to decrease the intrarenal accumulation of either inorganic mercury or methylmercury. They interpreted this as indicating that inorganic mercury or methylmercury that is bound to ligands other than MT in renal cells can be secreted readily into the tubular lumen with intracellular GSH. Other studies [41,42,182,225] have documented that induction of renal MT is associated with increased intrarenal accumulation of mercury and decreased severity of the nephropathy induced by either organic or inorganic mercury. Thus, it appears that there is a complex interplay between protein and nonprotein thiols in the renal disposition of mercury.

2.6.2 Extracellular Thiols and the Renal Accumulation and Toxicity of Mercury

While manipulation of intracellular thiols is sometimes used therapeutically to alter the accumulation of mercury and to modulate the effects of mercury once it enters target sites, administration of thiol-containing compounds can be applied prior to, or simultaneously with, mercury-containing compounds to alter the pharmacokinetics and pharmacodynamics of mercury. Both DMPS and DMSA are becoming two of the more commonly used metal chelators employed as antidotes for mercury poisoning, and their chemical and pharmacological properties have been reviewed by Aposhian [145,147]. Examples of some of their most distinguishing features are that in contrast to the earlier chelator dimercaprol (also known as British Anti-Lewisite or BAL), DMPS and DMSA are fairly nontoxic, very water soluble, not very lipid soluble, and effective if administered orally. The two compounds are quite versatile, being capable of chelating arsenic, lead, cadmium, and mercury. However, they differ in potency and specificity; for example, DMPS is generally more effective of the two in chelating inorganic forms of mercury [149,226]. Additional extracellular thiol reagents that have been used clinically for the removal of methylmercury are D-penicillamine and *N*-acetyl-DL-penicillamine [145]. Some of the reported variability in effectiveness and potency of the various chelators of mercury may be attributed to species differences, routes of administration, and doses of chelators given.

Zalups et al. [167] demonstrated dose-dependent protection with DMPS in rats from the nephropathy induced by inorganic mercury. Their data suggest that the protective effects of DMPS are attributed to decreases in the renal burden of mercury and increases in the urinary excretion of mercury. Furthermore, Maiorino et al. [227] demonstrated a high correlation between effectiveness of DMPS and urinary excretion of both inorganic mercury and DMPS in humans. In a recent study [149], the same dose of DMPS or DMSA, when administered to rats 24 h after the animals had received an intravenous nontoxic dose of mercuric chloride, was shown to reduce the renal burden of mercury significantly during the subsequent 24 h after treatment with the respective chelator. Treatment with DMPS caused a reduction in the renal burden of mercury by more than 80%, while DMSA caused a reduction in the renal burden of mercury by about 50%. These findings indicate that DMPS is more effective (on a per mole basis) in reducing the renal burden of mercury when given after exposure to inorganic mercury. The kinetics involved in the rapid reduction of the renal burden of mercury, after treatment with DMPS or DMSA, appear to indicate that transport of both these chelating agents by epithelial cells along the proximal tubule is involved in the reduction of the renal tubular burden of mercury. It is well established that proximal tubular epithelial cells transport organic anions, such as certain sulfonates and dicarboxylic acids.

In a recent mechanistic study using isolated perfused proximal tubular segments, Zalups et al. [85] provided data indicating that DMPS is taken up rapidly at the basolateral membrane by the PAH-dependent organic anion transport system. Figure 2.5 summarizes these data. The findings also show that once inorganic mercury binds to DMPS, the mercuric conjugates are not taken up readily at either the luminal or basolateral membranes. These particular findings are contrary to the commonly held presumption that mercuric conjugates of DMPS might be transported by the organic anion transport system [149]. Perhaps the most important findings from this study are those

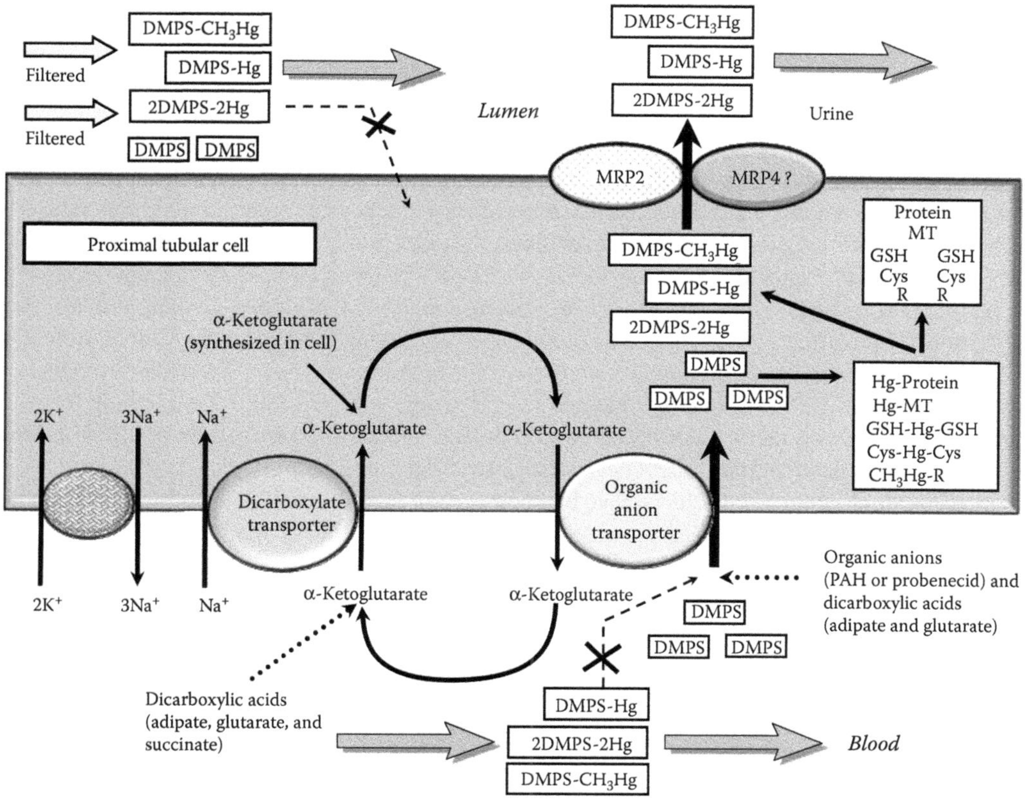

FIGURE 2.5 1,2-Dimercaptopropane-1-sulfonate (DMPS) is a very effective chelating agent for the removal of mercuric ions from the kidneys. The mechanisms by which DMPS reduces the renal tubular burden of mercury have been elucidated only recently. This figure shows the role of the organic anion transporters (OATs) in the basolateral membrane and multiple resistance proteins (MRP2) in the luminal membrane in proximal tubular epithelial cells in the manner by which DMPS extracts mercuric ions from within the cytosolic compartment of proximal tubular epithelial cells into the luminal compartment. Following systemic treatment with DMPS, both DMPS and mercuric conjugates of DMPS (DMPS-Hg, 2DMPS-2Hg, and/or other forms) are present in the blood. These compounds are filtered readily at the glomerulus and delivered to the luminal compartment of proximal tubules. They are also delivered to the basal compartment of proximal tubules via blood flow. Because of the polar negative charge associated with the sulfonate group of DMPS, neither DMPS nor mercuric conjugates of DMPS are absorbed readily at the luminal plasma membrane of proximal tubular cells. Interestingly, despite the fact that reduced and oxidized forms of DMPS are readily transported into proximal tubular cells, isolated perfused tubule data indicate that mercuric conjugates of DMPS are not transportable species of mercury at the basolateral membrane. Current evidence indicates that the therapeutic actions of DMPS involve the following steps: (1) DMPS (in a reduced and/or oxidized state) is taken up avidly at the basolateral membrane by one or more transport systems. (2) Intracellularly, reduced DMPS interacts with, competes for, and then extracts mercuric ions bonded to a host of molecules, including structurally and enzymatic proteins, low-molecular-weight proteins such as the MTs, and small nonprotein thiols such as GSH, Cys, and others (R-Hg-R). (3) Once a mercuric ion becomes bonded to one or two DMPS molecules, and a sufficient intracellular concentration of mercuric conjugates of DMPS has formed to generate a gradient favoring the outward movement of these complexes, DMPS *S*-conjugates are transported out of the proximal tubular epithelial cells by select luminal transport proteins, such as MRP2 and MRP2. (4) Finally, the conjugates are excreted into the urine because they cannot be taken up by any segment of the nephron or collecting duct due to the polar nature of the complexes.

indicating that DMPS can extract accumulated inorganic mercury from proximal tubular cells while it is being transported in a secretory manner from the basolateral to the luminal side of proximal tubular epithelial cells. Figure 2.5 illustrates the mechanisms involved in the renal cellular transport of DMPS and the mechanisms by which DMPS reduces the renal tubular burden of mercury. Additional support for the hypothesis that transport of DMPS and intracellular chelation of mercury occur along segments of the proximal tubule after treatment with DMPS comes from the study by Klotzbach and Diamond [228]. Using isolated perfused kidneys from male Long-Evans rats, they showed that DMPS undergoes net tubular secretion by a kinetically saturable process that is inhibited by PAH and probenecid. They also found that DMPS produced a dose-dependent decrease in the retention of inorganic mercury and an increase in the urinary excretion of inorganic mercury. Furthermore, both effects were blocked by probenecid, suggesting that the mechanism of protection by DMPS is by chelation of inorganic mercury within proximal tubular cells. Many investigators have observed that DMPS is readily oxidized in perfusates or in plasma to the disulfide form. To enable interaction with metals, DMPS is reduced back to the dithiol form within proximal tubular cells by a GSH-dependent thiol-disulfide exchange reaction [228,229].

Other low-molecular-weight thiols have been employed experimentally to modulate the nephrotoxicity of mercury. Because of its prominence as the primary intracellular, nonprotein thiol, exogenous GSH is a logical choice as a modulatory agent. Work by Jones and colleagues [230] has demonstrated that oral administration of GSH can significantly increase the content of GSH in the lung, kidney, heart, brain, small intestine, and skin, but not in the liver, under conditions where GSH is depleted. This suggests that GSH taken orally may supplement cellular GSH in some tissues under certain toxicological or pathological conditions. A large body of data from both *in vivo* and *in vitro* systems indicates that exogenous GSH can protect against mercury-induced renal injury. Zalups et al. [73] perfused isolated rabbit proximal tubules with 18.4 μM mercuric chloride and various thiols, including GSH or Cys. Both thiols, when present in the perfusate at a fourfold higher concentration than inorganic mercury, either prevented or significantly decreased the extent of acute tubular injury induced by unbound mercuric ions. An ultrafiltrate of rabbit plasma was similarly protective. The mechanism of protection by GSH, Cys, or plasma ultrafiltrate appeared to involve decreased uptake of inorganic mercury across the luminal membrane and subsequent accumulation. Houser and Berndt [165] administered GSH monoethyl ester to rats and found that both the renal cortical accumulation of inorganic mercury and the severity of mercury-induced renal injury were diminished.

The protective effects of exogenous GSH and DMPS have also been demonstrated in suspensions of isolated proximal tubular cells from rats [16]. Proximal tubular cells were first incubated for 15 min in an extracellular buffer containing bovine serum albumin and various concentrations of GSH or DMPS. They were then incubated for an additional hour in the presence of 250 μM mercuric chloride, which was found to be the threshold concentration of inorganic mercury that produced cellular injury under the incubation conditions being studied. GSH provided concentration-dependent protection from mercury-induced cytotoxicity, as assessed by decreases in the activity of total cellular lactate dehydrogenase. A GSH concentration of 500 μM, or twice that of inorganic mercury, was required to completely protect proximal tubular epithelial cells. DMPS, in contrast, provided complete protection against 250 μM mercuric chloride at a concentration (175 μM) that was less than that of inorganic mercury. Differences in the level of protection afforded by GSH and DMPS likely arise from differences in the chemistry and renal handling of the two compounds. Additional findings obtained from isolated proximal tubular epithelial cells from both normal and uninephrectomized rats have recently confirmed the protective effects of both GSH and DMPS against the cytotoxic effects of inorganic mercury *in vitro* [17].

In contrast to the *in vitro* data described above, Tanaka et al. [86] found that coadministration of GSH and mercuric chloride to mice caused the renal content of mercury to increase relative to that in mice that were given mercuric chloride alone. These investigators concluded that transport of inorganic mercury to the kidney may occur as a mercury–GSH complex and that the simultaneous presence of GSH enhances the uptake of mercury. Zalups and Barfuss [68,69,231] observed similar effects in rats coadministered a nontoxic dose of inorganic mercury with GSH or Cys. Consistent with these findings, Miller and Woods [231] showed recently that complexes of GSH and Hg^{2+} or GSSG and Hg^{+} promoted uroporphyrinogen oxidation and catalyzed the decomposition of hydrogen peroxide, indicating that mercury–GSH (or other thiol) complexes likely contribute to mercury-induced toxicity. Some of these results have also been confirmed in rats by Zalups (unpublished findings). Zalups and Barfuss [99] have very recent data from rats indicating that when a toxic 2.0 μmol kg^{-1} dose of mercuric chloride is coadministered with Cys, the nephropathy induced by the inorganic mercury becomes more severe. Resolution of the marked contrast between these findings and the *in vitro* findings described above will require a detailed mechanistic description of the renal transport of inorganic mercury. Although advances have been made in understanding the mechanisms of the renal transport of mercury, the role of thiols in the renal cellular uptake of mercury is still somewhat unclear.

In contrast to the highly effective protective effects of DMPS and DMSA against mercury-induced renal cellular injury, less definitive results have been obtained with two other dithiols, such as dithioerythritol and dithiothreitol. On the one hand, Barnes et al. [232] observed, in rats, evidence of protection against morphologic lesions and losses of activities of key marker enzymes for plasma membrane and mitochondria induced by mercury with dithiothreitol. Weinberg et al. [190] provided evidence of protection for isolated renal mitochondria from mercuric chloride-induced dysfunction by dithioerythritol, but only if the dithiol was added *in vitro* simultaneously with mercuric chloride; when the dithiol agent was added *in vitro* after the rats had been treated with mercuric chloride *in vivo*, no protection or reversal of toxicity was observed. To complicate further the understanding of how dithiols interact with mercury-containing compounds in biological systems, Chavez and Holguin [191] reported that the addition of dithiothreitol to renal mitochondria isolated from the rat that had been treated with inorganic mercury actually increased the degree of mitochondrial injury induced by mercury. They suggested that the dithiol made additional SH-sensitive sites available for interaction with mercury, thereby enhancing the toxic response. In the same study, the investigators also reported that the monothiol 2-mercaptoethanol also enhanced mercuric chloride-induced mitochondrial injury; however, higher concentrations than those of the dithiol were required to reproduce the effect.

Chavez et al. [192] also reported that the angiotensin converting enzyme inhibitor captopril [1-(3-mercapto-2-methyl-1-oxopropyl)-1-proline] was an effective protective agent both *in vivo* and *in vitro* against mercuric chloride-induced mitochondrial injury and morphological damage.

Besides low-molecular-weight thiols, administration of inorganic mercury complexed to the small SH-containing protein MT, although not providing protection against the toxicity induced by inorganic mercury, altered the renal site of injury [233]. Whereas the primary target of renal injury induced by mercuric chloride is the *pars recta* (S2 and S3 segments) of the proximal tubule, the primary target of renal injury induced by mercury-MT appears to be the *pars convoluta* and early *pars recta* (S1 and S2 segments) of the proximal tubule. Intrarenal accumulation and urinary excretion of inorganic mercury in rats were also greater when mercury was administered with MT than when mercury was administered by itself [173].

2.6.3 Reductions in Functioning Nephrons and Compensatory Tubular Hypertrophy and the Renal Disposition and Toxicity of Mercury

Reduction in the number of functioning nephrons, which can occur as a consequence of aging, renal disease, or surgical removal of renal tissue, has profound effects on renal cellular function and, consequently, on the renal handling of exogenous chemicals and on the susceptibility of renal tissue to chemically induced injury. Following a significant loss of renal mass, the remnant renal tissue undergoes compensatory growth, which is predominantly due (i.e., >85%) to cellular hypertrophy (rather than cellular hyperplasia), particularly in segments of the proximal tubule. One of the more prominent changes in renal function that occurs as a result of compensatory renal growth includes marked increases in mitochondrial metabolism, which may lead to an enhanced susceptibility of renal tissue to oxidative stress [234].

Numerous animal studies have shown that rats that have undergone a significant reduction in renal mass, such as unilateral nephrectomy, are more susceptible to the nephropathy induced by inorganic mercury than rats with two normal kidneys [11,12,16,165,168]. The biochemical changes that occur as a consequence of reduced renal mass and compensatory renal growth are retained *in vitro* when proximal tubular cells are isolated from rats [16]. Furthermore, the enhanced susceptibility of hypertrophied proximal tubular cells to the toxic effects of inorganic mercury is also retained *in vitro*. In the absence of exogenous thiols in the extracellular incubation medium, proximal tubular cells isolated from unilaterally nephrectomized (NPX) rats, in which compensatory renal growth had occurred, exhibited irreversible cellular injury at significantly lower concentrations of mercuric chloride than proximal tubular cells isolated from sham-operated rats.

Although the mechanism(s) for the enhanced susceptibility of proximal tubular cells from NPX rats to injury induced by mercury is/are not well characterized, it appears that enhanced accumulation of mercury is a contributing factor. Findings from studies with both mercuric chloride [10,12,168] and methylmercuric chloride [40] indicate that greater amounts of mercury, on a per gram tissue basis, accumulate in the remnant kidney of NPX rats than in the kidneys of sham-operated or control rats. Moreover, the findings indicate that the greatest increase in the accumulation of mercury occurs in the outer stripe of the outer medulla, and specifically in *pars recta* segments of proximal tubules [10], which coincides with the site where the toxicity of mercury is expressed in the kidney. Other factors, such as changes in intrarenal handling of mercury, are also probably involved in changing the cellular response to mercury exposure. Some of the altered accumulation of mercury that occurs in the remnant kidney is probably related to alteration in the renal concentrations of intracellular thiols. Recent findings show that the intracellular metabolisms of both GSH [12] and MT [41,42,235] are altered significantly after the renal mass is reduced following unilateral nephrectomy and compensatory renal growth. Zalups and Lash [12] showed that the cellular content of GSH in the remnant kidney increases following uninephrectomy, especially in the outer stripe of the outer medulla. This increase in renal cellular GSH has recently been shown to be linked to the increased activity of GCS [184], which is the rate-limiting enzyme involved in the intracellular synthesis of GSH. With respect to MT, recent molecular biological data indicate clearly that the increased renal cellular contents of MT that occur following uninephrectomy are linked directly to the increased transcription of the genes for MT-1 and MT-2 [235].

Despite the significant progress that has been made in defining the biochemical and physiological changes that occur during compensatory renal growth, much more research is needed to understand the precise mechanisms responsible for the increased proximal tubular uptake of, and susceptibility of renal injury to, inorganic mercury that occurs when renal mass has been reduced significantly.

ACKNOWLEDGMENT

Support for this review comes in part from grant ES05980 awarded by the National Institute of Environmental Health Science.

REFERENCES

1. ATSDR. Toxicological profile for mercury. (United States: U.S. Department of Health and Human Services, Public Health Service, Agency for Toxic Substance and Disease Registry), 2007.
2. Clarkson, T. W. The pharmacology of mercury compounds. *Annu Rev Pharmacol* 12, 375, 1972.
3. Zalups, R. K. and Lash, L. H. Advances in understanding the renal transport and toxicity of mercury. *J Toxicol Environ Health* 42, 1, 1994.
4. Zalups, R. K. Molecular interactions with mercury in the kidney. *Pharmacol Rev* 52, 113, 2000.

5. Rabenstein, D. L. Metal complexes of glutathione and their biological significance. In: D. Dolphin, O. Auramovbic, and R. Poulson (eds), *Glutathione: Chemical, Biochemical and Medical Aspects*, Vol. 3, p. 147. New York: Wiley Press, 1989.
6. Cember, H., Gallagher, P., and Faulkner, A. Distribution of mercury among blood fractions and serum proteins. *Am Ind Hyg Assoc J* 29, 233, 1968.
7. Friedman, H. L. Relationship between chemical structure and biological activity in mercurial compounds. *Ann N Y Acad Sci* 65, 461, 1957.
8. Lau, S. and Sarkar, B. Inorganic mercury(II)-binding components in normal human blood serum. *J Toxicol Environ Health* 5, 907, 1979.
9. Mussini, E. [Distribution of mercurial diuretics in the organism]. *Boll Soc Ital Biol Sper* 34, 1487, 1958.
10. Zalups, R. K. Renal accumulation and intrarenal distribution of inorganic mercury in the rabbit: Effect of unilateral nephrectomy and dose of mercuric chloride. *J Toxicol Environ Health* 33, 213, 1991.
11. Zalups, R. K. and Diamond, G. L. Mercuric chloride-induced nephrotoxicity in the rat following unilateral nephrectomy and compensatory renal growth. *Virchows Arch B Cell Pathol Incl Mol Pathol* 53, 336, 1987.
12. Zalups, R. K. and Lash, L. H. Effects of uninephrectomy and mercuric chloride on renal glutathione homeostasis. *J Pharmacol Exp Ther* 254, 962, 1990.
13. Ruegg, C. E., et al. Differential patterns of injury to the proximal tubule of renal cortical slices following *in vitro* exposure to mercuric chloride, potassium dichromate, or hypoxic conditions. *Toxicol Appl Pharmacol* 90, 261, 1987.
14. Barfuss, D. W., Robinson, M. K., and Zalups, R. K. Inorganic mercury transport in the proximal tubule of the rabbit. *J Am Soc Nephrol* 1, 910, 1990.
15. Zalups, R. K., Knutson, K. L., and Schnellmann, R. G. *In vitro* analysis of the accumulation and toxicity of inorganic mercury in segments of the proximal tubule isolated from the rabbit kidney. *Toxicol Appl Pharmacol* 119, 221, 1993.
16. Lash, L. H. and Zalups, R. K. Mercuric chloride-induced cytotoxicity and compensatory hypertrophy in rat kidney proximal tubular cells. *J Pharmacol Exp Ther* 261, 819, 1992.
17. Lash, L. H., Putt, D. A., and Zalups, R. K. Influence of exogenous thiols on inorganic mercury-induced injury in renal proximal and distal tubular cells from normal and uninephrectomized rats. *J Pharmacol Exp Ther* 291, 492, 1999.
18. Smith, M. W., et al. $HgCl_2$-induced changes in cytosolic Ca^{2+} of cultured rabbit renal tubular cells. *Biochim Biophys Acta* 931, 130, 1987.
19. Zalups, R. K. and Barfuss, D. Accumulation of inorganic mercury along the renal proximal tubule of the rabbit. *Toxicol Appl Pharmacol* 106, 245, 1990.
20. Adam, K. R. The effects of dithiols on the distribution of mercury in rabbits. *Br J Pharmacol Chemother* 6, 483, 1951.
21. Ashe, W. F., et al. Behavior of mercury in the animal organism following inhalation. *A M A Arch Ind Hyg Occup Med* 7, 19, 1953.
22. Berlin, M. and Gibson, S. Renal uptake, excretion, and retention of mercury. I. A study in the rabbit during infusion of mercuric chloride. *Arch Environ Health* 6, 617, 1963.
23. Cherian, M. G. and Clarkson, T. W. Biochemical changes in rat kidney on exposure to elemental mercury vapor: Effect on biosynthesis of metallothionein. *Chem Biol Interact* 12, 109, 1976.
24. Clarkson, T. W. and Magos, L. The effect of sodium maleate on the renal deposition and excretion of mercury. *Br J Pharmacol Chemother* 31, 560, 1967.
25. Friberg, L. Studies on the accumulation, metabolism and excretion of inorganic mercury (Hg203) after prolonged subcutaneous administration to rats. *Acta Pharmacol Toxicol (Copenh)* 12, 411, 1956.
26. Friberg, L. Studies on the metabolism of mercuric chloride and methylmercury dicyandiamide; experiments on rats given subcutaneous injections with radioactive mercury (Hg203). *AMA Arch Ind Health* 20, 42, 1959.
27. Hahn, L. J., et al. Dental "silver" tooth fillings: A source of mercury exposure revealed by whole-body image scan and tissue analysis. *FASEB J* 3, 2641, 1989.
28. Hahn, L. J., et al. Whole-body imaging of the distribution of mercury released from dental fillings into monkey tissues. *FASEB J* 4, 3256, 1990.
29. Zalups, R. K. Method for studying the *in vivo* accumulation of inorganic mercury in segments of the nephron in the kidneys of rats treated with mercuric chloride. *J Pharmacol Methods* 26, 89, 1991.
30. Zalups, R. K. Autometallographic localization of inorganic mercury in the kidneys of rats: Effect of unilateral nephrectomy and compensatory renal growth. *Exp Mol Pathol* 54, 10, 1991.

31. Zalups, R. K. and Diamond, G. L. Intrarenal distribution of mercury in the rat: Effect of administered dose of mercuric chloride. *Bull Environ Contam Toxicol* 38, 67, 1987.
32. Norseth, T. and Clarkson, T. W. Studies on the biotransformation of 203Hg-labeled methylmercury chloride in rats. *Arch Environ Health* 21, 717, 1970.
33. Norseth, T. and Clarkson, T. W. Biotransformation of methylmercury salts in the rat studied by specific determination of inorganic mercury. *Biochem Pharmacol* 19, 2775, 1970.
34. Magos, L. and Butler, W. H. The kinetics of methylmercury administered repeatedly to rats. *Arch Toxicol* 35, 25, 1976.
35. Magos, L., et al. Comparative study of the sensitivity of male and female rats to methylmercury. *Arch Toxicol* 48, 11, 1981.
36. Magos, L., et al. The comparative toxicology of ethyl- and methylmercury. *Arch Toxicol* 57, 260, 1985.
37. McNeil, S. I., Bhatnagar, M. K., and Turner, C. J. Combined toxicity of ethanol and methylmercury in rat. *Toxicology* 53, 345, 1988.
38. Prickett, C. S., Laug, E. P., and Kunze, F. M. Distribution of mercury in rats following oral and intravenous administration of mercuric acetate and phenylmercuric acetate. *Proc Soc Exp Biol Med* 73, 585, 1950.
39. Zalups, R. K., Gelein, R. M., and Cherian, M. G. Shifts in the dose–effect relationship for the nephropathy induced by cadmium-metallothionein in rats after a reduction in renal mass. *J Pharmacol Exp Ther* 262, 1256, 1992.
40. Zalups, R. K., Barfuss, D. W., and Kostyniak, P. J. Altered intrarenal accumulation of mercury in uninephrectomized rats treated with methylmercury chloride. *Toxicol Appl Pharmacol* 115, 174, 1992.
41. Zalups, R. K. and Cherian, M. G. Renal metallothionein metabolism after a reduction of renal mass. I. Effect of unilateral nephrectomy and compensatory renal growth on basal and metal-induced renal metallothionein metabolism. *Toxicology* 71, 83, 1992.
42. Zalups, R. K. and Cherian, M. G. Renal metallothionein metabolism after a reduction of renal mass. II. Effect of zinc pretreatment on the renal toxicity and intrarenal accumulation of inorganic mercury. *Toxicology* 71, 103, 1992.
43. Bergstrand, A., Friberg, L., and Odeblad, E. Localization of mercury in the kidneys after subcutaneous administration: A study on rabbits with mercuric chloride and phenylmercuric acetate. *AMA Arch Ind Health* 17, 253, 1958.
44. Berlin, M. Renal uptake, excretion, and retention of mercury. II. A study in the rabbit during infusion of methyl- and phenylmercuric compounds. *Arch Environ Health* 6, 626, 1963.
45. Berlin, M. and Ullberg, S. Accumulation and retention of mercury in the mouse. III. An autoradiographic comparison of methylmercuric dicyandiamide with inorganic mercury. *Arch Environ Health* 6, 610, 1963.
46. Berlin, M. and Ullberg, S. Accumulation and retention of mercury in the mouse. II. An autoradiographic comparison of phenylmercuric acetate with inorganic mercury. *Arch Environ Health* 6, 602, 1963.
47. Berlin, M. and Ullberg, S. Accumulation and retention of mercury in the mouse. I. An autoradiographic study after a single intravenous injection of mercuric chloride. *Arch Environ Health* 6, 589, 1963.
48. Berlin, M. and Ullrebg, S. Increased uptake of mercury in mouse brain caused by 2,3-dimercaptopropanol. *Nature* 197, 84, 1963.
49. Berlin, M. On estimating threshold limits for mercury in biological material. *Acta Med Scand Suppl* 396, 1, 1963.
50. Friberg, L., Odeblad, E., and Forssman, S. Distribution of two mercury compounds in rabbits after a single subcutaneous injection; a radiometric and autoradiographic study of the distribution of mercuric chloride and phenylmercuric acetate. *AMA Arch Ind Health* 16, 163, 1957.
51. Taugner, R., Winkel, K. Z., and Iravani, J. [On the localization of mercuric chloride concentration in the rat kidney]. *Virchows Arch Pathol Anat Physiol Klin Med* 340, 369, 1966.
52. Magos, L., Cikrt, M., and Snowden, R. The dependence of biliary methylmercury secretion on liver GSH and ligandin. *Biochem Pharmacol* 34, 301, 1985.
53. Hultman, P. and Enestrom, S. Localization of mercury in the kidney during experimental acute tubular necrosis studied by the cytochemical silver amplification method. *Br J Exp Pathol* 67, 493, 1986.
54. Hultman, P., et al. Murine susceptibility to mercury. I. Autoantibody profiles and systemic immune deposits in inbred, congenic, and intra-H-2 recombinant strains. *Clin Immunol Immunopathol* 65, 98, 1992.
55. Hultman, P. and Enestrom, S. Dose–response studies in murine mercury-induced autoimmunity and immune-complex disease. *Toxicol Appl Pharmacol* 113, 199, 1992.

56. Rodier, P. M. and Kates, B. Histological localization of methylmercury in mouse brain and kidney by emulsion autoradiography of 203Hg. *Toxicol Appl Pharmacol* 92, 224, 1988.
57. Rodier, P. M., Kates, B., and Simons, R. Mercury localization in mouse kidney over time: Autoradiography versus silver staining. *Toxicol Appl Pharmacol* 92, 235, 1988.
58. Danscher, G., Horsted-Bindslev, P., and Rungby, J. Traces of mercury in organs from primates with amalgam fillings. *Exp Mol Pathol* 52, 291, 1990.
59. Gage, J. C. Distribution and excretion of methyl and phenyl mercury salts. *Br J Ind Med* 21, 197, 1964.
60. Omata, S., et al. Time-dependent accumulation of inorganic mercury in subcellular fractions of kidney, liver, and brain of rats exposed to methylmercury. *Arch Toxicol* 44, 231, 1980.
61. Dunn, J. D. and Clarkson, T. W. Does mercury exhalation signal demethylation of methylmercury? *Health Phys* 38, 411, 1980.
62. Madsen, K. M. Mercury accumulation in kidney lysosomes or proteinuric rats. *Kidney Int* 18, 445, 1980.
63. Zalups, R. K. and Barfuss, D. W. Intrarenal distribution of inorganic mercury and albumin after coadministration. *J Toxicol Environ Health* 40, 77, 1993.
64. Mussini, E. [Bonds of mercurial diuretics to blood proteins]. *Boll Soc Ital Biol Sper* 34, 1588, 1958.
65. Cannon, V. T., Barfuss, D. W., and Zalups, R. K. Molecular homology and the luminal transport of Hg^{2+} in the renal proximal tubule. *J Am Soc Nephrol* 11, 394, 2000.
66. Cannon, V. T., Zalups, R. K., and Barfuss, D. W. Amino acid transporters involved in luminal transport of mercuric conjugates of cysteine in rabbit proximal tubule. *J Pharmacol Exp Ther* 298, 780, 2001.
67. Zalups, R. K. and Ahmad, S. Transport of *N*-acetylcysteine *S*-conjugates of methylmercury in Madin-Darby canine kidney cells stably transfected with human isoform of organic anion transporter 1. *J Pharmacol Exp Ther* 314, 1158, 2005.
68. Zalups, R. K. and Barfuss, D. W. Renal disposition of mercury in rats after intravenous injection of inorganic mercury and cysteine. *J Toxicol Environ Health* 44, 401, 1995.
69. Zalups, R. K. and Barfuss, D. W. Accumulation and handling of inorganic mercury in the kidney after coadministration with glutathione. *J Toxicol Environ Health* 44, 385, 1995.
70. Zalups, R. K. and Barfuss, D. W. Participation of mercuric conjugates of cysteine, homocysteine, and *N*-acetylcysteine in mechanisms involved in the renal tubular uptake of inorganic mercury. *J Am Soc Nephrol* 9, 551, 1998.
71. Zalups, R. K. and Lash, L. H. Binding of mercury in renal brush-border and basolateral membrane-vesicles. *Biochem Pharmacol* 53, 1889, 1997.
72. Zalups, R. K. and Minor, K. H. Luminal and basolateral mechanisms involved in the renal tubular uptake of inorganic mercury. *J Toxicol Environ Health* 46, 73, 1995.
73. Zalups, R. K., Robinson, M. K., and Barfuss, D. W. Factors affecting inorganic mercury transport and toxicity in the isolated perfused proximal tubule. *J Am Soc Nephrol* 2, 866, 1991.
74. Aslamkhan, A. G., et al. Human renal organic anion transporter 1-dependent uptake and toxicity of mercuric–thiol conjugates in Madin-Darby canine kidney cells. *Mol Pharmacol* 63, 590, 2003.
75. Zalups, R. K. Organic anion transport and action of gamma-glutamyl transpeptidase in kidney linked mechanistically to renal tubular uptake of inorganic mercury. *Toxicol Appl Pharmacol* 132, 289, 1995.
76. Zalups, R. K. Basolateral uptake of mercuric conjugates of *N*-acetylcysteine and cysteine in the kidney involves the organic anion transport system. *J Toxicol Environ Health A* 55, 13, 1998.
77. Zalups, R. K. Basolateral uptake of inorganic mercury in the kidney. *Toxicol Appl Pharmacol* 151, 192, 1998.
78. Zalups, R. K. and Ahmad, S. Homocysteine and the renal epithelial transport and toxicity of inorganic mercury: Role of basolateral transporter organic anion transporter 1. *J Am Soc Nephrol* 15, 2023, 2004.
79. Zalups, R. K. and Ahmad, S. Handling of cysteine *S*-conjugates of methylmercury in MDCK cells expressing human OAT1. *Kidney Int* 68, 1684, 2005.
80. Zalups, R. K. and Ahmad, S. Handling of the homocysteine *S*-conjugate of methylmercury by renal epithelial cells: Role of organic anion transporter 1 and amino acid transporters. *J Pharmacol Exp Ther* 315, 896, 2005.
81. Zalups, R. K., Aslamkhan, A. G., and Ahmad, S. Human organic anion transporter 1 mediates cellular uptake of cysteine-*S* conjugates of inorganic mercury. *Kidney Int* 66, 251, 2004.
82. Zalups, R. K. and Barfuss, D. W. Pretreatment with *p*-aminohippurate inhibits the renal uptake and accumulation of injected inorganic mercury in the rat. *Toxicology* 103, 23, 1995.
83. Zalups, R. K. and Barfuss, D. W. Small aliphatic dicarboxylic acids inhibit renal uptake of administered mercury. *Toxicol Appl Pharmacol* 148, 183, 1998.

84. Zalups, R. K. and Barfuss, D. W. Renal organic anion transport system: A mechanism for the basolateral uptake of mercury–thiol conjugates along the pars recta of the proximal tubule. *Toxicol Appl Pharmacol* 182, 234, 2002.
85. Zalups, R. K., et al. Mechanisms of action of 2,3-dimercaptopropane-1-sulfonate and the transport, disposition, and toxicity of inorganic mercury in isolated perfused segments of rabbit proximal tubules. *Mol Pharmacol* 54, 353, 1998.
86. Tanaka, T., Naganuma, A., and Imura, N. Role of gamma-glutamyltranspeptidase in renal uptake and toxicity of inorganic mercury in mice. *Toxicology* 60, 187, 1990.
87. Tanaka-Kagawa, T., Naganuma, A., and Imura, N. Tubular secretion and reabsorption of mercury compounds in mouse kidney. *J Pharmacol Exp Ther* 264, 776, 1993.
88. Berndt, W. O., et al. Renal glutathione and mercury uptake by kidney. *Fundam Appl Toxicol* 5, 832, 1985.
89. de Ceaurriz, J., et al. Role of extracellular glutathione and gamma-glutamyltranspeptidase in the disposition and kidney toxicity of inorganic mercury in rats. *J Appl Toxicol* 14, 201, 1994.
90. Kim, C. Y., Watanabe, C., and Satoh, H. Effects of buthionine sulfoximine (BSO) on mercury distribution after Hg(o) exposure. *Toxicology* 98, 67, 1995.
91. Kim, C. Y., et al. Inhibition of gamma-glutamyltranspeptidase decreases renal deposition of mercury after mercury vapor exposure. *Arch Toxicol* 69, 722, 1995.
92. Lash, L. H. and Jones, D. P. Distribution of oxidized and reduced forms of glutathione and cysteine in rat plasma. *Arch Biochem Biophys* 240, 583, 1985.
93. Parks, L. D., Zalups, R. K., and Barfuss, D. W. Heterogeneity of glutathione synthesis and secretion in the proximal tubule of the rabbit. *Am J Physiol* 274, F924, 1998.
94. Parks, L. D., Zalups, R. K., and Barfuss, D. W. Luminal and basolateral membrane transport of glutathione in isolated perfused S(1), S(2), and S(3) segments of the rabbit proximal tubule. *J Am Soc Nephrol* 11, 1008, 2000.
95. Schaub, T. P., et al. Expression of the conjugate export pump encoded by the mrp2 gene in the apical membrane of kidney proximal tubules. *J Am Soc Nephrol* 8, 1213, 1997.
96. Schaub, T. P., et al. Expression of the MRP2 gene-encoded conjugate export pump in human kidney proximal tubules and in renal cell carcinoma. *J Am Soc Nephrol* 10, 1159, 1999.
97. Keppler, D., et al. ATP-dependent transport of glutathione *S*-conjugates by the multidrug resistance protein MRP1 and its apical isoform MRP2. *Chem Biol Interact* 111–112, 153, 1998.
98. Naganuma, A., et al. Possible role of hepatic glutathione in transport of methylmercury into mouse kidney. *Biochem Pharmacol* 37, 291, 1988.
99. Zalups, R. K. and Barfuss, D. W. Nephrotoxicity of inorganic mercury co-administrated with L-cysteine. *Toxicology* 109, 15, 1996.
100. Schafer, J. A. and Watkins, M. L. Transport of L-cystine in isolated perfused proximal straight tubules. *Pflugers Arch* 401, 143, 1984.
101. Selhub, J. Homocysteine metabolism. *Annu Rev Nutr* 19, 217, 1999.
102. Clarke, R., et al. Hyperhomocysteinemia: An independent risk factor for vascular disease. *N Engl J Med* 324, 1149, 1991.
103. Dudman, N. P., et al. Human homocysteine catabolism: Three major pathways and their relevance to development of arterial occlusive disease. *J Nutr* 126, 1295S, 1996.
104. House, J. D., Brosnan, M. E., and Brosnan, J. T. Characterization of homocysteine metabolism in the rat kidney. *Biochem J* 328 (Pt 1), 287, 1997.
105. House, J. D., Brosnan, M. E., and Brosnan, J. T. Renal homocysteine metabolism. *Contrib Nephrol* 121, 79, 1997.
106. Brosnan, J. T., et al. Renal metabolism of homocysteine *in vivo*. *Biochem Soc Trans* 23, 470S, 1995.
107. Bostom, A., et al. Net uptake of plasma homocysteine by the rat kidney *in vivo*. *Atherosclerosis* 116, 59, 1995.
108. Friedman, A. N., et al. The kidney and homocysteine metabolism. *J Am Soc Nephrol* 12, 2181, 2001.
109. Bridges, C. C., et al. Mercuric conjugates of cysteine are transported by the amino acid transporter system b(0, +): Implications of molecular mimicry. *J Am Soc Nephrol* 15, 663, 2004.
110. Kanai, Y., et al. Expression of mRNA (D2) encoding a protein involved in amino acid transport in S3 proximal tubule. *Am J Physiol* 263, F1087, 1992.
111. Palacin, M., et al. Molecular biology of mammalian plasma membrane amino acid transporters. *Physiol Rev* 78, 969, 1998.
112. Palacin, M., et al. The amino acid transport system b(0, +) and cystinuria. *Mol Membr Biol* 18, 21, 2001.

113. Foreman, J. W., et al. Homocystine uptake in isolated rat renal cortical tubules. *Metabolism* 31, 613, 1982.
114. Naggar, H., et al. Regulation of reduced-folate transporter-1 (RFT-1) by homocysteine and identity of transport systems for homocysteine uptake in retinal pigment epithelial (RPE) cells. *Exp Eye Res* 77, 687, 2003.
115. Bridges, C. C. and Zalups, R. K. Homocysteine, system $b^{0,+}$ and the renal epithelial transport and toxicity of inorganic mercury. *Am J Pathol* 165, 1385, 2004.
116. Furriols, M., et al. rBAT, related to L-cysteine transport, is localized to the microvilli of proximal straight tubules, and its expression is regulated in kidney by development. *J Biol Chem* 268, 27060, 1993.
117. Pfeiffer, R., et al. Luminal heterodimeric amino acid transporter defective in cystinuria. *Mol Biol Cell* 10, 4135, 1999.
118. Clarkson, T. W. Molecular and ionic mimicry of toxic metals. *Annu Rev Pharmacol Toxicol* 33, 545, 1993.
119. Aschner, M. and Clarkson, T. W. Methylmercury uptake across bovine brain capillary endothelial cells *in vitro*: The role of amino acids. *Pharmacol Toxicol* 64, 293, 1989.
120. Aschner, M., et al. Methylmercury uptake in rat primary astrocyte cultures: The role of the neutral amino acid transport system. *Brain Res* 521, 221, 1990.
121. Aschner, M. and Aschner, J. L. Mercury neurotoxicity: Mechanisms of blood–brain barrier transport. *Neurosci Biobehav Rev* 14, 169, 1990.
122. Kerper, L. E., Ballatori, N., and Clarkson, T. W. Methylmercury transport across the blood–brain barrier by an amino acid carrier. *Am J Physiol* 262, R761, 1992.
123. Zalups, R. K. Early aspects of the intrarenal distribution of mercury after the intravenous administration of mercuric chloride. *Toxicology* 79, 215, 1993.
124. Zalups, R. K. and Lash, L. H. Cystine alters the renal and hepatic disposition of inorganic mercury and plasma thiol status. *Toxicol Appl Pharmacol* 214, 88, 2006.
125. Shimomura, A., Chonko, A. M., and Grantham, J. J. Basis for heterogeneity of para-aminohippurate secretion in rabbit proximal tubules. *Am J Physiol* 240, F430, 1981.
126. Ferrier, B., Martin, M., and Roch-Ramel, F. Effects of *p*-aminohippurate and pyrazinoate on the renal excretion of salicylate in the rat: A micropuncture study. *J Pharmacol Exp Ther* 224, 451, 1983.
127. Ullrich, K. J., Rumrich, G., and Kloss, S. Contraluminal *para*-aminohippurate transport in the proximal tubule of the rat kidney. III. Specificity: Monocarboxylic acids. *Pflugers Arch* 409, 547, 1987.
128. Ullrich, K. J., et al. Contraluminal para-aminohippurate (PAH) transport in the proximal tubule of the rat kidney. I. Kinetics, influence of cations, anions, and capillary preperfusion. *Pflugers Arch* 409, 229, 1987.
129. Ullrich, K. J., et al. Contraluminal *para*-aminohippurate (PAH) transport in the proximal tubule of the rat kidney. II. Specificity: Aliphatic dicarboxylic acids. *Pflugers Arch* 408, 38, 1987.
130. Pritchard, J. B. Coupled transport of *p*-aminohippurate by rat kidney basolateral membrane vesicles. *Am J Physiol* 255, F597, 1988.
131. Tanaka, T., Naganuma, A., and Imura, N. Routes for renal transport of methylmercury in mice. *Eur J Pharmacol* 228, 9, 1992.
132. Ban, M. and de Ceaurriz, J. Probenecid-induced protection against acute hexachloro-1,3-butadiene and methylmercury toxicity to the mouse kidney. *Toxicol Lett* 40, 71, 1988.
133. Rogulski, J. and Angielski, S. Effect of maleic acid on the kidney. IV. Synthesis of amino acids in the kidney of maleate-treated rats. *Acta Biochim Pol* 10, 133, 1963.
134. Rogulski, J. and Angielski, S. Effect of maleic acid on the kidney. III. Succinate as hydrogen donor in the reductive amination in rat kidney. *Acta Biochim Pol* 10, 125, 1963.
135. Rogulski, J. and Angielski, S. Effect of maleic acid on the kidney. II. Hydrogen donors in the reductive amination of alpha-ketoglutarate in rat kidney and liver. *Acta Biochim Pol* 10, 1, 1963.
136. Pritchard, J. B. and Miller, D. S. Mechanisms mediating renal secretion of organic anions and cations. *Physiol Rev* 73, 765, 1993.
137. Dantzler, W. H. and Evans, K. K. Effect of alpha-KG in lumen on PAH transport by isolated perfused rabbit renal proximal tubules. *Am J Physiol* 271, F521, 1996.
138. Dantzler, W. H. Comparative aspects of renal urate transport. *Kidney Int* 49, 1549, 1996.
139. Kim, Y. K. and Dantzler, W. H. Specificity of basolateral organic cation transport in snake renal proximal tubules. *Am J Physiol* 270, R1025, 1996.
140. Lash, L. H. and Jones, D. P. Renal glutathione transport. Characteristics of the sodium-dependent system in the basal-lateral membrane. *J Biol Chem* 259, 14508, 1984.
141. Lash, L. H. and Jones, D. P. Transport of glutathione by renal basal-lateral membrane vesicles. *Biochem Biophys Res Commun* 112, 55, 1983.

142. Lash, L. H. and Jones, D. P. Uptake of the glutathione conjugate *S*-(1,2-dichlorovinyl)glutathione by renal basal-lateral membrane vesicles and isolated kidney cells. *Mol Pharmacol* 28, 278, 1985.
143. Lash, L. H. and Anders, M. W. Uptake of nephrotoxic *S*-conjugates by isolated rat renal proximal tubular cells. *J Pharmacol Exp Ther* 248, 531, 1989.
144. Dantzler, W. H., Evans, K. K., and Wright, S. H. Kinetics of interactions of para-aminohippurate, probenecid, cysteine conjugates and *N*-acetyl cysteine conjugates with basolateral organic anion transporter in isolated rabbit proximal renal tubules. *J Pharmacol Exp Ther* 272, 663, 1995.
145. Aposhian, H. V. DMSA and DMPS—water soluble antidotes for heavy metal poisoning. *Annu Rev Pharmacol Toxicol* 23, 193, 1983.
146. Maiorino, R. M., et al. Determination and metabolism of dithiol chelating agents: X. In humans, meso-2,3-dimercaptosuccinic acid is bound to plasma proteins via mixed disulfide formation. *J Pharmacol Exp Ther* 254, 570, 1990.
147. Aposhian, H. V. and Aposhian, M. M. *meso*-2,3-Dimercaptosuccinic acid: Chemical, pharmacological and toxicological properties of an orally effective metal chelating agent. *Annu Rev Pharmacol Toxicol* 30, 279, 1990.
148. Aposhian, H. V., et al. Human studies with the chelating agents, DMPS and DMSA. *J Toxicol Clin Toxicol* 30, 505, 1992.
149. Zalups, R. K. Influence of 2,3-dimercaptopropane-1-sulfonate (DMPS) and meso-2,3-dimercaptosuccinic acid (DMSA) on the renal disposition of mercury in normal and uninephrectomized rats exposed to inorganic mercury. *J Pharmacol Exp Ther* 267, 791, 1993.
150. Bahn, A., et al. Interaction of the metal chelator 2,3-dimercapto-1-propanesulfonate with the rabbit multispecific organic anion transporter 1 (rbOAT1). *Mol Pharmacol* 62, 1128, 2002.
151. Islinger, F., Gekle, M., and Wright, S. H. Interaction of 2,3-dimercapto-1-propane sulfonate with the human organic anion transporter hOAT1. *J Pharmacol Exp Ther* 299, 741, 2001.
152. Burckhardt, B. C., et al. The renal Na(+)-dependent dicarboxylate transporter, NaDC-3, translocates dimethyl- and disulfhydryl-compounds and contributes to renal heavy metal detoxification. *J Am Soc Nephrol* 13, 2628, 2002.
153. Sugawara, N., et al. Decreased hepatobiliary secretion of inorganic mercury, its deposition and toxicity in the Eisai hyperbilirubinemic rat with no hepatic canalicular organic anion transporter. *Toxicology* 126, 23, 1998.
154. Zalups, R. K. and Bridges, C. C. MRP2 involvement in renal proximal tubular elimination of methylmercury mediated by DMPS or DMSA. *Toxicol Appl Pharmacol* 2008.
155. Bridges, C. C., Joshee, L., and Zalups, R. K. MRP2 and the DMPS- and DMSA-mediated elimination of mercury in TR(−) and control rats exposed to thiol *S*-conjugates of inorganic mercury. *Toxicol Sci* 105, 211, 2008.
156. Bridges, C. C., Joshee, L., and Zalups, R. K. Multidrug resistance proteins and the renal elimination of inorganic mercury mediated by 2,3-dimercaptopropane-1-sulfonic acid and meso-2,3-dimercaptosuccinic acid. *J Pharmacol Exp Ther* 324, 383, 2008.
157. van Aubel, R. A., et al. The MRP4/ABCC4 gene encodes a novel apical organic anion transporter in human kidney proximal tubules: Putative efflux pump for urinary cAMP and cGMP. *J Am Soc Nephrol* 13, 595, 2002.
158. Leslie, E. M., Deeley, R. G., and Cole, S. P. Multidrug resistance proteins: Role of P-glycoprotein, MRP1, MRP2, and BCRP (ABCG2) in tissue defense. *Toxicol Appl Pharmacol* 204, 216, 2005.
159. Madejczyk, M. S., et al. Accelerated urinary excretion of methylmercury following administration of its antidote *N*-acetylcysteine requires Mrp2/Abcc2, the apical multidrug resistance-associated protein. *J Pharmacol Exp Ther* 322, 378, 2007.
160. Zalups, R. K. and Barfuss, D. W. Diversion or prevention of biliary outflow from the liver diminishes the renal uptake of injected inorganic mercury. *Drug Metab Dispos* 24, 480, 1996.
161. Zalups, R. K. Intestinal handling of mercury in the rat: Implications of intestinal secretion of inorganic mercury following biliary ligation or cannulation. *J Toxicol Environ Health A* 53, 615, 1998.
162. Zalups, R. K., Barfuss, D. W., and Lash, L. H. Relationships between alterations in glutathione metabolism and the disposition of inorganic mercury in rats: Effects of biliary ligation and chemically induced modulation of glutathione status. *Chem Biol Interact* 123, 171, 1999.
163. Zalups, R. K., Barfuss, D. W., and Lash, L. H. Disposition of inorganic mercury following biliary obstruction and chemically induced glutathione depletion: Dispositional changes one hour after the intravenous administration of mercuric chloride. *Toxicol Appl Pharmacol* 154, 135, 1999.
164. Baggett, J. M. and Berndt, W. O. The effect of potassium dichromate and mercuric chloride on urinary excretion and organ and subcellular distribution of [203Hg] mercuric chloride in rats. *Toxicol Lett* 29, 115, 1985.

165. Houser, M. T. and Berndt, W. O. Unilateral nephrectomy in the rat: Effects on mercury handling and renal cortical subcellular distribution. *Toxicol Appl Pharmacol* 93, 187, 1988.
166. Madsen, K. M. and Hansen, J. C. Subcellular distribution of mercury in the rat kidney cortex after exposure to mercuric chloride. *Toxicol Appl Pharmacol* 54, 443, 1980.
167. Zalups, R. K., Gelein, R. M., and Cernichiari, E. DMPS as a rescue agent for the nephropathy induced by mercuric chloride. *J Pharmacol Exp Ther* 256, 1, 1991.
168. Zalups, R. K., Klotzbach, J. M., and Diamond, G. L. Enhanced accumulation of injected inorganic mercury in renal outer medulla after unilateral nephrectomy. *Toxicol Appl Pharmacol* 89, 226, 1987.
169. Smith, J.C., et al. The kinetics of intravenously administered methylmercury in man. *Toxicol Appl Pharmacol* 128, 251, 1994.
170. Mambourg, A. M. and Raynaud, C. [Study, using radioactive isotopes, of the mechanism of urinary excretion of mercury in the rabbit]. *Rev Fr Etud Clin Biol* 10, 414, 1965.
171. Vostal, J. and Heller, J. [The possibilities of studying transtubular transport in the kidney]. *Cesk Fysiol* 15, 178, 1966.
172. Baggett, J. M. and Berndt, W. O. The effect of depletion of nonprotein sulfhydryls by diethyl maleate plus buthionine sulfoximine on renal uptake of mercury in the rat. *Toxicol Appl Pharmacol* 83, 556, 1986.
173. Zalups, R. K., Cherian, M. G., and Barfuss, D. W. Mercury-metallothionein and the renal accumulation and handling of mercury. *Toxicology* 83, 61, 1993.
174. Magos, L. and Stoytchev, T. Combined effect of sodium maleate and some thiol compounds on mercury excretion and redistribution in rats. *Br J Pharmacol* 35, 121, 1969.
175. Trojanowska, B., Piotrowski, J. K., and Szendzikowski, S. The influence of thioacetamide on the excretion of mercury in rats. *Toxicol Appl Pharmacol* 18, 374, 1971.
176. Foulkes, E. C. Excretion and retention of cadmium, zinc, and mercury by rabbit kidney. *Am J Physiol* 227, 1356, 1974.
177. Zalups, R. K. Reductions in renal mass and the nephropathy induced by mercury. *Toxicol Appl Pharmacol* 143, 366, 1997.
178. Piotrowski, J. K., et al. Mercury binding in the kidney and liver of rats repeatedly exposed to mercuric chloride: Induction of metallothionein by mercury and cadmium. *Toxicol Appl Pharmacol* 27, 11, 1974.
179. Zalups, R. K. and Koropatnick, J. Temporal changes in metallothionein gene transcription in rat kidney and liver: Relationship to content of mercury and metallothionein protein. *J Pharmacol Exp Ther* 295, 74, 2000.
180. Woods, J. S., Davis, H. A., and Baer, R. P. Enhancement of gamma-glutamylcysteine synthetase mRNA in rat kidney by methylmercury. *Arch Biochem Biophys* 296, 350, 1992.
181. Chung, A. S., Maines, M. D., and Reynolds, W. A. Inhibition of the enzymes of glutathione metabolism by mercuric chloride in the rat kidney: Reversal by selenium. *Biochem Pharmacol* 31, 3093, 1982.
182. Fukino, H., et al. Effect of zinc pretreatment on mercuric chloride-induced lipid peroxidation in the rat kidney. *Toxicol Appl Pharmacol* 73, 395, 1984.
183. Siegers, C. P., Schenke, M., and Younes, M. Influence of cadmium chloride, mercuric chloride, and sodium vanadate on the glutathione-conjugating enzyme system in liver, kidney, and brain of mice. *J Toxicol Environ Health* 22, 141, 1987.
184. Lash, L. H. and Zalups, R. K. Activities of enzymes involved in renal cellular glutathione metabolism after uninephrectomy in the rat. *Arch Biochem Biophys* 309, 129, 1994.
185. Addya, S., et al. Effects of mercuric chloride on several scavenging enzymes in rat kidney and influence of vitamin E supplementation. *Acta Vitaminol Enzymol* 6, 103, 1984.
186. Gstraunthaler, G., Pfaller, W., and Kotanko, P. Glutathione depletion and *in vitro* lipid peroxidation in mercury or maleate induced acute renal failure. *Biochem Pharmacol* 32, 2969, 1983.
187. Lund, B. O., Miller, D. M., and Woods, J. S. Mercury-induced H_2O_2 production and lipid peroxidation *in vitro* in rat kidney mitochondria. *Biochem Pharmacol* 42 Suppl, S181, 1991.
188. Lund, B. O., Miller, D. M., and Woods, J. S. Studies on Hg(II)-induced H_2O_2 formation and oxidative stress *in vivo* and *in vitro* in rat kidney mitochondria. *Biochem Pharmacol* 45, 2017, 1993.
189. Weinberg, J. M., Harding, P. G., and Humes, H. D. Mitochondrial bioenergetics during the initiation of mercuric chloride-induced renal injury. II. Functional alterations of renal cortical mitochondria isolated after mercuric chloride treatment. *J Biol Chem* 257, 68, 1982.
190. Weinberg, J. M., Harding, P. G., and Humes, H. D. Mitochondrial bioenergetics during the initiation of mercuric chloride-induced renal injury. I. Direct effects of *in vitro* mercuric chloride on renal mitochondrial function. *J Biol Chem* 257, 60, 1982.
191. Chavez, E. and Holguin, J. A. Mitochondrial calcium release as induced by Hg^{2+}. *J Biol Chem* 263, 3582, 1988.

192. Chavez, E., et al. Protective behavior of captopril on Hg(++)-induced toxicity on kidney mitochondria. *In vivo* and *in vitro* experiments. *J Pharmacol Exp Ther* 256, 385, 1991.
193. Jung, K. Y., Uchida, S., and Endou, H. Nephrotoxicity assessment by measuring cellular ATP content. I. Substrate specificities in the maintenance of ATP content in isolated rat nephron segments. *Toxicol Appl Pharmacol* 100, 369, 1989.
194. Hieb, W. F. and Rothstein, M. Isolation from liver of a heat-stable requirement for reproduction of a free-living nematode. *Arch Biochem Biophys* 136, 576, 1970.
195. Anner, B. M. and Moosmayer, M. Preparation of Na,K-ATPase-containing liposomes with predictable transport properties by a procedure relating the Na,K-transport capacity to the ATPase activity. *J Biochem Biophys Methods* 5, 299, 1982.
196. Imesch, E., Moosmayer, M., and Anner, B. M. Mercury weakens membrane anchoring of Na-K-ATPase. *Am J Physiol* 262, F837, 1992.
197. Anner, B. M., Moosmayer, M., and Imesch, E. Mercury blocks Na-K-ATPase by a ligand-dependent and reversible mechanism. *Am J Physiol* 262, F830, 1992.
198. Verkman, A. S. Water channels in cell membranes. *Annu Rev Physiol* 54, 97, 1992.
199. Verkman, A. S. Lessons on renal physiology from transgenic mice lacking aquaporin water channels. *J Am Soc Nephrol* 10, 1126, 1999.
200. Levy, R. I., Weiner, I. M., and Mudge, G. H. The effects of acid–base balance on the diuresis produced by organic and inorganic mercurials. *J Clin Invest* 37, 1016, 1958.
201. Kuwahara, M., et al. Mercury-sensitive residues and pore site in AQP3 water channel. *Biochemistry* 36, 13973, 1997.
202. Woods, J. S., et al. Stimulation of porphyrinogen oxidation by mercuric ion. I. Evidence of free radical formation in the presence of thiols and hydrogen peroxide. *Mol Pharmacol* 38, 253, 1990.
203. Woods, J. S., Calas, C. A., and Aicher, L. D. Stimulation of porphyrinogen oxidation by mercuric ion. II. Promotion of oxidation from the interaction of mercuric ion, glutathione, and mitochondria-generated hydrogen peroxide. *Mol Pharmacol* 38, 261, 1990.
204. Bowers, M. A., et al. Quantitative determination of porphyrins in rat and human urine and evaluation of urinary porphyrin profiles during mercury and lead exposures. *J Lab Clin Med* 120, 272, 1992.
205. Woods, J. S., et al. Urinary porphyrin profiles as a biomarker of mercury exposure: Studies on dentists with occupational exposure to mercury vapor. *J Toxicol Environ Health* 40, 235, 1993.
206. Goering, P. L., et al. Relationship between stress protein induction in rat kidney by mercuric chloride and nephrotoxicity. *Toxicol Appl Pharmacol* 113, 184, 1992.
207. Hernadez-Pando, R., et al. Histological and subcellular distribution of 65 and 70 kD heat shock proteins in experimental nephrotoxic injury. *Exp Toxicol Pathol* 47, 501, 1995.
208. Miura, K., Inokawa, M., and Imura, N. Effects of methylmercury and some metal ions on microtubule networks in mouse glioma cells and *in vitro* tubulin polymerization. *Toxicol Appl Pharmacol* 73, 218, 1984.
209. Sager, P. R. and Syversen, T. L. Differential responses to methylmercury exposure and recovery in neuroblastoma and glioma cells and fibroblasts. *Exp Neurol* 85, 371, 1984.
210. Brown, D. L., et al. Effects of methylmercury on the microtubule system of mouse lymphocytes. *Toxicol Appl Pharmacol* 94, 66, 1988.
211. Vogel, D. G., Margolis, R. L., and Mottet, N. K. The effects of methylmercury binding to microtubules. *Toxicol Appl Pharmacol* 80, 473, 1985.
212. Roman-Franco, A. A., et al. Anti-basement membrane antibodies and antigen—antibody complexes in rabbits injected with mercuric chloride. *Clin Immunol Immunopathol* 9, 464, 1978.
213. Druet, P., et al. Immune type glomerulonephritis induced by HgCl2 in the Brown Norway rat. *Ann Immunol (Paris)* 129C, 777, 1978.
214. Henry, G. A., et al. Mercury-induced renal autoimmunity in the MAXX rat. *Clin Immunol Immunopathol* 49, 187, 1988.
215. Bigazzi, P. E. Autoimmunity induced by chemicals. *J Toxicol Clin Toxicol* 26, 125, 1988.
216. Bigazzi, P. E. Lessons from animal models: The scope of mercury-induced autoimmunity. *Clin Immunol Immunopathol* 65, 81, 1992.
217. Sapin, C., Druet, E., and Druet, P. Induction of anti-glomerular basement membrane antibodies in the Brown-Norway rat by mercuric chloride. *Clin Exp Immunol* 28, 173, 1977.
218. Enestrom, S. and Hultman, P. Immune-mediated glomerulonephritis induced by mercuric chloride in mice. *Experientia* 40, 1234, 1984.
219. Weening, J. J., Hoedemaeker, P. J., and Bakker, W. W. Immunoregulation and anti-nuclear antibodies in mercury-induced glomerulopathy in the rat. *Clin Exp Immunol* 45, 64, 1981.

220. Koropatnick, J. and Zalups, R. K. Effect of non-toxic mercury, zinc or cadmium pretreatment on the capacity of human monocytes to undergo lipopolysaccharide-induced activation. *Br J Pharmacol* 120, 797, 1997.
221. Johnson, D. R. Role of renal cortical sulfhydryl groups in development of mercury-induced renal toxicity. *J Toxicol Environ Health* 9, 119, 1982.
222. Baggett, J. M. and Berndt, W. O. Renal function in the rat after treatment with mercuric chloride plus potassium dichromate. *Fundam Appl Toxicol* 6, 98, 1986.
223. Zalups, R. K. and Lash, L. H. Depletion of glutathione in the kidney and the renal disposition of administered inorganic mercury. *Drug Metab Dispos* 25, 516, 1997.
224. Girardi, G. and Elias, M. M. Effectiveness of *N*-acetylcysteine in protecting against mercuric chloride-induced nephrotoxicity. *Toxicology* 67, 155, 1991.
225. Fukino, H., et al. Mechanism of protection by zinc against mercuric chloride toxicity in rats: Effects of zinc and mercury on glutathionine metabolism. *J Toxicol Environ Health* 19, 75, 1986.
226. Planas-Bohne, F. The effect of 2,3-dimercaptorpropane-1-sulfonate and dimercaptosuccinic acid on the distribution and excretion of mercuric chloride in rats. *Toxicology* 19, 275, 1981.
227. Maiorino, R. M., et al. Determination and metabolism of dithiol chelating agents. XII. Metabolism and pharmacokinetics of sodium 2,3-dimercaptopropane-1-sulfonate in humans. *J Pharmacol Exp Ther* 259, 808, 1991.
228. Klotzbach, J. M. and Diamond, G. L. Complexing activity and excretion of 2,3-dimercapto-1-propane sulfonate in rat kidney. *Am J Physiol* 254, F871, 1988.
229. Stewart, J. R. and Diamond, G. L. *In vivo* renal tubular secretion and metabolism of the disulfide of 2,3-dimercaptopropane-1-sulfonate. *Drug Metab Dispos* 16, 189, 1988.
230. Aw, T. Y., Wierzbicka, G., and Jones, D. P. Oral glutathione increases tissue glutathione *in vivo. Chem Biol Interact* 80, 89, 1991.
231. Miller, D. M. and Woods, J. S. Redox activities of mercury–thiol complexes: Implications for mercury-induced porphyria and toxicity. *Chem Biol Interact* 88, 23, 1993.
232. Barnes, J. L., et al. Studies on the pathophysiology of acute renal failure. IV. Protective effect of dithiothreitol following administration of mercuric chloride in the rat. *Virchows Arch B Cell Pathol Incl Mol Pathol* 32, 201, 1980.
233. Chan, H. M., et al. Exogenous metallothionein and renal toxicity of cadmium and mercury in rats. *Toxicology* 76, 15, 1992.
234. Nath, K. A., Croatt, A. J., and Hostetter, T. H. Oxygen consumption and oxidant stress in surviving nephrons. *Am J Physiol* 258, F1354, 1990.
235. Zalups, R. K., Fraser, J., and Koropatnick, J. Enhanced transcription of metallothionein genes in rat kidney: Effect of uninephrectomy and compensatory renal growth. *Am J Physiol* 268, F643, 1995.

3 Essential and Toxic Metal Transport in the Liver

Nazzareno Ballatori, David A. Aremu, and Michael S. Madejczyk

CONTENTS

3.1 Introduction ... 80
3.2 Models Used to Study Hepatic Transport ... 81
3.3 Determinants of Hepatic Metal Transport ... 82
3.3.1 Hepatic Microanatomy ... 82
3.3.2 Functional Polarity ... 83
3.3.3 Bile Secretion and Biliary-Hepatic Cycle ... 84
3.3.4 Paracellular Pathway for Biliary Excretion ... 85
3.3.5 Plasma and Intracellular Protein Binding ... 86
3.4 Sinusoidal Uptake of Metals from Blood Plasma, and Secretion Back into the Bloodstream ... 86
3.4.1 Transport during Membrane Recycling: Roles of Transferrin and Ceruloplasmin ... 87
3.4.2 Sinusoidal Metal Transporters and Protein Modifiers of Transport ... 89
3.4.2.1 Uptake on Phosphate or Sulfate Transporters ... 89
3.4.2.2 CTR1 and CTR2, the Copper Uptake Transporters ... 89
3.4.2.3 DMT1/SLC11A2, a Multispecific Metal Transporter of the Nramp Family ... 92
3.4.2.4 ZIP/SLC39 and ZNT/SLC30 in Zinc Uptake and Efflux, Respectively ... 92
3.4.2.5 Ferroportin, Hephaestin, and Hepcidin in Hepatic Iron Export ... 93
3.4.2.6 Organic Solute Carriers: OATPs, OCTs, NTCP, and Amino Acid and Peptide Transporters ... 94
3.4.3 Uptake and Efflux on Sinusoidal Membrane Channels and Pumps ... 94
3.4.4 Permeation as Lipid-Soluble Forms or by Convection ... 95
3.5 Intracellular Binding, Sequestration, and Metabolism ... 96
3.5.1 Intracellular Ligands ... 96
3.5.2 Metal Uptake and Release from Endosomal/Lysosomal Compartments: ATP7A, ATP7B, DMT1, and ZNT2 ... 96
3.5.3 Nuclear Uptake of Metals ... 98
3.5.4 Mitochondrial Metal Uptake ... 98
3.6 Transport Across the Canalicular Membrane into Bile ... 99
3.6.1 Transport during Membrane Recycling: Role of the Gene Products Defective in Wilson's Disease (ATP7B) and Menkes' Syndrome (ATP7A) ... 99
3.6.2 Canalicular Membrane Transporters and Pumps ... 99
3.6.2.1 MRP2-Mediated Excretion of GSH Complexes and Other Complexes ... 99
3.6.2.2 Additional Organic Solute Carriers: MDR1, MDR3, and BSEP ... 102
3.6.2.3 Sulfate Exchanger ... 102

3.7 Summary 102
Acknowledgments 102
References 103

3.1 INTRODUCTION

The liver plays a central role in the whole-body homeostasis of essential trace metals and in the disposition of toxic heavy metals. A major hepatic function involves uptake of metals from portal blood, their intracellular metabolism, storage, incorporation into apometalloproteins, and distribution of the resulting metal complexes either into the bloodstream for delivery to other tissues, or by excretion into bile. Biliary excretion is a principal pathway for the elimination of a number of metals from the body, including copper, manganese, lead, mercury, silver, and arsenic.

Significant progress has now been made in the identification of the molecular mechanisms by which metals traverse the liver cell sinusoidal (basolateral) and canalicular (apical) plasma membranes. For the sinusoidal membrane, several candidate metal transporters have been described, including the high-affinity copper transporter CTR1, the zinc transporter ZIP2, and a multispecific divalent metal ion transporter DMT1, a member of the natural-resistance-associated macrophage protein (Nramp) family of membrane proteins. DMT1, a divalent cation transporter with unusually broad metal substrate specificity, is expressed in rat liver in moderate amounts, although expression is low when compared to that of intestine or kidney. In contrast, *CTR1*, the human gene for a high-affinity copper uptake transporter, is expressed at its highest levels in the liver. Another key sinusoidal metal uptake system is the transferrin-receptor-mediated endocytosis of iron and possibly of other metals that bind to transferrin. Two transferrin receptors have now been identified (TfR1 and TfR2). Hepatic metal uptake also occurs by other receptor-mediated pathways (e.g., ferritin and asialoglycoprotein), and by adsorptive and fluid-phase endocytosis. Metals that are present as oxyanions, such as arsenate, vanadate, chromate, and selenate, are transported in large part by the phosphate or sulfate transporters, whereas transport of organometallic complexes may occur via some of the sinusoidal organic solute carriers (the OATPs, OATs, OCTs, etc.).

The release of metals from liver cells into blood plasma or bile is also mediated by both metal-selective membrane transport proteins and membrane recycling. Important insights into mechanisms for cellular export of metals were provided by the identification and characterization of several putative metal transporters, including the Wilson's disease and Menkes' syndrome genes, some members of the cation diffusion facilitator (CDF) family of transporters (e.g., ZNT1–ZNT9), and of some of the canalicular organic solute efflux pumps (of the MDR and MRP families). The Wilson and Menkes disease genes encode transmembrane P-type ATPases that are required for copper export. ZNT1, a member of the CDF family, appears to localize to the sinusoidal membrane where it may function to release zinc into the extracellular space. Additional mechanisms for the export of metals from liver cells into bile involve the canalicular ATP-dependent, multispecific organic solute transporters MDR1 and MRP2. These multispecific transport proteins are members of the ATP-binding cassette (ABC) superfamily of proteins, and are expressed at high levels on the liver cell canalicular membrane. Although their preferred substrates are amphiphatic organic molecules, they also transport organometallic complexes. MRP2 mediates biliary transport of metals complexed with glutathione (GSH), including arsenic and cisplatin, and possibly copper, cadmium, and mercury. Cells overexpressing MDR1 are more resistant to cationic lipophilic metal complexes, indicating transport of the metal complexes on MDR1.

The present discussion will provide an overview of the mechanisms that the liver cell utilizes to transport metals to and from the bloodstream, and into bile, with an emphasis on recent work on the characterization of metal-selective transporters. Previous studies of hepatic uptake and biliary excretion of specific metals are described in several comprehensive reviews (Ballatori, 1991a, 1991b,

1994, 2002; Ballatori and Madejczyk, 2006; Klaassen, 1976; Klaassen and Watkins, 1984), and are only discussed briefly in the present report.

3.2 MODELS USED TO STUDY HEPATIC TRANSPORT

The liver's complex microanatomy and reticuloendothelial system create a specialized epithelial barrier separating blood from bile. This complexity poses a special challenge in the design of experiments to investigate mechanisms of hepatobiliary transport of solutes, including metals. The intact animal and the isolated perfused liver model are the only systems that permit the simultaneous measurement of sinusoidal transport and biliary excretion; however, it is often difficult to obtain detailed kinetic data using these model systems. Other *in vitro* models are most often chosen; however, all *in vitro* models share a major limitation, namely the inability to reproduce the ionic and ligand composition of blood and blood plasma, a critical factor in regulating metal uptake. Metals differ from most other transported solutes in that they generally do not exist in their "free" form in biological fluids and tissues, but are normally complexed with amino acids, peptides, proteins, phospholipids, and other tissue constituents. As a consequence, their disposition, including their membrane transport, is regulated by the availability and relative concentrations of competing ligands (Ballatori, 1991a, 1994).

Despite the limitations of *in vitro* systems, isolated hepatocytes in culture have provided invaluable information on the characteristics of uptake of a variety of metals. Isolated hepatocytes have been used to determine the apparent K_m and V_{max} values and the driving forces for metal uptake, as well as the effects of individual ligands, metals, and other drugs on uptake. In contrast, isolated hepatocytes have not been as useful for studying the export of metals and other solutes from the cell. Mammalian hepatocytes rapidly lose their transport polarity upon isolation, which makes it difficult to distinguish excretion via the sinusoidal versus the canalicular membranes. Although several hepatoma and immortalized hepatocyte cell lines have been developed that display some features of the adult polarized hepatocyte phenotype, the conservation of the hepatocyte differentiation program is incomplete in these cell lines (Ballatori et al., 2006). The hepatocyte couplet system obviates some of these difficulties in that it retains a bile canalicular space (Graf et al., 1984), but has additional limitations, including the relative inaccessibility of the small bile canalicular space. Collagen entrapment, coculture of hepatocytes with extracellular matrix, and spherical multicellular aggregates (spheroids) are some of the techniques that have been used to improve the performance of hepatocytes in primary culture. Among these, hepatocyte spheroids have been shown to exhibit extended functional activity, indicating a close link between cellular architecture and tissue-specific functions (Ambrosino et al., 2005). A three-dimensional (3D) culture of hepatocytes is therefore essential for the reconstruction of functional hepatic tissues *in vitro*. Another method for 3D culture uses scaffolds, such as paper (Mizuguchi et al., 2000), poly-L-lactic-acid (Jiang et al., 2002; Mooney et al., 1995), or 3D-perfused micro-array bioreactors (Powers et al., 2002). The hepatocytes in the scaffolds tend to form spheroids. More recently, Sudo et al. (2005) has reported using a stack of cells composed of rat small hepatocytes (SHs; hepatic progenitor cells) created by inverting one membrane of an SH monolayer on top of another to form an SH bilayer. The cells within the bilayer exhibited mRNA transcription of the hepatic-differentiation markers. Transmission electron microscopy also revealed that the SHs of the upper and lower layers adhered to one another, and that bile canaliculi formed between them. Furthermore, bile canalicular proteins were localized to the lumina of the tubular structures. Although 3D cultures may be promising reconstructions of tissue-engineered livers, additional research is needed to establish the usefulness of this model to study hepatobiliary transport, and metal transport in particular.

In the early 1980s, another useful model system was introduced, a technique for the separation of plasma membrane vesicles derived from the liver cell sinusoidal and canalicular membrane domains (Meier et al., 1984). This made it possible to examine transport at both poles of the cell separately, without interference from cell metabolic functions. With this system, the electrogenic features, energy dependence, and ion requirements could be determined with relative ease. The limitations of this system include cross-contamination with other intracellular or plasma membranes, orientation of the vesicles, functional integrity of the transport proteins or the vesicles themselves, and the relatively high degree of nonspecific binding that is often observed with metals (Ballatori, 1994).

However, the most significant progress in the characterization of hepatic transport systems, including metal transporters, has come from studies that utilized molecular biological approaches. In particular, the use of complementation strategies in bacteria and yeast and the use of the *Xenopus laevis* oocyte expression system have allowed for the identification of several candidate hepatic metal transport proteins.

It is becoming increasingly apparent that most cell functions, including membrane transport functions, are highly conserved throughout evolution. Several mammalian transporters have recently been identified after structurally homologous proteins were first characterized and identified in either bacteria or yeast, as described in more detail below. For example, the human gene for the high-affinity copper uptake transporter CTR1 was identified by functional complementation in yeast (Zhou and Gitschier, 1997). A second strategy involves the use of *Xenopus* oocytes, which have become a standard tool for the cloning and functional expression of integral membrane proteins (Sigel, 1990; Soreq and Seidman, 1992). Gunshin et al. (1997) used this strategy to clone the cDNA for Dmt1, a multispecific divalent cation transporter of the Nramp family. *Xenopus* oocytes allow for the study of biogenesis, functional structure, modulation, and expression of plasma membrane proteins (Sigel, 1990; Soreq and Seidman, 1992). Oocytes efficiently translate exogenous mRNA injected into their cytoplasm, and are generally able to correctly post-translationally modify and target heterologously expressed proteins to the appropriate cellular compartment. This system has been used to clone and express membrane proteins of several classes, including transporters, channels, receptors, and enzymes (Sigel, 1990; Soreq and Seidman, 1992).

3.3 DETERMINANTS OF HEPATIC METAL TRANSPORT

3.3.1 Hepatic Microanatomy

Mammalian hepatocytes are arranged in plates or sheets that are only one cell thick, and are surrounded by capillaries (sinusoids), thus maximizing the surface area for exchange. The sinusoidal endothelium is highly fenestrated and has little if any basement membrane (Disse space). The sieve-like plates in the endothelial cells permit solutes up to the size of chylomicrons to pass from blood into the Disse space. As a result, metals associated with particles of this size or smaller in blood plasma have direct access to the basolateral surface of hepatocytes.

Three distinct plasma membrane domains are present on hepatocytes: the sinusoidal, lateral, and canalicular membranes. The sinusoidal or "basal" surface is covered with numerous irregular microvilli that increase the surface membrane area, thereby facilitating hepatic uptake. The lateral domain is a relatively smooth and straight extension of the sinusoidal membrane that projects between neighboring hepatocytes and lines the intercellular cleft. It contains gap junctions, desmosomes, and tight junctions, but no microvilli. In analogy with other epithelial cells, the sinusoidal and lateral domains together form the basolateral membrane. The apical or canalicular membrane differs histochemically and biochemically from the basolateral surface. Although the lumen of the canalicular space is only ~1 μm in diameter, the surface area is enlarged by an extensive network of microvilli, which often fill the lumen. The canalicular membrane constitutes approximately 13% of the plasma membrane surface.

Separation of blood spaces from the lumen of biliary spaces is accomplished largely by two belt-like junctional complexes that encircle each hepatocyte, enclosing the canaliculi and linking adjacent hepatocytes within the hepatic plate. The canalicular wall is therefore a specialized part of the hepatocellular plasma membrane. The tight junctions that separate bile from plasma were initially considered to be impermeable; however, it is now accepted that the paracellular pathway is the major route for the movement of certain solutes and water between plasma and bile. Studies in perfused rat liver demonstrate that greater than 75% of biliary Na^+, K^+, Cl^-, and water come directly from plasma via the paracellular pathway (Graf et al., 1984, 1987). Hepatocyte junctional complexes are of low electrical resistance and somewhat cation selective, suggesting the presence of negative charges. Studies by Graf et al. (1987) in isolated rat hepatocyte couplets indicate that the cation selectivity of the paracellular pathway is relatively weak. Paracellular permeability appears to be regulated by hormones, Ca^{2+}, and other endogenous factors and is altered by some drugs or chemicals or during hepatic injury. The role of this pathway in the hepatobiliary transport of metals remains obscure.

3.3.2 Functional Polarity

Differences in lipid and protein composition among the three plasma membrane domains generate the hepatocyte's functional polarity (Meier, 1988). The sinusoidal membrane is rich in receptors, enzymes, and transport proteins. Receptors for hormones and transmitters such as insulin, glucagon, vasopressin, and α-adrenergic agonists, and for metabolites, such as lipoproteins, asialoglycoproteins, transferrin, and IgA, are concentrated on this membrane domain. Hormone-sensitive adenylate cyclase and Na^+, K^+-ATPase enzyme activity are also localized to the sinusoidal membrane. Its high phagocytic and endocytic properties are revealed by the presence of clathrin-coated and noncoated vesicles.

Transport proteins are also abundant on the sinusoidal membrane. These proteins facilitate the uptake of ions, nutrients, bile acids, and organic anions and cations, including certain drugs and xenobiotics from blood plasma (Figure 3.1). Hepatic uptake of organic solutes across the sinusoidal membrane is mediated by multiple transport systems, and several classes of transporters have now been identified, including the following: the Na^+-coupled bile acid transporter NTCP (SLC10A1) (Hagenbuch et al., 1991), the electrogenic cation transporters OCT1-3 (SLC22A1-3), the cation and carnitine transporters OCTN1 (SLC22A4), OCTN2 (SLC22A5), and OCT6 (SLC22A16), and the proton/cation antiporters MATE1, MATE2-K, and MATE2-B (Grundemann et al., 1994; Koepsell et al., 2007), as well as the OATP (SLCO) multispecific transporters (Jacquemin et al., 1994; Kullak-Ublick et al., 1994; Noe et al., 1997). The driving force for uptake of rat Oatp1 has been suggested to be the GSH electrochemical gradient (Li et al., 1998). In contrast to this rat Oatp, human OATP1B3/OATP-8 and OATP1B1/OATP-C most likely function as bidirectional facilitated diffusion transporters and GSH is not a substrate or activator of their transport activity (Mahagita et al., 2007). The sinusoidal membrane also has transport systems that facilitate efflux of H^+, Ca^{2+}, amino acids, and possibly other solutes from liver cells back into blood plasma, and some of these are illustrated in Figure 3.1.

The lateral membrane participates in cell–cell interaction, but has not yet been shown to contain specific transport processes. Desmosomes and tight junctions within lateral membranes link adjacent hepatocytes and create the canalicular space, whereas its gap junctions are specialized for cell–cell communication.

On the other hand, the canalicular membrane is replete with membrane transport proteins. Some of these are illustrated in Figure 3.2. Transport systems for organic anions (MRP2/ABCC2), organic cations (MDR1/ABCB1), phospholipids (MDR3/ABCB4), and bile acids (BSEP/ABCB11), along with high specific activities of the enzymes γ-glutamyl transpeptidase, leucylnaphthylamidase, Mg^{2+}-ATPase, and alkaline phosphodiesterase, are localized to the canalicular membrane. MDR1 and MRP2 have been implicated in biliary excretion of metals complexes, as discussed further below. A Cl^-/HCO_3^- exchanger and a Cl^--selective channel have also been identified (Meier, 1988). Lysosomes and endocytic vesicles fuse with the canalicular membrane and discharge their contents (including metals) into bile (Renston et al., 1980).

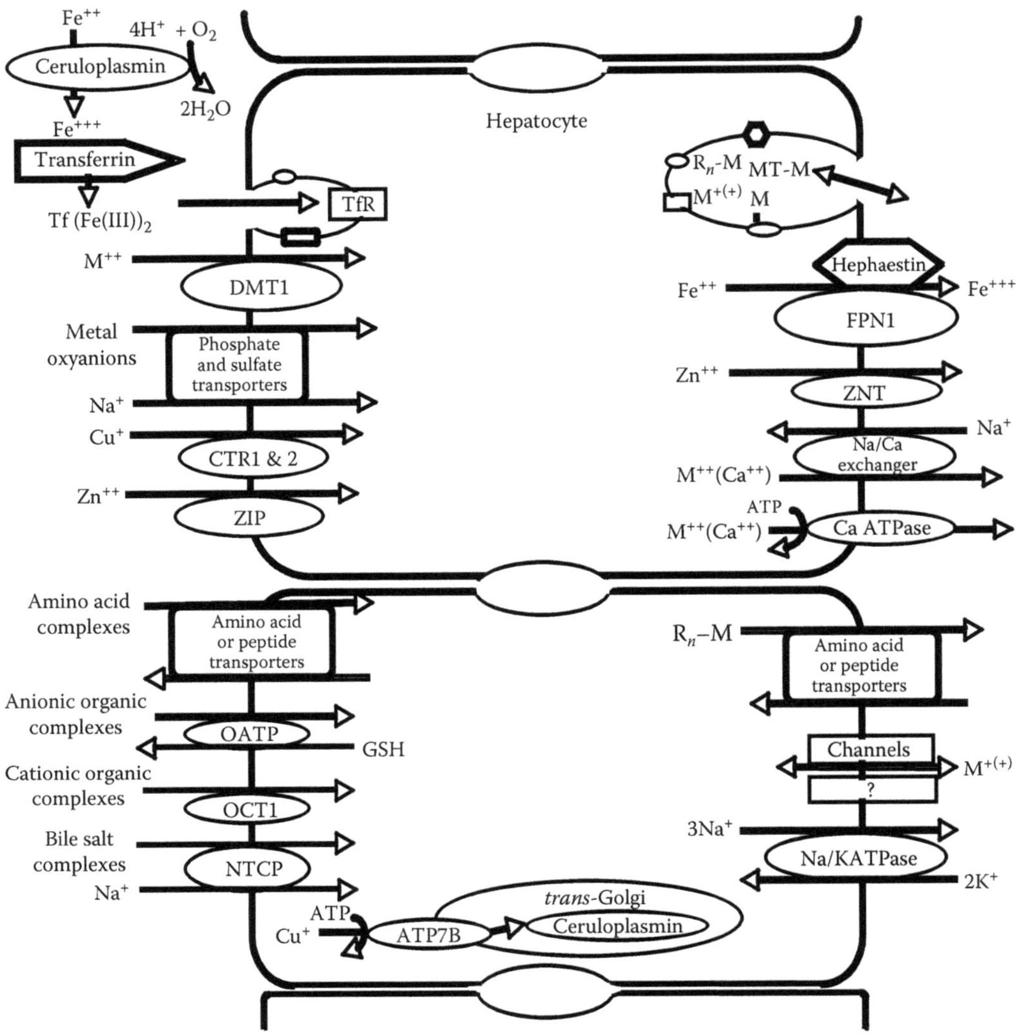

FIGURE 3.1 Putative hepatic sinusoidal metal uptake and efflux pathways.

3.3.3 Bile Secretion and Biliary-Hepatic Cycle

Hepatic bile is an isosmotic electrolyte solution containing mixed lipid micelles of bile acids, cholesterol, and phospholipids, along with porphyrins, proteins, peptides, amino acids, metals, and a variety of other endogenous and exogenous compounds (Boyer, 1986; Klaassen and Watkins, 1984). Bile is secreted largely in response to the osmotic gradient between bile, liver tissue, and plasma, generated by active secretion of impermeant solutes into the canalicular spaces. Water, electrolytes, and other solutes move across the canalicular membrane as well as through the tight junctions to dissipate the osmotic gradient created by active secretion of bile acids (Sperber, 1959), glutathione (Ballatori and Truong, 1989, 1992), and possibly other solutes.

After its secretion, canalicular bile winds its way through the labyrinth of canaliculi toward the periphery of the hepatic lobule, and drains into the hepatic ductules and ducts, and finally into the common duct and gallbladder. At each of these sites, canalicular bile may be modified by both absorptive and secretory processes, and some of its components metabolized by membrane-bound ectoproteins (Ballatori et al., 1986b, 1988), demonstrating that biliary epithelia are not inert conduits

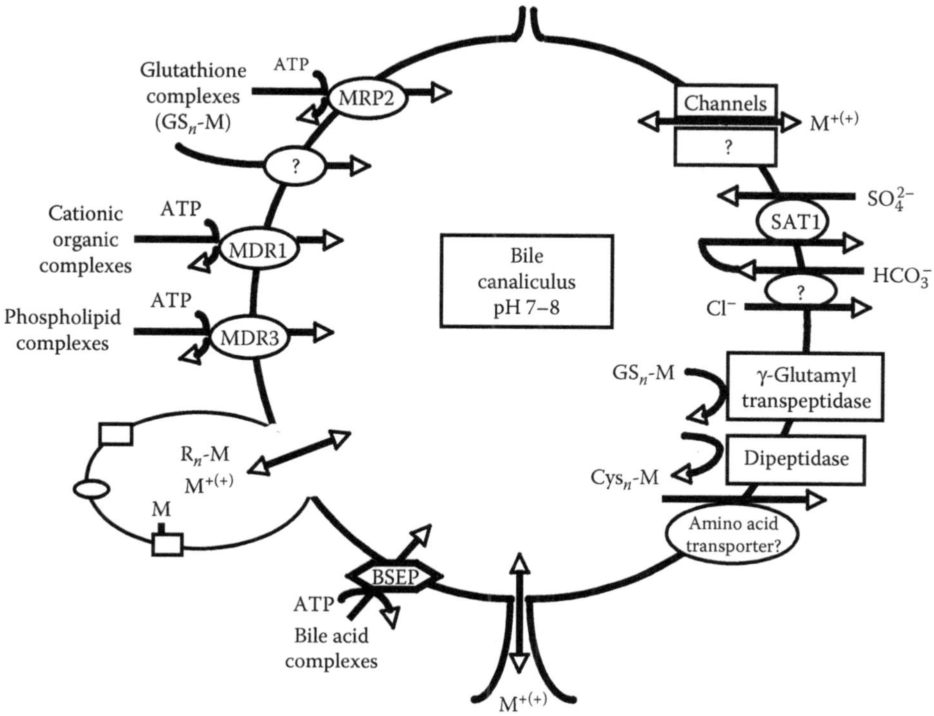

FIGURE 3.2 Hepatocyte bile canalicular membrane metal transport pathways.

for bile, but are actively involved in its postsecretory processing. In analogy with the enterohepatic circulation, there is a biliary-hepatic cycle for amino acids (Ballatori et al., 1986a, 1988; Moseley et al., 1988), and an organometallic compound, methylmercury (Dutczak and Ballatori, 1991). Thus, metals that are secreted by hepatocytes into bile canaliculi may be reabsorbed at more distal sites within the biliary tree (Ballatori, 1994).

3.3.4 Paracellular Pathway for Biliary Excretion

Rather than traversing the hepatocyte basolateral and canalicular membranes, metals may also enter bile via either the paracellular pathway (Figure 3.2) or the "transhepatocellular" pathway (i.e., a transepithelial route across bile ductular epithelia). However, because bile cannot be sampled simultaneously at its source (canaliculus) and within bile ductules, it is difficult to quantitate the role of the ductular epithelium in solute transport into bile or solute reabsorption from bile.

Movement of substances between blood plasma and bile is restricted and regulated primarily by the hepatocellular junctional complex. This junction is normally impermeable to solutes as large as proteins, but as already indicated, it appears to be the predominant route for entry of small ions and water into bile (Graf et al., 1984, 1987). Hepatic junctional complexes are considered relatively "leaky" since the shunt conductance accounts for a large fraction of the total tissue conductance. In freeze-fracture replicas, tight junctions normally contain 3–5 parallel strands that line the perimeters of the canaliculi, but the strand number may vary between 1 and 8. The junctional complexes are also relatively cation selective, suggesting a negatively charged barrier. Thus, movement of negatively charged compounds is impeded relative to cationic compounds of similar size; however, both readily cross the junction. Studies in rat hepatocyte couplets indicate a relatively low ion selectivity of the paracellular pathway (Graf et al., 1987).

Of importance to the present discussion, the Ca^{2+} and Mg^{2+} concentrations in bile are similar to those in blood plasma; these cations apparently enter bile via the paracellular pathway (Boyer, 1986).

It is reasonable to expect that other free divalent metal cations in plasma could also use the shunt pathway, but there is no direct evidence for this hypothesis. For most heavy metals, the role of the paracellular pathway appears minimal, probably because these metals are bound to plasma constituents that are normally excluded from the paracellular pathway. On the other hand, if the junctional complexes that line the paracellular pathway are damaged, paracellular flux might increase. By analogy, massive renal Mg^{2+} and Ca^{2+} wasting is observed in a hereditary disease called hypomagnesemia with hypercalciuria and nephrocalcinosis (FHHNC), and this is suggested to originate from a defect in paracellular permeation in the kidneys (Schmitz et al., 2007). The gene responsible for this disorder was identified by positional cloning and named paracellin-1 (PCLN-1) (Simon et al., 1999).

3.3.5 Plasma and Intracellular Protein Binding

Because formation of metal complexes is highly favored thermodynamically, most heavy metals are present in biological tissues and fluids as complexes, rather than as free cations. Although the thermodynamic stability of coordinate-covalent bonds is typically quite high, they are kinetically labile, so that a given metal may rapidly exchange from one ligand to another. This kinetic lability has proven to be the greatest stumbling block to the isolation and identification of metal complexes in biological fluids and tissues. For example, during tissue homogenization or chromatographic separation, additional binding sites may be exposed (or some eliminated), thus altering the distribution of the metal among various ligands. Reactivity varies between metals and is influenced by the nature of the ligand, whether mono- or multi-dentate, and the pH and ionic strength of the media. Copper for example forms relatively low-affinity complexes with albumin or amino acids, but is tightly bound to ceruloplasmin. Similarly, mercury and cadmium form kinetically labile complexes with amino acids, glutathione, or albumin, but more stable complexes with metallothionein (MT). Chelating agents normally display low specificity for metals: they bind a wide range of metals and differ in their efficacy for various forms of the same metal.

For the heavy metals, detoxification or protection from toxicity usually involves binding to specific proteins, including the following: MTs, which form complexes with copper, zinc, cadmium, mercury, and other metals; ferritin, transferrin, and hemosiderin, which are predominantly iron-binding proteins, but also have some affinity for other metals; and ceruloplasmin, which chelates copper and possibly other metals. Similarly, the biological activity of the essential trace metals is due to their ability to attach to specific prosthetic groups on proteins and other biological molecules. Manganese may be an exception to this generalization, since at least some of its biological functions are related to the free divalent metal.

Conversely, metal-induced toxicity is usually attributed to the reactivity of the "free" metal and is most often observed in tissues involved in their transport, such as intestine, liver, and kidney. Toxic metals, or an excess of trace metals, cause injury by binding to and perturbing the functions of cellular ligands. For example, mercury binds avidly to reduced sulfhydryl groups on cysteine moieties, and interferes in the function of all sulfhydryl-dependent proteins that have been examined (Clarkson, 1972). Because of the shared physicochemical properties between Ca^{2+}, Pb^{2+}, and Cd^{2+}, lead and cadmium probably exert some of their toxic effects by replacing Ca^{2+} on essential proteins or lipids. Similarly, vanadate and arsenate are structurally similar to phosphate and are able to displace phosphate from critical binding sites.

3.4 SINUSOIDAL UPTAKE OF METALS FROM BLOOD PLASMA, AND SECRETION BACK INTO THE BLOODSTREAM

As with other cell types, the major determinants of the chemical interaction that will ensue following exposure of a liver cell to a given metal are the type and speciation of the metal in the extracellular medium (blood plasma), and the presence and activation state of a given metal transport pathway.

Metal binding to the membrane and penetration into the cells are regulated by the affinity and accessibility of a specific chemical form to various membrane ligands, including metal-selective transport systems.

It is becoming increasingly apparent that hepatocytes and other cells possess efficient uptake systems for procuring essential metals, and that these uptake systems are under tight regulatory control (Eide, 1997; Thiele, 2003). Recent evidence indicates the presence of several distinct families of metal transport proteins (see below). The nonessential metals frequently mimic the essential metals and gain access via the same pathways, or enter via less selective pathways (Ballatori, 1994, 2002; Dawson and Ballatori, 1995).

3.4.1 Transport during Membrane Recycling: Roles of Transferrin and Ceruloplasmin

Endocytotic and exocytotic mechanisms play a critical role in hepatic metal homeostasis. The importance of receptor-mediated endocytosis in the transport of iron and copper has long been recognized, and many of the individual steps in this process have been characterized at the molecular level (Aisen, 1998). However, the role of receptor-mediated endocytosis in the transport of other metals and the general roles of fluid-phase and adsorptive endocytosis in metal transport have not been examined to a significant extent.

It is likely that the overall contribution of membrane recycling to metal transport has been underestimated. In rat hepatocytes, the volume of fluid taken up (and released) via the fluid phase pathway(s) is quite large, and far greater than by receptor-mediated pathways (Oka et al., 1989). The total rate of fluid internalization in rat liver cells is between 1 and 6 pL/cell/h (Oka et al., 1989; Scharschmidt et al., 1986), or greater than 20% of their volume each hour. From this rate it can be estimated that between 5 and 50 times the hepatocytes' plasma membrane is endocytosed each hour (Blomhoff et al., 1989; Oka et al., 1989; Scharschmidt et al., 1986). This phenomenal rate of plasma membrane and fluid internalization has obvious implications for the transport of substances across the membrane, yet little information is available on its role in hepatic metal transport. The rapid rate of membrane and fluid transport may be particularly important in facilitating movement of metals that either have a high affinity for plasma membrane binding sites or are bound to ligands that are selectively cleared via their hepatic receptors. Because the process is bidirectional, membrane recycling functions to transport metals to and from blood plasma (Figure 3.1) and to and from bile (Figure 3.2). A recent report has demonstrated endocytotic uptake of MT by HepG2 (human hepatocellular carcinoma) cells via a lipid raft-dependent endocytosis (Hao et al., 2007). Subcellular fractionation after MT uptake revealed significant amounts of MT in vesicular fractions, including lysosomes but virtually no MT in the cytosol. However, metals bound to MT are released into the cytosol. By implication, the findings may define an endocytotic pathway for cellular metal acquisition in hepatocytes *in vivo.*

Internalized endocytotic vesicles move rapidly (within minutes) to lysosomes and the pericanalicular Golgi area. The entrapped metal complexes may be metabolized, released to other intracellular compartments, or transferred in coated and noncoated vesicles to bile canaliculi. As in other epithelia, canalicular membrane biogenesis first requires insertion of integral membrane proteins into the basolateral plasma membrane followed by their selective endocytosis and delivery to the canalicular domain (Bartles et al., 1987). A tubulovesicular transport process has been suggested to function in the transcellular movement of certain proteins into bile (Coleman, 1987; Sakisaka et al., 1988). Although vesicular exocytosis into bile makes only a minimal contribution to the total volume of fluid secreted into bile, <4%, it is the predominant mechanism for biliary excretion of proteins and certain other macromolecules. Similar endocytotic/exocytotic mechanisms are probably quite important in hepatic metal transport, particularly for metals that are complexed with high-molecular-weight plasma constituents that are cleared by the liver. Even if only a small fraction of the metal in plasma were bound to such ligands, the efficiency of the process could make it the predominant pathway. Additional studies are needed to assess the relative contributions of these ubiquitous transport pathways.

The transferrin endocytotic process is relatively well characterized. Most investigators agree that the predominant mechanism of iron transport from blood plasma into hepatocytes is via the transferrin receptors (Graham et al., 2007; Morgan and Baker, 1988), as illustrated in Figure 3.1. Nearly all of the iron in plasma (~99%) is normally associated with transferrin, a protein that binds iron in the Fe(III) oxidation state. Oxidation of Fe(II) to Fe(III) is catalyzed by the copper-containing enzyme ceruloplasmin (Figure 3.1).

The interaction of transferrin with its receptors promotes iron uptake by two mechanisms. First, receptor-mediated endocytosis leads to the internalization of diferric transferrin, followed by release of iron within acidic vesicles, and extrusion of iron-depleted transferrin (apotransferrin). The mechanism by which the released iron is subsequently transferred from the endosome/lysosome to the cytosol is discussed in Section 3.5.2 (namely, it appears to be mediated by DMT1). A second and more speculative uptake mechanism involves the possibility that iron is released at the plasma membrane without internalization of the transferrin–receptor complex (Jordan and Kaplan, 1994; Thorstensen and Romslo, 1988; Thorstensen et al., 1995). In this model, transferrin-bound ferric iron or non-transferrin-bound iron is reduced to ferrous iron extracellularly, removed from the transferrin molecule, and transported into the cell via DMT1 or possibly one of the ZIP proteins (e.g., ZIP14) (Figure 3.1).

In addition to transferrin receptors, ferritin receptors play a major role in hepatic iron uptake (Mack et al., 1983; Osterloh and Aisen, 1989). Kupffer cells release a substantial fraction of the iron acquired by erythrophagocytosis in the form of ferritin, which is efficiently internalized by hepatocytes, via their ferritin receptors. It has been suggested that ferritin may be more important than transferrin in mediating hepatic iron uptake at physiological concentrations of these ligands (Aisen, 1998; Fisher et al., 2007). This issue remains to be resolved.

A third receptor-mediated endocytotic mechanism involves the asialoglycoprotein receptors (ASGP-R) (Oka and Weigel, 1983; Regoeczi et al., 1984; Stockert et al., 1980; Young et al. 1983). Galactose-specific ASGP-R are unique to hepatocytes, and have been identified in every mammalian species examined. One substrate for these receptors is asialotransferrin. Sinusoidal endothelial cells appear to be able to desialate transferrin (Tavassoli, 1988); the iron carried by these sialic acid-depleted molecules can be rapidly cleared from the blood by receptor-mediated endocytosis. Hepatic iron uptake is faster from asialotransferrin than from fully sialylated transferrin (Rudolph et al., 1986). Prata et al. (2006) demonstrated the hepatic receptor-mediated endocytotic uptake of the lanthanide(III) chelates of DOTA-type glycoconjugates in HepG2 cells and in rats. *In vitro* studies show high uptake of radiolabeled [^{153}Sm]-DOTAGal$_2$ by HepG2 cells containing the ASGP-R, which is decreased to less than 50% by the presence of its high-affinity ligand asialofetuin. Similarly, *in vivo* pharmacokinetic studies on Wistar rats using the [^{153}Sm]$^{3+}$-labeled glycoconjugates show a high uptake in the receptor-rich organ liver of the radiolabeled compounds containing terminal galactosyl groups, but very little uptake for those compounds with terminal glycosyl groups. Blocking the receptor *in vivo* reduced liver uptake by 90%, strongly suggesting that the liver uptake of these compounds is via receptor-mediated endocytosis by their binding to the ASGP-R (Prata et al., 2006).

Although the relative contributions of these auxiliary pathways of iron uptake are unknown, it is becoming quite clear that these pathways are important not only in iron-overload states, but work in parallel with the transferrin receptors to mediate normal hepatic iron uptake (Morley and Bezkorovainy, 1985).

Interestingly, iron-binding proteins may also be involved in the transmembrane movement of other metal cations, including manganese, zinc, and vanadium. Most of the manganese in plasma is bound to transferrin, a finding that may explain the rapid hepatic clearance of Mn from plasma. Similarly, a substantial fraction of vanadium in rat plasma is associated with transferrin. Vanadium accumulates preferentially in tissues that are also abundant in iron (liver, spleen, and kidney). Chromatographic separation of vanadium in liver homogenates reveals that the metal coelutes with fractions corresponding to transferrin and ferritin. It has been speculated that ferritin may serve as a general metal detoxicant because of its ability to bind to a variety of metal cations, including Cd, Zn, Be, and Al (Joshi et al., 1989). Metals also bind to albumin, and this complex may also be transported by vesicular mechanisms (Tibaduiza and Bobilya, 1996).

The role of membrane turnover in biliary excretion of metals is discussed in Section 3.6.

3.4.2 Sinusoidal Metal Transporters and Protein Modifiers of Transport

3.4.2.1 Uptake on Phosphate or Sulfate Transporters

One of the earliest metal transport mechanisms to be described was the transport of metal oxyanions on phosphate and sulfate carriers (Clarkson, 1993; van Veen et al., 1994; Wetterhahn-Jennette, 1981). Vanadate and arsenate are structurally similar to phosphate and can compete with phosphate for transport, as well as intracellular binding sites. Indeed, their toxicity is thought to be directly related to this competition. Similarly, chromate, selenate, and molybdate are structurally similar to sulfate, and are substrates for sulfate transporters.

Hepatic phosphate or sulfate transporters have been characterized functionally, and some of the key proteins have been identified at the molecular level. Hepatic phosphate uptake is mediated in part by a Na^+-dependent mechanism (Ghishan et al., 1993), and genes have been identified that mediate Na^+-dependent phosphate uptake when expressed in *Xenopus* oocytes (Chong et al., 1993; Kavanaugh and Kabat, 1996; Miyamoto et al., 1995; Norbis et al., 1997; Virkki et al., 2007). Two unrelated families of Na^+-dependent P_i transporters transport different P_i species in vertebrates: the type II Na^+/P_i cotransporters (SCL34) prefer divalent HPO_4^{2-}, and the type III Na^+/P_i cotransporters (SLC20) transport monovalent $H_2PO_4^-$. The SCL34 family comprises both electrogenic and electroneutral members that are expressed in various epithelia and other polarized cells. These proteins play a role in modulating the P_i content of luminal fluids. The two SLC20 family members PiT-1 and PiT-2 are electrogenic and ubiquitously expressed and may serve a housekeeping role for cell P_i homeostasis (Virkki et al., 2007). Phosphate transporter protein expression was recently confirmed by Western blotting in basolateral and canalicular membranes and by immunofluorescence in rat intact liver (Frei et al., 2005). Transport studies in canalicular membrane vesicles demonstrated sodium-dependent P_i uptake and this was compatible with transport characteristics of sodium-phosphate cotransporters NaPi-IIb, PiT-1, and PiT-2, of which the mRNAs were detected in rat liver (Frei et al., 2005). On the protein level, NaPi-IIb was detected at the canalicular membrane of hepatocytes and at the brush-border membrane of cholangiocytes. In contrast, PiT-1 and PiT-2 were detected at the basolateral membrane of hepatocytes. In addition to these plasma membrane transporters, phosphate transporters are also present in intracellular organelles, including mitochondria (Pratt et al., 1991). Arsenate and vanadate most likely enter hepatocytes and their organelles on these phosphate transporters (Wetterhahn-Jennette, 1981).

Hepatic sinusoidal sulfate uptake also appears to be mediated by a Na^+-dependent mechanism, and Na^+-sulfate cotransporters (the SLC13 transporters) have now been identified at the molecular level (Markovich and Murer, 2004; Pajor, 2006). The SLC13 gene family consists of five members in humans: two of the transporters, NaS1 and NaS2, carry substrates such as sulfate, selenate, and thiosulfate, and the other members of the family (NaDC1, NaDC3, and NaCT) are transporters for di- and tri-carboxylates, including succinate, citrate, and α-ketoglutarate. Functional properties of NaS2 have been characterized in human and rat as well as mouse (Markovich et al., 2005; Dawson et al., 2005, 2006). These sulfate transporters likely mediate hepatic uptake of chromate, selenate, and molybdate; however, additional research is needed to establish this possibility.

3.4.2.2 CTR1 and CTR2, the Copper Uptake Transporters

Although it has long been recognized that the liver is the central organ of copper homeostasis, the molecular basis for hepatic copper uptake was not identified until the mid-1990s (Bull and Cox, 1994; Vulpe et al., 1993; Vulpe and Packman, 1995; Yamaguchi et al., 1996). Important insights into this process were provided by studies in several laboratories, many of which took advantage of the remarkable similarities between yeast and mammalian cells in terms of copper and iron metabolism, and the relative ease with which genetic manipulations can be carried out in yeast. A great deal is now known about how lower eukaryotes, such as yeast, and prokaryotes defend themselves against toxic metals, whereas relatively little is known about defense strategies utilized by mammalian cells. In yeast, resistance to metals is usually due to genes whose products function to sequester metals in intracellular compartments, whereas bacterial defenses are mainly attributed to plasmid-encoded genes whose products function to export metals out of the cell (Silver, 1996; Solioz et al., 1994; Tsai et al., 1997).

Models illustrating the normal pathways of copper metabolism in hepatocytes are provided in Figure 3.3, and those of the yeast *Saccharomyces cerevisiae* in Figure 3.4. As depicted in these figures, there is a remarkable evolutionary conservation in the overall mechanisms of copper homeostasis from yeast to humans (Kaplan and O'Halloran, 1996; Kim et al., 2008; Perego and Howell, 1997; Solioz et al., 1994; Valentine and Grella, 1997).

S. cerevisiae acquire copper as Cu(I) through the action of a plasma membrane-associated Cu(II)–Fe(III) reductase (Fre1), two high-affinity copper transport proteins (Ctrl and Ctr3), and a low-affinity copper transporter (Ctr2) (Figure 3.4). Of the three transporters, only the Ctrl protein is indispensable for high-affinity copper uptake (Dancis et al., 1994). Once within the yeast cell, Cu(I) is transferred to small cytoplasmic proteins that function as intracellular transporters or chaperones. These intracellular copper-binding proteins deliver the copper to selected intracellular proteins or organelles (Figure 3.4). The Atx1 protein, for example, delivers copper to Ccc2, a P-type ATPase that pumps copper into endosomal/lysosomal compartments. Ccc2 is structurally and functionally similar to the gene products defective in Wilson's disease (ATP7B) and Menke's syndrome (ATP7A), and is localized to the same intracellular compartment (Figures 3.3 and 3.4). Ccc2 functions to deliver copper into this space for storage, as well as for insertion into Fet3, a multicopper oxidase that is similar to ceruloplasmin (Yuan et al., 1995).

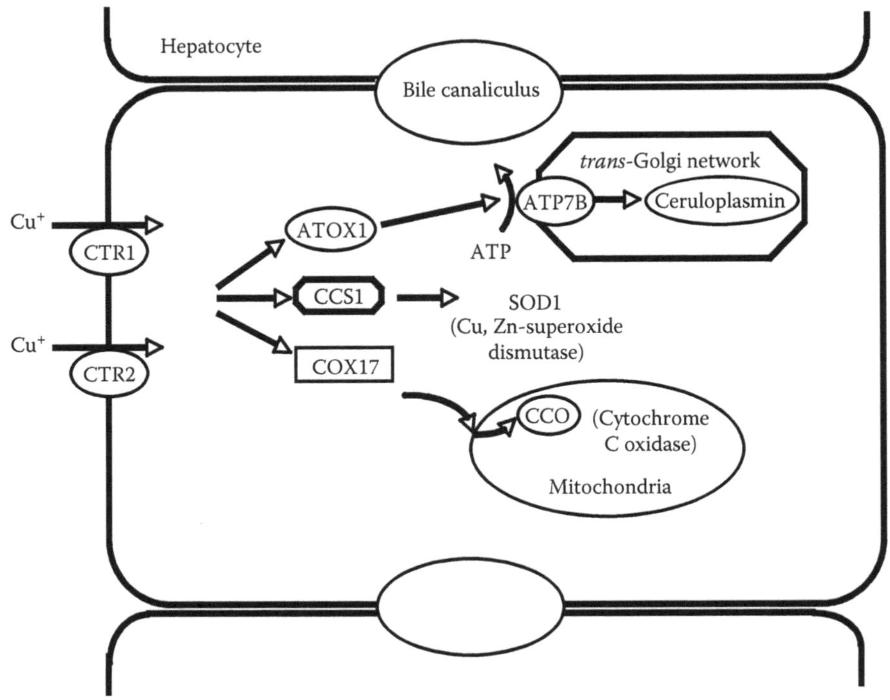

FIGURE 3.3 Model of hepatocellular copper uptake and utilization.

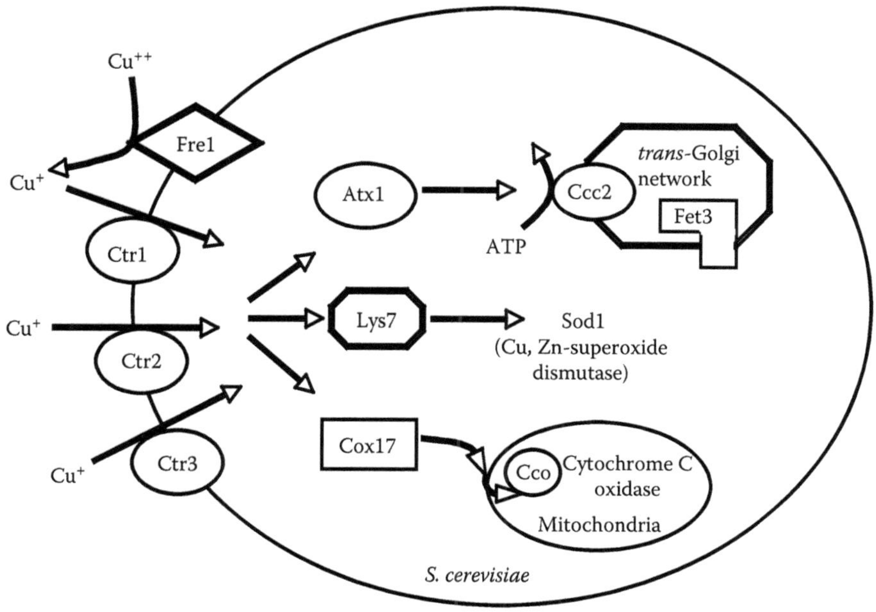

FIGURE 3.4 Yeast copper uptake mechanisms, and intracellular distribution to various copper acceptors.

The similarity between yeast and mammalian copper homeostasis is quite striking, and extends to the membrane transporters, the cytoplasmic chaperones, and the terminal copper acceptors (Figures 3.3 and 3.4). Of significance, Zhou and Gitschier (1997) isolated a human gene involved in copper uptake by complementation of the yeast high-affinity copper uptake mutant, *Ctrl*. The human gene product (CTR1; Figure 3.3) exhibits 29% amino acid identity with yeast Ctrl. A database search by Zhou and Gitschier (1997) revealed an additional human gene that was named *CTR2*. By Northern blot analysis, *CTR1* and *CTR2* were expressed in all the tissues examined, but the liver exhibited the highest level of expression. *Ctrl* mRNA levels in cells and animal models do not appear to change as a function of Cu status. Interestingly, an induction in trafficking of a functional epitope-tagged version of Ctrl from the plasma membrane to endosomal compartments has been demonstrated in response to elevated Cu levels (Petris et al., 2003), suggesting a potential mechanism for regulating high-affinity Cu^{1+} uptake. Ctrl is found both at the plasma membrane and in association with intracellular vesicles in some cultured cell lines and mouse tissues (Lee et al., 2002; Klomp et al., 2002). Although the mechanism for Ctrl trafficking elicited by increasing Cu concentrations is still unclear, it may be that Ctrl may function in Cu uptake at both the plasma membrane and from intracellular compartments (Kim et al., 2008). Western immunoblot analyses, using immunopurified antibody, detected monomeric (23 kDa) and oligomeric forms of Ctrl in the membrane fraction of several mouse organs, and immunohistochemical analyses detected abundant Ctrl protein in liver canaliculi (Kuo et al., 2006).

Once internalized, some of the copper is incorporated into ceruloplasmin, a protein that plays a dual role in facilitating iron uptake into hepatocytes. In addition to catalyzing the oxidation of Fe(II) to Fe(III) for incorporation into transferrin, ceruloplasmin stimulates iron uptake by a transferrin-independent mechanism (Mukhopadhyay et al., 1998). Mukhopadhyay et al. (1998) reported that although ceruloplasmin had no effect on iron uptake or release in iron-replete HepG2 cells, it stimulated iron uptake in iron-depleted cells by increasing the apparent affinity of iron for a trivalent cation-specific transporter (Attieh et al., 1999). Ceruloplasmin-stimulated iron uptake was completely blocked by unlabeled Fe^{3+} and by other trivalent cations including Al^{3+}, Ga^{3+}, and Cr^{3+}, but not by divalent cations (Attieh et al., 1999). The induction of ceruloplasmin-stimulated iron uptake by iron deficiency was also blocked by actinomycin D and cycloheximide, consistent with a transcriptionally induced or regulated transporter. Transcriptional activation of Cp by iron deficiency was later identified to be by hypoxia-inducible factor-1 (HIF-1) that binds to the hypoxia-responsive elements (HREs) at the 5′-flanking region of the human ceruloplasmin gene (Mukhopadhyay et al., 2000).

This requirement of a copper-containing protein for iron transport in yeast and mammalian cells provides at least a partial explanation for the observation that animals that have low ceruloplasmin levels due to either copper deficiency or the inability to synthesize the holoprotein are unable to mobilize iron from tissues (Kaplan and O'Halloran, 1996). Delivery of hepatocellular copper to apoceruloplasmin requires the endosomal transporter ATP7B (Figure 3.3), the protein that is defective in Wilson's disease. Other tissues utilize the homologous transporter ATP7A, the protein defective in Menkes' syndrome, to transport copper into intracellular compartments. Since the liver is a major site of ceruloplasmin synthesis, most Wilson's disease patients have diminished ceruloplasmin levels, yet these levels are nevertheless adequate to maintain iron homeostasis.

Interestingly, CTR1 may mediate cellular uptake of cisplatin (DDP), carboplatin (CBDCA), and oxaliplatin (L-OHP), whereas the efflux of these drugs may be mediated by the two Cu efflux transporters ATP7A and ATP7B (Safaei and Howell, 2005).

3.4.2.3 DMT1/SLC11A2, a Multispecific Metal Transporter of the Nramp Family

An additional mechanism for cellular uptake of ferrous iron and of other divalent metal ions was described by Gunshin et al. (1997). These investigators cloned and characterized a mammalian iron and divalent cation transporter that is expressed in a number of tissues, including liver. DMT1 (NRAMP2, DCT1, SCL11A2), a member of the Nramp family of membrane proteins, is able to transport a variety of divalent metal cations (Fe, Zn, Mn, Co, Cd, Cu, Ni, and Pb), by a proton-coupled and membrane potential-dependent mechanism. Although DMT1 likely contributes to the uptake of toxic metals, more work is needed to establish this possibility. Suzuki et al. (2008) reported that cadmium uptake in Dmt1-deficient mice is similar to control mice, indicating that other transport pathways are mediating cadmium transport.

The Nramp family of membrane-associated proteins displays a high sequence conservation from yeast to humans, with many species expressing at least two discrete gene copies (Cellier et al., 1995). In the intestine, this protein mediates metal uptake at the lumenal surface by cotransport with protons (with a stoichiometry of $1H^+:1M^{2+}$), although uncoupled divalent metal transport has also been demonstrated *in vitro* (Mackenzie et al., 2006). Insights into the role of Dmt1 in iron and manganese homeostasis were obtained from studies in the microcytic anemia (mk) mouse and the phenotypically similar Belgrade (b) rat (Chua and Morgan, 1997; Fleming et al., 1997, 1998), which have orthologous mutations (glycine 185 to arginine) in the *Dmt1* gene. In the liver, Dmt1 appears to be localized to the sinusoidal membrane of rat hepatocytes (Trinder et al., 2000), where it is likely to function as an uptake mechanism for non-transferrin-bound iron, manganese, and other divalent metals. There are four predicted splice variants of *DMT1*, which result from two alternative first exons (1A and 1B), combined with one of two 3′ UTR variations with or without an iron-response element (IRE) (Hubert and Hentze, 2002; Lee et al., 1998). The predominant form present in the liver is the 1B+IRE, although a small amount of the 1B–IRE form is also present. Studies examining the different isoforms indicated that the +IRE isoform is predominantly localized to the plasma membrane, has a relatively slower rate of internalization compared to the –IRE isoform, and is targeted to lysosomes, while the –IRE isoform is more efficiently endocytosed and targeted to recycling endosomes (Lam-Yuk-Tseung et al., 2005, 2006).

3.4.2.4 ZIP/SLC39 and ZNT/SLC30 in Zinc Uptake and Efflux, Respectively

Cellular zinc uptake is mediated in part by members of the Zrt/Irt-like protein (ZIP/SLC39) superfamily of metal ion transporters in mammals (Eide, 2006; Kambe et al., 2004, 2008). There are at least 14 ZIP transporters in human cells. The first member of this family to be identified from *Arabidopsis* was iron-regulated transporter-1 (Irt1), which preferentially imports iron, but can also transport manganese, cadmium, and zinc (Dufner-Beattie et al., 2003). Two other members of this superfamily (zinc-regulated transporter-1 [Zrt1] and Zrt2) were subsequently identified in yeast. Zinc uptake by *S. cerevisiae* is mediated by a high-affinity system that is required for growth in zinc-limited medium (Zrt1) and a lower-affinity system that is active in zinc-replete cells (Zrt2) (Eide, 1997; Zhao and Eide, 1996). One feature shared by all zinc transporters is a histidine-rich domain that probably functions as a coordination site for zinc (Eide, 1997).

Zinc efflux from cells is mediated in part by the ZNT/SLC30 proteins (Liuzzi and Cousins, 2004; Palmiter and Huang, 2004), which belong to the CDF family of metal ion transporters (Paulsen and Saier, 1997; Silver, 1996). There are at least nine ZNT transporters in human cells, and these play critical roles in maintaining zinc homeostasis through efflux and sequestration mechanisms during zinc excesses (Kirschke and Huang, 2003).

Palmiter and Findley (1995) isolated a cDNA encoding a zinc transporter (Znt1) from a rat kidney expression library by complementation of a mutated, zinc-sensitive baby hamster kidney (BHK) cell line. The transporter was localized to the plasma membrane of these cells and was found to mediate both uptake and efflux of zinc, although efflux was considered the physiological direction of transport. The energetics of transport was not identified: zinc transport was unaffected by metabolic poisons, or by the ionic composition of the culture medium (Palmiter and Findley, 1995).

ZnT1, ZnT4, and ZnT5 have been reported to be ubiquitously expressed, whereas ZnT2 is limited to small intestine, kidney, and placenta (Kambe et al., 2002; Liuzzi et al., 2001; McMahon and Cousins, 1998). In a study on the expression of intestinal and hepatic ZnT1 in an intact animal model, McMahon and Cousins (1998) demonstrated that intestinal ZnT1 was most abundant at the basolateral surface of enterocytes lining the villi of the duodenum and jejunum. Dietary zinc supplementation elevated the level of intestinal *ZnT1* mRNA and protein but had no effect in the liver. In response to an acute oral zinc dose, the level of intestinal *ZnT1* mRNA was increased eightfold, without a corresponding increase in ZnT1 protein. Conversely, the acute oral dose did not affect liver *ZnT1* mRNA, but resulted in a fivefold increase in liver ZnT1 protein. Thus, despite the zinc responsiveness of the *ZnT1* gene, additional factors may be regulating the steady-state level of ZnT1 transporter protein.

Palmiter et al. (1996a) have also identified a related transporter that is localized to intracellular endosomal/lysosomal vesicles, ZnT2. This transporter is relatively selective for zinc, but is not expressed in the liver (Palmiter et al., 1996a). This same group has also identified a related protein, ZnT3 whose expression is restricted to brain and testes (Palmiter et al., 1996b). Immunohistochemical analysis of murine brain suggests localization of ZnT3 to synaptic vesicles (Palmiter et al., 1996b). Huang and Gitschier (1997) identified another member of this group of transporters, ZnT4, as the protein that is defective in the inherited zinc deficiency in the lethal milk mouse. ZnT4 was shown to confer zinc resistance on a zinc-sensitive yeast strain, and was abundantly expressed in mammary epithelia and brain (Huang and Gitschier, 1997). Other ZnT proteins have been recently identified, including ZnT5–ZnT9. Human ZNT5 immunoreactivity was found to be associated with secretory granules by electron microscopy and was shown to play an important role in transporting zinc into secretory granules in pancreatic beta cells (Kambe et al., 2002). ZnT6 was predominantly detected in the chief cell of the stomach, columnar epithelial cells of the jejunum, cecum, colon, and rectum, whereas ZnT7 was observed in all epithelia of the mouse gastrointestinal tract with the highest expression in the small intestine (Yu et al., 2007). Expression of ZnT proteins in the absorptive epithelial cell of the gastrointestinal tract suggests that ZnT proteins may play important roles in zinc absorption and endogenous zinc secretion.

3.4.2.5 Ferroportin, Hephaestin, and Hepcidin in Hepatic Iron Export

As illustrated in Figure 3.1, cellular export of iron appears to be mediated by the complementary action of two proteins: a ferroxidase enzyme called hephaestin and a transporter called ferroportin (FPN1, SLC40A1, MTP1, IREG1) (Anderson et al., 2007; Andrews and Schmidt, 2007; Deugnier et al., 2008; Frazer et al., 2002; Graham et al., 2007; Nemeth, 2008). Ferroportin was reported independently by three groups (Abboud and Haile, 2000; Donovan et al., 2000; McKie et al., 2000) and appears to be the sole mediator of iron release from hepatocytes (Donovan et al., 2005). Ferroportin appears to transport ferrous iron, although this has not been shown directly, nor has the driving force for transport been identified (Graham et al., 2007).

Hephaestin is a transmembrane-bound ceruloplasmin homologue that functions as a multicopper ferroxidase (McKie et al., 2001). All residues involved in copper binding and disulfide bond formation in ceruloplasmin are conserved in hephaestin; however, unlike ceruloplasmin, hephaestin is an integral membrane protein with a single transmembrane domain. Under normal conditions, iron is taken up in the small intestine by DMT1 and is either stored bound to ferritin or is transported to the systemic circulation by ferroportin, while being oxidized by hephaestin to be incorporated into transferrin. Hephaestin is mutated in the sex-linked anemic mouse (*sla* mouse), is highly expressed in the intestine, and is necessary for iron egress from intestinal enterocytes into the circulation (Vulpe et al., 1999). Both ferroportin and hephaestin are expressed in hepatocytes at the mRNA level (Zhang et al., 2004). The specific mechanism by which hephaestin and ferroportin interact to mediate cellular metal export has not yet been defined.

Ferrportin function is also regulated by hepcidin, a small peptide synthesized by the liver that can sense iron stores, and inhibits iron export by inducing ferroportin degradation (Mena et al., 2008; Nemeth et al., 2004; Nemeth, 2008). In enterocytes, hepcidin may also inhibit apical iron uptake by decreasing *DMT1* transcription (Mena et al., 2008).

3.4.2.6 Organic Solute Carriers: OATPs, OCTs, NTCP, and Amino Acid and Peptide Transporters

Sinusoidal uptake and efflux of metals may also occur on organic solute carriers, although to date there is relatively little evidence for this possibility. As already mentioned, metals are normally present in biological fluids as complexes with endogenous molecules such as amino acids, peptides, or phospholipids, substances that are themselves transported across cell membranes by carrier proteins. These proteins also appear to be able to transport metal–substrate complexes (Ballatori, 1994; Dutczak and Ballatori, 1994; Kerper et al., 1992). Organic solute carriers are generally multispecific, that is, they accept substrates that differ considerably in their chemical structures, and most are not able to discriminate between substrates whose only modification is the presence of a metal ion. This mode of transport appears to be particularly important for toxic metals, given the absence of selective transport systems for these metals.

Several organic solute transporters have been identified on the sinusoidal membrane of hepatocytes, including the organic anion transporting polypeptides OATP (Jacquemin et al., 1994; Kullak-Ublick et al., 1994; Noe et al., 1997), the organic cation transporter OCT (Grundemann et al., 1994) and the Na^+-taurocholate (bile acid) cotransporting polypeptide NTCP (Hagenbuch et al., 1991) (Figure 3.1). Amino acid transporters (Malandro and Kilberg, 1996) and peptide transporters (Fei et al., 1994) have also been identified that could function both for hepatic uptake and release of these organic solutes. Theoretically, these sinusoidal carriers may facilitate transport of metals bound to their respective substrates (Figure 3.1). For example, an OATP protein appears to mediate hepatic uptake of an MRI contrast agent, the gadolinium complex Gd-B 20790 (Jigorel et al., 2005; Pascolo et al., 1999). Several groups have provided evidence for cellular copper or zinc uptake as histidine complexes (Aiken et al., 1992; Harris, 1993; Horn et al., 1995; Horn and Thomas, 1996), and our laboratory has demonstrated transport of methylmercury as a cysteine complex (Ballatori, 1994; Ballatori and Truong, 1995b; Dutczak and Ballatori, 1994; Kerper et al., 1992; Mokrzan et al., 1995). In erythrocytes, zinc also appears to be taken up as an anionic complex ($[Zn(HCO_3)_2Cl]^-$) through the anion exchanger (Torrubia and Garay, 1989).

3.4.3 Uptake and Efflux on Sinusoidal Membrane Channels and Pumps

The role of ion channels and primary active pumps in facilitating hepatic sinusoidal metal transport appears to be minimal, although these pathways may be important for metal transport in other cell types. For example, in excitable tissues Cd^{2+} and Pb^{2+} may enter cells via voltage-sensitive channels (Hinkle et al., 1987; Reuter, 1983; Simons and Pocock, 1987). Studies by Hinkle et al. (1987) in a pituitary cell line demonstrate that one route of cadmium uptake in these cells is via voltage-gated dihydropyridine-sensitive calcium channels. The voltage-gated calcium channels also admit Ba^{2+} and Sr^{2+} and are inhibited by a number of divalent metal cations.

In hepatocytes however, voltage-sensitive calcium channels have not been identified (Hughes and Barritt, 1989; Kawanishi et al., 1989). Calcium uptake by hepatocytes occurs largely through receptor-activated calcium channels, which could theoretically allow other divalent cations to enter the cell. Support for this possibility is provided by the observation that hepatocyte receptor-activated calcium channels are inhibited by Zn, Cd, Ni, Co, and Mn (Hughes and Barritt, 1989); however, the nature of the inhibition by the metals is unknown. Crofts and Barritt (1990) indicate that Mn^{2+} can move into hepatocytes through the receptor-activated Ca^{2+} inflow system, identifying a potential regulated mechanism for hepatic manganese uptake.

Several other interactions between hepatic uptake of calcium and other metals have been described. Studies *in vivo* (Tang et al., 1987) and with isolated rat hepatocytes (Sorensen, 1988) demonstrate higher cadmium uptake as the concentration of calcium is lowered. However, this effect is probably indirect since cadmium has no effect on ^{45}Ca influx (Sorensen, 1988). This investigator describes several mechanisms whereby changes in Ca^{2+} may alter the plasma membrane's physicochemical properties, and indirectly influence its transport properties. In contrast to these findings in intact hepatocytes, passive Ca^{2+} influx in isolated plasma membrane vesicles is inhibited by Zn^{2+}, Cu^{2+}, Ni^{2+}, Co^{2+}, Hg^{2+}, and *p*-chloromercuribenzoate (Bygrave et al., 1989). The involvement of sulfhydryl groups in this interaction has been suggested by Bygrave et al. (1989).

Metals also interact with plasma membrane Ca^{2+}-ATPases, the predominant proteins for hepatic Ca^{2+} efflux (Blitzer et al., 1989; Pavoine et al., 1987; Suzuki and Kawakita, 1993), raising the possibility that other metals may utilize these ATPases to exit the liver cell and enter blood plasma (Figure 3.1). Distinct Ca^{2+}-ATPases may be present on the canalicular and basolateral membranes (Blitzer et al., 1989). However, the existence of a Na^+-dependent hepatic Ca^{2+} efflux system (i.e., Na^+/Ca^{2+} exchanger) has not been established, although its existence and a hypothetical role in metal transport remain (Figure 3.1).

3.4.4 Permeation as Lipid-Soluble Forms or by Convection

Although often invoked to explain the "nonsaturable" component of metal transport in *in vitro* hepatocyte systems, it is unlikely that simple diffusion and convection play a role in metal transport across the plasma membrane. Diffusion denotes the nonmediated movement of a solute down its electrochemical gradient, leading to the dissipation of free energy in that gradient. Convection or bulk flow is the vectorial movement of an assembly of molecules under an imposed external potential such as hydrostatic pressure.

Most metals and metal complexes are polar, hydrophilic, and quite often charged, properties that would exclude them from the hydrophobic interior of the plasma membrane. Simple diffusion through a membrane protein matrix or ion channel is also highly unlikely. The reason for the preponderance of "nonsaturable" components of transport in *in vitro* systems is unclear. It is most likely an artifact of either the cell or tissue isolation procedure, or of the experimental conditions used to examine transport. Kinetics consistent with simple diffusion may be obtained if (1) the concentration of metal substrate is below the K_m of the transport system. The effective metal substrate concentration in the incubation media is often artificially lowered by extensive binding to nonsubstrate ligands; (2) there is a large component of nonspecific high-affinity (or essentially irreversible) binding to cellular constituents; (3) transport occurs across a cell membrane damaged either during cell isolation or by increasing concentrations of the metal itself. For example, protease contaminants in the collagenase used for hepatocyte isolation can degrade transferrin receptors, thereby inhibiting the predominant iron uptake mechanism and accelerating other nonspecific transport pathways

(Morgan and Baker, 1986). Finally, (4) uptake occurs via vesicular endocytosis/exocytosis. In this case, the metal would be transported nonspecifically along with the entrapped fluid and/or membrane vesicle components. All of these mechanisms need to be considered before attributing a particular component of transport to "simple diffusion."

Studies of metal transport in isolated hepatocyte systems are particularly susceptible to these artifacts. First, since a given metal normally binds to many endogenous ligands, some of which may be unidentified, it is difficult to reproduce *in vivo* conditions in a cell culture system. As discussed above, metals are normally complexed with amino acids, peptides, proteins, phospholipids, and other tissue constituents, and are distributed among all competing ligands on the basis of concentrations and relative affinities of available ligands. Because it is difficult to reproduce *in vivo* conditions in a culture dish, interpretation of transport studies performed *in vitro* should be performed with care.

Second, the propensity of metals to bind to things such as culture dishes, filters, components of the culture media, or more importantly cell membranes, creates both a practical and theoretical dilemma. This "nonspecific" attachment is difficult to eliminate and even more difficult to quantitate. When metal salts are added to protein- and amino acid-free culture media, the metal frequently associates rapidly with the cells. This rapid uptake is not due to the presence of an efficient "transport system" as often suggested, but may simply reflect the absence of competing ligands.

Third, as already indicated, the loss of hepatocyte polarity, the dedifferentiation, and the damage that may occur during cell isolation and culture may alter the hepatocyte's basic transport mechanisms. Many investigators using isolated hepatocytes tacitly assume that the transport that they observe is localized to the basolateral membrane. While this is usually a good approximation, there are exceptions. Loss of cell polarity in isolated hepatocytes implies that all plasma membrane domains contribute to the transport processes.

3.5 INTRACELLULAR BINDING, SEQUESTRATION, AND METABOLISM

3.5.1 Intracellular Ligands

Intracellular sequestration of essential metals is a regulated process. As discussed above for copper (Figure 3.3), there are specific metal-binding ligands that direct essential intracellular metals to their appropriate destination within the cell, or that control the reactivity (toxicity) of both essential and nonessential metals. However, only a few of these intracellular metal chaperones or transporters have been characterized at the molecular level (Aballay et al., 1996; Cousins, 1985; Koch et al., 1997; MacKenzie et al., 2008). In addition to these relatively selective ligands, there are a number of other compounds that play a more general role in metal disposition, including MT, ferritin, citrate, ascorbate, GSH, and amino acids. The relative abundance and affinity of these ligands are an important determinant of the transport of metals to various intracellular organelles, or back into the extracellular space (bile or blood plasma).

3.5.2 Metal Uptake and Release from Endosomal/Lysosomal Compartments: ATP7A, ATP7B, DMT1, and ZNT2

As already indicated, endosomal and lysosomal vesicles are key intracellular organelles for the storage and metabolism of copper, iron, and possibly other metals (Kim et al., 2008). For example, normal hepatocytes transport copper into the *trans*-Golgi network using the ATP-dependent transporter ATP7B, the transporter that is defective in Wilson's disease (Figure 3.5). ATOX1 delivers bound Cu^{+} to ATP7B at the secretory compartment, where it is then transferred from the chaperone to an N-terminal binding domain on the transporter, although it is still unclear how the metal is subsequently transferred to the transmembrane domain for movement across the membrane (Anastassopoulou et al., 2004; Banci et al., 2007; González-Guerrero et al., 2008; Pufahl et al., 1997; Wernimont et al., 2000). This same mechanism is also utilized in other cells for ATP7A

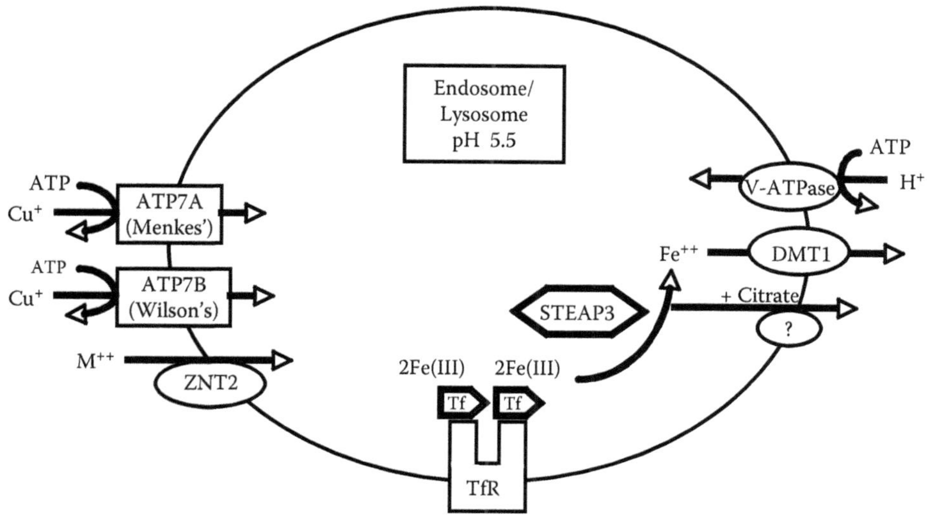

FIGURE 3.5 Possible mechanisms for metal sequestration and release from endosomal/lysosomal compartments.

(Kim et al., 2008). Within this intracellular compartment, copper is then used for the synthesis of copper-containing proteins such as ceruloplasmin, or it is stored for subsequent excretion (Davis et al., 1996). Some of the intravesicular copper is presumably sorted into vesicles destined for the lysosomal–biliary excretory pathway. According to this model, fusion of exocytic vesicles with the canalicular membrane delivers copper into bile (Figure 3.2); however, neither the site nor the mechanism by which this vesicular sorting occurs is known.

In tissues other than the liver, a comparable P-type ATPase (ATP7A) pumps copper into endosomal/lysosomal compartments. The Wilson and Menkes disease genes have 56% overall identity. The Menkes gene (*ATP7A*) is ubiquitously expressed in adult tissues, with little or no expression in the liver (Kuo et al., 1997), whereas the Wilson disease gene is expressed in only a few cell types, notably liver and certain neuronal cells. Mutations in *ATP7A* leads to significant copper accumulation in intestinal mucosa, kidney, and selected other tissues. This inability to deliver copper from sites of its absorption and storage results in a systemic copper insufficiency in Menkes patients. In contrast with the copper insufficiency of Menkes patients, Wilson's disease patients accumulate excess copper in many tissues (e.g., liver, brain, kidney, and cornea) due to the inability to excrete copper into bile, the main route of its elimination. At the cellular level, copper accumulates in the cytosol and the cell eventually succumbs to copper toxicity.

Cells utilize different mechanisms to transfer Fe to and from intracellular vesicles (Figure 3.5). Transferrin receptor-mediated endocytosis leads to the internalization of ferric iron, and acidification of the organelle by V-ATPases promotes the dissociation of ferric iron from transferrin and the dissociation of transferrin from its receptor (Figure 3.5). The mechanism utilized to transport iron and other metals from these compartments back into the cytosol has only recently been defined. Namely, DMT1 appears to be a major contributor to metal efflux from the endosomal compartments. DMT1 colocalizes with TFR1 and cycles through the endosomal compartment, appearing in acidic endosomes (Canonne-Hergaux et al., 2001; Su et al., 1998; Tabuchi et al., 2002; Touret et al., 2003). In addition, DMT1 transports iron optimally at pH 5.5 (Garrick et al., 2006), consistent with its presence in acidic endosomes and suggesting that the energy for iron transport may be provided by a proton gradient. Furthermore, work with the Belgrade rat, which is deficient in Dmt1, has implicated Dmt1 in endosomal Fe transport (Fleming et al., 1998; Garrick et al., 1993). Recently, the gene *STEAP3* has been identified as a candidate vesicular ferrireductase (Ohgami et al., 2005). *STEAP3* ("sixtrans-membrane epithelial antigen of the prostate 3") is one of four genes wholly or partially deleted in the *nm1054* iron deficiency anemia mouse. It is highly expressed in the liver and colocalizes in endosomes with TFR1 and DMT1. Of interest, iron uptake and ferrireductase activity were reduced in reticulocytes obtained from *nm1054* and *steap3* knockout mice. In addition, overexpression in HEK293T cells resulted in increased ferrireductase activity. The other three members of the Steap family (STEAP1, 2 and 4) have also been shown to be ferri- and cupric-reductases (Ohgami et al., 2006). The four genes are ubiquitously expressed; however, adult liver expresses predominantly STEAP3 with a small amount of STEAP1 (Ohgami et al., 2005, 2006). Like STEAP3, the other Steap proteins colocalize, at least partially, in an endosomal compartment with transferrin and TFR1.

An additional metal transporter that is localized to these intracellular vesicles is Znt2 (Palmiter et al., 1996a), a member of the CDF family of metal transporters (Figure 3.5). Palmiter et al. (1996a) reported that BHK cells that express Znt2 are able to accumulate zinc to levels that are much higher than are tolerated by cells that lack this gene product. Znt2 was able to confer resistance on zinc, but not to cadmium, cobalt, or copper, indicating that it is relatively specific for zinc. Zinc accumulation into endosomal/lysosomal vesicles was unaffected by bafilomycin, an inhibitor of the vacuolar-type H^+-ATPase, indicating that zinc uptake does not depend on the proton gradient. As with Znt1, Znt2 does not have recognizable nucleotide-binding domains, indicating that ATP is not directly required for transport function. Znt2 is expressed in many tissues, including kidney, intestine, seminal vesicles, and testes, but not liver (Palmiter et al., 1996a). Other speculative possibilities for lysosomal export of metals include transport as citrate complexes (Figure 3.5).

3.5.3 Nuclear Uptake of Metals

Nuclear metal homeostasis is probably also regulated by transport pathways, but few studies have examined this issue. Gurgueira and Meneghini (1996) demonstrated the presence of an ATP-dependent iron uptake system in isolated rat liver nuclei. Nuclear iron accumulation was mediated by an ATP-dependent mechanism when nuclei were incubated with either Fe(III)-citrate or Fe(III)-ATP, two physiologically important iron complexes. The characteristics of uptake indicate that a P-type ATPase mediates iron uptake, and that this ATPase is distinct from the nuclear membrane Ca^{2+}-ATPase (Gurgueira and Meneghini, 1996).

3.5.4 Mitochondrial Metal Uptake

Mitochondria have regulated calcium transporters that are important for Ca^{2+} homeostasis; however, the molecular mechanisms for transport of other metals in mitochondria have not been elucidated. One possibility is that the calcium transporters participate in mitochondrial uptake of heavy metals, but this has not been tested directly. A number of metals can be taken up via the mitochondrial Ca^{2+} uniporter and can thus be sequestered in mitochondria ($Sr^{2+} > Mn^{2+} > Ba^{2+} \gg La^{3+}$) (Gavin et al., 1999). Ca^{2+} efflux occurs via both Na^+-dependent and Na^+-independent mechanisms. However, Mn^{2+} transport out of the mitochondria is a much slower process, and occurs via the Na-independent efflux mechanism, which is the most important mechanism in liver mitochondria. In addition, intramitochondrial Mn^{2+} inhibits Ca^{2+} efflux via both the Na^+-dependent and Na^+-independent mechanisms (Gavin et al., 1999). The existence of a Na^+-dependent hepatic Ca^{2+} efflux system (i.e., Na^+/Ca^{2+} exchanger) has been reported, although the rate of transport in the liver mitochondria is much lower compared to heart mitochondria (Chiesi et al., 1987; Goldstone and Crompton, 1982).

A cobalt transporter (COT1) was identified in mitochondria from the yeast *S. cerevisiae* (Conklin et al., 1994). COT1 confers resistance on cobalt, and to a lesser extent on zinc, presumably by sequestering the metal in the mitochondria. COT1 is homologous to yeast ZRC1 and mammalian ZNT transporters, proteins that confer resistance on zinc. Each of these proteins has six membrane-spanning domains, a large histidine-rich intracellular loop, and long C-terminal tails; however, the energetics of metal ion transport have not been established for any of these proteins (Conklin et al., 1994; Palmiter et al., 1996a, 1996b).

3.6 TRANSPORT ACROSS THE CANALICULAR MEMBRANE INTO BILE

A model illustrating the mechanisms of metal transport at the liver cell bile canaliculus is provided in Figure 3.2, and the individual transport systems are described below.

3.6.1 Transport during Membrane Recycling: Role of the Gene Products Defective in Wilson's Disease (ATP7B) and Menkes' Syndrome (ATP7A)

Biliary excretion is the most important route for the elimination of copper, and three mechanisms for copper transport across the canalicular membrane have been identified: direct transport via ATP7B across the canalicular membrane, ATP7B-mediated transport into lysosomes followed by exocytosis into bile, and transport as a GSH complex (Ballatori, 1994; Deugnier et al., 2008; Graham et al., 2007; Hernandez et al. 2008; Kim et al., 2008; La Fontaine and Mercer, 2007; Lutsenko et al., 2007; Roelofsen et al., 2000). In copper overload conditions, the ATP7B protein that is normally retained in the *trans*-Golgi network is inserted into the bile canalicular membrane for direct transport of copper into bile, and there is also an increase in lysosomal exocytosis into bile (Hernandez et al., 2008; La Fontaine and Mercer, 2007). Sternlieb et al. (1973) found that the specific activity of copper in hepatic lysosomal fractions from a patient with Wilson's disease was similar to that in bile, suggesting that lysosomes might be the source of biliary copper. However, Gross et al. (1989) suggest that this process may not be as important when hepatocytes contain a normal amount of copper. Nevertheless, copper is normally present in relatively high concentrations in lysosomes and could contribute to its basal excretion.

Evidence in support of this mechanism was provided by the demonstration that the gene products defective in Wilson's disease (ATP7B) and Menkes' syndrome (ATP7A) are localized largely to intracellular vesicles (Dierick et al., 1997; Nagano et al., 1998; Yang et al., 1997), but traffick to the plasma membrane in the presence of excess copper (Hernandez et al., 2008; La Fontaine and Mercer, 2007). ATP7A, ATP7B, and other ATP-driven heavy metal pumps appear to form a separate branch of the ion-transporting P-type ATPases, the CPx-type ATPases, based on a common conserved intramembraneous cysteine–proline–cysteine or cysteine–proline–histidine motif (Solioz and Vulpe, 1996). The major defect in Wilson's disease is the inability to excrete copper into bile, which causes an accumulation of copper in various tissues, including liver and brain. Most Wilson's disease patients also have a reduction in serum ceruloplasmin activity. The Wilson's disease gene is expressed predominantly in the liver, whereas the Menkes' syndrome gene is expressed at low levels in the liver and higher levels in many other tissues. Menkes' syndrome patients accumulate copper at high levels in many tissues, but liver copper concentrations are normal.

The third mechanism of copper export involves transport as a GSH complex on the ATP-dependent canalicular membrane pump MRP2, or possibly on a related transporter (Figure 3.2; Ballatori, 1994), as discussed further below. Interestingly, patients with Wilson's disease have markedly lower hepatic concentrations of GSH (Summer and Eisenburg, 1985). Whether this is a consequence or a cause of hepatic copper accumulation is unknown.

Vesicular exocytosis may also be involved in biliary excretion of iron and possibly other metals (LeSage et al., 1986; Ramm et al., 1994; Regoeczi and Chindemi, 1995). Regoeczi and Chindemi (1995) measured the translocation of different forms of transferrin from blood to bile in the rat, and found that only a small fraction of the metal-containing protein is excreted in bile.

3.6.2 Canalicular Membrane Transporters and Pumps

3.6.2.1 MRP2-Mediated Excretion of GSH Complexes and Other Complexes

Among the metal-binding ligands, GSH is one of the most versatile and pervasive (Ballatori, 1994). GSH is the most abundant nonprotein sulfhydryl-containing compound within cells, at concentrations of 1–10 mM. The sulfhydryl group on the cysteine moiety has a high affinity for metals, forming thermodynamically stable but kinetically labile mercaptides with a number of metals including mercury, silver, cadmium, arsenic, lead, gold, zinc, and copper. Studies have demonstrated that the abundance of GSH, along with the high turnover rate for this tripeptide, and its adducts via both intra- and interorgan cycles play key roles in metal transport and metabolism (Ballatori, 1994; Ballatori and Clarkson, 1982, 1983). Of significance, membrane transport of several metals is accomplished by efflux of the corresponding GSH mercaptides on membrane proteins that normally transport either glutathione *S*-conjugates or GSH itself (Ballatori, 1994; Ballatori and Truong, 1995a; Dijkstra et al., 1995; Dutczak and Ballatori, 1994). The transporters for the glutathione *S*-conjugates have been identified at the molecular level (Cole et al., 1994; Deeley and Cole, 1997; Ito et al., 1997; Keppler et al., 1997; Kool et al., 1997; Paulusma et al., 1996, 1997), whereas GSH transporters remain poorly characterized (Ballatori et al., 2005).

GSH export is required for the delivery of its constituent amino acids to other tissues, detoxification of drugs, metals, and other reactive compounds of both endogenous and exogenous origin, protection against oxidant stress, and secretion of hepatic bile. Recent studies indicate that some members of the multidrug resistance-associated protein (MRP/CFTR or ABCC) family of ABC proteins, as well as some members of the organic anion transporting polypeptide (OATP or SLC0) family of transporters, contribute to this process (Ballatori et al., 2005). The ABCC family has 12 functional members, including nine *MRP* genes in humans (*MRP1–MRP9*), and much progress has been made in the functional characterization of these transporters (Bera et al., 2002; Borst et al., 1997; Buchler et al., 1996; Cole et al., 1994; Deeley and Cole, 1997; Hopper et al., 2001; Ito et al., 1997; Keppler et al., 1997; Kool et al., 1997, 1999; Paulusma et al., 1996, 1997; Tammur et al., 2001; Wijnholds et al., 1997). Five of the 12 members of this family appear to mediate GSH export from cells, namely, MRP1, MRP2, MRP4, MRP5, and CFTR. Additionally, two members of the OATP family, rat Oatp1 and Oatp2, have been identified as GSH transporters (Li et al., 1998, 2000). For the Oatp1 transporter, efflux of GSH may provide the driving force for the uptake of extracellular substrates. In humans, OATP-B and OATP8 do not appear to transport GSH (Mahagita et al., 2007); however, other members of this family have yet to be characterized in regard to GSH transport. Because transport is a key step in GSH homeostasis and is intimately linked to its biological functions, GSH export proteins are likely to modulate essential cellular functions.

MRP1 is found on the plasma membrane of many cell types, but not in the liver (Kool et al., 1997), whereas MRP2, which is also referred to as cMOAT (canalicular multispecific organic anion transporter) or cMRP (canalicular MRP), is selectively localized to the apical plasma membrane of specific transporting epithelia, including the hepatocyte canalicular membrane (Figure 3.2), and the apical membrane of kidney proximal tubules. Both MRP1 and MRP2 have a broad specificity for conjugates formed from GSH, glucuronides, or sulfate. Staining for MRP3 in normal human liver revealed high levels in the bile duct epithelial cells and lower levels in the basolateral membranes of hepatocytes (Kool et al., 1997). Although levels in the liver appear to be low across species, MRP4/Mrp4 has been localized to the basolateral membrane of human, mouse, and rat hepatocytes, and of HepG2 cells, providing an alternative pathway for these cells to transport GSH and monoanionic bile salts across the sinusoidal membrane into the blood (Rius et al., 2003).

There is now considerable evidence that MRP1, MRP2, and related membrane proteins also mediate cellular efflux of glutathione–metal complexes. First, cells overexpressing MRP1 are more resistant to arsenite, antimonite, and cisplatin, consistent with accelerated metal efflux, although resistance to the metals is relatively low (Borst et al., 1997; Chen et al., 1997; Deeley and Cole, 1997). These metals can form divalent and trivalent complexes with GSH *in vitro*, and presumably are forming these GSH complexes *in vivo* as well. Cells expressing mammalian MRP2 are also more resistant to arsenic toxicity (Kojima et al., 2006; Lee et al., 2006; Liu et al., 2002). Zaman et al. (1995) demonstrated MRP-mediated extrusion of [^{35}S]cysteine-containing compounds after addition of arsenite, suggesting coupled efflux. Second, Ishikawa et al. (1996, 1997) demonstrated enhanced transport of a glutathione–platinum complex in membrane vesicles prepared from a tumor cell line overexpressing MRP. Third, mutant rats that lack Mrp2 activity (TR$^-$, GY, or EHBR rats) exhibit impaired ability to transport several metals into bile, including zinc, excess copper, silver, cadmium, and methylmercury (Ballatori et al., 1995; Dijkstra et al., 1996, 1997; Houwen et al., 1990; Madejczyk et al., 2007; Sugawara et al., 1997). In contrast, basal excretion of endogenous copper is unaffected in these mutant rats (Houwen et al., 1990). Fourth, the predicted MRP1 protein shows 43% amino acid identity with the YCF1 protein of *S. cerevisiae*, a protein that confers cadmium resistance by transporting and sequestering the glutathione–cadmium complex [$Cd(GS)_2$] in the yeast vacuole (Li et al., 1996, 1997; Szczypka et al., 1994; Wemmie et al., 1994). Of significance, human *MRP1* cDNA can complement the loss of cadmium resistance in *YCF1*-disrupted yeast cells (Tommasini et al., 1996). The predicted MRP1 protein sequence also shows 34% amino acid identity with the *Leishmania* PgpA, ltpgpA (Cole et al., 1994), a transporter involved in resistance to arsenite and antimonite (Callahan and Beverley, 1991; Dey et al., 1996). LtpgpA confers resistance to arsenite and trivalent antimonials, but not to pentavalent antimonials, zinc, cadmium, or the typical MDR substrates vinblastine and puromycin (Callahan and Beverley, 1991). Dey et al. (1996) demonstrated ATP-dependent transport of the As(III)–glutathione complex [$As(GS)_3$] in membrane vesicles of *Leishmania tarentolae*. In addition, a family of four MRP-related proteins has been identified in the nematode *Caenorhabditis elegans*, one of which (ceMRP1) confers resistance to cadmium and arsenite (Broeks et al., 1996). *C. elegans* MRP1 is 47% identical to human MRP1 at the predicted amino acid level (Broeks et al., 1996).

3.6.2.1.1 Intrabiliary Degradation of GSH–Metal Complexes

Once secreted across the canalicular membrane into bile, GSH, glutathione *S*-conjugates, as well as glutathione complexes are degraded by γ-glutamyl transpeptidase and dipeptidase activities on the canalicular membrane of hepatocytes and the luminal surface of bile ductular epithelia (Figure 3.2; Ballatori et al., 1986a; Hinchman et al., 1991; Hinchman and Ballatori, 1994; Simmons et al., 1991), and the resulting products are partially reabsorbed from bile back to liver (Ballatori et al., 1986b, 1988; Hinchman et al., 1991; Moseley et al., 1988; Simmons et al., 1992). Species differences in biliary glutathione degradation result from differences among species in γ-glutamyl transpeptidase activity and localization within the liver (Ballatori et al., 1988; Hinchman and Ballatori, 1990).

Studies in γ-glutamyl transpeptidase-deficient mice provide direct evidence for a major role of this enzyme in regulating the metabolism, tissue distribution, and elimination of methylmercury and inorganic mercury (Ballatori et al., 1998a). These findings provide direct evidence for a major role of γ-glutamyl transpeptidase in regulating the tissue distribution and elimination of methylmercury and inorganic mercury (Ballatori et al., 1998a), and provide strong support for the observation that *N*-acetylcysteine may be an excellent antidote in methylmercury poisoning (Aremu et al., 2008; Ballatori et al., 1998b).

3.6.2.1.2 Reabsorption of Metals from Bile

The methylmercury that is secreted into bile is extensively reabsorbed from the gallbladder (Dutczak et al., 1991), and from intrahepatic biliary spaces (Dutczak and Ballatori, 1992). Reabsorption diminishes the amount excreted in bile, the main route of methylmercury elimination, and may account for the relatively long biological half-life and toxicity of this metal. Methylmercury reabsorption from guinea-pig biliary spaces back into the liver is facilitated by the γ-glutamyl transpeptidase-mediated conversion of the GSH complex to the cysteine complex (Figure 3.2). The guinea-pig liver has relatively high γ-glutamyl transpeptidase activity (Hinchman and Ballatori 1990), such that nearly all of the GSH (Ballatori et al., 1988) and GSH conjugates (Hinchman et al., 1991) secreted into bile are catabolized within the biliary tree. Dutczak and Ballatori (1992) demonstrated that γ-glutamyl transpeptidase also catalyzes the initial step in the conversion of the biliary methylmercury–glutathione complex to the cysteine complex and that in the latter form methylmercury is readily reabsorbed from the biliary tree.

Similarly, methylmercury reabsorption from guinea-pig gallbladder was more extensive when the metal was given as a cysteine or GSH complex, as compared with an albumin complex (Dutczak et al.,

1991). Because the gallbladder also has γ-glutamyl transpeptidase activity (Hinchman and Ballatori, 1990), a similar mechanism may be involved; however, there is as yet no evidence for this hypothesis.

3.6.2.2 Additional Organic Solute Carriers: MDR1, MDR3, and BSEP

MDR are constitutively expressed in several epithelial and endothelial tissues, including the liver cell bile canalicular membrane. Bile canalicular membranes contain MDR1, which transports cationic drugs into bile, and MDR3, which transports phospholipids into bile (Oude Elferink and Paulusma, 2007; Smit et al., 1993) (Figure 3.2). Sharma et al. (1996) demonstrated that cationic organometallic complexes are also substrates for MDR1. These investigators demonstrated that human epidermal carcinoma KB cells overexpressing MDR1 P-glycoprotein are more resistant to the cytotoxic activities of Al(III), Fe(III), and Ga(III) complexes of (ethylenediamine)-*N,N′*-bis[propyl[(2-hydroxy-4,6-dimethoxybenzyl)-imino]]. Broeks et al. (1996) reported that the *C. elegans* equivalent MDR protein (cePGP) also contributes to heavy metal (cadmium and arsenite) resistance in this organism, but the mechanism involved is unknown. These observations raise the possibility that MDR proteins on the canalicular membrane may be involved in the transport of metal complexes into bile.

An additional canalicular transporter that might be involved in biliary metal excretion is the ATP-dependent bile acid transporter, BSEP (Figure 3.2), although its role is probably minimal. Bile acid–metal complexes may form under certain conditions, and these would most likely be substrates for the multispecific BSEP protein.

3.6.2.3 Sulfate Exchanger

As noted earlier, vanadate and arsenate are structurally similar to phosphate and can compete with phosphate for transport, whereas chromate, selenate, and molybdate are structurally similar to sulfate, and are substrates for sulfate transporters. A sulfate transporter has been identified on the liver cell canalicular membrane (Sat1; Bissig et al., 1994), providing a mechanism by which these oxyanions may be secreted into bile, although this has not been tested directly. Canalicular sulfate efflux on Sat1 is driven by exchange with bicarbonate; the latter is secreted into bile by a chloride-bicarbonate exchanger (Figure 3.2).

3.7 SUMMARY

Several metal-selective transport mechanisms on hepatocyte cell membranes allows the liver to efficiently extract metals from plasma, metabolize, store, and redistribute them in various forms either back into the bloodstream or into bile. Bidirectional transport across the sinusoidal plasma membrane allows the liver to regulate plasma concentrations, and therefore availability to other tissues. Transport across the canalicular membrane is largely but not exclusively unidirectional, and is a major excretory pathway for excess metals and/or potentially toxic metals.

Some of the key transport mechanisms have now been identified at the molecular level, providing an unprecedented opportunity to examine their structure, function, and regulation. Additional studies are required to define the cellular and subcellular localization of key transport proteins in the liver, their functional orientation in the membrane (uptake or efflux), driving force, substrate selectivity, or regulation under physiological and pathophysiological conditions. It is likely that additional families of metal-specific transporters may also be present, but have not yet been identified. This more complete picture of metal transport systems will provide a solid framework for understanding metal homeostasis in both health and disease.

ACKNOWLEDGMENTS

Preparation of this overview was supported in part by National Institute of Health Grants DK48823 and DK067214, NIEHS Center Grants ES01247 and ES03828, and NIEHS Training Grant ES07026.

REFERENCES

Aballay, A., Arenas, G.N., and Mayorga, L.S. Calcium- and zinc-binding proteins in intracellular transport, *Biocell*, 20, 339, 1996.

Abboud, S. and Haile, D.J. A novel mammalian iron-regulated protein involved in intracellular iron metabolism, *J. Biol. Chem.*, 275, 19906, 2000.

Aiken, S.P., Horn, N.M., and Saunders, N.R. Effects of amino acids on zinc transport in rat erythrocytes, *J. Physiol.*, 445, 69, 1992.

Aisen, P. Transferrin, the transferrin receptor, and the uptake of iron by cells, *Metal Ions Biol. Sys.*, 35, 585, 1998.

Ambrosino, G., et al. Isolated hepatocytes versus hepatocyte spheroids: *In vitro* culture of rat hepatocytes, *Cell Transplant.*, 14, 397, 2005.

Anastassopoulou, I., et al. Solution structure of the apo and copper(I)-loaded human metallochaperone HAH1, *Biochemistry*, 43, 13046, 2004.

Anderson, G.J., Darshan, D., Wilkins, S.J., and Frazer, D.M. Regulation of systemic iron homeostasis: How the body responds to changes in iron demand, *Biometals*, 20, 665, 2007.

Andrews, N.C. and Schmidt, P.J. Iron homeostasis. *Annu. Rev. Physiol.*, 69, 69, 2007.

Aremu, D.A., Madejczyk, M.S., and Ballatori, N. *N*-Acetylcysteine as a potential antidote and biomonitoring agent of methylmercury exposure, *Environ. Health Perspect.*, 116, 26, 2008.

Attieh, Z.K., et al. Ceruloplasmin ferroxidase activity stimulates cellular iron uptake by a trivalent cation-specific transport mechanism, *J. Biol. Chem.*, 274, 1116, 1999.

Ballatori, N. Mechanisms of metal transport across liver cell plasma membranes, *Drug Metab. Rev.*, 23, 83, 1991a.

Ballatori, N. Biliary excretion of metals, in *Biliary Excretion of Drugs and Other Chemicals*, Siegers, C.-P. and Watkins III, J.B. (Eds), G. Fischer Verlag, Stuttgart-New York, Chapter 3.7, pp. 283–306, 1991b.

Ballatori, N. Glutathione mercaptides as transport forms of metals, *Adv. Pharm.*, 27, 271, 1994.

Ballatori, N. Transport of toxic metals by molecular mimicry, *Environ. Health Perspect.*, 110, 689, 2002.

Ballatori, N. and Clarkson, T.W. Developmental changes in the biliary excretion of methylmercury and glutathione, *Science*, 216, 61, 1982.

Ballatori, N. and Clarkson, T.W. Biliary transport of glutathione and methylmercury, *Am. J. Physiol.*, 244, G435, 1983.

Ballatori, N. and Dutczak, W.J. Identification and characterization of high and low affinity transport systems for reduced glutathione in liver cell canalicular membranes, *J. Biol. Chem.*, 269, 19731, 1994.

Ballatori, N., Gatmaitan, Z., and Truong A.T. Impaired biliary excretion and whole body elimination of methylmercury in rats with congenital defect in biliary glutathione excretion, *Hepatology*, 22, 1469, 1995.

Ballatori, N., Jacob, R., and Boyer, J.L. Intrabiliary glutathione hydrolysis—a source of glutamate in bile, *J. Biol. Chem.*, 261, 7860, 1986a.

Ballatori, N., Lieberman, M.W., and Wang W. *N*-Acetylcysteine as an antidote in methylmercury poisoning, *Environ. Health Perspect.*, 106, 267, 1998b.

Ballatori, N. and Madejczyk, M.S. Transport of nonessential metals across mammalian cell membranes. *Topics Curr. Genet.*, 14, 455, 2006.

Ballatori, N., Moseley, R.H., and Boyer, J.L. Sodium gradient-dependent L-glutamate transport is localized to the canalicular domain of liver plasma membranes, *J. Biol. Chem.*, 261, 6216, 1986b.

Ballatori, N. and Truong, A.T. Relation between biliary glutathione excretion and bile acid-independent bile flow, *Am. J. Physiol.*, 256, G22, 1989.

Ballatori, N. and Truong, A.T. Glutathione as a primary osmotic driving force in hepatic bile formation, *Am. J. Physiol.*, 263, G617, 1992.

Ballatori, N. and Truong, A.T. Multiple canalicular transport mechanisms for glutathione *S*-conjugates. Transport on both ATP- and voltage-dependent carriers, *J. Biol. Chem.*, 270, 3594, 1995a.

Ballatori, N. and Truong, A.T. Mechanisms of hepatic methylmercury uptake, *J. Toxicol. Environ. Health*, 46, 343, 1995b.

Ballatori, N., Wang, W., and Lieberman, M.W. Accelerated methylmercury elimination in gamma-glutamyl transpeptidase-deficient mice, *Am. J. Pathol.*, 152, 1049, 1998a.

Ballatori, N., et al., Biliary catabolism of glutathione and differential reabsorption of its amino acid constituents, *Am. J. Physiol.*, 254, G1, 1988.

Ballatori, N., et al. Molecular mechanisms of reduced glutathione transport: Role of the MRP/CFTR/ABCC and OATP/SLC21A families of membrane proteins, *Toxicol. Appl. Pharmacol.*, 204, 238, 2005.

Ballatori, N., et al. Retention of structural and functional polarity in cultured skate hepatocytes undergoing *in vitro* morphogenesis, *Comp. Biochem. Physiol. B, Biochem. Mol. Biol.*, 144, 167, 2006.

Banci, L., et al. Interaction of the two soluble metal-binding domains of yeast Ccc2 with copper(I)-Atx1, *Biochem. Biophys. Res. Commun.*, 364, 645, 2007.
Bartles, J.R., et al. Biogenesis of the rat hepatocyte plasma membrane *in vivo*: Comparison of the pathways taken by apical and basolateral proteins using subcellular fractionation, *J. Cell Biol.*, 105, 1241, 1987.
Bera, T.K., et al. MRP9, an unusual truncated member of the ABC transporter superfamily, is highly expressed in breast cancer, *Proc. Natl. Acad. Sci. USA*, 99, 6997, 2002.
Bingham, M.J., et al. ATP-dependent copper transporter in the Golgi apparatus of rat hepatocytes transports Cu(II) not Cu(I), *Am. J. Physiol.*, 271, G741, 1996.
Bissig, M., et al. Functional expression cloning of the canalicular sulfate transport system of rat hepatocytes, *J. Biol. Chem.*, 269, 3017, 1994.
Blitzer, B.L., Hostetler, B.R., and Scott, K.A. Hepatic adenosine triphosphate-dependent Ca^{2+} transport is mediated by distinct carriers on rat basolateral and canalicular membranes, *J. Clin. Invest.*, 83, 1319, 1989.
Blomhoff, R., et al. A multicompartmental model of fluid-phase endocytosis in rabbit liver parenchymal cells, *Biochem. J.*, 262, 605, 1989.
Borst, P., Kool, M., and Evers, R. Do CMOAT (MRP2), other MRP homologues, and LRP play a role in MDR, *Semin, Cancer Biol.*, 8, 205, 1997.
Boyer, J.L. Mechanisms of bile secretion and cholestasis, in *Physiology of Membrane Disorders*, Andreoli, T.E., Hoffman, J.F., Fanestil, D.D., and Schultz, S.G. (Eds), Plenum Publishing, New York, p. 609, 1986.
Broeks, A., et al. Homologues of the human multidrug resistance genes MRP and MDR contribute to heavy metal resistance in the soil nematode *Caenorhabditis elegans*, *EMBO J.*, 15, 6132, 1996.
Buchler, M., et al. cDNA cloning of the hepatocyte canalicular isoform of the multidrug resistance protein, cMrp, reveals a novel conjugate export pump deficient in hyperbilirubinemic mutant rats, *J. Biol. Chem.*, 271, 15091, 1996.
Bull, P.C. and Cox, D.W. Wilson disease and Menkes disease: New handles on heavy-metal transport, *Trends Genet.*, 10, 246, 1994.
Bygrave, F.L., Karjalainen, A., and Altin, J.G. Passive calcium influx by plasma membrane vesicles isolated from rat liver, *Cell Calcium*, 10, 235, 1989.
Callahan, H.L. and Beverley, S.M. Heavy metal resistance: A new role for P-glycoprotein in *Leishmania*, *J. Biol. Chem.*, 266, 18427, 1991.
Canonne-Hergaux, F., et al. Expression of the DMT1 (NRAMP2/DCT1) iron transporter in mice with genetic iron overload disorders, *Blood*, 97, 1138, 2001.
Cellier, M., et al. Nramp defines a family of membrane proteins. *Proc. Natl. Acad. Sci. USA*, 92, 10089, 1995.
Chen, Z.-S., et al. Reversal of heavy metal resistance in multidrug-resistant human KB carcinoma cells, *Biochem. Biophys. Res. Commun.*, 236, 586, 1997.
Chiesi, M., et al. Stereospecific action of diltiazem on the mitochondrial Na–Ca exchange system and on sarcolemmal Ca-channels, *Biochem. Pharmacol.*, 36, 2735, 1987.
Chong, S.S., et al. Molecular cloning of the cDNA encoding a human renal sodium phosphate transport protein and its assignment to chromosome 6p21.3-p23, *Genomics*, 18, 355, 1993.
Chua, A.C. and Morgan, E.H. Manganese metabolism is impaired in the Belgrade laboratory rat, *J. Comp. Physiol.*, 167, 361, 1997.
Clarkson, T.W. The pharmacology of mercury compounds, *Annu. Rev. Pharmacol.*, 12, 375, 1972.
Clarkson, T.W. Molecular and ionic mimicry of toxic metals, *Annu. Rev. Pharmacol. Toxicol.*, 33, 545, 1993.
Cole, S.P., et al. Pharmacological characterization of multidrug resistant MRP-transfected human tumor cells, *Cancer Res.*, 54, 5902, 1994.
Coleman, R. Biochemistry of bile secretion, *Biochem. J.*, 244, 249, 1987.
Conklin, D.S., Culbertson, M.R., and Kung, C. Interactions between gene products involved in divalent cation transport in *Saccharomyces cerevisiae*, *Mol. Genet.*, 244, 303, 1994.
Cousins, R.J. Absorption, transport, and hepatic metabolism of copper and zinc: Special reference to metallothionein and ceruloplasmin, *Physiol. Rev.*, 65, 238, 1985.
Crofts, J.N. and Barritt, G.J. The liver cell plasma membrane Ca^{++} inflow systems exhibit a broad specificity for divalent metal ions, *Biochem. J.*, 269, 579, 1990.
Dancis, A., et al. The *Saccharomyces cerevisiae* copper transport protein Ctr1p–biochemical characterization, regulation by copper, and physiologic role in copper uptake, *J. Biol. Chem.*, 269, 25660, 1994.
Davis, W., Chowrimootoo, G.F., and Seymour, C.A. Defective biliary copper excretion in Wilson's disease: The role of ceruloplasmin, *Eur. J. Clin. Invest.*, 26, 893, 1996.
Dawson, D.C. and Ballatori, N. Membrane transporters as sites of action and routes of entry for toxic metals, *Handbook Exp. Pharmacol.*, 115, 53, 1995.

Dawson, P.A., et al. The rat Na^+-sulfate cotransporter $rNaS_2$: Functional characterization, tissue distribution, and gene (slc13a4) structure, *Pflugers Arch.*, 450, 262, 2005.
Dawson, P.A., et al. Molecular cloning and characterization of the mouse Na^+ sulfate cotransporter gene (Slc13a4): Structure and expression, *Genes Genet. Syst.*, 81, 265, 2006.
Deeley, R.G. and Cole, S.P. Function, evolution and structure of multidrug resistance protein (MRP), *Semin. Cancer Biol.*, 8, 193, 1997.
Deugnier, Y., Brissot, P., and Loréal, O. Iron and the liver: Update 2008, *J. Hepatol.* 48 (Suppl. 1), S113, 2008.
Dey, S., et al. An ATP-dependent As(III)-glutathione transport system in membrane vesicles of *Leishmania tarentolae*, *Proc. Natl. Acad. Sci. USA*, 93, 2192, 1996.
Dierick, H.A., et al. Immunocytochemical localization of the Menkes copper transport protein (ATP7A) to the *trans*-Golgi network, *Hum. Mol. Genet.*, 6, 409, 1997.
Dijkstra, M., et al. Adenosine triphosphate-dependent copper transport in isolated rat liver plasma membranes, *J. Clin. Invest.*, 95, 412, 1995.
Dijkstra, M., et al. Bile secretion of cadmium, silver, zinc and copper in the rat. Involvement of various tranport systems, *Life Sci.*, 59, 1237, 1996.
Dijkstra, M., et al. Differences in hepatic processing of dietary and intravenously administered copper in rats, *Hepatology*, 26, 962, 1997.
Donovan, A., et al. Positional cloning of zebrafish ferroportin1 identifies a conserved vertebrate iron exporter, *Nature*, 403, 776, 2000.
Donovan, A., et al. The iron exporter ferroportin/Slc40a1 is essential for iron homeostasis, *Cell Metab.*, 1, 191, 2005.
Dufner-Beattie, J., et al. Structure, function, and regulation of a subfamily of mouse zinc transporter genes, *J. Biol. Chem.*, 278, 50142, 2003.
Dutczak, W.J. and Ballatori, N. Gamma-glutamyltransferase-dependent biliary-hepatic recycling of methylmercury in the guinea pig, *J. Pharmacol. Exp. Ther.*, 262, 619, 1992.
Dutczak, W.J. and Ballatori, N. Transport of the glutathione–methylmercury complex across liver canalicular membranes on reduced glutathione carriers, *J. Biol. Chem.*, 269, 9746, 1994.
Dutczak, W.J., Clarkson, T.W., and Ballatori N. Biliary-hepatic recycling of a xenobiotic: Gallbladder absorption of methylmercury, *Am. J. Physiol.*, 260, G873, 1991.
Eide, D. Molecular biology of iron and zinc uptake in eukaryotes, *Curr. Opin. Cell Biol.*, 9, 573, 1997.
Eide, D.J. Zinc transporters and the cellular trafficking of zinc. *Biochim. Biophys. Acta*, 1763, 711, 2006.
Fei, Y.J., et al. Expression cloning of a mammalian proton-coupled oligopeptide transporter, *Nature*, 368, 563, 1994.
Fisher, J., Devraj, K., Ingram, J., Slagle-Webb, B., Madhankumar, A.B., Liu, X., Klinger, M., Simpson, I.A., and Connor, J.R. Ferritin: A novel mechanism for delivery of iron to the brain and other organs, *Am. J. Physiol. Cell Physiol.*, 293, C641, 2007.
Fleming, M.D., et al. Microcytic anaemia mice have a mutation in Nramp2, a candidate iron transporter gene, *Nat. Genet.*, 16, 383, 1997.
Fleming, M.D., et al. Nramp2 is mutated in the anemic Belgrade (b) rat: Evidence of a role for Nramp2 in endosomal iron transport, *Proc. Natl. Acad. Sci. USA*, 95, 1148, 1998.
Frazer, D.M., Wilkins, S.J., Becker, E.M., Vulpe, C.D., McKie, A.T., Trinder, D., and Anderson, G.J. Hepcidin expression inversely correlates with the expression of duodenal iron transporters and iron absorption in rats, *Gastroenterology*, 123, 835, 2002.
Frei, P., et al. Identification and localization of sodium-phosphate cotransporters in hepatocytes and cholangiocytes of rat liver, *Am. J. Physiol. Gastrointest. Liver Physiol.*, 288, G771, 2005.
Garrick, M.D., et al. Transferrin and the transferrin cycle in Belgrade rat reticulocytes, *J. Biol. Chem.*, 268, 14867, 1993.
Garrick, M.D., et al. Comparison of mammalian cell lines expressing distinct isoforms of divalent metal transporter 1 in a tetracycline-regulated fashion, *Biochem. J.*, 398, 539, 2006.
Gavin, C.E., Gunter, K.K., and Gunter, T.E. Manganese and calcium transport in mitochondria: Implications for manganese toxicity, *Neurotoxicology*, 20, 445, 1999.
Ghishan, F.K., et al. Characterization and expression of a novel Na(+)-inorganic phosphate transporter at the liver plasma membrane of the rat, *Gastroenterology*, 105, 519, 1993.
Goldstone, T.P. and Crompton, M. Evidence for beta-adrenergic activation of Na^+-dependent efflux of Ca^{2+} from isolated liver mitochondria, *Biochem. J.*, 204, 369, 1982.
González-Guerrero, M. and Argüello, J.M. Mechanism of Cu^+-transporting ATPases: Soluble Cu^+ chaperones directly transfer Cu^+ to transmembrane transport sites. *Proc. Natl. Acad. Sci. USA*, 105, 5992, 2008.

Graf, J., Gautam, A., and Boyer, J.L. Isolated rat hepatocyte couplets: A primary secretory unit for electrophysiologic studies of bile secretory function, *Proc. Natl. Acad. Sci. USA*, 81, 6516, 1984.
Graf, J., et al. Cell membrane and transepithelial voltages and resistances in isolated rat hepatocyte couplets, *J. Membr. Biol.*, 95, 241, 1987.
Graham, R.M., Chua, A.C., Herbison, C.E., Olynyk, J.K., and Trinder, D. Liver iron transport, *World J. Gastroenterol.*, 13, 4725, 2007.
Gross, J.B., Jr., et al. Biliary copper excretion by hepatocyte lysosomes in the rat. Major excretory pathway in experimental copper overload, *J. Clin. Invest.*, 83, 30, 1989.
Grundemann, D., et al. Drug excretion mediated by a new prototype of polyspecific transporter, *Nature*, 372, 549, 1994.
Gunshin, H., et al. Cloning and characterization of a mammalian proton-coupled metal-ion transporter, *Nature*, 388, 482, 1997.
Gurgueira, S.A. and Meneghini, R. An ATP-dependent iron transport system in isolated rat liver nuclei, *J. Biol. Chem.*, 271, 13616, 1996.
Hagenbuch, B., et al. Functional expression cloning and characterization of the hepatocyte Na^+/bile acid cotransport system, *Proc. Natl. Acad. Sci. USA*, 88, 10629, 1991.
Hao, Q., Hong, S.H., and Maret, W. Lipid raft-dependent endocytosis of metallothionein in HepG2 cells, *J. Cell. Physiol.*, 210, 428, 2007.
Harris, E.D. The transport of copper, *Prog. Clin. Biol. Res.*, 380, 163, 1993.
Hernandez, S., et al. ATP7B copper-regulated traffic and association with the tight junctions: Copper excretion into the bile, *Gastroenterology*, 134, 1215, 2008.
Hinchman, C.A. and Ballatori, N. Glutathione-degrading capacities of liver and kidney in different species, *Biochem. Pharmacol.*, 40, 1131, 1990.
Hinchman, C.A. and Ballatori, N. Glutathione conjugation and conversion to mercapturic acids can occur as an intrahepatic process, *J. Toxicol. Environ. Health*, 41, 387, 1994.
Hinchman, C.A., et al. Intrahepatic conversion of a glutathione conjugate to its mercapturic acid, *J. Biol. Chem.*, 266, 22179, 1991.
Hinkle, P.M., Kinsella, P.A., and Osterhoudt, K.C. Cadmium uptake and toxicity via voltage-sensitive calcium channels, *J. Biol. Chem.*, 262, 16333, 1987.
Hopper, E., et al. Analysis of the structure and expression pattern of MRP7 (ABCC10), a new member of the MRP subfamily, *Cancer Lett.*, 162, 181, 2001.
Horn, N.M. and Thomas, A.L. Interactions between the histidine stimulation of cadmium and zinc influx into human erythrocytes, *J. Physiol.*, 496, 711, 1996.
Horn, N.M., Thomas, A.L., and Tompkins, J.D. The effect of histidine and cysteine on zinc influx into rat and human erythrocytes, *J. Physiol.*, 489, 73, 1995.
Houwen, R., et al. Two pathways for biliary copper excretion in the rat. The role of glutathione, *Biochem. Pharmacol.*, 39, 1039, 1990.
Huang, L.P. and Gitschier, J. A novel gene involved in zinc transport is deficient in the lethal milk mouse, *Nat. Genet.*, 17, 292, 1997.
Hubert, N. and Hentze, M.W. Previously uncharacterized isoforms of divalent metal transporter (DMT)-1: Implications for regulation and cellular function, *Proc. Natl. Acad. Sci. USA*, 99, 12345, 2002.
Hughes, B.P. and Barritt, G.J. Inhibition of the liver cell receptor-activated Ca^{2+} inflow system by metal ion inhibitors of voltage-operated Ca^{2+} channels but not by other inhibitors of Ca^{2+} inflow, *Biochim. Biophys. Acta*, 1013, 197, 1989.
Ishikawa, T., et al. Coordinated induction of MRP/GS-X pump and gamma-glutamylcysteine synthetase by heavy metals in human leukemia cells, *J. Biol. Chem.*, 271, 14981, 1996.
Ishikawa, T., et al. The GS-X pump in plant, yeast, and animal cells: Structure, function, and gene expression, *Biosci. Rep.*, 17, 189, 1997.
Ito, K., et al. Molecular cloning of canalicular multispecific organic anion transporter defective in EHBR, *Am. J. Physiol.*, 272, G16, 1997.
Jacquemin, E., et al. Expression cloning of a rat liver Na(+)-independent organic anion transporter, *Proc. Natl. Acad. Sci. USA*, 91, 133, 1994.
Jiang, J., et al. Cultivation of fetal liver cells in a three-dimensional poly-L-lactic acid scaffold in the presence of oncostatin M, *Cell Transplant.*, 11, 403, 2002.
Jigorel, E., et al. Functional expression of sinusoidal drug transporters in primary human and rat hepatocytes, *Drug Metab. Dispos.*, 33, 1418, 2005.
Jordan, I. and Kaplan, J. The mammalian transferrin-independent iron transport system may involve a surface ferrireductase activity, *Biochem. J.*, 302, 875, 1994.

Joshi, J.G., Sczekan, S.R., and Fleming, J.T. Ferritin—a general metal detoxicant, *Biol. Trace Elem. Res.*, 21, 105, 1989.

Kambe, T., et al. Cloning and characterization of a novel mammalian zinc transporter, zinc transporter 5, abundantly expressed in pancreatic beta cells, *J. Biol. Chem.*, 277, 19049, 2002.

Kambe, T., Geiser, J., Lahner, B., Salt, D.E., and Andrews, G.K. Slc39a1 to 3 (subfamily II) Zip genes in mice have unique cell-specific functions during adaptation to zinc deficiency, *Am. J. Physiol. Regul. Integr. Comp. Physiol.*, 294, R1474, 2008.

Kambe, T., Yamaguchi-Iwai, Y., Sasaki, R., and Nagao, M. Overview of mammalian zinc transporters, *Cell. Mol. Life Sci.*, 61, 49, 2004.

Kaplan, J. and O'Halloran, T.V. Iron metabolism in eukaryotes: Mars and Venus at it again, *Science*, 271, 1510, 1996.

Kavanaugh, M.P. and Kabat, D. Identification and characterization of a widely expressed phosphate transporter/retrovirus receptor family, *Kidney Int.*, 49, 959, 1996.

Kawanishi, T., et al. Ca^{2+} oscillations induced by hormonal stimulation of individual fura-2-loaded hepatocytes, *J. Biol. Chem.*, 264, 12859, 1989.

Keppler, D., Leier, I., and Jedlitschky, G. Transport of glutathione conjugates and glucuronides by the multidrug resistance proteins MRP1 and MRP2, *Biol. Chem.*, 378, 787, 1997.

Kerper, L.E., Ballatori, N., and Clarkson, T.W. Methylmercury transport across the blood–brain barrier by an amino acid carrier, *Am. J. Physiol.*, 262, R761, 1992.

Kim, B-E., Nevitt, T., and Thiele, D.J. Mechanisms for copper acquisition, distribution and regulation, *Nat. Chem. Biol.*, 4, 176, 2008.

Kirschke, C.P. and Huang, L. ZnT7, a novel mammalian zinc transporter, accumulates zinc in the Golgi apparatus, *J. Biol. Chem.*, 278, 4096, 2003.

Klaassen, C.D. Biliary excretion of metals, *Drug Metab. Rev.*, 5, 165, 1976.

Klaassen, C.D. and Watkins, III, J.B. Mechanisms of bile formation, hepatic uptake, and biliary excretion, *Pharmacol. Rev.*, 36, 1, 1984.

Klomp, A.E., et al. Biochemical characterization and subcellular localization of human copper transporter 1 (hCTR1), *Biochem. J.*, 364, 497, 2002.

Koch, K.A., Pena, M.M., and Thiele, D.J. Copper-binding motifs in catalysis, transport, detoxification and signaling, *Chem. Biol.*, 4, 549, 1997.

Koepsell, H., Lips, K., and Volk, C. Polyspecific organic cation transporters: Structure, function, physiological roles, and biopharmaceutical implications, *Pharm. Res.*, 24, 1227, 2007.

Kojima, C., et al. Chronic exposure to methylated arsenicals stimulates arsenic excretion pathways and induces arsenic tolerance in rat liver cells, *Toxicol. Sci.*, 91, 70, 2006.

Kool, M., et al. Analysis of expression of cMOAT (MRP2), MRP 3, MRP 4, and MRP 5, homologues of the multidrug resistance-associated protein gene (MRP 1), in human cancer cell lines, *Cancer Res.*, 57, 3537, 1997.

Kool, M., et al. Expression of human MRP6, a homologue of the multidrug resistance protein gene MRP1, in tissues and cancer cells, *Cancer Res.*, 59, 175, 1999.

Kullak-Ublick, G.A., et al. Functional characterization of the basolateral rat liver organic anion transporting polypeptide, *Hepatology*, 20, 411, 1994.

Kuo, Y.M., et al. Copper transport protein (Ctr1) levels in mice are tissue specific and dependent on copper status, *J. Nutr.*, 136, 21, 2006.

Kuo, Y.M., Gitschier, J., and Packman, S. Developmental expression of the mouse mottled and toxic milk genes suggests distinct functions for the Menkes and Wilson disease copper transporters, *Hum. Mol. Genet.*, 6, 1043, 1997.

La Fontaine, S. and Mercer, J.F. Trafficking of the copper-ATPases, ATP7A and ATP7B: Role in copper homeostasis, *Arch. Biochem. Biophys.*, 463, 149, 2007.

Lam-Yuk-Tseung, S., et al. Carboxylterminus determinants of the iron transporter DMT1/SLC11A2 isoform II (-IRE/1B) mediate internalization from the plasma membrane into recycling endosomes, *Biochemistry*, 44, 12149, 2005.

Lam-Yuk-Tseung, S. and Gros, P. Distinct targeting and recycling properties of two isoforms of the iron transporter DMT1 (NRAMP2, Slc11A2), *Biochemistry*, 45, 2294, 2006.

Lee, J., Petris, M.J., and Thiele, D.J. Characterization of mouse embryonic cells deficient in the ctr1 high affinity copper transporter, Identification of a Ctr1-independent copper transport system, *J. Biol. Chem.*, 277, 40253, 2002.

Lee, P.L., et al. The human Nramp2 gene: Characterization of the gene structure, alternative splicing, promoter region and polymorphisms, *Blood Cells Mol. Dis.*, 24, 199, 1998.

Lee, T.C., et al. Enhanced expression of multidrug resistance-associated protein 2 and reduced expression of aquaglyceroporin 3 in an arsenic-resistant human cell line, *J. Biol. Chem.*, 281, 18401, 2006. Erratum in: *J. Biol. Chem.*, 282, 11612, 2007.

LeSage, G.D., et al. Biliary excretion of iron from hepatocyte lysosomes in the rat. A major excretory pathway in experimental iron overload, *J. Clin. Invest.*, 77, 90, 1986.

Li, L., et al. Identification of glutathione as a driving force and leukotriene C_4 as a substrate for Oatp1, the hepatic sinusoidal organic solute transporter, *J. Biol. Chem.*, 273, 16184, 1998.

Li, L., Meier, P.J., and Ballatori, N. Oatp2 mediates bidirectional organic solute transport: A role for intracellular glutathione, *Mol. Pharmacol.*, 58, 335, 2000.

Li, Z.S., et al. The yeast cadmium factor protein (YCF1) is a vacuolar glutathione *S*-conjugate pump, *J. Biol. Chem.*, 271, 6509, 1996.

Li, Z.S., et al. A new pathway for vacuolar cadmium sequestration in *Saccharomyces cerevisiae*: YCF1-catalyzed transport of bis(glutathionato)cadmium, *Proc. Natl. Acad. Sci. USA*, 94, 42, 1997.

Liu, J., et al. Multidrug-resistance mdr1a/1b double knockout mice are more sensitive than wildtype mice to acute arsenic toxicity, with higher arsenic accumulation in tissues, *Toxicology*, 170, 55, 2002.

Liuzzi, J.P., Blanchard, R.K., and Cousins, R.J. Differential regulation of zinc transporter 1, 2, and 4 mRNA expression by dietary zinc in rats, *J. Nutr.*, 131, 46, 2001.

Liuzzi, J.P. and Cousins, R.J. Mammalian zinc transporters, *Annu. Rev. Nutr.*, 24, 151–172, 2004.

Lutsenko, S., Barnes, N.L., Bartee, M.Y., and Dmitriev, O.Y. Function and regulation of human copper-transporting ATPases, *Physiol. Rev.*, 87, 1011, 2007.

Mack, U., Powell, L.W., and Halliday, J.W. Detection and isolation of a hepatic membrane receptor for ferritin, *J. Biol. Chem.*, 258, 4672, 1983.

Mackenzie, B., et al. Divalent metal-ion transporter DMT1 mediates both H^+-coupled Fe^{2+} transport and uncoupled fluxes, *Pflugers Arch.*, 451, 544, 2006.

MacKenzie, E.L., Iwasaki, K., and Tsuji, Y. Intracellular iron transport and storage: From molecular mechanisms to health implications, *Antioxid. Redox. Signal.*, 10, 997–1030, 2008.

Madejczyk, M.S., et al. Accelerated urinary excretion of methylmercury following administration of its antidote *N*-acetylcysteine requires Mrp2/Abcc2, the apical multidrug resistance-associated protein, *J. Pharmacol. Exp. Ther.*, 322, 378, 2007.

Mahagita, C., et al. Human organic anion transporter 1B1 and 1B3 function as bidirectional carriers and do not mediate GSH-bile acid cotransport, *Am. J. Physiol. Gastrointest. Liver Physiol.*, 293, G271, 2007.

Malandro, M.S. and Kilberg, M.S. Molecular biology of mammalian amino acid transporters, *Annu. Rev. Biochem.*, 65, 305, 1996.

Markovich, D., et al. Functional characterization and genomic organization of the human Na(+)-sulfate cotransporter hNaS2 gene (SLC13A4), *Biochem. Biophys. Res. Commun.*, 326, 729, 2005.

Markovich, D. and Murer, H. The SLC13 gene family of sodium sulphate/carboxylate cotransporters, *Pflugers Arch.*, 447, 594, 2004.

McKie, A.T., et al. A novel duodenal iron-regulated transporter, IREG1, implicated in the basolateral transfer of iron to the circulation, *Mol. Cell*, 5, 299, 2000.

McKie, A.T., et al. An iron-regulated ferric reductase associated with the absorption of dietary iron, *Science*, 291, 1755, 2001.

McMahon, R.J. and Cousins, R.J. Regulation of the zinc transporter ZnT-1 by dietary zinc, *Proc. Natl. Acad. Sci. USA*, 95, 4841, 1998.

Meier, P.J. Transport polarity of hepatocytes, *Semin. Liver Dis.*, 8, 293, 1988.

Meier, P.J., et al. Structural and functional polarity of canalicular and basolateral plasma membrane vesicles isolated in high yield from rat liver, *J. Cell Biol.*, 98, 991, 1984.

Mena, N.P., Esparza, A., Tapia, V., Valdés, P., and Núñez, M.T. Hepcidin inhibits apical iron uptake in intestinal cells, *Am. J. Physiol. Gastrointest. Liver Physiol.*, 294, G192, 2008.

Miyamoto, K., et al. Cloning and functional expression of a Na^+-dependent phosphate co-transporter from human kidney: CDNA cloning and functional expression, *Biochem. J.*, 305, 81, 1995.

Mizuguchi, T., et al. Paper is a compatible bed for rat hepatocytes, *Artif. Organs*, 24, 271, 2000.

Mokrzan, E.M., et al. Methylmercury-thiol uptake into cultured brain capillary endothelial cells on amino acid system L, *J. Pharmacol. Exp. Ther.*, 272, 1277, 1995.

Mooney, D.J., et al. Biodegradable sponges for hepatocyte transplantation, *J. Biomed. Mater. Res.*, 29, 959, 1995.

Morgan, E.H. and Baker, E. Iron uptake and metabolism by hepatocytes, *Fed. Proc.*, 45, 2810, 1986.

Morgan, E.H. and Baker, E. Role of transferrin receptors and endocytosis in iron uptake by hepatic and erythroid cells, *Ann. N. Y. Acad. Sci.*, 526, 65, 1988.

Morley, C.G. and Bezkorovainy, A. Cellular iron uptake from transferrin: Is endocytosis the only mechanism?, *Int. J. Biochem.*, 17, 553, 1985.
Moseley, R.H., Ballatori, N., and Murphy, S.M. Na^+-Glycine cotransport in canalicular liver plasma membrane vesicles, *Am. J. Physiol.*, 255, G253, 1988.
Mukhopadhyay, C.K., Attieh, Z.K., and Fox, P.L. Role of ceruloplasmin in cellular iron uptake, *Science*, 279, 714, 1998.
Mukhopadhyay, C.K., Mazumder, B., and Fox, P.L. Role of hypoxia-inducible factor-1 in transcriptional activation of ceruloplasmin by iron deficiency, *J. Biol. Chem.*, 275, 21048, 2000.
Nagano, K., et al. Intracellular distribution of the Wilson's disease gene product (ATPase7B) after *in vitro* and *in vivo* exogenous expression in hepatocytes from the LEC rat, an animal model of Wilson's disease, *Hepatology*, 27, 799, 1998.
Nemeth, E., Tuttle, M.S., Powelson, J., Vaughn, M.B., Donovan, A., Ward, D.M., Ganz, T., and Kaplan, J. Hepcidin regulates cellular iron efflux by binding to ferroportin and inducing its internalization, *Science*, 306, 2090, 2004.
Nemeth, E. Iron regulation and erythropoiesis, *Curr. Opin. Hematol.*, 15, 169, 2008.
Noé, B., et al. Isolation of a multispecific organic anion and cardiac glycoside transporter from rat brain, *Proc. Natl. Acad. Sci. USA*, 94, 10346, 1997.
Norbis, F., et al. Identification of a cDNA/protein leading to an increased Pi-uptake in *Xenopus laevis* oocytes, *J. Membr. Biol.*, 156, 19, 1997.
Ohgami, R.S., et al. Identification of a ferrireductase required for efficient transferrin-dependent iron uptake in erythroid cells, *Nat. Genet.*, 37, 1264, 2005.
Ohgami, R.S., et al. The Steap proteins are metalloreductases, *Blood*, 108, 1388, 2006.
Oka, J.A. and Weigel, P.H. Recycling of the asialoglycoprotein receptor in isolated rat hepatocytes. Dissociation of internalized ligand from receptor occurs in two kinetically and thermally distinguishable compartments, *J. Biol. Chem.*, 258, 10253, 1983.
Oka, J.A., Christensen, M.D., and Weigel, P.H. Hyperosmolarity inhibits galactosyl receptor-mediated but not fluid phase endocytosis in isolated rat hepatocytes, *J. Biol. Chem.*, 264, 12016, 1989.
Osterloh, K. and Aisen, P. Pathways in the binding and uptake of ferritin by hepatocytes, *Biochim. Biophys. Acta*, 1011, 40, 1989.
Oude Elferink, R.P. and Paulusma, C.C. Function and pathophysiological importance of ABCB4 (MDR3 P-glycoprotein), *Pflugers Arch.*, 453, 601, 2007.
Pajor, A.M. Molecular properties of the SLC13 family of dicarboxylate and sulfate transporters, *Pflugers Arch.*, 451, 597, 2006.
Palmiter, R.D., Cole, T.B., and Findley, S.D. ZnT-2, a mammalian protein that confers resistance to zinc by facilitating vesicular sequestration, *EMBO J.*, 15, 1784, 1996a.
Palmiter, R.D., et al., ZnT-3, a putative transporter of zinc into synaptic vesicles, *Proc. Natl. Acad. Sci. USA*, 93, 14934, 1996b.
Palmiter, R.D. and Findley, S.D., Cloning and functional characterization of a mammalian zinc transporter that confers resistance to zinc, *EMBO J.*, 14, 639, 1995.
Palmiter, R.D. and Huang, L. Efflux and compartmentalization of zinc by members of the SLC30 family of solute carriers, *Pflugers Arch.*, 447, 744, 2004.
Pascolo, L., et al. Molecular mechanisms for the hepatic uptake of magnetic resonance imaging contrast agents, *Biochem. Biophys. Res. Commun.*, 257, 746, 1999.
Paulsen, I.T. and Saier, M.H., Jr. A novel family of ubiquitous heavy metal ion transport proteins, *J. Membr. Biol.*, 156, 99, 1997.
Paulusma, C., et al. Congenital jaundice in rats with a mutation in a multidrug resistance-associated protein gene, *Science*, 271, 1126, 1996.
Paulusma, C.C. and Oude Elferink, R.P.J. The canalicular multispecific organic anion transporter and conjugated hyperbilirubinemia in rat and man, *J. Mol. Med.*, 75, 420, 1997.
Pavoine, C., et al. The high affinity (Ca2 + -Mg2 +)-ATPase in liver plasma membranes is a Ca^{2+} pump. Reconstitution of the purified enzyme into phospholipid vesicles, *J. Biol. Chem.*, 262, 5113, 1987.
Perego, P. and Howell, S.B. Molecular mechanisms controlling sensitivity to toxic metal ions in yeast, *Toxicol. Appl. Pharmacol.*, 147, 312, 1997.
Petris, M.J., et al. Copper-stimulated endocytosis and degradation of the human copper transporter, hCtr1, *J. Biol. Chem.*, 278, 9639, 2003.
Powers, M.J., et al. Functional behavior of primary rat liver cells in a three-dimensional perfused microarray bioreactor, *Tissue Eng.*, 8, 499, 2002.

Prata, M.I., et al. Targeting of lanthanide(III) chelates of DOTA-type glycoconjugates to the hepatic asyaloglycoprotein receptor: Cell internalization and animal imaging studies, *Contrast Media Mol. Imaging*, 1, 246, 2006.
Pratt, R.D., Ferreira, G.C., and Pedersen, P.L. Mitochondrial phosphate transport. Import of the H^+/Pi symporter and role of the presequence, *J. Biol. Chem.*, 266, 1276, 1991.
Pufahl, R.A., et al. Metal ion chaperone function of the soluble Cu(I) receptor Atx1, *Science*, 278, 853, 1997.
Ramm, G.A., Powell, L.W., and Halliday, J.W. Pathways of intracellular trafficking and release of ferritin by the liver in vivo: The effect of chloroquine and cytochalasin D, *Hepatology*, 19, 504, 1994.
Regoeczi, E. and Chindemi, P.A. Translocation of different forms of transferrin from blood to bile in the rat, *Hepatology*, 21, 1063, 1995.
Regoeczi, E., Chindemi, P.A., and Debanne, M.T. Partial resialylation of human asialotransferrin types 1 and 2 in the rat, *Can. J. Biochem. Cell Biol.*, 62, 853, 1984.
Renston, R.H., et al. Evidence for a vesicular transport mechanism in hepatocytes for biliary secretion of immunoglobulin A, *Science*, 208, 1276, 1980.
Reuter, H. Calcium channel modulation by neurotransmitters, enzymes and drugs, *Nature*, 301, 569, 1983.
Rius, M., et al. Cotransport of reduced glutathione with bile salts by MRP4 (ABCC4) localized to the basolateral hepatocyte membrane, *Hepatology*, 38, 374, 2003.
Roelofsen, H., Wolters, H., Van Luyn, M.J., Miura, N., Kuipers, F., and Vonk, R.J. Copper-induced apical trafficking of ATP7B in polarized hepatoma cells provides a mechanism for biliary copper excretion, *Gastroenterology*, 119, 782, 2000.
Rudolph, J.R., et al. Preferential hepatic uptake of iron from rat asialotransferrin: Possible engagement of two receptors, *Am. J. Physiol.*, 251, G398, 1986.
Safaei, R. and Howell, S.B. Copper transporters regulate the cellular pharmacology and sensitivity to Pt drugs, *Crit. Rev. Oncol. Hematol.*, 53, 13, 2005.
Sakisaka, S., Ng, O.C., and Boyer, J.L. Tubulovesicular transcytotic pathway in isolated rat hepatocyte couplets in culture. Effect of colchicine and taurocholate, *Gastroenterology*, 95, 793, 1988.
Scharschmidt, B.F., et al. Fluid phase endocytosis by cultured rat hepatocytes and perfused rat liver: Implications for plasma membrane turnover and vesicular trafficking of fluid phase markers, *Proc. Natl. Acad. Sci. USA*, 83, 9488, 1986.
Schmitz, C., Deason, F., and Perraud, A.L. Molecular components of vertebrate Mg^{2+}-homeostasis regulation, *Magnes. Res.*, 20, 6, 2007.
Sharma, V., Crankshaw, C.L., and Piwnica-Worms, D. Effects of multidrug resistance (MDR1) P-glycoprotein expression levels and coordination metal on the cytotoxic potency of multidentate (N4O2) (ethylenediamine)bis[propyl(*R*-benzylimino)]metal(III) cations, *J. Med. Chem.*, 39, 3483, 1996.
Sigel, E. Use of *Xenopus* oocytes for the functional expression of plasma membrane proteins, *J. Membr. Biol.*, 117, 201, 1990.
Silver, S. Bacterial resistances to toxic metal ions—a review, *Gene*, 179, 9, 1996.
Simmons, T.W., Anders, M.W., and Ballatori, N. Cysteine and *S*-(1,2-dichlorovinyl)-L-cysteine transport in rat liver canalicular membrane vesicles: Potential reabsorption mechanisms for biliary metabolites of glutathione and its *S*-conjugates, *J. Pharmacol. Exp. Ther.*, 262, 1182, 1992.
Simmons, T.W., Hinchman, C.A., and Ballatori, N. Polarity of hepatic glutathione and glutathione *S*-conjugate efflux, and intraorgan mercapturic acid formation in the skate, *Biochem. Pharmacol.*, 42, 2221, 1991.
Simon, D.B., et al. Paracellin-1, a renal tight junction protein required for paracellular Mg^{2+} resorption, *Science*, 285, 103, 1999.
Simons, T.J. and Pocock, G. Lead enters bovine adrenal medullary cells through calcium channels, *J. Neurochem.*, 48, 383, 1987.
Smit, J.J.M., et al. Homogeneous disruption of the murine mdr2 P-glycoprotein gene leads to a complete absence of phospholipid from bile and to liver disease, *Cell*, 75, 451, 1993.
Solioz, M., Odermatt, A., and Krapf, R. Copper pumping ATPases: Common concepts in bacteria and man, *FEBS Lett.*, 346, 44, 1994.
Solioz, M. and Vulpe, C. CPx-type ATPases: A class of P-type ATPases that pump heavy metals, *Trends Biochem. Sci.*, 21, 237, 1996.
Sorensen, E.M. Modulatory effect of calcium on the influx and binding of cadmium in primary cultures of neonatal rat hepatocytes, *Toxicol. Lett.*, 41, 39, 1988.
Soreq, H. and Seidman, S. *Xenopus* oocyte microinjection: From gene to protein, *Meth. Enzymol.*, 207, 225, 1992.
Sperber, I. Secretion of organic anions in the formation of urine and bile, *Pharmacol. Rev.*, 11, 109, 1959.

Sternlieb, I., et al. Lysosomal defect of hepatic copper excretion in Wilson's disease (hepatolenticular degeneration), *Gastroenterology*, 64, 99, 1973.
Stockert, R.J., et al. Endocytosis of asialoglycoprotein-enzyme conjugates by hepatocytes, *Lab. Invest.*, 43, 556, 1980.
Su, M.A., et al. The G185R mutation disrupts function of the iron transporter Nramp2, *Blood*, 92, 2157, 1998.
Sudo, R., et al. Reconstruction of 3D stacked-up structures by rat small hepatocytes on microporous membranes, *FASEB J.*, 19, 1695, 2005.
Sugawara, N., et al. Lack of biliary excretion of Cd linked to an inherent defect of the canalicular isoform of multidrug resistance protein (cMrp) does not abnormally stimulate accumulation of Cd in the Eisai hyperbilirubinemic (EHB) rat liver, *Arch. Toxicol.*, 71, 336, 1997.
Summer, K.H. and Eisenburg, J. Low content of hepatic reduced glutathione in patients with Wilson's disease, *Biochem. Med.*, 34, 107, 1985.
Suzuki, T. and Kawakita, M. Uncoupling of ATP splitting from Ca(2+)-transport in Ca(2+)-transporting ATPase of the sarcoplasmic reticulum as a result of modification by *N*-(3-pyrene)maleimide: Activation of a channel with a specificity for alkaline earth metal ions, *J. Biochem.*, 114, 203, 1993.
Suzuki, T., Momoi, K., Hosoyamada, M., Kimura, M., and Shibasaki, T. Normal cadmium uptake in microcytic anemia mk/mk mice suggests that DMT1 is not the only cadmium transporter *in vivo*, *Toxicol. Appl. Pharmacol.*, 227, 462, 2008.
Szczypka, M.S., et al. A yeast metal resistance protein similar to human cystic fibrosis transmembrane conductance regulator (CFTR) and multidrug resistance-associated protein, *J. Biol. Chem.*, 269, 22853, 1994.
Tabuchi, M., et al. Alternative splicing regulates the subcellular localization of divalent metal transporter 1 isoforms, *Mol. Biol. Cell*, 13, 4371, 2002.
Tammur, J., et al. Two new genes from the human ATP-binding cassette transporter superfamily, ABCC11 and ABCC12, tandemly duplicated on chromosome 16q12, *Gene*, 273, 89, 2001.
Tang, D.T., Zhang, S., and Zhong, X. Studies on the influence of calcium on cadmium absorbed by rat liver and the resistivity against the toxicity of cadmium in rat, *Bull. Environ. Contam. Toxicol.*, 39, 453, 1987.
Tavassoli, M. The role of liver endothelium in the transfer of iron from transferrin to the hepatocyte, *Ann. N. Y. Acad. Sci.*, 526, 83, 1988.
Thiele, D.J. Integrating trace element metabolism from the cell to the whole organism, *J. Nutr.,* 133, 1579S, 2003.
Thorstensen, K., et al. Uptake of iron from N-terminal half-transferrin by isolated rat hepatocytes. Evidence of transferrin-receptor-independent iron uptake, *Eur. J. Biochem.*, 232, 129, 1995.
Thorstensen, K. and Romslo, I. Uptake of iron from transferrin by isolated rat hepatocytes. A redox-mediated plasma membrane process?, *J. Biol. Chem.*, 263, 8844, 1988.
Tibaduiza, E.C. and Bobilya, D.J. Zinc transport across an endothelium includes vesicular cotransport with albumin, *J. Cell. Physiol.*, 167, 539, 1996.
Tommasini, R., et al. The human multidrug resistance-associated protein functionally complements the yeast cadmium resistance factor 1, *Proc. Natl. Acad. Sci. USA*, 93, 6743, 1996.
Torrubia, J.O.A. and Garay, R. Evidence for a major route for zinc uptake in human red blood cells: $[Zn(HCO_3)_2Cl]^-$ influx through the $[Cl^-/HCO_3^-]$ anion exchanger, *J. Cell. Physiol.*, 138, 316, 1989.
Touret, N., et al. Dynamic traffic through the recycling compartment couples the metal transporter Nramp2 (DMT1) with the transferrin receptor, *J. Biol. Chem.*, 278, 25548, 2003.
Trinder, D., Oates, P.S., Thomas, C., Sadleir, J., and Morgan, E.H. Localisation of divalent metal transporter 1 (DMT1) to the microvillus membrane of rat duodenal enterocytes in iron deficiency, but to hepatocytes in iron overload, *Gut*, 46, 270, 2000.
Tsai, K.-J., Hsu, C.-M., and Rosen, B.P. Efflux mechanisms of resistance to cadmium, arsenic and antimony in prokaryotes and eukaryotes, *Zool. Stud.*, 36, 1, 1997.
van Veen, H.W., et al. Translocation of metal phosphate via the phosphate inorganic transport system of *Escherichia coli*, *Biochemistry*, 33, 1766, 1994.
Valentine, J.S. and Gralla, E.B. Delivering copper inside yeast and human cells, *Science*, 278, 817, 1997.
Virkki, L.V., Biber, J., Murer, H., and Forster, I.C. Phosphate transporters: A tale of two solute carrier families, *Am. J. Physiol. Renal Physiol.*, 293, F643, 2007.
Vulpe, C., et al. Isolation of a candidate gene for Menkes disease and evidence that it encodes a copper-transporting ATPase, *Nat. Genet.*, 3, 7, 1993.
Vulpe, C.D., et al. Hephaestin, a ceruloplasmin homologue implicated in intestinal iron transport, is defective in the sla mouse, *Nat. Genet.*, 21, 195, 1999.
Vulpe, C.D. and Packman, S. Cellular copper transport, *Annu. Rev. Nutr.*, 15, 293, 1995.

Wemmie, J.A., et al. Cadmium tolerance mediated by the yeast AP-1 protein requires the presence of an ATP-binding cassette transporter-encoding gene, YCF1, *J. Biol. Chem.*, 269, 32592, 1994.
Wernimont, A.K., et al. Structural basis for copper transfer by the metallochaperone for the Menkes/Wilson disease proteins, *Nat. Struct. Biol.*, 7, 766, 2000.
Wetterhahn-Jennette, K. The role of metals in carcinogenesis: Biochemistry and metabolism, *Environ. Health Perspect.*, 40, 233, 1981.
Wijnholds, J., et al. Increased sensitivity to anticancer drugs and decreased inflammatory response in mice lacking the multidrug resistance-associated protein, *Nat. Med.*, 3, 1275, 1997.
Yamaguchi, Y., et al. Biochemical characterization and intracellular localization of the Menkes disease protein, *Proc. Natl. Acad. Sci. USA*, 93, 14030, 1996.
Yang, X.L., et al. Two forms of Wilson disease protein produced by alternative splicing are localized in distinct cellular compartments, *Biochem. J.*, 326, 897, 1997.
Young, S.P., Bomford, A.B., and Williams, R. Dual pathways for the uptake of rat asialotransferrin by rat hepatocytes, *J. Biol. Chem.*, 258, 4972, 1983.
Yu, Y.Y., Kirschke, C.P., and Huang, L. Immunohistochemical analysis of ZnT1, 4, 5, 6, and 7 in the mouse gastrointestinal tract, *J. Histochem. Cytochem.*, 55, 223, 2007.
Yuan, D.S., et al. The Menkes/Wilson disease gene homologue in yeast provides copper to a ceruloplasmin-like oxidase required for iron uptake, *Proc. Natl. Acad. Sci. USA*, 92, 2632, 1995.
Zaman, G.J.R., et al. Role of glutathione in the export of compounds from cells by the multidrug-resistance-associated protein, *Proc. Natl. Acad. Sci. USA*, 92, 7690, 1995.
Zhang, A.S., et al. Localization of iron metabolism-related mRNAs in the rat liver indicate that HFE is expressed predominantly in hepatocytes, *Blood*, 103, 1509, 2004.
Zhao, H. and Eide, D. The yeast ZRT1 gene encodes the zinc transporter protein of a high-affinity uptake system induced by zinc limitation, *Proc. Natl. Acad. Sci. USA*, 93, 2454, 1996.
Zhou, B. and Gitschier, J. hCTR1: A human gene for copper uptake identified by complementation in yeast, *Proc. Natl. Acad. Sci. USA*, 94, 7481, 1997.

4 Molecular and Cell Biology of Lead

Bruce A. Fowler

CONTENTS

4.1 Introduction 114
4.2 Lead Transport and Cellular Uptake 114
4.3 Cellular Uptake 114
4.4 Intracellular Binding 114
4.4.1 Delta-Aminolevulinic Acid Dehydratase 115
4.5 Target Organs 115
4.6 Nervous System 115
4.7 Kidneys 115
4.7.1 Proximal Tubule Cells 116
4.7.1.1 Mitochondria 116
4.7.1.2 Nuclei 116
4.7.1.3 Cytoplasm 116
4.7.1.4 Glomerular and Interstitial Capillaries 116
4.8 Cardiovascular System 116
4.9 Lead-Induced Alterations of Cellular Signaling Pathways 117
4.10 Lead and the Immune System 117
4.10.1 Humoral Immune Effects 117
4.10.2 Cellular Immune Effects 117
4.10.3 Hematopoietic System/Heme Biosynthetic Pathway 118
4.10.3.1 ALA Synthetase 118
4.10.3.2 ALA Dehydratase 118
4.10.3.3 Uroporphyrinogen Decarboxylase 118
4.10.3.4 Coproporphyrin Oxidase 118
4.11 Reproductive System 119
4.12 Lead and Cancer 119
4.13 Mitigating Factors 119
4.14 Pharmacological Considerations 120
4.14.1 Dose and Duration of Exposure Level 120
4.14.2 Diet and Nutrition 120
4.14.2.1 Age 120
4.14.2.2 Gender 120
4.14.2.3 Genetics 120
4.14.2.4 Mixtures Exposures 120
4.14.2.5 Pb × As Interaction Studies 121
4.14.2.6 Pb × Cd 121
4.14.2.7 Pb × Cd × As 121
4.15 Nanomaterials 121
References 122

4.1 INTRODUCTION

Lead is a ubiquitous toxic metal found in virtually all living organisms including man. The toxic properties of this element from elevated exposures have been known for several thousand years. In recent decades, the ability of lead to produce toxic effects in a number of target organ systems, and essential cellular and molecular pathways at low dose levels has been increasingly appreciated as a result of the advent of modern and more sensitive and specific toxicological tools derived from basic cell and molecular biology. Molecular (omic) biomarkers are excellent examples of these modern tools, which provide insights into the impact of low-dose lead exposures on basic biological processes and also the influence of factors such as mixture exposures and genetic inheritance on health outcomes and delineation of populations at special risk. The ability to detect and interpret low-dose lead effects is of increasing importance since lead is now being incorporated into nanomaterials [1–3] that can be expected to increase biological exposures in new ways. These new uses of lead means that there will be an increasing need for evaluation of how and why this metal may exert effects on target cell populations in essential organ systems and delineation of key-sensitive molecular pathways.

This chapter will attempt to provide a concise overview of current understandings regarding how lead is taken up into cells, molecular binding patterns of lead in the circulation and in target cell populations in relation to known effects of lead on sensitive molecular systems in those cells such as the heme biosynthetic pathway, intracellular signaling systems, alterations in gene expression patterns, and mechanisms of cell death and replacement. The role of genetic inheritance as a factor in mediating sensitivity to lead will be taken up in relation to defining populations at special risk for low-dose lead toxicity.

4.2 LEAD TRANSPORT AND CELLULAR UPTAKE

The heme biosynthetic pathway enzyme, delta-aminolevulinic acid dehydratase (ALAD), which is found in red blood cells, has been shown [4] to be the major carrier protein for lead in blood. As discussed below, this enzyme exists as two polymorphic forms that vary in their affinity for lead. In addition to lead bound to ALAD, there is a smaller "diffusible" fraction of lead that appears to be important in mediating the bioavailability of lead to cells in target organ systems.

4.3 CELLULAR UPTAKE

Uptake of lead from the circulation may occur as a result of ionic lead binding to the surface coat proteins of target organ cells such as kidney proximal tubule cells [5] with subsequent intake via endocytosis secondary to cell membrane turnover. Other investigators [6] reported that lead as well as a number of other nephrotoxic metals damaged the integrity of isolated renal brush-border and basolateral membrane vesicles during *in vitro* exposure studies. No active transport mechanisms for lead at the cell membrane have been demonstrated. A second possible mechanism involves endocytosis of lead bound to proteins or low-molecular-weight molecules with subsequent lysosomal hydrolysis of the protein moiety and release of lead into the cytoplasm with subsequent rebinding to proteins such as ALAD, or low-molecular-weight proteins (PbBPs) as discussed in Section 4.4.

4.4 INTRACELLULAR BINDING

The intracellular binding patterns of lead vary by tissue and dose level of lead exposure. At high dose levels of lead, lead inclusion bodies that are pathognomonic indicators of lead poisoning and contain the highest intracellular concentrations of lead are formed in target organs such as the brain [7] and kidney [8–11]. At low-dose or environmental exposure levels of lead, soluble low-molecular-weight proteins (PbBPs) are the major binding sites for lead in target organs such as the brain [12–14] and kidney [12–18]. Studies on lead binding indicate that these low molecular weight proteins demonstrate consistent dissociation constants (Kds) of 10^{-8} M for lead across species, apparently as a result of highly conserved non-SH-based binding sites that normally also bind zinc. These proteins are hence distinct from metallothionein, which does not bind lead to any appreciable degree but binds zinc via SH–multidentate binding sites. Both the PbBPs and metallothioneins are capable of donating zinc to zinc-dependent proteins such as ALAD, as discussed below.

4.4.1 Delta-Aminolevulinic Acid Dehydratase

As noted earlier, the heme biosynthetic pathway enzyme ALAD is the major binding component for lead in blood and it exists as two polymorphic alleles (1 and 2) that also vary in their binding affinity/stability such that general terms, and depending on lead exposure level, persons possessing the ALAD 2 allele experience higher blood lead values than those possessing the ALAD 1 allele [19]. This genetic difference in lead-binding patterns has important implications for the target organ cellular toxicology of lead. In addition, both PbBPs and metallothionein discussed earlier may influence lead binding to ALAD by donating zinc to this zinc-dependent/zinc-activated enzyme [20–25].

4.5 TARGET ORGANS

It has been appreciated for many years that lead is a broad-spectrum toxic agent that affects a number of major organ systems in different ways and at different dose levels. Further, lead may selectively damage specific cell types in these organs as a function of a number of different mitigating factors as discussed below. Brief summaries of lead toxicity to a number of these major organ systems at the cellular and molecular levels of biological organization are provided latter and will hopefully be of value to the reader in accessing the voluminous literature on lead toxicity in a focused manner.

4.6 NERVOUS SYSTEM

The central nervous system (CNS) and the peripheral nervous system (PNS) components of the nervous system may each be affected by lead toxicity as a function of dose level and duration of exposure. In general, the known mechanisms of lead toxicity to specific cell types in these individual nervous system components will vary in part on the basis of the normal biological functions conducted by these cell types.

In the CNS, loss of cognitive abilities (e.g., IQ) may occur in some individuals at low lead dose levels and amplify at higher dose levels and extended periods of exposure to overt clinical neurological disease [26–28]. The effects of lead on neurogenesis have been extensively studied since the developing nervous system is particularly sensitive to lead, and children are generally regarded as being a population at special risk. At the cellular level, numerous studies conducted in experimental systems have suggested lead effects on a number of nerve cell developmental processes [29–33] but with variable demonstration of linkages to cognitive development depending on the behavioral test method system utilized. A number of molecular processes including alterations in various signaling pathways [34], changes in neuronal cytoskeletal proteins [35,36] and cell adhesion molecules [37,38]. Alterations in nervous cell signaling pathways may influence neuronal cell growth [39–42] as well as changes in gene/protein expression [42–44]. These various effects may overlap and will vary as a function of specific cell type, lead dose level, and duration of exposure.

In the PNS, decrements in nerve conduction velocities have been reported at lower lead dose levels while persons with high-dose/long-term exposures may experience overt neuromuscular symptoms such as “wrist drop” or “ankle drop” [45].

4.7 KIDNEYS

There are several specific cell types in the kidney that may be damaged by lead exposure including the cells of the proximal tubules and interstitial capillaries. Damage to these cells may contribute to changes in a number of other organ systems such as the cardiovascular system with regard to blood pressure.

4.7.1 Proximal Tubule Cells

The epithelial cells of the renal proximal tubules are a major target cell population for lead toxicity and the effects on several organelle and molecular systems have been demonstrated by a number of investigators [7–11].

4.7.1.1 Mitochondria

Lead disturbances of mitochondrial energy production [9] and inhibition of mitochondrial heme pathway enzymes [9] are among the important metabolic disturbances produced by lead in this organelle.

4.7.1.2 Nuclei

The formation of lead intranuclear inclusion bodies as noted earlier is pathognomonic of lead poisoning and is associated with alterations in DNA and protein synthesis patterns. The low-molecular-weight lead-binding proteins discussed below appear to play an important role in mediating the intranuclear movement of lead and binding to chromatin at low dose levels. These alterations in nuclear function have been associated with increased cellular division of renal proximal tubule cells in rodents [46] and may be related to the development of renal adenocarcinomas in rodents following chronic lead exposure. Other *in vitro* studies [15,16] showed that the cytosolic low-molecular-weight PbBPs were capable of translocating into purified isolated nuclei and binding to chromatin. It was observed that cadmium and zinc were more effective than cold lead in displacing radioactive lead from these proteins *in vitro.*

Such data are consistent with *in vivo* lead–cadmium–arsenic interaction studies described below that demonstrated that concomitant exposure to cadmium attenuated the formation of lead intranuclear inclusion bodies.

4.7.1.3 Cytoplasm

There are a number of cytoplasmic enzymes (e.g., ALAD) that have been shown to be altered by lead exposure both *in vivo* [9–11] and *in vitro* [20–25]. Data from these studies point to a major influence of cytosolic PbBPs and metallothionein in mediating the interaction of lead with these enzymes.

4.7.1.4 Glomerular and Interstitial Capillaries

Fibrosis of the glomerular and interstitial capillaries has extensively documented in both workers with high-dose lead exposure [45] and in experimental animals (7). These fibrotic changes undoubtedly play a role in the development of increased BUN and creatinine in lead-exposed workers [45]. Other studies on the molecular level [47] have reported induction of the stress protein HSPA5 (GRP78) following *in vitro* incubation of rat glomerular mesangial cells (rGMCs) with lead at 1 or 10 M Pb, indicating that this class of stress protein is activated by lead exposure perhaps secondary to lead-induced oxidative stress [48] that may play an important role in lead-induced glomerular/interstitial fibrosis [49].

4.8 CARDIOVASCULAR SYSTEM

Lead exposure has been extensively linked to the development of hypertension in humans [19,40,50–53]. The underlying mechanisms for the effects of lead on blood pressure are complex and appear to involve lead-induced reactive oxygen species (ROS) and nitric oxide (NO) signaling [50,52,53]. Biphasic alterations of angiotension converting enzyme (ACE) activity in rats as a function of duration of exposure have also been reported [40]. In addition, the activation of antioxidant enzyme systems such as superoxide dismutase (SOD), catalase (CAT), and glutathione peroxidase (GPX) have also been reported by several groups of investigators [54–56] and the interplay of these antioxidant systems may well provide an explanation for variability in the degree to which lead induces hypertension.

Other effects such as reduced β-adrenergic receptors and plasma catecholamine levels [57,58], endothelial cell damage, and atherosclerosis have also been reported in relation to lead exposure in a variety of clinical and experimental reports [51]. These effects appear to be mediated at least in part by lead binding to ALAD in blood with differences noted between ALAD 1 and 2 isoforms [19]. There is an ongoing discussion as to whether the hypertensive effects of lead are secondary to lead effects in the kidney noted earlier, a result of lead interaction with signaling pathways that regulate blood pressure as discussed below or some combination of these effects. Further research is clearly needed to delineate the complex underlying mechanisms by which produces hypertension.

4.9 LEAD-INDUCED ALTERATIONS OF CELLULAR SIGNALING PATHWAYS

As noted earlier, lead is a broad-spectrum toxic agent that produces effects on a number of organ, cell, and molecular systems. More recent studies have identified that lead is capable of producing changes in important cellular signaling pathways such as the nitric oxide synthase (NOS) system [50,53] molecular signaling pathways. Some of these changes appear to be related to increased production of ROS via inhibition of mitochondrial respiration [9]. It is clear that further research is needed to understand the interrelationships between lead effects observed in these complex pathways, alterations in other organelle or molecular systems, and the mediating factors discussed below.

4.10 LEAD AND THE IMMUNE SYSTEM

There is an extensive and well-documented literature concerning the effects of lead on the immune system both in terms of humoral [59–61] and cellular [62–70] aspects of immune function. In general, lead exposure, both *in vivo* and *in vitro*, has been shown to exert a variety of immunomodulating effects on various aspects of immune function suggesting multiple points of interaction with cellular machinery [69]. The mechanisms underlying these effects appear to involve the same organelle and basic molecular systems discussed earlier, but the health consequences in terms of susceptibility to infection, autoimmune reactions, and cancer are clearly important.

4.10.1 Humoral Immune Effects

Lead exposure has been shown to alter production of circulating antibodies to a number of antigens and microorganisms. The obvious potential health risks to lead-exposed organisms for infection or increased susceptibility to cancer cannot be understated. Miller et al. [60] reported differential effects of oral lead exposure in Fischer 344 rats with increases in several humoral parameters but depression of cell-mediated immune responses. Ercal et al. [61] demonstrated a linkage between lead-induced oxidative damage and suppression of serum levels of IgA, IgM, and IgG in Fischer 344 rats.

4.10.2 Cellular Immune Effects

Lead-induced suppression of cellular immunity is also an important aspect of lead toxicity since it can also have important impacts on both the ability of fight infection and immune surveillance for cancerous cells in major organ systems. McCabe et al. [59] reported findings of studies on lead-induced increases of major histocompatibility complex (MHC) class II molecules (Ia) on the surface of mouse B lymphocytes. These alterations appeared to be related to either increased translation or post-translational events since changes in messenger RNA for these proteins were not observed. Miller et al. [60] showed that developmental exposure of Fischer 344 rats produced chronic immune alterations in type 1 and type 2 immune responses and suggested that these effects may be related to a lead-induced shift in T helper cell development or functioning. Subsequent studies by Bunn et al. [63,64] reported gender differences as a result of embryonic exposure of Sprague-Dawley rats to lead [63] or as a result of oral lead exposure of juveniles or adult Fischer 344 rats [64] to lead in drinking water. In general, females were found to be more sensitive than males with regard to delayed-type hypersensitivity (DTH) responses. McCabe et al. [65] and Shen et al. [66] reported on the *in vitro* effects of lead on CD4(+) T cells. McCabe et al. [65] showed that lead stimulated proliferation and expansion of alloantigen-reactive T helper cells at culture lead concentrations down to 0.1 μM Pb. Shen et al. [66] reported that lead-induced cytokine responses at concentrations that did not affect T cell proliferation. Data from these studies indicate that immune response profiles are lead concentration dependent. Subsequent studies by Farrer and coworkers [68] led enhanced CD4+ proliferation by indirectly affecting antigen presenting cells and antigen interaction. These investigators recently showed that myeloid suppressor cell-produced NO provides at least one mechanism for producing the observed CD4+ T cell proliferative response. Lynes and coworkers [69] highlighted the importance of lead-induced alterations in gene expression as an overarching set of interrelated mechanisms that regulate observed changes in the various aspects of lead-induced changes in immune function.

4.10.3 Hematopoietic System/Heme Biosynthetic Pathway

Lead-induced inhibition of a number of enzymes in the heme biosynthetic pathway has been documented in a number of major organ systems and at high-dose exposures to produce anemia as a result of suppression of the hematopoietic system [26,45]. The most well-studied enzymes in the pathway documented to show inhibition are discussed below.

4.10.3.1 ALA Synthetase

ALA synthetase (ALAS) is the first and rate-limiting enzyme in the heme biosynthetic pathway. It is located loosely bound to the mitochondrial inner membrane and has been known for many years to be sensitive to the inhibitory effects of a number of metals including lead [9]. This enzyme is also known to be inducible by decreases in cellular heme levels and so may exhibit differential responses to lead inhibition depending on dose and duration of exposure [71].

4.10.3.2 ALA Dehydratase

ALA dehydratase (ALAD) is the second enzyme in the heme biosynthetic pathway and is located in the cellular cytoplasm. It is a zinc-dependent enzyme and regarded as the most sensitive enzyme to lead inhibition with an IC_{50} on the order of 10^{-7} M [20–25]. As noted earlier, this enzyme exists as two isoforms (1 and 2) that vary in their binding affinity for lead with isoform 2 having the stronger binding affinity for lead. Lead inhibition of ALAD and the increased presence of ALA in the urine of exposed individuals have been used for many as sensitive and relatively specific biomarkers for elevated lead exposure.

4.10.3.3 Uroporphyrinogen Decarboxylase

Uroporphyrinogen decarboxylase (UROD) is located at the fourth step in the heme biosynthetic pathway and is sensitive to oxidative stress from a number of trace elements [72]. This inhibition results in the increased presence of uroporphyrin in urine of lead-exposed individuals. Lead-induced oxidative stress may be the underlying mechanism for inhibition of this cytosolic enzyme.

4.10.3.4 Coproporphyrin Oxidase

Coproporphyrin oxidase (CO) is a mitochondrial enzyme, which is the penultimate enzyme in the heme biosynthetic pathway. It is also sensitive to lead inhibition [72] and results in the increased presence of coproporphyrin in the urine of lead-exposed individuals perhaps in concert with lead inhibition of ferrochelatase (FC) as discussed below.

4.10.3.4.1 Ferrochelatase

FC is the final enzyme in the heme biosynthetic pathway that catalyzes the insertion of ferrous (Fe^{2+}) iron into the protoporphyrin ring to form heme. This enzyme that is located as an integral protein in the mitochondrial inner membrane is also sensitive to lead inhibition [73–75]. The mechanism of this that may be secondary to lead inhibition of mitochondrial respiration and energy-linked mitochondrial transformations [76] leading to a decrement in the reduction of Fe^{3+} to Fe^{2+} during transport across the mitochondrial inner membrane and prior to insertion of Fe^{2+} into the protoporphyrin ring to form heme. If Fe^{2+} is not inserted into the porphyrin ring, then Zn^{2+} is inserted leading to an increase in Zn-protoporphyrin (ZPP) concentration. Increased ZPP in circulating red blood cells has also been used as a biomarker for lead toxicity and decreased iron intake [75,77].

4.11 REPRODUCTIVE SYSTEM

As noted earlier, lead is a broad-spectrum toxic agent that exerts toxic manifestations in a number of organ systems including the reproductive system. Lead is capable of crossing the placenta and exerting deleterious effects on the developing embryo leading to increased risk of spontaneous abortion [78,79] and has been demonstrated to decrease sperm counts and alter sperm morphology/motility in males with moderate to elevated lead exposures [80–92]. The mechanisms underlying these effects also appear to be complex and may involve the interplay of a number of cellular and molecular biochemical systems. Lead-induced oxidative stress has been reported by a number of investigators [93–95] as an early and consistent finding in both lead-exposed experimental animals and humans. Other investigators have reported alterations in endocrine functions affecting the testes [96–98], altered DNA/protein synthesis [90], and DNA protamine binding [99,100]. Finally, other investigators have reported direct lead-induced cytotoxicity to male germinal cells in mice [101], decreased testes weights associated with decreased sperm concentrations [102] in lead-treated rats, and that neonatal lead exposure reduced fertility in adult males, which was associated with increased numbers of apoptotic cells and decreased numbers of testicular macrophages. As with the other organ systems reviewed in this chapter, it is clear that lead may exert a broad spectrum of cellular effects related to oxidative stress and which appear to be mediated by a complex interplay of molecular systems.

4.12 LEAD AND CANCER

Chronic exposure of rodents to lead in drinking water has been demonstrated to produce elevated rates of renal cancer. Epidemiological studies in workers with elevated occupational exposures to lead have produced inconsistent results with regard to cancer rates [103]. Results from experimental cellular/molecular systems have demonstrated evidence of DNA damage that may be secondary to lead-induced oxidative stress [104,105] produced by ROS. International Agency for Research against Cancer (IARC) has classified lead as a possible human carcinogen (Class 2B) based on these lines of evidence. Further research is clearly needed to determine what molecular mechanisms are underlying the observed positive effects in rodents and more equivocal findings in humans and the interactive effects of DNA repair mechanisms such as ERK1/2 [106] metallothionein expression [107] and expression of the ALAD2 allele [108].

There are several possible interrelated mechanisms by which lead can elicit the carcinogenic response. One involves the known binding of lead to DNA transcription factor zinc finger proteins [109–111] that can alter normal DNA replication. Lead induced oxidative stress [112–114], which could produce direct DNA damage, and lead induced apoptosis [115,116], which could produce a cell's death and replacement scenario that may give rise to damaged replacement cells. Finally, it has been hypothesized [117] that the renal lead-binding proteins (PbBPs) act in a manner similar to hormonal receptors. They normally bind/transport zinc into target cell nuclei, bind to chromatin and alter gene expression patterns leading to cell proliferation.

4.13 MITIGATING FACTORS

As with many toxic agents, the mechanistic effects of lead on the various organs and biochemical systems discussed earlier will vary in part as a result of a number of mitigating factors such as those discussed below.

4.14 PHARMACOLOGICAL CONSIDERATIONS

4.14.1 Dose and Duration of Exposure Level

The various effects of lead will vary as a function of dose and duration of exposure with some toxic manifestations occurring at chronic low dose levels and others occurring at elevated and more acute or intermediate exposures [118]. This also means that the operative underlying mechanisms may also vary in prominence under a given set of exposure conditions. It is also important to note that lead is a cumulative toxic agent that is predominately deposited in the skeleton and it may be redistributed from this depot under conditions that produce demineralization/or mobilization of calcium from the skeleton, such as osteoporosis or pregnancy, respectively [119].

4.14.2 Diet and Nutrition

It has been appreciated for many years that diet and nutritional status may exert pronounced effects on susceptibility to lead toxicity [120]. The underlying factors are related to influencing the uptake of lead from the gastrointestinal tract and influencing the bioavailability of essential metals such as zinc and iron to lead-sensitive enzymatic processes such as those found in the heme biosynthetic pathway. Treatment of animals with zinc [23–25] has been shown to attenuate the inhibitory effects of lead on ALAD and iron deficiency has been shown to increase the formation of ZPP [75]. More recent studies [67] have reported that *in vivo* administration of alpha-tocopherol protects against the immunotoxic effects of lead in mice apparently via antioxidative effects.

4.14.2.1 Age

Age is a well-known factor that influences susceptibility to lead toxicity; the very young developing organisms and older individuals are at special risk [26]. As noted earlier, lead is capable of crossing the placenta and producing effects in developing offspring. In addition, lead may be mobilized from the skeletal stores in older individuals such that lead accumulation from a lifetime may be returned to an individual in later years from this internal compartment.

4.14.2.2 Gender

Males and females may vary in susceptibility to lead as a function of the factors discussed earlier with different organ systems manifesting responses. The issue of osteoporosis for females and the possible effects of lead on this process have been extensively debated [119]. For males, the issue of decreased fertility due to alterations in spermatogenesis is also an important consideration in the gender-specific issue.

4.14.2.3 Genetics

It has been appreciated for many years that individuals with similar lead exposure and other equivalent risk factors will vary in susceptibility to lead toxicity suggesting that genetic factors may also play an important role. The most well-studied of these genetic risk factors for susceptibility to lead toxicity involves the role of ALAD polymorphisms in mediating lead bioavailability to sensitive biochemical processes. As noted earlier, ALAD in humans exists as 2 alleles (1 and 2) such that individuals may have 1, 1–2, or 2 polymorphic forms of this enzyme. In most populations, ALAD1 is the most prevalent isoform (~85%) while ALAD-1–2 and ALAD2 together comprise ~15%). These latter individuals will, depending on exposure level, have generally higher blood lead values due to the higher binding affinity of lead to the ALAD2 allele. The dynamic interaction of lead from this carrier protein in blood with other molecular processes is a subject of ongoing research [19].

4.14.2.4 Mixtures Exposures

Lead exposure in combination with other toxic trace elements such as cadmium and arsenic is a common phenomenon under a number of occupational and environmental situations (e.g., Superfund sites). The issue of lead interactions in these mixture situations is a major concern for risk assessment purposes since additive, synergistic, or antagonistic interactions from combined exposures may greatly influence the shape of the dose–response curve and expected toxic outcome measures. The following brief review of the published literature on interactions between lead and other common toxic elements such as arsenic and cadmium is intended to only act as a stimulus for research in this important area. The issue of lead-based nanomaterials will be introduced since these materials are only now being produced for commercial applications, but it is clear that the small size of these materials may greatly influence the absorption and distribution of lead in humans.

4.14.2.5 Pb × As Interaction Studies

The early Pb × Cd × As factorial design interaction studies by Mahaffey and coworkers [121–123] using stressor dose levels of these elements demonstrated an additive interaction for a number of parameters in rats exposed to these elements in a semipurified diet at 10 weeks. More recent studies [124] also showed synergistic interactive immunotoxic effects of concomitant lead and arsenic treatments in mice following acute administration of these elements.

4.14.2.6 Pb × Cd

The studies by Mahaffey and coworkers noted earlier, also examined Pb × Cd interactions and found that concomitant exposure to Cd eliminated formation of the pathognomonic lead intranuclear inclusion bodies in kidneys, which was associated with an approximate 60% reduction in total renal lead concentrations but additive increases in urinary porphyrin excretion patterns [122,123]. More recent studies [125] demonstrated a positive lead–cadmium interactive effect on the responsiveness of rat mesenteric blood vessels to norepinephrine and angiotension II. These data indicate that the blood vasculature effects related to lead exposure discussed earlier may also be altered by concomitant exposure to cadmium, which is also well known to produce vascular alterations.

4.14.2.7 Pb × Cd × As

The above studies [121–123] included a Pb × Cd × As exposure group, which also showed elimination of the renal lead intranuclear inclusion bodies, marked reduction of total renal lead burden, and further additive effects on urinary porphyrin excretion patterns. More recent Pb × Cd × As studies using LOEL dose levels of these elements in drinking water [126] demonstrated similar effects on the heme biosynthetic pathway, which showed a marked time-course pattern with more pronounced effects at a 30-day time point, attenuation of the effects at 90 days apparently as a result of induction of antioxidant systems, and the appearance of oxidative damage markers at 180 days apparently as a result of prolonged oxidative stress. Taken together, data from these studies demonstrate the value of biomarker endpoints in elucidating interactions between lead and other common toxic trace elements and the alterations that may occur between tissue concentrations of lead and expected biological effects under mixture exposure conditions.

4.15 NANOMATERIALS

A rapidly expanding area to toxicological concern involves the use of toxic metals such as lead in nanomaterials. At present, little is understood about the toxic properties of materials such as Pb-based nanomaterials [1–3], but it is clear that the synthesis of these materials may easily positively alter expected relationships between inhalation or ingestion exposures to a given concentration of lead and uptake of this element. These concerns clearly support the need for further research into the bioavailability of Pb from such materials and a rigorous examination of health risks from both the particle size and combined elemental exposures perspective.

REFERENCES

1. Hur, S.G., et al., Influences of alkaline earth metal substitution on the crystal structure and physical properties of magnetic RuSr1.9A0.1GdCu2O8 (A=Ca, Sr, and Ba) superconductors, *J. Phys. Chem. B*, 109, 21694, 2005.
2. Csanády, A., et al., Nanocomposite bulk of mechanically milled Al-Pb samples consolidated pore-free by the high-energy rate forming technique, *J. Nanosci. Nanotechnol.*, 5 869, 2005.
3. Zhou, G., et al., Controlled synthesis of high-quality PbS star-shaped dendrites, multipods, truncated nanocubes, and nanocubes and their shape evolution process, *J. Phys. Chem. B*, 110, 6543, 2006.
4. Bergdahl, I.A., et al., Plasma and blood lead in humans: Capacity-limited binding to delta-aminolevulinic acid dehydratase and other lead-binding components, *Toxicol. Sci.*, 46, 247, 1998.
5. Victery, W.W., Miller, C.R., and Fowler, B.A., Lead accumulation by rat renal brush border membrane vesicles, *J. Pharmacol. Exp. Therap.*, 231, 589, 1984.
6. Herak-Kramberger, C.M. and Sabolic, I., The integrity of renal cortical brush-border and basolateral membrane vesicles is damaged *in vitro* by nephrotoxic heavy metals, *Toxicology*, 156, 139, 2001.
7. Goyer, R.S. and Rhyne, B.C., Pathological effects of lead, *Int. Rev. Exp. Pathol.*, 12, 1, 1973.
8. Goyer, R.A., et al., Lead dosage and the role of the intranuclear inclusion body: An experimental study, *Arch. Environ. Health*, 20, 704, 1971.
9. Fowler, B.A., et al., Chronic low level lead toxicity in the rat III. An integrated toxicological assessment with special reference to the kidney, *Toxicol. Appl. Pharmacol.*, 56, 59, 1980.
10. Oskarsson, A. and Fowler, B.A., Effects of lead on the heme biosynthetic pathway in rat kidney, *Exp. Mol. Pathol.*, 43, 397, 1985.
11. Oskarsson, A. and Fowler, B.A., Effects of lead inclusion on subcellular distribution of lead in rat kidney: The relationship to mitochondrial function, *Exp. Mol. Pathol.*, 43, 409, 1985.
12. Oskarsson, A., Squibb, K.S., and Fowler, B.A., Intracellular binding of lead in the kidney: Partial isolation and characterization of post-mitochondrial supernatant lead-binding components, *Biochem. Biophys. Res. Commun.*, 104, 290, 1982.
13. DuVal, G.E. and Fowler, B.A., Preliminary purification and characterization studies of a low molecular weight high affinity cytosolic lead-binding protein in rat brain, *Biochem. Biophys. Res. Commun.*, 159, 177, 1989.
14. Quintanilla-Vega, B., et al., Lead-binding proteins in brain tissue of environmentally-lead exposed humans, *Chem. Biol. Interact.*, 98, 193, 1995.
15. Mistry, P., Lucier, G.W., and Fowler, B.A., High affinity lead-binding proteins from rat kidney cytosol: Mediate cell-free nuclear translocation of lead, *J. Pharmacol. Exp. Therap.*, 232, 462–469, 1985.
16. Mistry, P., Mastri, C., and Fowler, B.A., Influence of metal ions on renal cytosolic lead-binding proteins and nuclear uptake of lead in the kidney, *Biochem. Pharmacol.*, 35, 711, 1986.
17. Fowler, B.A., et al., Implications of lead-binding proteins for risk assessment of lead exposure, *J. Expo. Anal. Environ. Epidemiol.*, 3, 441, 1993.
18. Smith, D.R., et al., High affinity renal lead-binding proteins in environmentally exposed humans, *Chem. Biol. Interact.*, 115, 39, 1998.
19. Scinicariello, F., et al., Lead and δ-aminolevulinic acid dehydratase polymorphism: Where does it lead? A meta-analysis, *Environ. Health Perspect.*, 115, 35, 2007.
20. Goering, P.L. and Fowler, B.A., Regulation of lead inhibition of δ-aminolevulinic dehydratase by a high affinity renal lead-binding protein, *J. Pharmacol. Exp. Therap.*, 231, 66, 1984.
21. Goering, P.L. and Fowler, B.A., Mechanisms of renal lead-binding protein protection against lead-inhibition of δ-aminolevulinic acid dehydratase, *J. Pharmacol. Exp. Therap.*, 234, 365, 1985.
22. Goering, P.L., Mistry, P., and Fowler, B.A., A high affinity lead-binding protein in brain attenuates lead inhibition of δ-aminolevulinic acid dehydratase: Comparison with a renal lead-binding protein, *J. Pharmacol. Exp. Therap.*, 237, 220, 1986.
23. Goering, P.L. and Fowler, B.A., Mechanism of kidney metallothionein reversal of lead inhibition of δ-aminolevulinic acid dehydratase, *Arch. Biochem. Biophys.*, 253, 48, 1987.
24. Goering, P.L. and Fowler, B.A., Metal constitution of metallothionein influences inhibition of δ-aminolevulinic acid dehydratase (porphobilinogen synthase) by lead, *Biochem. J.*, 245, 339, 1987.
25. Goering, P.L. and Fowler, B.A., Regulatory roles of high affinity metal binding proteins in mediating lead effects on δ-aminolevulinic acid dehydratase, *Ann. N. Y. Acad. Sci.*, 514, 235, 1987.
26. NAS/NRC, *Measuring Lead Exposure in Infants, Children, and Other Sensitive Populations* (p. 337), National Academies Press, Washington, DC, 1993.
27. Guilarte, T.R., et al., Environmental enrichment reverses cognitive and molecular deficits induced by developmental lead exposure, *Ann. Neurol.*, 53, 50, 2003.

28. White, L.D., et al., New and evolving concepts in the neurotoxicology of lead, *Toxicol. Appl. Pharmacol.*, 225, 1, 2007.
29. Chang, W., et al., Effects of Brn-3a protein and RNA expression in rat brain following low-level lead exposure during development on spatial learning and memory, *Toxicol. Lett.*, 164, 63, 2006.
30. Davidovics, Z. and DiCicco-Bloom, E., Moderate lead exposure elicits neurotrophic effects in cerebral cortical precursor cells in culture, *J. Neurosci. Res.*, 80, 817, 2005.
31. Gilbert, M.E., et al., Chronic developmental lead exposure reduces neurogenesis in adult rat hippocampus but does not impair spatial learning, *Toxicol. Sci.*, 86, 211, 2005.
32. Heidmets, L.T., et al., Early post-natal, low-level lead exposure increases the number of PSA-NCAM expressing cells in the dentate gyrus of adult rat hippocampus, *Neurotoxicology*, 27, 39, 2006.
33. Kermani, S., et al., Effect of lead on proliferational and neural differentiation of bone marrow-mesenchymal stem cells, *Toxicol. In Vitro*, 22, 995, 2008.
34. Xiao, C., et al., Pb^{2+} impairs GABAergic synaptic transmission in rat hippocampal slices: A possible involvement of presynaptic calcium channels, *Brain Res.*, 1088, 93, 2006.
35. Harry, G.J., et al., Lead-induced alterations of glial fibrillary acidic protein (GFAP) in the developing rat brain, *Toxicol. Appl. Pharmacol.*, 139, 84, 1996.
36. Scortegagna, M., Chikhale, E. and Hanbauer, I., Effect of lead on cytoskeletal proteins expressed in E14 mesencephalic primary cultures, *Neurochem. Int.*, 32, 353, 1998.
37. Prozialeck, W.C., et al., Cadherins and NCAM as potential targets in metal toxicity, *Toxicol. Appl. Pharmacol.*, 183, 255, 2002.
38. Hu, Q., et al., Maternal low-level lead exposure reduces the expression of PSA-NCAM and the activity of sialyltransferase in the hippocampi of neonatal rat pups, *Neurotoxicology*, 29, 675, 2008.
39. Crumpton, T., et al., Lead exposure in pheochromocytoma (PC12) cells alters neural differentiation and Sp1 DNA-binding, *Neurotoxicology*, 22, 49, 2001.
40. Sharifi, A.M., et al., Investigation of circulatory and tissue ACE activity during development of lead-induced hypertension, *Toxicol. Lett.*, 153, 233, 2004.
41. Williams, T.M., Lead enhances NGF-induced neurite outgrowth in PC12 cells by potentiating ERK/MAPK activation, *Neurotoxicology*, 21(6), 1081–1089, 2000.
42. Schneider, J.S., Huang, F.N., Vemuri, M.C. Effects of low-level lead exposure on cell survival and neurite length in primary mesencephalic cultures. *Neurotoxicology and Teratology*. 25: 555–559, 2003.
43. Guilarte, T.R., McGlothan, J.L., and Nihei, M.K., Hippocampal expression of *N*-methyl-D-aspartate receptor (NMDAR1) subunit splice variant mRNA is altered by developmental exposure to Pb(2+), *Brain Res. Mol. Brain Res.*, 76, 299, 2000.
44. Qian, Y., et al., Lead targets GRP78, a molecular chaperone, in C6 rat glioma cells, *Toxicol. Appl. Pharmacol.*, 163, 260, 2000.
45. Skerfving, S., Bergdahl, I.A., Lead, In: G.F. Nordberg, B.A. Fowler, M. Nordberg, and L. Friberg (Eds), *Handbook of Toxicology of Metals* (3rd edition, pp. 599–643), Elsevier Publishers, Amsterdam, 2007.
46. Choie, D.D. and Richter, G.W., Cell proliferation in rat kidney induced by lead acetate and effects of urinephrectomy on the proliferation, *Am. J. Pathol.*, 66, 265, 1972.
47. Falahatpisheh, H., et al., Activation profiles of HSPA5 during the glomerular mesangial cell stress response to chemical injury, *Cell Stress Chaperones*, 12, 209, 2007.
48. Daggett, D.A., et al., Effects of lead on rat kidney and liver: GST expression and oxidative stress, *Toxicology*, 128, 191, 1998.
49. Rodríguez-Iturbe, B., et al., Chronic exposure to low doses of lead results in renal infiltration of immune cells, NF-kappaB activation, and overexpression of tubulointerstitial angiotensin II, *Antioxid. Redox Signal*, 7, 1269, 2005.
50. Vaziri, N.D. and Sica, D.A., Lead-induced hypertension: Role of oxidative stress, *Curr. Hypertens. Rep.*, 6, 314, 2004.
51. Vaziri, N.D., Mechanisms of lead-induced hypertension and cardiovascular disease, *Am. J. Physiol. Heart Circ. Physiol.*, 295, H454, 2008.
52. Dursun, N., et al., Blood pressure relationship to nitric oxide, lipid peroxidation, renal function, and renal blood flow in rats exposed to low lead levels, *Biol. Trace Elem. Res.*, 104, 141, 2005.
53. Robles, H.V., et al., Lead exposure effect on angiotensin II renal vasoconstriction, *Hum Exp. Toxicol.*, 26, 499, 2007.
54. Farmand, F., et al., Lead-induced dysregulation of superoxide dismutases, catalase, glutathione peroxidase, and guanylate cyclase, *Environ. Res.*, 98, 33, 2005.
55. Vaziri, N.D., et al., Superoxide dismutase, catalase, glutathione peroxidase and NADPH oxidase in lead-induced hypertension, *Kidney Int.*, 63, 186, 2003.

56. Zhan, C.D., et al., Superoxide dismutase, catalase and glutathione peroxidase in the spontaneously hypertensive rat kidney: Effect of antioxidant-rich diet, *J. Hypertens.*, 22, 2025, 2004.
57. Chang, H.R., et al., Reduced vascular beta-adrenergic receptors and catecholamine response in rats with lead induced hypertension, *Arch. Toxicol.*, 71, 778, 1997.
58. Chang, H.R., et al., The change of beta-adrenergic system after cessation of lead exposure, *Toxicology*, 207, 73, 2005.
59. McCabe, M.J., Jr., Dias, J.A., and Lawrence, D.A., Lead influences translational or posttranslational regulation of Ia expression and increases invariant chain expression in mouse B cells, *J. Biochem. Toxicol.*, 6, 269, 1991.
60. Miller, T.E., et al., Developmental exposure to lead causes persistent immunotoxicity in Fischer 344 Rats, *Toxicol Sci.*, 42, 129, 1998.
61. Ercal, N., et al., A role for oxidative stress in suppressing serum immunoglobulin levels in lead-exposed Fischer 344 rats, *Arch. Environ. Contam. Toxicol.*, 39, 251, 2000.
62. Pyatt, D.W., et al., Inorganic lead activates NF-kappa B in primary human CD4+ T lymphocytes, *Biochem. Biophys. Res. Commun.*, 227, 380, 1996.
63. Bunn, T.L., et al., Exposure to lead during critical windows of embryonic development: Differential immunotoxic outcome based on stage of exposure and gender, *Toxicol. Sci.*, 64, 57, 2001.
64. Bunn, T.L., et al., Gender-based profiles of developmental immunotoxicity to lead in the rat: Assessment in juveniles and adults, *J. Toxicol. Environ. Health A*, 64, 223, 2001.
65. McCabe, M.J., Jr., Singh, K.P., and Reiners, J.J., Jr., Low level lead exposure *in vitro* stimulates the proliferation and expansion of alloantigen-reactive CD4(high) *T* cells, *Toxicol. Appl. Pharmacol.*, 177, 219, 2001.
66. Shen, X., Lee, K., and König, R., Effects of heavy metal ions on resting and antigen-activated CD4(+) T cells, *Toxicology*, 169, 67, 2001.
67. Fernandez-Cabezudo, M.J., et al., Alpha tocopherol protects against immunosuppressive and immunotoxic effects of lead, *Free Radic. Res.*, 37, 437, 2003.
68. Farrer, D.G., Hueber, S.M., and McCabe, M.J., Jr., Lead enhances CD4+ T cell proliferation indirectly by targeting antigen presenting cells and modulating antigen-specific interactions, *Toxicol. Appl. Pharmacol.*, 207, 125, 2005.
69. Lynes, M.A., et al., Gene expression influences on metal immunomodulation, *Toxicol. Appl. Pharmacol.*, 210, 9, 2006.
70. Farrer, D.G., et al., Reduction of myeloid suppressor cell derived nitric oxide provides a mechanistic basis of lead enhancement of alloreactive CD4(+) T cell proliferation, *Toxicol. Appl. Pharmacol.*, 229, 135, 2008.
71. Moore, M.R., Goldberg A., and Yeung-Laiwah, A.A.C., Lead effects on the heme biosynthetic pathway: Relationship to toxicity, *Ann. N. Y. Acad. Sci.*, 514, 191, 1987.
72. Woods, J.S. and Fowler, B.A., Metal alterations of uroporphyrinogen decarboxylase and coproporphyrinogen oxidase, *Ann. N. Y. Acad. Sci.*, 514, 55, 1987.
73. Dailey, H.A., Metal inhibition of ferrochelatase, *Ann. N. Y. Acad. Sci.*, 514, 81, 1987.
74. Oskarsson, A. and Fowler, B.A., Alterations in renal heme biosynthesis during metal nephrotoxicity, *Ann. N. Y. Acad. Sci.*, 514, 268, 1987.
75. Piomelli, S., Seaman, C., and Kapoor, S., Lead-induced abnormalities of porphyrin metabolism, *Ann. N. Y. Acad. Sci.*, 514, 278, 1987.
76. Fowler, B.A., Oskarsson, A., and Woods, J.S., Metal and metalloid-induced porphyrinurias: Relationships to cell injury, *Ann. N. Y. Acad. Sci.*, 514, 172, 1987.
77. Labbe, R.F., et al., Zinc protoporphyrin, past, present and future, *Ann. N. Y. Acad. Sci.*, 514, 7, 1987.
78. Borja-Aburto, V.H., et al., Blood lead levels measured prospectively and risk of spontaneous abortion, *Am. J. Epidemiol.*, 150, 590, 1999.
79. Hertz-Picciotto, I., The evidence that lead increases the risk for spontaneous abortion, *Am. J. Ind. Med.*, 38, 300, 2000.
80. Hsu, P.C., et al., Lead-induced changes in spermatozoa function and metabolism, *J. Toxicol. Environ. Health A*, 55, 45, 1998.
81. Apostoli, P., et al., Male reproductive toxicity of lead in animals and humans, ASCLEPIOS Study Group, *Occup. Environ. Med.*, 55, 364, 1998.
82. Telisman, S., et al., Semen quality and reproductive endocrine function in relation to biomarkers of lead, cadmium, zinc, and copper in men, *Environ. Health Perspect.*, 108, 45, 2000.
83. Telisman, S., et al., Reproductive toxicity of low-level lead exposure in men, *Environ. Res.*, 105, 256, 2007.

84. Huang, Y.L., Tseng, W.C., and Lin, T.H., *In vitro* effects of metal ions (Fe^{2+}, Mn^{2+}, Pb^{2+}) on sperm motility and lipid peroxidation in human semen, *J. Toxicol. Environ. Health A*, 62, 259, 2001.
85. Bonde, J.P., et al., Sperm count and chromatin structure in men exposed to inorganic lead: Lowest adverse effect levels, *Occup. Environ. Med.*, 59, 234, 2002.
86. Eibensteiner, L. et al., Lead exposure and semen quality among traffic police in Arequipa, Peru, *Int. J. Occup. Environ. Health*, 11, 161, 2005.
87. Hernández-Ochoa, I., et al., Low lead environmental exposure alters semen quality and sperm chromatin condensation in northern Mexico, *Reprod. Toxicol.*, 20, 221, 2005.
88. Naha, N., et al., Structural alteration of spermatozoa in the persons employed in lead acid battery factory, *Indian J. Physiol. Pharmacol.*, 49, 153, 2005.
89. Naha, N. and Chowdhury, A.R., Inorganic lead exposure in battery and paint factory: Effect on human sperm structure and functional activity, *J. U.O.E.H.*, 28, 157, 2006.
90. Chowdhuri, D.K., Narayan, R., and Saxena, D.K., Effect of lead and chromium on nucleic acid and protein synthesis during sperm–zona binding in mice, *Toxicol. In Vitro*, 15, 605, 2001.
91. Naha, N. and Manna, B., Mechanism of lead induced effects on human spermatozoa after occupational exposure, *Kathmandu Univ. Med. J.*, 5, 85, 2007.
92. Oliveira, H., et al., Lead chloride affects sperm motility and acrosome reaction in mice: Lead affects mice sperm motility and acrosome reaction, *Cell Biol. Toxicol.*, 4, 341–353, 2009.
93. Hsu, P.C., et al., Lead exposure causes generation of reactive oxygen species and functional impairment in rat sperm, *Toxicology*, 122, 133, 1997.
94. Marchlewicz, M., Michalska, T., and Wiszniewska, B., Detection of lead-induced oxidative stress in the rat epididymis by chemiluminescence, *Chemosphere*, 57, 1553, 2004.
95. Kasperczyk, A., et al., Assessment of semen function and lipid peroxidation among lead exposed men, *Toxicol. Appl. Pharmacol.*, 228, 378, 2008.
96. Thoreux-Manlay, A., et al., Impairment of testicular endocrine function after lead intoxication in the adult rat, *Toxicology*, 100, 101, 1995.
97. Ronis, M.J., Gandy, J., and Badger, T., Endocrine mechanisms underlying reproductive toxicity in the developing rat chronically exposed to dietary lead, *J. Toxicol. Environ. Health A*, 54, 77, 1998.
98. Mahmoud, A., Is inhibin B involved in the toxic effect of lead on male reproduction? *Int. J. Androl.*, 28, 150, 2005.
99. Quintanilla-Vega, B., et al., Lead effects on protamine–DNA binding, *Am. J. Ind. Med.*, 38, 324, 2000.
100. Quintanilla-Vega, B., et al., Lead interaction with human protamine (HP2) as a mechanism of male reproductive Toxicity, *Chem. Res. Toxicol.*, 13, 594, 2000.
101. Acharya, U.R., Acharya, S., and Mishra, M., Lead acetate induced cytotoxicity in male germinal cells of Swiss mice, *Ind. Health*, 41, 291, 2003.
102. Wang, L., et al., Effects of lead exposure on sperm concentrations and testes weight in male rats: A meta-regression analysis, *J. Toxicol. Environ. Health A*, 71, 454, 2008.
103. Steenland, K. and Boffetta, P., Lead and cancer in humans: Where are we now? *Am. J. Ind. Med.*, 38, 295, 2000.
104. Fracasso, M.E., et al., Lead induced DNA strand breaks in lymphocytes of exposed workers: Role of reactive oxygen species and protein kinase C, *Mutat. Res.*, 515, 159, 2002.
105. Wozniak, K. and Blasiak, J., *In vitro* genotoxicity of lead acetate: Induction of single and double DNA strand breaks and DNA–protein cross-links, *Mutat. Res.*, 535, 127, 2003.
106. Lin, Y.W., Chuang, S.M., and Yang, J.L., Persistent activation of ERK1/2 by lead acetate increases nucleotide excision repair synthesis and confers anti-cytotoxicity and anti-mutagenicity, *Carcinogenesis*, 24, 53, 2003.
107. Waalkes, M.P., et al., Metallothionein-I/II double knockout mice are hypersensitive to lead-induced kidney carcinogenesis: Role of inclusion body formation, *Cancer Res.*, 64, 7766, 2004.
108. Rajaraman, P., et al., Lead, genetic susceptibility, and risk of adult brain tumors, *Cancer Epidemiol. Biomarkers Prev.*, 15, 2514, 2006.
109. Hanas, J.S., et al., Lead inhibition of DNA-binding mechanism of Cys(2)His(2) zinc finger proteins, *Mol. Pharmacol.*, 56, 982, 1999.
110. Zawia, N.H., et al., Disruption of the zinc finger domain: A common target that underlies many of the effects of lead, *Neurotoxicology*, 21, 1069, 2000.
111. Jarzecki, A.A., Lead-poisoned zinc fingers: Quantum mechanical exploration of structure, coordination, and electronic excitations, *Inorg. Chem.*, 46, 7509, 2007.
112. Aykin-Burns, N., et al., Oxidative effects of lead in young and adult Fischer 344 rats, *Arch. Environ. Contam. Toxicol.*, 44, 417, 2003.

113. Garçon, G., et al., Biologic markers of oxidative stress and nephrotoxicity as studied in biomonitoring of adverse effects of occupational exposure to lead and cadmium, *J. Occup. Environ. Med.*, 46, 1180, 2004.
114. Jurczuk, M., Moniuszko-Jakoniuk, J., and Brzóska, M.M., Involvement of some low-molecular thiols in the peroxidative mechanisms of lead and ethanol action on rat liver and kidney, *Toxicology*, 219, 11, 2006.
115. He, L., et al., Lead and calcium produce rod photoreceptor cell apoptosis by opening the mitochondrial permeability transition pore, *J. Biol. Chem.*, 275, 12175, 2000.
116. Xu, J., Ji, L.D., and Xu, L.H., Lead-induced apoptosis in PC 12 cells: Involvement of p53, Bcl-2 family and caspase-3, *Toxicol. Lett.*, 166, 60, 2006.
117. Fowler, B.A., Kahng, M.W., and Smith, D.R., Role of lead binding proteins in renal cancer, *Environ. Health Perspect.*, 102, 115, 1994.
118. Moffett, D., El-Masri, H., and Fowler, B.A., General considerations of dose-effect and dose-response relationships, In: G.F. Nordberg, B.A. Fowler, M. Nordberg, and L. Friberg (Eds), *Handbook of Toxicology of Metals* (3rd edition, pp. 101–115), Elsevier Publishers, Amsterdam, 2007.
119. Nordberg, G.F., Mahaffey, K.R., and Fowler, B.A., Introduction and summary. International workshop on lead in bone: Implications for dosimetry and toxicology, Columbia, Maryland, March, 1991, *Environ. Health Perspect.*, 91, 3, 1991.
120. Nordberg, G.F., et al., Interactions in metal toxicology, In: G.F. Nordberg, B.A. Fowler, M. Nordberg, and L. Friberg, (Eds), *Handbook of Toxicology of Metals* (3rd edition, pp. 117–145), Elsevier Publishers, Amsterdam, 2007.
121. Mahaffey, K.R. and Fowler, B.A., Effects of concurrent administration of diet of lead, cadmium, and arsenic in the rat, *Environ. Health Perspect.*, 19, 165, 1977.
122. Mahaffey, K.R., et al., Concurrent exposure to lead, cadmium and arsenic: Effects on toxicity and tissue metal concentrations in the rat, *J. Lab. Clin. Med.*, 98, 463, 1981.
123. Fowler, B.A. and Mahaffey, K.R., Interaction between lead, cadmium and arsenic in relation to porphyrin excretion patterns, *Environ. Health Perspect.*, 25, 87, 1978.
124. Bishayi, B. and Sengupta, M., Synergism in immunotoxicological effects due to repeated combined administration of arsenic and lead in mice, *Int. Immunopharmacol.*, 6, 454, 2006.
125. Skoczyńska, A., Wróbel, J., and Andrzejak, R., Lead–cadmium interaction effect on the responsiveness of rat mesenteric vessels to norepinephrine and angiotensin II, *Toxicology*, 162, 157, 2001.
126. Fowler, B.A., et al., Oxidative stress induced by lead, cadmium and arsenic mixtures: 30-day, 90-day, and 180-day drinking water studies in rats: An overview, *Biometals*, 17, 567, 2004.

5 Transport and Biological Impact of Manganese

Keith M. Erikson, Khristy J. Thompson, and Michael Aschner

CONTENTS

5.1 Overview ... 127
5.2 Mn: Essentiality and Toxicity ... 127
5.3 Mn Transport in the Lungs ... 129
5.4 Mn Transport and Biology in the Brain ... 130
5.4.1 Tf/Tf Receptor in Mn Transport ... 130
5.4.2 DMT1 Is an Mn Transporter ... 131
5.4.3 Olfactory Bulb ... 131
5.4.4 Other Potential Transporters of Brain Mn ... 133
5.4.5 ZIP Transporter Proteins ... 133
5.4.6 Dopamine Transporter (DAT): Putative Mn Transporter? ... 133
5.4.7 Astrocytes Regulate Brain Mn Concentrations ... 134
5.5 Biological Consequences of Abnormal Mn Levels ... 134
5.5.1 Neurochemical Alterations Associated with Mn Toxicity ... 134
5.5.2 Mitochondria and Oxidative Stress in PD and Mn-Induced DAergic Neurodegeneration ... 135
5.6 Conclusions ... 136
Acknowledgments ... 136
References ... 137

5.1 OVERVIEW

Manganese (Mn) is an essential nutrient and, unlike other trace elements (e.g., iron), toxicity is more prevalent than dietary deficiency. This chapter will commence with a discussion on the essentiality of Mn and its general biological functions. We will then discuss putative Mn transport mechanisms with a particular emphasis on the lung and brain, the primary organs involved in the etiology of Mn neurotoxicity (manganism). We conclude the chapter with several sections focusing on the neurobiology of manganism. Special emphasis is placed on the neurochemical and biochemical aspects of Mn-induced neuropathology and the biochemical similarities it shares with Parkinson's disease (PD).

5.2 Mn: ESSENTIALITY AND TOXICITY

Mn is found in many foods (e.g., legumes and green leafy vegetables) and is required for proper growth and development. Current guidelines suggest that adults consume 2–5 mg Mn/day.[1] Mn is an important component of many key enzymes involved in the detoxification of reactive oxygen species referred to as ROS (Mn superoxide dismutase), amino acid synthesis (glutamine synthetase,

GS), the urea cycle (arginase), and energy production (pyruvate carboxylase).[2–4] Both deficiency and excess of this metal result in health problems, although Mn toxicity is more common. Too little dietary Mn can lead to growth retardation, alterations in glucose and high-density lipid (HDL) cholesterol levels, and even reproductive failure.[5] On the other hand, excess Mn can lead to the onset of a movement disorder, termed "manganism," which is similar to PD.

Manganism is often associated with occupations such as Mn mining,[6–8] steel production, welding,[9–12] and the dry cell battery industry. Indeed, the characteristics of Mn neurotoxicity are best studied in persons from these groups. In general, symptoms of Mn poisoning can be broken down into three stages. The first stage involves behavioral changes, such as increased anxiety or nervousness, apathy, loss of memory, and/or decreases in concentration. In the second phase, parkinsonian features (an expressionless face, impaired writing ability, and difficulty in walking backwards) become apparent. Finally, dystonia of the trunk and extremities, expressed characteristically as "cock-walk" gait, is observed. Interestingly, tremor, a typical feature of PD, is not commonly associated with manganism.[13,14] However, once the disease progresses to the latter stages, manganism is generally irreversible even if the patient is removed from the source of Mn.

Ingested Mn is efficiently cleared from the body through the bile, resulting in the subsequent excretion of approximately 95–97% of the metal. Clearance is much lower if liver function is impaired, such as in chronic liver disease or cirrhosis, or if Mn is injected directly into the blood stream, as in parenteral nutrition. Thus, persons in either group are at an increased risk for Mn intoxication. Following absorption in the intestines, Mn is transported throughout the body bound to various proteins (transferrin, Tf), plasma macroglobulin, (and/or albumin) and small molecules (citrate).[15–19] As Mn and iron share transporter systems (Tf/Tf receptor and divalent metal transporter-1 [DMT1]), recent studies have examined whether iron deficiency and anemia could also be risk factors for Mn neurotoxicity.[20]

Few reports exist on brain Mn concentrations upon Mn intoxication. The concentrations of Mn in human striatum and globus pallidus (GP) (primary target areas) are unknown. Cerebellar concentrations of Mn are available from Japanese accident victims, and at the time of autopsy they ranged between 9 and 10 μM.[21] However, these values are likely lower than those in the striatum and GP, both of which are known to accumulate more Mn than the cerebellum. A study performed almost three decades ago examined Mn concentration in the striatum and GP (two regions in the nonhuman primate known to accumulate Mn) of monkeys dosed for three months with Mn dioxide.[22] Striatal Mn concentration reached 264 μM, while Mn concentration in GP peaked at 334 μM. In rats, Mn concentrations can reach up to 200 μM depending on the brain region and dosing regimen.[23–25] Thus, during Mn toxicity, it is possible for brain levels to exceed 350 μM. Mn concentrations in various parts of the rat striatum have been reported to range from 4.4 to 18 μM, and in exposed rats, the levels of Mn in the striatum increase to 23–70 μM.[23–25] Notably, these studies suggest that both in primates and rodents, homeostatic control of brain Mn concentrations is tight, because even in conditions of relatively high exposures, levels of brain Mn increase only several (3–4) fold.

The Mn fuel additive, methylcyclopentadienyl manganese tricarbonyl (MMT), has antiknock properties. When combusted, the Mn from MMT is in the form of several aerosolized salts, the most abundant species being manganese phosphate and manganese sulfate.[26,27] Pharmacokinetic studies have shown that the salt characteristics will determine the rate of transport into the brain with the following rank order: $MnCl_2 > MnSO_4 > MnPO_4$.[28,29] Nonetheless, it is important to note, that regardless of exposure, once absorbed and within biological media (e.g., blood and cerebrospinal fluid), Mn would be expected to bind to the same ligands and behave in an analogous pharmacokinetic fashion irrespective of portal of entry. Thus, the physical and chemical properties of these aerosolized salts will only govern their absorption and elimination properties, but overall the tissue distribution and their mechanisms of toxicity would be expected to be similar, yet on a differential temporal scale (e.g., attainment of toxic concentrations of Mn in target tissue would be reached faster for the more soluble Mn salts).

Mn neurotoxicity (manganism) shares neurological symptoms with several clinical disorders commonly described as "extrapyramidal motor system dysfunction," and in particular, PD.[30] Evidence for the involvement of oxidative stress in the etiology of PD exists, supported by the following observations: (1) Monoamine oxidase (MAO) activity, which catabolizes intraneuronal dopamine and yields hydrogen peroxide (H_2O_2), increases with age.[31] (2) Pharmacological manipulations that enhance dopamine turnover cause an increase in oxidized glutathione (GSH), which can be suppressed by simultaneous treatment with the MAO inhibitors clorgyline and deprenyl, indicating that MAO activity is a source of oxidative stress.[32] (3) In postmortem studies of PD brains, the activity of glutathione peroxidase and the amounts of GSH in the substantia nigra are reduced.[33] (4) Oxidative stress is one of the proximate causes of 1-methly-4-phenyl-1,2,3,6-tetrahyrdopyridine (MPTP)-induced dopaminergic (DAergic) neuronal degeneration, and transgenic mice overexpressing copper/zinc superoxide dismutase are resistant to the neurotoxic action of MPTP.[34] Common features in the etiologies of manganism-like PD and idiopathic PD are discussed in Section 5.5.

5.3 Mn TRANSPORT IN THE LUNGS

The respiratory system is the first entry point for inhaled Mn. Deposition of Mn particles is dependent on particle size where larger particles are deposited in the upper airway and smaller size particles reach the lower airways.[35,36] Upper airway particles are more readily cleared through mucociliary actions unlike the lower airways. The biological availability of the Mn is then dependent on the form of Mn, exposure time, and liberation of the Mn following engulfment by alveolar macrophages.[37] The lung facilitates two exit mechanisms for Mn: the mucociliary elevator clearance mechanism to the gastrointestinal system or absorption via alveolar epithelial cells directly to the blood stream. Clearance of Mn from the lung through the mucociliary elevator mechanism results in the same fate as dietary Mn, which is rapidly cleared and nontoxic. Absorption/transport into circulation across the alveolar epithelium plays a more significant role in Mn neurotoxicity.

The mechanisms involved in Mn absorption by the lung are unclear. Uptake of Mn by alveolar epithelial cells, more specifically Type II cells, is believed to occur through the Tf/Tf receptor/DMT1 mechanism or through Tf-independent pathways.[38] In either case, lung Mn absorption and clearance have been shown to be altered by manipulating iron status. Pharmacokinetic analyses found that iron deficiency by diet or phlebotomy altered blood clearance on Mn following intratracheal instillation,[39,40] whereas lung iron loading either by direct instillation of iron oxide particles or through a diet of 1% carbonyl iron or decreased transport from the lung to circulation.[39,41] In these studies, neither DMT1 nor Tf receptor mRNA levels were altered in the lung despite the change in iron status. Studies using the Belgrade rat found that DMT1 does not play a role in lung Mn transport, leading to a search for other transport mechanisms for Mn to the circulatory systems outside of the Tf/TfR/DMT1 pathway.[42]

There is increasing evidence that nonselective calcium channels transport divalent Mn in the heart, liver, and brain.[42] Mn^{2+} uptake by cardiomyocytes *in vitro*[43] and uptake into the liver[44] have been shown to be altered by L-type voltage-gated channels. Analysis of isolated blood–brain-barrier-derived cells and *in situ* brain perfusion assays indicate regulation of Mn^{2+} uptake by store-operated calcium channels.[45] Several members of the transient receptor potential (TRP) family of receptor- and store-operated calcium channels have also been implicated in the transport of Mn.[46,47] Uptake studies of ^{54}Mn by A549 cells suggest that metal uptake by Type II alveolar epithelial cells is associated with both L-type Ca^{2+} channels and TRPM7, a member of the TRP melastatin subfamily. Mn seems to be absorbed by the pulmonary epithelium through different pathways. A putative role for nonselective calcium channels in lung metal clearance is rapidly emerging.[42]

Mn can also cause damage to the lung tissue itself, perhaps due to the reactive oxygen species generated.[48] The subsequent events may lead to epithelial damage, increases in permeability, and inflammatory responses leading to decreased lung function.[49–51] Vascular endothelial growth factor

(VEGF), regulated through the hypoxia inducible factor-1/hypoxia-response element mechanism, has recently been shown to increase in response to just short-term Mn exposure.[52] It is suggested that alternation of pulmonary gene expression will lead to increased susceptibility of the lung to respiratory disease. A study examining school-age children who had been exposed to airborne metal particles found a correlation between exposure to Mn-containing particles and a decline in peak expiratory flow rate (PEFR).[53] Long-term exposure of miners to Mn has been implicated in decreased pulmonary function; however, the compounding factor of active smoking by the workers presents some difficulty in the interpretation of the results.[54]

5.4 Mn TRANSPORT AND BIOLOGY IN THE BRAIN

5.4.1 Tf/Tf Receptor in Mn Transport

A series of studies has shown that Mn influx to the brain, but not efflux, is carrier mediated.[55,56] In plasma, Mn is bound to small molecular weight carriers such as albumin and citrate or proteins like Tf. Tf is the primary iron binding/transport protein in plasma and is also known to bind Mn. Mn does not avidly complex with sulfhydryl (-SH) groups or amines, and it shows little variation in its stability constants for endogenous complexing ligands such as glycine, cysteine, riboflavin, and guanosine. In the absence of Fe, the binding sites of Tf can accommodate a number of other metals raising the possibility that Tf functions *in vivo* as a transport agent for many of these metals. Mn binding to Tf is time dependent.[15,57,58] When complexed with Tf, Mn is exclusively present in the trivalent oxidation state, with two metal ions tightly bound to each Tf molecule.[59] At normal plasma iron (Fe) concentrations (0.9–2.8 μg/mL), normal Fe-binding capacity (2.5–4 μg/mL), and normal Tf concentration in plasma, namely 3 mg/mL, with two metal-ion-binding sites per molecule (M_r 77,000) of which only 30% are occupied by Fe^{3+}, Tf has available 50 micromoles of unoccupied Mn^{3+}-binding sites per liter.[15]

Since Tf receptors are present on the *surface* of the cerebral capillaries[60–62] and endocytosis of Tf is known to occur in these capillaries,[62] it has been suggested that Mn (in the trivalent oxidation state) enters the endothelial cells complexed with Tf. Mn is then released from the complex in the endothelial cell interior by endosomal acidification and the apo-Tf/TfR complex is returned to the luminal surface[63,64] without the assistance of DMT1.[65] Mn released within the endothelial cells is subsequently transferred to the abluminal cell surface for release into the extracellular fluid. The endothelial Mn is delivered to brain-derived Tf for extracellular transport and the Mn is subsequently taken up by neurons that possess both Tf receptors and DMT1.[66] A recent study by Moos et al.[65] indicates that Tf-bound metal gains access to the brain through the blood–CSF barrier where the metal then diffuses into neurons from the ventricles and subarachnoid space. Support for receptor-mediated endocytosis of an Mn–Tf complex in cultured neuroblastoma cells (SH-SY5Y) was recently demonstrated by Suarez and Eriksson.[67] Sloot and Gramsbergen have demonstrated anterograde axonal transport of ^{54}Mn in both nigrostriatal and striatonigral pathways.[68] Furthermore, *in vivo*, intravenous administration of the ferric-hydroxide dextran complex significantly inhibits Mn brain uptake, and high Fe intake reduces central nervous system (CNS) Mn concentrations, corroborating a relationship between Fe and Mn transport.[15,69]

The basal ganglia; including pallidum, thalamic nuclei, and substantia nigra, contains the highest Mn concentrations.[70] Interestingly, Fe concentrations in these structures are the highest as well.[71] Although the areas with dense Tf distribution[72] do not correspond to the distribution of Mn (or Fe), the fact that Mn-accumulating areas are efferent to areas of high Tf receptor density suggests that these sites may accumulate Mn through neuronal transport.[68] For example, the Mn-rich areas of the ventral pallidum, GP, and substantia nigra receive input from the nucleus accumbens and the caudate-putamen, two areas abundantly rich in Tf receptors.[73,74]

Experiments examining Tf's role in Mn transport are often conducted with Mn-bound Tf complexes. However, experiments using the hypotransferrinemic mouse, which lacks functional Tf, have shown that Mn transport to the brain is not affected, indicating that Tf is not necessary for brain Mn acquisition across the blood–brain barrier.[75] Tf gains access to the cell by binding to its receptor, the TfR, where the complex is endocytosed. DMT1 is a component of the endosome and functions to acidify the endosome to allow for release of Tf-bound metal and then functions to pump the metal into the cytosol. It is presumed that Mn also is released from Tf–Mn complexes by the same mechanism[76,77]; however, more studies are needed. This mechanism was elucidated in reticulocytes where TfR is abundant, and Tf–Fe complexes (and perhaps Tf–Mn complexes) were able to enter cells, but the metal could not be released into the cytosol.[78] Several *in vitro* studies have shown that Mn^{3+} does not accumulate intracellularly to any significant amount to cause neurotoxicity. Gunter et al. found no evidence of stabilization of Mn^{3+} complexes in human neuroteratocarcinoma (NT2) cells[79] and primarily rat astrocyte cultures or in nerve growth factor-treated PC12 cells.[80] The cause of Mn neurotoxicity is suggested to be due to Mn^{2+} inhibition of Ca^{2+} activation and control of ATP production.[77]

5.4.2 DMT1 Is an Mn Transporter

A prominent Mn transporter is the DMT1 [also known as NRAMP-2 and divalent cation transporter 1 (DCT-1)]. It is best known for its role as an intestinal interluminal protein responsible for Fe regulation.[81,82] Indeed, gene transcription for this protein is actually regulated by Fe concentration via an iron-response element (IRE) located on the mRNA.[78] There is growing evidence that DMT1 is involved in brain Mn delivery. It has been well established that DMT1 has an affinity for Mn.[76] In the microcytic anemia (mk) mouse and the phenotypically similar Belgrade (b) rat,[83,84] orthologous mutations (glycine 185 to arginine) in the DMT1 gene result in significantly reduced dietary iron (Fe) absorption. The role of the defective DMT1 allele in the transport of Mn across the BBB has been evaluated in homozygous Belgrade (b/b) rats that exhibit hypochromic anemia and in heterozygous (+/b) Belgrade rats.[85] Plasma clearance and uptake by the CNS after intravenous injection of radioactive ^{54}Mn bound to Tf or mixed with serum have demonstrated that plasma clearance of Mn–Tf was much slower than Mn serum, but both were faster than the clearance of Fe–Tf. Uptake of ^{54}Mn as well as ^{59}Fe by the brain was less in b/b compared with +/b rats, suggesting that the defective DMT1 allele affects the metabolism of both metals and that Mn and Fe might share DMT1 transporters in the BBB.[85] Other *in vitro* works also strongly support a role for this transporter in Mn movement across the BBB. Interestingly, DMT1 activity is pH dependent, and likely Fe dependent as well. Additionally, because metal transport via DMT1 is an active process, it is likely to be temperature dependent. This is interesting because data suggest that Mn transport across rat brain endothelial (RBE4) cells in culture is, among other things, temperature, pH, and Fe dependent.[4] While the mammalian data are probably more germane to human Mn transport, it is also known that mutations in NRAMP-2, the DMT1 homologue in bacteria and yeast, disrupt Mn transport.[86–88] However, it is clear that DMT1 is not the sole transporter of Mn across the BBB, since in the Belgrade rat Mn is efficiently transported into the CNS.[18] This is not altogether surprising since many metals and other molecules are dependent on numerous transporters and other proteins to facilitate their movement across various barrier systems. For an extensive review on the role of DMT1 in mammalian transport of multiple metals, refer to Garrick et al.[78]

5.4.3 Olfactory Bulb

The nasal passage component of the respiratory system offers another uptake pathway for Mn via the olfactory neurons of the olfactory pathway. In the nose, the olfactory epithelium is constituted by primary olfactory neurons, sustentacular (supporting) cells, and Bowman's glands that provide mucus to cover the epithelial surface. Axons of the olfactory neurons extend through the cribiform plate of the ethmoid bone to project into the olfactory bulb. There they form synaptic connections with secondary olfactory neurons called mitral and tufted cells. Axons from the latter neurons project into the olfactory cortex, which include the anterior olfactory nucleus, the olfactory tubercle, and the pyriform, amygdaloid, and entorhinal cortices. The structural architecture and neuronal connections of the olfactory system are reviewed by Shipley and colleagues (Figure 5.1).[90]

While the brain is largely protected by the blood–brain barrier, the presence of primary olfactory neurons that are in close contact with the external environment allows the nasal cavity to provide a unique window to the CNS. Axonal transport in the olfactory system occurs with solvents, metals, and even larger particles.[91–94] Sunderman has argued for the importance of understanding rhinotoxicity in studies of metal exposure.[91] Over the past decade, strong evidence has been accumulating that inhaled Mn can be directly transported from the nose to the brain via the olfactory tract. Studies of olfactory transport of Mn in fish[95] and rats[96] show that ^{54}Mn travels rapidly (~2 mm/h) along primary olfactory neurons to reach regions of the olfactory cortex. One proposal has been that Mn(II) acts as a calcium ortholog to enter neurons through channels and become packaged into vesicles that travel anterograde and in a microtubule-dependent (colchicine-sensitive) manner.[68] After release from presynaptic neurons, Mn would cross the synapse to be taken up by postsynaptic neurons and thus be distributed throughout interconnecting regions of the brain.[97]

These early studies mapping olfactory metal uptake relied on autoradiographic imaging of brain tissue slices from ^{54}Mn-instilled animals.[68,96,98] However, more recent efforts have established that the paramagnetic features of Mn(II) can be used for *in vivo* imaging. Paramagnetic Mn(II) ions shorten T1 relaxation of water protons to increase T1-weighted MRI signals.[99] This technique has been exploited to map the mouse hippocampus after intracerebral injection of $MnCl_2$,[100] to study the neuronal connections in the basal ganglia of monkeys,[101] and to follow odorant stimulation of the

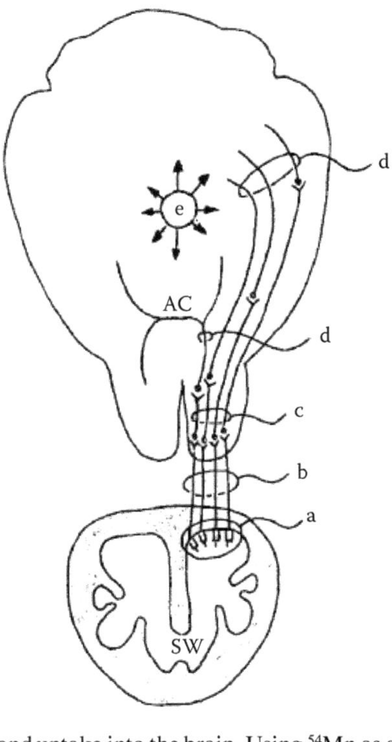

FIGURE 5.1 The olfactory tract and uptake into the brain. Using ^{54}Mn as a tracer, Henriksson et al.[89] mapped the following route: uptake from the nasal cavity by primary olfactory neurons (a), transport along these neurons (b), uptake and transport by secondary (mitral and tufted) neurons in the olfactory bulb (c), uptake and transport by tertiary neurons of the olfactory cortex (d), and into the rest of the brain (e). At increasing doses, radioisotopes crossed the septal window (SW) into the chamber opposite (contralateral) to the instillation chamber (ipsilateral).

mouse olfactory system.[102] More recently, Cross and colleagues have used this approach to map rat olfactory connections,[103] and the results of their study agree remarkably well with early work by Tjalve and coworkers.[94–96] Recently, Thompson et al. have identified that DMT1 is expressed in the olfactory bulb and is present in the sustentacular cells of the olfactory epithelium.[104] The levels of DMT1 increased in response to anemia. Consistent with this increased expression of DMT1 was an increase in the amount of Mn transported to the brain following intranasal instillation of Mn including the key areas of the basal ganglia and hippocampus. Transport was significantly reduced also in Belgrade rats intranasally instilled with Mn. The results suggest that the neurotoxicity of Mn can be modified by iron status due to the iron-responsive regulation of DMT1.

5.4.4 Other Potential Transporters of Brain Mn

As mentioned above, the mechanism by which Mn is transported across the BBB has recently come to light. It appears that facilitated diffusion,[105] active transport,[17,105,106] DMT1-mediated transport,[20,78] ZIP8-, store-operated calcium channels, as well as Tf-dependent transport mechanisms are all involved in shuttling Mn across the BBB.[17] Although non-protein-bound Mn enters the brain more rapidly than Tf-bound Mn,[105,106] it is unclear as to which form represents the predominant mechanism of transport *in situ*.

A small fraction of Mn is found in plasma as Mn citrate.[19,55] The same authors have suggested that an Mn citrate tridentate complex with a noncoordinated central carboxylate recognition moiety is likely a substrate for the organic anion transporter or a monocarboxylate transporter (MCT). Candidates for transport of Mn citrate may include MCT and/or members of the organic anion transporter polypeptide (OATP) or ATP-binding cassette (ABC) superfamilies.

5.4.5 ZIP Transporter Proteins

Recent evidence implies a role for the zrt-irt-like (ZIP) transporter proteins in Mn transport, although whether it is functional at the BBB is yet to be determined. The ZIP transporter proteins, members of the solute-carrier-39 (SLC39) metal-transporter family, having 14 members—highly conserved orthologs—between mouse and human.[107] Zip proteins have been originally divided into two subfamilies, referred to as subfamily I, consisting mostly of fungal and plant sequences, and subfamily II, which consists of insect, nematode, and mammalian sequences. More recently, PSI-BLAST (position-specific iterative-basic local alignment search tool) analysis uncovered two additional subfamilies, referred to as gufA and the LIV-1 or LIV-1 subfamily of ZIP transporters (LZT). Zip4–8, Zip10, and Zip12–14 belong to the LZT subfamily. Zip1–3 are from the Zip II subfamily; Zip9 is from the Zip I subfamily; and Zip11 clusters within the gufA subfamily. In general, Zip proteins contain eight transmembrane domains (TMDs), yet LIV-1 has only six. The fourth and fifth TMDs are highly conserved along the phylogenetic chain and likely form the pore through which metals pass. In plants, several ZIP proteins have been implicated in divalent metal transport, including Zn, Fe, and Mn.

A recent *in vitro* study in mouse fetal fibroblast cultures established that ZIP8 has a high affinity for Mn.[108] The K_m for Mn^{2+} of 2.2 μM is close to physiological concentrations and within the same range determined in many cell lines or tissues. However, whether Slc39 and organic transporters function in Mn transport remains to be examined under physiologically relevant conditions. Definitive studies to assess other protein functions (Slc39, MCT) in physiological roles are needed.

5.4.6 Dopamine Transporter (DAT): Putative Mn Transporter?

Why and by what means Mn specifically accumulates in brain regions rich in dopamine remains unclear. The neurotoxins 1-methyl-4-phenylpyridium (MPP^+) and paraquat are known to be selectively lethal to DAergic neurons due to the fact that they are transported by the DAT.[109,110]

Ingersoll et al. examined the effect on Mn uptake by inhibiting DAT.[23] Cocaine, a DAT inhibitor, and reserpine, which depletes extracellular dopamine concentrations by inhibiting vesicular reuptake in the synapse, were administered to rats injected with Mn intrathecally. A significant decrease in Mn accumulation in the ventral pallidum was seen in rats treated with both. This suggests a potential role for the DAT and dopamine metabolism in the transport of Mn. However, both cocaine and reserpine are known to affect other neurotransmitters, most notably serotonin and norepinephrine.[111] Therefore, it was unclear if this decrease in ventral pallidum Mn concentrations is directly related to the DAT being inhibited or is related to the inhibition of the serotonin transporter and/or the norepinephrine transporter.

A recent study by this laboratory examined the potential role of the DAT in Mn accumulation in the brain using a knockout mouse model.[112] Homozygous DAT-KO and WT mice were exposed to an acute dose of 50 mg/kg of $MnCl_2$ via intraperitoneal injection. After 15 min, brains were removed for metal analysis. No difference in basal Mn levels was seen between the DAT-KO mice and the WT mice. However, there was a significant 40% decrease in Mn accumulation in the striatum of the DAT-KO mice receiving Mn injection compared to the WT mice receiving Mn injection. These data were followed up in rats enduring chronic Mn exposure; upon blockade of the DAT with a specific inhibitor (GBR 12909), Mn uptake was attenuated in the GP.[113] This suggests that the DAT may not play a central role in normal Mn transport in the brain, but may become relevant in a toxicological paradigm in terms of Mn exposure.

5.4.7 Astrocytes Regulate Brain Mn Concentrations

TfR and DMT1 are both associated with the foot processes of astrocytes and play an integral role in maintaining extracellular concentrations of Mn in the brain.[38,75,76,114] Astrocytes maintain brain homeostasis through precise regulation of extracellular constituents. Astrocytes perform several functions that are essential for normal neuronal activity, including glutamate uptake, glutamate release, ionic buffering, and water transport. Mn is one such constituent in which astrocytic control is critical for normal neurological function. The astrocyte efficiently transports Mn, and an astrocyte-specific manganoprotein critical for ammonia metabolism, GS, accounts for about 80% of the total brain Mn.[115]

5.5 BIOLOGICAL CONSEQUENCES OF ABNORMAL Mn LEVELS

The first reported cases of Mn deficiency were in chickens suffering from perosis, which later was discovered to be due to inadequate glycosyltransferase activity causing the malformed bones. This symptom of Mn deficiency manifests itself during the formation and growth of bones and connective tissues during development. In adult animals, including humans, Mn deficiency, which is extremely rare, is characterized by weight loss and blood clotting problems. Therefore, in humans, pathologies associated with abnormal Mn biology revolve around exposure to excessive Mn and not its deficiency. The next section will focus on the consequences of Mn overexposure on neurochemistry and oxidative stress-related pathologies.

5.5.1 Neurochemical Alterations Associated with Mn Toxicity

γ-Aminobutyric acid (GABA) is the most abundant inhibitory neurotransmitter in the adult brain.[116,117] Glutamate is converted to GABA by decarboxylation via glutamate decarboxylase (GAD) and is degraded via GABA-transaminase. Cortical glutamate afferents project into the striatum where, in concert with GABA and dopamine, motor behaviors are controlled.[118] Reports of GABA concentrations in the rat brain upon Mn exposure are inconsistent. For example, exposure to 6 mg Mn/kg/day led to a significant increase in brain Mn concentrations and significant decrease in

GABA concentrations.[119] Another report showed that rats exposed to 20 mg Mn/kg/day had significantly increased brain Mn and GABA concentrations.[120] Gwiazda et al. found a significant 16% increase in striatal GABA concentrations resulting from cumulative low-dose Mn exposure in a pre-Parkinsonian rat model.[121] Consequently, it appears that a relationship exists between the severity of Mn exposure and GABA concentrations, with lower Mn exposure leading to decreased GABA and high Mn exposure leading to increased GABA concentrations.

Recent studies suggest that changes in GABA metabolism precipitate dopamine-related changes due to Mn exposure.[121,122] These changes in GABA metabolism due to accumulating Mn in GABA-rich brain regions are caused by altered reuptake of GABA.[122] The motor pathologies and characteristic behaviors associated with manganism revolve around the consequences of altered GABAergic inhibitory firing in the basal ganglia.

Glutamate is the most prevalent excitatory neurotransmitter,[123] whereas GABA is the most abundant inhibitory neurotransmitter.[117] Cortical glutamate afferents project into the striatum where, in concert with GABA and dopamine, motor behaviors are controlled.[118] Glutamate is converted to GABA by decarboxylation via GAD and is degraded via GABA-transaminase. Altered glutamatergic and GABAergic functioning can contribute to altered striatal dopamine metabolism.[124,125] Therefore, we have postulated that the neurotoxic effects of Mn on striatal dopamine may be indirectly mediated via abnormal striatal glutamate and/or GABA metabolism, and that temporally, changes in areas that are known to avidly accumulate Mn precede the well-described effects of Mn on DAergic function. Specifically, it is hypothesized that Mn accumulation in the GP causes decreased GABAergic efferent firing from the GP into the subthalamic nuclei. Consequently, glutamatergic projections into the substantia nigra that originate from the subthalamic nuclei will fire in an uncontrolled manner, causing dysregulation of DAergic output into the striatum from the substantia nigra (Figure 5.2).

5.5.2 Mitochondria and Oxidative Stress in PD and Mn-Induced DAergic Neurodegeneration

Several hypotheses on the loss of DAergic neurons in PD suggest that mitochondrial damage is a primary cause of cell death. It has been postulated that (1) mitochondria of DAergic neurons are selectively vulnerable to a number of environmental contaminants that can lead to mitochondrial dysfunction; (2) DAergic neurons can, under certain circumstances, produce endogenous mitochondrial toxin; or (3) mitochondria possess various enzyme defects (complex-I, for example) that may result in impaired energy metabolism. The centrality of mitochondria in these postulates arises from observations that mitochondrial toxins (MPP^+ ions, rotenone, 6-hydroxydopamine (6-OHDA), etc.) can mimic a parkinsonian-like syndrome in multiple animal species. They are potent inhibitors of mitochondrial respiration and model many of the features inherent to human PD.

The mitochondrial complex-I inhibitors, MPTP and rotenone (model toxins for PD), damage nigral neurons by mechanisms involving oxidation.[126] Oxidative damage also plays a significant role in 6-OHDA-induced DAergic neuronal cell death. H_2O_2, superoxide ions, and hydroxyl radicals[127] are generated by nonenzymatic breakdown of 6-OHDA and the direct inhibition of complex-I activity, leading to lipid peroxidation, protein denaturation, and a decrease in GSH, all hallmark features of postmortem PD.[128–130]

There are many parallels between the aforementioned features of PD and Mn-induced neurotoxicity. Intracellular Mn^{2+} is sequestered by mitochondria via the Ca^{2+} uniporter, and intrastriatal Mn injections result in loss of DAergic neurons, resembling toxicity caused by the mitochondrial poisons, aminooxyacetic acid, and MPP^+. MPTP-induced DAergic neurodegeneration involves glutamate-mediated toxicity; noncompetitive or competitive *N*-methyl D-aspartate (NMDA) antagonists protect nigral neurons from this effect.[131] Consistent with increased production of ROS, Mn inhibits mitochondrial complex-I, a feature inherent to PD and its experimental models (MPP^+, 6-OHDA, rotenone, and paraquat).[132]

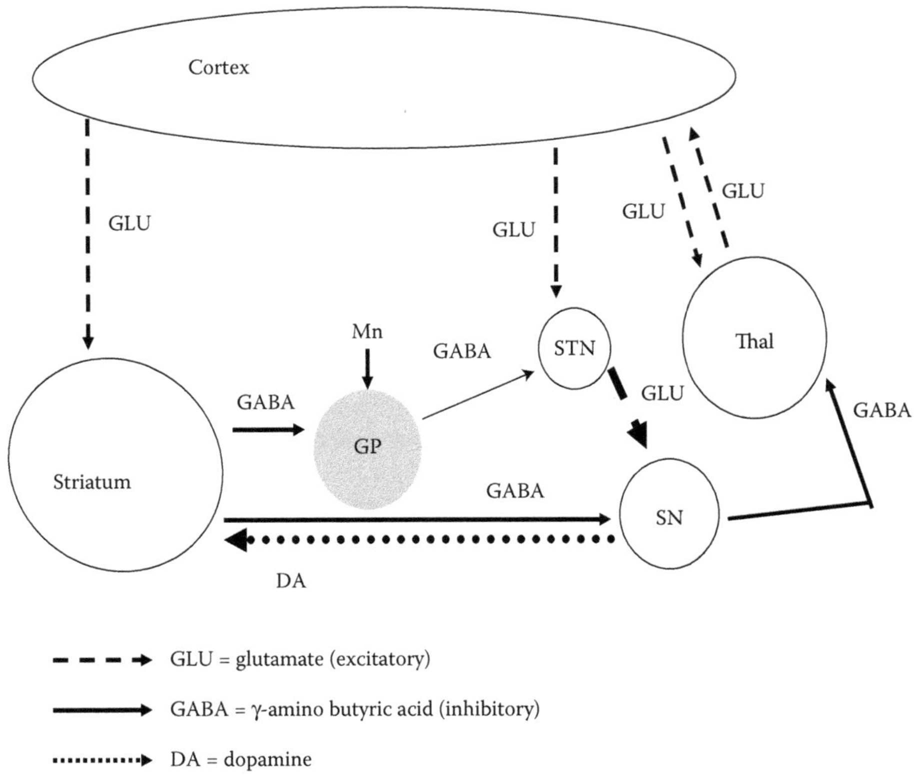

FIGURE 5.2 During manganism, it is hypothesized that Mn accumulates primarily in the GP. In turn, increased Mn concentrations in the GP cause neurodegeneration leading to decreased GABAergic output into the subthalamic nuclei. Glutamatergic input into the substantia nigra from the subthalamic nuclei is unregulated leading to DAergic dysfunction in the striatum. The thickness of the lines illustrates the strength of neurotransmission. GP: globus pallidus, STN: subthalamic nuclei, Thal: thalamus, and SN: substantia nigra. (Adapted from Graybiel, A., *Trends Neurosci.*, 13, 244–254, 1990 and Verity, M.A., *Neurotoxicology*, 20, 489–497, 1999.)

5.6 CONCLUSIONS

Mn-induced neurotoxicity arises over a prolonged period of time and when Mn intake via either portal (lung and GI tract) exceeds its elimination. While concentrations of Mn (1 mg Mn/m^3) in air that are associated with overt Mn neurotoxicity are rarely, if ever, found in the modern workplace, subtle neurological effects in people exposed to significantly lower levels, whether from air or food sources, are likely to continue to occur. Studies over the last few decades have not only greatly improved the understanding of the health risks associated with exposure to Mn and its symptoms, but have also refined the understanding of Mn transport and neurotoxicity mechanisms. Specifically, much has been learned about Mn transporters such as DMT1 and Tf, the relationship between Mn and Fe, as well as shared mechanisms of neurotoxicity between Mn and several mitochondrial poisons. Combined, these studies have just begun to scratch the surface on the relationship between Mn and other divalent metals and the primary biological mechanisms implicated in its toxic effects on select brain regions. It seems that future experimental emphasis would be most profitable if directed at a better understanding of the many facets of Mn homeostasis, both in health and disease.

ACKNOWLEDGMENTS

This review was partially supported by NIEHS grant no. 10563 and DoD grant no. W81XWH-05-1-0239 (to M.A.).

REFERENCES

1. Gregor, J.L., Dietary standards for manganese: Overlap between nutritional and toxicological studies. *J. Nutr.*, 128, 368S–371S, 1998.
2. Hurley, L.S. and Keen, C.L., Manganese, In: E. Underwood and W. Mertz (eds), *Trace Elements in Human Health and Animal Nutrition*, pp. 185–223. New York: Academic Press, 1987.
3. Erikson, K.M. and Aschner, M., Manganese neurotoxicity and glutamate–GABA interaction. *Neurochem. Internat.*, 43, 475–480, 2003.
4. Fitsanakis, V.A., Garcia, S.J., and Aschner, M., Manganese dynamics, distribution and neurotoxicity. In: M. Aschner and L.G. Costa (eds), *The Role of Glia in Neurotoxicity*, pp. 395–415. Boca Rotan, FL: CRC Press, 2005.
5. Gregor, J.L., Nutrition versus toxicology of manganese in humans: Evaluation of potential biomarkers. *Neurotoxicology*, 20, 205–212, 1999.
6. Garcia-Avila, M. and Penalver-Ballina, R., Manganese poisoning in the mines of Cuba. *Ind. Med. Surg.*, 22, 220–221, 1953.
7. Rodier, J., Manganese poisoning in Moroccan miners. *Br. J. Ind. Med.*, 12, 21–35, 1955.
8. Myers, J.E., et al., Nervous system effects of occupational manganese exposure on South African manganese mineworkers. *Neurotoxicology*, 24, 649–656, 2003.
9. Racette, B.A., et al., Welding-related parkinsonism: Clinical features, treatment, and pathophysiology. *Neurology*, 56, 8–13, 2001.
10. Racette, B.A., et al., [^{18}F]FDOPA PET and clinical features in parkinsonism due to manganism. *Mov. Disord.*, 20, 492–496, 2005.
11. Sadek, A.H., Rauch, R., and Schulz, P.E., Parkinsonism due to manganism in a welder. *Int. J. Toxicol.*, 22, 393–401, 2003.
12. Levy, B.S. and Nassetta, W.J., Neurologic effects of manganese in humans: A review. *Int. J. Occup. Environ. Health*, 9, 153–163, 2003.
13. Olanow, C.W., Manganese-induced parkinsonism and Parkinson's disease. *Ann. NY Acad. Sci.*, 1012, 209–223, 2004.
14. Pal, P.K., Samii, A., and Calne, D.B., Manganese neurotoxicity: A review of clinical features, imaging and pathology. *Neurotoxicology*, 20, 227–238, 1999.
15. Aschner, M. and Aschner, J.L., Manganese transport across the blood–brain barrier: Relationship to iron homeostasis. *Brain. Res. Bull.*, 24, 857–860, 1990.
16. Aschner, M. and Aschner, J.L., Manganese neurotoxicity: Cellular effects and blood–brain barrier transport. *Neurosci. Biobehav. Rev.*, 15, 333–340, 1991.
17. Aschner, M. and Gannon, M., Manganese (Mn) transport across the rat blood–brain barrier: Saturable and transferrin dependent transport mechanisms. *Brain Res. Bull.*, 33, 345–349, 1994.
18. Crossgrove, J.S. and Yokel, R.A., Manganese distribution across the blood–brain barrier III: The divalent metal transporter-1 is not the major mechanism mediating brain manganese uptake. *Neurotoxicology*, 25, 451–460, 2004.
19. Yokel, R.A. and Crossgrove, J.S., Manganese toxicokinetics at the blood–brain barrier. *Res. Rep. Health Eff. Inst.*, 119, 7–73, 2004.
20. Erikson, K.M., et al., Globus pallidus: A target brain region for divalent metal accumulation associated with dietary iron deficiency. *J. Nutr. Biochem.*, 15, 335–341, 2004.
21. Sumino, K., et al., Heavy metals in normal Japanese tissues. *Arch. Environ. Health*, 30, 487–494, 1975.
22. Suzuki, Y., et al., Study of substance toxicity of manganese dioxide in monkeys. *Tokushima J. Exp. Med.*, 22, 5–10, 1975.
23. Ingersoll, R.T., Montgomery, E.B., and Aposhian, H.V., Central nervous system toxicity of manganese. II: Cocaine or reserpine inhibit manganese concentration in the rat brain. *Neurotoxicology*, 20, 467–476, 1999.
24. Lai, J.C., et al., Manganese mineral interactions in brain. *Neurotoxicology*, 20, 433–444, 1999.
25. Roels, H., et al., Influence of the route of administration and the chemical form ($MnCl_2$, MnO_2) on the absorption and cerebral distribution of manganese in rats. *Arch. Toxicol.*, 71, 223–230, 1997.
26. Lynam, D.R., et al., Environmental effects and exposures to manganese from use of methylcyclopentadienyl manganese tricarbonyl (MMT) in gasoline. *Neurotoxicology*, 20, 145–150, 1999.
27. Zayed, J., Hong, B., and L'Esperance, G., Characterization of manganese-containing particles collected form the exhaust emissions of automobiles running with MMT additive. *Environ. Sci. Technol.*, 33, 3341–3346, 1999.

28. Dorman, D.C., et al., Influence of particle solubility on the delivery of inhaled manganese to the rat brain: Manganese sulfate and manganese tetroxide pharmacokinetics following repeated (14-day) exposure. *Toxicol. Appl. Pharmacol.*, 170:79–87, 2001.
29. Drown, D.B., Oberg, S.G., and Sharma, S.P., Pulmonary clearance of soluble and insoluble forms of manganese. *J. Toxicol. Environ. Health*, 17, 201–212, 1986.
30. Calne, D.B., et al., Manganism and idiopathic parkinsonism: Similarities and difference. *Neurology*, 44, 1583–1586, 1994.
31. Fahn, S. and Cohen, G., The oxidant stress hypothesis in Parkinson's disease: Evidence supporting it. *Ann. Neurol.*, 32, 804–812, 1992.
32. Cohen, G. and Spina, M.B., Deprenyl suppresses the oxidant stress associated with dopamine turnover. *Ann. Neurol.*, 26, 689–690, 1989.
33. Riederer, P., et al., Transition metals, ferritin, glutathione, and ascorbic acid in parkinsonian brains. *J. Neurochem.*, 52, 515–520, 1989.
34. Przedborski, S., et al., Transgenic mice with increased Cu/Zn-superoxide dismutase activity are resistant to *N*-methyl-4-phenyl-1,2,3,6-tetrahydropyridine-induced neurotoxicity. *J. Neurosci.*, 12, 1658–1667, 1992.
35. Pityn, P., et al., The topography of particle deposition in the human lung. *Respir. Physiol.*, 78, 19–29, 1989.
36. Yu, I.J., et al., Pattern of deposition of stainless steel welding fume particles inhaled into the respiratory systems of Sprague-Dawley rats exposed to a novel welding fume generating system. *Toxicol. Lett.*, 116, 103–111, 2000.
37. Roth, J.A., Homeostatic and toxic mechanisms regulating manganese uptake, retention, and elimination. *Biol. Res.*, 39, 45–57, 2006.
38. Zheng, W., Aschner, M., and Ghersi-Egea, J.F., Brain barrier systems: A new frontier in metal neurotoxicological research. *Toxicol. Appl. Pharmacol.*, 192, 1–11, 2003.
39. Brain, J.D., et al., Effects of iron status on transpulmonary transport and tissue distribution of Mn and Fe. *Am. J. Respir. Cell Mol. Biol.*, 34, 330–337, 2006.
40. Heilig, E., et al., Pharmacokinetics of pulmonary manganese absorption: Evidence for increased susceptibility to manganese loading in iron-deficient rats. *Am. J. Physiol. Lung Cell Mol. Physiol.*, 288, L887–L893, 2005.
41. Thompson, K., et al., The influence of high iron diet on rat lung manganese absorption. *Toxicol. Appl. Pharmacol.*, 210, 17–23, 2006.
42. Heilig, E.A., et al., Manganese and iron transport across pulmonary epithelium. *Am. J. Physiol. Lung Cell Mol. Physiol.*, 290, L1247–L1259, 2006.
43. Tsushima, R.G., et al., Modulation of iron uptake in heart by L-type Ca^{2+} channel modifiers: Possible implications in iron overload. *Circ. Res.*, 84, 1302–1309, 1999.
44. Colet, J.M., Vander Elst, L., et al., Dynamic evaluation of the hepatic uptake and clearance of manganese-based MRI contrast agents: A ^{31}P NMR study on the isolated and perfused rat liver. *J. Magn. Reson. Imaging*, 8, 663–669, 1998.
45. Crossgrove, J.S. and Yokel, R.A., Manganese distribution across the blood–brain barrier. IV. Evidence for brain influx through store-operated calcium channels. *Neurotoxicology*, 26, 297–307, 2005.
46. Chen, J. and Barritt, G.J., Evidence that TRPC1 (transient receptor potential canonical 1) forms a Ca(2 +)-permeable channel linked to the regulation of cell volume in liver cells obtained using small interfering RNA targeted against TRPC1. *Biochem. J.*, 373, 327–336, 2003.
47. Hofmann, T., et al., Direct activation of human TRPC6 and TRPC3 channels by diacylglycerol. *Nature*, 397, 259–263, 1999.
48. Ghio, A.J. and Cohen, M.D., Disruption of iron homeostasis as a mechanism of biologic effect by ambient air pollution particles. *Inhal. Toxicol.*, 17, 709–716, 2005.
49. Prahalad, A.K., et al., Enhancement of 2′-deoxyguanosine hydroxylation and DNA damage by coal and oil fly ash in relation to particulate metal content and availability. *Chem. Res. Toxicol.*, 13, 1011–1019, 2000.
50. Prahalad, A.K., et al., Air pollution particles mediated oxidative DNA base damage in a cell free system and in human airway epithelial cells in relation to particulate metal content and bioreactivity. *Chem. Res. Toxicol.*, 14, 879–887, 2001.
51. Roemer, W., et al., Pollution effects on asthmatic children in Europe, the PEACE study. *Clin. Exp. Allergy*, 30, 1067–1075, 2000.
52. Bredow, S., et al., Subchronic inhalation of soluble manganese induces expression of hypoxia-associated angiogenic genes in adult mouse lungs. *Toxicol. Appl. Pharmacol.*, 221, 148–157, 2007.

53. Hong, Y.C., et al., Metals in particulate pollutants affect peak expiratory flow of schoolchildren. *Environ. Health Perspect.*, 115, 430–434, 2007.
54. Boojar, M.M. and Goodarzi, F., A longitudinal follow-up of pulmonary function and respiratory symptoms in workers exposed to manganese. *J. Occup. Environ. Med.*, 44, 282–290, 2004.
55. Crossgrove, J.S., et al., Manganese distribution across the blood–brain barrier. I. Evidence for carrier-mediated influx of manganese citrate as well as manganese and manganese transferrin. *Neurotoxicology*, 24, 3–13, 2003.
56. Yokel, R.A., Crossgrove, J.S., and Bukaveckas, B.L., Manganese distribution across the blood–brain barrier. II. Manganese efflux from the brain does not appear to be carrier mediated. *Neurotoxicology*, 24, 15–22, 2003.
57. Keefer, R.C., Barak, A.J., and Boyett, J.D., Binding of manganese and transferring in rat serum. *Biochim. Biophys. Acta*, 221, 390–393, 1970.
58. Scheuhammer, A.M. and Cherian, M.G., Binding of manganese in human and rat plasma. *Biochim. Biophys. Acta*, 840, 163–169, 1985.
59. Aisen, P., Aasa, R., and Redfield, A.G., The chromium, manganese, and cobalt complexes of transferrin. *J. Biol. Chem.*, 244, 4628–4633, 1969.
60. Fishman, J.B., et al., Receptor-mediated transcytosis of transferrin across the blood–brain barrier. *J. Cell. Biol.*, 101, 423A, 1985.
61. Jeffries, W.A., et al., Transferrin receptor on endothelium of brain capillaries. *Nature*, 132, 162–163, 1984.
62. Partridge, W.M., Eisenberg, J., and Yang, J., Human blood–brain barrier transferrin receptor. *Metabolism*, 36, 892–895, 1987.
63. Morris, C.M., et al., Uptake and distribution of iron and transferrin in the adult brain. *J. Neurochem.*, 59, 300–306, 1992.
64. Morris, C.M., et al., Brain iron homeostasis. *J. Inorg. Biochem.*, 47, 257–265, 1992.
65. Moos, T., et al., Brain capillary endothelial cells mediate iron transport into the brain by segregating iron from transferrin without the involvement of divalent metal transporter 1. *J. Neurochem.*, 98, 1946–1958, 2006.
66. Moos, T. and Morgan, E.H., A morphological study of the developmentally regulated transport of iron into the brain. *Dev. Neurosci.*, 24, 99–105, 2002.
67. Suarez, N. and Eriksson, H., Receptor-mediated endocytosis of a manganese complex of transferrin into neuroblastoma (SHSY5Y) cells in culture. *J. Neurochem.*, 61, 127–131, 1993.
68. Sloot, W.N. and Gramsbergen, J.B., Axonal transport of manganese and its relevance to selective neurotoxicity in the rat basal ganglia. *Brain Res.*, 657, 124–132, 1994.
69. Diez-Ewald, M., Weintraub, L.R., and Crosby, W.H., Inter relationship of iron and manganese metabolism. *Proc. Soc. Exp. Biol. Med.*, 129, 448–451, 1968.
70. Barbeau, A., Inoué, N., and Cloutier, T., Role of manganese in dystonia. *Adv. Neurol.*, 14, 339–352, 1976.
71. Hill, J.M. and Switzer, R.C., III, The regional distribution and cellular localization of iron in the rat brain. *Neuroscience*, 11, 595–603, 1984.
72. Hill, J.M., Ruff, M.R., and Weber, R.J., Transferrin receptors in rat brain: Neuropeptide-like pattern and relationship to iron distribution. *Proc. Natl. Acad. Sci. USA*, 82, 4553–4557, 1985.
73. Walaas, I. and Fonnum, F., The distribution and origin of glutamate decarboxylase and choline acetyltransferase in ventral pallidum and other basal forebrain regions. *Brain Res.*, 177, 325–336, 1979.
74. Nagy, J.I., Carter, D.A., and Fibiger, H.C., Evidence for a GABA-containing projection from the enopenduncular nucleus to the lateral habenula in the rat. *Brain Res.*, 145, 360–364, 1978.
75. Malecki, E.A., et al., Transferrin is required for normal distribution of ^{59}Fe and ^{54}Mn in mouse brain. *J. Neurol. Sci.*, 170, 112–118, 1999.
76. Roth, J.A. and Garrick, M.D., Iron interactions and other biological reactions mediating the physiological and toxic actions of manganese. *Biochem. Pharmacol.*, 66, 1–13, 2003.
77. Gunter, T.E., et al., Speciation of manganese in cells and mitochondria: A search for the proximal cause of manganese neurotoxicity. *Neurotoxicology*, 27, 765–776, 2006.
78. Garrick, M., et al., DMT1: A mammalian transporter for multiple metals. *Biometals*, 16, 41–54, 2003.
79. Gunter, K.K., et al., Determining the oxidation states of manganese in NT2 cells and cultured astrocytes. *Neurobiol. Aging*, 27, 1816–1826, 2005.
80. Gunter, K.K., et al., Determining the oxidation states of manganese in PC12 and nerve growth factor-induced PC12 cells. *Free Radic. Biol. Med.*, 39, 164–181, 2005.
81. Conrad, M.E. and Umbreit, J.N., Pathways of iron absorption. *Blood Cells Mol. Dis.*, 29, 336–355, 2002.

82. Mackenzie, B. and Hediger, M., Slc11 family of H+-coupled metal-ion transporters NRAMP1 and DMT. *Pflug. Arch.*, 447, 571–579, 2004.
83. Fleming, M.D., et al., Nramp-2 is mutated in the anemic Belgrade (b) rat: Evidence of a role for Nramp-2 in endosomal iron transport. *Proc. Natl. Acad. Sci. USA*, 95, 1148–1153, 1998.
84. Fleming, M.D., et al., Microcytic anemia mice have a mutation in Nramp2, a candidate iron transporter gene. *Nat. Genet.*, 16, 383–386, 1997.
85. Chua, A.C. and Morgan, E.H., Manganese metabolism is impaired in the Belgrade laboratory rat. *J. Comp. Physiol.*, 167, 361–369, 1997.
86. Portnoy, M., Jensen, L., and Culotta, V., The distinct methods by which manganese and iron regulate the NRAMP transporters in yeast. *Biochem. J.*, 362, 110–124, 2002.
87. Rosakis, A. and Koster, W., Transition metal transport in the green microalga *Chlamydomonas reinhardtii*-genomic sequence analysis. *Res. Microbiol.*, 155, 201–210, 2000.
88. Rosakis, A. and Koster, W., Divalent metal transport in the green microalga *Chlamydomonas reinhardtii* is mediated by a protein similar to prokaryotic nramp homologues. *Biometals*, 18, 107–120, 2005.
89. Henriksson, J., Tallkvist, J., and Tjalve, H., Transport of manganese via the olfactory pathway in rats: Dosage dependency of the uptake and subcellular distribution of the metal in the olfactory epithelium and the brain. *Toxicol. Appl. Pharmacol.*, 156, 119–128, 1999.
90. Shipley, M.T., Puche, A.C., and Ennis, M., *The Olfactory System*, G. Paxinos (ed.), New York: Academic Press, 2003.
91. Sunderman, F.W., Jr., Nasal toxicity, carcinogenicity, and olfactory uptake of metals. *Ann. Clin. Lab. Sci.*, 31, 3–24, 2001.
92. Lee, B.J., et al., Spread of bovine herpesvirus type 5 (BHV-5) in the rabbit brain after intranasal inoculation. *J. Neurovirol.*, 5, 474–484, 1999.
93. Oberdorster, G., et al., Translocation of inhaled ultrafine particles to the brain. *Inhal. Toxicol.*, 16, 437–445, 2004.
94. Tjalve, H. and Henriksson, J., Uptake of metals in the brain via olfactory pathways. *Neurotoxicology*, 20, 181–195, 1999.
95. Tjalve, H., Mejare, C., and Borg-Neczak, K., Uptake and transport of manganese in primary and secondary olfactory neurones in pike. *Pharmacol. Toxicol.*, 77, 23–31, 1995.
96. Tjalve, H., et al., Uptake of manganese and cadmium from the nasal mucosa into the central nervous system via olfactory pathways in rats. *Pharmacol. Toxicol.*, 79, 347–356, 1996.
97. Takeda, A., et al., Manganese transport in the neural circuit of rat CNS. *Brain Res. Bull.*, 45, 149–152, 1998.
98. Brenneman, K.A., et al., Direct olfactory transport of inhaled manganese ((54)MnCl(2)) to the rat brain: Toxicokinetic investigations in a unilateral nasal occlusion model. *Toxicol. Appl. Pharmacol.*, 169, 238–248, 2000.
99. Lin, Y.J. and Koretsky, A.P., Manganese ion enhances T1-weighted MRI during brain activation: An approach to direct imaging of brain function. *Magn. Reson. Med.*, 38, 378–388, 1997.
100. Watanabe, T., et al., *In vivo* 3D MRI staining of the mouse hippocampal system using intracerebral injection of $MnCl_2$. *Neuroimage*, 22, 860–867, 2004.
101. Saleem, K.S., et al., Magnetic resonance imaging of neuronal connections in the macaque monkey. *Neuron*, 34, 685–700, 2002.
102. Pautler, R.G. and Koretsky, K.P., Tracing odor-induced activation in the olfactory bulbs of mice using manganese-enhanced magnetic resonance imaging. *Neuroimage*, 16, 441–448, 2002.
103. Cross, D.J., et al., Statistical mapping of functional olfactory connections of the rat brain *in vivo*. *Neuroimage*, 23, 1326–1335, 2004.
104. Thompson, K., et al., Olfactory uptake of manganese requires Divalent Metal Transporter-1 and is enhanced by anemia. *FASEB J.*, 21, 223–230, 2007.
105. Rabin, O., et al., Rapid brain uptake of manganese(II) across the blood–brain barrier. *J. Neurochem.*, 61, 509–517, 1993.
106. Murphy, V.A., et al., Saturable transport of manganese (II) across the rat blood–brain barrier. *J. Neurochem.*, 57, 948–954, 1991.
107. Eide, D.J. The SLC39 family of metal ion transporters. *Pflueg. Arch. Eur. J. Physiol.*, 447, 796–800, 2004.
108. He, L., et al., ZIP8, Member of the solute-carrier-39 (SLC39) metal-transporter family: Characterization of transporter properties. *Mol. Pharmacol.*, 70, 171–180, 2006.
109. Petzinger, G.M., et al., Behavioral motor recovery in the 1-methyl-4-phenyl-1,2,3,6-tetrahydropyridine-lesioned squirrel monkey (*Saimiri sciureus*): Changes in striatal dopamine and expression of tyrosine hydroxylase and dopamine transporter proteins. *J. Neurosci. Res.*, 83, 332–347, 2006.

110. Yang, W. and Tiffany-Castiglioni, E., The bipyridyl herbicide paraquat produces oxidative stress-mediated toxicity in human neuroblastoma SH-SY5Y cells: Relevance to the dopaminergic pathogenesis. *J. Toxicol. Environ. Health*, 68, 1939–1961, 2005.
111. Ritz, M.C., et al., ^{3}H-WIN 35, 065-2: A ligand for cocaine receptors in striatum. *J. Neurochem.*, 55, 1556–1562, 1990.
112. Erikson, K.M., et al., Manganese accumulation in striatum of mice exposed to toxic doses is dependent upon a functional dopamine transporter. *Environ. Toxicol. Pharmacol.*, 20, 390–394, 2005.
113. Anderson, J.G., Cooney, P.T., and Erikson, K.M., Inhibition of DAT function attenuates manganese accumulation in the globus pallidus. *Environ. Toxicol. Pharmacol.*, 23, 179–184, 2007.
114. Burdo, J.R., et al., Cellular distribution of iron in the brain of the Belgrade rat. *Neuroscience*, 93, 1189–1196, 1999.
115. Wedler, F.C. and Denman, R.B., Glutamine synthetase: The major Mn(II) enzyme in mammalian brain. *Curr. Top. Cell Regul.*, 24, 153–169, 1984.
116. Beleboni, R.O., et al., Pharmacological and biochemical aspects of GABAergic neurotransmission: Pathological and neuropsychobiological relationships. *Cell Mol. Neurobiol.*, 24, 707–728, 2004.
117. Olsen, R.W. and DeLorey, T.M. GABA and glycine, In: G.J. Siegel, B.W. Agranoff, R.W. Albers, S.K. Fisher, and M.D. Uhler (eds), *Basic Neurochemistry*, pp. 335–346. Philadelphia: Lipincott Williams and Wilkins, 1999.
118. Carlsson, M. and Carlsson, A., Interactions between glutamatergic and monoaminergic systems within the basal ganglia-implications for schizophrenia and Parkinson's disease. *Trends Neurosci.*, 13, 272–276, 1990.
119. Chandra, S.V., Malhotra, K.M., and Shukla, G.S., GABAergic neurochemistry in manganese exposed rats. *Acta Pharmacol. Toxicol.*, 51, 456–458, 1982.
120. Lipe, G.W., et al., Effect of manganese on the concentration of amino acids in different regions of the rat brain. *J. Environ. Sci. Health B*, 34, 119–132, 1999.
121. Gwiazda, R.H., et al., Sub-chronic low cumulative exposure to manganese affects striatal GABA but not dopamine. *Neurotoxicology*, 23, 69–76, 2002.
122. Anderson, J.G., Cooney, P.T., and Erikson, K.M., Brain manganese accumulation is inversely related to GABA uptake in male and female rats. *Toxicol. Sci.*, 95, 188–195, 2007.
123. Cotman, C.W., Foster, A.C., and Lanhorn, T., An overview of glutamate as a neurotransmitter. *Adv. Biochem. Psychopharmacol.*, 271, 1981.
124. Page, G., et al., Modulation of the neuronal dopamine transporter activity by metabotropic glutamate receptor mGluR5 in rat striatal synaptosomes through phosphorylation mediated processes. *J. Neurochem*, 76, 1282–1290, 2001.
125. Castro, S.L. and Zigmond, M.J., Stress-induced increase in extracellular dopamine in striatum glutamatergic action via *N*-methyl-D-aspartate receptors in substantia nigra. *Brain Res.*, 901, 47–54, 2001.
126. Chiueh, C.C., et al., *In vivo* generation of hydroxyl radicals and MPTP-induced dopaminergic toxicity in the basal ganglia. *Ann. N. Y. Acad. Sci.*, 738, 25–36, 1994.
127. De Iuliis, A., et al., A proteomic approach in the study of an animal model of Parkinson's disease. *Clin. Chim. Acta*, 357, 202–209, 2005.
128. Jenner, P., Oxidative mechanisms in nigral cell death in Parkinson's disease. *Mov. Disord.*, 13, 24–34, 1998.
129. Perier, C., et al., Complex I deficiency primes Bax-dependent neuronal apoptosis through mitochondrial oxidative damage. *Proc. Natl. Acad. Sci. USA*, 102, 19126–19131, 2005.
130. Sherer, T.B., et al., Mechanism of toxicity in rotenone models of Parkinson's disease. *J. Neurosci.*, 23, 10756–10764, 2003.
131. Takeda, A., Sotogaku, N., and Oku, N., Influence of manganese on the release of neurotransmitters in rat striatum. *Brain Res.*, 965, 279–282, 2003.
132. Stredrick, D.L., et al., Manganese-induced cytotoxicity in dopamine-producing cells. *Neurotoxicology*, 25, 543–553, 2004.
133. Graybiel, A., Neurotransmitters and neuromodulators in the basal ganglia. *Trends Neurosci.*, 13, 244–254, 1990.
134. Verity, M.A., Manganese neurotoxicity: A mechanistic hypothesis. *Neurotoxicology*, 20, 489–497, 1999.

6 Metallothionein and Metal Homeostasis

James Koropatnick

CONTENTS

6.1 Introduction 143
6.2 MT Proteins 144
6.3 MT Gene Structure and Function 145
6.3.1 MT Gene Promoter Elements 146
6.3.1.1 Basal MT Gene Transcription 146
6.3.1.2 Induced MT Gene Transcription 146
6.3.1.3 Metal Response Element-Binding Transcription Factor-1 (MTF-1) 147
6.3.1.4 Oxidative Stress Induction 149
6.3.1.5 Glucocorticoid Induction 149
6.3.1.6 Cytokine and Stress-Mediated Induction 149
6.3.1.7 Epigenetic Events: Chromatin Structure and DNA Methylation 150
6.3.1.8 Post-Transcriptional Regulation 150
6.3.1.9 Developmental and Tissue-Specific MT Expression 151
6.4 MT Function 151
6.4.1 MTs as Zinc Buffers 151
6.5 MT in Health and Disease 153
6.5.1 MT in Immune Function 153
6.5.2 MT and Cancer 153
6.5.3 MT and Cardiac Toxicity 155
6.5.3.1 MT, Diabetes, and Cardiac Toxicity 155
6.5.4 MT and Neurological Function 156
6.6 Summary and Perspective 157
Acknowledgment 157
References 158

6.1 INTRODUCTION

Every environmental and biological niche on earth is accessible to heavy metals. They enter into, and accumulate in, many of those compartments as the most long-lived product of both natural and man-made sources [1]. Metal ions are undoubtedly toxic to cells and organs in varying degree, but they are also required for viability by all living systems. For example, copper is essential for the function of oxidative enzymes (including catalase, peroxidase, cytochrome oxides, and others) but is also a cellular toxicant through its capacity to generate oxygen and other radicals that react with and diminish or destroy the function of essential cellular macromolecules [2]. The dual capacity to both induce damage and mediate essential physiological events is reflected in the development of cellular mechanisms to (a) transport metals into cells and subcellular compartments, (b) secrete metals into the extracellular environment, and (c) associate metal ions with chaperone macromolecules that aid in those transport processes, control or abrogate their toxicity, and regulate metal ion availability for physiological processes. These events have been assumed to function to control metal toxicity, but recent evidence indicates that they also regulate the function of proteins that associate with metals (because those proteins depend on metal ions for function and/or they act as metal ion sensors important in regulating cellular events triggered by high or low metal ion levels in cells and cellular compartments). For metal ions that are essential and with low capacity to induce damage under normal circumstances (e.g., zinc) this may be especially true. More than 3% of the human genome encodes proteins with zinc-binding domains important for function and the fraction may be as high as 10% [3,4]. The large fraction of the human genome devoted to proteins encoding zinc-binding proteins is reflected in the absolute requirement of zinc for life in a broad range of organisms. Like other Group IIB metal ions, zinc is capable of mediating redox reactions for biological purposes. It has the advantage of being relatively (but not completely) nontoxic compared to other metal ions, and that advantage is reflected in the many proteins that utilize zinc to coordinate protein–protein and protein–nucleic acid interactions (zinc fingers and zinc clusters, and nuclear hormone receptors) [5,6]. Severe zinc deficiency, although rare in humans, contributes to growth retardation, hypogonadism, dermatitis, and immune dysfunction [7]. Mild zinc deficiency, which has been suggested to be a health crisis of global proportions [8], has been linked to poor neuropsychological performance, abnormal fetal development (low birth weight and increased incidence of childhood disease in developing countries [9]), increased cancer risk [10–12], and overall increases in disease and resulting mortality [13,14]. Zinc availability, and zinc participation in physiological events mediated by zinc-binding proteins, is homeostatically regulated in organisms by controlled and mutually responsive interactions among an array of transporters that efflux and import zinc (reviewed by Haase and Maret, this volume), transcription factors (including metal-responsive transcription factor-1 [MTF-1]) that both require zinc for activity and mediate cellular response to alterations in levels of biologically available zinc, and zinc storage vesicles (zincosomes) [15]. Mutations in zinc transporter genes aggravate the consequences of dietary zinc deficiency, or lead to pathological conditions arising from inadequate zinc regulation, independent of the amount of zinc in diet. For example, *Acrodermatitis enteropathica* is a rare condition involving zinc deficiency and is due to the presence of a mutated human Zip4 (SLC39A4) zinc importer [16,17]. Transient neonatal zinc deficiency (caused by low zinc levels in maternal milk) is caused by a mutation in the human ZnT2 zinc exporter [18]. Transport systems shuttle zinc into and out of cells and organs at a relatively high rate: in humans, approximately 1% of zinc in the body is replaced daily by the zinc obtained from dietary sources [19]. The primary site of zinc uptake is through intestinal enterocytes, while zinc excretion is by pancreatic acinar cells and other intestinal secretions. The expression of specific zinc transporters (and the rate of zinc uptake and excretion) is responsive to dietary zinc levels and is coupled to regulated expression and action of intracellular proteins that associate with zinc, particularly metallothioneins (MTs). MTs have been implicated in a variety of functions, foremost of which is homeostatic regulation of zinc. In this chapter, we review the function and regulation of expression of MTs in the context of other proteins important in zinc regulation and the evidence for a homeostatic role for MTs in zinc metabolism.

6.2 MT PROTEINS

MTs collectively comprise a family of small (typically less than 10 kDa), cysteine-rich, metal-binding proteins found in all eukaryotes. All MTs contain 61–62 amino acids organized into single α and β domains (one α and one β domain per MT molecule). The high number of cysteine residues (11 in the C-terminal α domain and 9 in the N-terminal β domain) are responsible for a high capacity of MTs to coordinate covalent binding of divalent metal cations in metal/thiolate clusters. For example, MT molecules can bind up to a total of seven zinc or cadmium ions, or up to 12 copper ions, in aggregate between the two domains. MTs can bind a variety of metals but (under normal conditions and in the absence of exposure to unusually high levels of other metal ions in the environment or diet) they are bound primarily to zinc and secondarily to copper [20]. Although MTs from different species differ in some respects, they retain a high degree of similarity at the protein level [21] and that similarity implies conserved and important biological function. Essential functions for the MT family of proteins have not yet been identified, but a number of roles in response to cellular and tissue stress, and in physiological function under normal conditions, have been attributed to MTs. These include detoxification of heavy metals, homeostatic regulation of metals essential for life (primarily zinc and copper) and important in growth, differentiation, and function of a variety of specialized cells and organs, and protection of tissues against various forms of oxidative injury [22].

Four different MT isoforms (MT-1, MT-2, MT-3, and MT-4) have been described in mammals. MT-1 and MT-2 are present in virtually all mammalian cells and organs, with the highest expression in liver tissue. MT-3 and MT-4 are limited to specific cells and tissues. MT-3 was first identified as a molecule that suppressed growth of rat neuronal cells in culture [23] and later as an MT [24]. MT-3 is expressed in brain and primarily in glutaminergic neurons, with additional very low levels reported in pancreas and intestine [25]. MT-4 is expressed, apparently exclusively, in squamous epithelial cells in skin and tongue [26]. MT-3 and MT-4 are constitutively expressed in the organs where they are expressed but do not appear to be inducible by exogenous agents (although MT-4 may increase, to a limited extent, in response to stress [27]). MT-1 and MT-2, on the other hand, are both constitutively expressed and highly inducible, and events and agents capable of enhancing MT-1 and MT-2 expression generally do not increase MT-3 and MT-4. Increased levels of different MT isoforms are associated with proliferation of human tumor cells in culture [28,29] and MT is transiently increased in the nuclei of rapidly proliferating cells during G_1-to-S phase transition and after cell injury, possibly to meet a high intranuclear demand for zinc for DNA repair and synthesis by zinc-associated MT [30]. Furthermore, in mice with genetically ablated MT-1 and MT-2 genes, proliferation of liver cells after chemical injury using thioacetamide (a liver toxin) or partial hepatectomy was reduced by approximately 50% [31]. In rats undergoing compensatory renal growth in response to uninephrectomy, transcription of MT-1 and MT-2 genes (particularly in the cortex and the outer stripe of the outer medulla) and the amount of MT protein (in renal cortex and whole kidney) were increased [32]. MTs are therefore associated with cellular proliferation and hypertrophy *in vitro* and *in vivo* and may mediate components of those events. The levels of MT isoforms vary and intracellular localization of MTs changes during differentiation [27,33–37] and during monocyte differentiation and macrophage activation [38–41]. Regulation of genes responsible for different MT isoforms, leading to variations in MT levels and differential cellular localization, is associated with a broad range of physiological events.

Among the many conspicuous features of MTs, the broad range of agents and conditions capable of inducing MT-1 and MT-2 genes is among the most remarkable. MT biosynthesis is regulated primarily at the level of transcription through the action of transcription factors (directly or indirectly responsive to zinc) on regulatory DNA elements linked to MT gene open reading frames (ORFs). The many factors and physiological and stress events that directly or indirectly induce MT genes provide insight into the potential physiological roles of MT genes.

6.3 MT GENE STRUCTURE AND FUNCTION

Although only four MT proteins have been identified in mammals (including humans), at least 11 MT-1 genes have been identified in humans (MT1A, B, E, F, G, H, I, J, K, L, and X), some of which appear to encode RNAs incapable of directing the production of MT protein. The 11 MT-1 genes are clustered within the q13 region of chromosome 16, along with one MT-2 gene (also known as MT-2A), one MT-3 gene, and one MT-4 gene. An MTL-5 gene (MT-like, directing the production of tesmin protein and a marker of early male germ cell differentiation) has been reported on chromosome 11 (in the q13.2–q13.3 region) [42]. Human MT-M and MT-E genes (chromosomal location unknown) have been reported by direct submission to the National Center for Biotechnology

Information (NCBI). In mice, one copy each of MT-1, MT-2, MT-3, and MT-4 are clustered in a 60 kb region of chromosome 8. A mouse MTL-5 gene on chromosome 19 is, like MT-1 and MT-2, induced by heavy metals [43]; and a gene sequence termed "MT-1 activator" (MT-1A, on mouse chromosome 15, and similar to mouse MT-1) has been reported to the NCBI.

Multiple genes in the MT family, multiple genes in the MT-1 subfamily in humans, and the possible existence of novel additional MT genes (MT-M, MT-E) hint at complexity in terms of function and expression. Detailed comparative analysis has not been undertaken and remains an area with great potential to illuminate MT protein function. To date, the study of transcriptional control of MT genes has focused on MT-2 in humans and MT-1 and MT-2 in mice.

6.3.1 MT Gene Promoter Elements

6.3.1.1 Basal MT Gene Transcription

Multiple signals regulate both low-level, basal MT expression and MT induced by developmental and environmental signals. MT-1 and MT-2 genes contain classical TATA boxes and initiator regions (InR) that recruit transcription factor IID (TFIID) as the initial step in gene transcription, and both are required for basal and inducible MT gene transcription. In addition, a series of metal regulatory elements (MREs; with the consensus sequence CTNTGC(G/A)CNCGGCCC, and present in non-identical copies MREa to MREg), initially described as required for induction by metals, are also essential for MT expression in the absence of exogenous metal ions [44]. The MREs act in conjunction with the zinc-dependent, zinc-responsive transcription factor MTF-1. Therefore, MREs, MTF-1, and the essential zinc ions that associate with MTF-1, all contribute to both basal and inducible MT expression. Basal expression, in the absence of added MT inducers, is not uniform in all cells and varies considerably among cultured mammalian cell lines: those levels correlate well with differences in basal MTF-1 activity [45].

MREs associating with MTF-1, although important, are not the only *cis* elements responsible for basal MT transcription. MREc′ (MREc plus an additional contiguous 11 base pairs) binds C′-binding proteins 1 and 2 (C′BP-1 and C′ BP-2, closely related to the C/EBP family of CCAAT/enhancer binding proteins) and may partially mediate basal transcription [46]. Further, mouse and hamster MT-1 genes contain an MLTF (adeno*ma*jor *l*ate *t*ranscription *f*actor) binding site overlapping an *a*ntioxidant *r*esponse *e*lement (ARE) necessary for both basal and cadmium-induced MT-1 gene expression [47,48]. Many MT gene promoters contain a perfect consensus ARE in addition to the hybrid MLTF/ARE present in mouse MT-1 genes [49]. AREs are found in genes responsive to free radicals (redox-cycling xenobiotics and H_2O_2) and are stimulated by interaction with NF-E2-related factor 2 (Nrf-2) [50]. Although Nrf-2 enhances transcription, Fos and Fra-1 transcription factors can also interact with AREs to repress transcription. The combination of control by both positive (activator) and negative (repressor) interactions between DNA and regulatory proteins appears to contribute to basal MT gene transcription in the absence of inducers.

6.3.1.2 Induced MT Gene Transcription

MT genes are induced by exogenous inorganic agents, including transition metal ions and other stress-inducing, redox-active species, and by organic molecules that transmit receptor-mediated signals (cytokines; steroid hormones including glucocorticoid, progesterone, and vitamin D_3; bacterial lipopolysaccharide [LPS]; and others). Many heavy metals (including zinc, cadmium, mercury, copper, and bismuth; all of which bind to MT) induce MT-1 and MT-2 (but not MT-3 and MT-4) in mammalian cells [22]. Nickel and cobalt ions also induce MT gene transcription, but do not bind to the resulting protein. The potency of induction of MT gene transcription and protein synthesis varies among inducing metals [51], but zinc and cadmium are among the most potent. MREs are essential for MT gene inducibility (as they are for basal MT transcription) and are present in all mammalian MT genes regardless of their inducibility by metals. MREd is the most powerful mediator of metal induction [52] but all six MREs are fully engaged with regulatory proteins during metal-induced induction of transcription [53]. Although functional MREs are essential for MT gene transcription, their distribution and orientation in promoter/enhancer gene regions are of little or no consequence [54].

6.3.1.3 Metal Response Element-Binding Transcription Factor-1 (MTF-1)

MTF-1 is an MRE-binding protein that binds optimally to DNA and stimulates MT gene transcription when available zinc levels are high [55]. MTF-1 binding to DNA is reversibly induced by direct binding to zinc, and only zinc. It is expressed in all mammalian cells and is required for both basal and induced MT expressions [56,57]. MTF-1 gene knockout in mice is lethal, and mouse embryos lacking MTF-1 expression do not survive beyond the embryo stage: the protein is essential for liver development and adequate response to stress [57,58]. In humans, the MTF-1 gene is located on chromosome 1 (1p33) [59] and the 753-amino-acid protein it encodes contains, in the N-terminal region, six Cys_2His_2 (cysteine, histidine) zinc finger regions that mediate binding to DNA. Although the majority of zinc fingers are constitutively associated with zinc, they do not appear to modulate their DNA-binding capacity in response to variable intracellular zinc levels. At least one, however, has characteristics (including substantially lower affinity for zinc than is typical for canonical Cys_2His_2 zinc fingers [6]) consistent with a novel ability to modulate the transcription-inducing activity of MTF-1 as available zinc levels increase or decrease [57]. Although attention has focused on this region with respect to responsiveness to variations in zinc, there are differences in metal ion homeostasis and stress response among species in which MTF-1 has been identified that may be attributable to additional, as yet uncharacterized, regions in MTF-1 orthologues that also participate in responses to zinc or other heavy metals [6]. Three distinct transcription activation domains (acidic; proline-rich; and serine/threonine-rich) exist in the MTF-1 C-terminal region. Interaction among all three is required for metal-inducible MT induction.

The capacity of MTF-1 to mediate MT gene transcription in response to metals other than zinc, and to induction by agents and events other than exposure to exogenous metals, suggests that an indirect zinc redistribution mechanism (ultimately leading to increased zinc association with MTF-1) may be responsible for that versatility in responsiveness. Although MT-mediated association of zinc with an inhibitor molecule leading to release of the inhibitor and subsequent activation of MTF-1 has been suggested [60], demonstration of a zinc-sensitive inhibitor has not been reported. Direct activation of MTF-1 by zinc, or by introduction of zinc to MTF-1 by a chaperone molecule or molecules, remains the strongest possibility.

MTs themselves have been suggested to be regulatory molecules (chaperones) that provide zinc (and only zinc) directly or indirectly to MTF-1: an attractive concept, since MTs exhibit a hierarchy of metal-binding affinities. The affinity of MT for copper and cadmium is approximately 100-fold greater than for zinc [61]. MTs are predominantly associated with zinc in cells and tissues not exposed to high levels of other metals for which MTs have high affinity, and that difference in affinity leads to displacement of zinc ions from MT in the presence of added copper or cadmium, leading to local release of zinc [62]. Released zinc would then be available for binding to, and activation of, MTF-1, particularly if the released zinc were present in the nucleus. In fact, MT (which is preferentially located in the cytoplasm in resting mammalian cells) [30,33] has been localized to the nucleus under conditions of toxic metal overload [63], during mammalian development [64], and in some human tumors [65,66]. Furthermore, cadmium-induced transcription of MT-1 genes *in vitro*, mediated by MTF-1, requires the presence of zinc in the form of zinc-loaded MT-1 protein (zn7-MT), not free zinc [67], suggesting a role for MT as an MTF-1-regulating metallochaperone. On the other hand, such a role might not be universal for all MTs regardless of species. In *Drosophila*, for example, MT proteins are not necessary for copper-induced activation of MT promoters *in vivo* [68]. Nevertheless, current evidence favors the concept that regulation of zinc transfer to MTF-1 could be regulated by a zinc chaperone. MTs are candidates for zinc chaperone status in mammals and might be particularly important under conditions where zinc availability is limited (i.e., when zinc is likely to be preferentially associated with cellular molecules with particularly high affinity for zinc, including zinc finger- and zinc cluster-containing transcription factors, and MT).

Although MTF-1 regulates basal and inducible expression of MT genes, it also integrates multiple environmental signals and indirectly or directly modifies transcription of a large number of genes [69,70]. MTF-1 directly associates or cooperates with multiple factors associated with cellular response to stress and resistance to toxicity. These include gamma glutamyl-cysteine synthetase heavy chain (γGSC_{hc}), a key enzyme for synthesis of glutathione and implicated in resistance to a number of chemotherapeutic agents [14]. In addition, genes encoding the transcription factor C/EBPα (induced during acute phase response [71,72] in response to stress), embryonic alfafetoprotein (AFP; which, in addition to other functions, scavenges heavy metal ions and reactive oxygen [73]), the zinc transporter ZnT-1, and placental growth factor (PlGF; a member of the vascular endothelial growth factor [VEGF] family of angiogenic factors [74] the gene for which is, in addition, regulated by nuclear factor κB [NFκB] [75] [see below]) are likely targets of MTF-1 [76]. In addition, MTF-1 appears to be involved in tumor development and modulation of malignant characteristics. Loss of MTF-1 leads to increased activity of transforming growth factor β1 (TGF-β1) and tissue transglutaminase (tTG), both of which are involved in production of extracellular matrix capable of enhancing tumor growth, and delayed tumor growth and reduced vascular density in mouse embryonic fibroblasts transformed with *ras* [77].

MTF-1, in addition to directly controlling the expression of target genes, also interacts with other transcription factors or cooperates with them to modulate their activity in controlling gene transcription. MTF-1 has been reported to interact with the p65 subunit of NFκB (which interacts with the PlGF promoter to enhance transcription) and, under hypoxic conditions stimulating translocation of p65 to the nucleus, to stimulate transcription of PlGF [75]. On the other hand, hypoxia leads to suppression rather than stimulation of PlGF expression in some cells and, in that situation, decreased MTF-1 levels and decreased MTF-1 association with the PlGF promoter occur [78]. Thus, increased and decreased MTF-1 levels, and MTF-1 association with promoter regions of regulated genes, correlate with up- and down-regulation of PlGF expression (respectively) under hypoxic conditions in different cell types, strengthening the causal connection of MTF-1 with transcription of stress genes other than those directing production of MTs.

The connection between NFκB and MTF-1 can be extended to a connection between NFκB and MT. NFκB transcription factors are dimers, with the p50/p65 complex being the most abundant form in mammals [79]. NFκB interacts with an inhibitory IκB family member protein in the cytoplasm [80] or nucleus [81] under conditions when it is required to be inactive. A variety of stimuli, including cellular stress, stimulate degradation of IκBs, release of NFκB and its entry into the nucleus, binding to target gene promoter/enhancers, and regulation of genes involved in immune response, growth, and apoptosis [79]. Active, nuclear NFκB activates antiapoptotic genes and protects cells from death induced by tumor necrosis factor (TNF) and other stimulators of apoptosis [38,82,83]. MTs, like NFκB, are antiapoptotic: antisense downregulation of MT increases apoptosis [82]. Literature reports are contradictory with respect to the role of MT in NFκB (reviewed in reference [84]). However, clonal fibroblastic cell lines from mice with genetically ablated MT-1 and MT-2 genes (MT knockout cells) have lower levels of NFκB subunit p65 (the same subunit reported to interact with MTF-1 [75]) than wild-type mouse fibroblasts, lower levels of NFκB-induced reporter gene transcription, and increased sensitivity to apoptosis. Reconstitution of MT by stable incorporation of mouse MT-1 expression in multiple clonal population of MT knockout cells increased NFκB subunit p65, NFκB-dependent reporter activity, and resistance to apoptosis [84]. Association of both MTF-1 and MT with NFκB, combined with the requirement of MTF-1 for zinc-associated MT to mediate its activity under certain conditions, raises the intriguing possibility that the three molecules/molecular complexes associate to coordinately regulate zinc availability to NFκB to regulate its activity under stress conditions.

6.3.1.4 Oxidative Stress Induction

Chemical agents (e.g., diethyl maleate, paraquat, and menadione) that mediate the production of free radicals (reactive oxygen species [ROS]) have been reported to induce rodent liver MT [85,86]. Mouse Hepa cells exposed to generators of hydroxyl radicals or superoxide anions (hydrogen peroxide, *tert*-butyl hydroquinone [tBHQ], or menadione), at concentrations that do not reduce viability, have increased expression of MT-1 and heme oxygenase. Mitochondria-specific ROS generators (antimycin A and 2,4-dinitrophenol) also induce MT [87]. Oxidative stress of various types induces MT.

Both MREs and the hybrid MLTF(USF)/ARE (described above) mediate MT gene oxidative stress response. Mouse MREd responds to metals (zinc and cadmium) and ROS generators (hydrogen peroxide and tBHQ) [47]. On the other hand, the MLTF/ARE motif is responsive to cadmium and hydrogen peroxide, but not zinc or tBHQ [88]. Optimal induction of highly redox-active species (cadmium or hydrogen peroxide) appears to depend on both MRE and MLTF/ARE elements in the mouse MT-1 promoter, with induction with relatively redox-inactive zinc, and tBHQ, requiring only MREs [56]. Induction with zinc, hydrogen peroxide, or tBHQ led to association of *trans*-acting factors over all six MREs, although induction of transcription by hydrogen peroxide was transient (30 min or less) compared to more sustained transcription in response to zinc or tBHQ (several hours).

6.3.1.5 Glucocorticoid Induction

Glucocorticoid hormones (GCs) induce MT [89]. GC receptors in the cytoplasm are associated with heat shock proteins (HSPs) to suppress glucocorticoid receptor (GR) activity. GC binding to the inactive GR monomer induces dissociation of HSPs, binding of monomers into active GR homodimers, and translocation of the active GR dimer into the nucleus. GR dimers bind to glucocorticoid response elements (GREs) adjacent to MT-1 and MT-2 gene ORFs to activate transcription [90]. GREs are present as single copies upstream of the transcription start site in human MT-2 genes, and as two tandem copies in mouse MT-1 and MT-2 genes. Where tandem copies are present, each GRE can independently (alone and without the second GRE) mediate transcription in response to endogenous glucocorticoids (e.g., hydrocortisone) or synthetic glucocorticoid agonists such as dexamethasone (dex). The GR receptor antagonist RU-486 abolishes GC-induced MT expression. Glucocorticoid inductions of MT-1 and MT-2 in mouse cells are approximately equal, while MT-2 is more strongly induced than MT-1 [61]. The GRE(s) upstream of human and rodent MT-2 genes (upstream of MT-1 genes) appear to mediate responsiveness of both MT-1 and MT-2 to glucocorticoids [90]. Considering the role of glucocorticoids in inflammatory response and infectious disease, and the contribution of dysregulation of glucocorticoid action in the development of autoimmune and inflammatory disease [91], GC induction of MT expression and the potential capacity of MTs to mediate GR activity (Section 6.4.1) and immune cell activity (Section 6.5.1), further investigation of the connection between MTs and GR activity is warranted.

6.3.1.6 Cytokine and Stress-Mediated Induction

Inflammation, bacterial infection, and physical/physiological stress strongly induce intracellular MT protein in liver [92], similar to induction of other acute phase proteins (including C-reactive protein, fibrinogen, and mannan-binding lectin) that increase in serum during infection or inflammatory reaction [93]. Induction is mediated by glucocorticoids and cytokines. Acute phase proteins (including MTs) are induced in hepatocytes by interleukin-6 (IL-6), TNFα, and IL-1α and IL-1β secreted by macrophages, of which IL-6 is the most potent [94,95]. Secreted IL-6 associates with IL-6 receptor on hepatocytes and induces receptor clustering to activate Janus kinases (JAK). Activated JAK phosphorylates signal transducer and activator of transcription (STAT) kinases that then form homodimers, translocate to the nucleus, and induce transcription by binding to STAT response elements [96]. Restraint stress can also induce both MT-1 and MT-2 in mice [97] and it is clear that both glucocorticoids (elevated by activation of the hypothalamus–pituitary–adrenal gland axis) and IL-6 are responsible, since either treatment with the glucocorticoid antagonist RU-486

[98] or assessment in mice with genetically ablated IL-6 genes [99] leads to deficient MT induction in response to restraint. Overlapping IL-6 response elements (IL-6-REs; which bind to STAT transcription factors) are present in the mouse MT-1 promoter and these, in close proximity with GREs and other *cis* elements responsive to STAT3 (also called APRF), are responsible for synergistic induction of MT by glucocorticoid and IL-6 [100]. It appears, therefore, that physical interaction between STAT3 and glucocorticoids, either directly or through intermediate protein(s), can enhance MT gene transcription in response to stress.

IL-1α, IL-1β, TNFα, IFNβ, and bacterial LPS also induce MT in liver in a concentration-dependent manner [92]. Direct treatment of human fibroblasts with TNFα and IFNβ also induces MT-2, indicating that combined treatment with other cytokines or signaling molecules from specialized nonfibroblasts is not needed for them to induce MTs [101].

6.3.1.7 Epigenetic Events: Chromatin Structure and DNA Methylation

Chemical agents that induce changes in chromatin organization can also alter MT gene transcription. Rodent tumor cells treated *in vitro* with sodium butyrate (which relaxes tight chromatin conformation by inhibiting deacetylation of chromatin-associated histones [102,103]) have increased basal MT gene transcription, and increased sensitivity to induction with heavy metals [104]. MT-1 is also developmentally regulated in a pattern that correlates well with the conformation of chromatin in which the gene resides: MT-1 and alfafetoprotein genes undergo reversible changes in chromatin structure in step with changes in the degree of developmentally regulated expression [34], possibly rendering MT genes available to transcription factors under inducing conditions.

The role of DNA methylation and chromatin structure in inhibition of MT expression has been investigated. CpG islands (methylated in higher eukaryotes as a mechanism for gene silencing in a tissue-specific manner [105]) are present in rodent MT genes, and they are highly methylated and transcriptionally repressed in rodent tumor cell lines derived from liver and thymus tumors. Normal, nontumor liver and thymus cells with low methylation levels had elevated levels of MT transcription [106]. Overall, these data suggest a role for DNA methylation in regulating the sensitivity of MT genes to induction by metals and other agents and conditions.

DNA methylation can lead to recruitment of HDACS (corepressor histone deacetylase complexes) that deacetylate histones in chromatin and promote close order packing that inhibits access by transcription factors. Hypoacetylation of histones leads to silencing of MT-1 and MT-2 in mouse lymphosarcoma cells [107], supporting a model of transcriptional control of MT genes where DNA hypermethylation leads to loss of acetylation in histones to favor close order chromatin packing and transcriptional silencing. Developmental or other events would lead to inhibition of histone deacetylases and/or activation of histone acetylases, enhanced histone acetylation, and relaxed chromatin structure. Subsequent demethylation would lead to loss of association of methylated CpG binding proteins and sets the stage for association of transcription factors to induce or repress MT gene transcription. In this scenario, regulation of MT gene transcription would be a layered process where changes in methylation status and chromatin structure would be prerequisites to allow the action of transcription factor mediation of transcription.

6.3.1.8 Post-Transcriptional Regulation

The preponderance of evidence indicates that MTs are transcriptionally regulated, but post-transcriptional events may also mediate expression under certain circumstances. Transcription of MT genes and MT protein levels in mice in response to treatment with inorganic mercury varies between organs. In liver, the rate of MT gene transcription and MT protein levels are closely correlated with each other and the content of mercury. In kidney, however, those correlations were weak or absent. Overall, it appears that hepatic levels of MT protein in response to mercury induction are determined primarily by MT gene transcription, but post-transcriptional events are important in determining the renal content of MT protein after mercury treatment [108]. Furthermore, while transcription of human MT-1 and MT-2 genes, and mRNA accumulation, are similar after cadmium or zinc induction of human cells *in vitro*, the stability of MT mRNA differs considerably [109]. In rats treated with cadmium or copper, MT-1 mRNA was recruited into translational polyribosomes in parallel with the increase in MT-1 gene transcription; however, MT-2 mRNA was recruited into polyribosomes at a rate lower than the increase in MT-2 mRNA [110]. This suggests that MT mRNA translation is regulated and that post-transcriptional events can determine MT protein levels under certain circumstances. Although post-transcriptional events may mediate MT levels to some degree, their nature remains largely unknown and requires further investigation.

6.3.1.9 Developmental and Tissue-Specific MT Expression

MT transcription and protein accumulation are regulated according to differentiation state, and cell and tissue type. Cell-specific activation of MT genes occurs during mammalian embryo development [111], and MT genes are constitutively and coordinately expressed in embryos prior to implantation. MT-1 genes in early mouse embryos prior to implantation are not responsive to metal induction and become so only at later stages [37]. MT is expressed at high levels in endoderm of yolk sac early in development and, when metal responsiveness develops, is regulated by MTF-1 and MLTF/USF I response to zinc [111]. Before birth, MT mRNA and protein levels in many tissues are higher than in adults, including lung, spleen, pancreas, kidney, heart, and testes, although the primary site of MT synthesis is in prenatal and perinatal liver [112]. There is a rapid decline in tissue MT levels following birth, falling to adult levels within two weeks *postpartum*. Maternal liver, kidney, and duodenum MT levels, however, remain elevated during lactation in the absence of external induction [113].

MTs are also differentially expressed in different cell types of varying physiological function. They are highly expressed during growth and proliferation in a human prostate tumor cell line [28], in rat kidney undergoing compensatory hypertrophy [32], and in cells on the proliferating edge of epithelial cells *in situ* [114]. The intracellular location of MTs depends on the stage of tissue development, and MTs are located in the nuclei of parenchymal cells in fetal and newborn liver but primarily in cytoplasm in adult liver [33]. Mouse MT isoforms change in skin layers during differentiation. As inner layer keratinocytes in stratified epithelium differentiate into outer stratum spinosum and switch keratin isoform production, they also switch from MT-1 to MT-4 gene transcription and mRNA accumulation [27]. MTs are therefore associated with proliferation and hypertrophy, differentiation, and activation. The complete range of molecular signaling events responsible for MT gene expression associated with proliferation and differentiation is not clearly delineated, but the broad range of cells with different proliferation and replication status, and MT expression, provides useful models for exploration.

6.4 MT FUNCTION

6.4.1 MTs as Zinc Buffers

Interestingly, apometallothioneins with reduced thiolate residues capable of binding zinc or other metals, but with at least some available thiolates remaining unassociated with those metals (apoMT in the form of "reduced thionein" or T_R), have been reported *in vivo* and in tumor cell lines [115,116] and have been shown to be metal ion donors and acceptors that regulate the *in vitro* activity of zinc-dependent proteins [117–119]. The donor/acceptor capacity of MTs has been suggested to be active in living cells and implicates MTs as proteins that link zinc availability and redox metabolism. In addition to T_R, oxidized apoMT (thionin or T_O) is generated and zinc is released when MTs, when reacted with oxidants, present cells and tissues undergoing oxidative stress. The dynamic interplay between MT proteins associated with zinc; MT proteins unassociated with zinc due to oxidation of zinc-bearing MTs (T_O); MT proteins unassociated with zinc but capable of sequestering zinc ions (T_R); and zinc unassociated with MT protein ("free" or "available" zinc) has been suggested to constitute a zinc buffering system that homeostatically regulates zinc availability to zinc-requiring and zinc-sensing proteins [120]. The regulation of intracellular free/available zinc to within a relatively small range may be important in maintaining normal physiological function and avoiding pathological conditions, and MTs may be central to that process.

There is evidence supporting the concept that MT (associated with metals and conceivably in the form of T_O and T_R) can mediate zinc availability to zinc-binding cellular macromolecules. DNA metabolizing proteins and signal transduction molecules require zinc [121]. GRs and ERs (estrogen receptors), respectively require zinc for stability and activity [122]. Although MTs have not been shown to be required for hormone receptor function *in vivo*, MT genes are induced by glucocorticoids and their analogs [89,123]. The cellular level of Zn–MT is correlated with GR activity in rodent cells [20] and MTs donate and sequester zinc to and from ER *in vitro* [22].

With respect to MT structure and its capacity to mediate zinc transfer, MT-1 and MT-2 bind seven zinc atoms and the majority are deep within α and β domains and unavailable for exchange with other molecules without disruption of zinc–thiol coordinate covalent bonds. However, two zinc-binding sites allow relatively simple transfer without destruction of MT protein [124]. Evidence suggesting that MTs with an incomplete complement of zinc (apoMT, or T_R) may sequester zinc when it is scarce, lowering its availability to other proteins—although moderate overexpression of MT-2 in a human breast cancer-derived cell line (MCF-7) enhanced transcriptional activity of the zinc-dependent transcription factor p53, expression at higher levels suppressed p53 activity [125]. Cultured mouse fibroblast cells selected for survival under low zinc conditions have high MT expression, but transcription of a transfected reporter gene construct responsive to activation by MTF-1 was suppressed, consistent with the concept that MT sequestered scarce zinc ions away from zinc-requiring/sensing MTF-1 [126]. MT with an incomplete complement of zinc has also been reported to selectively activate enzymes that are inhibited by zinc association, implying that zinc sequestration by MTs can also enhance activity of target proteins [117].

In addition to the capacity of glucocorticoids to induce MTs (as mentioned above), the role of MTs in regulating the activity of the zinc-dependent GR itself has been investigated. *In vitro* treatment of mouse mammary tumor 2305 cells harboring a GR-responsive reporter gene with zinc, mercury, or heat (to induce heat shock) enhanced dex-induced reporter transcription. This vector responds to the artificial glucocorticoid dex by translocating to the cell nucleus to mediate target gene transcription. The level of enhancement correlated, not with the absolute level of MT protein, but with the amount of MT associated with zinc [20,127]. For example, treatment of 2305 cells with cadmium or copper increased CdMT and CuMT but had no effect on dex-induced reporter gene transcription. Heat shock did not increase absolute levels of MT protein, but increased both the amount of MT associated with zinc and dex responsiveness of the reporter. Together, these data strengthen the concept that MT mediates GR activity by regulating zinc.

These reports are consistent with the hypothesis that, as zinc levels decrease, zinc-dependent transcription factor activity may become increasingly dependent on zinc-associated MTs capable of transferring metal ions to those transcription factors. Conversely, zinc-activated protein function may decrease under conditions where apoMT is present and free zinc is low.

A critical question with respect to zinc availability to cellular macromolecules, and a potential role for MT as a donor/acceptor of zinc, is with respect to the direct availability of zinc to cellular macromolecules without the need for intermediate "chaperone" donors/acceptors. For such intermediates to be needed, the availability of zinc in cells must necessarily be limited. If there were a "free" zinc pool readily available for all proteins requiring them, strategies to control access to zinc ions would be circumvented. However, virtually all cellular zinc is associated with macromolecules and "free" zinc is consequently limited [121]. Although the large number of cellular and extracellular species associated with zinc exhibit a broad range of affinities for zinc, where the supply of dietary zinc is limited and suboptimal, zinc would presumably be preferentially associated with species with the highest affinity. In that scenario, storage of zinc ions in high-affinity molecules, coupled with direct or indirect transfer from those species to zinc-requiring (and possibly zinc-regulated) molecules, would be a strategy to homeostatically regulate zinc availability, particularly where zinc supplies are limited. Chaperone-mediated regulation of other metals has been reported, suggesting that a similar regulatory strategy could apply to zinc. Copper, for example, is regulated by transporters and chaperones first described in yeast, but which imply analogous metal ion regulation in higher eukaryotes including humans [128]. MTs (and thioneins partially or completely unassociated with zinc) may be intracellular zinc chaperones capable of regulating zinc availability to proteins that mediate physiological events and cellular response to stress.

6.5 MT IN HEALTH AND DISEASE

6.5.1 MT in Immune Function

Zinc and other metal ions (selenium, copper, and iron among them) are leading elements in maintenance of immune function in mammals [129–131]. Subclinical and overt zinc deficiency reduces the functions of macrophages (including phagocytosis and capacity to kill target cells), neutrophils (chemotaxis and respiratory burst), NK cells, and complement [129]. Zinc deficiency in patients with acrodermatitis enterohepatica (mediated by a genetic defect resulting in abnormal zinc transport) is associated with atrophy of the thymus, poor lymphocyte proliferation in response to mitogens, poor thymulin activity, altered CD4 + /CD8+ T cell ratios, and decreased NK and monocyte activity, all of which are ameliorated by zinc supplementation [132]. Zinc is required for proteins mediating activation of monocyte/macrophages [133]. Activation of monocytes by a variety signaling molecules proceeds through a G-protein pathway, with phosphorylation of protein kinases leading to cytokine production and arachidonate release. Protein kinase C (PKC) and inositol 1,4,5-triphosphate (IP_3) are activated, and active PKC leads to degradation of IκBα and release of NFκB. Released NFκB translocates to the nucleus to control transcription of multiple activation genes, including IL-1α, IL-1B, and TNFα [134,135]. Evidence exists implicating MTs in innate immunity (including monocyte/macrophage and other innate immune activities). Mice with genetically ablated MT-1 and MT-2 genes (MT knockout mice) are protected against TNF-induced systemic inflammatory response syndrome (SIRS) and MT-1-overexpressing mice are sensitized to the lethal effects of TNF [136,137], consistent with the hypothesis that MTs may mediate zinc-requiring proteins required for monocyte activation.

MT is expressed at a high level in primary human monocytes and is increased in a human monocytic leukemia cell line (THP-1) after activation [41]; and administration of antisense RNA to suppress basal, cadmium, and LPS-induced MT in THP-1 cells abolishes subsequent activation by LPS and GM-CSF without reducing viability [40]. Loss of activation potential as a result of MT downregulation suggests a role for MT in monocyte activation. In addition, it has been proposed that increased MT in lymphocytes from aged human subjects and Down syndrome patients may be responsible for decreased immune cell function associated with age (immunosenescence), because of sequestration of zinc in an unavailable form in MT through mechanisms that have not yet been described [138].

MT-1/MT-2 knockout (MT KO) mice provide a model to study the role of MT in immune cells. MT KO mice are phenotypically normal under nonstress conditions, but are sensitive to cadmium and oxidant insult and have reduced zinc uptake, suggesting that zinc homeostasis is disturbed [139]. Innate immune function (monocyte/macrophage activation) is compromised in MT KO mice, consistent with reduced sensitivity to SIRS [39], and MT KO mouse peritoneal macrophages have been reported to have reduced response to activation [140] due to a lack of MT interaction with NFκB [38]. Characterization of the effects of MT on innate immune cell function, with an emphasis on a potential role in zinc homeostasis, will be important areas for future investigation of MT.

6.5.2 MT and Cancer

In humans, MTs (MT-1 and MT-2) have been reported to have aberrant expression in a number of human tumors. For those arising in breast, kidney, lung, salivary gland, ovary, testes, leukemia, urinary bladder, nasopharyngeal tissue, and non-Hodgkin lymphoma, an association with increased MT levels (compared to nontumor surrounding tissue) has been reported [141–143]. In view of increased MT levels in rapidly proliferating cells (see Section 6.2) and the association between MT and resistance to toxicity-induced cell death [22,57,58], these data have been suggested to support a role for MTs in tumor cell proliferation and resistance to cytotoxicity induced by radiation or chemotherapeutic drugs. On the other hand, decreased levels of MT-1 and MT-2 in human liver, prostate, thyroid, central nervous system, and testicular tumors have been linked to increased malignancy and poorer prognosis [66,141,144–149] and, in acute lymphoblastic leukemia, lack of change in MT level associated with transformation has been reported [149]. The overall role of MTs in tumors, therefore, is not completely clear.

MTs may contribute to tumor cell treatment resistance by directly sequestering chemotherapeutic drugs to prevent their cytotoxicity. For example, platinum-containing *cis*-diamminedichloroplatin(II) (CDDP or cisplatin) and its metabolites can bind to MT-1 or MT-2 [141,146,150], thereby preventing CDDP association with DNA and contribution to production of reactive oxygen intermediates, and resulting cell death. Indirect binding of reactive oxygen species to MTs [151] may also reduce chemotherapy-induced oxidative stress and tumor cell death. Direct evidence supporting a role for MTs in tumor cell resistance to therapy has been reported, at least in mice: MT knockout mice and cell lines derived from them are sensitive to death induced by cisplatin, bleomycin, melphalan, ara-c, and cytarabin [143,152]. In human tumors, an inverse relationship between MT-1 and MT-2 levels and sensitivity to chemotherapeutic agents (including carboplatin, etoposide, doxorubicin, methotrexate, and adriamycin) and to radiotherapy-induced damage has been reported [146,147,153–155]. Overall, targeted reduction in MT level is associated with increased susceptibility to toxins, and increased MT, as a single factor, is correlated with resistance to damage. In spite of the potential role that MTs might play in mediating resistance to treatment, MT levels in human tumors are not a reliable clinical prognostic tool [142]. For example, higher (not lower) MT protein level correlated with improved response to cisplatin treatment in testicular germ cell tumors [144]. In addition, *in vitro* analysis of cisplatin sensitivity in a panel of cells from human pancreatic tumors did not reveal any correlation between MT level and drug resistance [156], and other factors (including glutathione-*S*-transferase-II, *c*-glutamylcysteine synthetase, and topoisomerase-I and II) may be more important than MTs in mediating cisplatin resistance [142]. In fact, cellular characteristics other than differences in MT levels may be paramount in mediating resistance to a variety of toxic agents in mammalian cells. MT knockout cells are, paradoxically, less sensitive to toxicity mediated by zinc, cadmium, or cisplatin than wild-type cells [157], although zinc induction of MT in wild-type cells, but not in MT knockout cells, increased the resistance to these agents (indicating an increasing role for higher level, induced MTs, but not low-level, basal MT content).

The potential for MTs to mediate zinc availability to and from proteins that are important in malignancy has been supported by several lines of evidence. For example, mutations in the p53 gene leading to inactivation of zinc-containing p53 protein have been reported in a majority of human tumors [158]. There is a strong association between increased MT and mutated p53 in human tumors [159] and MT knockout cells have higher levels of functional p53 than wild-type cells [152]. ApoMT (T_R) has been reported to remove zinc from p53 and inactivate it, and direct interaction of apoMT (and not zinc-containing MT) with p53 has been reported in human breast tumor cells [160]. These data support the possibility that MTs may reduce p53 function by sequestering zinc, leading to reduced p53-mediated transcription of genes mediating reduced malignancy and apoptosis. MTs may be oncogenic factors in this scenario. On the other hand, there is no obvious correlation between MT protein levels and p53 levels in malignant carcinoma [161] and, in p53-positive cells, MTs are inducible to higher levels than in p53-negative cells [162]. Overall, it is likely that the potential for MTs to mediate zinc availability to multiple proteins in addition to p53, which mediate a broad range of functions (e.g., glucocorticoid receptor and $NF_{\kappa}B$ subunits), the possibility (still unresolved) for zinc-containing MT to increase (not decrease) p53 activity, and the potential for differences in levels of MT isoforms and subcellular location to introduce variation into the ultimate physiological effects of MTs in tumor cells lead either to decreased or increased malignancy and is not clearly predictable given current information.

6.5.3 MT and Cardiac Toxicity

Cardiac tissue is particularly sensitive to oxidative stress due to low antioxidant activity in that organ. Rats fed a copper-deficient diet to induce oxidative stress-induced lipid peroxidation selectively in liver, without altered peroxidation in liver, indicating organ-selective heart damage. In addition, the basal activities of enzymes protective against active oxygen (superoxide dismutase, catalase, and glutathione peroxidase) were lower in heart than in liver, suggesting that a weak antioxidant defense system is responsible for the relatively high degree of oxidative damage in heart, at least by copper deficiency [163]. Anthracyclines (daunorubicin, doxorubicin, epirubicin, and others) are widely used in the treatment of a variety of human tumors, including leukemia, lymphomas, and breast, uterine, ovarian, and lung cancers (including pediatric tumors), but irreversible cardiotoxicity in patients treated with them has limited their effectiveness [164].

Transgenic mice overexpressing human MT-2A [165] or catalase [166] in heart tissue, both of which mediate antioxidant protection, have been generated. Transgenic MT or catalase expression was targeted to the heart and localized exclusively in cardiomyocytes [167]. Both suppressed acute and chronic anthracycline (doxorubicin)-induced cardiotoxicity, as evidenced by fewer pathological changes in histology, ultrastructure, and function [165,166,168,169]. Although transgenic MT had a direct antioxidant effect, the authors suggest that MT regulation of zinc availability to mitochondrial aconitase (an enzyme with reduced activity under conditions of oxidative stress) and post-translational modification of CCO-Va (the cytochrome *c* oxidase subunit Va, decreased under oxidation conditions) by MT may also play a role. Overall, the induction and expression of MTs in the heart may be important in reducing cardiac toxicity induced by therapeutic agents, and possibly under other circumstances (infection, exposure to environmental toxins, diabetes, and aging).

6.5.3.1 MT, Diabetes, and Cardiac Toxicity

Cardiovascular disease, and particularly congestive heart failure, poses a significantly higher risk for patients with type 2 diabetes mellitus than for those without the disease [170]. Hyperglycemia in diabetic patients is associated with myocardial changes characteristic of diabetic cardiomyopathy and heart failure. In studies with diabetic mice (induced by streptozotocin) exposed to high levels of glucose, hyperglycemia directly induced apoptotic cell death in the myocardium *in vivo*. The apoptosis was mediated, in part, by cytochrome *c*-activated caspase-3, potentially triggered by oxidative stress concomitant with high glucose exposure [171]. Increased oxidative stress results in increased lipid peroxidation, F2-isoprostanes, nitrotyrosine, and DNA damage and (in diabetes) release of superoxide anions. Supplementation with antioxidants (α-tocopherol, α-lipoate, and ascorbic acid) has been shown to be beneficial in diabetic patients and antioxidant therapy may be a productive approach to treatment. Nevertheless, experimental approaches have been inconsistent due to difficulty in maintaining therapeutically effective antioxidant levels, both in circulation and in tissues at risk for oxidative damage, and sufficiently effective, nontoxic exogenous antioxidants [172,173]. Endogenous antioxidants (including MTs) have been proposed as appropriate agents to reduce oxidative damage in heart. MT has been shown to reduce diabetic cardiomyopathy (manifested by protein nitration, lipid peroxidation, and increased mitochondrial mixed glutathione in conjunction with decreased mitochondrial reduced glutathione) in streptozotocin-induced type 1 diabetes in wild-type mice, but not in transgenic mice expressing high levels of human MT2A in cardiac tissue [174]. Furthermore, treatment with zinc salts that induced increased MT in the cardiac tissue of wild-type mice ameliorated cardiomyopathy (morphological abnormalities, fibrosis, LV systolic and diastolic dysfunction, and catecholamine desensitization) [175]. The protective effect of MT was applicable to other mouse diabetic models. Diabetic mice (OVE26, a transgenic mouse developing severe early-onset type 1 diabetes) that were mated with transgenic mice specifically expressing elevated MT in the heart generated diabetic offspring with elevated cardiac MT (OVE26MT mice). OVE26MT mice exhibited cardiomyopathy, reduced morphological abnormalities, and reduced impairment of ischemic contractility exhibited in diabetic mouse hearts [176,177], suggesting a protective effect of MT on heart tissue in this diabetic model.

In addition to cardiac dysfunction, renal pathologies are also associated with diabetes and are a significant factor in morbidity and mortality. OVE26 transgenic mice that were mated with Nmt transgenic mice specifically expressing MT in podocytes (critical cells integrated into the selective filtration barrier of the glomerulus and susceptible to oxidative damage) resulted in double-transgenic offspring (OVE26Nmt) that overexpressed MT specifically in podocytes under conditions modeling human diabetes. OVE26Nmt mice developed diabetes similar to OVE26 mice, but the enhanced expression of MT in renal cells reduced podocyte damage. OVE26Nmt mice had more podocytes, less glomerular cell death, and a higher density of podocyte foot processes than OVE26 mice without enhanced podocyte MT levels. Moreover, expansion of glomerular and mesangial volume was significantly less in OVE26Nmt mice than in OVE26 animals. Thus, MT can protect glomerular podocytes in diabetic mice and reduce or delay diabetic nephropathy [178]. The observation that both constitutive high-level MT in mouse cardiac and kidney tissue, and zinc-induced MT in heart, attenuates impaired cardiac and kidney function in two models of diabetes suggests that enhanced MT expression (likely by nontoxic treatment with essential metals such as zinc) may be an area for therapeutic development in diabetes, and further work in this important area is warranted.

6.5.4 MT and Neurological Function

MTs are closely associated with astrocyte response to central nervous system (CNS) injury. The most abundant MT isoforms in brain (and primarily in astrocytes, with possible trace amounts in some neurons) are MT-1 and MT-2 [179]. MT-3 is found almost exclusively in the brain, with much lower levels in other tissues, and MT-4 is not found in the brain at all. Both MT-1 and MT-2 are correlated with reactive astrogliosis [180]: a state entered into by resident astrocytes after neuronal damage (induced by stroke, physical trauma, and neurodegenerative disease) involving cellular hypertrophy, proliferation, and alterations in gene expression [181]. Both MT-1 and MT-2 are also increased in neurodegenerative diseases [179,180].

MTs are clearly involved in protecting against CNS injury. MT-1/MT-2 knockout mice recover poorly from cortical cryolesion wounding, have elevated levels of neuronal apoptosis [182], and have impaired response to a wide range of other experimental CNS damage, including kainic acid lesions [183] and focal ischemia [184]. Transgenic mice overexpressing MT-1 (TgMT1*) respond better to cortical cryolesion, with reduced oxidative stress and neuronal apoptosis [185]. Remarkably, exogenous administration of MT-2A (injected intraperitoneally into both wild-type and MT-1/MT-2 knockout mice) had a similar effect. Direct application of recombinant human MT-2A in the brain after focal cortical lesion to the adult rat neocortex similarly enhanced regeneration of axons [186]. Thus, in spite of relatively little evidence that endogenous MTs are localized extracellularly (except under conditions where dying cells may release MTs bound to toxic metals such as cadmium [187]), MT-1 and/or MT-2 applied outside cells can have beneficial effects similar to those exerted by endogenous, intracellular MT. There is growing evidence to support a role for extracellular MT. For example, exogenous MT-1 and MT-2 applied *in vitro* to dopaminergic and hippocampal neurons [188] and cortical neurons [186] promote neurite outgrowth. Retinal ganglion cells are similarly affected [189] and MT-1/MT-2 protect dopaminergic neurons from the toxic effects of 6-hydroxydopamine, and hippocampal neurons from β amyloid peptide-induced neurotoxicity [188]. Although extracellular MT could, conceivably, mediate events leading to increased *de novo* production of endogenous MT to exert such effects, an obvious possibility is that extracellular MT is capable of acting as a ligand for cell surface molecules (MT "receptors") to carry out protective functions.

In fact, MT-1 and MT-2 have been reported to interact specifically with members of the low-density lipoprotein (LDL) family of receptors in neurons, specifically LRP1 and LRP2 (megalin) [190], to activate mitogen-activated protein kinase (MAPK) and PI3 kinase/Akt signaling. Both these pathways are well recognized as mediators of neuronal survival and neurite outgrowth in response to cyclic adenosine monophosphate (cAMP) response element binding protein (CREB) [191]. CREB signaling is also important in overriding the inhibition of neuron regeneration induced by myelin-associated glycoprotein (MAG) and myelin [192], and MT treatment boosts axon regeneration in injured adult cortex where MAG levels are elevated [186]. Taken together, these results raise the possibility that MT might induce axonal regeneration by activating CREB to overcome MAG inhibition by activating CREB [193,194].

Cellular uptake of MT associated with LDL receptors in the brain is an unresolved possibility, although it is well established that LDL receptors participate in classical endocytosis of extracellular ligands. Specifically, LRP2 can mediate uptake of cadmium-MT in the kidney [195], although it is not clear whether a similar uptake of zinc-MT by LRP2 occurs in neurons. Considering the neuroactive roles of zinc (a signaling ion released by neural activity at excitatory synapses, and a modulator of the neuronal death associated with transient global ischemia and sustained seizures [196]), it is an attractive hypothesis that LRP2 acts to increase intracellular zinc levels by mediating uptake of extracellular zinc-MT. A model incorporating this concept has been proposed [197] in which MT-1 and MT-2 induced in astrocytes by stress or inflammation would be actively or passively released by those cells, taken up by LDL receptors on neurons, which, in turn, activate PI3 kinase/Akt, MAPK, and CREB to enhance neuronal survival and neurite outgrowth [193]. Overall, the role of MTs (and, in particular, extracellular MTs) appears to have potential as therapeutic proteins in CNS damage and recovery. In particular, peptides modeled on motifs recognized in linker domains as responsible for MT binding to megalin (so-called Emtin peptides, C19 KCK [190] or S_{24}CKKSCC [198]) may be a major step in identifying MT-based therapeutic molecules to treat CNS injury and neurodegenerative diseases.

6.6 SUMMARY AND PERSPECTIVE

Although many metal ions, including zinc, have toxic effects that diminish cell viability, they are also essential for hundreds of cellular macromolecules. These macromolecules carry out a myriad of metal-associated functions, including sensing the presence of metals to homeostatically regulate them for functions essential for life or for appropriate responses to external stimuli, resistance to damage, or differentiation. MTs are candidates as modulators of the function of metal-dependent and metal-responsive transcription factors, enzymes, and structural proteins through regulation of the availability of zinc, especially under conditions of restricted access to zinc because of low dietary or extracellular zinc levels, or binding of zinc to molecules of relatively high affinity. Although the viability and apparent normality of MT knockout mice under low stress conditions indicates that not all metal-requiring proteins are obliged to be regulated by MT under all circumstances, the impaired response of MT knockout cells and animals under stress conditions (including stresses imposed in models of chronic human diseases including diabetes, cardiac deficiencies induced by stress, cancer, and CNS damage trauma-associated or not) and improved response when MTs are endogenously upregulated (or, remarkably, applied extracellularly) suggest protective roles for MTs in circumstances of particular importance in the health of humans. The potential for MTs to be therapeutic agents is emerging as an exciting area for development.

ACKNOWLEDGMENT

Support for this review came, in part, from a grant to J.K. from the Canadian Institutes for Health Research (MOP 82720).

REFERENCES

1. Goyer, R. A. 1997. Toxic and essential metal interactions. *Annu Rev Nutr* **17**, 37–50.
2. Brewer, G. J. 2008. The risks of free copper in the body and the development of useful anticopper drugs. *Curr Opin Clin Nutr Metab Care* **11**, 727–732.
3. Andreini, C., Banci, L., Bertini, I., and Rosato, A. 2006. Zinc through the three domains of life. *J Proteome Res* **5**, 3173–3178.
4. Blasie, C. A. and Berg, J. M. 2002. Structure-based thermodynamic analysis of a coupled metal binding-protein folding reaction involving a zinc finger peptide. *Biochemistry* **41**, 15068–15073.
5. Laity, J. H., Lee, B. M., and Wright, P. E. 2001. Zinc finger proteins: New insights into structural and functional diversity. *Curr Opin Struct Biol* **11**, 39–46.
6. Laity, J. H. and Andrews, G. K. 2007. Understanding the mechanisms of zinc-sensing by metal-response element binding transcription factor-1 (MTF-1). *Arch Biochem Biophys* **463**, 201–210.
7. Hambidge, M. 2000. Human zinc deficiency. *J Nutr* **130**, 1344S–1349S.
8. Prasad, A. S. 2003. Zinc deficiency. *BMJ* **326**, 409–410.
9. Hambidge, K. M. and Krebs, N. F. 2003. Zinc, low birth weight, and breastfeeding. *Pediatrics* **112**, 1419–1420.
10. Fong, L. Y., Jiang, Y., and Farber, J. L. 2006. Zinc deficiency potentiates induction and progression of lingual and esophageal tumors in p53-deficient mice. *Carcinogenesis* **27**, 1489–1496.
11. Frederickson, C. J., Koh, J. Y., and Bush, A. I. 2005. The neurobiology of zinc in health and disease. *Nat Rev Neurosci* **6**, 449–462.
12. Sandstead, H. H., Frederickson, C. J., and Penland, J. G. 2000. History of zinc as related to brain function. *J Nutr* **130**, 496S–502S.
13. Black, R. E. 2003. Zinc deficiency, infectious disease and mortality in the developing world. *J Nutr* **133**, 1485S–1489S.
14. Lichtlen, P. and Schaffner, W. 2001. The "metal transcription factor" MTF-1: Biological facts and medical implications. *Swiss Med Wkly* **131**, 647–652.
15. Wellenreuther, G., Cianci, M., Tucoulou, R., Meyer-Klaucke, W., and Haase, H. 2009. The ligand environment of zinc stored in vesicles. *Biochem Biophys Res Commun* **380**, 198–203.
16. Kury, S., Dreno, B., Bezieau, S., Giraudet, S., Kharfi, M., Kamoun, R., and Moisan, J. P. 2002. Identification of SLC39A4, a gene involved in acrodermatitis enteropathica. *Nat Genet* **31**, 239–240.
17. Wang, K., Zhou, B., Kuo, Y. M., Zemansky, J., and Gitschier, J. 2002. A novel member of a zinc transporter family is defective in acrodermatitis enteropathica. *Am J Hum Genet* **71**, 66–73.
18. Chowanadisai, W., Lonnerdal, B., and Kelleher, S. L. 2006. Identification of a mutation in SLC30A2 (ZnT-2) in women with low milk zinc concentration that results in transient neonatal zinc deficiency. *J Biol Chem* **281**, 39699–39707.
19. Cousins, R. J., Liuzzi, J. P., and Lichten, L. A. 2006. Mammalian zinc transport, trafficking, and signals. *J Biol Chem* **281**, 24085–24089.
20. DeMoor, J. M., Kennette, W. A., Collins, O. M., and Koropatnick, J. 2001. Zinc-metallothionein levels are correlated with enhanced glucocorticoid responsiveness in mouse cells exposed to ZnCl(2), HgCl(2), and heat shock. *Toxicol Sci* **64**, 67–76.
21. Miles, A. T., Hawksworth, G. M., Beattie, J. H., and Rodilla, V. 2000. Induction, regulation, degradation, and biological significance of mammalian metallothioneins. *Crit Rev Biochem Mol Biol* **35**, 35–70.
22. Haq, F., Mahoney, M., and Koropatnick, J. 2003. Signaling events for metallothionein induction. *Mutat Res* **533**, 211–226.
23. Uchida, Y., Takio, K., Titani, K., Ihara, Y., and Tomonaga, M. 1991. The growth inhibitory factor that is deficient in the Alzheimer's disease brain is a 68 amino acid metallothionein-like protein. *Neuron* **7**, 337–347.
24. Palmiter, R. D., Findley, S. D., Whitmore, T. E., and Durnam, D. M. 1992. MT-III, a brain-specific member of the metallothionein gene family. *Proc Natl Acad Sci USA* **89**, 6333–6337.
25. Ebadi, M., Iversen, P. L., Hao, R., Cerutis, D. R., Rojas, P., Happe, H. K., Murrin, L. C., and Pfeiffer, R. F. 1995. Expression and regulation of brain metallothionein. *Neurochem Int* **27**, 1–22.
26. Meloni, G., Zovo, K., Kazantseva, J., Palumaa, P., and Vasak, M. 2006. Organization and assembly of metal-thiolate clusters in epithelium-specific metallothionein-4. *J Biol Chem* **281**, 14588–14595.
27. Quaife, C. J., Findley, S. D., Erickson, J. C., Froelick, G. J., Kelly, E. J., Zambrowicz, B. P., and Palmiter, R. D. 1994. Induction of a new metallothionein isoform (MT-IV) occurs during differentiation of stratified squamous epithelia. *Biochemistry* **33**, 7250–7259.

28. Koropatnick, J., Kloth, D. M., Kadhim, S., Chin, J. L., and Cherian, M. G. 1995. Metallothionein expression and resistance to cisplatin in a human germ cell tumor cell line. *J Pharmacol Exp Ther* **275**, 1681–1687.
29. Nagel, W. W. and Vallee, B. L. 1995. Cell cycle regulation of metallothionein in human colonic cancer cells. *Proc Natl Acad Sci USA* **92**, 579–583.
30. Cherian, M. G. and Apostolova, M. D. 2000. Nuclear localization of metallothionein during cell proliferation and differentiation. *Cell Mol Biol (Noisy-le-grand)* **46**, 347–356.
31. Cherian, M. G. and Kang, Y. J. 2006. Metallothionein and liver cell regeneration. *Exp Biol Med (Maywood)* **231**, 138–144.
32. Zalups, R. K., Fraser, J., and Koropatnick, J. 1995. Enhanced transcription of metallothionein genes in rat kidney: Effect of uninephrectomy and compensatory renal growth. *Am J Physiol* **268**, F643–650.
33. Cherian, M. G. 1994. The significance of the nuclear and cytoplasmic localization of metallothionein in human liver and tumor cells. *Environ Health Perspect* **102**(Suppl. 3), 131–135.
34. Koropatnick, J. and Duerksen, J. D. 1987. Nuclease sensitivity of alpha-fetoprotein, metallothionein-1, and immunoglobulin gene sequences in mouse during development. *Dev Biol* **122**, 1–10.
35. Nartey, N. O., Banerjee, D., and Cherian, M. G. 1987. Immunohistochemical localization of metallothionein in cell nucleus and cytoplasm of fetal human liver and kidney and its changes during development. *Pathology* **19**, 233–238.
36. Panemangalore, M., Banerjee, D., Onosaka, S., and Cherian, M. G. 1983. Changes in the intracellular accumulation and distribution of metallothionein in rat liver and kidney during postnatal development. *Dev Biol* **97**, 95–102.
37. Andrews, G. K., Huet-Hudson, Y. M., Paria, B. C., McMaster, M. T., De, S. K., and Dey, S. K. 1991. Metallothionein gene expression and metal regulation during preimplantation mouse embryo development (MT mRNA during early development). *Dev Biol* **145**, 13–27.
38. Kanekiyo, M., Itoh, N., Kawasaki, A., Matsuyama, A., Matsuda, K., Nakanishi, T., and Tanaka, K. 2002. Metallothionein modulates lipopolysaccharide-stimulated tumour necrosis factor expression in mouse peritoneal macrophages. *Biochem J* **361**, 363–369.
39. Koropatnick, J. and Zalups, R. K. 1997. Effect of non-toxic mercury, zinc or cadmium pretreatment on the capacity of human monocytes to undergo lipopolysaccharide-induced activation. *Br J Pharmacol* **120**, 797–806.
40. Leibbrandt, M. E., Khokha, R., and Koropatnick, J. 1994. Antisense down-regulation of metallothionein in a human monocytic cell line alters adherence, invasion, and the respiratory burst. *Cell Growth Differ* **5**, 17–25.
41. Leibbrandt, M. E. and Koropatnick, J. 1994. Activation of human monocytes with lipopolysaccharide induces metallothionein expression and is diminished by zinc. *Toxicol Appl Pharmacol* **124**, 72–81.
42. Sugihara, T., Wadhwa, R., Kaul, S. C., and Mitsui, Y. 1999. A novel testis-specific metallothionein-like protein, tesmin, is an early marker of male germ cell differentiation. *Genomics* **57**, 130–136.
43. Matsuura, T., Kawasaki, Y., Miwa, K., Sutou, S., Ohinata, Y., Yoshida, F., and Mitsui, Y. 2002. Germ cell-specific nucleocytoplasmic shuttling protein, tesmin, responsive to heavy metal stress in mouse testes. *J Inorg Biochem* **88**, 183–191.
44. Samson, S. L. and Gedamu, L. 1998. Molecular analyses of metallothionein gene regulation. *Prog Nucleic Acid Res Mol Biol* **59**, 257–288.
45. Ghoshal, K. and Jacob, S. T. 2001. Regulation of metallothionein gene expression. *Prog Nucleic Acid Res Mol Biol* **66**, 357–384.
46. Aniskovitch, L. P. and Jacob, S. T. 1998. Distinct rat proteins can recognize CCAAT-homologous sequences of the metallothionein promoter and trans-activate this promoter. *Oncogene* **16**, 1475–1486.
47. Dalton, T. P., Li, Q., Bittel, D., Liang, L., and Andrews, G. K. 1996. Oxidative stress activates metal-responsive transcription factor-1 binding activity. Occupancy *in vivo* of metal response elements in the metallothionein-I gene promoter. *J Biol Chem* **271**, 26233–26241.
48. Li, Q., Hu, N., Daggett, M. A., Chu, W. A., Bittel, D., Johnson, J. A., and Andrews, G. K. 1998. Participation of upstream stimulator factor (USF) in cadmium-induction of the mouse metallothionein-I gene. *Nucleic Acids Res* **26**, 5182–5189.
49. Dalton, T., Palmiter, R. D., and Andrews, G. K. 1994. Transcriptional induction of the mouse metallothionein-I gene in hydrogen peroxide-treated Hepa cells involves a composite major late transcription factor/antioxidant response element and metal response promoter elements. *Nucleic Acids Res* **22**, 5016–5023.

50. Nguyen, T., Sherratt, P. J., and Pickett, C. B. 2003. Regulatory mechanisms controlling gene expression mediated by the antioxidant response element. *Annu Rev Pharmacol Toxicol* **43**, 233–260.
51. Murata, M., Gong, P., Suzuki, K., and Koizumi, S. 1999. Differential metal response and regulation of human heavy metal-inducible genes. *J Cell Physiol* **180**, 105–113.
52. Palmiter, R. D. 1998. The elusive function of metallothioneins. *Proc Natl Acad Sci USA* **95**, 8428–8430.
53. Gronostajski, R. M. 2000. Roles of the NFI/CTF gene family in transcription and development. *Gene* **249**, 31–45.
54. Koizumi, S., Suzuki, K., Ogra, Y., Yamada, H., and Otsuka, F. 1999. Transcriptional activity and regulatory protein binding of metal-responsive elements of the human metallothionein-IIA gene. *Eur J Biochem* **259**, 635–642.
55. Westin, G. and Schaffner, W. 1988. A zinc-responsive factor interacts with a metal-regulated enhancer element (MRE) of the mouse metallothionein-I gene. *EMBO J* **7**, 3763–3770.
56. Andrews, G. K. 2000. Regulation of metallothionein gene expression by oxidative stress and metal ions. *Biochem Pharmacol* **59**, 95–104.
57. Heuchel, R., Radtke, F., Georgiev, O., Stark, G., Aguet, M., and Schaffner, W. 1994. The transcription factor MTF-1 is essential for basal and heavy metal-induced metallothionein gene expression. *EMBO J.* **13**, 2870–2875.
58. Radtke, F., Heuchel, R., Georgiev, O., Hergersberg, M., Gariglio, M., Dembic, Z., and Schaffner, W. 1993. Cloned transcription factor MTF-1 activates the mouse metallothionein I promoter. *EMBO J* **12**, 1355–1362.
59. Brugnera, E., Georgiev, O., Radtke, F., Heuchel, R., Baker, E., Sutherland, G. R., and Schaffner, W. 1994. Cloning, chromosomal mapping and characterization of the human metal-regulatory transcription factor MTF-1. *Nucleic Acids Res* **22**, 3167–3173.
60. Palmiter, R. D. 1994. Regulation of metallothionein genes by heavy metals appears to be mediated by a zinc-sensitive inhibitor that interacts with a constitutively active transcription factor, MTF-1. *Proc Natl Acad Sci USA* **91**, 1219–1223.
61. Hamer, D. H. 1985. Metallothionein. *Annu Rev Biochem* **55**, 913–951.
62. Suzuki, C. A., Ohta, H., Albores, A., Koropatnick, J. and Cherian, M. G. 1990. Induction of metallothionein synthesis by zinc in cadmium pretreated rats. *Toxicology* **63**, 273–284.
63. Deng, D. X., Ono, S., Koropatnick, J. and Cherian, M. G. 1998. Metallothionein and apoptosis in the toxic milk mutant mouse. *Lab Invest* **78**, 175–183.
64. Oliver, J. R., Mara, T. W. and Cherian, M. G. 2005. Impaired hepatic regeneration in metallothionein-I/II knockout mice after partial hepatectomy. *Exp Biol Med (Maywood)* **230**, 61–67.
65. Surowiak, P., Materna, V., Maciejczyk, A., Pudelko, M., Markwitz, E., Spaczynski, M., Dietel, M., Zabel, M. and Lage, H. 2007. Nuclear metallothionein expression correlates with cisplatin resistance of ovarian cancer cells and poor clinical outcome. *Virchows Arch* **450**, 279–285.
66. Pedersen, M. O., Larsen, A., Stoltenberg, M., and Penkowa, M. 2009. The role of metallothionein in oncogenesis and cancer prognosis. *Prog Histochem Cytochem* **44**, 29–64.
67. Zhang, B., Georgiev, O., Hagmann, M., Gunes, C., Cramer, M., Faller, P., Vasak, M., and Schaffner, W. 2003. Activity of metal-responsive transcription factor 1 by toxic heavy metals and H_2O_2 *in vitro* is modulated by metallothionein. *Mol Cell Biol* **23**, 8471–8485.
68. Egli, D., Yepiskoposyan, H., Selvaraj, A., Balamurugan, K., Rajaram, R., Simons, A., Multhaup, G., et al. 2006. A family knockout of all four *Drosophila* metallothioneins reveals a central role in copper homeostasis and detoxification. *Mol Cell Biol* **26**, 2286–2296.
69. Kindermann, B., Doring, F., Budczies, J., and Daniel, H. 2005. Zinc-sensitive genes as potential new target genes of the metal transcription factor-1 (MTF-1). *Biochem Cell Biol* **83**, 221–229.
70. Marr, M. T., 2nd, Isogai, Y., Wright, K. J., and Tjian, R. 2006. Coactivator cross-talk specifies transcriptional output. *Genes Dev* **20**, 1458–1469.
71. Burgess-Beusse, B. L. and Darlington, G. J. 1998. C/EBPalpha is critical for the neonatal acute-phase response to inflammation. *Mol Cell Biol* **18**, 7269–7277.
72. Wang, N. D., Finegold, M. J., Bradley, A., Ou, C. N., Abdelsayed, S. V., Wilde, M. D., Taylor, L. R., Wilson, D. R., and Darlington, G. J. 1995. Impaired energy homeostasis in C/EBP alpha knockout mice. *Science* **269**, 1108–1112.
73. Mizejewski, G. J. 1997. Alpha-fetoprotein as a biologic response modifier: Relevance to domain and subdomain structure. *Proc Soc Exp Biol Med* **215**, 333–362.
74. Green, C. J., Lichtlen, P., Huynh, N. T., Yanovsky, M., Laderoute, K. R., Schaffner, W., and Murphy, B. J. 2001. Placenta growth factor gene expression is induced by hypoxia in fibroblasts: A central role for metal transcription factor-1. *Cancer Res* **61**, 2696–2703.

75. Cramer, M., Nagy, I., Murphy, B. J., Gassmann, M., Hottiger, M. O., Georgiev, O., and Schaffner, W. 2005. NF-kappaB contributes to transcription of placenta growth factor and interacts with metal responsive transcription factor-1 in hypoxic human cells. *Biol Chem* **386**, 865–872.
76. Lichtlen, P., Wang, Y., Belser, T., Georgiev, O., Certa, U., Sack, R., and Schaffner, W. 2001. Target gene search for the metal-responsive transcription factor MTF-1. *Nucleic Acids Res* **29**, 1514–1523.
77. Haroon, Z. A., Amin, K., Lichtlen, P., Sato, B., Huynh, N. T., Wang, Z., Schaffner, W., and Murphy, B. J. 2004. Loss of metal transcription factor-1 suppresses tumor growth through enhanced matrix deposition. *FASEB J* **18**, 1176–1184.
78. Nishimoto, F., Sakata, M., Minekawa, R., Okamoto, Y., Miyake, A., Isobe, A., Yamamoto, T., et al. 2009. Metal transcription factor-1 is involved in hypoxia-dependent regulation of placenta growth factor in trophoblast-derived cells. *Endocrinology* **150**, 1801–1808.
79. Siebenlist, U., Franzoso, G., and Brown, K. 1994. Structure, regulation and function of NF-kappa B. *Annu Rev Cell Biol* **10**, 405–455.
80. Karin, M. 1999. The beginning of the end: IkappaB kinase (IKK) and NF-kappaB activation. *J Biol Chem* **274**, 27339–27342.
81. Nelson, G., Paraoan, L., Spiller, D. G., Wilde, G. J., Browne, M. A., Djali, P. K., Unitt, J. F., Sullivan, E., Floettmann, E., and White, M. R. 2002. Multi-parameter analysis of the kinetics of NF-kappaB signalling and transcription in single living cells. *J Cell Sci* **115**, 1137–1148.
82. Abdel-Mageed, A. B. and Agrawal, K. C. 1998. Activation of nuclear factor kappaB: Potential role in metallothionein-mediated mitogenic response. *Cancer Res* **58**, 2335–2338.
83. Kanekiyo, M., Itoh, N., Kawasaki, A., Tanaka, J., Nakanishi, T., and Tanaka, K. 2001. Zinc-induced activation of the human cytomegalovirus major immediate-early promoter is mediated by metallothionein and nuclear factor-kappaB. *Toxicol Appl Pharmacol* **173**, 146–153.
84. Butcher, H. L., Kennette, W. A., Collins, O., Zalups, R. K., and Koropatnick, J. 2004. Metallothionein mediates the level and activity of nuclear factor kappa *B* in murine fibroblasts. *J Pharmacol Exp Ther* **310**, 589–598.
85. Bauman, J. W., Liu, J., Liu, Y. P., and Klaassen, C. D. 1991. Increase in metallothionein produced by chemicals that induce oxidative stress. *Toxicol Appl Pharmacol* **110**, 347–354.
86. Min, K. S., Terano, Y., Onosaka, S., and Tanaka, K. 1991. Induction of hepatic metallothionein by nonmetallic compounds associated with acute-phase response in inflammation. *Toxicol Appl Pharmacol* **111**, 152–162.
87. Kondoh, M., Inoue, Y., Atagi, S., Futakawa, N., Higashimoto, M., and Sato, M. 2001. Specific induction of metallothionein synthesis by mitochondrial oxidative stress. *Life Sci* **69**, 2137–2146.
88. Favreau, L. V. and Pickett, C. B. 1993. Transcriptional regulation of the rat NAD(P)H: Quinone reductase gene. Characterization of a DNA-protein interaction at the antioxidant responsive element and induction by 12-O-tetradecanoylphorbol 13-acetate. *J Biol Chem* **268**, 19875–19881.
89. Karin, M., Andersen, R. D., Slater, E., Smith, K., and Herschman, H. R. 1980. Metallothionein mRNA induction in HeLa cells in response to zinc or dexamethasone is a primary induction response. *Nature* **286**, 295–297.
90. Collingwood, T. N., Urnov, F. D. and Wolffe, A. P. 1999. Nuclear receptors: Coactivators, corepressors and chromatin remodeling in the control of transcription. *J Mol Endocrinol* **23**, 255–275.
91. Tait, A. S., Butts, C. L., and Sternberg, E. M. 2008. The role of glucocorticoids and progestins in inflammatory, autoimmune, and infectious disease. *J Leukoc Biol* **84**, 924–931.
92. De, S. K., McMaster, M. T., and Andrews, G. K. 1990. Endotoxin induction of murine metallothionein gene expression. *J Biol Chem* **265**, 15267–15274.
93. Ceciliani, F., Giordano, A., and Spagnolo, V. 2002. The systemic reaction during inflammation: The acute-phase proteins. *Protein Pept Lett* **9**, 211–223.
94. Schroeder, J. J. and Cousins, R. J. 1990. Interleukin 6 regulates metallothionein gene expression and zinc metabolism in hepatocyte monolayer cultures. *Proc Natl Acad Sci USA* **87**, 3137–3141.
95. Itoh, N., Kasutani, K., Muto, N., Otaki, N., Kimura, M., and Tanaka, K. 1996. Blocking effect of anti-mouse interleukin-6 monoclonal antibody and glucocorticoid receptor antagonist, RU38486, on metallothionein-inducing activity of serum from lipopolysaccharide-treated mice. *Toxicology* **112**, 29–36.
96. Boudny, V. and Kovarik, J. 2002. JAK/STAT signaling pathways and cancer. Janus kinases/signal transducers and activators of transcription. *Neoplasma* **49**, 349–355.
97. Ghoshal, K., Wang, Y., Sheridan, J. F., and Jacob, S. T. 1998. Metallothionein induction in response to restraint stress. Transcriptional control, adaptation to stress, and role of glucocorticoid. *J Biol Chem* **273**, 27904–27910.

98. Hernandez, J., Carrasco, J., Belloso, E., Giralt, M., Bluethmann, H., Kee Lee, D., Andrews, G. K., and Hidalgo, J. 2000. Metallothionein induction by restraint stress: Role of glucocorticoids and IL-6. *Cytokine* **12**, 791–796.
99. Molotkov, A., Nishimura, N., Satoh, M., and Tohyama, C. 2000. Role of IL-6 in the induction of hepatic metallothionein in mice after partial hepatectomy. *Life Sci* **66**, 963–970.
100. Kasutani, K., Itoh, N., Kanekiyo, M., Muto, N., and Tanaka, K. 1998. Requirement for cooperative interaction of interleukin-6 responsive element type 2 and glucocorticoid responsive element in the synergistic activation of mouse metallothionein-I gene by interleukin-6 and glucocorticoid. *Toxicol Appl Pharmacol* **151**, 143–151.
101. Sciavolino, P. J. and Vilcek, J. 1995. Regulation of metallothionein gene expression by TNF-alpha and IFN-beta in human fibroblasts. *Cytokine* **7**, 242–250.
102. Klehr, D., Schlake, T., Maass, K., and Bode, J. 1992. Scaffold-attached regions (SAR elements) mediate transcriptional effects due to butyrate. *Biochemistry* **31**, 3222–3229.
103. Walia, H., Chen, H. Y., Sun, J. M., Holth, L. T., and Davie, J. R. 1998. Histone acetylation is required to maintain the unfolded nucleosome structure associated with transcribing DNA. *J Biol Chem* **273**, 14516–14522.
104. Birren, B. W. and Herschman, H. R. 1986. Regulation of the rat metallothionein-I gene by sodium butyrate. *Nucleic Acids Res* **14**, 853–867.
105. Wolffe, A. P. and Matzke, M. A. 1999. Epigenetics: Regulation through repression. *Science* **286**, 481–486.
106. Lieberman, M. W., Beach, L. R., and Palmiter, R. D. 1983. Ultraviolet radiation-induced metallothionein-I gene activation is associated with extensive DNA demethylation. *Cell* **35**, 207–214.
107. Ghoshal, K., Datta, J., Majumder, S., Bai, S., Dong, X., Parthun, M., and Jacob, S. T. 2002. Inhibitors of histone deacetylase and DNA methyltransferase synergistically activate the methylated metallothionein I promoter by activating the transcription factor MTF-1 and forming an open chromatin structure. *Mol Cell Biol* **22**, 8302–8319.
108. Zalups, R. K. and Koropatnick, J. 2000. Temporal changes in metallothionein gene transcription in rat kidney and liver: Relationship to content of mercury and metallothionein protein. *J Pharmacol Exp Ther* **295**, 74–82.
109. Sadhu, C. and Gedamu, L. 1989. Metal-specific posttranscriptional control of human metallothionein genes. *Mol Cell Biol* **9**, 5738–5741.
110. Vasconcelos, M. H., Tam, S. C., Beattie, J. H., and Hesketh, J. E. 1996. Evidence for differences in the post-transcriptional regulation of rat metallothionein isoforms. *Biochem J* **315**(Pt 2), 665–671.
111. Andrews, G. K., Lee, D. K., Ravindra, R., Lichtlen, P., Sirito, M., Sawadogo, M., and Schaffner, W. 2001. The transcription factors MTF-1 and USF1 cooperate to regulate mouse metallothionein-I expression in response to the essential metal zinc in visceral endoderm cells during early development. *EMBO J* **20**, 1114–1122.
112. Andrews, G. K., Adamson, E. D., and Gedamu, L. 1984. The ontogeny of expression of murine metallothionein: Comparison with the alpha-fetoprotein gene. *Dev Biol* **103**, 294–303.
113. Solaiman, D., Jonah, M. M., Miyazaki, W., Ho, G., and Bhattacharyya, M. H. 2001. Increased metallothionein in mouse liver, kidneys, and duodenum during lactation. *Toxicol Sci* **60**, 184–192.
114. Kontozoglou, T. E., Banerjee, D., and Cherian, M. G. 1989. Immunohistochemical localization of metallothionein in human testicular embryonal carcinoma cells. *Virchows Arch A Pathol Anat Histopathol* **415**, 545–549.
115. Krezel, A. and Maret, W. 2007. Different redox states of metallothionein/thionein in biological tissue. *Biochem J* **402**, 551–558.
116. Pattanaik, A., Shaw, C. F., 3rd, Petering, D. H., Garvey, J., and Kraker, A. J. 1994. Basal metallothionein in tumors: Widespread presence of apoprotein. *J Inorg Biochem* **54**, 91–105.
117. Maret, W., Jacob, C., Vallee, B. L., and Fischer, E. H. 1999. Inhibitory sites in enzymes: Zinc removal and reactivation by thionein. *Proc Natl Acad Sci USA* **96**, 1936–1940.
118. Ye, B., Maret, W., and Vallee, B. L. 2001. Zinc metallothionein imported into liver mitochondria modulates respiration. *Proc Natl Acad Sci USA* **98**, 2317–2322.
119. Zeng, J., Heuchel, R., Schaffner, W., and Kagi, J. H. 1991. Thionein (apometallothionein) can modulate DNA binding and transcription activation by zinc finger containing factor Sp1. *FEBS Lett* **279**, 310–312.
120. Maret, W. and Krezel, A. 2007. Cellular zinc and redox buffering capacity of metallothionein/thionein in health and disease. *Mol Med* **13**, 371–375.
121. Vallee, B. L. and Falchuk, K. H. 1993. The biochemical basis of zinc physiology. *Physiol Rev* **73**, 79–118.

122. Evans, R. M. 1988. The steroid and thyroid hormone receptor superfamily. *Science* **240**, 889–895.
123. Hager, L. J. and Palmiter, R. D. 1981. Transcriptional regulation of mouse liver metallothionein-I gene by glucocorticoids. *Nature* **291**, 340–342.
124. Maret, W. 2003. Cellular zinc and redox states converge in the metallothionein/thionein pair. *J Nutr* **133**, 1460S–1462S.
125. Meplan, C., Richard, M. J., and Hainaut, P. 2000. Metalloregulation of the tumor suppressor protein p53: Zinc mediates the renaturation of p53 after exposure to metal chelators *in vitro* and in intact cells. *Oncogene* **19**, 5227–5236.
126. Suhy, D. A., Simon, K. D., Linzer, D. I., and O'Halloran, T. V. 1999. Metallothionein is part of a zinc-scavenging mechanism for cell survival under conditions of extreme zinc deprivation. *J Biol Chem* **274**, 9183–9192.
127. DeMoor, J. M. and Koropatnick, D. J. 2000. Metals and cellular signaling in mammalian cells. *Cell Mol Biol (Noisy-le-grand)* **46**, 367–381.
128. Rees, E. M. and Thiele, D. J. 2004. From aging to virulence: Forging connections through the study of copper homeostasis in eukaryotic microorganisms. *Curr Opin Microbiol* **7**, 175–184.
129. Ibs, K. H. and Rink, L. 2003. Zinc-altered immune function. *J Nutr* **133**, 1452S–1456S.
130. Fraker, P. J. and King, L. E. 2004. Reprogramming of the immune system during zinc deficiency. *Annu Rev Nutr* **24**, 277–298.
131. Wintergerst, E. S., Maggini, S., and Hornig, D. H. 2007. Contribution of selected vitamins and trace elements to immune function. *Ann Nutr Metab* **51**, 301–323.
132. Prasad, A. S. 2000. Effects of zinc deficiency on Th1 and Th2 cytokine shifts. *J Infect Dis* **182**(Suppl 1), S62–68.
133. Szallasi, Z., Bogi, K., Gohari, S., Biro, T., Acs, P., and Blumberg, P. M. 1996. Non-equivalent roles for the first and second zinc fingers of protein kinase Cdelta. Effect of their mutation on phorbol ester-induced translocation in NIH 3T3 cells. *J Biol Chem* **271**, 18299–18301.
134. Hancock, W. W., Grey, S. T., Hau, L., Akalin, E., Orthner, C., Sayegh, M. H., and Salem, H. H. 1995. Binding of activated protein *C* to a specific receptor on human mononuclear phagocytes inhibits intracellular calcium signaling and monocyte-dependent proliferative responses. *Transplantation* **60**, 1525–1532.
135. Meier, R. W., Niklaus, G., Dewald, B., Fey, M. F., and Tobler, A. 1995. Inhibition of the arachidonic acid pathway prevents induction of IL-8 mRNA by phorbol ester and changes the release of IL-8 from HL 60 cells: Differential inhibition of induced expression of IL-8, TNF-alpha, IL-1 alpha, and IL-1 beta. *J Cell Physiol* **165**, 62–70.
136. Waelput, W., Broekaert, D., Vandekerckhove, J., Brouckaert, P., Tavernier, J., and Libert, C. 2001. A mediator role for metallothionein in tumor necrosis factor-induced lethal shock. *J Exp Med* **194**, 1617–1624.
137. Waelput, W., Brouckaert, P., Broekaert, D., and Tavernier, J. 2006. A role for leptin in the systemic inflammatory response syndrome (SIRS) and in immune response, an update. *Curr Med Chem* **13**, 465–475.
138. Mocchegiani, E., Costarelli, L., Giacconi, R., Cipriano, C., Muti, E., and Malavolta, M. 2006. Zinc-binding proteins (metallothionein and alpha-2 macroglobulin) and immunosenescence. *Exp Gerontol* **41**, 1094–1107.
139. Coyle, P., Philcox, J. C., and Rofe, A. M. 1995. Hepatic zinc in metallothionein-null mice following zinc challenge: *In vivo* and *in vitro* studies. *Biochem J* **309**(Part 1), 25–31.
140. Sugiura, T., Kuroda, E., and Yamashita, U. 2004. Dysfunction of macrophages in metallothionein-knock out mice. *J UOEH* **26**, 193–205.
141. Cherian, M. G., Jayasurya, A., and Bay, B. H. 2003. Metallothioneins in human tumors and potential roles in carcinogenesis. *Mutat Res* **533**, 201–209.
142. Theocharis, S. E., Margeli, A. P., Klijanienko, J. T., and Kouraklis, G. P. 2004. Metallothionein expression in human neoplasia. *Histopathology* **45**, 103–118.
143. Thirumoorthy, N., Manisenthil Kumar, K. T., Shyam Sundar, A., Panayappan, L., and Chatterjee, M. 2007. Metallothionein: An overview. *World J Gastroenterol* **13**, 993–996.
144. Meijer, C., Timmer, A., De Vries, E. G., Groten, J. P., Knol, A., Zwart, N., Dam, W. A., Sleijfer, D. T., and Mulder, N. H. 2000. Role of metallothionein in cisplatin sensitivity of germ-cell tumors. *Int J Cancer* **85**, 777–781.
145. Huang, Y., de la Chapelle, A., and Pellegata, N. S. 2003. Hypermethylation, but not LOH, is associated with the low expression of MT1G and CRABP1 in papillary thyroid carcinoma. *Int J Cancer* **104**, 735–744.
146. Dziegiel, P. 2004. Expression of metallothioneins in tumor cells. *Pol J Pathol* **55**, 3–12.

147. Endo, T., Yoshikawa, M., Ebara, M., Kato, K., Sunaga, M., Fukuda, H., Hayasaka, A., Kondo, F., Sugiura, N., and Saisho, H. 2004. Immunohistochemical metallothionein expression in hepatocellular carcinoma: Relation to tumor progression and chemoresistance to platinum agents. *J Gastroenterol* **39**, 1196–1201.
148. Tao, X., Zheng, J. M., Xu, A. M., Chen, X. F., and Zhang, S. H. 2007. Downregulated expression of metallothionein and its clinicopathological significance in hepatocellular carcinoma. *Hepatol Res* **37**, 820–827.
149. Ferrario, C., Lavagni, P., Gariboldi, M., Miranda, C., Losa, M., Cleris, L., Formelli, F., Pilotti, S., Pierotti, M. A., and Greco, A. 2008. Metallothionein 1G acts as an oncosupressor in papillary thyroid carcinoma. *Lab Invest* **88**, 474–481.
150. Hagrman, D., Goodisman, J., Dabrowiak, J. C., and Souid, A. K. 2003. Kinetic study on the reaction of cisplatin with metallothionein. *Drug Metab Dispos* **31**, 916–923.
151. Liu, J., Qu, W. and Kadiiska, M. B. 2009. Role of oxidative stress in cadmium toxicity and carcinogenesis. *Toxicol Appl Pharmacol* **238**, 209–214.
152. Kondo, Y., Rusnak, J. M., Hoyt, D. G., Settineri, C. E., Pitt, B. R., and Lazo, J. S. 1997. Enhanced apoptosis in metallothionein null cells. *Mol Pharmacol* **52**, 195–201.
153. Shimoda, R., Achanzar, W. E., Qu, W., Nagamine, T., Takagi, H., Mori, M., and Waalkes, M. P. 2003. Metallothionein is a potential negative regulator of apoptosis. *Toxicol Sci* **73**, 294–300.
154. Surowiak, P., Materna, V., Kaplenko, I., Spaczynski, M., Dietel, M., Lage, H., and Zabel, M. 2005. Augmented expression of metallothionein and glutathione S transferase pi as unfavourable prognostic factors in cisplatin-treated ovarian cancer patients. *Virchows Arch* **447**, 626–633.
155. Dutsch-Wicherek, M., Popiela, T. J., Klimek, M., Rudnicka-Sosin, L., Wicherek, L., Oudinet, J. P., Skladzien, J., and Tomaszewska, R. 2005. Metallothionein stroma reaction in tumor adjacent healthy tissue in head and neck squamous cell carcinoma and breast adenocarcinoma. *Neuro Endocrinol Lett* **26**, 567–574.
156. Ohshio, G., Imamura, T., Okada, N., Wang, Z. H., Yamaki, K., Kyogoku, T., Suwa, H., Yamabe, H., and Imamura, M. 1996. Immunohistochemical study of metallothionein in pancreatic carcinomas. *J Cancer Res Clin Oncol* **122**, 351–355.
157. Kennette, W., Collins, O. M., Zalups, R. K., and Koropatnick, J. 2005. Basal and zinc-induced metallothionein in resistance to cadmium, cisplatin, zinc, and tertbutyl hydroperoxide: Studies using MT knockout and antisense-downregulated MT in mammalian cells. *Toxicol Sci* **88**, 602–613.
158. Lu, C. and El-Deiry, W. S. 2009. Targeting p53 for enhanced radio- and chemosensitivity. *Apoptosis* **14**, 597–606.
159. Formigari, A., Irato, P., and Santon, A. 2007. Zinc, antioxidant systems and metallothionein in metal mediated-apoptosis: Biochemical and cytochemical aspects. *Comp Biochem Physiol C Toxicol Pharmacol* **146**, 443–459.
160. Ostrakhovitch, E. A., Olsson, P. E., Jiang, S., and Cherian, M. G. 2006. Interaction of metallothionein with tumor suppressor p53 protein. *FEBS Lett* **580**, 1235–1238.
161. Pastuszewski, W., Dziegiel, P., Krecicki, T., Podhorska-Okolow, M., Ciesielska, U., Gorzynska, E., and Zabel, M. 2007. Prognostic significance of metallothionein, p53 protein and Ki-67 antigen expression in laryngeal cancer. *Anticancer Res* **27**, 335–342.
162. Fan, L. Z. and Cherian, M. G. 2002. Potential role of p53 on metallothionein induction in human epithelial breast cancer cells. *Br J Cancer* **87**, 1019–1026.
163. Chen, Y., Saari, J. T., and Kang, Y. J. 1994. Weak antioxidant defenses make the heart a target for damage in copper-deficient rats. *Free Radic Biol Med* **17**, 529–536.
164. Cardinale, D., Colombo, A., and Cipolla, C. M. 2008. Prevention and treatment of cardiomyopathy and heart failure in patients receiving cancer chemotherapy. *Curr Treat Options Cardiovasc Med* **10**, 486–495.
165. Kang, Y. J., Chen, Y., Yu, A., Voss-McCowan, M., and Epstein, P. N. 1997. Overexpression of metallothionein in the heart of transgenic mice suppresses doxorubicin cardiotoxicity. *J Clin Invest* **100**, 1501–1506.
166. Kang, Y. J., Chen, Y., and Epstein, P. N. 1996. Suppression of doxorubicin cardiotoxicity by overexpression of catalase in the heart of transgenic mice. *J Biol Chem* **271**, 12610–12616.
167. Zhou, Z. and Kang, Y. J. 2000. Immunocytochemical localization of metallothionein and its relation to doxorubicin toxicity in transgenic mouse heart. *Am J Pathol* **156**, 1653–1662.
168. Sun, X., Zhou, Z., and Kang, Y. J. 2001. Attenuation of doxorubicin chronic toxicity in metallothionein-overexpressing transgenic mouse heart. *Cancer Res* **61**, 3382–3387.
169. Kang, Y. J., Sun, X., Chen, Y., and Zhou, Z. 2002. Inhibition of doxorubicin chronic toxicity in catalase-overexpressing transgenic mouse hearts. *Chem Res Toxicol* **15**, 1–6.

170. Giles, T. D. 2003. The patient with diabetes mellitus and heart failure: At-risk issues. *Am J Med* **115(Suppl 8A)**, 107S–110S.
171. Trost, S. and LeWinter, M. 2001. Diabetic cardiomyopathy. *Curr Treat Options Cardiovasc Med* **3**, 481–492.
172. Wiernsperger, N. F. 2003. Oxidative stress as a therapeutic target in diabetes: Revisiting the controversy. *Diabetes Metab* **29**, 579–585.
173. Vega-Lopez, S., Devaraj, S., and Jialal, I. 2004. Oxidative stress and antioxidant supplementation in the management of diabetic cardiovascular disease. *J Investig Med* **52**, 24–32.
174. Cai, L., Wang, Y., Zhou, G., Chen, T., Song, Y., Li, X., and Kang, Y. J. 2006. Attenuation by metallothionein of early cardiac cell death via suppression of mitochondrial oxidative stress results in a prevention of diabetic cardiomyopathy. *J Am Coll Cardiol* **48**, 1688–1697.
175. Wang, J., Song, Y., Elsherif, L., Song, Z., Zhou, G., Prabhu, S. D., Saari, J. T., and Cai, L. 2006. Cardiac metallothionein induction plays the major role in the prevention of diabetic cardiomyopathy by zinc supplementation. *Circulation* **113**, 544–554.
176. Liang, Q., Carlson, E. C., Donthi, R. V., Kralik, P. M., Shen, X., and Epstein, P. N. 2002. Overexpression of metallothionein reduces diabetic cardiomyopathy. *Diabetes* **51**, 174–181.
177. Ye, G., Metreveli, N. S., Ren, J., and Epstein, P. N. 2003. Metallothionein prevents diabetes-induced deficits in cardiomyocytes by inhibiting reactive oxygen species production. *Diabetes* **52**, 777–783.
178. Burney, B. O., Kalaitzidis, R. G., and Bakris, G. L. 2009. Novel therapies of diabetic nephropathy. *Curr Opin Nephrol Hypertens* **18**, 107–111.
179. Hidalgo, J., Aschner, M., Zatta, P., and Vasak, M. 2001. Roles of the metallothionein family of proteins in the central nervous system. *Brain Res Bull* **55**, 133–145.
180. West, A. K., Chuah, M. I., Vickers, J. C., and Chung, R. S. 2004. Protective role of metallothioneins in the injured mammalian brain. *Rev Neurosci* **15**, 157–166.
181. Sofroniew, M. V. 2005. Reactive astrocytes in neural repair and protection. *Neuroscientist* **11**, 400–407.
182. Penkowa, M., Moos, T., Carrasco, J., Hadberg, H., Molinero, A., Bluethmann, H., and Hidalgo, J. 1999. Strongly compromised inflammatory response to brain injury in interleukin-6-deficient mice. *Glia* **25**, 343–357.
183. Carrasco, J., Hernandez, J., Bluethmann, H., and Hidalgo, J. 1998. Interleukin-6 and tumor necrosis factor-alpha type 1 receptor deficient mice reveal a role of IL- 6 and TNF-alpha on brain metallothionein-I and -III regulation. *Brain Res Mol Brain Res* **57**, 221–234.
184. Trendelenburg, G., Prass, K., Priller, J., Kapinya, K., Polley, A., Muselmann, C., et al. 2002. Serial analysis of gene expression identifies metallothionein-II as major neuroprotective gene in mouse focal cerebral ischemia. *J Neurosci.* **22**, 5879–5888.
185. Giralt, M., Penkowa, M., Lago, N., Molinero, A., and Hidalgo, J. 2002. Metallothionein-1 + 2 protect the CNS after a focal brain injury. *Exp Neurol* **173**, 114–128.
186. Chung, R. S., Vickers, J. C., Chuah, M. I., and West, A. K. 2003. Metallothionein-IIA promotes initial neurite elongation and postinjury reactive neurite growth and facilitates healing after focal cortical brain injury. *J Neurosci* **23**, 3336–3342.
187. Nomiyama, K., Nomiyama, H., and Kameda, N. 1998. Plasma cadmium-metallothionein, a biological exposure index for cadmium-induced renal dysfunction, based on the mechanism of its action. *Toxicology* **129**, 157–168.
188. Kohler, L. B., Berezin, V., Bock, E., and Penkowa, M. 2003. The role of metallothionein II in neuronal differentiation and survival. *Brain Res* **992**, 128–136.
189. Fitzgerald, M., Nairn, P., Bartlett, C. A., Chung, R. S., West, A. K., and Beazley, L. D. 2007. Metallothionein-IIA promotes neurite growth via the megalin receptor. *Exp Brain Res* **183**, 171–180.
190. Ambjorn, M., Asmussen, J. W., Lindstam, M., Gotfryd, K., Jacobsen, C., Kiselyov, V. V., Moestrup, S. K., Penkowa, M., Bock, E. and Berezin, V. 2008. Metallothionein and a peptide modeled after metallothionein, EmtinB, induce neuronal differentiation and survival through binding to receptors of the low-density lipoprotein receptor family. *J Neurochem* **104**, 21–37.
191. Chang, L. and Karin, M. 2001. Mammalian MAP kinase signalling cascades. *Nature* **410**, 37–40.
192. Hannila, S. S. and Filbin, M. T. 2008. The role of cyclic AMP signaling in promoting axonal regeneration after spinal cord injury. *Exp Neurol* **209**, 321–332.
193. Chung, R. S., Hidalgo, J., and West, A. K. 2008. New insight into the molecular pathways of metallothionein-mediated neuroprotection and regeneration. *J Neurochem* **104**, 14–20.
194. Chung, R. S., Penkowa, M., Dittmann, J., King, C. E., Bartlett, C., Asmussen, J. W., et al. 2008. Redefining the role of metallothionein within the injured brain: Extracellular metallothioneins play an important role in the astrocyte-neuron response to injury. *J Biol Chem.* **283**, 15349–15358.

195. Wolff, N. A., Abouhamed, M., Verroust, P. J., and Thevenod, F. 2006. Megalin-dependent internalization of cadmium-metallothionein and cytotoxicity in cultured renal proximal tubule cells. *J Pharmacol Exp Ther* **318**, 782–791.
196. Choi, D. W. and Koh, J. Y. 1998. Zinc and brain injury. *Annu Rev Neurosci* **21**, 347–375.
197. Chung, R. S. and West, A. K. 2004. A role for extracellular metallothioneins in CNS injury and repair. *Neuroscience* **123**, 595–599.
198. Klassen, R. B., Crenshaw, K., Kozyraki, R., Verroust, P. J., Tio, L., Atrian, S., Allen, P. L., and Hammond, T. G. 2004. Megalin mediates renal uptake of heavy metal metallothionein complexes. *Am J Physiol Renal Physiol* **287**, F393–F403.

7 Cellular and Molecular Biology of Iron-Binding Proteins

Danuta S. Kalinowski and Des R. Richardson

CONTENTS

7.1 Introduction ... 167
7.2 Iron Metabolism ... 168
7.2.1 Dietary Iron Uptake ... 168
7.2.1.1 Inorganic Iron ... 168
7.2.1.2 Heme ... 169
7.2.1.3 The Labile Iron Pool (LIP) ... 169
7.2.1.4 Iron Export from Enterocytes ... 169
7.2.2 Cellular Iron Uptake ... 169
7.2.2.1 Transferrin ... 169
7.2.2.2 Iron Export from the Endosome ... 170
7.2.2.3 Endosomal Ferrireductase ... 170
7.2.2.4 Endosomal Docking to the Mitochondrion, Sec15l1 ... 171
7.2.3 The Iron Storage Protein, Ferritin ... 171
7.3 Mitochondrial Iron Metabolism ... 172
7.3.1 Iron Transport into the Mitochondrion ... 172
7.3.2 Heme Transport from the Mitochondrion ... 172
7.3.3 Frataxin ... 173
7.3.4 Export of [Fe–S] Clusters from the Mitochondrion ... 173
7.4 Iron Homeostasis ... 173
7.4.1 Regulation of Cellular Iron Metabolism: The IRPs ... 174
7.4.2 Regulation of Systemic Iron Metabolism: The Iron Regulatory Hormone, Hepcidin ... 175
7.5 Conclusion ... 176
Acknowledgments ... 176
References ... 176

7.1 INTRODUCTION

Iron (Fe) is an essential element involved in a variety of cellular processes critical for life, including oxygen transport, energy generation, and DNA synthesis.[1–3] This is due to the unique redox activity of iron, which is able to act as an electron donor or acceptor by cycling between the ferric (Fe^{3+}) and ferrous (Fe^{2+}) states. Importantly, iron can participate in the oxidation–reduction process known as the Haber–Weiss reaction where hydrogen peroxide reacts to form the hydroxide anion and the reactive oxygen species (ROS), the hydroxyl radical.[4]

The properties that make iron intrinsic for life also make it potentially deleterious. Excess free iron participating in the Haber–Weiss reaction can generate the highly reactive hydroxyl radical with toxic side effects.[4,5] This ROS is able to induce oxidative cell damage to a number of significant biomolecules, resulting in lipid peroxidation, mitochondrial damage, and DNA oxidation.[5,6] In fact, such oxidative reactions lead to the impairment of cellular functions, resulting in tissue and organ damage evident in the iron-loading diseases, β-thalassemia and Friedreich's ataxia (FA).[7–10] Consequently, to prevent the toxic side effects of the Haber–Weiss reaction, an intricate system of proteins involved in iron transport, storage, and homeostasis has evolved.

7.2 IRON METABOLISM

7.2.1 Dietary Iron Uptake

7.2.1.1 Inorganic Iron

Iron absorption in the gut from dietary sources occurs predominantly within the duodenum of the small intestine. Enterocytes take up iron from two sources, namely inorganic iron and heme.[11] Inorganic iron must be in the ferrous (Fe^{2+}) state to be absorbed (Figure 7.1). Thus, any Fe^{3+} in the diet must be reduced to the Fe^{2+} state at the apical surface of enterocytes before absorption. This role was thought to be carried out by the membrane-bound duodenal cytochrome b (Dcytb) protein that possesses ferrireductase activity.[12] However, no iron-deficient phenotype was observed after the ablation of the murine Dcytb homolog, *Cybrd1*.[13] These results suggest that Dcytb is not critical for dietary iron uptake in mice and that another undiscovered ferrireductase may be responsible.[13] Irrespective of the ferriductase involved, Fe^{2+} is then transported across the enterocyte cell membrane by the divalent metal transporter 1 (DMT1) that is found to be crucial for intestinal inorganic iron uptake (Figure 7.1).[14–16] The expression of DMT1 is regulated by intracellular iron levels through the binding of iron regulatory proteins (IRPs) to an iron-response element (IRE) in its 3′-untranslated region (UTR) (see Section 7.4.2).

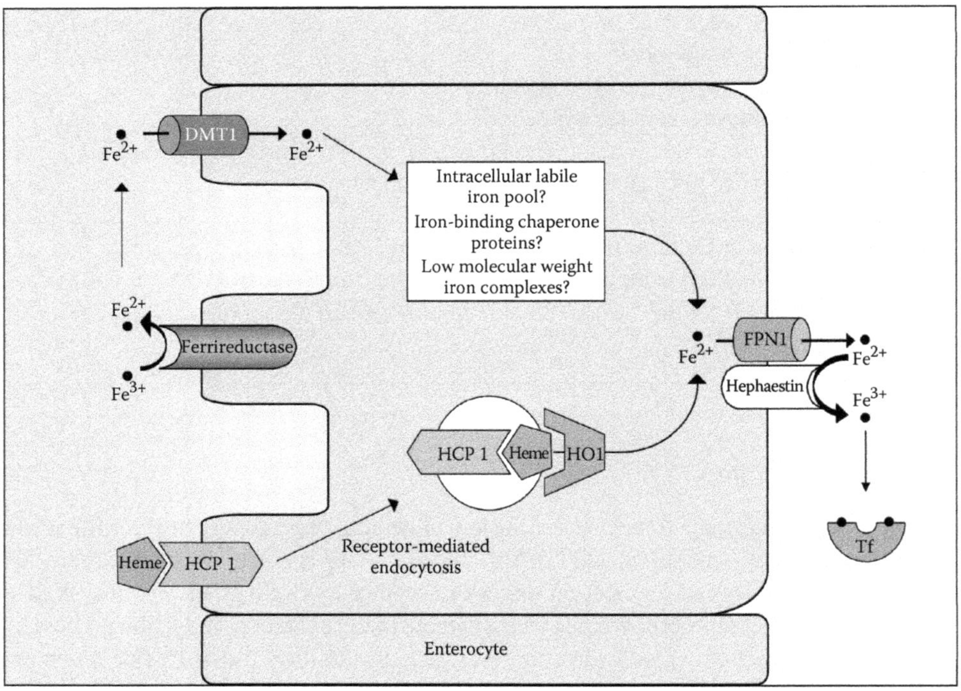

FIGURE 7.1 Diagram of iron uptake from dietary sources: Inorganic ferric iron is reduced to the ferrous state before being transported into the enterocyte by DMT1. Dietary heme is thought to be transported into the enterocyte by the HCP1 and is internalized into an endosome. However, the identity of HCP1 as a physiological heme transporter remains controversial. Iron in heme is released by HO1. Once liberated, iron can enter the putative LIP or be exported from the enterocyte in the ferrous state by FPN1 located in the basolateral membrane. This ferrous iron is then thought to be oxidized by hephaestin, which can then be bound by Tf and is transported through the circulation.

7.2.1.2 Heme

The other major source of dietary iron is heme, a protoporphyrin ring that binds iron and is derived from the breakdown of hemoglobin and myoglobin of meat products. Heme is thought to be internalized by the receptor, heme carrier protein-1 (HCP1), which is expressed at high levels in the duodenum.[17] Once heme binds to HCP1 on the cell surface, the complex is internalized via receptor-mediated endocytosis (Figure 7.1). This endosome is thought to migrate to the endoplasmic reticulum, where heme is metabolized by heme oxygenase-1 (HO1), located on the reticulum surface, to liberate iron (Figure 7.1).[17]

Studies examining HCP1 also demonstrated that heme transport is regulated post-transcriptionally by iron stores.[17] How iron levels regulate heme transport remains unclear. However, during iron depletion HCP1 is localized to the apical membrane, while in iron-replete mice HCP1 is located within endosomal vesicles.[17] Although HCP1 appears to transport heme, there is no strong evidence as yet that this is the physiologically relevant mechanism. In addition, a latter study demonstrated that HCP1 was a folate transporter,[18] questioning its role in heme metabolism[19] or suggesting that it transports both heme and folate and potentially other lipophilic substrates.

7.2.1.3 The Labile Iron Pool (LIP)

Once dietary iron has been imported into the enterocyte, it is then thought to enter the hypothesized LIP (Figure 7.1). The precise character of the LIP is unknown, but may comprise small-molecular-weight iron complexes or high-molecular-weight intermediates. While the exact identity of these components is elusive, it was once believed to consist of iron bound to amino acids, nucleotides, and sugars.[20] Clearly, the presence of low-molecular-weight iron within the cell would have disastrous consequences due to the generation of ROS. However, low-molecular-weight iron is present at exceedingly low levels in the cell, suggesting that if such a pool exists it must be in very low concentrations or may not be present.[21,22] Other models of intracellular iron transport do not involve low-molecular-weight iron complexes and propose that iron is transported within the cell without entering a chelatable LIP.[21,23]

7.2.1.4 Iron Export from Enterocytes

The exact proteins, compartments, or mechanisms of iron transport within the enterocyte to the basolateral membrane remain unknown. Iron in the Fe^{2+} state is exported by ferroportin 1 (FPN1) through the basolateral membrane to the interstitial space and may be oxidized by hephaestin, a transmembrane ferrioxidase (Figure 7.1).[24,25]

Hepcidin, a peptide hormone secreted by the liver, acts as an iron-regulatory hormone and negatively regulates FPN1 expression that controls mammalian systemic Fe metabolism.[26] Under conditions of iron overload, hepcidin is expressed at high levels in the liver.[27] Hepcidin in the circulation can bind to FPN1 on the cell membrane, causing FPN1 to be internalized and degraded. This leads to a reduction of iron efflux from enterocytes, thus regulating iron levels.[26]

FPN1 is critical in cellular iron export and was demonstrated to be the only known mechanism for iron efflux as found in transgenic mice.[28] In fact, deletion of FPN1 was embryonically lethal.[28] Once exported by FPN1, iron is then bound with high affinity by the serum iron-transport protein, transferrin (Tf), and then carried in the circulation (Figure 7.1).

7.2.2 Cellular Iron Uptake

7.2.2.1 Transferrin

In the serum, iron is found in a highly soluble form, bound to Tf to prevent its deleterious interactions with hydrogen peroxide. This transport molecule can bind two iron ions in their ferric state with high affinity at iron-binding sites located in its N- and C-termini (Figure 7.2).[29] Two molecules of diferric Tf are then bound to its cell surface receptor, transferrin receptor 1 (TfR1; Figure 7.2). A second receptor, transferrin receptor 2 (TfR2), has also been identified, although its precise

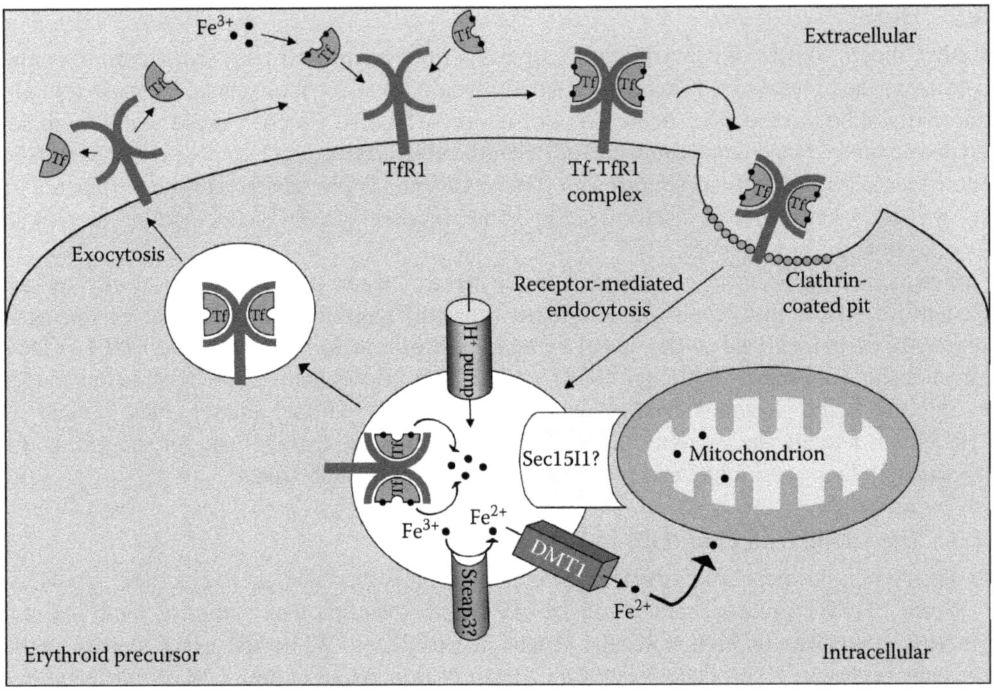

FIGURE 7.2 Erythroid iron uptake: Diferric Tf binds to its receptor, the TfR1, and the resulting complex is internalized into an endosome. A proton pump-mediated endosomal pH decrease releases ferric iron from Tf. In erythroid precursors, iron is then reduced by the six-transmembrane epithelial antigen of the prostate-3 (Steap3) in the endosome before being exported by DMT1. It is thought that the Sec15l1 protein plays a role in Tf-containing endosome cycling and may also dock to mitochondria to directly deliver iron to this important organelle. Apo-transferrin and TfR1 are then returned to the cell surface.

role in iron metabolism and uptake remains unclear. In fact, TfR2 does not appear to play a role in quantitative iron uptake[30,31] and appears to be involved in regulating iron homeostasis.

This Tf–TfR1 complex undergoes receptor-mediated endocytosis and is internalized into an endosome (Figure 7.2).[29,32] Iron is subsequently released from Tf following a decrease in endosomal pH, mediated by a membrane-bound proton pump (Figure 7.2).[29]

7.2.2.2 Iron Export from the Endosome

It has been observed that Tf undergoes a conformational change after binding to TfR1, where each lobe opens and may possibly allow other proteins, such as endosomal DMT1, to access the bound iron within the vesicle.[33,34] If such a direct iron transfer occurs from protein to protein, this would prevent the precipitation of iron and allow highly efficient Fe transport from the endosome. Indeed, studies using the mk/mk mouse model that suffers hypochromic microcytic anemia have demonstrated that DMT1 is the physiologically relevant Fe transporter.[15,35]

7.2.2.3 Endosomal Ferrireductase

While Tf binds iron in the ferric state, it is released from the endosome by DMT1 in the ferrous state. It is unclear whether the ferric iron is reduced prior to or after its release from Tf, and this suggested that a ferrireductase must exist within the endosomal vesicle.[36] The identity of this ferrireductase was recently determined through the characterization of the *nm1054* mutant mouse model. This mouse model of iron deficiency has a phenotype of hypochromic microcytic anemia, where there is impaired hemoglobin synthesis due to decreased iron uptake.[36] Using this model, the recently reported six-transmembrane epithelial antigen of the prostate-3 (Steap3) was determined to be the endosomal ferrireductase responsible for iron reduction in the endosomes of erythroid precursors for iron utilization (Figure 7.2).[36] Steap3 is highly expressed on erythroid cells and is localized to Tf–TfR1-containing endosomes.[36] However, Steap3 is not needed for efficient iron uptake in other cell types. In light of the discovery of Steap3 and the potential role of Dcytb in iron absorption in the gut, it is possible that several ferrireductase systems may be utilized in iron metabolism, some of which are probably yet to be identified.[13,36]

Once iron is released into the hypothesized LIP, the endosome undergoes exocytosis returning the TfR1 to the cell surface and releasing apo-Tf back into the serum to further facilitate cellular iron uptake (Figure 7.2).[29,32]

7.2.2.4 Endosomal Docking to the Mitochondrion, Sec15l1

Another mouse model of hypochromic microcytic anemia enabled an increased understanding of Tf cycling.[37,38] A mutation of the *Sec15l1* gene in hemoglobin-deficit (hbd) mice was found to contribute to their anemic state, whereby the mice appeared to lack effective cycling of Tf-containing endosomes in erythroid cells.[37–39] The Sec15l1 gene product, which is part of the mammalian exocyst complex, is thought to play a role not only in the cycling of Tf-containing endosomes, but also in the docking of endosomes to the mitochondrion for the direct delivery of iron to this compartment (Figure 7.2).[39] However, it has yet to be demonstrated whether there is reduced iron levels in the mitochondria of reticulocytes in *hbd* mice. This system would enable iron released from the endosomal vesicle to be directly delivered to the mitochondrion and is consistent with reticulocyte studies suggesting that iron is transported directly from protein to protein and/or compartment to compartment.[21] These models, which bypass the LIP, question the existence of this compartment. However, its existence in other cell types, such as hepatocytes and macrophages, cannot be ruled out. It is speculated that iron in the LIP is distributed to specific organelles within the cell by an iron-trafficking system to be incorporated into the active sites of proteins or for storage in the iron storage protein, ferritin.[40]

7.2.3 The Iron Storage Protein, Ferritin

Iron is stored within cells in the storage protein, ferritin, which is composed of two types of subunits, namely the heavy (H) and light (L) chains.[41] This protein acts both as an iron storage and detoxification system, sequestering free iron and preventing it from reacting deleteriously in the Haber–Weiss reaction.[41] Under conditions of excess iron, ferritin synthesis is upregulated, while ferritin levels are reduced when iron availability is low, contributing toward overall iron homeostasis.[6]

All ferritins have 24 subunits, with varying H-to-L ratios found in different species.[41] These are arranged to form a hollow protein shell that can accommodate over 4500 atoms of iron in the ferric state as inorganic complexes in the core.[42,43] Importantly, the H-chain is associated with the oxidation of ferrous iron to the ferric state through a ferrioxidase site composed of seven conserved amino acid residues.[41,44] The L-chain contains a nucleation site that is thought to assist in core formation.[41] It is believed that iron is taken up through channels that transverse the ferritin protein shell[45–51] and is initially oxidized at specific sites to the ferric state, being laid down as inorganic iron (ferrihydrite) in a preformed cavity.[41]

Little is understood regarding the mechanisms involved in the release of iron from ferritin. Traditionally, the mobilization of iron from ferritin is believed to occur via two possible mechanisms. Present knowledge favors the lysosomal degradation pathway for the reutilization of ferritin iron.[52,53] Iron release from ferritin has been suggested to occur through the degradation of ferritin within lysosomes[52–54] and *in vitro* studies have demonstrated this.[53,55] On the other hand, studies *in vitro* have shown that biological reductants, such as superoxide,[56] ascorbate,[57] and flavins[58] can promote iron release from ferritin. In fact, it has been speculated that structural fluctuations in ferritin may allow for the access of small reductants to enter the protein and directly interact with its iron core.[58] After being released from ferritin, the stored iron can then be incorporated into iron-containing proteins including those located within the mitochondrion.

7.3 MITOCHONDRIAL IRON METABOLISM

7.3.1 Iron Transport into the Mitochondrion

The mitochondrion is a critical compartment in iron metabolism that is responsible for heme synthesis and iron–sulfur ([Fe–S]) cluster biosynthesis.[2] It has recently been suggested that iron is transported into the mitochondrion of murine erythroblasts by the iron transporter, mitoferrin1 (Figure 7.3).[59] Mutations in murine *mitoferrin1* and its zebrafish homolog, *frascati*, was found to result in the impairment of heme synthesis due to defects in mitochondrial iron uptake.[59,60] Also, the mutations in the yeast homologs, MRS3 and MRS4, showed impaired heme and [Fe–S] cluster biosynthesis.[61] Given that mutations in mitoferrin1 do not lead to a deleterious phenotype, it can be speculated that other mechanisms of mitochondrial iron transport must also exist (Figure 7.3).

7.3.2 Heme Transport from the Mitochondrion

Once in the mitochondrion, iron can be utilized for a wide range of metabolic processes, including heme biosynthesis (Figure 7.3). Newly synthesized heme is transported out of the mitochondrion for incorporation into proteins, but the mechanism by which it is exported remains elusive. A number of protein candidates have been identified as possible mitochondrial heme exporters. The breast cancer resistance protein (ABCG2)[62] and the ABC-mitochondrial erythroid (ABC-me) transporter,[63] members of the ATP-binding cassette superfamily of membrane transporters, are thought to be candidates in the trafficking of heme from the mitochondrion (Figure 7.3). Ablation of ABCG2 in mice studies was found to lead to an accumulation of protoporphyrin IX, a heme synthesis intermediate.[62] This result indicated that ABCG2 may have a role in the export of heme from the mitochondrion to be

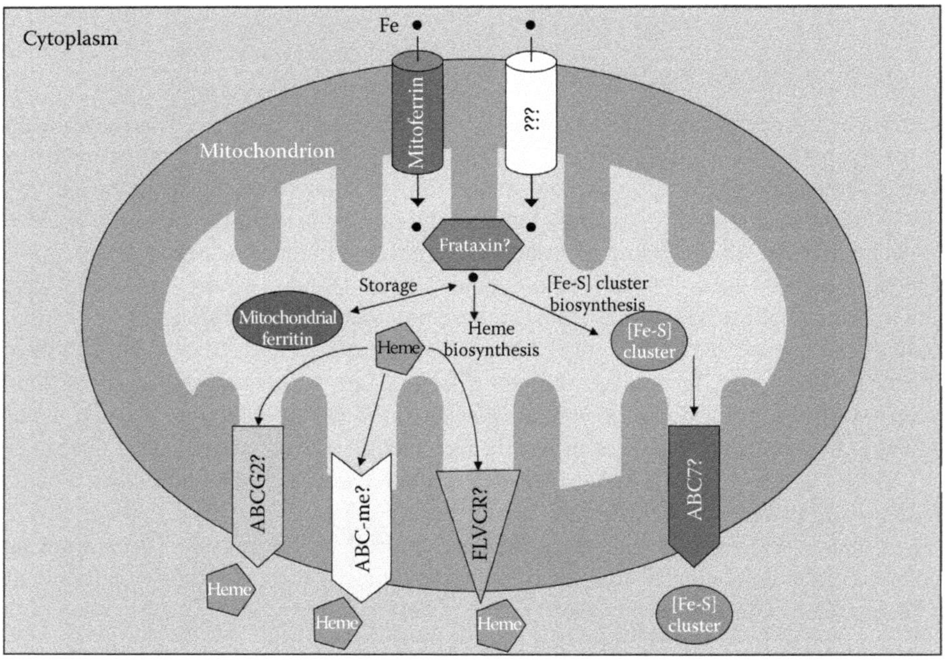

FIGURE 7.3 Iron metabolism in the mitochondrion: Iron is transported into the mitochondrion by mitoferrin1 in erythroid cells and potentially by other iron transporters. Frataxin, an inner mitochondrial protein, may play a role in determining whether this iron is utilized for heme and/or [Fe-S] cluster biosynthesis. Alternatively, iron can also be stored by mitochondrial ferritin. Heme is hypothesized to be exported from the mitochondrion by ABCG2, ABC-me, or FLVCR, while [Fe-S] clusters are thought to be exported by ABC7 for incorporation into proteins.

incorporated in proteins. Additionally, ABC-me overexpression was found to enhance hemoglobin synthesis in erythroleukemia cells and, thus, is speculated to facilitate mitochondrial transport functions that are related to heme biosynthesis (Figure 7.3).[63] The feline leukemia virus subgroup C receptor (FLVCR) is also speculated to be another protein involved in the transport of heme.[64] Studies using human FLVCR demonstrated that it was involved in the export of cytoplasmic heme and was hypothesized to protect developing erythroid cells from heme toxicity.[64] More recently, ablation of FLVCR in mice was found to lead to a lack of definitive erythropoiesis with these animals dying in mid-gestation, further demonstrating the possible role FLVCR plays in heme transport and iron homeostasis (Figure 7.3).[65] However, the precise roles of ABC-me and FLVCR in mitochondrial heme trafficking remain uncertain.

7.3.3 Frataxin

The role of other proteins involved in the trafficking of iron in the mitochondrion has been identified through a number of disease states. For example, the expression of frataxin (Figure 7.3), an inner mitochondrial protein, is decreased in the disease, FA, which leads to iron loading in the mitochondrion.[66,67] The decrease in frataxin expression correlates with iron deposits seen in the mitochondria of the liver, heart, and spleen of FA patients[68] and decreased respiratory chain activity of the [Fe–S] cluster-containing complexes I, II, and III.[69] While the form of this excess mitochondrial iron is unknown, it may be an unbound macromolecular precipitate, or less likely, stored in mitochondrial ferritin (Figure 7.3). Frataxin is believed to play a role in the regulation of mitochondrial iron utilization and was hypothesized to act as a switch between heme synthesis and the production of [Fe–S] clusters (Figure 7.3).[70] However, the exact role that frataxin plays in mitochondrial iron trafficking and metabolism has yet to be determined.

A recent investigation has examined the physiological role of frataxin by utilizing the muscle creatine kinase (MCK) conditional frataxin knockout (mutant) mouse model that reproduces the classical traits associated with cardiomyopathy in FA.[67] This study examined the mechanisms responsible for the increased cardiac mitochondrial iron loading in mutants. These experiments showed that increased mitochondrial iron in the myocardium of mutants was due to marked transferrin–iron uptake, which was the result of enhanced TfR1 expression.[71] In contrast to the mitochondrion, cytosolic ferritin and FPN1 expression as well as the proportion of cytosolic Fe were decreased in mutant mice, indicating cytosolic iron deprivation and markedly increased mitochondrial iron targeting.[71] These studies demonstrated that loss of frataxin alters cardiac iron metabolism due to pronounced changes in iron trafficking away from the cytosol to the mitochondrion. Hence, frataxin deficiency leads to decreased [Fe–S] synthesis, which results in a compensatory increase in iron transport to the mitochondrion.[71] This response may be an attempt to rescue the decreased [Fe–S] synthesis that is vital for energy generation. A model summarizing the altered iron trafficking observed is presented in Figure 7.4.

7.3.4 Export of [Fe–S] Clusters from the Mitochondrion

A mutation in the membrane transporter, ABC7, found in X-linked sideroblastic anemia with ataxia, demonstrated its function in the maturation of cytosolic [Fe–S] cluster-containing proteins.[72] Expression of the mutated ABC7 protein product was found to lead to a low efficiency of cytosolic [Fe-S] protein maturation and thus the functional ABC7 protein was hypothesized to be involved in the transfer of [Fe–S] clusters from the mitochondrion to the cytoplasm (Figure 7.3).[72] The resulting iron accumulation in the mitochondria of neural cells in patients affected with X-linked sideroblastic anemia leads to cellular damage and eventually death.[72]

7.4 IRON HOMEOSTASIS

The regulation of Fe metabolism is complex and can be divided into those mechanisms that occur at the cellular and systemic levels.

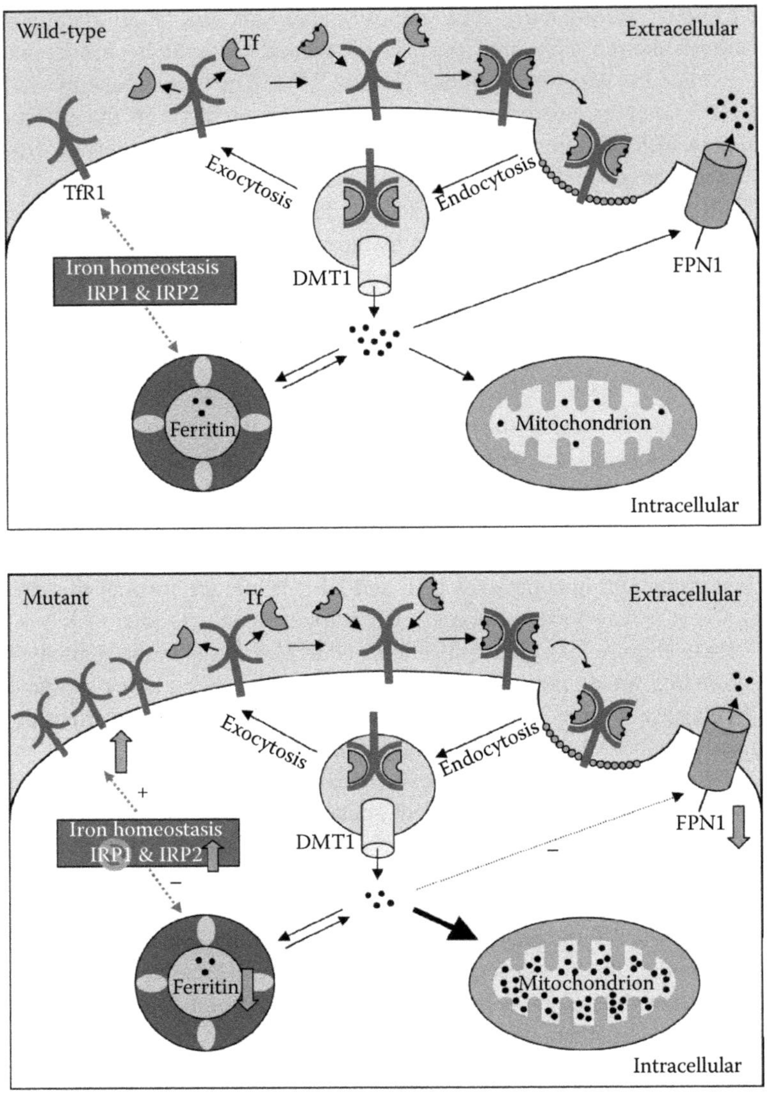

FIGURE 7.4 Alterations in iron trafficking in the MCK conditional frataxin knockout (mutant) mouse: Ablation of frataxin causes decreased [Fe–S] synthesis that is critical for energy generation. Increased TfR1 and decreased ferritin and FPN1 levels are evident in the mutant compared to the wild-type mice, which are thought to be mediated by increased IRP2–RNA binding. There is a cytosolic iron deficiency in mutant animals and increased trafficking of iron to the mitochondrion leading to iron-loading of this organelle. (Adapted from Whitnall et al. *Proc. Natl. Acad. Sci. USA*, 105, 9757, 2009.)

7.4.1 Regulation of Cellular Iron Metabolism: The IRPs

The uptake, metabolism, and storage of iron in cells are regulated, at least in part, by IRP1 and IRP2 (Figure 7.5). These cytosolic proteins have been identified as key iron sensors that form a post-transcriptional regulatory network by which iron homeostasis is controlled.[73] Both IRP1 and IRP2 are able to recognize and bind in a structure- and sequence-specific manner to the IRE, a highly conserved 28-nucleotide sequence motif present in the UTR of a range of mRNAs encoding proteins involved in iron metabolism. These IRE-containing mRNAs include the *TfR1*, *ferritin*, *DMT1*, and *FPN1* (Figure 7.5).[73,74] As described below, the IRP–IRE mechanism plays an important role in regulating the expression of these molecules via sensing intracellular iron levels in almost all cells.[75]

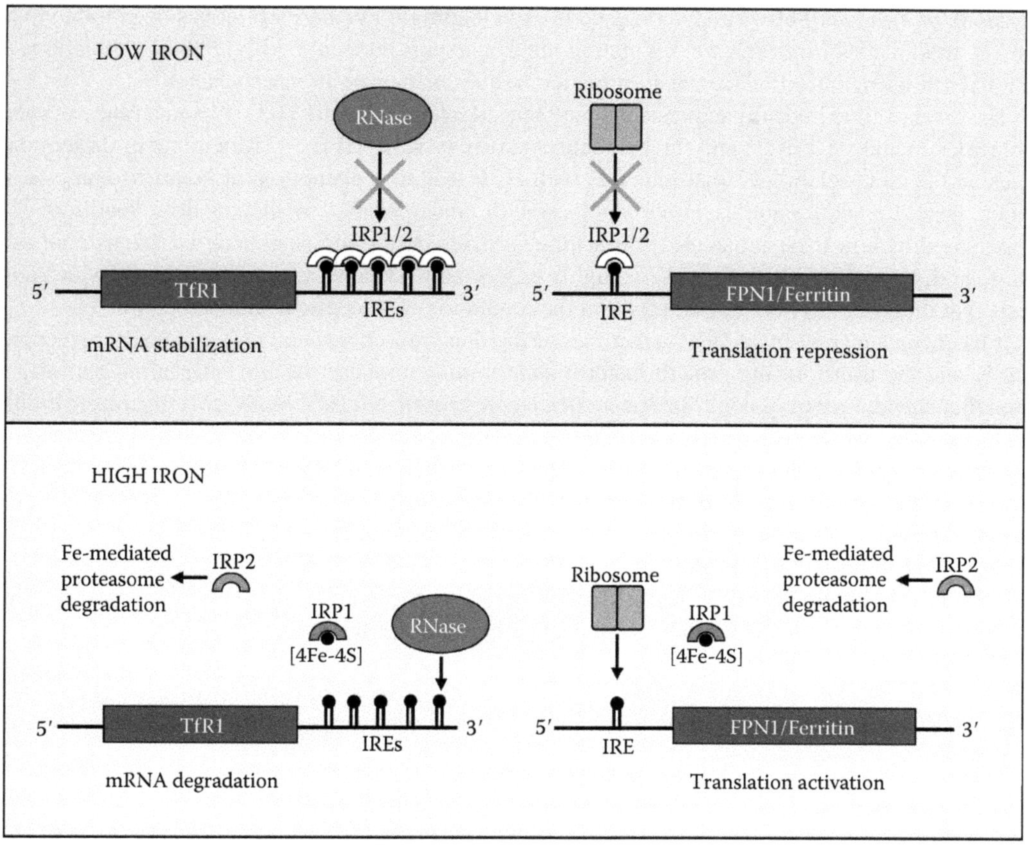

FIGURE 7.5 The role of IRP1 and IRP2 in iron homeostasis: Intracellular iron levels regulate the binding of IRP1 and IRP2 to their IREs. Under conditions of low intracellular iron levels, IRP1 and IRP2 are able to bind to the five IREs at the 3′ end of *TfR1* mRNA, resulting in mRNA stabilization and increased TfR1 levels. Simultaneously, the translation of *FPN1* and *ferritin* is repressed by the binding of IRP1 and IRP2 to the 5′ IRE, leading to decreased iron export and storage. Under high iron levels, the formation of a [4Fe-4S] cluster prevents the binding of IRP1 to IREs, while IRP2 is targeted for proteasome degradation. Hence, *TfR1* mRNA undergoes degradation, leading to decreased iron uptake, while *FPN1* and *ferritin* mRNA are translated, resulting in increased export and storage.

Both IRP1 and IRP2 bind to the five IREs in the 3′ UTR of *TfR1* mRNA under conditions of low iron levels resulting in mRNA stabilization (Figure 7.5). This leads to the subsequent translation of the mRNA. When iron levels are high, there is a loss of the mRNA-binding activity of IRPs, which leads to mRNA degradation. As the IRE-containing *DMT1* transcript also contains a single 3′ IRE, a similar mechanism is also involved in its regulation by iron.[35,76] While IRP1 and IRP2 play similar roles in post-transcriptional regulation of DMT1 expression, these two proteins act through different mechanisms.[73,77] In fact, in the case of IRP1, IRE binding is regulated by the presence of an [Fe–S] cluster that forms only when iron is abundant,[75] whereas IRP2 is regulated via iron-mediated degradation by the proteasome (Figure 7.5).[78]

7.4.2 Regulation of Systemic Iron Metabolism: The Iron Regulatory Hormone, Hepcidin

At the systemic level, regulation of iron metabolism is achieved by the peptide, hepcidin, that is primarily synthesized by the liver[79] and maintains iron homeostasis.[80] Indeed, hepcidin is the *hormone of iron metabolism* and is induced by iron loading.[81] Furthermore, it has been found to be the key modulator of iron release, regulating the iron export molecule, FPN1.[82] It also functions as a signal inhibiting intestinal iron absorption and sequestering iron in macrophages.[83]

The level of liver hepcidin expression is influenced by hemojuvelin (HJV)[84] and a range of other molecules including TfR2[85] and the hemochromatosis protein, HFE.[86,87] Mutations in these molecules and β_2-microglobulin[88] that interacts with HFE lead to a phenotype of hemochromatosis, in which hepcidin expression is either decreased or unresponsive to dietary iron loading.[85,88,89] Understanding how these molecules interact may provide important insights into a spectrum of iron overload diseases. Significantly, HJV is highly expressed in skeletal muscle and heart,[90] which suggests that these tissues may be important in the regulation of systemic iron metabolism.

It has been suggested that HJV, a member of the bone morphogenetic protein (BMP) coreceptor family and the transforming growth factor-β superfamily, may use the BMP signaling pathway to upregulate hepcidin expression.[86] In this model, it is proposed that HJV binds BMP ligands to induce BMP signaling that in turn activates the SMAD signaling pathway to directly increase hepcidin gene expression.[86] Indeed, mice whose hepatocytes were deficient in SMAD4 had reduced hepcidin expression and a phenotype of iron overload.[91] This pathway is not used by HFE or TFR2[92] and the precise mechanism of these upstream regulators remains unclear. HJV is typically membrane bound by a glycosylphosphatidylinositol anchor, although a soluble HJV (sHJV) form does exist.[93] Interestingly, it is hypothesized that sHJV competes with membrane-bound HJV for receptor binding on the cell surface, with sHJV inhibiting the signaling cascade that induces hepcidin expression.[93]

Hepcidin also acts as an acute phase protein and its expression can be markedly induced by interleukin-6 (IL-6).[81] IL-6 is a myokine, that is, a cytokine secreted from active skeletal muscle.[94] Thus, in summary, increased HJV or increased IL-6 leads to increased expression of the iron metabolism hormone, hepcidin.

7.5 CONCLUSION

Although iron is essential for life owing to its role in a variety of critical cellular processes, excess iron can have severe health consequences. Thus, tight regulation of iron levels through the implementation of a vast array of transport, storage, and regulatory proteins is necessary. Recently, a wide range of new molecules involved in iron metabolism and homeostasis have been discovered, many of their roles being highlighted through their involvement in a number of disease states. These new findings have answered a number of questions regarding the regulation and transport of iron at a cellular level and have ultimately altered our previous perceptions of iron metabolism. Although many gaps in our knowledge have been recently filled, numerous facets of iron metabolism have yet to be fully characterized. Only through further investigation can we begin to complete the picture of iron metabolism and homeostasis.

ACKNOWLEDGMENTS

This work was supported by grants and fellowship support (to D.R.R.) from the National Health and Medical Research Council of Australia, Australian Research Council, Muscular Dystrophy Association USA, and Friedreich's Ataxia Research Alliance of Australia and USA. We thank Ms Zaklina Kovacevic and Ms Yu Yu for proofreading the manuscript prior to submission.

REFERENCES

1. Andrews, N.C. Disorders of iron metabolism. *N. Engl. J. Med.*, 341, 1986, 1999.
2. Hentze, M.W., Muckenthaler, M.U., and Andrews, N.C. Balancing acts: Molecular control of mammalian iron metabolism. *Cell*, 117, 285, 2004.

3. Richardson, D.R. and Ponka, P. The molecular mechanisms of the metabolism and transport of iron in normal and neoplastic cells. *Biochim. Biophys. Acta*, 1331, 1, 1997.
4. Halliwell, B. and Gutteridge, J.M. Oxygen free radicals and iron in relation to biology and medicine: Some problems and concepts. *Arch. Biochem. Biophys.*, 246, 501, 1986.
5. Barnham, K.J., Masters, C.L., and Bush, A.I. Neurodegenerative diseases and oxidative stress. *Nat. Rev. Drug Discov.*, 3, 205, 2004.
6. Lieu, P.T., et al. The roles of iron in health and disease. *Mol. Aspects Med.*, 22, 1, 2001.
7. Shinar, E. and Rachmilewitz, E.A. Oxidative denaturation of red blood cells in thalassemia. *Semin. Hematol.*, 27, 70, 1990.
8. Wong, A., et al. The Friedreich's ataxia mutation confers cellular sensitivity to oxidant stress which is rescued by chelators of iron and calcium and inhibitors of apoptosis. *Hum. Mol. Genet.*, 8, 425, 1999.
9. Alper, G. and Narayanan, V. Friedreich's ataxia. *Pediatr. Neurol.*, 28, 335, 2003.
10. Schrier, S.L., et al. The role of oxidant injury in the pathophysiology of human thalassemias. *Redox. Rep.*, 8, 241, 2003.
11. Conrad, M.E. and Umbreit, J.N. Pathways of iron absorption. *Blood Cells Mol. Dis.*, 29, 336, 2002.
12. McKie, A.T., et al. An iron-regulated ferric reductase associated with the absorption of dietary iron. *Science*, 291, 1755, 2001.
13. Gunshin, H., et al. Cybrd1 (duodenal cytochrome b) is not necessary for dietary iron absorption in mice. *Blood*, 106, 2879, 2005.
14. Fleming, M.D., et al. Nramp2 is mutated in the anemic Belgrade (b) rat: Evidence of a role for Nramp2 in endosomal iron transport. *Proc. Natl. Acad. Sci. USA*, 95, 1148, 1998.
15. Fleming, M.D., et al. Microcytic anaemia mice have a mutation in Nramp2, a candidate iron transporter gene. *Nat. Genet.*, 16, 383, 1997.
16. Gunshin, H., et al. Slc11a2 is required for intestinal iron absorption and erythropoiesis but dispensable in placenta and liver. *J. Clin. Invest.*, 115, 1258, 2005.
17. Shayeghi, M., et al. Identification of an intestinal heme transporter. *Cell*, 122, 789, 2005.
18. Qiu, A., et al. Identification of an intestinal folate transporter and the molecular basis for hereditary folate malabsorption. *Cell*, 127, 917, 2006.
19. Andrews, N.C. When is a heme transporter not a heme transporter? When it's a folate transporter. *Cell Metab.*, 5, 5, 2007.
20. Jacobs, A. Low molecular weight intracellular iron transport compounds. *Blood*, 50, 433, 1977.
21. Richardson, D.R., Ponka, P., and Vyoral, D. Distribution of iron in reticulocytes after inhibition of heme synthesis with succinylacetone: Examination of the intermediates involved in iron metabolism. *Blood*, 87, 3477, 1996.
22. Vyoral, D., Hradilek, A., and Neuwirt, J. Transferrin and iron distribution in subcellular fractions of K562 cells in the early stages of transferrin endocytosis. *Biochim. Biophys. Acta*, 1137, 148, 1992.
23. Zhang, A.S., Sheftel, A.D., and Ponka, P. Intracellular kinetics of iron in reticulocytes: Evidence for endosome involvement in iron targeting to mitochondria. *Blood*, 105, 368, 2005.
24. Chen, H., et al. Hephaestin is a ferroxidase that maintains partial activity in sex-linked anemia mice. *Blood*, 103, 3933, 2004.
25. Vulpe, C.D., et al. Hephaestin, a ceruloplasmin homologue implicated in intestinal iron transport, is defective in the sla mouse. *Nat. Genet.*, 21, 195, 1999.
26. Nemeth, E., et al. Hepcidin regulates cellular iron efflux by binding to ferroportin and inducing its internalization. *Science*, 306, 2090, 2004.
27. Pigeon, C., et al. A new mouse liver-specific gene, encoding a protein homologous to human antimicrobial peptide hepcidin, is overexpressed during iron overload. *J. Biol. Chem.*, 276, 7811, 2001.
28. Donovan, A., et al. The iron exporter ferroportin/Slc40a1 is essential for iron homeostasis. *Cell Metab.*, 1, 191, 2005.
29. Morgan, E.H. Transferrin biochemistry, physiology and clinical significance. *Mol. Aspects Med.*, 4, 1, 1981.
30. Kawabata, H., et al. Transferrin receptor 2-alpha supports cell growth both in iron-chelated cultured cells and *in vivo*. *J. Biol. Chem.*, 275, 16618, 2000.
31. Kawabata, H., et al. Molecular cloning of transferrin receptor 2. A new member of the transferrin receptor-like family. *J. Biol. Chem.*, 274, 20826, 1999.
32. Klausner, R.D., et al. Binding of apotransferrin to K562 cells: Explanation of the transferrin cycle. *Proc. Natl. Acad. Sci. USA*, 80, 2263, 1983.

33. Cheng, Y., et al. Structure of the human transferrin receptor–transferrin complex. *Cell*, 116, 565, 2004.
34. Richardson, D.R. Mysteries of the transferrin–transferrin receptor 1 interaction uncovered. *Cell*, 116, 483, 2004.
35. Gunshin, H., et al. Cloning and characterization of a mammalian proton-coupled metal–ion transporter. *Nature*, 388, 482, 1997.
36. Ohgami, R.S., et al. Identification of a ferrireductase required for efficient transferrin-dependent iron uptake in erythroid cells. *Nat. Genet.*, 37, 1264, 2005.
37. Lim, J.E., et al. A mutation in Sec15l1 causes anemia in hemoglobin deficit (hbd) mice. *Nat. Genet.*, 37, 1270, 2005.
38. White, R.A., et al. Iron metabolism mutant hbd mice have a deletion in Sec15l1, which has homology to a yeast gene for vesicle docking. *Genomics*, 86, 668, 2005.
39. Zhang, A.S., Sheftel, A.D., and Ponka, P. The anemia of "haemoglobin-deficit" (hbd/hbd) mice is caused by a defect in transferrin cycling. *Exp. Hematol.*, 34, 593, 2006.
40. Kalinowski, D.S. and Richardson, D.R. The evolution of iron chelators for the treatment of iron overload disease and cancer. *Pharmacol. Rev.*, 57, 547, 2005.
41. Harrison, P.M. and Arosio, P. The ferritins: Molecular properties, iron storage function and cellular regulation. *Biochim. Biophys. Acta*, 1275, 161, 1996.
42. Ford, G.C., et al. Ferritin: Design and formation of an iron-storage molecule. *Philos. Trans. R. Soc. Lond. B. Biol. Sci.*, 304, 551, 1984.
43. Harrison, P.M., et al. Probing structure–function relations in ferritin and bacterioferritin. *Adv. Inorg. Chem.*, 36, 449, 1991.
44. Xu, X., Persson, H.L., and Richardson, D.R. Molecular pharmacology of the interaction of anthracyclines with iron. *Mol. Pharmacol.*, 68, 261, 2005.
45. Desideri, A., et al. Iron entry route in horse spleen apoferritin. Involvement of the three-fold channels as probed by selective reaction of cysteine-126 with the spin label 4-maleimido-tempo. *FEBS Lett.*, 287, 10, 1991.
46. Harrison, P.M., Treffry, A., and Lilley, T.H. Ferritin as an iron-storage protein: Mechanisms of iron uptake. *J. Inorg. Biochem.*, 27, 287, 1986.
47. Stefanini, S., et al. Identification of the iron entry channels in apoferritin. Chemical modification and spectroscopic studies. *Biochemistry*, 28, 378, 1989.
48. Treffry, A., et al. Defining the roles of the threefold channels in iron uptake, iron oxidation and iron-core formation in ferritin: A study aided by site-directed mutagenesis. *Biochem. J.*, 296, 721, 1993.
49. Treffry, A., et al. Recombinant H-chain ferritins: Effects of changes in the 3-fold channels. *FEBS Lett.*, 247, 268, 1989.
50. Wardeska, J.G., Viglione, B., and Chasteen, N.D. Metal ion complexes of apoferritin. Evidence for initial binding in the hydrophilic channels. *J. Biol. Chem.*, 261, 6677, 1986.
51. Yablonski, M.J. and Theil, E.C. A possible role for the conserved trimer interface of ferritin in iron incorporation. *Biochemistry*, 31, 9680, 1992.
52. Persson, H.L., Nilsson, K.J., and Brunk, U.T. Novel cellular defenses against iron and oxidation: Ferritin and autophagocytosis preserve lysosomal stability in airway epithelium. *Redox. Rep.*, 6, 57, 2001.
53. Radisky, D.C. and Kaplan, J. Iron in cytosolic ferritin can be recycled through lysosomal degradation in human fibroblasts. *Biochem. J.*, 336, 201, 1998.
54. Roberts, S. and Bomford, A. Ferritin iron kinetics and protein turnover in K562 cells. *J. Biol. Chem.*, 263, 19181, 1988.
55. Hernandez-Yago, J., et al. Autophagy of ferritin incorporated into the cytosol of HeLa cells by liposomes. *Cell Tissue Res.*, 205, 303, 1980.
56. Bolann, B.J. and Ulvik, R.J. Release of iron from ferritin by xanthine oxidase. Role of the superoxide radical. *Biochem. J.*, 243, 55, 1987.
57. Bienfait, H.F. and Van den Briel, M.L. Rapid mobilization of ferritin iron by ascorbate in the presence of oxygen. *Biochim. Biophys. Acta*, 631, 507, 1980.
58. Funk, F., et al. Reductive mobilisation of ferritin iron. *Eur. J. Biochem.*, 152, 167, 1985.
59. Shaw, G.C., et al. Mitoferrin is essential for erythroid iron assimilation. *Nature*, 440, 96, 2006.
60. Ransom, D.G., et al. Characterization of zebrafish mutants with defects in embryonic hematopoiesis. *Development*, 123, 311, 1996.
61. Zhang, Y., et al. Mrs3p, Mrs4p, and frataxin provide iron for Fe-S cluster synthesis in mitochondria. *J. Biol. Chem.*, 281, 22493, 2006.

62. Jonker, J.W., et al. The breast cancer resistance protein protects against a major chlorophyll-derived dietary phototoxin and protoporphyria. *Proc. Natl. Acad. Sci. USA*, 99, 15649, 2002.
63. Shirihai, O.S., et al. ABC-me: A novel mitochondrial transporter induced by GATA-1 during erythroid differentiation. *EMBO J.*, 19, 2492, 2000.
64. Quigley, J.G., et al. Identification of a human heme exporter that is essential for erythropoiesis. *Cell*, 118, 757, 2004.
65. Keel, S.B., et al. A heme export protein is required for red blood cell differentiation and iron homeostasis. *Science*, 319, 825, 2008.
66. Cossee, M., et al. Inactivation of the Friedreich ataxia mouse gene leads to early embryonic lethality without iron accumulation. *Hum. Mol. Genet.*, 9, 1219, 2000.
67. Puccio, H., et al. Mouse models for Friedreich ataxia exhibit cardiomyopathy, sensory nerve defect and Fe–S enzyme deficiency followed by intramitochondrial iron deposits. *Nat. Genet.*, 27, 181, 2001.
68. Sanchez-Casis, G., Cote, M., and Barbeau, A. Pathology of the heart in Friedreich's ataxia: Review of the literature and report of one case. *Can. J. Neurol. Sci.*, 3, 349, 1976.
69. Rotig, A., et al. Aconitase and mitochondrial iron–sulphur protein deficiency in Friedreich ataxia. *Nat. Genet.*, 17, 215, 1997.
70. Becker, E.M., et al. Erythroid differentiation and protoporphyrin IX down-regulate frataxin expression in Friend cells: Characterization of frataxin expression compared to molecules involved in iron metabolism and hemoglobinization. *Blood*, 99, 3813, 2002.
71. Whitnall, M., et al. The MCK mouse heart model of Friedreich's ataxia: Alterations in iron-regulated proteins and iron chelation limits cardiac hypertrophy. *Proc. Natl. Acad. Sci. USA*, 105, 9757, 2009.
72. Bekri, S., et al. Human ABC7 transporter: Gene structure and mutation causing X-linked sideroblastic anemia with ataxia with disruption of cytosolic iron–sulfur protein maturation. *Blood*, 96, 3256, 2000.
73. Wallander, M.L., Leibold, E.A., and Eisenstein, R.S. Molecular control of vertebrate iron homeostasis by iron regulatory proteins. *Biochim. Biophys. Acta*, 1763, 668, 2006.
74. Wang, J. and Pantopoulos, K. The pathway for IRP2 degradation involving 2-oxoglutarate-dependent oxygenase(s) does not require the E3 ubiquitin ligase activity of pVHL. *Biochim. Biophys. Acta*, 1743, 79, 2005.
75. Hentze, M.W. and Kuhn, L.C. Molecular control of vertebrate iron metabolism: mRNA-based regulatory circuits operated by iron, nitric oxide, and oxidative stress. *Proc. Natl. Acad. Sci. USA*, 93, 8175, 1996.
76. Gunshin, H., et al. Iron-dependent regulation of the divalent metal ion transporter. *FEBS Lett.*, 509, 309, 2001.
77. Zoller, H., et al. Mechanisms of iron mediated regulation of the duodenal iron transporters divalent metal transporter 1 and ferroportin 1. *Blood Cells Mol. Dis.*, 29, 488, 2002.
78. Guo, B., et al. Iron regulates the intracellular degradation of iron regulatory protein 2 by the proteasome. *J. Biol. Chem.*, 270, 21645, 1995.
79. Park, C.H., et al. Hepcidin, a urinary antimicrobial peptide synthesized in the liver. *J. Biol. Chem.*, 276, 7806, 2001.
80. Nicolas, G., et al. Hepcidin, a new iron regulatory peptide. *Blood Cells Mol. Dis.*, 29, 327, 2002.
81. Nemeth, E., et al. Hepcidin, a putative mediator of anemia of inflammation, is a type II acute-phase protein. *Blood*, 101, 2461, 2003.
82. Ganz, T., Molecular control of iron transport. *J. Am. Soc. Nephrol.*, 18, 394, 2007.
83. Fleming, R.E. and Sly, W.S. Hepcidin: A putative iron-regulatory hormone relevant to hereditary hemochromatosis and the anemia of chronic disease. *Proc. Natl. Acad. Sci. USA*, 98, 8160, 2001.
84. Huang, F.W., et al. A mouse model of juvenile hemochromatosis. *J. Clin. Invest.*, 115, 2187, 2005.
85. Wallace, D.F., et al. First phenotypic description of transferrin receptor 2 knockout mouse, and the role of hepcidin. *Gut*, 54, 980, 2005.
86. Babitt, J.L., et al. Bone morphogenetic protein signaling by hemojuvelin regulates hepcidin expression. *Nat. Genet.*, 38, 531, 2006.
87. Zhou, X.Y., et al. HFE gene knockout produces mouse model of hereditary hemochromatosis. *Proc. Natl. Acad. Sci. USA*, 95, 2492, 1998.
88. Muckenthaler, M.U., et al. Molecular analysis of iron overload in beta2-microglobulin-deficient mice. *Blood Cells Mol. Dis.*, 33, 125, 2004.
89. Niederkofler, V., Salie, R., and Arber, S. Hemojuvelin is essential for dietary iron sensing, and its mutation leads to severe iron overload. *J. Clin. Invest.*, 115, 2180, 2005.

90. Papanikolaou, G., et al. Mutations in HFE2 cause iron overload in chromosome 1q-linked juvenile hemochromatosis. *Nat. Genet.*, 36, 77, 2004.
91. Wang, R.H., et al. A role of SMAD4 in iron metabolism through the positive regulation of hepcidin expression. *Cell Metab.*, 2, 399, 2005.
92. Truksa, J., et al. Bone morphogenetic proteins 2, 4, and 9 stimulate murine hepcidin 1 expression independently of Hfe, transferrin receptor 2 (Tfr2), and IL-6. *Proc. Natl. Acad. Sci. USA*, 103, 10289, 2006.
93. Lin, L., Goldberg, Y.P., and Ganz, T. Competitive regulation of hepcidin mRNA by soluble and cell-associated hemojuvelin. *Blood*, 106, 2884, 2005.
94. Febbraio, M.A., et al. Interleukin-6 is a novel factor mediating glucose homeostasis during skeletal muscle contraction. *Diabetes*, 53, 1643, 2004.

8 The Regulatory and Signaling Functions of Zinc Ions in Human Cellular Physiology

Hajo Haase and Wolfgang Maret

CONTENTS

8.1 Introduction 182
8.2 Zinc Proteins and the Zinc Proteome 182
8.3 Zinc Homeostatic Proteins and Cellular Zinc Homeostasis 183
8.3.1 Zinc Transporters 185
8.3.2 Zinc Sensors 185
8.3.3 Metallothioneins 186
8.3.3.1 Regulation of MT 188
8.3.3.2 MT Trafficking 188
8.4 Zinc in Intracellular Signal Transduction 189
8.4.1 Generation and Occurrence of a Zinc Signal 189
8.4.2 Involvement of the Zinc Signal in a Biological Process 189
8.4.3 Action of the Zinc Signal via Signal Transduction Pathways 189
8.4.4 Identification of the Molecular Targets of the Zinc Signal 189
8.4.5 Effect of Zinc on Signal Transduction Pathways 191
8.4.5.1 Insulin Signaling 191
8.4.5.2 EGF Signaling 191
8.4.5.3 Mitogen-Activated Protein Kinase 191
8.4.5.4 Protein Kinase C 192
8.4.6 Molecular Targets of Zinc in Signal Transduction 192
8.4.6.1 Transient Binding of Zinc to Structural Sites 192
8.4.6.2 Transient Binding of Zinc to Sites that Affect Enzymatic Activity 192
8.4.6.3 Zinc in the Assembly of Signaling Complexes 193
8.5 Zinc in Intercellular Signal Transduction 194
8.5.1 Specialized Cells that Secrete Zinc 194
8.5.1.1 Prostate 195
8.5.1.2 Pancreas 196
8.5.1.3 Neurons 197
8.5.1.4 Osteoblasts 198
8.5.1.5 Paneth Cells 198
8.5.1.6 Eye 198
8.5.1.7 Mammary Gland Epithelial Cells 198
8.5.1.8 Intestine 198
8.5.1.9 Skin 199
8.6 Zinc Signaling in the Immune System 199
8.6.1 Cells of the Adaptive Immune System 199
8.6.2 Cells of the Innate Immune System 200

8.6.3 Cytokine Production 200
8.7 Systemic Zinc Homeostasis 201
8.8 Conclusion 202
Acknowledgment 202
References 203

8.1 INTRODUCTION

Zinc is important for virtually all aspects of cellular functions. Two recent developments significantly advanced our understanding of human zinc biology: the recognition that zinc is a constituent of several thousand proteins and the discoveries that cellular zinc homeostasis requires at least three dozen proteins and a high degree of regulation and integration into cellular signaling networks. This chapter focuses on yet another development, namely emerging general and cell-specific functions of zinc ions beyond their roles in established zinc proteins. How zinc is controlled in human cells will be the basis for discussing the functions of zinc ions in cellular regulation and signal transduction, especially in phosphorylation cascades, and their transient interactions with some proteins that are generally not recognized as zinc proteins. Comparative and specific aspects of the zinc biology in plants, single-cell eukaryotes, or microorganisms are beyond the scope of this chapter. New functions of zinc can be summarized under a role of zinc ions as intercellular and intracellular signals, although they embody quite different principles of action. In essence, these functions establish additional biological activities and a much broader role of zinc in biology than previously acknowledged.

8.2 ZINC PROTEINS AND THE ZINC PROTEOME

At least 10% of the human genes encode proteins with zinc-binding motifs [1]. Among the approximately 3000 putative human zinc proteins, about 1000 are enzymes with catalytic functions of zinc and the rest are proteins in which zinc primarily has a structural function in domains that interact with other proteins, nucleic acids, or lipids. The count of zinc proteins includes 397 hydrolases; 302 ligases; 167 transferases; 43 oxidoreductases; 24 lyases/isomerases; 957 transcription factors; 221 signaling proteins; 141 transport/storage proteins; 53 proteins with structural zinc sites; 19 proteins involved in DNA repair, replication, and translation; 427 zinc finger proteins of unknown function; and 456 additional proteins of unknown function. This estimate of the human zinc proteome is based on ligand signatures for zinc-binding sites in protein/nucleotide sequences and the use of these signatures for homology searches in databases [1,2]. However, the actual number of zinc-dependent proteins could be considerably larger because this approach fails to account for some types of zinc–protein interactions [3]. The estimates do not include proteins to which zinc ions bind transiently and which have not been recognized as zinc proteins because zinc is not present when the proteins are isolated and characterized. A distinction was made between metalloproteins that bind metals tightly, and hence are isolated with the metal, and metal-activated proteins that bind metals less tightly and hence lose their metal during isolation [4]. For zinc, it is known that this metal inhibits the activity of a number of proteins by binding with relatively high (nanomolar) affinity. These proteins have to be isolated in the presence of a chelating agent in order to remain active, and therefore their nature as potential zinc metalloproteins is not noticed. Another example is zinc binding at protein interfaces in homologous and heterologous protein–protein interactions [5]. In this case, the individual isolated proteins lack zinc, but their assembly to complexes depends on zinc binding.

With few exceptions, the building blocks of zinc coordination environments in proteins are the side chains of histidine, cysteine, and glutamate/aspartate. However, permutations of these ligands and the use of ligand bridges in multizinc sites provide a remarkably rich variation of zinc-binding sites [5]. The use of zinc in so many proteins raises several issues. How does the cell ensure that all these proteins obtain the zinc they need for their function? How does it redistribute zinc without interfering with the metabolism of the other essential metal ions? How does it avoid zinc overload that could cause unspecific interactions and side effects? How can it maintain regulatory and signaling functions of zinc in addition to supplying zinc to zinc-requiring proteins? Answers to these questions are forthcoming from studies of the proteins that control cellular zinc homeostasis.

8.3 ZINC HOMEOSTATIC PROTEINS AND CELLULAR ZINC HOMEOSTASIS

Both thermodynamics and kinetics have a role in the way zinc is redistributed in the cell and in the body but the relative contributions of each are a matter of debate. If one considers thermodynamics only, every metal ion is available at concentrations commensurate with its binding affinity to metalloproteins [6]. When isolated, most eukaryotic zinc proteins contain zinc and no other metal ions in the zinc-binding site. Thus, rarely is there any promiscuity of metal binding in eukaryotic proteins, indicating high specificity in providing a metalloprotein with the correct metal ion. For example, in a protein that requires different metal ions, such as cytosolic superoxide dismutase, zinc is always in the structural site, while copper is always in the catalytic site. A specific copper chaperone, the copper chaperone for superoxide dismutase (CCS) protein, inserts copper into the catalytic site [7]. However, a specific chaperone for zinc is not known. One can argue that too many chaperones would be needed to supply the structurally quite variable sites in zinc proteins with zinc and that control of zinc ion insertion into proteins requires different mechanisms. Such mechanisms can entail controlling the insertion of other metals via specific chaperones and insertases, but making a sufficiently high concentration of zinc ions available for zinc sites in proteins and controlling their availability tightly. Minimally, control of cellular zinc involves the coordination properties of zinc-binding sites in proteins, membrane transporters for zinc, and proteins that bind and distribute zinc within the cell. Insights into how zinc is handled within the cell began to emerge with the identification of proteins involved in cellular zinc homeostasis and trafficking. The very existence of these proteins and a zinc homeostatic system that is at least as complex as those for iron or copper attests to the importance that the cell places on controlling zinc. Two dozen mammalian zinc transport proteins have been described so far. They are differentially localized within cells and mediate uptake and intracellular distribution of zinc. Zinc sensors, such as metal-response element-binding transcription factor-1 (MTF-1), and at least a dozen metallothioneins (MTs) complement these zinc transporters (Figure 8.1). Regulation of these zinc homeostatic proteins by various mechanisms affords a high degree of integration into cellular signaling networks, suggesting that control of zinc ion concentrations and zinc distribution is critical for many cellular functions. Compromised functions of these proteins in the presence of sufficient zinc perturb zinc metabolism and are a cause for diseases. A mutation in the *Zip4* (*SLC39A4*) gene causes acrodermatitis enteropathica, a rare disorder that is based on impaired intestinal zinc absorption and results in severe zinc deficiency [8,9]. In transient neonatal zinc deficiency, low zinc concentrations in the mother's milk are due to a mutation in the *ZnT-2* (*SLC30A2*) gene [10]. Individuals with a mutation in the pancreatic β-cell-specific zinc transporter protein ZnT-8 (SLC30A8) have a higher risk for type 2 diabetes [11–15]. A polymorphism in the *MT-1A* gene also has been associated with diabetes and its cardiovascular complications [16,17].

A critical question in zinc biology is how cellular zinc ions are distributed, specifically whether they are transferred between proteins by protein–protein interactions without ever being free (associative mechanism) or whether zinc dissociates first and then associates with another protein (dissociative mechanism). An associative mechanism works well for zinc exchange reactions between proteins in which zinc is bound to sulfur ligands [18]. In a dissociative mechanism, however, the metal is free, at least for a limited time. Measuring cellular free zinc ion concentrations and determining whether they are sufficient for providing zinc to zinc-requiring proteins is one way of testing whether a dissociative mechanism could be operative. In prokaryotic organisms, there is virtually no free zinc [19]. While certainly true in the sense of the word that zinc ions without any ligands do not exist in the biological milieu, the conditions are different in eukaryotic cells. Estimates

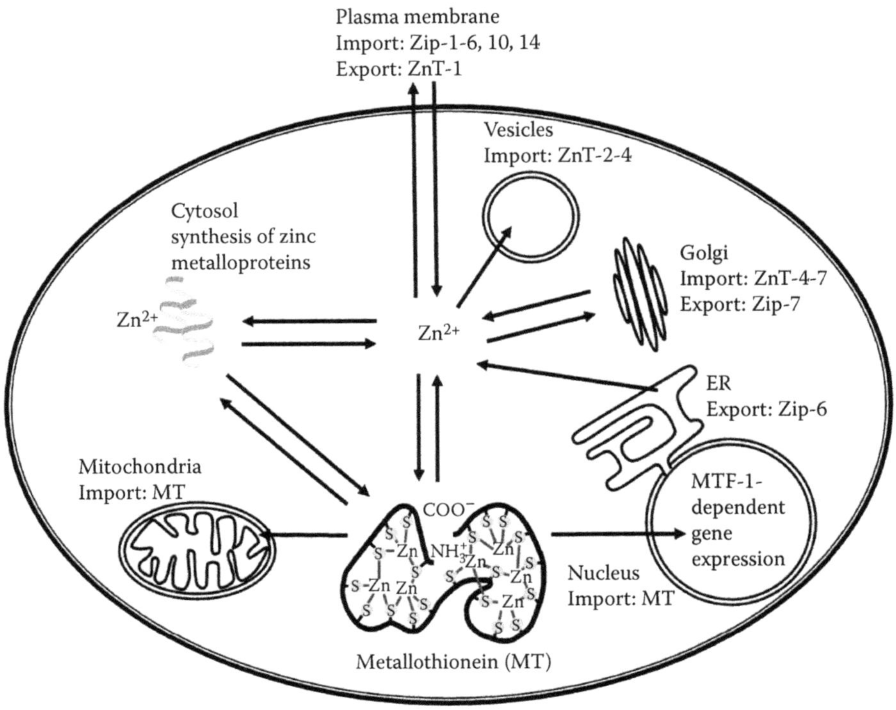

FIGURE 8.1 Cellular zinc homeostasis. 14 Zip (Zrt, Irt-like proteins, SLC39 gene family) and 10 ZnT proteins (SLC30 gene family) control zinc transfer across cellular membranes into or out of the cytosol. MTs sequester zinc in two zinc/thiolate clusters and supply zinc to zinc-requiring apoproteins. MT also transports zinc into the nucleus and into the intermembrane space of mitochondria. Zinc activates the zinc sensor MTF-1, which translocates to the nucleus and activates transcription of its target genes. All these proteins control free zinc ion concentrations that are critical for general as well as specific cellular, subcellular, and extracellular functions.

of the concentrations of free zinc ions under steady-state conditions are in the range of hundreds of picomolars in rabbit skeletal muscle, 24 pM in erythrocytes, and 500 pM in leukemia and neuroblastoma cells [20–23]. More recent estimates for eukaryotic cells are 5–10 pM for pheochromocytoma (PC12) cells, 170 pM and 350 pM for primary human monocytes and lymphocytes, respectively, between 625 pM and 1.25 nM for human colon cancer (HT29) cells depending on the state of the cells, and 1.07 nM for cortical neurons [24–27]. These concentrations are at least six orders of magnitude lower than the total cellular zinc concentration, which is about 200–300 μM. At which picomolar concentrations zinc ions are available in the cell will depend on the binding constants of the proteins that bind zinc tightly and the concentrations of any additional ligands that buffer zinc. For cytosolic zinc enzymes, the affinity for zinc is in the picomolar range, for example, carbonic anhydrase, superoxide dismutase, and sorbitol dehydrogenase [28]. The affinities of secreted zinc enzymes for zinc are also in the picomolar range [3]. Epithelial cells have an intracellular zinc buffering capacity of 20–30 μM with picomolar affinity for zinc [27]. Clearly, the overall zinc buffering capacity is much higher, but only the capacity that determines physiological zinc concentrations is important in this regard. How chemical buffering and mechanisms other than physicochemical buffering, referred to as muffling [29], interact determines the amplitudes of zinc ion fluctuations that can be used for cellular functions. Free zinc ion concentrations can change from picomolar to single-digit nanomolar under a variety of conditions, prompting the question of what the targets of such zinc ion fluctuations are [30–37]. One target of increased free zinc ion concentrations is the cellular zinc sensor, MTF-1, which is essential for basal and zinc-induced expression of MT [38–40]. Other targets will be discussed later in this chapter.

8.3.1 Zinc Transporters

Zinc transporters belong to two major protein families. So far, there are 14 Zip (Zrt/Irt-like) proteins, designated as SLC39A1-A14, which transport zinc into the cytosol, and 10 ZnT transporters, designated as SLC30A1-A10, which generally transport zinc out of the cytosol. Some of these proteins also transport metals such as manganese and iron, and therefore, it is not clear whether or not all of them are bona fide zinc transporters.

The regulation and the driving forces by which the different transporters maintain zinc homeostasis remain poorly characterized. Potential mechanisms include the passive facilitation of diffusion through membranes following a concentration gradient, active transport driven by hydrolysis of ATP through an ion pump, or secondary active transport coupled to the concentration gradient of another ion.

ZnT proteins are the mammalian subgroup of the cation diffusion facilitators (CDF). The name suggests a passive mechanism of transport. However, cation diffusion cannot be the sole driving force for transport, because in some cases zinc is transported against remarkable concentration gradients. Thus, zinc accumulates to millimolar concentrations in vesicles in the presence of only pico- to nanomolar concentrations in the cytosol. No ATP-binding motifs have been identified in the sequences of either vesicular importers ZnT-2 or -3 or in Zip proteins [41–43]. While zinc-transporting *P*-type ATPases exist in other organisms such as ZntA in *Escherichia coli* and HMA2 and HMA4 in *Arabidopsis thaliana*, none have been reported for mammalian cells [44–46].

In other CDF proteins from nonmammalian sources, namely the CzcD protein in *Bacillus subtilis*, ZitB and YiiP in *E. coli*, and Zrc1 in *Saccharomyces cerevisiae*, antiport with H^+ or K^+ is the driving force for zinc transport [47–50]. Thus, ZnT proteins can also function as active, secondary transporters. For some Zip proteins, it was already shown that they transport zinc ions together with hydrogen carbonate [51,52]. In addition, evidence for a Na^+/Zn^{2+} exchange activity mediating active zinc export on the plasma membrane of mammalian cells has been presented [53].

A crystal structure of the bacterial YiiP Zn^{2+}/H^+ exchanger provides a first glimpse at possible mechanisms of zinc transport through membranes [54]. Zinc ions may be delivered by a putative zinc metallochaperone to the C-terminal domain and then transported through the intracellular cavity, perhaps along a thermodynamic gradient provided by different binding sites, to the periplasmic space.

There is considerable cell-specific expression of some of the transporters in response to zinc status and endocrine signaling. Their regulation and their changing protein levels seem to be critical for zinc-dependent functions of a cell. Regulation includes heterodimerization, alternative splicing and different localization of the proteins, and control of their RNA stability [55,56]. While membrane zinc transporters are discussed in a separate chapter in this book, the following discussion will focus on the events in the soluble fractions of the cell.

8.3.2 Zinc Sensors

The only cytosolic proteins so far identified in controlling cellular zinc are MTs and the zinc sensor MTF-1. MTF-1 senses high zinc ion availability and induces the expression of thionein (T), the apo-form of MT, and ZnT-1 to bind and export zinc. Sensing is based on the coordination chemistry of at least two of its six zinc fingers with a Cys_2His_2 coordination motif. The affinities of the fingers for zinc are thought to vary about 10–50-fold even though the donor atoms in the fingers are identical [57]. The affinities are in the nanomolar range, which is lower than in "classic" zinc fingers that typically have picomolar affinities [40]. It is worth noting that these affinities are in the range where zinc sensing is expected based on the observed cellular free zinc ion concentrations.

On the other hand, zinc deprivation can also change gene expression, including differential expression of MTs and zinc transporters, as well as hundreds of genes that are not directly related to zinc homeostasis [58–60]. How such a reduced intracellular availability of zinc ions is sensed is not known.

8.3.3 Metallothioneins

MTs bind zinc, and under specific circumstances, they bind copper in addition to zinc. In humans, they are a family of at least 10 proteins, which are classified into MT-1 with multiple forms, and single MT-2, MT-3, and MT-4 proteins [61]. Zinc ions in MT are bound exclusively to the sulfur donors of cysteines in Zn_3S_9 and Zn_4S_{11} clusters that have characteristic thiolate ligand bridges between the zinc ions [62,63]. This structure of MT, in which 20 reduced cysteinyl residues bind the seven zinc ions buried inside the protein, has been the accepted model in discussions of the functions of MT. However, this model does not represent the physiologically important state, because MT occurs in the holoform, "MT," in the apo-form, "thionein," and in oxidized forms collectively called "thionin" in tissues and cells [64,65]. Under physiological conditions, MT neither is fully loaded with zinc ions nor has fully reduced cysteines. In this regard, it differs from many zinc proteins that are usually saturated with zinc. In rat liver, 20% of MT is in the apo-form and 7% is oxidized. The distribution of these forms changes when cells are incubated with zinc ions or when the cellular thiol/disulfide redox balance is perturbed [64]. This variation of the structure of MT is at the center of its mechanism of action.

Although all seven zinc ions are formally in similar tetrathiolate (Cys_4) coordination environments in the two clusters, the affinities of the sites for zinc differ by four orders of magnitude [66]. Four zinc ions are bound tightly (log $K = 11.8$), two are bound with intermediate strength (log $K \sim 10$), and one is bound relatively weakly (log $K = 7.7$). Based on these data, MT is described with Zn_4T, Zn_5T, Zn_6T, and Zn_7T species. Such a description has significant implications for the functions of MT. It suggests that the species Zn_7T does not exist under physiological conditions, because there are not enough free zinc ions available in the cell to saturate the weak binding site. With these properties, MT can participate actively in zinc metabolism rather than trapping zinc ions in a thermodynamically stable complex. Its coordination environments with fine-tuned affinities for zinc make picomolar to nanomolar concentrations of free zinc ions available.

A new aspect of zinc biology became evident from the studies of MT. A characteristic chemical feature of MT and many intracellular zinc proteins is the interaction of zinc with sulfur ligands [3]. For one, according to Pearson's Hard Soft Acid Base (HSAB) principle, zinc ions are at the borderline between hard and soft cations and therefore can accommodate both hard donors, such as oxygen, and soft donors, such as sulfur. The sulfur coordination environments provide thermodynamic stability for zinc, but kinetically they allow fast ligand exchange. In addition, the sulfur donors from the cysteines permit reversible redox reactions with concomitant release and binding of zinc [67]. In this way, the availability of zinc ions can be controlled by using metabolic energy in the form of redox changes to mobilize zinc from its tight binding sites under oxidizing conditions and to bind zinc under reducing conditions [68]. This redox chemistry of the ligands of redox-inert zinc differs from that of copper and iron, where the central atom is redox-active and changes its valence state. Coupled redox reactions establish a redox cycle for MT, in which the protein can be either a zinc donor or a zinc acceptor (Figure 8.2) [69].

The reactivity of the sulfur ligands in MT demonstrates how reactive compounds can affect zinc metabolism. Chemically different compounds react with the sulfur donors in zinc proteins and release zinc ions [67]. For example, MT can be nitrosylated *in vitro* or in cultured cells by agents that elevate cytoplasmic calcium and thereby activate calmodulin-dependent nitric oxide (NO) synthase [70,71]. In this process, zinc released from MT activates MTF-1-dependent gene transcription [72]. This Ca^{2+}–NO–Zn^{2+} pathway is cytoprotective in the lung, but can be cytotoxic in the brain when calcium influx into neurons activates the synthesis of NO and superoxide, which combine to form peroxynitrite and release cytotoxic levels of zinc ions [73,74]. The threshold of free zinc ion concentrations that separates pathology from physiology depends on both the redox and zinc buffering capacities and has not been defined clearly.

Another group of compounds that react with MT and other proteins with zinc–sulfur coordination environments is reactive carbonyls. Aldehydes introduced through environmental exposures, or

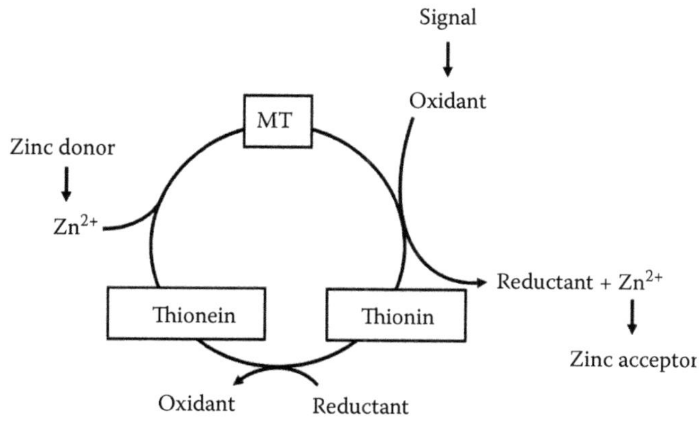

FIGURE 8.2 MT redox cycle in cellular signaling pathways. One redox couple controls zinc release from MT, the zinc-loaded state of the protein, to a partially oxidized and partially zinc-depleted state (thionin). Another redox couple controls the equilibrium between this state and a fully reduced and partially zinc-depleted state (thionein). Zinc binding to this state completes the cycle. In the presence of catalytic selenium compounds, the redox cycle of MT couples with the glutathione/glutathione disulfide redox pair (Data from Chen, Y. and Maret, W. *Eur. J. Biochem.*, 268, 3346, 2001). In this way, zinc can be transferred from zinc donors to zinc acceptors and redox reactions can control free zinc ion concentrations.

formed under a variety of conditions, such as oxidative stress, lipid peroxidation, hyperglycemia-induced glycations, and ethanol metabolism, modify the cysteine ligands in MT and release its zinc ions [75].

Thionylation of MT, such as glutathionylation or homocysteinylation, releases zinc ions [76,77]. The consequences of these reactions are at least twofold. First, the released zinc activates gene transcription [76]. Second, the modified cysteines in MT can no longer scavenge reactive species, thus lowering cellular antioxidant capacity. Given that some cells have micromolar concentrations of MT and that MT is a highly inducible protein, the reducing capacity of the 20 thiols in MT is significant. A function of MT as an antioxidant and scavenger of reactive species is acknowledged [78,79]. As a zinc-dependent redox system, MT can complement other cellular thiol/disulfide redox systems as has been demonstrated for a coupling between thionein, thionin, thioredoxin, and methionine sulfoxide reductase [80]. However, when such functions are discussed, attention is rarely given to the fact that the zinc ions released in these reactions are potent cellular effectors. Perturbation of cellular free zinc ion concentrations is necessarily linked to antioxidant functions of MT and scavenging of reactive species. The capacity of zinc to protect the cell against oxidative damage has been recognized for many years and has led to the notion that zinc itself is an antioxidant [81,82]. Chemically, this concept is not well founded, because zinc is redox-inert in biology. An antioxidant function must be indirect, that is, "pro-antioxidant" by a mechanism that links zinc and biochemical processes with an antioxidant potential [83]. One basis for such a function of zinc is that part of the cellular zinc buffering capacity depends on thiols [84]. These thiols are protected against oxidation when they bind zinc, but unprotected when there is not enough zinc, such as in zinc deficiency, which is a pro-oxidant condition [85–87]. Pro-antioxidant functions of zinc involve the activation of MTF-1, which induces thionein with high reducing capacity owing to its 20 cysteines. If the cellular zinc buffering capacity is exhausted at high concentrations of zinc, concentrations of free zinc ions will increase and zinc will bind to other targets and also generate a pro-oxidant condition by depleting cellular energy and increasing mitochondrial production of reactive oxygen species [88]. Thus, whether zinc is cytotoxic or cytoprotective is a matter of the concentrations of zinc ions, the state of MT, and the cellular zinc and redox buffering capacity [89,90].

8.3.3.1 Regulation of MT

Not only the expression of the zinc transporters, but also that of the T genes is tightly controlled by a host of chemical and physical events [91,92]. In this way, many cellular pathways can affect zinc availability and distribution. The total amount of MT varies over 400-fold in different cell lines [93]. T is induced by increased concentrations of zinc, various forms of stress, the acute phase response, and inflammation, with isoform- and tissue-specific expressions [94]. Interferons, cytokines, growth factors, and agents that act on nuclear hormone receptors or through the adenylate cyclase, phospholipase C, and Jak/Stat pathways all exert some control over MT [95,96]. Control at the gene and protein levels establishes that the zinc and redox functions of MT and T are an integral part of cellular biology to control the availability of zinc ions for regulatory functions. At low cellular free zinc ion concentrations, any newly synthesized T will equilibrate with any MT already existing and it can acquire additional zinc only if zinc is available. Therefore, the functions of these inducers can relate primarily to the properties of T as a zinc acceptor and reductant.

Zinc ion fluctuations must be tightly controlled to avoid unspecific reactions. For example, proteins must fold in an environment of low zinc availability, because zinc ion concentrations in the nanomolar range can lead to misfolding [97]. The different species of MT are ideal candidates for such a control, because they can be either zinc donors or zinc acceptors and, in this way, control a variety of zinc-dependent biological processes. MT can activate the apo-form of sorbitol dehydrogenase, which has a log K of 11.2 for zinc, or reactivate zinc-inhibited protein tyrosine phosphatase 1B (PTP1B), which has a log K of 7.8 for zinc [28]. Whether or not these cytosolic enzymes with zinc affinities that differ by at least three orders of magnitude receive zinc from MT or MT receives zinc from them depends on the cellular redox poise and the metal load of MT, which is described by the T/(MT + T) molar ratio. At ratios between 0.08 and 0.31 prevailing in tissues and cells, picomolar concentrations of free zinc ions are available from MT for reconstituting apo-enzymes with zinc but such ratios will not allow zinc inhibition of PTP1B. Under conditions of decreased ratios, such as those found during oxidative stress, nanomolar concentrations of free zinc ions become available from MT/T and affect enzymes that are not zinc metalloenzymes, such as PTP1B.

8.3.3.2 MT Trafficking

MTs are both extracellular and intracellular proteins, are distributed among different compartments in the cell, and interact with other proteins. The extensive trafficking of MT provides ways of distributing zinc. The driving forces for this trafficking are not known, however.

Within the cell, MT is imported into the intermembrane space of liver mitochondria. MT inhibits mitochondrial respiration in a zinc-dependent manner, while T activates respiration by removing zinc from the inhibitory site(s) [98]. Translocation of MT into the nucleus requires energy (ATP) and depends on phosphorylation signaling, the state of the cell, and the oxidation of a cytosolic factor [99–103].

Cells secrete MT and take it up from the extracellular space [104,105]. Secretion of MT from the exocrine pancreas does not seem to occur via a classic vesicular secretory pathway [106]. Uptake was first shown for kidney cell lines as an endocytotic process that involves the scavenger receptors megalin/cubilin and receptor-associated protein [107–109]. In hepatocarcinoma (HepG2) cells, uptake is cholesterol dependent, and the metals, but not the protein, are translocated to the cytoplasm [110]. MT-1 and MT-2 bind megalin and the lipoprotein receptor-related protein-1 on cerebellar granule neurons, suggesting that MT secreted from astrocytes affects neurons by promoting neurite outgrowth and survival [111–113]. In this regard, the biological activities of MT-1 and MT-2 are opposite to the one of MT-3, which prevents neurite outgrowth [114].

MT-3 has been isolated from brain tissue [115]. Extracellular MT-3 is a growth inhibitory factor (GIF) for neurons [115]. This activity is specific for the MT-3 isoform and has been mapped to a characteristic ^{6}CPCP motif in the N-terminal β-domain [116]. The molecular basis for the growth inhibitory activity of MT-3 is unknown. MT-3 is released from cultured astrocytes and functions in a pathway that loads neuronal synaptic vesicles with zinc ions [114,117]. In this process, MT-3 interacts with the small GTPase Rab3a, which in turn interacts with docking and trafficking proteins in the exo-endocytotic cycle of synaptic vesicles [118].

In the remaining discussion, the cell will continue to be the focus because zinc is basically an intracellular ion and used for cell-specific functions. The cytoplasm and organelles of hepatocytes contain 50% of cellular zinc, 30–40% are in the nucleus, and the remainder is in the microsomal fraction. The lowest concentrations are in mitochondria [119]. In general, not much is known about subcellular concentrations of total and free zinc. Given the intracellular distribution of transporters, the concentrations are likely to be different. Specific vesicles, such as lysosomes and zincosomes, are believed to have millimolar concentrations of zinc. Studies with zinc radioisotopes demonstrated uptake of the metal ion in the cytosol, mitochondria, and nuclei [120,121]. In several types of cells, the resting levels of free zinc ions are significantly lower in the nucleus than in the cytosol [122,123]. However, nuclear and mitochondrial zinc transporters are not known. MT translocates to both mitochondria and the nucleus and can serve as a zinc shuttle for these cellular compartments. In mitochondria, MT localizes to the intermembrane space [98]. Zinc transported into the mitochondrial matrix by the calcium uniporter is a very potent inhibitor of redox enzymes [124]. Whether or not zinc is required for physiological processes in the matrix is not known.

8.4 ZINC IN INTRACELLULAR SIGNAL TRANSDUCTION

The biology of calcium has shown that even the smallest chemical entity, namely an ion, can be an efficient regulator of cellular signal transduction. A similar function has been proposed for zinc [125]. For zinc ions to qualify as carriers of information, certain criteria should be met. Four conditions seem to be both sufficient and necessary.

8.4.1 Generation and Occurrence of a Zinc Signal

As the most basic condition, the concentrations of cytosolic free zinc ions should change in response to a stimulus. A cellular activator would have to trigger such a change either by an exchange of zinc with the extracellular environment or by release of zinc from an intracellular store. Alternatively, a translocation of free zinc ions between different cellular compartments could establish a subcellularly restricted signal.

8.4.2 Involvement of the Zinc Signal in a Biological Process

A change in the concentrations of free zinc ions should elicit a biological response to the stimulus that triggered it. The response does not need to elicit a global change of cellular function. Rather, it can be limited to just one specific aspect of metabolism or signaling, or it can establish crosstalk between pathways by modulating the effect of a different stimulus on another pathway. Specific chelation of zinc ions should inhibit such a signal, while increases in zinc ions should amplify it.

8.4.3 Action of the Zinc Signal via Signal Transduction Pathways

Zinc is a permanent structural and enzymatic cofactor in a significant number of proteins, most of which are not involved in signal transduction. The zinc activation of such proteins can be mistaken for a biological signal. Therefore, zinc should either be part of, or modulate, a signaling pathway.

8.4.4 Identification of the Molecular Targets of the Zinc Signal

The final proof of principle is the identification of interactions of zinc with specific molecular elements of a signaling pathway. Specificity, a hallmark of signaling, can be provided by the coordination chemical features of a zinc-binding site. At least three modes of actions of zinc have been observed (Figure 8.3). Zinc stabilizes structural zinc-binding motifs that influence the protein's biological activity, such as DNA binding and transcriptional activation (example A); zinc modulates the enzymatic activity of a signaling protein, such as inhibition of PTPs (example B); zinc affects the assembly of signaling complexes through binding at interfaces between proteins (example C).

Without being comprehensive, the following selected examples provide strong evidence that the first two conditions for zinc ions being biological signals are fulfilled. With regard to the generation and occurrence of zinc signals, it is recognized that zinc coordination environments in proteins with cysteine ligands can transduce redox signals into zinc signals [126]. Specifically, reactive species and oxidizing agents can react with the sulfur ligands of zinc in proteins and release zinc ions. For example, nitric monoxide (NO) reacts with zinc-bound thiols to form *S*-nitrosothiols or disulfides with concomitant release of zinc [36,71,127]. When cytokines stimulate endothelial cells, MT translocates to the nucleus and releases its zinc in an NO-dependent reaction, generating a transient nuclear zinc signal [35]. Furthermore, stimulation of protein kinase C (PKC) signaling causes a cellular redistribution of zinc and a rise of intracellular free zinc ions [128,129]. Another way of generating zinc signals is the release of zinc ions from intracellular compartments [130,131]. While relatively fast zinc signals can be generated by transducing redox signals or releasing compartmentalized zinc ions, different cellular free zinc ion concentrations can also result from altered expression of MTs and zinc transporters, a process that is much slower. In conclusion, a distinction is made

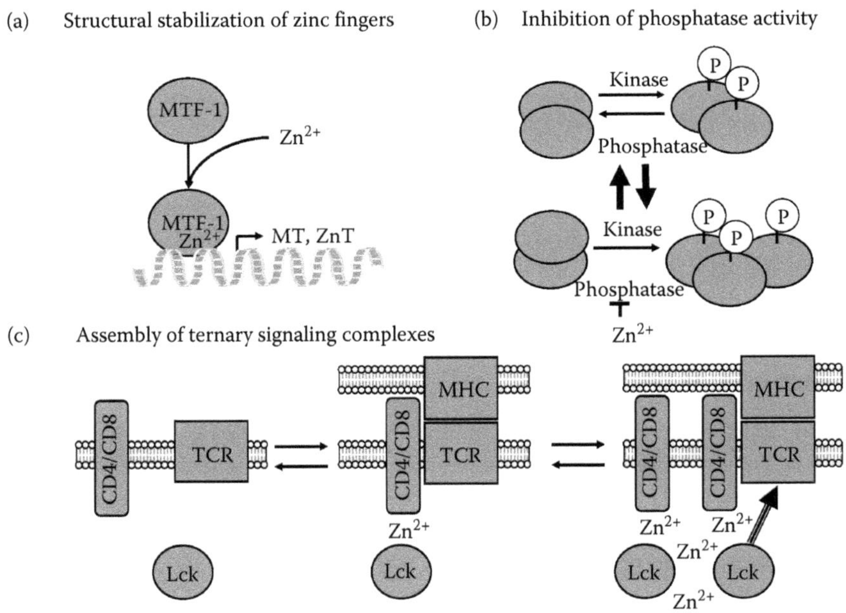

FIGURE 8.3 Mechanisms of intracellular zinc signaling. Three molecular mechanisms of how low nanomolar concentrations of zinc ions reversibly affect cellular signaling are known. (a) Structural stabilization. Zinc ions bind to the zinc fingers of the sensor MTF-1, which is translocated to the nucleus and activates gene transcription. (b) Regulation of enzymatic activity. Zinc ions inhibit PTPs and affect phosphorylation cascades. (c) Assembly of ternary complexes. Two zinc-dependent steps are essential for early TCR activation. First, upon contact with the MHC on an antigen-presenting cell, the kinase Lck is recruited to the complex of the TCR by binding to the costimulatory molecules CD4 or CD8 via a zinc ion at the interface of the two proteins. Second, a zinc-dependent dimerization through the interfaces between the SH3 domains brings Lck in close proximity, thereby promoting autocatalytic activation. Only after these two steps involving bridging zinc ions have occurred, Lck phosphorylates tyrosine residues in the CD3 signal transduction component of the TCR and zeta-chain (TCR) associated protein 70 kDa (ZAP70) kinase, thus initiating downstream TCR signaling.

here between stimuli that trigger zinc signals and stimuli that lead to adaptations of a cell to different levels of free zinc ions and concomitant changes of the functions of zinc-sensitive proteins.

The involvement of zinc signals in biological processes is treated in a rather extensive body of literature on the role of zinc in cellular proliferation, differentiation, and apoptosis [132]. Proliferation of baby hamster kidney cells depends on uptake of zinc, which is required for signals that lead to DNA synthesis [133]. In addition, zinc is required for creatine kinase expression during differentiation of myoblasts [134]. These effects of zinc may be mediated by MT, which undergoes nuclear translocation during proliferation of hepatocytes and during differentiation of myoblasts and adipocytes [99,100,135]. The basis for these growth factor-like activities of zinc ions will be explored further in the next section.

The two remaining conditions are similarly strongly supported by the effects of zinc ions on phosphorylation cascades, although the evidence is less widely appreciated.

8.4.5 Effect of Zinc on Signal Transduction Pathways

8.4.5.1 Insulin Signaling

It has been known for 70 years that the zinc content of the diabetic pancreas is lower than that of the normal pancreas and it has been known for over 40 years that zinc has a role in insulin signaling, because zinc-deficient laboratory animals are much less sensitive to insulin [136,137]. Zinc is insulin-mimetic, meaning that incubation of cells or tissues with zinc ions has the same effects on lipid and glucose metabolism as insulin itself [138,139]. The insulin-mimetic properties of zinc ions are clearly seen in cell cultures where zinc ions can completely replace insulin [140]. In the insulin signaling pathway, zinc ions affect the phosphorylation of the insulin receptor (IR) β subunit at the activating tyrosines 1158, 1162, and 1163 [141,142]. Because the effect is all the way upstream in the pathway, proteins downstream in the pathway are affected, as well. This point is important because the actions of zinc ions may appear pleiotropic, while in fact only a small number of upstream targets or even a single target may be affected. Thus, zinc ions influence the phosphorylation of tyrosine 856 on the IR substrate-1 protein, phosphatidyl inositol-3-kinase (PI3K)-mediated Akt phosphorylation, and hence glucose transport, and the activating phosphorylation of p21-activated kinases 1 and 2 and extracellular signal-regulated kinase (ERK) 1/2 [130,142].

8.4.5.2 EGF Signaling

Another receptor tyrosine kinase affected by zinc is the epidermal growth factor receptor (EGFR). Exposing human bronchial epithelial cells to zinc ions enhances mitogen-activated protein kinase (MAPK)-mediated cytokine expression by causing a dimerization-independent phosphorylation of EGFR at tyrosines 845, 1068, and 1173, which is mediated by the kinase Src [143,144]. The phosphorylation activates downstream signaling in the Ras/MEK/ERK and PI3K/Akt pathways by promoting the degradation of the negative regulator, the phosphatase and tensin homolog (PTEN) [145,146].

8.4.5.3 Mitogen-Activated Protein Kinase

Signaling pathways originating from membrane receptors other than the IR and the EGFR also involve MAPKs and are affected by zinc ions. Thus, ERK was activated when the growth factor-like effects of zinc ions on fibroblasts were examined [147]. Likewise, MEK-mediated phosphorylation of ERK was observed in the zinc-induced expression of the zinc finger transcription factor Egr-1 in neurons, and in the zinc-induced activation of the p70S6 kinase in neuroblastoma cells [148,149]. ERK is also activated when peroxynitrite releases zinc in oligodendrocytes [150]. However, from these experiments, one should not infer that zinc invariably stimulates MAPKs. In rat glioma cells, only low concentrations of zinc activate ERK. Higher concentrations are inhibitory [130]. In *Caenorhabditis elegans*, the ZnT-1 homolog CDF-1 is a positive regulator of EGFR signaling. Here, CDF-1-mediated zinc efflux activates the Ras/Raf/MEK/ERK pathway [151]. A potential molecular mechanism involves an effect of zinc on the phosphorylation of KSR (kinase suppressor of Ras), a scaffolding protein that stabilizes the interaction of Ras, Raf, and MEK [152]. A similar mechanism may occur in higher eukaryotes, but only at zinc concentrations that are significantly higher than the ones that activate MAPKs.

8.4.5.4 Protein Kinase C

PKC is part of signaling pathways that are activated by many different receptors. Chelation of free zinc ions inhibits the activation of PKC by different stimuli, such as activation of the NMDA receptor, the calcium ionophore A23187, phorbol esters, or antigen stimulation of lymphocytes [153–155]. One step in the activation of PKC, which can be induced by zinc ions, is its translocation to the plasma membrane [156]. Zinc ions increase the interaction of PKC with actin filaments [157,158]. In the zinc finger domain of PKC, zinc is required for the binding of phorbol esters [154,155].

8.4.6 Molecular Targets of Zinc in Signal Transduction

There are at least three different paradigms for the targets of zinc signals (Figure 8.3). In all three instances, free zinc ions must be tightly controlled to achieve sufficient biological specificity. Only selected examples are discussed because the extent to which zinc ions are employed in this capacity is unknown. Observations of zinc ion fluctuations in the heart support the notion that intracellular zinc signals are used more widely. When subcellular distribution of zinc in cardiomyocytes was investigated, transverse striations of millimolar concentrations of zinc (and iron) were found with a periodicity of sarcomers [159]. It was suggested that these intracellular zinc stores can be released into the cytosol within seconds [159].

8.4.6.1 Transient Binding of Zinc to Structural Sites

MTF-1 is a cellular zinc ion sensor [40,160]. MTF-1 controls the expression of proteins that are essential for zinc homeostasis, such as MTs, in particular the isoforms MT-1 and MT-2, the transporter ZnT-1, and at least 40 additional proteins that are not directly involved in zinc metabolism [39,161,162]. Increased zinc ion availability effects its nuclear translocation and DNA binding, and activation of gene expression. MTF-1 binds with six highly conserved Cys_2His_2 zinc finger domains to the promoters of several genes that contain metal-response elements (MREs) with a core consensus sequence TGCRCnC [163]. Interaction with DNA is mediated by reversible binding of zinc to some of these structural motifs and stabilization of the zinc finger domains. Which zinc finger(s) bind zinc constitutively, which ones are sensitive to changes in free zinc ion concentrations, and which other structural motifs are involved in fine-tuning zinc binding to MTF-1 is widely discussed [40,57,164–166].

While zinc binding controls the interaction between MTF-1 and DNA, zinc also influences MTF-1-mediated transcription by an indirect mechanism that involves phosphorylation of MTF-1 at serine and tyrosine residues by PKC, c-Jun N-terminal kinase (JNK), PI3K, casein kinase II, tyrosine kinases, and calcium signaling [167,168]. Reactive oxygen species can also activate MTF-1-dependent transcription. In this case, zinc ions are the mediators of redox signals. In cell-free experiments, hydrogen peroxide releases zinc from MT, which then binds to MTF-1 and activates transcription [169]. In intact cells, NO has the same effect [72].

How many other proteins bind zinc reversibly in zinc finger domains with consequences for their functions is not known.

8.4.6.2 Transient Binding of Zinc to Sites that Affect Enzymatic Activity

8.4.6.2.1 Zinc and PTP Activity

Zinc ions inhibit many enzymes, in particular those with cysteines in the active site, such as PTPs [170]. Zinc inhibition of PTP1B, the main phosphatase controlling tyrosine phosphorylation of the

IR, enhances phosphorylation signaling and demonstrates how zinc ions can be insulin-mimetic and have growth factor-like activities. Chelation of free zinc ions inhibits insulin- and insulin-like growth factor-1 (IGF-1)-dependent intracellular signals, an effect that has been attributed to a loss of PTP inhibition by zinc [130,141].

Many reported zinc inhibition constants for proteins are not within the physiological range of cytosolic free zinc ion concentrations. However, low nanomolar concentrations of zinc inhibit several enzymes [171]. Measurement of such low inhibition constants is not straightforward because enzymes must be free of chelating agents, such as DTT and EDTA, and they need to be active in the absence of these agents. Furthermore, low free zinc ion concentrations must be controlled by appropriate metal ion buffers. Employing such controlled conditions, a K_i value of 15 nM for zinc inhibition of PTP1B was measured [28,141]. The zinc-binding site has not yet been identified, but the strong zinc inhibition of the enzymatic activity of the truncated catalytic domain of PTP1C (=SHP-1) indicates that it is on this domain and likely involves the catalytic cysteine [130].

8.4.6.2.2 Zinc and Phosphodiesterase (PDE) Activity

Several isoforms of cyclic nucleotide PDEs are activated at low zinc concentrations, because they require a catalytic zinc ion [172]. On the other hand, slightly higher concentrations of zinc ions than those necessary for activation inhibit PDEs [173,174]. For example, zinc inhibition of PDEs causes an increase in cyclic nucleotides and subsequent PKA-mediated inhibition of proinflammatory cytokine signaling in monocytes [175,176].

8.4.6.2.3 Zinc and Mitogen-Activated Protein (MAP) Kinase Phosphatase (MKP) Activity

A third group of phosphatases that are sensitive to zinc ions is MKP. They are dual specificity phosphatases that dephosphorylate both tyrosine and threonine residues in activated MAPKs. Zinc inhibition of MKPs is one mechanism of how zinc stimulates the expression of interleukin-8 (IL-8) in epithelial cells [177]. However, not all of the zinc effects on MAPK phosphorylation and activation can be explained only by zinc inhibition of MKP, because zinc also affects targets upstream of the MAPKs. For example, a combined effect of zinc on PTP and MKP is thought to modulate IL-8 expression in airway epithelial cells [177].

Inhibition of phosphatases is not a classic signal because it complements kinase activity. Zinc inhibition of a phosphatase sustains increased phosphorylation initiated by activation of a receptor tyrosine kinase. In this way, variations of the cellular zinc status can influence the intensity and duration of phosphorylation signaling and control the responsiveness of a cell toward growth factors. A loss of zinc modulation results in constitutively active phosphatases and compromises phosphorylation signaling. This interaction between zinc and phosphorylation signaling is an important aspect of zinc signaling and may explain the association between zinc deficiency and loss of growth and differentiation.

8.4.6.3 Zinc in the Assembly of Signaling Complexes

The T cell receptor (TCR) has no intrinsic kinase activity and depends on a Src-family tyrosine kinase, Lck, for signal transduction. Signal transduction of the TCR is initiated by the specific interaction with an antigen-loaded major histocompatibility complex (MHC) molecule on the surface of a neighboring antigen-presenting cell, followed by the assembly of a multiprotein signaling complex. The costimulatory membrane proteins CD4 or CD8 also bind to MHC with their extracellular domains and are recruited to the vicinity of the TCR signaling complex. Lck is bound to the intracellular domains of CD4 or CD8 in a complex that is stabilized by a zinc ion bound to two cysteine residues from each protein at the interface site between Lck and CD4/CD8 [178]. In addition, the homodimerization of the SH3 domains of Lck is also zinc-dependent. In this case, two zinc ions at the dimer interface of the two SH3 domains stabilize the complex, and zinc coordination does not involve cysteines [179]. The estimated K_d value for zinc is ≤100 nM. The activation of Lck is a complex event that involves autocatalytic transphosphorylation of the activating tyrosine 394 between neighboring Lck molecules. Zinc-induced homodimerization may bring two Lck molecules into close proximity, thereby facilitating its full activation. This mechanism is supported by the observation that zinc increases phosphorylation of purified Lck [180]. It is worth noting that the overall effect of zinc ions in this system is similar to the ones described for phosphatases, namely establishing and sustaining a phosphorylation signal.

8.5 ZINC IN INTERCELLULAR SIGNAL TRANSDUCTION

Although this chapter started with a discussion of the concept of zinc ions as an intracellular signal, secretion of zinc ions from neuronal vesicles as an intercellular signal was critical in developing the concept of zinc signaling (Figure 8.4).

8.5.1 SPECIALIZED CELLS THAT SECRETE ZINC

The total cellular zinc concentration of a few hundred micromolar is an average value that includes all cellular compartments. Significant variations in zinc concentrations of tissues have been reported, indicating that zinc is not distributed randomly. The evidence presented in this chapter suggests that zinc is not only a static component of 10% of the proteome, but rather has multiple regulatory functions that require dynamic changes in its concentrations, distribution, and interactions with proteins.

A special aspect of the cell biology of zinc is the role of zinc ions that are not tightly bound to proteins and can be detected with chromophoric and fluorescent chelating agents. As discussed in Section 8.2, free zinc ions are maintained at very low concentrations in the cytosol to avoid unspecific interactions with proteins. Thus, zinc signaling in the cytosol must occur with tight control of

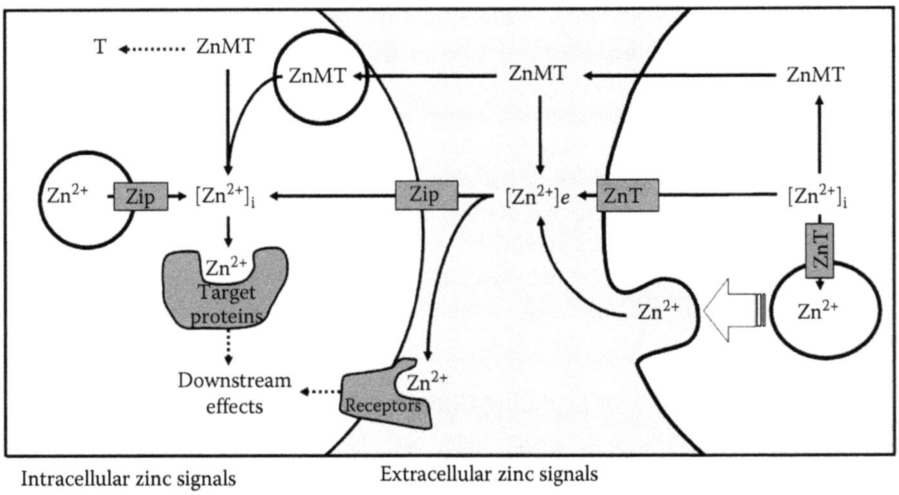

FIGURE 8.4 Intra- and intercellular zinc signaling. Aspects of intracellular zinc signaling are illustrated for the cell shown on the left, where $[Zn^{2+}]_i$ increases due to release of zinc from vesicles, transport into the cell, or release from cellular or endocytosed MT. The increase of $[Zn^{2+}]_i$ is an effector of many proteins (Figure 8.3). The generation of extracellular zinc signals is illustrated for the cell shown on the right, where $[Zn^{2+}]_i$ decreases due to sequestration of zinc in vesicles, vesicular secretion, MT exocytosis, or transport of zinc out of the cell. The extracellular free zinc, $[Zn^{2+}]_e$, is a paracrine messenger and binds to proteins on the cytoplasmic membrane of neighboring cells and triggers intracellular signaling events. For simplicity, processes in the cell on the left are shown only with the involvement of Zip transporters, while processes in the cell on the right are shown only with the involvement of ZnT transporters. Clearly, both types of transporters are active in each cell.

low concentrations of locally generated zinc signals. However, in certain cellular compartments, such as secretory vesicles, lysosomes, and vesicles that may or may not be identical to lysosomes and have been called "zincosomes," the concentrations of free zinc ions can be relatively high (estimated to be millimolar). Hence, zinc in these compartments can be readily visualized by autometallographic methods and fluorescent chelating agents. The vesicular sequestration occurs through transporters of the ZnT family. This vesicular sequestration of high concentrations of zinc ions and their release allows physiological actions of zinc ions at concentrations that are quite different from those in the cytosol. Free zinc ions are present in synaptic vesicles of zinc-enriched nerve terminals, secretory vesicles of somatotrophic cells in the pituitary gland, zymogene granules in pancreatic acinar cells, β-cells of the islets of Langerhans, Paneth cells of the crypts of Lieberkühn, and secretory cells of the tubuloacinar glands of the prostate, the epithelium of parts of the epididymal ducts, and osteoblasts [181]. Zinc ions can be released from these vesicles—or the vesicles released together with their zinc ions—to the extracellular space and serve a variety of functions. Zinc-secreting cells also include the submandibular salivary gland, mast cells, and granulocytes [182].

Like in the case of the chelatable iron pool, for which terms such as "labile iron," "intracellular transit iron," "free iron," "low molecular weight iron," "exchangeable iron," and "metabolically active iron" have been used, a similar terminology has been adopted to methodically define pools of zinc ions [183]. Thus, depending on the method used, different results may be obtained. How the introduction of a chelating agent into the cell affects intracellular equilibria and which specific reagents are employed for visualization of zinc may affect the outcome of the measurements. To account for possible interference of the chelating agent, it has been suggested to perform measurements at different concentrations of the chelating agent and then extrapolate the data to a zero concentration of the agent [27]. A comparison of histochemical and histofluorimetric determination of chelatable zinc in the developing mouse embryo demonstrated general agreement of the methods but also revealed significant differences [184]. Chelatable zinc was found in many rapidly proliferating tissues with a broad distribution, suggesting a specific role of zinc in those tissues. Quantitative determinations are not possible with these histological methods. Thus, whether or not chelatable zinc ions correspond to the free zinc ions discussed above is an issue that needs further investigation.

8.5.1.1 Prostate

Prostate epithelial cells accumulate high levels of zinc, making this organ the one with the highest zinc concentration [9.2 mg/g (ash)] [185]. One physiological role of high zinc in the prostate is thought to be the inhibition of the Krebs cycle enzyme m-aconitase [186]. A consequence of this inhibition is the accumulation of citrate, the substrate of aconitase [187]. Zinc ions are localized to the secretory granules in the lateral lobe and secretory ducts of the epithelial cells. They are secreted into the prostatic fluid where their concentrations are millimolar, that is, at least about 100 times higher than in the blood. One of the functions of the secreted zinc may be to provide sufficiently high concentrations for the inhibition of proteolytic enzymes. Low micromolar concentrations of zinc inhibit proteinases that do not depend on zinc as the catalytic metal ion, such as the human kallikrein (hK) family of serine proteinases. The $K_{i(\mathrm{app})}$ values (μM) are 3–5 (pH 7.5) for hK2; 6 (pH 7.5) for hK3; 16 (pH 7, IC_{50}) for hK4; 8 (pH 7) for hK5; 10 (pH 7.5) for hK7; and 0.01 (pH 8) for hK14 [188–193]. The kallikrein hK4 binds zinc with a histidine ligand and a glutamate ligand [191]. In the rat kallikrein tonin, the zinc ligands are three histidines and a glutamate from a neighboring protein molecule. In hK5, two histidine ligands bind the zinc ions, and a third histidine ligand may also be recruited for zinc binding [192]. Since zinc concentrations in seminal plasma and prostate fluid are around 10 mM, the zinc inhibition is believed to be significant for those of the hKs that function in prostate physiology, such as hK3 [prostate-specific antigen (PSA)] and hK4 (prostase), and those that are found in seminal plasma (hK2, -3, -5, and -14). Activation of hKs in seminal plasma is thought to occur through a proteolytic cascade and to involve a drop in the available zinc following ejaculation and zinc chelation by semenogelins in the semen coagulum. Activation results in semen liquefaction and release of motile sperm cells [188].

8.5.1.2 Pancreas

Both the exocrine and endocrine pancreas need significant amounts of zinc for their functions (Figure 8.5). Autometallography provided information about pancreatic distribution of zinc. The highest vesicular zinc content is in the β-cells, but vesicular zinc is also present in the α-cells and in the zymogen granules of acinar cells [194]. The amount of zinc secreted from the exocrine pancreas corresponds to about the amount taken up by the intestine [195]. Zinc is needed for the many proteolytic enzymes that require zinc and are secreted from the pancreas. In addition, zinc inhibition may be a mechanism to control the activity of secreted pancreatic proteins. Thus, the zinc-proteinase carboxypeptidase A binds an additional, inhibitory zinc ion with a K_i value of 0.5 μM [196]. The only protein ligand involved in binding the inhibitory zinc is a glutamate, but the interaction is stabilized by a hydroxide ion that bridges the inhibitory zinc and the catalytic zinc [197].

The islets of Langerhans are in the endocrine component of the pancreas. Within these islets, β-cells produce insulin and release it via exocytosis in response to elevated blood glucose. The result is a stimulation of glucose uptake into insulin-sensitive cells in target tissues and thus a lowering of blood glucose. Inside the secretory vesicles, insulin is stored as a crystalline hexamer with two bound zinc ions and one bound calcium ion [198]. The concentrations of insulin and zinc in the β-cell granules are both millimolar, and zinc is present at a 1.5-fold molar excess over the amount required to saturate the binding sites in insulin [199]. Because of the roles of zinc in both the exocrine and endocrine pancreas, the pancreas has a very active zinc metabolism as reflected by the expression of a total of 16 ZnT and Zip zinc transporters [200]. Two zinc transporters, ZnT-5 and ZnT-8, are abundantly expressed, especially in β-cells, and seem to transport zinc into the secretory vesicles [201–203]. Zinc is released together with insulin, and the islet staining for zinc is reduced significantly after insulin secretion [204,205]. An average of 0.2 amol of free zinc ions per single event is released from one vesicle and generates a local concentration of 150 nM [204]. Hence, a considerable amount of zinc can be released during high secretory activity. With the release rates of insulin [206] and the stoichiometry of the zinc/insulin hexamer, a daily release of 0.265 μmol (17.2 μg) zinc from the endocrine pancreas is calculated for a 70 kg human. This amount is a minute

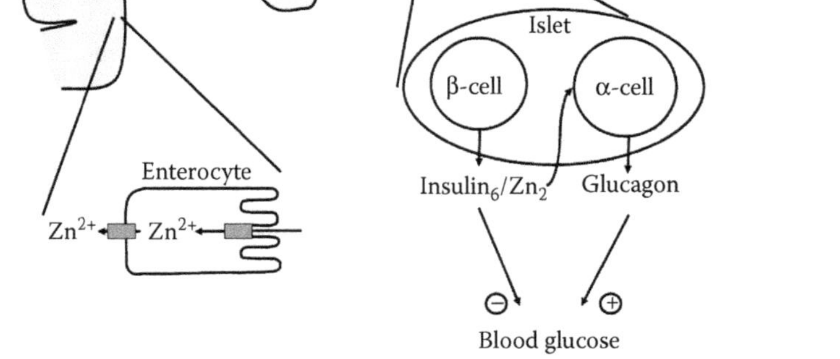

FIGURE 8.5 Zinc signaling of the pancreas. The exocrine pancreas stores zinc in acinar cells and secretes zinc-dependent proteases. In the endocrine pancreas (islets), insulin is stored as a crystalline Zn2–insulin hexamer in β-cells and secreted together with zinc. Secreted zinc ions inhibit the secretion of glucagon from α-cells. In enterocytes, zinc is taken up from the apical side by Zip4 and secreted by ZnT-1 at the basolateral membrane into the blood.

fraction of total plasma zinc and is certainly not high enough to be an endocrine signal. However, approximately 75% of insulin is secreted in pulses with a mean interval of 5 min [206]. During one pulse, 0.6 μmol (0.04 μg) zinc is released from the endocrine pancreas. This amount can elicit local oscillating concentrations within the islets, and potentially in the pancreas. The released zinc is thought to serve as a paracrine signal for islet α-cells that secrete the hormone glucagon, which antagonizes insulin actions by raising blood glucose levels. Glucagon release could be regulated directly by blood glucose levels, or β-cell activation could suppress α-cell activity by a paracrine mechanism. In the perfused rat pancreas, zinc, but not zinc-free insulin, mediates this paracrine effect of β-cells on inhibition of glucagon secretion [207,208]. However, the significance of this finding is not undisputed [209]. In rodent and human pancreatic islets, ATP-sensitive potassium channels (K_{ATP}) mediate inhibition of glucagon secretion in response to elevated glucose concentrations [210]. Because zinc affects K_{ATP} channels, the results are not mutually exclusive: A combination of glucose levels and zinc released from β-cells may regulate glucagon release from α-cells [208,211].

8.5.1.3 Neurons

Zinc was first visualized with a chromophoric chelating agent (dithizone) in the cornu ammonis (CA) region of the hippocampus [212]. Subsequently, it was localized to the presynaptic giant terminals ("boutons") of the hippocampal mossy fiber synapses by electron microscopy [213]. Total zinc concentrations of up to 15 μg/g have been measured in the human hippocampus [214]. The "zinc-containing neurons" featured prominently in developing the concept that zinc ions are stored in synaptic vesicles and released as neurotransmitters/neuromodulators [215]. Upon electric stimulation, the zinc-rich vesicles fuse with the cell membrane on the axonal terminals and release zinc ions into the synaptic cleft with subsequent effects on the dendritic spines of the postsynaptic neuron (Figure 8.6). One major target of zinc is the NR1-NR2A subunits of the NMDA receptor where zinc binds with high affinity (IC_{50} value of 10 nM) and inhibits calcium entry into the cell [216,217]. The mossy fibers are just one example of many cerebrocortical pathways where zinc is sequestered in the axon terminals [182]. Effects of zinc ions on purinergic, glycine, and nicotinic acetylcholine receptors have been observed [218–220]. Zinc is also translocated into the postsynaptic neuron [221]. In a neuronal activity-dependent way, zinc transactivates the neurotrophin receptor TrkB (tropomyosin-related kinase) signaling pathway. The protein tyrosine kinase Csk, which has an IC_{50} value of 500 nM for zinc, has been suggested to be the target of zinc [222,223]. Zinc inhibition of Csk in turn inhibits the Src kinase, which normally activates TrkB. Another possible site of action of zinc is the postsynaptic

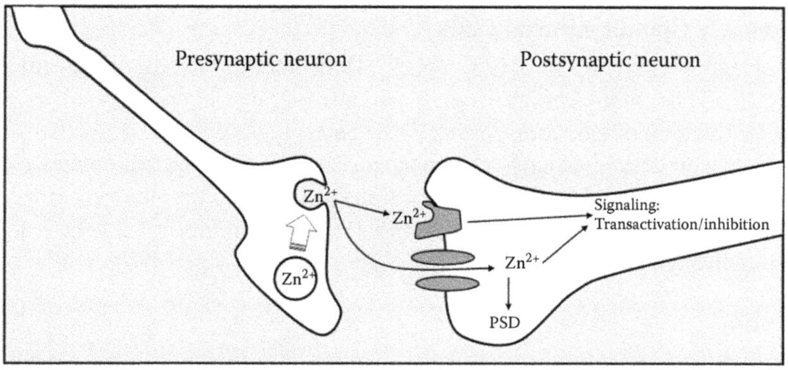

FIGURE 8.6 Intercellular signaling. Specialized neurons secrete zinc ions into the synaptic cleft (presynaptic neuron). The released zinc ions affect proteins on the cytoplasmic membrane of dendrites (postsynaptic neuron) and/or are translocated into the cytosol where they can affect gene transcription, signaling, and assembly of the scaffolding proteins of the postsynaptic density (PSD).

density. Zinc is coordinated tetrahedrally with two histidines, one glutamate, and a chloride ion between protomers of the scaffolding protein Shank3-SAM, whose fiber assembly and packaging are remarkably enhanced by zinc ions [224]. Zinc is also a constituent of cypin (cytosolic PSD-95 interactor), which promotes dendrite branching [225]. One of the functional consequences of released neuronal zinc is long-lasting synaptic modifications in the neural circuitry of learned behavior [226]. The biochemistry of zinc at the synapse requires many steps in addition to the ones shown in Figure 8.6. For example, the presynaptic vesicles have to be loaded with zinc. ZnT-3 and MT-3 are involved in this process. In addition, astrocytes secrete MT, which is taken up by neurons [113].

8.5.1.4 Osteoblasts

Bone contains 29% of human body zinc. Zinc is essential for formation of skeletal tissue, and it stimulates osteoblastic bone formation and inhibits osteoclastic bone resorption [227]. Extracellular matrix vesicles are the initial sites of mineral deposition in epiphyseal cartilage, dentine, and other mineralizing tissue [228]. Zinc is conspicuously high in matrix vesicles and regulates the onset of mineralization [228]. During bone growth in the epiphyseal growth plate, the chondrocytes differentiate, undergo spontaneous apoptosis, and release matrix vesicles [229].

8.5.1.5 Paneth Cells

Paneth cells are located at the bases of the crypts of Lieberkühn in the small intestine. Granular zinc in these cells is detectable with fluorescent chelating agents and secreted upon stimulation with a secretagogue, such as pilocarpine [230]. Paneth cells secrete microbicidal defensin peptides (cryptidins). The secretion of zinc is thought to be an "adjuvant" to this microbicidal function.

8.5.1.6 Eye

When compared with other tissues, the zinc concentration is particularly high in the retina (464 μg/g dry weight), retinal pigment epithelium, and choroid (472 μg/g dry weight) [231,232]. Zinc is essential for vision. Aside from its role in vitamin A metabolism, at least two functions that are specific for the retina have been identified. Among the three putative zinc ions of the rhodopsin monomer and an additional zinc ion at the interface between two monomers, the zinc-binding site in the transmembrane region is critical for the function and structure of the protein [233]. Zinc is released from photoreceptor terminals, where it is involved in neural processing at the first visual synapse in vertebrates [234].

Zinc ions are also secreted from other cells for purposes that are not linked to known signaling functions. Whether packaging of zinc ions into vesicles and vesicular exocytosis or export of nonvesicular zinc is the major route of secretion is not known.

8.5.1.7 Mammary Gland Epithelial Cells

The zinc concentration in human milk is 3–5 mg/L. Concentrations in the colostrum are about 3–5 times higher [235]. Thus, in a lactating woman, about 0.5–1 mg zinc per day are secreted into the milk [236]. Relocalization of zinc transporters in the cell is believed to be a mechanism for controlling zinc secretion, a process in which the hormone prolactin is involved. Mammary gland epithelial cells are highly specialized and employ Zip3 for transporting zinc into the cell. In addition, ZnT-2 has a major role in zinc secretion into the milk and a mutation in the ZnT-2 gene leads to transient neonatal zinc deficiency [10].

8.5.1.8 Intestine

Zip4 was identified as a major transporter responsible for intestinal uptake of zinc because a mutation in this gene causes acrodermatitis enteropathica, a genetic disease that is fatal if not treated with zinc [8,9,237]. Zip4 localizes to the luminal (apical) surface, while Zip5 is expressed on the basolateral surface of enterocytes, possibly mediating zinc uptake from the blood [238,239]. ZnT-1 also localizes to the basolateral membrane and is thought to participate in secretion of zinc into the blood, although it is likely that also other transporters participate in this process [240]. As discussed above, Paneth cells secrete zinc into the lumen.

8.5.1.9 Skin

The zinc physiology of skin demonstrates another principle, namely the control of extracellular zinc by chelation. The S100 protein psoriasin (S100A7) is secreted from keratinocytes and protects the human skin from *E. coli* infections. Its antimicrobial activity has been linked to zinc chelation [241]. In the crystal structure, the protein forms a dimer, in which a zinc ion is sandwiched between two molecules with an aspartate and a histidine from one molecule and two histidines from another [242]. An antimicrobial activity of zinc (and manganese) chelation has also been demonstrated for calprotectin (S100A8/S100A9). Neutrophil-derived calprotectin inhibits the growth of *Staphylococcus aureus* in tissue abscesses as a critical factor in the innate immune response [243].

8.6 ZINC SIGNALING IN THE IMMUNE SYSTEM

Zinc deficiency is the fifth leading health risk in developing countries with high mortality [244]. Characteristically, zinc-deficient individuals suffer from recurrent infections and parasitic diseases, highlighting the particular requirement of zinc for immune function [244]. In fact, the immune system critically depends on zinc [245–247]. Zinc deficiency increases the risk for several infectious diseases, such as diarrhea, pneumonia, and malaria [248].

The immune system employs mainly two mechanisms to combat an infection. The first one is innate immunity with an immediate response to conserved molecular structures of pathogens. The second one is adaptive immunity with a delayed, but more specific response to a particular pathogen. It is based on clonal expansion of antigen-specific cells and production of soluble factors, such as antibodies. Zinc homeostasis is crucial for the function of both kinds of immune cells, but the effects of zinc are specific for each cell type. In addition to its role in the generation and effectiveness of immune cells, zinc is also vital for the communication between them by soluble factors, such as cytokines [249]. Investigations of zinc-deficient mice and humans provided comprehensive insights into both intracellular and intercellular signaling functions of zinc in immunity.

8.6.1 Cells of the Adaptive Immune System

In zinc-deficient mice, there is atrophy of thymic and lymphoid tissue, resulting in depressed T and B lymphocyte reactions [250]. While zinc deficiency impairs lymphocyte development and augments apoptosis, it does not affect the relative composition of major lymphocyte subpopulations in the spleen [251–253].

T lymphocytes are highly sensitive to zinc status. T cell development takes place in the thymus, and thymic atrophy is a result of zinc deficiency in humans, for example, in acrodermatitis enteropathica [237]. Consequently, zinc deficiency is associated with reduced T cell activity and T cell numbers in the periphery [254]. These events may be mediated by thymulin, a thymic nonapeptide hormone that binds zinc with an affinity of $5 \pm 2 \times 10^{-7}$ M and regulates T cell formation and function [255]. Zinc supplementation restores thymulin activity in zinc deficiency, demonstrating that sufficient hormone is made but it is not active in the absence of zinc [256]. The zinc status also impacts T helper (TH) cell populations, skewing the balance between TH1 cells, which have a function in cell-mediated immunity, and TH2 cells, which are important for humoral immunity [257,258].

In vivo, zinc deficiency suppresses T cell functions. Zinc supplementation improves the delayed-type hypersensitivity reaction and stimulates the *in vitro* response to mitogens [259]. Improvement requires longer periods of time, usually several weeks, because new T cells need to be produced. However, a rapid zinc activation of T cells occurs *in vitro*. It depends on cytokines secreted by monocytes and is not mediated by a direct effect of zinc on T cells [247]. Higher, supraphysiological concentrations of zinc suppress T cell function. Zinc inhibition of the IL-1 type I receptor-associated kinase is involved in this suppression [260].

The zinc status also affects B lymphocytes. Zinc deficiency increases the rates of apoptosis in precursor and immature B cells and it reduces total B cell numbers, but apparently not their functions [245,261]. Thus, while B cells are not affected on a per cell basis, the production of antibodies can decrease as a consequence of reduced B cell numbers. Results of zinc supplementation on antibody production in response to vaccination are variable but indicate that an optimized supplementation protocol can improve the efficacy of vaccinations [259,262]. However, successful vaccination is a complex process, in which several types of immune cells participate. So far, there is no evidence that the functions of mature B cells depend on zinc. The observed effects on B cells could be based on zinc modulation of TH cell activity, because antibody production in response to T cell-dependent antigens is more affected than that in response to T cell-independent antigens [263,264].

8.6.2 Cells of the Innate Immune System

In natural killer (NK) cells, the interaction between the killer cell Ig-like receptors (KIR) and MHC class I depends on zinc [265–267]. In addition, the lytic activity of NK cells is lower in zinc-deficient individuals [251,268].

As mentioned above, mast cells contain granules that are rich in zinc [269]. These zinc stores can be depleted upon activation, indicating that mast cells can release significant amounts of zinc into their environment. The function of the released zinc is unknown. It may regulate activation of procaspases in the granules or may be a paracrine signal for other immune cells [270].

In zinc-deficient mice, myeloid precursor cells have a higher rate of proliferation and are found in higher absolute numbers, while the generation of lymphoid cells is reduced [271]. In contrast to adaptive immunity, innate immunity responds to zinc deficiency with increased production of new cells, pointing toward a "reprogramming" of the immune system rather than a general loss of function [245]. In addition to their effects on granulopoiesis, zinc ions at a concentration that is not realized *in vivo* (0.5 mM) have a chemotactic activity on granulocytes [272]. However, if zinc acts synergistically with chemokines, lower concentrations of zinc may be effective.

8.6.3 Cytokine Production

Zinc affects T cells by altering their ability to secrete cytokines. The balance of TH cells is maintained by a set of cytokines expressed in each cell type. Zinc deficiency reduces TH1 cytokines, mostly interferon-γ (IFN-γ) and IL-2, while TH2 cytokines are not affected [257,258]. IL-2 affects T cell growth, but also NK cell activity, while IFN-γ is important for macrophage activation. In this way, the action of zinc on T cells compromises the functions of other immune cells.

While the functions of most immune cells, such as T and NK cells, are suppressed in response to elevated zinc concentrations, monocytes/macrophages respond with elevated production and release of proinflammatory cytokines [247], and intracellular zinc signals are involved in the LPS-induced production of these cytokines [273]. However, this effect is concentration dependent as high zinc concentrations can inhibit lipopolysaccharide-induced production of the same cytokines [175].

Unlike in the case of monocytes, *in vitro* zinc supplementation does not stimulate secretion of cytokines from mast cells. However, cytokine secretion of mast cells in response to physiological stimuli seems to depend on zinc. Thus, crosslinking of the high-affinity immunoglobulin E receptor releases zinc from an intracellular store in mast cells. The process is dependent on calcium influx and MAPK/ERK activation. Chelation of zinc impairs nuclear factor κB (NF-κB) activation and abrogates proinflammatory cytokine production triggered by crosslinking F_C-receptors [270,274].

A comparable effect was observed in epithelial cells, which also produce cytokines. Here, chelation of zinc inhibits the production of several chemokines in response to tumor necrosis factor-α (TNF-α) stimulation [275].

In conclusion, many studies during zinc deficiency and zinc supplementation support the significance of zinc and zinc signaling for the immune system *in vivo*. They are supported by *in vitro* observations that demonstrate the requirement of zinc for the development and function of immune cells, and for a coordinated defense based on cellular communication via cytokines. The direct effects of zinc on the immune system influence the susceptibility toward infections, allergies, autoimmunity, and tumor surveillance, with serious consequences of zinc deficiency for an individual's health. In addition, the zinc status has poorly understood, long-term consequences on the immune system. Thus, gestational zinc deficiency results in measurable immunological changes in the offspring but also in subsequent generations, indicating that epigenetic effects of zinc on the immune system extend over several generations [276].

8.7 SYSTEMIC ZINC HOMEOSTASIS

This chapter would be incomplete without a discussion of systemic zinc homeostasis. The human body (70 kg) contains 2–3 g of zinc [277]. Zinc concentrations in tissues vary by at least one order of magnitude, from approximately 100 μg/g wet tissue in bone to approximately 10 μg/g in brain. Per gram of ash, there is also a 10-fold difference, with brain being lowest (0.8 mg/g) and prostate being highest (9.2 mg/g). In blood plasma, the zinc concentration is yet another order of magnitude lower, that is, 1 μg/g, 10–100 times lower than in most tissues, suggesting that an active transport process is necessary for maintaining higher cellular zinc concentrations. Together, skeletal muscle (57%) and bone (29%) account for 86% (2.3 g) of body zinc. Skin (6%) and liver (5%) add another 11%, leaving only about 3% for the remaining tissues [278]. The uneven zinc distribution in organs reflects the specialization of cells with regard to zinc.

Zinc is mainly an intracellular ion. Zinc in blood plasma accounts for less than 3 mg of body zinc, about 0.1%. The range that is considered normal is 12–18 μM (78–117 μg/dL). In zinc deficiency or in disease due to an acute phase response associated with infection, these values can be lower, but acute catabolism increases the plasma concentrations. In blood, 84% is bound to albumin, 15% to α_2-macroglobulin, and an estimated 1% to amino acids. Because total zinc in blood plasma represents only such a minor fraction of the total body zinc and cellular zinc is under the control of additional homeostatic mechanisms, measurements of serum or plasma zinc are not reliable indicators of the adequate zinc status in tissues.

At present, the recommended dietary allowance (RDA) of zinc is 11 mg for men and 8 mg for women (age 19–71+). With an absorption efficiency of less than 50%, the net requirement is about 3–4 mg [estimated average requirement (EAR)]. RDA and EAR are based on endogenous secretion, namely a linear correlation between excretion and uptake. Thus, only about 0.1% of body zinc, roughly the zinc in whole blood, is replenished every day and efficient homeostatic mechanisms control the maintenance of body zinc. Only approximately 24% of the zinc content of whole blood is found in the plasma. The major fraction (~74%) is in the erythrocytes, while leukocytes contribute only a minor fraction (~2.5%). The predominance of erythrocyte zinc results from the relative high numbers of these cells in blood. On a per cell basis, the zinc content of leukocytes is more than 20 times higher than that of erythrocytes [279].

Zinc from pancreatic and intestinal secretions and from digesting food is absorbed downstream in the jejunum, ileum, cecum, and perhaps ascending colon. The downstream absorption is crucial for the maintenance of zinc homeostasis. Absorption efficiency is the primary control of total body zinc [280]. This regulation allows the body to respond to relatively large fluctuations of zinc in the diet. However, indigestible zinc-binding ligands have a major effect on bioavailability.

During development, considerable redistribution of zinc occurs among bone, liver, and muscle. The liver is thought to be a central organ in zinc metabolism. Kinetically, there are at least two major metabolic pools, one with rapid turnover (12.5 days) and one with slow turnover (300 days) [280]. Kinetically fast tissues are primarily the liver, but also the kidney, pancreas, and spleen. When the sources for zinc under zinc deficiency were investigated in fully grown rats, the skeleton was identified as the major source [281]. Several phases were observed. First, fecal losses were counterbalanced by mobilization of zinc from the skeleton. Second, mobilization of zinc changed transiently to soft tissue (muscle and adipose tissue). Third, mobilization from the skeleton resumed until stores from which zinc can be mobilized are exhausted, resulting in severe zinc deficiency.

Little is known about hormonal regulation of zinc metabolism. Administration of adrenocorticotropic hormone (ACTH), which stimulates the production of glucocorticoids in the adrenal cortex, lowers serum zinc concentrations by 30–40% within 3–4 h with concomitant uptake in the liver [282]. An IL-6-mediated mechanism induces Zip14 on hepatocytes during the acute phase response [283]. A variety of other stresses lower plasma zinc, increase zinc uptake in tissues, and cause hypozincemia. During anabolism, tissues take up zinc, but during catabolism, zinc is lost from tissues and excreted in the urine. A possible relationship to intermediary metabolism is further indicated by a cyclic eating pattern in zinc deficiency accompanied by cyclic changes of zinc in blood plasma [284].

8.8 CONCLUSION

The science of zinc biology appears to undergo a paradigm shift. For many decades, the focus has been on the discovery of zinc proteins and the characterization of their structures and functions. The multiple consequences of zinc deficiency were considered to be the result of a loss of function of these zinc-binding proteins. However, symptoms of zinc deficiency often precede a measurable reduction of protein-bound zinc or even a reduction of plasma zinc levels. The idea is taking hold that the free zinc ion pool is not just needed for supplying apoproteins with zinc, but that it is tightly regulated and serves a role in intracellular and intercellular signaling in neurotransmission, immune responses, and cellular proliferation, differentiation, and death.

What are the next questions to be addressed? The molecular structures of regulatory zinc-binding sites are virtually unknown and so are the number of targets of zinc ions. The concentrations of signaling zinc ions are many orders of magnitude lower than the total concentrations of structural and catalytic zinc sites in proteins. Hence, their characterization continues to be a major analytical challenge, although exceptional progress has been made recently through the applications of fluorescent chelating agents that distinguish total from free zinc. It will be necessary to define further the amplitudes and durations of physiological zinc signals and the processes that trigger them. The significance of zinc signals will need to be demonstrated *in vivo*, because cultured cells encountering concentrations of several hundreds of micromolar zinc in the presence of only a fraction of the physiological concentrations of zinc-binding serum proteins are clearly not conditions that pertain *in vivo*. To what extent zinc ions entering cultured cells mimic zinc signals that are generated by physiological processes needs to be established. Yet, results from studies with cultured cells will continue to direct our attention to additional physiological roles of zinc ions. We now know that free zinc ions have a paracrine but likely not an endocrine effect. The interactions of zinc ions with many cellular signaling pathways indicate regulatory functions that critically contribute to the physiology in every cell of the human body. Perturbation of such a system is likely the cause for many diseases with at present poorly defined etiologies.

ACKNOWLEDGMENT

W.M. thanks Professors C.J. Frederickson and H.H. Sandstead for many stimulating discussions on various topics of zinc biology.

REFERENCES

1. Andreini, C., et al. Counting the zinc-proteins encoded in the human genome. *J. Proteome Res.*, 5, 196, 2006.
2. Vallee, B.L. and Auld, D.S. Zinc coordination, function, and structure of zinc enzymes and other proteins. *Biochemistry*, 29, 5647, 1990.
3. Maret, W. Zinc and sulfur: A critical biological partnership. *Biochemistry*, 43, 3301, 2004.
4. Vallee, B.L. and Wacker, W.E.C. *The Proteins* (2nd edition). Academic Press, New York, 1970.
5. Auld, D.S. Zinc coordination sphere in biochemical zinc sites. *Biometals*, 14, 271, 2001.
6. Williams, R.J.P. and da Silva, J.J.R.F. The distribution of elements in cells. *Coord. Chem. Rev.*, 200, 247, 2000.
7. O'Halloran, T.V. and Culotta, V.C. Metallochaperones, an intracellular shuttle service for metal ions. *J. Biol. Chem.*, 275, 25057, 2000.
8. Kury, S., et al. Identification of SLC39A4, a gene involved in acrodermatitis enteropathica. *Nat. Genet.*, 31, 239, 2002.
9. Wang, K., et al. A novel member of a zinc transporter family is defective in acrodermatitis enteropathica. *Am. J. Hum. Genet.*, 71, 66, 2002.
10. Chowanadisai, W., Lonnerdal, B., and Kelleher, S.L. Identification of a mutation in SLC30A2 (ZnT-2) in women with low milk zinc concentration that results in transient neonatal zinc deficiency. *J. Biol. Chem.*, 281, 39699, 2006.
11. Chimienti, F., Favier, A., and Seve, M. ZnT-8, a pancreatic beta-cell-specific zinc transporter. *Biometals*, 18, 313, 2005.
12. Saxena, R., et al. Genome-wide association analysis identifies loci for type 2 diabetes and triglyceride levels. *Science*, 316, 1331, 2007.
13. Scott, L.J., et al. A genome-wide association study of type 2 diabetes in Finns detects multiple susceptibility variants. *Science*, 316, 1341, 2007.
14. Sladek, R., et al. A genome-wide association study identifies novel risk loci for type 2 diabetes. *Nature*, 445, 881, 2007.
15. Zeggini, E., et al. Replication of genome-wide association signals in UK samples reveals risk loci for type 2 diabetes. *Science*, 316, 1336, 2007.
16. Giacconi, R., et al. +647 A/C and +1245 MT1A polymorphisms in the susceptibility of diabetes mellitus and cardiovascular complications. *Mol. Genet. Metab.*, 94, 98, 2008.
17. Yang, L., et al. Polymorphisms in metallothionein-1 and 2 genes associated with the risk of type 2 diabetes mellitus and its complications. *Am. J. Physiol. Endocrinol. Metab.*, 294, E987, 2008.
18. Maret, W., Larsen, K.S., and Vallee, B.L. Coordination dynamics of biological zinc "clusters" in metallothioneins and in the DNA-binding domain of the transcription factor Gal4. *Proc. Natl. Acad. Sci. USA*, 94, 2233, 1997.
19. Outten, C.E. and O'Halloran, T.V. Femtomolar sensitivity of metalloregulatory proteins controlling zinc homeostasis. *Science*, 292, 2488, 2001.
20. Adebodun, F. and Post, J.F. Role of intracellular free Ca(II) and Zn(II) in dexamethasone-induced apoptosis and dexamethasone resistance in human leukemic CEM cell lines. *J. Cell. Physiol.*, 163, 80, 1995.
21. Benters, J., et al. Study of the interactions of cadmium and zinc ions with cellular calcium homoeostasis using 19F-NMR spectroscopy. *Biochem. J.*, 322, 793, 1997.
22. Simons, T.J. Intracellular free zinc and zinc buffering in human red blood cells. *J. Membr. Biol.*, 123, 63, 1991.
23. Peck, E.J., Jr. and Ray, W.J., Jr. Metal complexes of phosphoglucomutase *in vivo*. Alterations induced by insulin. *J. Biol. Chem.*, 246, 1160, 1971.
24. Colvin, R.A., et al. Insights into Zn^{2+} homeostasis in neurons from experimental and modeling studies. *Am. J. Physiol. Cell Physiol.*, 294, C726, 2008.
25. Bozym, R.A., et al. Measuring picomolar intracellular exchangeable zinc in PC-12 cells using a ratiometric fluorescence biosensor. *ACS Chem. Biol.*, 1, 103, 2006.
26. Haase, H., et al. Flow cytometric measurement of labile zinc in peripheral blood mononuclear cells. *Anal. Biochem.*, 352, 222, 2006.
27. Krezel, A. and Maret, W. Zinc-buffering capacity of a eukaryotic cell at physiological pZn. *J. Biol. Inorg. Chem.*, 11, 1049, 2006.
28. Krezel, A. and Maret, W. Thionein/metallothionein control Zn(II) availability and the activity of enzymes. *J. Biol. Inorg. Chem.*, 13, 401, 2008.

29. Thomas, R.C., Coles, J.A., and Deitmer, J.W. Homeostatic muffling. *Nature*, 350, 564, 1991.
30. Aizenman, E., et al. Induction of neuronal apoptosis by thiol oxidation: Putative role of intracellular zinc release. *J. Neurochem.*, 75, 1878, 2000.
31. Atar, D., et al. Excitation–transcription coupling mediated by zinc influx through voltage-dependent calcium channels. *J. Biol. Chem.*, 270, 2473, 1995.
32. Choi, D.W. and Koh, J.Y. Zinc and brain injury. *Annu. Rev. Neurosci.*, 21, 347, 1998.
33. Sensi, S.L., et al. Modulation of mitochondrial function by endogenous Zn^{2+} pools. *Proc. Natl. Acad. Sci. USA*, 100, 6157, 2003.
34. Smith, P.J., et al. DNA damage-induced [Zn(2+)](i) transients: Correlation with cell cycle arrest and apoptosis in lymphoma cells. *Am. J. Physiol. Cell Physiol.*, 283, C609, 2002.
35. Spahl, D.U., et al. Regulation of zinc homeostasis by inducible NO synthase-derived NO: Nuclear metallothionein translocation and intranuclear Zn^{2+} release. *Proc. Natl. Acad. Sci. USA*, 100, 13952, 2003.
36. St Croix, C.M., et al. Nitric oxide-induced changes in intracellular zinc homeostasis are mediated by metallothionein/thionein. *Am. J. Physiol. Lung Cell Mol. Physiol.*, 282, L185, 2002.
37. Turan, B., Fliss, H., and Desilets, M. Oxidants increase intracellular free Zn^{2+} concentration in rabbit ventricular myocytes. *Am. J. Physiol.*, 272, H2095, 1997.
38. Giedroc, D.P., Chen, X., and Apuy, J.L. Metal response element (MRE)-binding transcription factor-1 (MTF-1): Structure, function, and regulation. *Antioxid. Redox Signal.*, 3, 577, 2001.
39. Heuchel, R., et al. The transcription factor MTF-1 is essential for basal and heavy metal-induced metallothionein gene expression. *EMBO J.*, 13, 2870, 1994.
40. Laity, J.H. and Andrews, G.K. Understanding the mechanisms of zinc-sensing by metal-response element binding transcription factor-1 (MTF-1). *Arch. Biochem. Biophys.*, 463, 201, 2007.
41. Palmiter, R.D., et al. ZnT-3, a putative transporter of zinc into synaptic vesicles. *Proc. Natl. Acad. Sci. USA*, 93, 14934, 1996.
42. Palmiter, R.D., Cole, T.B., and Findley, S.D. ZnT-2, a mammalian protein that confers resistance to zinc by facilitating vesicular sequestration. *EMBO J.*, 15, 1784, 1996.
43. Gaither, L.A. and Eide, D.J. The human ZIP1 transporter mediates zinc uptake in human K562 erythroleukemia cells. *J. Biol. Chem.*, 276, 22258, 2001.
44. Banci, L., et al. A new zinc–protein coordination site in intracellular metal trafficking: Solution structure of the Apo and Zn(II) forms of ZntA(46–118). *J. Mol. Biol.*, 323, 883, 2002.
45. Eren, E. and Arguello, J.M. Arabidopsis HMA2, a divalent heavy metal-transporting P(IB)-type ATPase, is involved in cytoplasmic Zn^{2+} homeostasis. *Plant Physiol.*, 136, 3712, 2004.
46. Hussain, D., et al. P-type ATPase heavy metal transporters with roles in essential zinc homeostasis in *Arabidopsis*. *Plant Cell*, 16, 1327, 2004.
47. Chao, Y. and Fu, D. Kinetic study of the antiport mechanism of an *Escherichia coli* zinc transporter, ZitB. *J. Biol. Chem.*, 279, 12043, 2004.
48. Grass, G., et al. FieF (YiiP) from *Escherichia coli* mediates decreased cellular accumulation of iron and relieves iron stress. *Arch. Microbiol.*, 183, 9, 2005.
49. Guffanti, A.A., et al. An antiport mechanism for a member of the cation diffusion facilitator family: Divalent cations efflux in exchange for K+ and H +. *Mol. Microbiol.*, 45, 145, 2002.
50. MacDiarmid, C.W., Milanick, M.A., and Eide, D.J. Biochemical properties of vacuolar zinc transport systems of *Saccharomyces cerevisiae*. *J. Biol. Chem.*, 277, 39187, 2002.
51. Gaither, L.A. and Eide, D.J. Functional expression of the human hZIP2 zinc transporter. *J. Biol. Chem.*, 275, 5560, 2000.
52. Girijashanker, K., et al. Slc39a14 gene encodes ZIP14, a metal/bicarbonate symporter: Similarities to the ZIP8 transporter. *Mol. Pharmacol.*, 73, 1413, 2008.
53. Ohana, E., et al. A sodium zinc exchange mechanism is mediating extrusion of zinc in mammalian cells. *J. Biol. Chem.*, 279, 4278, 2004.
54. Lu, M. and Fu, D. Structure of the zinc transporter YiiP. *Science*, 317, 1746, 2007.
55. Jackson, K.A., et al. Splice variants of the human zinc transporter ZnT5 (SLC30A5) are differentially localized and regulated by zinc through transcription and mRNA stability. *J. Biol. Chem.*, 282, 10423, 2007.
56. Ishihara, K., et al. Zinc transport complexes contribute to the homeostatic maintenance of secretory pathway function in vertebrate cells. *J. Biol. Chem.*, 281, 17743, 2006.
57. Potter, B.M., et al. The six zinc fingers of metal-responsive element binding transcription factor-1 form stable and quasi-ordered structures with relatively small differences in zinc affinities. *J. Biol. Chem.*, 280, 28529, 2005.
58. Cousins, R.J., et al. A global view of the selectivity of zinc deprivation and excess on genes expressed in human THP-1 mononuclear cells. *Proc. Natl. Acad. Sci. USA*, 100, 6952, 2003.

59. Daniel, H. and tom Dieck, H. Nutrient-gene interactions: A single nutrient and hundreds of target genes. *Biol. Chem.*, 385, 571, 2004.
60. Haase, H., et al. Differential gene expression after zinc supplementation and deprivation in human leukocyte subsets. *Mol. Med.*, 13, 362, 2007.
61. Li, Y. and Maret, W. Human metallothionein metallomics. *J. Anal. Atom. Spectrom.*, 23, 1055, 2008.
62. Arseniev, A., et al. Three-dimensional structure of rabbit liver [Cd7]metallothionein-2a in aqueous solution determined by nuclear magnetic resonance. *J. Mol. Biol.*, 201, 637, 1988.
63. Robbins, A.H., et al. Refined crystal structure of Cd, Zn metallothionein at 2.0 A resolution. *J. Mol. Biol.*, 221, 1269, 1991.
64. Krezel, A. and Maret, W. Different redox states of metallothionein/thionein in biological tissue. *Biochem. J.*, 402, 551, 2007.
65. Yang, Y., Maret, W., and Vallee, B.L. Differential fluorescence labeling of cysteinyl clusters uncovers high tissue levels of thionein. *Proc. Natl. Acad. Sci. USA*, 98, 5556, 2001.
66. Krezel, A. and Maret, W. Dual nanomolar and picomolar Zn(II) binding properties of metallothionein. *J. Am. Chem. Soc.*, 129, 10911, 2007.
67. Maret, W. and Vallee, B.L. Thiolate ligands in metallothionein confer redox activity on zinc clusters. *Proc. Natl. Acad. Sci. USA*, 95, 3478, 1998.
68. Maret, W., The function of zinc metallothionein: Alink between cellular zinc and redox state. *J. Nutr.*, 130, 1455S, 2000.
69. Chen, Y. and Maret, W. Catalytic selenols couple the redox cycles of metallothionein and glutathione. *Eur. J. Biochem.*, 268, 3346, 2001.
70. Pearce, L.L., et al. Metallothionein, nitric oxide and zinc homeostasis in vascular endothelial cells. *J. Nutr.*, 130, 1467S, 2000.
71. Kroncke, K.D., et al. Nitric oxide destroys zinc–sulfur clusters inducing zinc release from metallothionein and inhibition of the zinc finger-type yeast transcription activator LAC9. *Biochem. Biophys. Res. Commun.*, 200, 1105, 1994.
72. Stitt, M.S., et al. Nitric oxide-induced nuclear translocation of the metal responsive transcription factor, MTF-1 is mediated by zinc release from metallothionein. *Vascul. Pharmacol.*, 44, 149, 2006.
73. Bossy-Wetzel, E., et al. Crosstalk between nitric oxide and zinc pathways to neuronal cell death involving mitochondrial dysfunction and p38-activated K+ channels. *Neuron*, 41, 351, 2004.
74. St Croix, C.M., et al. Nitric oxide and zinc homeostasis in acute lung injury. *Proc. Am. Thorac. Soc.*, 2, 236, 2005.
75. Hao, Q. and Maret, W. Aldehydes release zinc from proteins. A pathway from oxidative stress/lipid peroxidation to cellular functions of zinc. *FEBS J.*, 273, 4300, 2006.
76. Barbato, J.C., et al. Targeting of metallothionein by L-homocysteine: A novel mechanism for disruption of zinc and redox homeostasis. *Arterioscler. Thromb. Vasc. Biol.*, 27, 49, 2007.
77. Maret, W., Oxidative metal release from metallothionein via zinc-thiol/disulfide interchange. *Proc. Natl. Acad. Sci. USA*, 91, 237, 1994.
78. Kang, Y.J., The antioxidant function of metallothionein in the heart. *Proc. Soc. Exp. Biol. Med.*, 222, 263, 1999.
79. Sato, M. and Bremner, I. Oxygen free radicals and metallothionein. *Free Radic. Biol. Med.*, 14, 325, 1993.
80. Sagher, D., et al. Thionein can serve as a reducing agent for the methionine sulfoxide reductases. *Proc. Natl. Acad. Sci. USA*, 103, 8656, 2006.
81. Bray, T.M. and Bettger, W.J. The physiological role of zinc as an antioxidant. *Free Radic. Biol. Med.*, 8, 281, 1990.
82. Powell, S.R., The antioxidant properties of zinc. *J. Nutr.*, 130, 1447S, 2000.
83. Hao, Q. and Maret, W. Imbalance between pro-oxidant and pro-antioxidant functions of zinc in disease. *J. Alzheimers Dis.*, 8, 161, 2005.
84. Krezel, A., Hao, Q., and Maret, W. The zinc/thiolate redox biochemistry of metallothionein and the control of zinc ion fluctuations in cell signaling. *Arch. Biochem. Biophys.*, 463, 188, 2007.
85. Hennig, B., et al. Antioxidant-like properties of zinc in activated endothelial cells. *J. Am. Coll. Nutr.*, 18, 152, 1999.
86. Ho, E., Courtemanche, C., and Ames, B.N. Zinc deficiency induces oxidative DNA damage and increases p53 expression in human lung fibroblasts. *J. Nutr.*, 133, 2543, 2003.
87. Oteiza, P.I., et al. Zinc deficiency induces oxidative stress and AP-1 activation in 3T3 cells. *Free Radic. Biol. Med.*, 28, 1091, 2000.
88. Dineley, K.E., Votyakova, T.V., and Reynolds, I.J. Zinc inhibition of cellular energy production: Implications for mitochondria and neurodegeneration. *J. Neurochem.*, 85, 563, 2003.

89. Maret, W. and Krezel, A. Cellular zinc and redox buffering capacity of metallothionein/thionein in health and disease. *Mol. Med.*, 13, 371, 2007.
90. Maret, W., Metallothionein redox biology in the cytoprotective and cytotoxic functions of zinc. *Exp. Gerontol.*, 43, 363, 2008.
91. Cousins, R.J., Liuzzi, J.P., and Lichten, L.A. Mammalian zinc transport, trafficking, and signals. *J. Biol. Chem.*, 281, 24085, 2006.
92. Haq, F., Mahoney, M., and Koropatnick, J. Signaling events for metallothionein induction. *Mutat. Res.*, 533, 211, 2003.
93. Woo, E.S., et al. Diversity of metallothionein content and subcellular localization in the National Cancer Institute tumor panel. *Cancer Chemother. Pharmacol.*, 41, 61, 1997.
94. Kägi, J.H.R., Evolution, structure and chemical activity of class I metallothioneins: An overview. In: *Metallothionein III*, K.T. Suzuki, N. Imura, and M. Kimura (eds), pp. 29–55. Birkhäuser, Basel, 1993.
95. Andrews, G.K., Regulation of metallothionein gene expression. *Prog. Food Nutr. Sci.*, 14, 193, 1990.
96. Ghoshal, K. and Jacob, S.T. Regulation of metallothionein gene expression. *Prog. Nucleic Acid Res. Mol. Biol.*, 66, 357, 2001.
97. Butler, J.S. and Loh, S.N. Zn(2+)-dependent misfolding of the p53 DNA binding domain. *Biochemistry*, 46, 2630, 2007.
98. Ye, B., Maret, W., and Vallee, B.L. Zinc metallothionein imported into liver mitochondria modulates respiration. *Proc. Natl. Acad. Sci. USA*, 98, 2317, 2001.
99. Apostolova, M.D., Ivanova, I.A., and Cherian, M.G. Signal transduction pathways, and nuclear translocation of zinc and metallothionein during differentiation of myoblasts. *Biochem. Cell Biol.*, 78, 27, 2000.
100. Schmidt, C. and Beyersmann, D. Transient peaks in zinc and metallothionein levels during differentiation of 3T3L1 cells. *Arch. Biochem. Biophys.*, 364, 91, 1999.
101. Takahashi, Y., Ogra, Y., and Suzuki, K.T. Nuclear trafficking of metallothionein requires oxidation of a cytosolic partner. *J. Cell. Physiol.*, 202, 563, 2005.
102. Tsujikawa, K., et al. Localization of metallothionein in nuclei of growing primary cultured adult rat hepatocytes. *FEBS Lett.*, 283, 239, 1991.
103. Woo, E.S., et al. Energy-dependent nuclear binding dictates metallothionein localization. *J. Cell. Physiol.*, 182, 69, 2000.
104. Moltedo, O., et al. Zinc transport and metallothionein secretion in the intestinal human cell line Caco-2. *J. Biol. Chem.*, 275, 31819, 2000.
105. Trayhurn, P., et al. Regulation of metallothionein gene expression and secretion in rat adipocytes differentiated from preadipocytes in primary culture. *Horm. Metab. Res.*, 32, 542, 2000.
106. De Lisle, R.C., et al. Metallothionein is a component of exocrine pancreas secretion: Implications for zinc homeostasis. *Am. J. Physiol.*, 271, C1103, 1996.
107. Erfurt, C., Roussa, E., and Thevenod, F. Apoptosis by Cd^{2+} or CdMT in proximal tubule cells: Different uptake routes and permissive role of endo/lysosomal CdMT uptake. *Am. J. Physiol. Cell Physiol.*, 285, C1367, 2003.
108. Klassen, R.B., et al. Megalin mediates renal uptake of heavy metal metallothionein complexes. *Am. J. Physiol. Renal Physiol.*, 287, F393, 2004.
109. Wolff, N.A., et al. Megalin-dependent internalization of cadmium-metallothionein and cytotoxicity in cultured renal proximal tubule cells. *J. Pharmacol. Exp. Ther.*, 318, 782, 2006.
110. Hao, Q., Hong, S.H., and Maret, W. Lipid raft-dependent endocytosis of metallothionein in HepG2 cells. *J. Cell. Physiol.*, 210, 428, 2007.
111. Ambjorn, M., et al. Metallothionein and a peptide modeled after metallothionein, EmtinB, induce neuronal differentiation and survival through binding to receptors of the low-density lipoprotein receptor family. *J. Neurochem.*, 104, 21, 2008.
112. Fitzgerald, M., et al. Metallothionein-IIA promotes neurite growth via the megalin receptor. *Exp. Brain Res.*, 183, 171, 2007.
113. Chung, R.S., et al. Redefining the role of metallothionein within the injured brain: Extracellular metallothioneins play an important role in the astrocyte-neuron response to injury. *J. Biol. Chem.*, 283, 15349, 2008.
114. Uchida, Y., et al. Growth inhibitory factor prevents neurite extension and the death of cortical neurons caused by high oxygen exposure through hydroxyl radical scavenging. *J. Biol. Chem.*, 277, 32353, 2002.
115. Uchida, Y., et al. The growth inhibitory factor that is deficient in the Alzheimer's disease brain is a 68 amino acid metallothionein-like protein. *Neuron*, 7, 337, 1991.

116. Sewell, A.K., et al. Bioactivity of metallothionein-3 correlates with its novel beta domain sequence rather than metal binding properties. *Biochemistry*, 34, 4740, 1995.
117. Cole, T.B., et al. Seizures and neuronal damage in mice lacking vesicular zinc. *Epilepsy Res.*, 39, 153, 2000.
118. Knipp, M., et al. Zn7metallothionein-3 and the synaptic vesicle cycle: Interaction of metallothionein-3 with the small GTPase Rab3A. *Biochemistry*, 44, 3159, 2005.
119. Thiers, R.E. and Vallee, B.L. Distribution of metals in subcellular fractions of rat liver. *J. Biol. Chem.*, 226, 911, 1957.
120. Bartholomew, M.E., Tupper, R., and Wormall, A. Incorporation of Zn-65 in the sub-cellular fractions of the liver and spontaneously occurring mammary tumours of mice after the injection of zinc-glycine containing Zn-65. *Biochem. J.*, 73, 256, 1959.
121. Weser, U. and Bischoff, E. Incorporation of 65Zn in rat liver nuclei. *Eur. J. Biochem.*, 12, 571, 1970.
122. Haase, H. and Beyersmann, D. Intracellular zinc distribution and transport in C6 rat glioma cells. *Biochem. Biophys. Res. Commun.*, 296, 923, 2002.
123. Zalewski, P.D., Forbes, I.J., and Betts, W.H. Correlation of apoptosis with change in intracellular labile Zn(II) using zinquin [(2-methyl-8-p-toluenesulphonamido-6-quinolyloxy)acetic acid], a new specific fluorescent probe for Zn(II). *Biochem. J.*, 296, 403, 1993.
124. Gazaryan, I.G., et al. Zinc irreversibly damages major enzymes of energy production and antioxidant defense prior to mitochondrial permeability transition. *J. Biol. Chem.*, 282, 24373, 2007.
125. Williams, R.J.P., Zinc: What is its role in biology? *Endeavour*, 8, 65, 1984.
126. Maret, W., Zinc coordination environments in proteins as redox sensors and signal transducers. *Antioxid. Redox Signal.*, 8, 1419, 2006.
127. Berendji, D., et al. Nitric oxide mediates intracytoplasmic and intranuclear zinc release. *FEBS Lett.*, 405, 37, 1997.
128. Korichneva, I., et al. Zinc release from protein kinase *C* as the common event during activation by lipid second messenger or reactive oxygen. *J. Biol. Chem.*, 277, 44327, 2002.
129. Csermely, P. and Somogyi, J. Zinc as a possible mediator of signal transduction in T lymphocytes. *Acta Physiol. Hung.*, 74, 195, 1989.
130. Haase, H. and Maret, W. Fluctuations of cellular, available zinc modulate insulin signaling via inhibition of protein tyrosine phosphatases. *J. Trace Elem. Med. Biol.*, 19, 37, 2005.
131. Yamasaki, S., et al. Zinc is a novel intracellular second messenger. *J. Cell Biol.*, 177, 637, 2007.
132. Beyersmann, D. and Haase, H. Functions of zinc in signaling, proliferation and differentiation of mammalian cells. *Biometals*, 14, 331, 2001.
133. Grummt, F., et al. Zinc as a second messenger of mitogenic induction. Effects on diadenosine tetraphosphate (Ap4A) and DNA synthesis. *Exp. Cell Res.*, 163, 191, 1986.
134. Petrie, L., Chesters, J.K., and Franklin, M. Inhibition of myoblast differentiation by lack of zinc. *Biochem. J.*, 276, 109, 1991.
135. Cherian, M.G. and Apostolova, M.D. Nuclear localization of metallothionein during cell proliferation and differentiation. *Cell. Mol. Biol. (Noisy-le-grand)*, 46, 347, 2000.
136. Quarterman, J., Mills, C.F., and Humphries, W.R. The reduced secretion of, and sensitivity to insulin in zinc-deficient rats. *Biochem. Biophys. Res. Commun.*, 25, 354, 1966.
137. Scott, D.A. and Fisher, A.M. The insulin and the zinc content of normal and diabetic pancreas. *J. Clin. Invest.*, 17, 725, 1938.
138. Coulston, L. and Dandona, P. Insulin-like effect of zinc on adipocytes. *Diabetes*, 29, 665, 1980.
139. May, J.M. and Contoreggi, C.S. The mechanism of the insulin-like effects of ionic zinc. *J. Biol. Chem.*, 257, 4362, 1982.
140. Wong, V.V., et al. Zinc as an insulin replacement in hybridoma cultures. *Biotechnol. Bioeng.*, 93, 553, 2006.
141. Haase, H. and Maret, W. Intracellular zinc fluctuations modulate protein tyrosine phosphatase activity in insulin/insulin-like growth factor-1 signaling. *Exp. Cell Res.*, 291, 289, 2003.
142. Tang, X. and Shay, N.F. Zinc has an insulin-like effect on glucose transport mediated by phosphoinositol-3-kinase and Akt in 3T3-L1 fibroblasts and adipocytes. *J. Nutr.*, 131, 1414, 2001.
143. Samet, J.M., et al. Mechanisms of Zn(2+)-induced signal initiation through the epidermal growth factor receptor. *Toxicol. Appl. Pharmacol.*, 191, 86, 2003.
144. Samet, J.M., et al. Activation of MAPKs in human bronchial epithelial cells exposed to metals. *Am. J. Physiol.*, 275, L551, 1998.
145. Wu, W., et al. Zinc-induced PTEN protein degradation through the proteasome pathway in human airway epithelial cells. *J. Biol. Chem.*, 278, 28258, 2003.

146. Wu, W., et al. Src-dependent phosphorylation of the epidermal growth factor receptor on tyrosine 845 is required for zinc-induced Ras activation. *J. Biol. Chem.*, 277, 24252, 2002.
147. Hansson, A., Extracellular zinc ions induces mitogen-activated protein kinase activity and protein tyrosine phosphorylation in bombesin-sensitive Swiss 3T3 fibroblasts. *Arch. Biochem. Biophys.*, 328, 233, 1996.
148. An, W.L., et al. Mechanism of zinc-induced phosphorylation of p70 S6 kinase and glycogen synthase kinase 3beta in SH-SY5Y neuroblastoma cells. *J. Neurochem.*, 92, 1104, 2005.
149. Park, J.A. and Koh, J.Y. Induction of an immediate early gene egr-1 by zinc through extracellular signal-regulated kinase activation in cortical culture: Its role in zinc-induced neuronal death. *J. Neurochem.*, 73, 450, 1999.
150. Zhang, Y., et al. Intracellular zinc release and ERK phosphorylation are required upstream of 12-lipoxygenase activation in peroxynitrite toxicity to mature rat oligodendrocytes. *J. Biol. Chem.*, 281, 9460, 2006.
151. Bruinsma, J.J., et al. Zinc ions and cation diffusion facilitator proteins regulate Ras-mediated signaling. *Dev. Cell.*, 2, 567, 2002.
152. Yoder, J.H., et al. Modulation of KSR activity in *Caenorhabditis elegans* by Zn ions, PAR-1 kinase and PP2A phosphatase. *EMBO J.*, 23, 111, 2004.
153. Baba, A., Etoh, S., and Iwata, H. Inhibition of NMDA-induced protein kinase C translocation by a Zn^{2+} chelator: Implication of intracellular Zn^{2+}. *Brain Res.*, 557, 103, 1991.
154. Csermely, P., et al. Zinc increases the affinity of phorbol ester receptor in T lymphocytes. *Biochem. Biophys. Res. Commun.*, 154, 578, 1988.
155. Forbes, I.J., et al. Interaction between protein kinase C and regulatory ligand is enhanced by a chelatable pool of cellular zinc. *Biochim. Biophys. Acta.*, 1053, 113, 1990.
156. Zalewski, P.D., et al. Synergy between zinc and phorbol ester in translocation of protein kinase C to cytoskeleton. *FEBS Lett.*, 273, 131, 1990.
157. Forbes, I.J., et al. Zinc induces specific association of PKC with membrane cytoskeleton. *Biochem. Int.*, 22, 741, 1990.
158. Zalewski, P.D., et al. Regulation of protein kinase C by Zn(2+)-dependent interaction with actin. *Biochem. Int.*, 24, 1103, 1991.
159. Palmer, B.M., et al. Intracellular distributions of essential elements in cardiomyocytes. *J. Struct. Biol.*, 155, 12, 2006.
160. Lichtlen, P. and Schaffner, W. The "metal transcription factor" MTF-1: Biological facts and medical implications. *Swiss Med. Wkly.*, 131, 647, 2001.
161. Lichtlen, P., et al. Target gene search for the metal-responsive transcription factor MTF-1. *Nucleic Acids Res.*, 29, 1514, 2001.
162. Langmade, S.J., et al. The transcription factor MTF-1 mediates metal regulation of the mouse ZnT1 gene. *J. Biol. Chem.*, 275, 34803, 2000.
163. Radtke, F., et al. Cloned transcription factor MTF-1 activates the mouse metallothionein I promoter. *EMBO J.*, 12, 1355, 1993.
164. Chen, X., Chu, M., and Giedroc, D.P. MRE-Binding transcription factor-1: Weak zinc-binding finger domains 5 and 6 modulate the structure, affinity, and specificity of the metal-response element complex. *Biochemistry*, 38, 12915, 1999.
165. Jiang, H., Daniels, P.J., and Andrews, G.K. Putative zinc-sensing zinc fingers of metal-response element-binding transcription factor-1 stabilize a metal-dependent chromatin complex on the endogenous metallothionein-I promoter. *J. Biol. Chem.*, 278, 30394, 2003.
166. Koizumi, S., et al. Roles of zinc fingers and other regions of the transcription factor human MTF-1 in zinc-regulated DNA binding. *J. Cell. Physiol.*, 185, 464, 2000.
167. LaRochelle, O., et al. Phosphorylation is involved in the activation of metal-regulatory transcription factor 1 in response to metal ions. *J. Biol. Chem.*, 276, 41879, 2001.
168. Saydam, N., et al. Regulation of metallothionein transcription by the metal-responsive transcription factor MTF-1: Identification of signal transduction cascades that control metal-inducible transcription. *J. Biol. Chem.*, 277, 20438, 2002.
169. Zhang, B., et al. Activity of metal-responsive transcription factor 1 by toxic heavy metals and H2O2 *in vitro* is modulated by metallothionein. *Mol. Cell. Biol.*, 23, 8471, 2003.
170. Brautigan, D.L., Bornstein, P., and Gallis, B. Phosphotyrosyl-protein phosphatase. Specific inhibition by Zn. *J. Biol. Chem.*, 256, 6519, 1981.
171. Maret, W., et al. Inhibitory sites in enzymes: Zinc removal and reactivation by thionein. *Proc. Natl. Acad. Sci. USA*, 96, 1936, 1999.
172. Conti, M. and Beavo, J. Biochemistry and physiology of cyclic nucleotide phosphodiesterases: Essential components in cyclic nucleotide signaling. *Annu. Rev. Biochem.*, 76, 481, 2007.

173. Percival, M.D., Yeh, B., and Falgueyret, J.P. Zinc dependent activation of cAMP-specific phosphodiesterase (PDE4A). *Biochem. Biophys. Res. Commun.*, 241, 175, 1997.
174. Pillai, R., Staub, S.F., and Colicelli, J. Mutational mapping of kinetic and pharmacological properties of a human cardiac cAMP phosphodiesterase. *J. Biol. Chem.*, 269, 30676, 1994.
175. von Bulow, V., Rink, L., and Haase, H. Zinc-mediated inhibition of cyclic nucleotide phosphodiesterase activity and expression suppresses TNF-alpha and IL-1 beta production in monocytes by elevation of guanosine 3′,5′-cyclic monophosphate. *J. Immunol.*, 175, 4697, 2005.
176. von Bulow, V., et al. Zinc-dependent suppression of TNF-alpha production is mediated by protein kinase A-induced inhibition of Raf-1, I kappa B kinase beta, and NF-kappa B. *J. Immunol.*, 179, 4180, 2007.
177. Kim, Y.M., et al. Zn^{2+}-induced IL-8 expression involves AP-1, JNK, and ERK activities in human airway epithelial cells. *Am. J. Physiol. Lung Cell Mol. Physiol.*, 290, L1028, 2006.
178. Kim, P.W., et al. A zinc clasp structure tethers Lck to T cell coreceptors CD4 and CD8. *Science*, 301, 1725, 2003.
179. Romir, J., et al. Crystal structure analysis and solution studies of human Lck-SH3; zinc-induced homodimerization competes with the binding of proline-rich motifs. *J. Mol. Biol.*, 365, 1417, 2007.
180. Pernelle, J.J., et al. Phosphorylation of the lymphoid cell kinase p56lck is stimulated by micromolar concentrations of Zn^{2+}. *FEBS Lett.*, 281, 278, 1991.
181. Danscher, G. and Stoltenberg, M. Zinc-specific autometallographic *in vivo* selenium methods: Tracing of zinc-enriched (ZEN) terminals, ZEN pathways, and pools of zinc ions in a multitude of other ZEN cells. *J. Histochem. Cytochem.*, 53, 141, 2005.
182. Frederickson, C.J., Koh, J.Y., and Bush, A.I. The neurobiology of zinc in health and disease. *Nat. Rev. Neurosci.*, 6, 449, 2005.
183. Petrat, F., et al. The chelatable iron pool in living cells: A methodically defined quantity. *Biol. Chem.*, 383, 489, 2002.
184. Nitzan, Y.B., Sekler, I., and Silverman, W.F. Histochemical and histofluorescence tracing of chelatable zinc in the developing mouse. *J. Histochem. Cytochem.*, 52, 529, 2004.
185. Schroeder, H.A., et al. Essential trace metals in man: Zinc. Relation to environmental cadmium. *J. Chronic. Dis.*, 20, 179, 1967.
186. Costello, L.C., et al. Zinc inhibition of mitochondrial aconitase and its importance in citrate metabolism of prostate epithelial cells. *J. Biol. Chem.*, 272, 28875, 1997.
187. Franklin, R.B. and Costello, L.C. Zinc as an anti-tumor agent in prostate cancer and in other cancers. *Arch. Biochem. Biophys.*, 463, 211, 2007.
188. Emami, N. and Diamandis, E.P. Human kallikrein-related peptidase 14 (KLK14) is a new activator component of the KLK proteolytic cascade. Possible function in seminal plasma and skin. *J. Biol. Chem.*, 283, 3031, 2008.
189. Lovgren, J., Airas, K., and Lilja, H. Enzymatic action of human glandular kallikrein 2 (hK2). Substrate specificity and regulation by Zn^{2+} and extracellular protease inhibitors. *Eur. J. Biochem.*, 262, 781, 1999.
190. Malm, J., et al. Enzymatic action of prostate-specific antigen (PSA or hK3): Substrate specificity and regulation by Zn(2+), a tight-binding inhibitor. *Prostate*, 45, 132, 2000.
191. Debela, M., et al. Crystal structures of human tissue kallikrein 4: Activity modulation by a specific zinc binding site. *J. Mol. Biol.*, 362, 1094, 2006.
192. Debela, M., et al. Structural basis of the zinc inhibition of human tissue kallikrein 5. *J. Mol. Biol.*, 373, 1017, 2007.
193. Debela, M., et al. Chymotryptic specificity determinants in the 1.0 A structure of the zinc-inhibited human tissue kallikrein 7. *Proc. Natl. Acad. Sci. USA*, 104, 16086, 2007.
194. Kristiansen, L.H., et al. Autometallography allows ultrastructural monitoring of zinc in the endocrine pancreas. *Histochem. Cell Biol.*, 115, 125, 2001.
195. Matseshe, J.W., et al. Recovery of dietary iron and zinc from the proximal intestine of healthy man: Studies of different meals and supplements. *Am. J. Clin. Nutr.*, 33, 1946, 1980.
196. Larsen, K.S. and Auld, D.S. Carboxypeptidase A: Mechanism of zinc inhibition. *Biochemistry*, 28, 9620, 1989.
197. Gomez-Ortiz, M., et al. Inhibition of carboxypeptidase A by excess zinc: Analysis of the structural determinants by X-ray crystallography. *FEBS Lett.*, 400, 336, 1997.
198. Dunn, M.F., Zinc-ligand interactions modulate assembly and stability of the insulin hexamer—a review. *Biometals*, 18, 295, 2005.
199. Foster, M.C., et al. Elemental composition of secretory granules in pancreatic islets of Langerhans. *Biophys. J.*, 64, 525, 1993.

200. Liuzzi, J.P., et al. Responsive transporter genes within the murine intestinal-pancreatic axis form a basis of zinc homeostasis. *Proc. Natl. Acad. Sci. USA*, 101, 14355, 2004.
201. Chimienti, F., et al. Identification and cloning of a beta-cell-specific zinc transporter, ZnT-8, localized into insulin secretory granules. *Diabetes*, 53, 2330, 2004.
202. Chimienti, F., et al. *In vivo* expression and functional characterization of the zinc transporter ZnT8 in glucose-induced insulin secretion. *J. Cell Sci.*, 119, 4199, 2006.
203. Kambe, T., et al. Cloning and characterization of a novel mammalian zinc transporter, zinc transporter 5, abundantly expressed in pancreatic beta cells. *J. Biol. Chem.*, 277, 19049, 2002.
204. Qian, W.J., Gee, K.R., and Kennedy, R.T. Imaging of Zn^{2+} release from pancreatic beta-cells at the level of single exocytotic events. *Anal. Chem.*, 75, 3468, 2003.
205. Zalewski, P.D., et al. Video image analysis of labile zinc in viable pancreatic islet cells using a specific fluorescent probe for zinc. *J. Histochem. Cytochem.*, 42, 877, 1994.
206. Porksen, N., et al. In humans at least 75% of insulin secretion arises from punctuated insulin secretory bursts. *Am. J. Physiol.*, 273, E908, 1997.
207. Ishihara, H., et al. Islet beta-cell secretion determines glucagon release from neighbouring alpha-cells. *Nat. Cell Biol.*, 5, 330, 2003.
208. Zhou, H., et al. Zinc, not insulin, regulates the rat alpha-cell response to hypoglycemia *in vivo. Diabetes*, 56, 1107, 2007.
209. Ravier, M.A. and Rutter, G.A. Glucose or insulin, but not zinc ions, inhibit glucagon secretion from mouse pancreatic alpha-cells. *Diabetes*, 54, 1789, 2005.
210. MacDonald, P.E., et al. A K ATP channel-dependent pathway within alpha cells regulates glucagon release from both rodent and human islets of Langerhans. *PLoS Biol.*, 5, e143, 2007.
211. Franklin, I., et al. Beta-cell secretory products activate alpha-cell ATP-dependent potassium channels to inhibit glucagon release. *Diabetes*, 54, 1808, 2005.
212. Maske, H., Über den topochemischen Nachweis von Zink in Ammonshorn verschiedener Säugetiere. *Naturwissenschaften*, 42, 424, 1955.
213. Haug, F.M., Electron microscopical localization of the zinc in hippocampal mossy fibre synapses by a modified sulfide silver procedure. *Histochemie*, 8, 355, 1967.
214. Dobrowolska, J., et al. Quantitative imaging of zinc, copper and lead in three distinct regions of the human brain by laser ablation inductively coupled plasma mass spectrometry. *Talanta*, 74, 717, 2008.
215. Frederickson, C.J. and Bush, A.I. Synaptically released zinc: Physiological functions and pathological effects. *Biometals*, 14, 353, 2001.
216. Paoletti, P., Ascher, P., and Neyton, J. High-affinity zinc inhibition of NMDA NR1-NR2A receptors. *J. Neurosci.*, 17, 5711, 1997.
217. Paoletti, P., et al. Zinc at glutamatergic synapses. *Neuroscience*, 158, 126, 2009.
218. Nagaya, N., et al. An intersubunit zinc binding site in rat P2X2 receptors. *J. Biol. Chem.*, 280, 25982, 2005.
219. Hsiao, B., et al. Determinants of zinc potentiation on the alpha4 subunit of neuronal nicotinic receptors. *Mol. Pharmacol.*, 69, 27, 2006.
220. Hirzel, K., et al. Hyperekplexia phenotype of glycine receptor alpha1 subunit mutant mice identifies Zn(2+) as an essential endogenous modulator of glycinergic neurotransmission. *Neuron*, 52, 679, 2006.
221. Li, Y., et al. Induction of mossy fiber–> Ca3 long-term potentiation requires translocation of synaptically released Zn^{2+}. *J. Neurosci.*, 21, 8015, 2001.
222. Huang, Y.Z., et al. Zinc-mediated transactivation of TrkB potentiates the hippocampal mossy fiber-CA3 pyramid synapse. *Neuron*, 57, 546, 2008.
223. Sun, G. and Budde, R.J. Substitution studies of the second divalent metal cation requirement of protein tyrosine kinase CSK. *Biochemistry*, 38, 5659, 1999.
224. Gundelfinger, E.D., et al. A role for zinc in postsynaptic density assembly and plasticity? *Trends Biochem. Sci.*, 31, 366, 2006.
225. Fernandez, J.R., Welsh, W.J., and Firestein, B.L. Structural characterization of the zinc binding domain in cytosolic PSD-95 interactor (cypin): Role of zinc binding in guanine deamination and dendrite branching. *Proteins*, 70, 873, 2008.
226. Kodirov, S.A., et al. Synaptically released zinc gates long-term potentiation in fear conditioning pathways. *Proc. Natl. Acad. Sci. USA*, 103, 15218, 2006.
227. Yamaguchi, M. and Fukagawa, M. Role of zinc in regulation of PTP activity in osteoblastic MC3T3-E1 cells: Zinc modulation of insulin-like growth factor-I's effect. *Calcif. Tissue Int.*, 76, 32, 2005.
228. Wuthier, R.E., Mechanism of Matrix Vesicle-Mediated Mineralization of Cartilage. *Isi Atlas Sci. Biochem.*, 1, 231, 1988.

229. Sauer, G.R., et al. Intracellular zinc fluxes associated with apoptosis in growth plate chondrocytes. *J. Cell. Biochem.*, 88, 954, 2003.
230. Giblin, L.J., et al. Zinc-secreting Paneth cells studied by ZP fluorescence. *J. Histochem. Cytochem.*, 54, 311, 2006.
231. Grahn, B.H., et al. Zinc and the eye. *J. Am. Coll. Nutr.*, 20, 106, 2001.
232. Ugarte, M. and Osborne, N.N. Zinc in the retina. *Prog. Neurobiol.*, 64, 219, 2001.
233. Stojanovic, A., Stitham, J., and Hwa, J. Critical role of transmembrane segment zinc binding in the structure and function of rhodopsin. *J. Biol. Chem.*, 279, 35932, 2004.
234. Redenti, S., Ripps, H., and Chappell, R.L. Zinc release at the synaptic terminals of rod photoreceptors. *Exp. Eye Res.*, 85, 580, 2007.
235. Underwood, E.J., *Trace Elements in Human and Animal Nutrition*, 4th ed. Academic Press, San Diego, 1977.
236. Kelleher, S.L. and Lonnerdal, B. Zip3 plays a major role in zinc uptake into mammary epithelial cells and is regulated by prolactin. *Am. J. Physiol. Cell Physiol.*, 288, C1042, 2005.
237. Endre, L., Katona, Z., and Gyurkovits, K. Letter: Zinc deficiency and cellular immune deficiency in acrodermatitis enteropathica. *Lancet*, 1, 1196, 1975.
238. Dufner-Beattie, J., et al. The adaptive response to dietary zinc in mice involves the differential cellular localization and zinc regulation of the zinc transporters ZIP4 and ZIP5. *J. Biol. Chem.*, 279, 49082, 2004.
239. Wang, F., et al. The mammalian Zip5 protein is a zinc transporter that localizes to the basolateral surface of polarized cells. *J. Biol. Chem.*, 279, 51433, 2004.
240. Kambe, T., Weaver, B.P., and Andrews, G.K. The genetics of essential metal homeostasis during development. *Genesis*, 46, 214, 2008.
241. Glaser, R., et al. Antimicrobial psoriasin (S100A7) protects human skin from Escherichia coli infection. *Nat. Immunol.*, 6, 57, 2005.
242. Brodersen, D.E., Nyborg, J., and Kjeldgaard, M. Zinc-binding site of an S100 protein revealed. Two crystal structures of Ca2 + -bound human psoriasin (S100A7) in the Zn^{2+}-loaded and Zn^{2+}-free states. *Biochemistry*, 38, 1695, 1999.
243. Corbin, B.D., et al. Metal chelation and inhibition of bacterial growth in tissue abscesses. *Science*, 319, 962, 2008.
244. World Health Organization. The World Health Report, p. 83, 2002.
245. Fraker, P.J. and King, L.E. Reprogramming of the immune system during zinc deficiency. *Annu. Rev. Nutr.*, 24, 277, 2004.
246. Shankar, A.H. and Prasad, A.S. Zinc and immune function: The biological basis of altered resistance to infection. *Am. J. Clin. Nutr.*, 68, 447S, 1998.
247. Wellinghausen, N., Kirchner, H., and Rink, L. The immunobiology of zinc. *Immunol. Today*, 18, 519, 1997.
248. Fischer Walker, C. and Black, R.E. Zinc and the risk for infectious disease. *Annu. Rev. Nutr.*, 24, 255, 2004.
249. Rink, L. and Haase, H. Zinc homeostasis and immunity. *Trends Immunol.*, 28, 1, 2007.
250. Prasad, A.S., Zinc and immunity. *Mol. Cell. Biochem.*, 188, 63, 1998.
251. Allen, J.I., et al. Alterations in human natural killer cell activity and monocyte cytotoxicity induced by zinc deficiency. *J. Lab. Clin. Med.*, 102, 577, 1983.
252. Fraker, P.J. and King, L.E. A distinct role for apoptosis in the changes in lymphopoiesis and myelopoiesis created by deficiencies in zinc. *FASEB J.*, 15, 2572, 2001.
253. King, L.E. and Fraker, P.J. Flow cytometric analysis of the phenotypic distribution of splenic lymphocytes in zinc-deficient adult mice. *J. Nutr.*, 121, 1433, 1991.
254. Overbeck, S., Rink, L., and Haase, H. Modulating the immune response by oral zinc supplementation: A single approach for multiple diseases. *Arch. Immunol. Ther. Exp. (Warsz.)*, 56, 15, 2008.
255. Gastinel, L.N., et al. Studies on the zinc binding site to the serum thymic factor. *Biochim. Biophys. Acta*, 797, 147, 1984.
256. Prasad, A.S., et al. Serum thymulin in human zinc deficiency. *J. Clin. Invest.*, 82, 1202, 1988.
257. Beck, F.W., et al. Changes in cytokine production and T cell subpopulations in experimentally induced zinc-deficient humans. *Am. J. Physiol.*, 272, E1002, 1997.
258. Cakman, I., et al. Dysregulation between TH1 and TH2 T cell subpopulations in the elderly. *Mech. Ageing Dev.*, 87, 197, 1996.
259. Duchateau, J., et al. Beneficial effects of oral zinc supplementation on the immune response of old people. *Am. J. Med.*, 70, 1001, 1981.
260. Wellinghausen, N., Martin, M., and Rink, L. Zinc inhibits interleukin-1-dependent T cell stimulation. *Eur. J. Immunol.*, 27, 2529, 1997.

261. Cook-Mills, J.M. and Fraker, P.J. Functional capacity of the residual lymphocytes from zinc-deficient adult mice. *Br. J. Nutr.*, 69, 835, 1993.
262. Haase, H., Overbeck, S., and Rink, L. Zinc supplementation for the treatment or prevention of disease: Current status and future perspectives. *Exp. Gerontol.*, 43, 394, 2008.
263. Fraker, P.J., et al. Regeneration of T-cell helper function in zinc-deficient adult mice. *Proc. Natl. Acad. Sci. USA*, 75, 5660, 1978.
264. Moulder, K. and Steward, M.W. Experimental zinc deficiency: Effects on cellular responses and the affinity of humoral antibody. *Clin. Exp. Immunol.*, 77, 269, 1989.
265. Rajagopalan, S., et al. The Ig-related killer cell inhibitory receptor binds zinc and requires zinc for recognition of HLA-C on target cells. *J. Immunol.*, 155, 4143, 1995.
266. Rajagopalan, S. and Long, E.O. Zinc bound to the killer cell-inhibitory receptor modulates the negative signal in human NK cells. *J. Immunol.*, 161, 1299, 1998.
267. Vales-Gomez, M., et al. The role of zinc in the binding of killer cell Ig-like receptors to class I MHC proteins. *Proc. Natl. Acad. Sci. USA*, 98, 1734, 2001.
268. Fernandes, G., et al. Impairment of cell-mediated immunity functions by dietary zinc deficiency in mice. *Proc. Natl. Acad. Sci. USA*, 76, 457, 1979.
269. Gustafson, G.T., Heavy metals in rat mast cell granules. *Lab. Invest.*, 17, 588, 1967.
270. Ho, L.H., et al. Labile zinc and zinc transporter ZnT4 in mast cell granules: Role in regulation of caspase activation and NF-kappaB translocation. *J. Immunol.*, 172, 7750, 2004.
271. King, L.E. and Fraker, P.J. Zinc deficiency in mice alters myelopoiesis and hematopoiesis. *J. Nutr.*, 132, 3301, 2002.
272. Hujanen, E.S., Seppa, S.T., and Virtanen, K. Polymorphonuclear leukocyte chemotaxis induced by zinc, copper and nickel *in vitro*. *Biochim. Biophys. Acta.*, 1245, 145, 1995.
273. Haase, H., et al. Zinc signals are essential for lipopolysaccharide-induced signal transduction in monocytes. *J. Immunol.*, 181, 6491, 2008.
274. Kabu, K., et al. Zinc is required for Fc epsilon RI-mediated mast cell activation. *J. Immunol.*, 177, 1296, 2006.
275. Richter, M., et al. Zinc chelators inhibit eotaxin, RANTES, and MCP-1 production in stimulated human airway epithelium and fibroblasts. *Am. J. Physiol. Lung Cell Mol. Physiol.*, 285, L719, 2003.
276. Beach, R.S., Gershwin, M.E., and Hurley, L.S. Gestational zinc deprivation in mice: Persistence of immunodeficiency for three generations. *Science*, 218, 469, 1982.
277. Widdowson, E.M., McCance, R.A., and Spray, C.M. The chemical composition of the human body. *Clin. Sci.*, 10, 113, 1951.
278. Jackson, M.J., Physiology of zinc: General aspects. In: *Zinc in Human Biology*, C.F. Mills (ed.), pp. 1–14. Springer, London, 1989.
279. Vallee, B.L. and Gibson, J.G. The zinc content of normal human whole blood, plasma, leukocytes, and erythrocytes. *J. Biol. Chem.*, 176, 445, 1948.
280. King, J.C. and Cousins, R.J. Zinc. In: *Modern Nutrition in Health and Disease*, M.E. Shils et al. (eds), pp. 271–285. Lippincott Williams & Wilkins, Philadelphia, 2006.
281. Windisch, W. Development of zinc deficiency in 65Zn labeled, fully grown rats as a model for adult individuals. *J. Trace Elem. Med. Biol.*, 17, 91, 2003.
282. Falchuk, K.H. Effect of acute disease and ACTH on serum zinc proteins. *N. Engl. J. Med.*, 296, 1129, 1977.
283. Liuzzi, J.P., et al. Interleukin-6 regulates the zinc transporter Zip14 in liver and contributes to the hypozincemia of the acute-phase response. *Proc. Natl. Acad. Sci. USA*, 102, 6843, 2005.
284. Chesters, J.K. Zinc. In: *Handbook of Nutritionally Essential Mineral Elements*, B. O'Dell and R. Sunde (eds), pp. 185–230. Marcel Dekker, New York, 1997.

9 Glutathione, Protein Thiols, and Metal Homeostasis

Lawrence H. Lash

CONTENTS

9.1 Introduction 213
9.2 Interorgan Metabolism of GSH and Metal Transport 214
9.2.1 GSH Synthesis and Degradation 214
9.2.2 Pathways for Interorgan Transport of GSH and GSH *S*-Conjugates 216
9.2.3 Hepatic GSH and GSH *S*-Conjugate Transport 218
9.2.4 Renal GSH and GSH *S*-Conjugate Transport 218
9.2.5 Metals as GSH *S*-Conjugates: Implications for Tissue Distribution 220
9.3 Modulation of Tissue Accumulation and Excretion of Metals by GSH and Protein Thiols 221
9.3.1 Hepatic and Biliary GSH and Metal Transport and Disposition 221
9.3.2 Plasma Thiols and Metal Disposition 221
9.3.3 Renal GSH and Transport, Disposition, and Toxicity of Mercury 222
9.3.3.1 Exogenous GSH 222
9.3.3.2 Uninephrectomy and Compensatory Renal Growth (NPX) Model 225
9.3.4 Cadmium Transport and Disposition 226
9.3.5 Arsenic Metabolism and Disposition 228
9.4 Metal-Induced Changes in GSH Status 229
9.4.1 Metal-Induced Changes in Cellular GSH Status 229
9.4.2 Regulation of Gene Expression by Metals 232
9.5 Summary and Conclusions 233
References 233

9.1 INTRODUCTION

The underlying chemical basis for much of the biological activity of heavy metals, such as mercury (Hg)*, cadmium (Cd), and arsenic (As), is their strong reactivity with sulfhydryl groups on both low-molecular-weight thiol compounds and the cysteinyl residues of proteins. Because of its predominance as the major intracellular nonprotein thiol compound, glutathione (GSH) is of particular importance in the disposition of many metals. Besides its presence in cells, GSH is also found extracellularly in plasma [1–7]. This presence suggests that metals are likely to be bound to GSH and translocated through the plasma to various tissues. Although GSH binds tightly to metals, Oram and colleagues [8] calculated that the formation constant of the 1:2 complex between Hg^{2+} and GSH is actually substantially smaller than what was originally thought. This has important implications for metal ion disposition, because it suggests that metals bound to GSH may exchange with other ligands. Such exchanges are, of course, necessary for the metal ions to interact with cellular molecules to produce their biological effects.

* When the precise speciation of the metal is not known or when a general reference is made to the metal without regard to its specific form, only the symbol (e.g., Hg, Cd, and As) is used.

Besides GSH, other low-molecular-weight thiols that are present in plasma, and thus may influence the interorgan distribution of metal ions such as Hg^{2+} and Cd^{2+}, include L-cysteine (L-Cys) and L-homocysteine [4,5,9]. By far, the most predominant thiol groups in plasma, however, are those present in albumin [10]. The ligand to which the metal ion is bound when it is presented to a target tissue is critical in determining the mechanism by which the metal ion enters cells. Thus, whereas metal ions bound to albumin typically enter target cells, such as hepatocytes or renal proximal tubular cells, by endocytosis or by exchanging with other ligands, metals ions bound to low-molecular-weight thiols can be transported directly into cells by specific membrane carrier proteins, as described later in this chapter and elsewhere in this book. For the kidneys as the target organ, the ligand to which the metal is bound will influence the proportion of exposure of the tissue that occurs by glomerular filtration as compared to that in renal plasma.

This chapter summarizes the current state of knowledge on the interactions between GSH and protein thiols with metals, focusing on Hg, Cd, As, and Se. The first three metals are nonessential metals to which humans and animals are typically exposed in environmental contaminations, although As is also a component in some therapeutic agents. Se is an essential trace metal that serves as a prosthetic group for a selected number of enzymes, is toxic in excess, and plays important roles in cellular responses to oxidative stress. After a review of how GSH and GSH *S*-conjugates are handled in the body and how complexes of GSH with metals mimic other GSH *S*-conjugates, the next section presents the interaction between metals and thiols, focusing on the influence of thiols on metal ion disposition and toxicity. This includes both the effect of exogenous GSH as well as the influence of different intracellular GSH concentrations on metal ion disposition and toxicity. The final section considers the influence of these metals on GSH and thiol status in cells, both in terms of effects on target tissue concentrations as well as effects on gene expression that lead to changes in cellular GSH status.

9.2 INTERORGAN METABOLISM OF GSH AND METAL TRANSPORT

9.2.1 GSH SYNTHESIS AND DEGRADATION

To understand the functions of GSH in metal ion disposition and modulation of metal ion toxicity, it is first important to understand how GSH concentrations and redox status are determined in cells and some of the basic functions of GSH. Accordingly, this section summarizes how GSH is synthesized and degraded, how these processes are regulated, how GSH and GSH *S*-conjugates are translocated from certain tissues (primarily the liver) and taken up by other tissues (primarily the kidneys), specific pathways for GSH and GSH *S*-conjugate transport in the two major target organs of interest in this chapter (viz., the liver and kidneys), and how complexes between metals and GSH are essentially handled like other GSH *S*-conjugates.

GSH is the predominant nonprotein thiol in most cells, being present at concentrations of 1–7 mM, depending on cell type and physiological state. As such, it plays numerous, critical roles in the metabolism of reactive electrophiles, peroxides, various hormones and prostaglandins, and in the thiol–disulfide status of cysteinyl sulfhydryl groups of proteins. GSH is a tripeptide containing the amino acids L-glutamate (L-Glu), L-Cys, and glycine (Gly) (Figure 9.1). The two key features of its structure are the thiol group on the cysteinyl residue and the linkage of the L-Glu and L-Cys residues through the γ-carboxyl group of L-Glu and the α-amino group of L-Cys. The first feature, the thiol group, is what gives GSH its biochemical functions, as it imparts on the molecule the properties of being a reductant and nucleophile. The second feature, the so-called γ-glutamyl or isopeptide bond, stabilizes the GSH molecule, preventing it from being degraded by proteases. Moreover, the resistance to proteolysis or degradation by all but a specific enzyme, γ-glutamyltransferase (also called γ-glutamyltranspeptidase; GGT; EC 2.3.2.2), enables GSH to be secreted from tissues (primarily the liver) and travel via the circulation to other tissues for extraction (see below discussion on interorgan transport pathways).

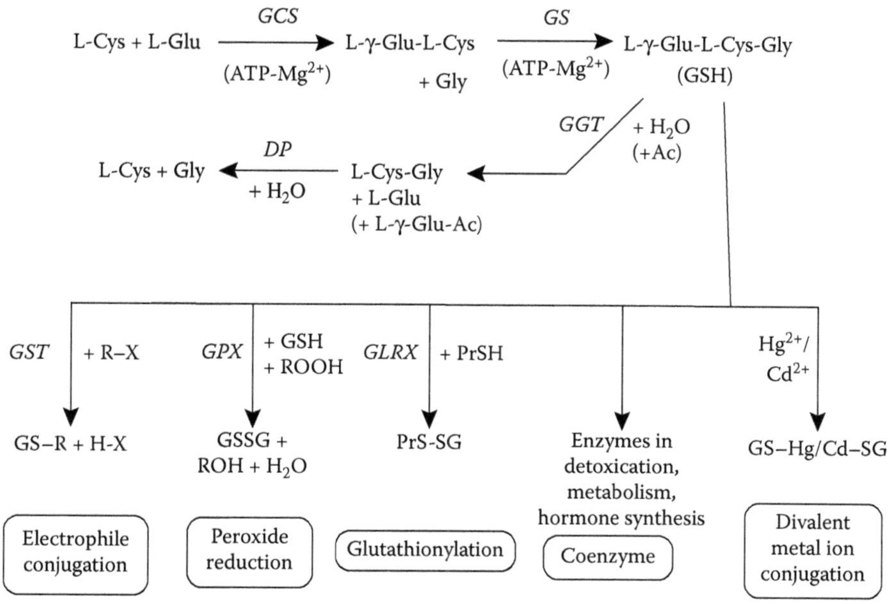

FIGURE 9.1 Summary scheme of GSH metabolism and GSH-dependent functions. GSH synthesis occurs by the sequence of two reactions on top, whereas degradation occurs by the reactions catalyzed by GGT and DP. Reactions of GSH can be divided into the five categories as shown in the figure. *Abbreviations:* Ac, acceptor for GGT transpeptidation reaction; DP, dipeptidase; GCS, γ-glutamylcysteine synthetase; GGT, γ-glutamyltransferase; GLRX, glutaredoxin; GPX, glutathione peroxidase; GS–Cd–SG, diglutathionyl complex with Cd^{2+}; GS–Hg–SG, diglutathionyl complex with Hg^{2+}; GS–R, GSH *S*-conjugate; GST, GSH *S*-transferase; PrSH, protein sulfhydryl; PrS–SG; glutathionylated protein; ROH, alcohol; and ROOH, organic peroxide.

GSH synthesis and degradation occur by two separate pathways, which enable more efficient control of the two processes and maintain tight control of GSH levels. GSH is synthesized from its precursor amino acids by two ATP-dependent reactions that occur predominantly, if not entirely, within the cytoplasm [11,12]. The first reaction is catalyzed by γ-glutamylcysteine synthetase (also called glutamate–cysteine ligase; GCS; EC 6.3.2.2), is the rate-limiting step in GSH synthesis, and produces L-γ-glutamyl-L-cysteine. The second step attaches Gly to a peptide bond to form GSH and is catalyzed by GSH synthetase (GS; EC 6.3.2.3). Because L-Glu and Gly are usually present at concentrations equal to or above the K_m values and are readily obtained by conversion from other amino acids, they are not limiting for GCS. In contrast, L-Cys is an essential amino acid that can only be obtained from dietary L-Cys or L-Met and is typically present in the cell at concentrations below the K_m value of 0.3 mM, making it limiting for GCS [13]. Thus, the two major factors that regulate GSH biosynthesis are the availability of L-Cys for GCS and feedback inhibition of GCS by GSH (K_i = 2.3 mM).

Modulation of GSH synthesis activity is an effective means of altering cellular GSH concentrations to investigate its role in a given process of interest. For example, a series of sulfoximine derivatives were developed by Meister and colleagues [14–16] as potent and selective inhibitors of GCS. Among these, buthionine sulfoximine (BSO) is the most potent and selective inhibitor and has become a standard reagent for the depletion of cellular GSH levels. Other ways of depleting cellular GSH include the addition of substrates for GSH *S*-transferases (GSTs), such as diethyl maleate (DEM) or *N*-ethylmaleimide (NEM). Because GCS expression and activity are tightly regulated, including the aforementioned feedback inhibition by GSH, approaches to increase cellular concentrations of GSH are more limited. As discussed below, addition of GSH or L-γ-glutamyl-L-cysteine can be used to increase GSH content of renal proximal tubular cells. Other cell types, such as hepatocytes or cardiac myocytes, for example, cannot use exogenous GSH because they lack transport systems for uptake of GSH [17].

The GCS enzyme from rat kidney is a heterodimer, with a heavy subunit of M_r 73 kDa and a light subunit of M_r 28 kDa. The heavy subunit (also called the glutamate–cysteine ligase catalytic or GCLC subunit) provides the catalytic function for the enzyme whereas the light subunit (also called the glutamate–cysteine ligase modulatory or GCLM subunit) modulates the kinetic properties of the catalytic subunit [18–21]. Although GCLM is catalytically inactive, it functions in a modulatory manner, as the name suggests, by lowering the K_m for L-Glu and by increasing the K_i for GSH. This results in increased catalytic efficiency and less feedback inhibition by GSH. GCLM, although important, however, is not essential as $Gclm^{-/-}$ knockout mice are viable but have markedly reduced levels of cellular GSH [22].

In contrast with synthesis, GSH degradation occurs exclusively extracellularly and is mediated by GGT and a membrane-bound dipeptidase (DP; cysteinylglycine dipeptidase—EC 3.4.13.6 or aminopeptidase M—EC 3.4.11.2). In the kidneys, GSH degradation occurs almost exclusively in the proximal tubules and is localized on the luminal or brush-border plasma membrane (BBM). GGT is an integral membrane glycoprotein that is constitutively expressed at high levels in adult animals in tissues such as kidneys, intestine, choroid plexus, and epididymus. Although GGT is highly expressed in perinatal liver, it is almost undetectable in adult liver. Conversely, a perinatal kidney exhibits very low levels of GGT, whereas an adult kidney expresses very high levels of GGT [23,24]. In fact, although the ratio of GGT activity in the kidneys to that in the liver varies considerably among species (ranging from ~800:1 in rats to ~400:1 in rabbits to ~100:1 in pigs and humans), the kidneys have by far the highest GGT activity of any tissue. As described below, this fact plays a major role in determining patterns of interorgan translocation of GSH and GSH *S*-conjugates.

GGT cleaves the γ-glutamyl peptide bond of GSH, transferring the glutamyl residue either to water (hydrolysis reaction) or to an amino acid or peptide acceptor (transpeptidation reaction) (Figure 9.1). Some dipeptides and most neutral amino acids are good acceptors, whereas branched-chain, acidic, and basic amino acids are poor substrates and D-amino acids and L-proline are inactive. Besides GSH, various *S*-substituted derivatives of GSH, including GSH *S*-conjugates of reactive electrophiles, GSH complexes with metal ions, and various γ-glutamyl amino acids and peptides, are also substrates. The importance of GGT in overall maintenance of GSH and sulfur amino acid homeostasis is highlighted by the observations that both treatment of animals with a potent GGT inhibitor and a patient with a severe GGT deficiency exhibited profound glutathionuria, L-γ-glutamyl-L-cysteinuria, and cysteinuria [25,26].

The five major categories of GSH-dependent functions are also shown in Figure 9.1, and they include detoxication of reactive electrophiles, as mediated by GSTs, peroxide reduction, as mediated by GSH peroxidase (GPX; EC 1.6.4.2), glutathionylation of proteins, as mediated by enzymes such as glutaredoxin (GLRX), function as a coenzyme in various drug metabolism and hormone biosynthetic reactions, and in conjugation or complexation of divalent metal ions.

9.2.2 Pathways for Interorgan Transport of GSH and GSH *S*-Conjugates

Being degraded only by GGT and being resistant to other proteases, GSH and GSH *S*-conjugates can translocate as the intact tripeptide or conjugate through the circulation. For GSH, this capability allows it to serve as a reservoir for L-Cys for protein synthesis or resynthesis of GSH. For GSH *S*-conjugates, this capability enables movement from the primary site of synthesis to the primary site of further metabolism and/or excretion. Because of its high biosynthetic capacity and high activity of GST, liver is the primary source of extracellular GSH and GSH *S*-conjugates, respectively. In turn, because it is the predominant site of GGT activity in the body, kidneys are the primary tissue that extracts GSH and GSH *S*-conjugates from the circulation. As shown in Figure 9.2, there are both enterohepatic and renal–hepatic circulations that are involved in the turnover and interorgan

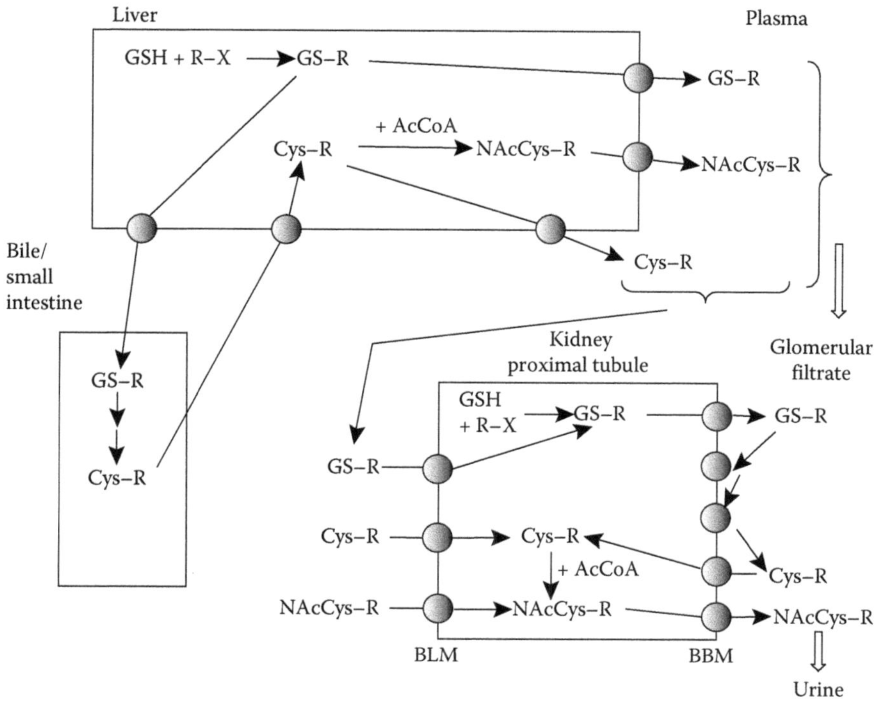

FIGURE 9.2 Interorgan pathways for transport and metabolism of GSH *S*-conjugates. The scheme summarizes the interorgan flux, via enterohepatic and renal–hepatic circulations, and metabolism of GSH *S*-conjugates (GS–R). *Abbreviations:* AcCoA, acetyl-CoA; BBM, brush-border membrane; BLM, basolateral membrane; Cys–R, cysteine *S*-conjugate; NAcCys–R, *N*-acetyl-L-cysteine *S*-conjugate; and R-X, electrophilic substrate for GST.

distribution of GSH *S*-conjugates. Thus, GSH *S*-conjugate that is formed in the liver is exported across the canalicular plasma membrane (CPM) to either bile or plasma; the former is degraded by biliary and/or small-intestinal GGT and DP to yield the corresponding cysteine *S*-conjugate, which returns to the liver. Some of the hepatic cysteine *S*-conjugate is converted to the mercapturate (i.e., *N*-acetylcysteine *S*-conjugate) and these two forms, along with any GSH *S*-conjugate, can be exported across the sinusoidal plasma membrane (SPM). These three forms enter the renal circulation where a portion (~30% during a single pass through the renal circulation) is filtered by the glomerulus and enters the tubular lumen. Luminal GSH *S*-conjugate is degraded by GGT and DP on the BBM to the cysteine *S*-conjugate, which is taken up into the proximal tubular cell for further metabolism. The nonfiltered fraction can be taken up by specific transporters on the basolateral plasma membrane (BLM). Intracellular GSH *S*-conjugate is readily exported into the lumen by transport across the BBM.

The classic view of GSH and GSH *S*-conjugate interorgan metabolism included efflux from the liver and renal extraction by glomerular filtration, degradation on the BBM by GGT and DP, and uptake into proximal tubular cells of the constituent amino acids and the corresponding cysteine *S*-conjugate [13,27–31]. The concept that GSH or GSH *S*-conjugates were transported into cells as the intact molecules was considered an artifact [32–34] despite the observation that most of the renal extraction occurs at the BLM and not at the BBM. However, studies in the isolated perfused rat kidney [35–38], isolated proximal tubular cells from rat kidney [39], isolated proximal tubules from rabbit kidney [40], and isolated BLM vesicles from rat renal cortex [41,42] provided the definitive proof that GSH could indeed be transported across the BLM from renal plasma into the proximal tubular cell as the intact tripeptide.

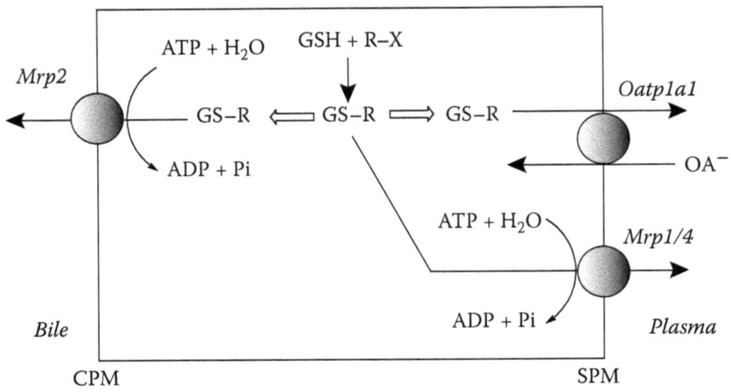

FIGURE 9.3 Hepatic transport of GSH *S*-conjugates. The scheme summarizes the known pathways for transport of GSH *S*-conjugates (GS–R) across the CPM and SPM. GS–R are transported across the CPM by the ATP-dependent Mrp2 and are transported across the SPM by both Mrp1 or Mrp4 and the organic anion polypeptide transporting protein 1a1 (Oatp1a1).

9.2.3 Hepatic GSH and GSH *S*-Conjugate Transport

Liver is the predominant site of GSH conjugation as up to 5% of total cytoplasmic protein is GST [43]. Although GSH conjugation can occur in several other tissues, including the kidneys (cf. Figure 9.2), the primary flux of GSH *S*-conjugate is from the liver to the kidneys, either by direct transport across the SPM into plasma or by transport across the CPM into bile, subsequent enterohepatic circulation, and efflux into plasma. The membrane carriers identified as transporting GSH *S*-conjugates across the SPM and CPM are illustrated in Figure 9.3. On the CPM, one member of the multidrug resistance-associated protein (Mrp) family of transporters, Mrp2 (*Abcc2*), can transport GSH and GSH *S*-conjugates [44,45]. On the SPM, the organic anion polypeptide transporter 1a1 (Oatp1a1; *Slco1a1*) has been shown to transport both GSH and GSH *S*-conjugates in exchange for other organic anions [44–46]. There are also two Mrp proteins, Mrp1 and Mrp4 (*Abcc1* and *Abcc4*, respectively), that may function in GSH *S*-conjugate efflux.

Despite the presence of several putative transporters on the SPM, most of the flux of GSH *S*-conjugates of halogenated solvents from hepatocytes appears to be across the CPM into bile, except at relatively high concentrations [47]. What is clear, however, is that under all physiological conditions, there is a very specific directionality of flux of GSH or GSH *S*-conjugates from the liver [48–50]. It is not clear, however, whether the flux of GSH complexes with metal ions exhibits a similar directionality with flux predominantly into bile. It is likely, however, that even if such complexes are handled similarly as those with halogenated solvents, metabolism and enterohepatic and renal–hepatic circulation would end up delivering the metal ion to the kidneys in the form of complexes with either GSH, L-Cys, or *N*-acetyl-L-cysteine (NAC) for ultimate excretion.

9.2.4 Renal GSH and GSH *S*-Conjugate Transport

In the case of kidneys, most of our attention focuses on the proximal tubules, because this cell population is the primary target for nephrotoxic GSH *S*-conjugates and heavy metals. Additionally, the proximal tubule is also the cell population along the nephron with the most highly developed array of membrane transporters for uptake and efflux of a broad range of chemicals, including organic anions and cations and amino acids [17,51].

As indicated in Figure 9.4, there are potentially three carriers on the BLM that may mediate uptake of GSH *S*-conjugates. These include Oat1 (*Slc22a6*), Oat3 (*Slc22a8*), and the sodium-dicarboxylate 2 carrier (NaC2; *Slc13a3*). Evidence is only correlative for the specific function of these carriers in the transport of GSH *S*-conjugates and direct evidence is limited for their function

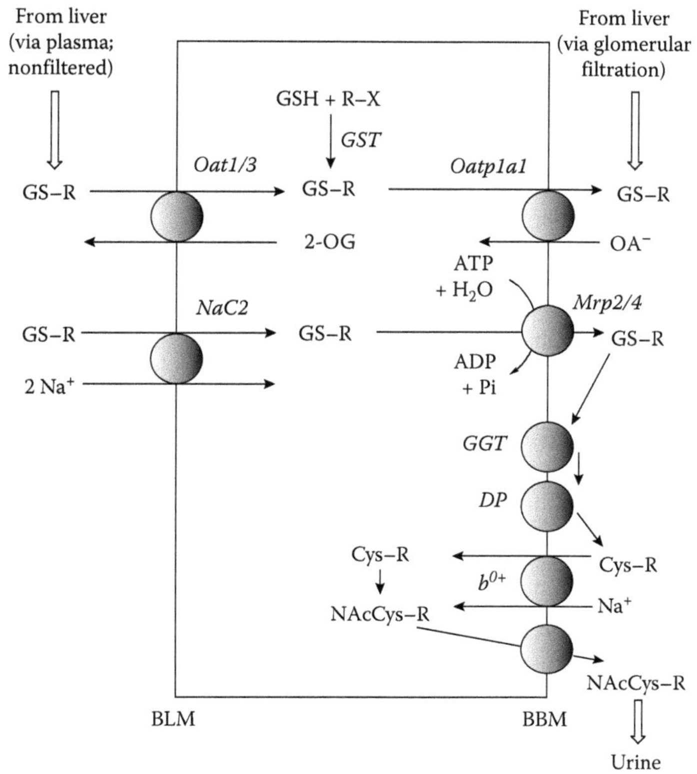

FIGURE 9.4 Renal proximal tubular transport and metabolism of GSH *S*-conjugates. Renal proximal tubular cells are exposed to GSH *S*-conjugates (GS–R) either at the BBM by glomerular filtration or at the BLM from the renal plasma. In the tubular lumen, GS–R are sequentially degraded by GGT and DP activity to yield the corresponding cysteine *S*-conjugate (Cys–R), which is subsequently taken up into the proximal tubular cell by Na^+-dependent amino acid transporters such as System b^{0+}. Once inside the proximal tubular cell, Cys–R are *N*-acetylated to form the mercapturate, which is generally readily excreted into the urine after efflux across the BBM. GS–R in the renal plasma is transported across the BLM by either the Oat1/3 in exchange for 2-oxoglutarate (2-OG) or by the sodium-dicarboxylate carrier 2 (NaC2) with two sodium ions in a cotransport process. Once inside the proximal tubular cell, GS–R is efficiently transported across the BBM into the tubular lumen by either the organic anion transporting polypeptide 1a1 (Oatp1a1) or the Mrp2/4, whereupon it is processed like the filtered GS–R.

in transporting GSH. Early studies in isolated BLM vesicles from rat kidney cortex [41,42] showed that GSH uptake was mediated by both Na^+-coupled and Na^+-independent processes and was inhibited by the nonspecific organic anion transport inhibitor probenecid. It was further shown in renal BLM vesicles and isolated proximal tubular cells from rat kidney [52] that a GSH *S*-conjugate competes with GSH for uptake and exhibits similar properties with respect to Na^+ ion dependence, energetics, and substrate specificity, suggesting that they are transported, at least in part, by the same carrier(s). Other studies in proximal tubular cells from rat kidneys [53,54] showed that in addition to probenecid, *p*-aminohippurate (PAH) and dimethylsuccinate inhibited GSH uptake. PAH inhibition implicates function of either Oat1 or Oat3 whereas dimethylsuccinate inhibition and the direct coupling to Na^+ ions implicate the additional function of NaC2. The only BLM carrier for which direct evidence of function in GSH transport exists is Oat3; bacterial expressed, purified, and reconstituted rat Oat3 transported GSH in exchange for both PAH and 2-oxoglutarate (2-OG) and the latter two in exchange for GSH [55].

Although GSH uptake into BBM vesicles of rat kidney cortex can be measured [56], the directionality of transport under physiological conditions is from cell to lumen rather than from lumen to cell [25,57]. Because of the selective localization of GSH uptake carriers on the renal BLM and the predominance of GGT on the renal BBM, GSH degradation of both filtered GSH and intracellular GSH is critical for GSH and sulfur amino acid homeostasis in the body [25]. The presence of the GGT active site facing into the lumen necessitates the need for efflux of intracellular GSH across the BBM but also prevents intracellular GSH from being degraded without the transport step. This is important both in terms of preserving intracellular GSH concentrations and also in separately regulating synthesis and degradation.

There are several putative carriers that may mediate GSH and GSH *S*-conjugate efflux into the tubular lumen (cf. Figure 9.4). While there is no direct evidence for the function of any single carrier on the BBM, the best indirect evidence is for Oatp1a1; while its function in the renal proximal tubule has not been directly investigated, hepatic Oatp1a1, which is expressed on the SPM, does transport GSH in exchange for an organic anion (OA^-) [44–46], as noted above. The other two potential carriers are Mrp2 and Mrp4. Whereas Mrp2 is expressed on the luminal membrane (BBM in the proximal tubule and CPM in the hepatocyte) in both liver and kidney, Mrp4 is expressed on the BBM in kidney but on the SPM in liver.

After the GSH *S*-conjugate is pumped out into the tubular lumen or arrives there by glomerular filtration, it is readily degraded to the cysteine *S*-conjugate, which is then taken back up into the proximal tubular cell by amino acid transporters such as system b^{0+}. Inside the proximal tubular cells, cysteine *S*-conjugates are readily *N*-acetylated to form mercapturic acid, which is typically secreted back into the lumen for urinary excretion. This step thus completes the interorgan pathway by which GSH *S*-conjugate that is synthesized in the liver is metabolized in the biliary/small intestinal epithelium or kidneys and is excreted into the urine.

9.2.5 Metals as GSH *S*-Conjugates: Implications for Tissue Distribution

Although heavy metal ions such as Hg^{2+}, Cd^{2+}, and $As^{3+(5+)}$ are not metabolized *per se* and they do not exist to any appreciable extent as the free metal ion, they do undergo metabolism in the sense that they are bound to various ligands that affect their interorgan translocation and ultimate interaction with target tissues or excretion and the ligand can be metabolized. Because of the predominance of GSH and its high metal ion binding affinity, complexes of heavy metals with GSH should be readily formed and found in various locations throughout the body. As discussed above, GSH *S*-conjugates undergo interorgan metabolism involving enterohepatic and renal–hepatic circulations and several discrete transport steps. Although the coordination complexes formed between metal ions and GSH are not chemically identical to the thioether conjugates formed between GSH and reactive electrophiles of organic molecules, they appear to be largely functionally treated in the same manner [58]. This idea that complexes between GSH and metal ions are handled by the body in a manner analogous to that for other GSH *S*-conjugates because of structural similarities is expressed as part of the concept of "molecular and ionic mimicry" [59–61].

The molecular mimicry concept has other important implications for the mechanism by which GSH–metal ion complexes are distributed in the body. It suggests that GSH–metal ion complexes are ultimately metabolized to cysteine or *N*-acetylcysteine conjugates and are transported out of the liver and taken up by renal proximal tubular cells by the same carriers as those identified to function in the transport of other thioether *S*-conjugates, including Mrps, Oat1/3, and Oatp1a1. Another important consideration is that pathological or toxicological conditions that modify cellular GSH status, including oxidative stress or the presence of inducers of GSTs or other GSH-dependent enzymes, may alter the extent to which heavy metals can gain access to their target tissues. The extent of metabolism of the GSH moiety in the GSH–metal ion complex is important because cysteine and *N*-acetylcysteine *S*-conjugates for the most part use different transporters than GSH *S*-conjugates in both liver and kidneys. As discussed below, rates of metal ion conjugate transport also differ substantially as the thiol ligand varies.

9.3 MODULATION OF TISSUE ACCUMULATION AND EXCRETION OF METALS BY GSH AND PROTEIN THIOLS

9.3.1 Hepatic and Biliary GSH and Metal Transport and Disposition

Liver functions not only as a potential target organ for, but also as an intermediary depository of, heavy metals. Hence, there are transport mechanisms for both uptake and efflux of metals. Several studies by Ballatori and Clarkson [62–64] showed that hepatic secretion of both Hg^{2+} and methylmercury (MeHg) into bile was dependent on the presence of GSH, forming a thiol–Hg^{2+} or thiol–MeHg complex. They later showed that rats with a congenital defect in biliary GSH excretion also had impaired biliary excretion of MeHg [65]. Similarly, Sugawara et al. [66] found that the Eisai hyperbilirubinemic rat, which lacks hepatic canalicular organ anion transport, also exhibits markedly decreased rates of hepatobiliary secretion of Hg^{2+}. Dutczak and Ballatori [67] also found that isolated canalicular membrane vesicles from rat liver transported a GSH–MeHg complex on the same carrier(s) as GSH, providing further evidence that hepatic Hg^{2+} or MeHg transport is linked with that of GSH.

The importance of biliary efflux of Hg^{2+} in terms of overall disposition and the renal accumulation of Hg^{2+} was highlighted in studies by Zalups and colleagues [68,69]. In these studies, either diversion of biliary flow from the liver to the intestines or biliary ligation markedly reduced renal accumulation of Hg^{2+}, particularly in the cortex and outer stripe of the outer medulla. Furthermore, modulation of both hepatic and renal GSH by *in vivo* treatment with either L-buthionine-*S*,*R*-sulfoximine (BSO) or DEM to deplete GSH or L-(αS,5*S*)-α-amino-3-chloro-4,5-dihydro-5-isoxazole acetic acid (acivicin) to inhibit GGT, had varying effects on the renal and hepatic accumulation of Hg^{2+}. Modulation of GSH status in conjunction with biliary ligation had additive effects in causing reductions in net renal accumulation of Hg^{2+}. At the same time, such manipulations increased hepatic retention of Hg^{2+}. These results provide further support for the concept that some fraction of the renal accumulation of Hg^{2+} is linked mechanistically to the hepatobiliary and GSH systems.

Other potential thiol ligands for heavy metals, such as lipoic acid, also appear to influence biliary excretion, although their effect is likely through modulation of GSH status. For example, studies by Gregus et al. [70] found that lipoic acid reduced biliary efflux of MeHg, Cd, zinc, and copper by forming a dihydrolipoic acid–GSH mixed disulfide in the liver. The mixed disulfide thus formed prevented GSH from forming a complex with the metals, thereby inhibiting their biliary efflux. This further supports the importance of GSH in the hepatic efflux of these metals.

9.3.2 Plasma Thiols and Metal Disposition

Although the thiol–disulfide redox status in plasma is considerably more oxidized than that inside cells [5], there is a large pool of protein and nonprotein thiols/sulfhydryl groups to which metals such as Hg and Cd can bind. Quantitatively, the cysteinyl sulfhydryl groups on serum albumin represent the largest metal-binding entity, with albumin containing eight cysteinyl residues per molecule. With albumin being a large protein (M_r = 66 kDa), however, metal ions bound to it are not normally filtered at the glomerulus to a large extent and cannot be directly transported across cellular membranes. As noted in the studies by Oram et al. [8], however, the existence of high binding affinity does not preclude exchange reactions occurring between albumin and other ligands. Both GSH and L-Cys are present in plasma at concentrations ranging from 1 to 20 μM, whereas the respective disulfides, glutathione disulfide (GSSG), and L-cystine are present at similar to higher concentrations [1–7].

As described in Section 9.2.2 about interorgan metabolism of GSH and GSH *S*-conjugates, the presence of an isopeptide bond between the γ-carboxyl group of the glutamyl residue and the α-amino group of the cysteinyl residue is a critical feature of the GSH molecule, making it resistant to the various proteases in plasma and cells. Being degraded only by GGT, GSH and GSH *S*-conjugates, which derive primarily from the liver, can translocate as the intact molecule through circulation and move to other target tissues. Because the kidneys are the predominant site of GGT activity in the body, they are the primary tissue that extracts GSH and GSH *S*-conjugates from the circulation. These facts suggest that the major transportable forms of metals such as Hg or Cd in the plasma will be complexes with either GSH or L-Cys. Besides GSH and L-Cys, L-homocysteine is also involved in the translocation and renal cellular uptake of Hg^{2+} [71]. Because of the importance of plasma L-homocysteine in different pathophysiological states, variations in L-homocysteine may alter metal ion distribution and, therefore, susceptibility to metal-induced toxicity.

9.3.3 Renal GSH and Transport, Disposition, and Toxicity of Mercury

This section discusses how variations in GSH status can directly and markedly influence the renal transport, disposition, and toxicity of Hg. The first subsection reviews the data on how administration of exogenous GSH can directly impact the renal handling of Hg and the consequences of exposure to this metal. The second subsection focuses on how preexisting changes in renal GSH status impact Hg handling and toxicity in the kidneys. Although one would intuitively think that administration of GSH or the existence of a condition in which intracellular GSH concentrations are elevated would automatically result in protection from mercury-induced nephrotoxicity, this is not necessarily the case because of what can be seen as a dual role for GSH in the handling of Hg. As described below, it will become clear that under certain conditions, addition of GSH is protective by forming a complex with Hg, thereby preventing the metal from interacting with critical sulfhydryl groups on structural proteins or enzymes that could lead to their inactivation or altered structure and conformation. Under other conditions, however, formation of a GSH–Hg complex promotes the renal cellular uptake and accumulation of Hg, thereby enhancing cytotoxicity.

9.3.3.1 Exogenous GSH

There are four distinct mechanisms by which Hg and GSH have been reported to interact in the kidneys to explain the influence of exogenous GSH on Hg transport, disposition, and toxicity (Figure 9.5): (1) Exogenous GSH can increase intracellular GSH, which can then act as a sink for Hg, preventing it from interacting with other cellular thiols, such as those on critical enzymes and structural proteins, thereby diminishing toxicity; (2) extracellular GSH can increase intracellular GSH, which can then act as a sink for Hg, enhancing intracellular accumulation of the metal, thereby enhancing toxicity; (3) extracellular GSH can bind to Hg, preventing its interaction with membrane or intracellular protein thiol groups, thereby diminishing toxicity; and (4) extracellular GSH can bind Hg, forming a complex that serves as a transport form of Hg, thereby enhancing its intracellular accumulation and enhancing toxicity.

The key to understanding how exogenous GSH influences Hg accumulation and toxicity by such diverse mechanisms depends on a clear definition of the conditions of exposure of renal cells to Hg and GSH in terms of time and relative concentrations. The following will summarize the literature on the influence of exogenous GSH, acting both as an extracellular and intracellular ligand for Hg, focusing primarily on Hg^{2+}, because most of the available literature has studied this form rather than MeHg or other forms.

9.3.3.1.1 Exogenous GSH Influencing Intracellular GSH Levels

Berndt and colleagues [72,73] investigated the effects of modulation of renal GSH metabolism on the renal disposition of administered $HgCl_2$ in mice. Inhibition of GSH degradation with acivicin led to increased urinary excretion of both GSH and Hg^{2+} whereas GSH depletion with either DEM or BSO increased Hg^{2+}-induced nephrotoxicity but decreased renal accumulation of Hg^{2+}. Hepatic GSH was not similarly altered with the GSH-depleting agents. Thus, these data suggested that the relationship between renal accumulation of Hg^{2+} and Hg^{2+}-induced nephrotoxicity is not straightforward. Zalups and Lash [74] performed similar studies in rats, with the important difference being that the influence on Hg^{2+} disposition of modulation of renal and hepatic GSH status was studied

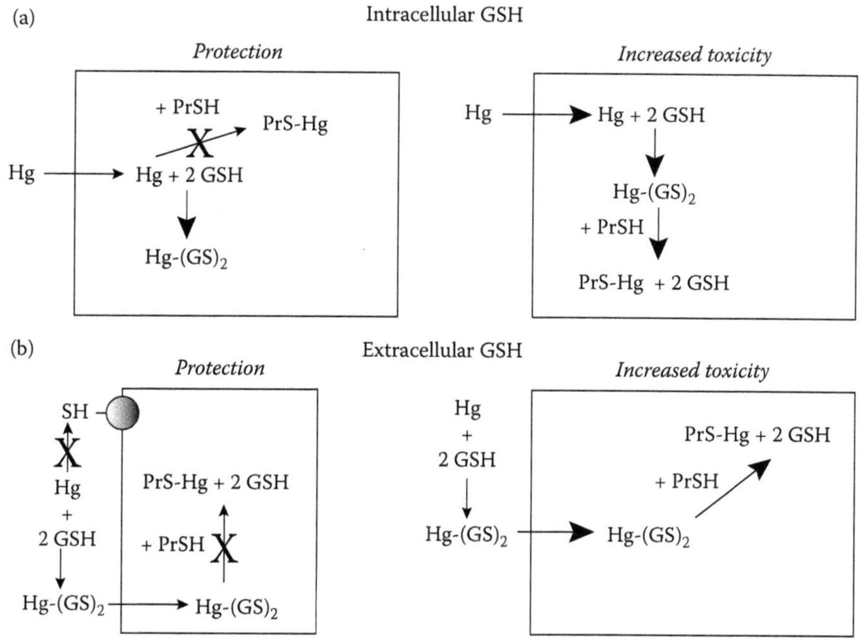

FIGURE 9.5 Potential modes of interaction of Hg with GSH in renal proximal tubular cells. The scheme illustrates four potential modes of interaction between Hg and GSH, two each for intracellular and extracellular GSH, respectively. (a) *Left panel*: For intracellular GSH, intracellular GSH may serve as a sink for Hg, resulting in the formation of a complex between Hg and GSH [Hg-$(GS)_2$], thereby preventing reaction of Hg (likely in the form of Hg^{2+} ions) with protein sulfhydryl (PrSH) groups, which would otherwise form a complex (PrS–Hg). This mode of interaction leads to cytoprotection. *Right panel*: Intracellular GSH forms a complex with Hg, leading to increased intracellular accumulation of Hg, which can then form complexes with PrSH, leading to increased cytotoxicity. (b) *Left panel*: Extracellular GSH forms a complex with Hg, thereby preventing Hg from interacting with membrane protein sulfhydryl groups. The Hg from the Hg–$(GS)_2$ complex thus formed may also be transported into the proximal tubular cell but would not appreciably interact with PrSH groups. This mode of interaction leads to cytoprotection. *Right panel*: Extracellular GSH forms a complex with Hg, thereby facilitating its uptake into the proximal tubular cell. Once inside the cell, Hg on the Hg–$(GS)_2$ complex can undergo an exchange reaction forming complexes with PrSH, thereby leading to cytotoxicity.

with a nontoxic dose of $HgCl_2$. Rats received an injection of 0.5 μmol $HgCl_2$/kg after pretreatment with either acivicin to inhibit GGT, BSO to inhibit GSH synthesis, or DEM to deplete GSH by conjugation. Whereas acivicin and DEM decreased renal GSH and renal Hg^{2+} content, BSO decreased renal GSH content but had no significant effect on renal Hg^{2+} content. Pretreatment with both BSO and DEM together significantly increased hepatic Hg^{2+} content, which was interpreted as being due to a decreased ability of hepatocytes to export Hg^{2+} as the Hg^{2+}–GSH complex.

Tanaka et al. [75] and Tanaka-Kagawa et al. [76] pretreated mice with acivicin and found that after administration of $HgCl_2$, renal Hg^{2+} content decreased, Hg^{2+} and GSH excretion in urine increased, and nephrotoxicity decreased. These findings suggested that renal Hg^{2+} uptake and accumulation is linked to GSH and is dependent on GGT activity. Similar findings were found for MeHg disposition [76]. de Ceaurriz et al. [77] similarly found that rats pretreated with acivicin exhibited decreased renal accumulation of Hg^{2+}, increased urinary Hg^{2+} excretion, and decreased $HgCl_2$-induced nephrotoxicity. However, GSH content was increased in both kidneys and plasma. Thus, despite the increase in renal GSH content, Hg^{2+} accumulation and nephrotoxicity were decreased, suggesting that GSH degradation was required for Hg^{2+} uptake. As discussed below, this is consistent with complexes of Hg^{2+} with either cysteinylglycine or L-Cys, and not GSH, being the primary transport forms for Hg^{2+}.

Both Girardi and Elias [78] and Johnson [79] similarly treated rats with DEM prior to treatment with $HgCl_2$, and found that this led to decreased renal GSH content. However, whereas Girardi and Elias [78] found that this produced increased renal accumulation of Hg^{2+} with decreased urinary Hg^{2+} excretion and increased renal toxicity, Johnson [79] found that the decrease in renal nonprotein sulfhydryl content was associated with a decrease in renal Hg^{2+} content and less of an impairment in renal function as compared with rats administered $HgCl_2$ alone. Consistent with their results with DEM, Girardi and Elias [78,80] found that pretreatment with NAC resulted in increases in renal nonprotein sulfhydryl content, increased urinary Hg^{2+} excretion, decreased renal Hg^{2+} accumulation, and decreased nephrotoxicity.

There have been fewer studies on the relationship between GSH status and MeHg disposition. Adachi et al. [81] administered three different diets to mice: a low-protein diet, a normal protein diet, and an amino acid-supplemented diet that contained sulfur amino acids. They found that the amino acid-supplemented diet resulted in increased levels of GSH, L-Cys, and L-homocysteine in plasma. This was accompanied by decreases in hepatic MeHg, marked decreases in renal MeHg, but increased uptake of MeHg in the brain.

9.3.3.1.2 Exogenous GSH as an Extracellular Ligand for Mercury and the Function of Mercury–GSH Complex as a Transport Form

A number of investigators have coadministered GSH or other thiols with Hg to assess the requirement for these thiols in transport and renal cellular accumulation. As with the studies described above on prior manipulations of renal GSH status, these studies often revealed differing results, depending on the particulars of the experimental design. In a second series of experiments, Tanaka et al. [75] coadministered GSH with $HgCl_2$ and found that renal content of Hg^{2+} was increased as compared to mice administered $HgCl_2$ alone. Their interpretation was that renal Hg^{2+} uptake occurs as a Hg^{2+}–GSH complex that is dependent on GGT activity. Zalups and Barfuss [82] conducted a similar study in rats, but with a nontoxic dose of $HgCl_2$ of 0.25 μmol/kg and a twofold higher amount of GSH. This coadministration of GSH and $HgCl_2$ resulted in increases in plasma and renal Hg^{2+} content as compared to administration of $HgCl_2$ alone. This was also interpreted as supporting a role for a Hg^{2+}–GSH complex as a transport form for renal accumulation of Hg^{2+}.

Several other studies have used other types of *in vivo* and *in vitro* renal models and have provided evidence that either a Hg^{2+}–GSH complex or a complex of Hg^{2+} with another low-molecular-weight thiol promotes the renal proximal tubular uptake and accumulation of Hg^{2+}. For example, in an *in vivo* study, Zalups [83] performed ureteral ligation on rats to reduce the glomerular filtration rate (GFR) to essentially zero, to examine the basolateral uptake of Hg^{2+} coupled to thiols. Hg^{2+} was administered as either $HgCl_2$ or as Hg^{2+}–GSH or Hg^{2+}–Cys complexes. Coadministration of Hg^{2+} with GSH or L-Cys resulted in significant increases in renal Hg^{2+} uptake. Pretreatment with the Oat inhibitor PAH decreased renal Hg^{2+} accumulation, consistent with Oat carrier proteins (Oat1 and/or Oat3) functioning in the renal proximal tubular uptake of both Hg^{2+}–GSH and Hg^{2+}–Cys complexes.

Other studies, however, are consistent with Hg^{2+}–GSH complexes playing at best a minor role in transport but Hg^{2+}–Cys complexes being the more likely transport forms. Studies in renal basolateral and brush-border membrane vesicles from rat kidneys showed that when compared to $HgCl_2$ alone, incubation of membrane vesicles with Hg^{2+} and L-Cys in a 1:3 ratio resulted in a significant increase in the binding and transport of Hg^{2+}, whereas incubation with Hg^{2+} and either GSH or 2,3-dimercapto-1-propanesulfonate (DMPS) in a 1:3 ratio resulted in only a minimal association of Hg^{2+} with the membranes [84]. Similarly, incubation of freshly isolated proximal tubular cells from rat kidney with either $HgCl_2$ or Hg^{2+} and different thiol-containing compounds (L-Cys, GSH, bovine serum albumin (BSA), or DMPS) in a 1 mol Hg^{2+} to 4 mol thiol ratio showed that cellular accumulation was found to be, in order of decreasing amount, Hg^{2+}–Cys > $HgCl_2$ > Hg^{2+}–GSH > Hg^{2+}–BSA, and Hg^{2+}–DMPS [85]. In another study with freshly isolated proximal tubular cells from rat kidneys, Lash et al. [86] compared the effects of preloading cells with GSH and coadministration with GSH; preloading GSH resulted in increased $HgCl_2$-induced cytotoxicity, consistent with the higher levels of intracellular GSH acting as a sink for Hg^{2+}; in contrast, coadministration of Hg^{2+} with GSH resulted in cytoprotection as compared with administration of $HgCl_2$. Studies with acivicin to inhibit GGT activity had no effect on cytotoxicity. Coadministration with BSA resulted in complete cytoprotection. Coadministration of Hg^{2+} with GSH in the renal proximal tubular cell line LLC-PK1 cells significantly protected against Hg^{2+}-induced cytotoxicity [87] as did coadministration of GSH monoethyl ester in rats [88]. Other *in vivo* studies in rats [89], *in vitro* studies in basolateral membrane vesicles from rat kidneys [90], and *in vitro* studies in rabbit proximal tubule fragments [91,92] provided evidence consistent with Hg^{2+}–Cys being the most active and, under most conditions, predominant transport form of Hg^{2+} across renal plasma membranes.

From the large number of studies described above, we are left with the question of the involvement of GSH in renal proximal tubular transport, accumulation, and toxicity of Hg^{2+} being unresolved. On the one hand, several studies have shown enhanced uptake and accumulation of Hg^{2+} with GSH coadministration. Other studies, however, have found little enhancement of uptake with GSH and much more enhancement with L-Cys. Some of the differences can be linked to the use of nontoxic concentrations of Hg^{2+} as compared with toxic concentrations of Hg^{2+}. Other differences, however, are not clearly rationalized. The influence of GSH and other thiols, therefore, appears to be highly complex and variable. It is thus likely that all four mechanisms illustrated in Figure 9.5 are operative under different conditions.

9.3.3.2 Uninephrectomy and Compensatory Renal Growth (NPX) Model

The NPX model of reduced renal mass illustrates further the complex relationship between renal cellular GSH status and the disposition and toxicity of Hg^{2+}. Within a short period of time after functional renal mass is reduced by uninephrectomy, the remnant renal tissue undergoes a series of compensatory physiological, morphological, and biochemical changes [93–95]. This compensatory response is predominantly a cellular hypertrophy and, among tubular epithelial cells, occurs primarily in the proximal tubules. At the tissue level, characteristics include increases in GFR, renal blood flow (RBF), and single-nephron GFR (SNGFR), while maintaining the tubuloglomerular feedback mechanism. The hypertrophied remnant kidney can thus be described as being in a state of hyperperfusion and hyperfiltration, which is often accompanied by some degree of hypertension [93,96–98]. At the cellular and biochemical levels, characteristics include increases in cellular content of protein, BBM and BLM membrane surface area, rates of cellular energy metabolism, activities of various enzymes, and activities of membrane transporters involved in the uptake and efflux of inorganic anions and cations, organic anions and cations, and various intermediary metabolites.

Although many processes or parameters increase in proportion to the increase in cellular protein, other processes increase disproportionately to protein, suggesting that they may have particular significance for cellular function in the hypertrophied state. One such parameter is cellular GSH content, which was shown in several studies to be significantly higher in renal homogenates from NPX rats, particularly in tissue from the outer stripe of the outer medulla, which is primarily composed of proximal straight tubules [99,100]. One might expect that this increased GSH content would diminish oxidative stress and susceptibility to various forms of chemically induced nephrotoxicity. This, however, is not the case as NPX rats *in vivo* and renal cells *in vitro* from NPX rats are actually more susceptible than normal rats or renal cells from normal rats to many toxicants, including $HgCl_2$. Some of these studies will be summarized here, with an emphasis on the relationship between renal GSH status and Hg^{2+} transport and toxicity.

Early studies comparing susceptibility of rats that had undergone uninephrectomy and compensatory renal growth (NPX rats) with that of control or sham-operated rats to Hg^{2+} universally found the NPX rats to be markedly more sensitive to Hg^{2+}-induced nephrotoxicity [101–104]. Moreover, the increased toxicity was clearly associated, at least at the moderately toxic doses, with increased renal accumulation of Hg^{2+}, particularly in the outer stripe of the outer medulla [100,104,105]. The same increase in susceptibility was retained in several *in vitro* models of proximal tubular cells from NPX rats [86,106,107]. The enhanced susceptibility to Hg^{2+}-induced nephrotoxicity and cytotoxicity is, as noted above, accompanied by increases in renal GSH content, especially in the nephron segment that primarily accumulates and becomes intoxicated by Hg^{2+}, namely the pars recta or straight (S2) segment of the proximal tubules. Besides increases in cellular GSH content in kidneys from NPX rats as compared to those from control or sham-operated rats, activities of several GSH-dependent enzymes, including GCS and GSSG reductase (GRD), are significantly increased [108–110]. Additionally, activities of several enzymes that are associated with mitochondrial energy metabolism and prominent, energy-dependent processes such as the $(Na^+ + K^+)$-stimulated ATPase, are also significantly elevated in proximal tubules from NPX rats as compared to those from control rats [106,108–110]. Hence, in addition to changes in GSH redox status, compensatory renal hypertrophy is associated with profound changes in cellular energetics, and has been interpreted as the hypertrophied proximal tubule existing in a hypermetabolic state [109–112].

Besides $HgCl_2$, the hypertrophied proximal tubule, relative to that from normal kidneys, also accumulates more MeHg [113] after exposure to methylmercuric chloride and more Cd^{2+} after exposure to either Cd-metallothionein (CdMT) [114] or $CdCl_2$ [115]. Thus, despite significant elevations in GSH content in the hypertrophied proximal tubule, both nephrotoxicity or *in vitro* cytotoxicity and accumulation of multiple metals, in multiple forms, are increased. This suggests that in this case, the enhanced intracellular GSH levels are functioning as a sink for the metals, thereby increasing their intracellular accumulation, as shown in the right-side panel of Figure 9.5a.

9.3.4 Cadmium Transport and Disposition

GSH has long been regarded as important in the cellular response to Cd, being considered a first line of defense [116]. As with Hg, there is a complex interplay between GSH and Cd that can vary considerably with concentration, whether GSH is coadministered with Cd, and tissue. Whereas the kidney, in particular the proximal tubules, is the primary target organ for Hg^{2+}, the kidney is only one of several major target organs for Cd^{2+}. As with Hg^{2+}, Cd^{2+} binds avidly to low-molecular-weight thiols and protein sulfhydryl groups. As one would expect, therefore, many of the specific molecular targets for Cd^{2+} that have been identified are sulfhydryl-containing proteins. For example, in a series of studies, Prozialeck and colleagues [117–120] identified intercellular junctions and E-cadherin in particular, as specific targets in renal and other epithelial cells. In part, the mechanism of action involves the displacement of endogenous Ca^{2+} ions by Cd^{2+}, but sulfhydryl groups exist on many of the proteins that comprise the cellular cytoskeleton.

Sabolic et al. [121,122] demonstrated that the very early effects of exposure of proximal tubules to Cd^{2+} or CdMT involve direct interactions with BBM transporters, shortening and loss of microvilli, and the loss of specific transporters. These changes were linked to alterations in cytoskeletal protein polymerization, losses of actin, megalin, and villin, and increases of α-tubulin, all of which contain cysteinyl residues that are essential for their structure and conformation. Consistent with these findings, Kinne et al. [123] showed a potent inhibition by $CdCl_2$ of Na^+-dependent uptake of L-Glu in rabbit renal BBM vesicles. Interestingly, the same potent effect was not elicited by CdMT. This may be explained by the suggested mechanism of CdMT-induced nephrotoxicity involving uptake of CdMT by endocytosis, accumulation within lysosomes, with subsequent intracellular release of Cd^{2+} ions into the cytoplasm [124].

Pearson and Prozialeck [125] went on to hypothesize further that besides directly causing disruption of cell–cell junctions and acute cytotoxicity, Cd^{2+}-induced disruption of E-cadherin-mediated cell adhesion and effects on cytoskeletal proteins can trigger the β-catenin signaling pathway, which can result in activation of oncogene expression in epithelial cells and an increase in the invasiveness of existing epithelial-derived cancers. In this manner, the authors proposed that Cd^{2+}-induced disruption of E-cadherin-dependent cell–cell junction may be the initial, pivotal step in induction of cancer and/or tumor promotion. Cd is a well-established human carcinogen [126], including the induction of malignant transformation of prostate epithelial cells [127].

Not surprisingly, therefore, coadministration of GSH or altered intracellular levels of GSH have been demonstrated to modulate proximal tubular accumulation and toxicity of Cd, although not always in predicted ways. Tang et al. [128] found that administration of either GSH or Zn^{2+} protected MT-null mice from CdMT-induced nephrotoxicity. The primary mechanism of protection was shown to involve a decrease in renal accumulation of Cd. Similarly, Wispriyono et al. [129] found that preincubation and coincubation of LLC-PK1 cells with NAC completely protected from $CdCl_2$-induced cytotoxicity, which was associated with a twofold increase in cellular content of GSH and a decrease in Cd^{2+} uptake. Addition of BSO to inhibit GSH synthesis, however, did not abolish the protection, suggesting that NAC-dependent protection was independent of GSH status.

In an analogy with studies on renal Hg^{2+} uptake, Zalups [130] coadministered Cd^{2+} with GSH or L-Cys in either a 1:4 or 1:2 ratio, and found that both thiols resulted in enhanced renal accumulation of Cd^{2+}. This suggested that complexes of Cd^{2+} with low-molecular-weight thiols are transport forms for Cd^{2+}, much in the same manner that mercuric complexes with GSH or L-Cys are transport forms of Hg^{2+}. Shaikh et al. [131], however, found distinctly different results exploring the relationship between GSH status (both extracellular and intracellular) and accumulation and nephrotoxicity of Cd in the form of CdMT. In their study, administration of BSO to deplete intracellular GSH enhanced CdMT-induced nephrotoxicity whereas administration of GSH protected against CdMT. In the case of BSO treatment, this was accompanied by no change in renal Cd accumulation, whereas the GSH-dependent protection was associated with decreased renal Cd content. These results were interpreted as indicating the existence of two GSH-dependent mechanisms of protection, one being intracellular and the other being extracellular. The intracellular mechanism would involve GSH acting as a sink for Cd^{2+}, preventing its interaction with cellular proteins, or acting as an antioxidant to detoxify Cd^{2+}-induced formation of reactive oxygen species (ROS). The extracellular mechanism would involve formation of a Cd^{2+}–GSH complex that would not be transported into the renal proximal tubule, thereby decreasing intracellular Cd^{2+} accumulation.

A recent study [132] on the protective effects of GSH against Cd^{2+}-induced cytotoxicity in a small-cell lung cancer-derived cell line (SR3A cells and SR3A cells that overexpress GCLC) provides a more detailed insight into the complex mechanisms that exist. While one mechanism of GSH-dependent protection is thought to involve GSH serving as a cofactor for MRP1-mediated Cd^{2+} efflux, this study found a more indirect effect relating to Cd^{2+} transport. These authors found that the SR3A-GCLC cells, which had markedly higher intracellular levels of GSH, were more resistant to Cd^{2+} than the parent cells. This resistance was due to decreased intracellular accumulation of Cd^{2+} due to a downregulation of the Cd^{2+} transporter ZIP8 (*Slc39a8*) rather than effects on Ca^{2+} channels. The *ZIP8* gene has a binding site for the Spl transcription factor in its proximal promoter region that is sensitive to GSH levels. Spl binding is decreased by increased levels of GSH.

The Spl transcription factor contains three zinc finger domains that are critical for DNA binding. Each zinc finger domain is composed of two cysteinyl and two histidinyl residues that are coordinated by zinc in a tetrahedral conformation. The structure of the zinc finger, particularly the cysteinyl residues, is very sensitive to redox status, with its stability being affected by intracellular thiol/disulfide pools such as the GSH/GSSG pool [133–135]. GSH has also been shown to suppress Spl binding to DNA promoters by inhibiting components of the mitogen-activated protein kinase (MAPK) pathway [136]. Thus, more complex effects of GSH that involve thiol-sensitive signaling pathways may also regulate Cd^{2+} disposition.

Certainly, as is the case with Hg^{2+}, the interaction between GSH and Cd^{2+} is multifaceted and can either be protective and sometimes associated with decreased renal Cd^{2+} accumulation, but can also serve as a means to transport Cd^{2+} into the renal proximal tubule. The existence of CdMT as a physiologically important form to which the kidneys or liver are exposed and some of the reported mechanistic differences between effects of Cd^{2+} when it is bound to MT as opposed to being either the free ion or complexed with a low-molecular-weight thiol, provides another layer of complexity in understanding the functions of GSH in Cd^{2+} transport, disposition, and toxicity. GSH-dependent modulation of signaling pathways may be another mechanism that can alter the disposition, and

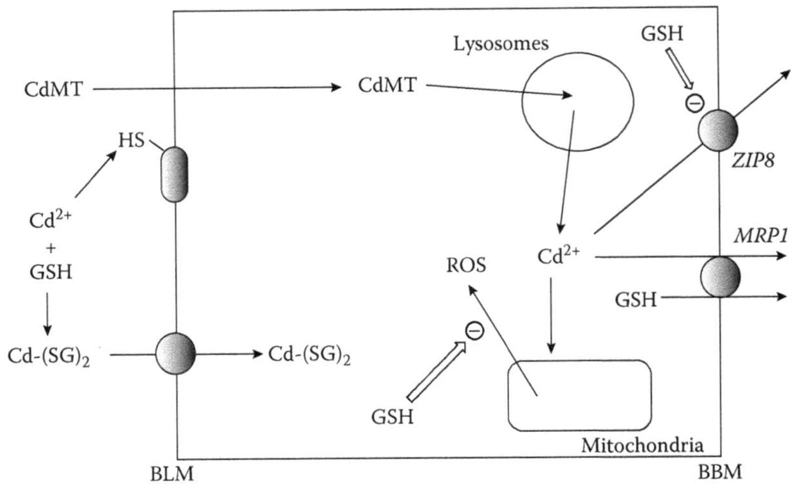

FIGURE 9.6 Modes of transport and cytotoxicity of Cd in renal proximal tubular cells. The renal proximal tubular cell is exposed to Cd^{2+} in the form of complexes with either the small, cysteine-rich protein metallothionein, forming CdMT, or as a complex with GSH [Cd-$(GS)_2$]. CdMT is taken up into the cell at the BLM primarily by endocytosis and accumulates in lysosomes, where the free Cd^{2+} ion is released. Intracellular Cd^{2+} can interact with mitochondria, producing reactive oxygen species (ROS), or can be transported across the BBM out into the tubular lumen by either MRP1 or ZIP8. An arrow with ⊖ shows an inhibitory effect of GSH. In the case of Cd^{2+}-induced ROS formation, GSH functions as an antioxidant to detoxify the ROS. In the case of ZIP8, GSH downregulates expression of the *ZIP8* gene.

hence toxicity, of Cd^{2+}. A summary scheme (Figure 9.6) is presented that illustrates these diverse and sometimes competing effects of GSH on Cd^{2+} disposition and toxicity. Further work is needed to define the precise conditions under which each pathway operates and to what extent each pathway functions in different target tissues.

9.3.5 Arsenic Metabolism and Disposition

Arsenic (As) is not only a metal contaminant of drinking water, but also a component of a limited number of therapeutic agents and traditional medicines [137]. It exists in both inorganic and organic forms with a valence charge of either +3 or +5. While As-V is well known to inhibit mitochondrial respiration, most of the toxic effects are associated with As-III. Like Hg and Cd, As also has a high affinity for thiols and sulfhydryl groups. The involvement of GSH in As metabolism and disposition, however, differs substantially from how it influences those processes for Hg and Cd. While there is evidence that complexes between As and GSH are transported by MRP1 [138,139], addition of GSH to incubations of rabbit renal cortical slices with sodium arsenite does not alter cellular As accumulation [140]. Thus, although complexes between As and GSH appear to be transport forms, modulation of GSH status does not directly influence As accumulation or toxicity in the same manner as it does for Hg and Cd.

One clear function for GSH is in As metabolism. As discussed in studies of As metabolism [141,142] and summarized in a recent review by Aposhian and Aposhian [143], GSH is involved in reduction of As-V compounds to As-III compounds and potentially in formation of several GSH conjugates. A scheme showing the known and proposed intermediates in the GSH-dependent metabolism of As compounds is shown in Figure 9.7. GSH is involved not only as a direct substrate in glutathionylation reactions, but also as a cofactor in several methylation reactions. The omega-class of GSTs (GSTO in humans, GSTω in rodents) has been identified as the specific isoform responsible for GSH-dependent As metabolism. hGSTO isoenzymes have thioltransferase and dehydroascorbate reductase activities, as well as reductase activity, toward arsenate, monomethylarsenate, and

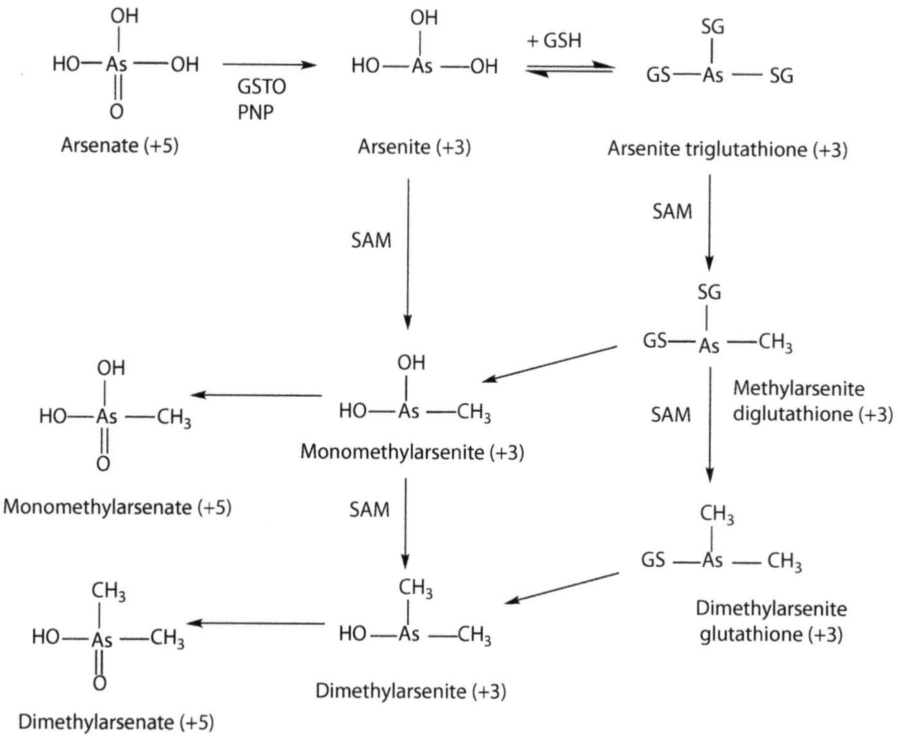

FIGURE 9.7 Metabolism of arsenic in mammalian cells. GSH is involved in the interconversion of various As-containing species. While some of the steps are well established, others, such as formation of various mono-, di- and tri-glutathionyl conjugates are hypothesized. GSH *S*-transferase omega (GSTO in humans, GSTω in rodents) is thought to be involved in several steps. Additionally, a purine nucleoside phosphorylase and methylation reactions involving *S*-adenosyl-L-methionine (SAM) are also involved in several of the reactions. Ultimately, As-containing species are oxidized to mono- and dimethylarsenate species (both +5 valence), which are excreted in the urine.

dimethylarsenate (which all contain As in the +5 valence form). Other methyltransferases are thought to be involved as well as a purine nucleoside phosphorylase, but a specific enzyme has not been unequivocally identified in most cases.

9.4 METAL-INDUCED CHANGES IN GSH STATUS

9.4.1 Metal-Induced Changes in Cellular GSH Status

Whereas the previous section considered how exogenous GSH or changes in intracellular GSH status modulate or regulate metal ion disposition and toxicity, this section summarizes our knowledge of how metals affect GSH status in target cells. The first subsection considers how exposure to metals alters intracellular GSH levels while the second subsection considers how metals can regulate the expression of GSH synthesizing enzymes and other GSH-dependent drug metabolizing enzymes.

Exposure of kidneys to a subtoxic dose of $HgCl_2$ was found to markedly increase intrarenal levels of GSH [100,144]. The increases were observed in both the cortex and outer stripe of the outer medulla, but were most prominent in the latter region. A composite of results from studies on the influence of a range of doses of $HgCl_2$ on renal levels of GSH and on activity of GCS in proximal tubular and distal tubular cells from rat kidneys is shown in Figure 9.8. Inasmuch as GCS is the rate-limiting enzyme for GSH synthesis, the approximately 100% increase in activity in proximal

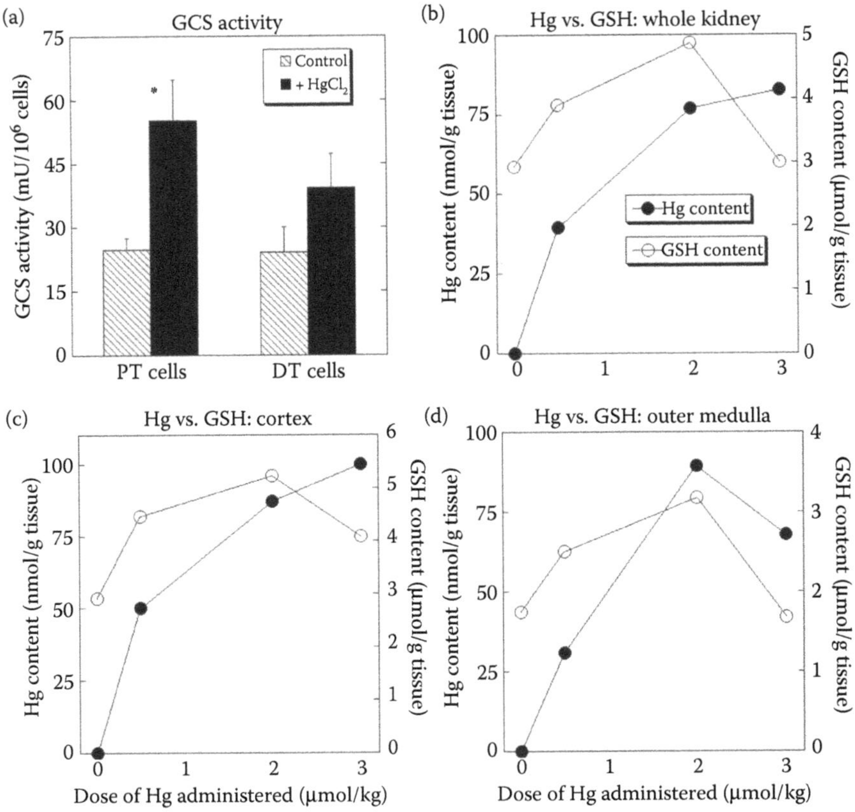

FIGURE 9.8 Effects of inorganic mercury on GSH synthesis rates and renal GSH contents. The data are derived from Zalups and Lash (Data from Zalups, R.K. and Lash, L.H., *J. Pharmacol. Exp. Ther.* 254, 962–970, 1990) and Lash and Zalups (Data from Lash, L.H. and Zalups, R.K., *J. Pharmacol. Exp. Ther.* 261, 819–829, 1992; Lash, L.H. and Zalups, R.K., *Arch. Biochem. Biophys.* 309, 129–138, 1994), and summarize studies on the influence of $HgCl_2$ incubated with either proximal tubular (PT) or distal tubular (DT) cells from rat kidneys on activity of γ-glutamylcysteine synthetase (GCS) activity (a) or of various doses of $HgCl_2$ injected in rats on GSH content in homogenates of whole kidney (b), renal cortex (c), or renal outer medulla (d). (a) Rats were given an i.p. dose of either saline (=Control) or 0.5 μmol Hg/kg body weight. After 24 h, PT and DT cells were isolated and GCS activity was measured. (b–d) Rats were given an i.p. dose of either saline (=0 Hg) or 0.5, 2, or 3 μmol $^{203}HgCl_2$/kg body weight. After 24 h, content of ^{203}Hg and GSH were measured in homogenates of whole kidney, renal cortex, and renal outer stripe of the outer medulla. Results for GCS activity are the means ± SEM of measurements from three separate cell preparations and for Hg and GSH, contents are the means of values from four separate animals.

tubular cells, normalized to cell number, indicates a dramatic upregulation of GSH synthesis. Moreover, it was found that even a moderately nephrotoxic dose of $HgCl_2$ (2.0 μmol/kg) caused a significant increase in GSH content. At a higher, more nephrotoxic dose (3.0 μmol/kg), however, GSH levels decreased due to massive tubular necrosis. Consistent with this high-dose effect, Addya et al. [145] found that chronic administration of a high dose of $HgCl_2$ decreased renal levels of GSH as well as activities of GPX and GST. Induction of acute renal failure in rats by administration of $HgCl_2$ increased lipid peroxidation and decreased renal GSH content in one study [146] and decreased renal nonprotein sulfhydryl content but increased GPX activity in direct proportion to decreases in renal Hg^{2+} accumulation [147].

The compensatory increase in renal GSH levels observed with low doses of Hg^{2+} has also been observed with MeHg [148,149]. Low doses of MeHg increased expression of GCS mRNA and increased intracellular GSH content in NRK-52E cells, a renal proximal tubular cell line. GSTα1/α2 expression is also induced by low-dose Hg^{2+} exposures in both mouse kidney [150] and rat kidney [151]. Thus, a compensatory response to low-dose exposures of either Hg^{2+} or MeHg occurs in which a battery of antioxidants, including several components of the GSH system, is induced.

Several studies [152–154] have implicated effects on the GSH redox system of divalent metal ions, including Hg^{2+}, Cd^{2+}, and arsenite, as being mediated through modulation of the nuclear factor-kappa B (NF-κB) transcription factor. NF-κB is a prosurvival, pleiotropic transcription factor that is a heterodimer, typically composed of p50 and p65 subunits. In its inactive form, NF-κB is found in the cytoplasm bound to a protein called inhibitory factor-κB (IκB). Upon stimulation by various agents, including diverse agents such as tumor necrosis factor-α, lipopolysaccharide, various oxidants, and heavy metals, IκB is released, which allows NF-κB to translocate to the nucleus where it can bind to various sites on the promoter regions of different genes, leading to increased expression of numerous genes. While ROS and oxidants promote activation and nuclear translocation of NF-κB, they inhibit binding to DNA; conversely, thiols and other reductants decrease activation and nuclear translocation of NF-κB but promote its binding to DNA [153]. Hg^{2+} has been shown to inhibit NF-κB activation and DNA binding [152,154], leading to apoptosis of renal proximal tubular cells [154].

Much in the same manner that subtoxic doses of $HgCl_2$ cause renal GSH concentrations to increase, injection of mice with a sublethal dose of $CdCl_2$ (3 mg/kg iv) produced modest increases in renal GSH levels, no change in hepatic GSH levels, but significant decreases in pulmonary and brain GSH [155]. High-dose Cd^{2+}, like high-dose Hg^{2+}, disrupts cellular thiol–disulfide homeostasis and causes oxidative stress and cell injury [156].

Similarities between effects on the GSH system elicited by Hg and other metals have been noted as well. For example, Di Simplicio et al. [157] noted similar effects of MeHg and sodium selenite on expression and activity of several GSH-dependent enzymes. Inhibition of enzymes of GSH metabolism by high doses of $HgCl_2$, however, can be reversed by selenium, which stimulates GSH synthesis at low concentrations [158]. Incubation of human keratinocytes for 24 h with subtoxic concentrations of As-III resulted in increases in intracellular GSH levels, which were associated with increases in activity and mRNA expression of GCS [159]. Activity and mRNA expression of GRD were also increased. The authors suggested that low levels of inorganic As-III elicit compensatory responses that are associated with its tumor-promoting effects; this is related mechanistically to As-III-induced increases in ROS, which result in activation of certain cell signaling and transcriptional pathways [160].

A final metal that influences GSH status, but that differs from those discussed above is Se. Unlike Hg, Cd, and As, Se is an essential macronutrient, although it does cause toxicity at high levels. Its biological functions are largely attributed to its presence in selenoproteins, for which 25 genes have been identified in humans [161]. Se is present in these proteins as selenocysteine, which is more reactive than L-Cys as it has a lower pK_a and stronger nucleophilicity. While biological functions are unclear for all the known selenoproteins, those that are known function as oxidoreductases, in redox signaling, antioxidant defense, thyroid hormone metabolism, and in immune response.

Important selenoproteins that utilize GSH or influence thiol redox status include GPX and thioredoxin reductase (TrxR). The so-called GSH redox cycle, involving GPX and GRD, is illustrated in Figure 9.9. Organic peroxides or H_2O_2, generated from lipid peroxidation or other forms of oxidative stress, are reduced by the action of GPX. The selenocysteinyl moiety of GPX plays a critical role in the catalytic cycle, with transient formation of a selenoglutathione intermediate. The glutathionyl moiety is eventually removed by reaction with another molecule of GSH to generate GSSG and regenerate the active GPX enzyme. GSH is then regenerated by GRD, which uses NADPH, itself generated from the activity of various substrate dehydrogenases. Transcription of several selenoproteins, including TrxR and GPX, is regulated by the redox-sensitive transcription factor NE-F2-related factor (Nrf2) and its modulator protein Kelch ECH Associating Protein 1 (Keapl). Because of the redox sensitivity of the Nrf2/Keapl system, other metals such as Hg^{2+} can alter the function of Nrf2 and thus influence expression levels of key redox proteins such as GPX and TrxR.

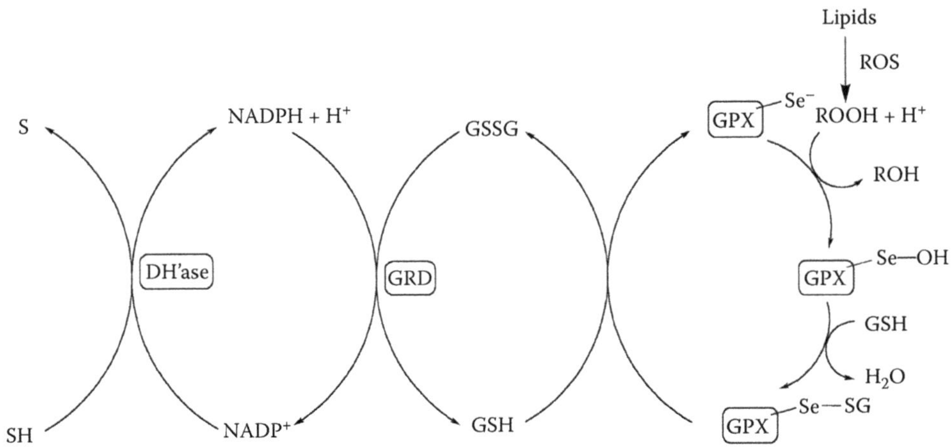

FIGURE 9.9 GSH redox cycle showing involvement of Se in the active site of GPX. Lipid peroxidation, induced by ROS, generates lipid peroxides (ROOH) that are reduced by GPX to the corresponding alcohol, ROH. The selenocysteine moiety of GPX initially forms a selenic acid (GPX-Se-OH) intermediate, which interacts with GSH to form a seleno-GSH conjugate intermediate. Reaction with an additional molecule of GSH generates GSSG and regenerates the active GPX enzyme. GSH is regenerated from GSSG by action of the NADPH-dependent GRD. NADPH levels are in turn maintained by the function of various substrate-level dehydrogenases (DH'ase).

9.4.2 Regulation of Gene Expression by Metals

Thus, as with the effects of GSH on disposition and toxicity of metal ions, the converse relationship, that of effects of metal ions on GSH status, is complex, dose-dependent, and involves both direct and indirect effects. An important aspect of the influence of metal ions on GSH and thiol–disulfide redox status that has only recently been investigated is the influence of metals on transcription factors. The next section briefly discusses the regulation of GSH synthesis and how metal ions can both directly and indirectly impact transcription.

GCLC and GCLM are encoded by two separate genes in most species, including humans and rodents, although yeast and bacteria have only a single gene [162]. Although there is often coordinated regulation of the two genes, they are regulated by distinct mechanisms [163,164]. The two GCL subunits are coordinately upregulated by many inducers of oxidative stress and treatment with antioxidants. There are several examples, however, where they are differentially regulated. Several transcription factor binding sites have been identified on both GCLC and GCLM promoters [164–167]. The most prominent feature on both genes is the presence of multiple antioxidant response elements (AREs; also called electrophile response element, EpRE). The importance of the ARE in modulation of enzyme expression is highlighted by three recent reviews [168–170]. Besides Gclc and Gclm, AREs are present in the promoters of numerous drug metabolism enzyme genes, such as those for GSTs, NAD(P)H: quinone oxidoreductase, superoxide dismutase 1, heme oxygenase-1, and aldo-keto reductases, among others.

One of the major differences between the promoter regions for Gclc and Gclm is that the former has a binding sequence for NF-κB (position –2391) but the latter does not. Cai et al. [163] showed that whereas transcription of both subunits is regulated by oxidative stress, only that of Gclc is mediated through NF-kB. Additionally, only Gclc is hormonally regulated (being responsive to insulin and hydrocortisone) and responsive to growth phase in cell culture. These differences are consistent with the "modulatory," although nonessential, function of GCLM.

The interaction between Hg^{2+} and $MeHg^{+}$ with NF-κB was noted above. Moreover, induction of oxidative stress by various forms of Hg and Cd, with the consequent increases in levels of ROS, can be associated with changes in a number of redox-sensitive transcription factors and transcription factor binding sites on the promoter region of Gclc, Gclm, and other genes. Thus, both direct and indirect alterations in transcription can occur due to exposure to metals.

9.5 SUMMARY AND CONCLUSIONS

This chapter has reviewed some of the key interactions between GSH, other selected thiols, and four metals: Hg, Cd, As, and Se. The first two metals, Hg and Cd, are environmental contaminants that can both modulate GSH status and have their disposition and toxicity modulated by GSH. Interactions between the metal ions and the GSH system are complex. The complexity arises from the multitude of interactions that occur, some of which produce opposing effects in terms of cellular accumulation and/or toxicity. Thus, while higher concentrations of GSH often result in cytoprotection *in vitro* or prevention of nephrotoxicity or other target-organ toxicity *in vivo*, there are other scenarios that result in increased metal ion accumulation and enhancement of toxicity. Similarly, exogenous GSH can have opposing effects, sometimes serving as a cytoprotective agent but other times, when it forms a complex or conjugate with the metal (particularly Hg or Cd), this can serve as a transport form that enhances renal cellular accumulation and toxicity of the metal. A central concept for understanding the disposition and toxicological consequences of formation of complexes between GSH and metals is that of "molecular mimicry." This chapter also summarized the other side of the interactions between metals and GSH, namely that of how metals affect GSH status. Earlier studies described the effect of exposures to metal ions at a range of doses, comprising both subtoxic to toxic doses, on levels of GSH and activity and/or expression of GSH-dependent enzymes. More recent studies have shown that metal ions can also affect GSH status by modulation of transcription factor activation and/or nuclear translocation. Such studies have revealed a complex array of effects that both illustrate the compensatory mechanisms that exist in cells in response to perturbations such as oxidative stress and suggest possible molecular targets for pharmacological agents in the design of new, therapeutic agents. The trace nutrient Se plays a significant role in modulation of GSH status and in the response of cells to oxidative stress. As a component of selenoproteins such as GPX and TrxR, selenium status plays a central role in the functions of GSH and related redox systems, such as the thioredoxin and GLRX pathways.

REFERENCES

1. Adams, J.D., Jr., Lauterburg, B.H., and Mitchell, J.R., Plasma glutathione and glutathione disulfide in the rat: Regulation and response to oxidative stress. *J. Pharmacol. Exp. Ther.* 227, 749–754, 1983.
2. Anderson, M.E. and Meister, A., Dynamic state of glutathione in blood plasma. *J. Biol. Chem.* 255, 9530–9533, 1980.
3. Hagenfeldt, L., Arvidsson, A., and Larsson, A., Glutathione and γ-glutamylcysteine in whole blood, plasma and erythrocytes. *Clin. Chim. Acta* 85, 167–173, 1978.
4. Kleinman, W.A. and Richie, J.P., Jr., Status of glutathione and other thiols and disulfides in human plasma. *Biochem. Pharmacol.* 60, 19–29, 2000.
5. Lash, L.H. and Jones, D.P., Distribution of oxidized and reduced forms of glutathione and cyst(e)ine in rat plasma. *Arch. Biochem. Biophys.* 240, 583–592, 1985.
6. Sies, H., Wahllander, A., Waydhas, C., Soboll, S., and Haberle, D., Functions of intracellular glutathione in hepatic hydroperoxide and drug metabolism and the role of extracellular glutathione. *Adv. Enzyme Regul.* 18, 303–320, 1979.
7. Wendel, A. and Cikryt, P., The level and half-life of glutathione in human plasma. *FEBS Lett.* 120, 209–211, 1980.
8. Oram, P.D., Fang, X., Fernando, Q., Letkeman, P., and Letkeman, D., The formation constants of mercury(II)–glutathione complexes. *Chem. Res. Toxicol.* 9, 709–712, 1996.
9. Andersson, A., Isaksson, A., Brattström, L., and Hultberg, B., Homocysteine and other thiols determined in plasma by HPLC and thiol-specific postcolumn derivatization. *Clin. Chem.* 39, 1590–1597, 1993.

10. Zalups, R.K. and Lash, L.H., Recent advances in understanding the renal transport and toxicity of mercury. *J. Toxicol. Environ. Health* 42, 1–44, 1994.
11. Griffith, O.W. and Meister, A., Origin and turnover of mitochondrial glutathione. *Proc. Natl. Acad. Sci. USA* 82, 4668–4672, 1985.
12. McKernan, T.B., Woods, E.B., and Lash, L.H., Uptake of glutathione by renal cortical mitochondria. *Arch. Biochem. Biophys.* 288, 653–663, 1991.
13. Meister, A. and Tate, S.S., Glutathione and related γ-glutamyl compounds: Biosynthesis and utilization. *Annu. Rev. Biochem.* 45, 559–604, 1976.
14. Griffith, O.W. and Meister, A., Differential inhibition of glutamine and γ-glutamylcysteine synthetases by α-alkyl analogs of methionine sulfoximine that induce convulsions. *J. Biol. Chem.* 253, 2333–2338, 1978.
15. Griffith, O.W. and Meister, A., Potent and specific inhibition of glutathione synthesis by buthionine sulfoximine (*S-n*-butyl homocysteine sulfoximine). *J. Biol. Chem.* 254, 7558–7560, 1979.
16. Griffith, O.W., Anderson, M.E., and Meister, A., Inhibition of glutathione biosynthesis by prothionine sulfoximine (*S-n*-propyl homocysteine sulfoximine), a selective inhibitor of γ-glutamylcysteine synthetase. *J. Biol. Chem.* 254, 1205–1210, 1979.
17. Lash, L.H., Role of glutathione transport processes in kidney function. *Toxicol. Appl. Pharmacol.* 204, 329–342, 2005.
18. Huang, C.-S., Anderson, M.E., and Meister, A., Amino acid sequence and function of the light subunit of rat kidney γ-glutamylcysteine synthetase. *J. Biol. Chem.* 268, 20578–20583, 1993.
19. Huang, C.-S., Chang, L.-S., Anderson, M.E., and Meister, A., Catalytic and regulatory properties of the heavy subunit of rat kidney γ-glutamylcysteine synthetase. *J. Biol. Chem.* 268, 19675–19680, 1993.
20. Lu, S.C., Regulation of glutathione synthesis. *Mol. Aspects Med.* 30, 42–59, 2009.
21. Yan, N. and Meister, A., Amino acid sequence of rat kidney γ-glutamylcysteine synthetase. *J. Biol. Chem.* 265, 1588–1593, 1990.
22. Yang, Y., Dieter, M.Z., Chen, Y., Shertzer, H.G., Nebert, D.W., and Dalton, T.P., Initial characterization of the glutamate–cysteine ligase modified subunit Gclm$^{-/-}$ knockout mouse: Novel model system for a severely compromised oxidative stress response. *J. Biol. Chem.* 277, 49446–49452, 2002.
23. Hinchman, C.A. and Ballatori, N., Glutathione-degrading capacities of liver and kidney in different species. *Biochem. Pharmacol.* 40, 1131–1135, 1990.
24. Ikeda, Y. and Taniguchi, N., Gene expression of γ-glutamyltranspeptidase. *Methods Enzymol.* 401, 408–425, 2005.
25. Griffith, O.W. and Meister, A., Translocation of intracellular glutathione to membrane-bound γ-glutamyl transpeptidase as a discrete step in the γ-glutamyl cycle: Glutathionuria after inhibition of transpeptidase. *Proc. Natl. Acad. Sci. USA* 76, 268–272, 1979.
26. Griffith, O.W. and Meister, A., Excretion of cysteine and γ-glutamylcysteine moieties in human and experimental animal γ-glutamyl transpeptidase deficiency. *Proc. Natl. Acad. Sci. USA* 77, 3384–3387, 1980.
27. Anderson, M.E. and Meister, A., Dynamic state of glutathione in blood plasma. *J. Biol. Chem.* 255, 9530–9533, 1980.
28. Griffith, O.W. and Meister, A., Glutathione: Interorgan translocation, turnover, and metabolism. *Proc. Natl. Acad. Sci. USA* 76, 5606–5610, 1979.
29. Häberle, D., Wahlländer, A., and Sies, H., Assessment of the kidney function in maintenance of plasma glutathione concentration and redox state in anesthetized rats. *FEBS Lett.* 108, 335–340, 1979.
30. McIntyre, T.M. and Curthoys, N.P., The interorgan metabolism of glutathione. *Int. J. Biochem.* 12, 545–551, 1980.
31. Wendel, A. and Cikryt, P., The level and half-life of glutathione in human plasma. *FEBS Lett.* 120, 209–211, 1980.
32. Abbott, W.A., Bridges, R.J., and Meister, A., Extracellular metabolism of glutathione accounts for its disappearance from the basolateral circulation of the kidneys. *J. Biol. Chem.* 259, 15393–15400, 1984.
33. Anderson, M.E., Bridges, R.J., and Meister, A., Direct evidence for interorgan transport of glutathione and that the nonfiltration mechanism for glutathione utilization involves γ-glutamyl transpeptidase. *Biochem. Biophys. Res. Commun.* 96, 848–863, 1980.
34. Inoue, M., Shinozuka, S., and Morino, Y., Mechanism of peritubular extraction of plasma glutathione: The catalytic activity of contralumenal γ-glutamyltransferase is prerequisite to the apparent peritubular extraction of plasma glutathione. *Eur. J. Biochem.* 157, 605–609, 1986.
35. Fonteles, M.C., Pillio, D.J., Jeske, A.H., and Leibach, F.H., Extraction of glutathione by the isolated perfused rabbit kidney. *J. Surg. Res.* 21, 169–174, 1976.

36. Ormstad, K., Lastbom, T., and Orrenius, S., Evidence for different localization of glutathione oxidase and γ-glutamyltransferase activities during extracellular glutathione metabolism in isolated perfused kidney. *Biochim. Biophys. Acta* 700, 148–153, 1982.
37. Rankin, B.B. and Curthoys, N.P., Evidence for renal paratubular transport of glutathione. *FEBS Lett.* 147, 193–196, 1982.
38. Rankin, B.B., Wells, W., and Curthoys, N.P., Rat renal peritubular transport and metabolism of plasma ^{35}S glutathione. *Am. J. Physiol.* 249, F198–F204, 1985.
39. Hagen, T.M., Aw, T.Y., and Jones, D.P., Glutathione uptake and protection against oxidative injury in isolated kidney cells. *Kidney Int.* 34, 74–81, 1988.
40. Parks, L.D., Zalups, R.K., and Barfuss, D.W., Heterogeneity of glutathione synthesis and secretion in the proximal tubule of the rabbit. *Am. J. Physiol.* 274, F924–F931, 1998.
41. Lash, L.H. and Jones, D.P., Transport of glutathione by renal basal-lateral membrane vesicles. *Biochem. Biophys. Res. Commun.* 112, 55–60, 1983.
42. Lash, L.H. and Jones, D.P., Renal glutathione transport: Characteristics of the sodium-dependent system in the basal-lateral membrane. *J. Biol. Chem.* 259, 14508–14514, 1984.
43. Hayes, J.D., Flanagan, J.U., and Jowsey, I.R., Glutathione transferases. *Annu. Rev. Pharmacol. Toxicol.* 45, 51–88, 2005.
44. Ballatori, N., Hammond, C.L., Cunningham, J.B., Krance, S.M., and Marchan, R., Molecular mechanisms of reduced glutathione transport: Role of the MRP/CTFR/ABCC and OATP/SLC21A families of membrane proteins. *Toxicol. Appl. Pharmacol.* 204, 238–255, 2005.
45. Ballatori, N., Krance, S.M., Marchan, R., and Hammond, C.L., Plasma membrane glutathione transporters and their roles in cell physiology and pathophysiology. *Mol. Aspects Med.* 30, 13–28, 2009.
46. Mittur, A., Wolkoff, A.W., and Kaplowitz, N., The thiol sensitivity of glutathione transport in sidedness-sorted basolateral liver plasma membrane and in Oatp1-expressing HeLa cell membrane. *Mol. Pharmacol.* 61, 425–435, 2002.
47. Gietl, Y.S. and Anders, M.W., Biosynthesis and biliary excretion of S-conjugates of hexachlorobuta-1,3-diene in the perfused rat liver. *Drug Metab. Dispos.* 19, 274–277, 1991.
48. Hinchman, C.A., Matsumoto, H., Simmons, T.W., and Ballatori, N., Intrahepatic conversion of a glutathione conjugate to its mercapturic acid: Metabolism of 1-chloro-2,4-dinitrobenzene in isolated perfused rat and guinea pig livers. *J. Biol. Chem.* 266, 22179–22185, 1991.
49. Hinchman, C.A., Truong, A.T., and Ballatori, N., Hepatic uptake of intact glutathione S-conjugate, inhibition by organic anions, and sinusoidal catabolism. *Am. J. Physiol.* 265, G547–G554, 1993.
50. Simmons, T.W., Hinchman, C.A., and Ballatori, N., Polarity of hepatic glutathione and glutathione S-conjugate efflux, and intraorgan mercapturic acid formation in the skate. *Biochem. Pharmacol.* 42, 2221–2228, 1991.
51. Lash, L.H., Putt, D.A., and Cai, H., Membrane transport function in primary cultures of human proximal tubular cells. *Toxicology* 228, 200–218, 2006.
52. Lash, L.H. and Jones, D.P., Uptake of the glutathione conjugate *S*-(1,2-dichlorovinyl)glutathione by renal basal-lateral membrane vesicles and isolated kidney cells. *Mol. Pharmacol.* 28, 278–282, 1985.
53. Lash, L.H. and Putt, D.A., Renal cellular transport of exogenous glutathione: Heterogeneity at physiological and pharmacological concentrations. *Biochem. Pharmacol.* 58, 897–907, 1999.
54. Visarius, T.M., Putt, D.A., Schare, J.M., Pegouske, D.M., and Lash, L.H., Pathways of glutathione metabolism and transport in isolated proximal tubular cells from rat kidney. *Biochem. Pharmacol.* 52, 259–272, 1996.
55. Lash, L.H., Putt, D.A., Xu, F., and Matherly, L.H., Role of rat organic anion transporter 3 (Oat3) in the renal basolateral transport of glutathione. *Chem. Biol. Interact.* 170, 124–134, 2007.
56. Inoue, M. and Morino, Y., Direct evidence for the role of the membrane potential in glutathione transport by renal brush-border membranes. *J. Biol. Chem.* 260, 326–331, 1985.
57. Scott, R.D. and Curthoys, N.P., Renal clearance of glutathione measured in rats pretreated with inhibitors of glutathione metabolism. *Am. J. Physiol.* 252, F877–F882, 1987.
58. Ballatori, N., Glutathione mercaptides as transport forms of metals. *Adv. Pharmacol.* 27, 271–298, 1994.
59. Clarkson, T.W., Molecular and ionic mimicry of toxic metals. *Annu. Rev. Pharmacol. Toxicol.* 32, 545–571, 1993.
60. Zalups, R.K., Molecular interactions with mercury in the kidney. *Pharmacol. Rev.* 52, 113–143, 2000.
61. Bridges, C. C. and Zalups, R.K., Molecular and ionic mimicry and the transport of toxic metals. *Toxicol. Appl. Pharmacol.* 204, 274–308, 2005.
62. Ballatori, N. and Clarkson, T.W., Biliary transport of glutathione and methylmercury. *Am. J. Physiol.* 244, G435–G441, 1983.

63. Ballatori, N. and Clarkson, T.W., Dependence of biliary secretion of inorganic mercury on the biliary transport of glutathione. *Biochem. Pharmacol.* 33, 1093–1098, 1984.
64. Ballatori, N. and Clarkson, T.W., Inorganic mercury secretion into bile as a low molecular weight complex. *Biochem. Pharmacol.* 33, 1087–1092, 1984.
65. Ballatori, N., Gatmaitan, Z., and Truong, A.T., Impaired biliary excretion and whole body elimination of methylmercury in rats with a congenital defect in biliary glutathione excretion. *Hepatology* 22, 1469–1473, 1995.
66. Sugawara, N., Lai, Y.-R., Sugaware, C., and Arizono, K., Decreased hepatobiliary secretion of inorganic mercury, its deposition and toxicity in the Eisai hyperbilirubinemic rat with no hepatic canalicular organic anion transporter. *Toxicology* 126, 23–31, 1998.
67. Dutczak, W.J. and Ballatori, N., Transport of the glutathione–methylmercury complex across liver canalicular membranes on reduced glutathione carriers. *J. Biol. Chem.* 269, 9746–9751, 1994.
68. Zalups, R.K. and Barfuss, D.W., Diversion or prevention of biliary outflow from the liver diminishes the renal uptake of injected inorganic mercury. *Drug Metab. Dispos.* 24, 480–486, 1996.
69. Zalups, R.K., Barfuss, D.W., and Lash, L.H., Disposition of inorganic mercury following biliary obstruction and chemically-induced glutathione depletion: Dispositional changes one hour after the intravenous administration of mercuric chloride. *Toxicol. Appl. Pharmacol.* 154, 135–144, 1999.
70. Gregus, Z., Stein, A.F., Varga, F., and Klaassen, C.D., Effect of lipoic acid on biliary excretion of glutathione and metals. *Toxicol. Appl. Pharmacol.* 114, 88–96, 1992.
71. Zalups, R.K. and Barfuss, D.W., Participation of mercuric conjugates of cysteine, homocysteine, and *N*-acetylcysteine in mechanisms involved in the renal tubular uptake of inorganic mercury. *J. Am. Soc. Nephrol.* 9, 551–561, 1998.
72. Baggett, J.M. and Berndt, W.O., The effect of depletion of nonprotein sulfhydryls by diethyl maleate plus buthionine sulfoximine on renal uptake of mercury in the rat. *Toxicol. Appl. Pharmacol.* 83, 556–562, 1986.
73. Berndt, W.O., Baggett, J.McC., Blacker, A., and Houser, M., Renal glutathione and mercury uptake by kidney. *Fundam. Appl. Toxicol.* 5, 832–839, 1985.
74. Zalups, R.K. and Lash, L.H., Depletion of glutathione in the kidney and the renal disposition of administered inorganic mercury. *Drug Metab. Dispos.* 25, 516–523, 1997.
75. Tanaka, T., Naganuma, A., and Imura, N., Role of γ-glutamyltranspeptidase in renal uptake and toxicity of inorganic mercury in mice. *Toxicology* 60, 187–198, 1990.
76. Tanaka-Kagawa, T., Naganuma, A., and Imura, N., Tubular secretion and reabsorption of mercury compounds in mouse kidney. *J. Pharmacol. Exp. Ther.* 264, 776–782, 1993.
77. de Ceaurriz, J., Payan, J.P., Morel, G., and Brondeau, M.T., Role of extracellular glutathione and γ-glutamyltranspeptidase in the disposition and kidney toxicity of inorganic mercury in rats. *J. Appl. Toxicol.* 14, 201–206, 1994.
78. Girardi, G. and Elias, M.M., Effect of different renal glutathione levels on renal mercury disposition and excretion in the rat. *Toxicology* 81, 57–67, 1993.
79. Johnson, D.R., Role of renal cortical sulfhydryl groups in development of mercury-induced renal toxicity. *J. Toxicol. Environ. Health* 9,119–126, 1982.
80. Girardi, G. and Elias, M.M., Effectiveness of *N*-acetylcysteine in protecting against mercuric chloride-induced nephrotoxicity. *Toxicology* 67, 155–164, 1991.
81. Adachi, T., Yasutake, A., and Hirayama, K., Influence of dietary levels of protein and sulfur amino acids on the fate of methylmercury in mice. *Toxicology* 93, 225–234, 1994.
82. Zalups, R.K. and Barfuss, D.W., Accumulation and handling of inorganic mercury in the kidney after coadministration with glutathione. *J. Toxicol. Environ. Health* 44, 385–399, 1995.
83. Zalups, R.K., Basolateral uptake of inorganic mercury in the kidney. *Toxicol. Appl. Pharmacol.* 151, 192–199, 1998.
84. Zalups, R.K. and Lash, L.H., Binding of mercury in renal brush-border and basolateral membrane-vesicles: Implication of a cysteine conjugate of mercury involved in the luminal uptake of inorganic mercury in the kidney. *Biochem. Pharmacol.* 53, 1889–1900, 1997.
85. Lash, L.H., Putt, D.A., and Zalups, R.K., Role of extracellular thiols in uptake and distribution of inorganic mercury in rat renal proximal and distal tubular cells. *J. Pharmacol. Exp. Ther.* 285, 1039–1050, 1998.
86. Lash, L.H., Putt, D.A., and Zalups, R.K., Influence of exogenous thiols on mercury-induced cellular injury in isolated renal proximal tubular and distal tubular cells from normal and uninephrectomized rats. *J. Pharmacol. Exp. Ther.* 291, 492–502, 1999.

87. Divine, K.K., Ayala-Fierro, F., Barber, D.S., and Carter, D.E., Glutathione, albumin, cysteine, and Cys–Gly effects on toxicity and accumulation of mercuric chloride in LLC-PK1 cells. *J. Toxicol. Environ. Health* 57, 489–505, 1999.
88. Houser, M.T., Milner, L.S., Kolbeck, P.C., Wei, S.H., and Stohs, S.J., Glutathione monoethyl ester moderates mercuric chloride-induced acute renal failure. *Nephron* 61, 449–455, 1992.
89. Zalups, R.K., Organic anion transport and action of γ-glutamyl transpeptidase in kidney linked mechanistically to renal tubular uptake of inorganic mercury. *Toxicol. Appl. Pharmacol.* 132, 289–298, 1995.
90. Lash, L.H., Hueni, S.E., Putt, D.A., and Zalups, R.K., Role of organic anion and amino acid carriers in transport of inorganic mercury in rat renal basolateral membrane vesicles: Influence of compensatory renal growth. *Toxicol. Sci.* 88, 630–644, 2005.
91. Zalups, R.K., Knutson, K.L., and Schnellmann, R.G., *In vitro* analysis of the accumulation and toxicity of inorganic mercury in segments of the proximal tubule isolated from the rabbit kidney. *Toxicol. Appl. Pharmacol.* 119, 221–227, 1993.
92. Wei, H., Qiu, L., Divine, K.K., Ashbaugh, M.D., McIntyre, L.C., Jr., Fernando, Q., and Gandolfi, A.J., Toxicity and transport of three synthesized mercury–thiol complexes in isolated rabbit renal proximal tubule suspensions. *Drug Chem. Toxicol.* 22, 323–341, 1999.
93. Fine, L., The biology of renal hypertrophy. *Kidney Int.* 29, 619–634, 1986.
94. Shirley, D.G. and Walter, S.J., Acute and chronic changes in renal function following unilateral nephrectomy. *Kidney Int.* 40, 62–68, 1991.
95. Meyer, T.W., Baboolal, K., and Brenner, B.M., Nephron adaptation to renal injury. In: B.M. Brenner (ed.), *The Kidney*, 5th edition, pp. 2011–2048, W.B. Saunders Company, Philadelphia, PA, 1996.
96. Hostetter, T.H., Progression of renal disease and renal hypertrophy. *Annu. Rev. Physiol.* 57, 263–278, 1995.
97. Gandhi, M., Olson, J.L., and Meyer, T.W., Contribution of tubular injury to loss of remnant kidney function. *Kidney Int.* 54, 1157–1165, 1998.
98. Kasiske, B.L., Ma, J.Z., Louis, T.A., and Swan, S.K., Long-term effects of reduced renal mass in humans. *Kidney Int.* 48, 814–819, 1995.
99. Zalups, R.K. and Veltman, J.C., Renal glutathione homeostasis in compensatory renal growth. *Life Sci.* 42, 2171–2176, 1988.
100. Zalups, R.K. and Lash, L.H., Effects of uninephrectomy and mercuric chloride on renal glutathione homeostasis. *J. Pharmacol. Exp. Ther.* 254, 962–970, 1990.
101. Houser, M.T. and Berndt, W.O., The effect of unilateral nephrectomy on the nephrotoxicity of mercuric chloride in the rat. *Toxicol. Appl. Pharmacol.* 83, 506–515, 1986.
102. Houser, M.T. and Berndt, W.O., Unilateral nephrectomy in the rat: Effects on mercury handling and renal cortical subcellular distribution. *Toxicol. Appl. Pharmacol.* 93, 187–194, 1988.
103. Zalups, R.K. and Diamond, G.L., Mercuric chloride-induced nephrotoxicity in the rat following unilateral nephrectomy and compensatory renal growth. *Virchows Arch. B* 53, 336–346, 1987.
104. Zalups, R.K., Klotzbach, J.M., and Diamond, G.L., Enhanced accumulation of injected inorganic mercury in renal outer medulla after unilateral nephrectomy. *Toxicol. Appl. Pharmacol.* 89, 226–236, 1987.
105. Zalups, R.K., Autometallographic localization of inorganic mercury in the kidneys of rats: Effect of unilateral nephrectomy and compensatory renal growth. *Exp. Mol. Pathol.* 54, 10–21, 1991.
106. Lash, L.H. and Zalups, R.K., Mercuric chloride-induced cytotoxicity and compensatory hypertrophy in rat kidney proximal tubular cells. *J. Pharmacol. Exp. Ther.* 261, 819–829, 1992.
107. Lash, L.H., Putt, D.A., and Zalups, R.K., Influence of compensatory renal growth on susceptibility of primary cultures of renal cells to chemically induced injury. *Toxicol. Sci.* 94, 417–427, 2006.
108. Lash, L.H. and Zalups, R.K., Activities of enzymes involved in renal cellular glutathione metabolism after uninephrectomy in the rat. *Arch. Biochem. Biophys.* 309, 129–138, 1994.
109. Lash, L.H., Putt, D.A., and Zalups, R.K., Biochemical and functional characteristics of cultured renal epithelial cells from uninephrectomized rats: Factors influencing nephrotoxicity. *J. Pharmacol. Exp. Ther.* 296, 243–251, 2001.
110. Lash, L.H., Putt, D.A., Horky, III, S.J., and Zalups, R.K., Functional and toxicological characteristics of isolated renal mitochondria: Impact of compensatory renal growth. *Biochem. Pharmacol.* 62, 383–395, 2001.
111. Harris, D.C.H., Chan, L., and Schrier, R.W., Remnant kidney hypermetabolism and progression of chronic renal failure. *Am. J. Physiol.* 254, F267–F276, 1988.
112. Shapiro, J.I., Elkins, N., Reiss, O.K., Suleymanlar, G., Jin, H., Schrier, R.W., and Chan, L., Energy metabolism following reduction of renal mass. *Kidney Int.* 45(Suppl. 45), S100–S105, 1994.

113. Zalups, R.K., Barfuss, D.W., and Kostyniak, P.J., Altered intrarenal accumulation of mercury in uninephrectomized rats treated with methylmercury chloride. *Toxicol. Appl. Pharmacol.* 115, 174–182, 1992.
114. Zalups, R.K., Gelein, R.M., and Cherian, M.G., Shifts in the dose–effect relationship for the nephropathy induced by cadmium-metallothionein in rats after a reduction in renal mass. *J. Pharmacol. Exp. Ther.* 262, 1256–1266, 1992.
115. Zalups, R.K., Influence of different degrees of reduced renal mass on the renal and hepatic disposition of administered cadmium. *J. Toxicol. Environ. Health* 51, 245–264, 1997.
116. Singhal, R.K., Anderson, M.E., and Meister, A., Glutathione, a first line of defense against cadmium toxicity. *FASEB J.* 1, 220–223, 1987.
117. Niewenhuis, R.J., Dimitriu, C., and Prozialeck, W.C., Ultrastructural characterization of the early changes in intercellular junctions in response to cadmium (Cd^{2+}) exposure in LLC-PK1 cells. *Toxicol. Appl. Pharmacol.* 142, 1–12, 1997.
118. Prozialeck, W.C. and Lamar, P.C., Comparison of the cytotoxic effects of cadmium (Cd^{2+}) in high and low resistance strains of MDCK cells that express different levels of E-cadherin. *Toxicol. In Vitro* 12, 633–647, 1998.
119. Prozialeck, W.C. and Lamar, P.C., Interaction of cadmium (Cd^{2+}) with a 13-residue polypeptide analog of a putative calcium-binding motif of E-cadherin. *Biochim. Biophys. Acta* 1451, 93–100,1999.
120. Prozialeck, W.C., Evidence that E-cadherin may be a target for cadmium toxicity in epithelial cells. *Toxicol. Appl. Pharmacol.* 164, 231–249, 2000.
121. Sabolic, I., Herak-Kramberger, C.M., and Brown D., Subchronic cadmium treatment affects the abundance and arrangement of cytoskeletal proteins in rat renal proximal tubular cells. *Toxicology* 165, 205–216, 2001.
122. Sabolic, I., Ljubojevic, M., Herak-Kramberger, C.M., and Brown, D., Cd-MT causes endocytosis of brush-border transporters in rat renal proximal tubules. *Am. J. Physiol.* 283, F1389–F1402, 2002.
123. Kinne, R.K.H., Schütz, H., and Kinne-Saffran, E., The effect of cadmium chloride *in vitro* on sodium-glutamate cotransport in brush border membrane vesicles isolated from rabbit kidney. *Toxicol. Appl. Pharmacol.* 135, 216–221, 1995.
124. Tang, W. and Shaikh, Z.A., Renal cortical mitochondrial dysfunction upon cadmium metallothionein administration to Sprague-Dawley rats. *J. Toxicol. Environ. Health* 63, 221–235, 2001.
125. Pearson, C.A. and Prozialeck, W.C., E-Cadherin, β-catenin and cadmium carcinogenesis. *Med. Hypothesis* 56, 573–581, 2001.
126. Waalkes, M.P., Cadmium carcinogenesis. *Mutat. Res.* 533, 107–120, 2003.
127. Achanzar, W.E., Diwan, B.A., Liu, J., Quader, S.T., Webber, M.M., and Waalkes, M.P., Cadmium-induced malignant transformation of human prostate epithelial cells. *Cancer Res.* 61, 455–458, 2001.
128. Tang, W., Sadovic, S., and Shaikh, Z.A., Nephrotoxicity of cadmium-metallothionein: Protection by zinc and role of glutathione. *Toxicol. Appl. Pharmacol.* 151, 276–282, 1998.
129. Wispriyono, B., Matsuoka, M., Igisu, H., and Matsuno, K., Protection from cadmium cytotoxicity by *N*-acetylcysteine in LLC-PK1 cells. *J. Pharmacol. Exp. Ther.* 287, 344–351, 1998.
130. Zalups, R.K., Evidence for basolateral uptake of cadmium in the kidneys of rats. *Toxicol. Appl. Pharmacol.* 164, 15–23, 2000.
131. Shaikh, Z.A., Northrup, J.B., and Vestergaard, P., Dependence of cadmium-metallothionein nephrotoxicity on glutathione. *J. Toxicol. Environ. Health* 56, 211–222, 1999.
132. Aiba, I., Hossain, A., and Kuo, M.T., Elevated GSH level increases cadmium resistance through down-regulation of Sp1-dependent expression of the cadmium transporter ZIP8. *Mol. Pharmacol.* 74, 823–833, 2008.
133. Dasari, A., Bartholemew, J.N., Volonte, D., and Galbiati, F., Oxidative stress induces premature senescence by stimulating caveolin-1 gene transcription through p38 mitogen-activated protein kinase/Sp1-mediated activation of two GC-rich promoter elements. *Cancer Res.* 66, 10805–10814, 2006.
134. Schafer, G., Cramer, T., Suske, G., Kemmer, W., Wiedenmann, B., and Höcker, M., Oxidative stress regulates vascular endothelial growth factor-A gene transcription through Sp1- and Sp3-dependent activation of two proximal GC-rich promoter elements. *J. Biol. Chem.* 278, 8190–8198, 2003.
135. Webster, K.A., Prentice, H., and Bishopric, N.H., Oxidation of zinc finger transcription factors: Physiological consequences. *Antiox. Redox Signal.* 3, 535–548, 2001.
136. Vayalil, P.K., Iles, K.E., Choi, J., Yi, A.K., Postlethwait, E.M., and Liu, R.M., Glutathione suppresses TGF-beta-induced PAI-1 expression by inhibiting p38 and JNK MAPK and binding of AP-1, SP-1, and Smad to the PAI-1 promoter. *Am. J. Physiol.* 293, L1281–L1292, 2007.
137. Liu, J., Lu, Y., Goyer, R.A., and Waalkes, M.P., Mineral arsenicals in traditional medicine: Orpiment, realgar, and arsenolite. *J. Pharmacol. Exp. Ther.* 326, 363–368, 2008.

138. Leslie, E.M., Haimeur, A., and Waalkes, M.P., Arsenic transport by the human multidrug resistance protein 1 (MRP1/ABCC1): Evidence that tri-glutathione conjugate is required. *J. Biol. Chem.* 279, 32700–32708, 2004.
139. Rappa, G., Lorico, A., Flavell, R.A., and Sartorelli, A.C., Evidence that the multidrug resistance protein (MRP) functions as a co-transporter of glutathione and natural product toxins. *Cancer Res.* 57, 5232–5237, 1997.
140. Burton, C.A., Hatlelid, K., Divine, K., Carter, D.E., Fernando, Q., Brendel, K., and Gandolfi, A.J., Glutathione effects on toxicity and uptake of mercuric chloride and sodium arsenite in rabbit renal cortical slices. *Environ. Health Perspect.* 103, 81–84, 1995.
141. Scott, N., Hatlelid, K.M., MacKenzie, N.E., and Carter, D.E., Reactions of arsenic(III) and arsenic(V) species with glutathione. *Chem. Res. Toxicol.* 6, 102–106, 1993.
142. Thompson, D.J., A chemical hypothesis for arsenic methylation in mammals. *Chem. Biol. Interact.* 88, 89–114, 1993.
143. Aposhian, H.V. and Aposhian, M.M., Arsenic toxicology: Five questions. *Chem. Res. Toxicol.* 19, 1–15, 2006.
144. Lash, L.H. and Zalups, R.K., Alterations in renal cellular glutathione metabolism after *in vivo* administration of a subtoxic dose of mercuric chloride. *J. Biochem. Toxicol.* 11, 1–9, 1996.
145. Addya, S., Chakravarti, K., Basu, A., Santra, M., Haldar, S., and Chatterjee, G.C., Effects of mercuric chloride on several scavenging enzymes in rat kidney and influence of vitamin E supplementation. *Acta Vitaminol. Enzymol.* 6, 103–107, 1984.
146. Gstraunthaler, G., Pfaller, W., and Kotanki, P., Glutathione depletion and *in vitro* lipid peroxidation in mercury or maleate induced acute renal failure. *Biochem. Pharmacol.* 32, 2969–2972, 1983.
147. Girardi, G. and Elias, M.M., Mercuric chloride effects on rat renal redox enzymes activities: SOD protection. *Free Radic. Biol. Med.* 18, 61–66, 1995.
148. Woods, J.S., Davis, H.A., and Baer, R.P., Enhancement of γ-glutamylcysteine synthetase mRNA in rat kidney by methylmercury. *Arch. Biochem. Biophys.* 296, 350–353, 1992.
149. Woods, J.S. and Ellis, M.E., Up-regulation of glutathione synthesis in rat kidney by methylmercury: Relationship to mercury-induced oxidative stress. *Biochem. Pharmacol.* 50, 1719–1724, 1995.
150. McGuire, S., Daggett, D.A., Bostad, E., Nelson, S., Wright, L.S., Siegel, F.L., and Kornguth, S., Increased levels of glutathione *S*-transferases and appearance of novel alpha class isoenzymes in kidneys of mice exposed to mercuric chloride. *Nephron* 77, 452–460, 1997.
151. Lash, L.H., Putt, D.A., Hueni, S.E., Payton, S.G., and Zwickl, J., Interactive toxicity of inorganic mercury and trichloroethylene in rat and human proximal tubules: Effects on apoptosis, necrosis, and glutathione status. *Toxicol. Appl. Pharmacol.* 221, 349–362, 2007.
152. Dieguez-Acuna, F.J., Ellis, M.E., Kushleika, J., and Woods, J.S., Mercuric ion attenuates nuclear factor-κB activation and DNA binding in normal rat kidney epithelial cells: Implications for mercury-induced nephrotoxicity. *Toxicol. Appl. Pharmacol.* 173, 176–187, 2001.
153. Shumilla, J.A., Wetterhahn, K.E., and Barchowsky, A., Inhibition of NF-κB binding to DNA by chromium, cadmium, mercury, zinc, and arsenite *in vitro*: Evidence of a thiol mechanism. *Arch. Biochem. Biophys.* 349, 356–362, 1998.
154. Woods, J.S., Dieguez-Acuna, F.J., Ellis, M.E., Kushleika, J., and Simmonds, P.L., Attenuation of nuclear factor kappa B (NF-κB) promotes apoptosis of kidney epithelial cells: A potential mechanism of mercury-induced nephrotoxicity. *Environ. Health Perspect.* 110(Suppl. 5), 819–822, 2002.
155. Siegers, C.-P., Schenke, M., and Younes, M., Influence of cadmium chloride, mercuric chloride and sodium vanadate on the glutathione-conjugating enzyme system in liver, kidney, and brain of mice. *J. Toxicol. Environ. Health* 22, 141–148, 1987.
156. Figueirido-Pereira, M.E., Yakushin, S., and Cohen, G., Disruption of the intracellular sulfhydryl homeostasis by cadmium-induced oxidative stress leads to protein thiolation and ubiquitination in neuronal cells. *J. Biol. Chem.* 273, 12703–12709, 1998.
157. Di Simplicio, P., Gorelli, M., Vignani, R., and Leonzio, C., The differential modulation of the enzymes of glutathione metabolism: Indication of overlapping effects of toxicity and repair in mouse liver and kidney after dietary treatment with methylmercury and sodium selenite. *Biol. Trace Elem. Res.* 36, 167–181, 1993.
158. Chung, A.-S., Maines, M.D., and Reynolds, W.A., Inhibition of the enzymes of glutathione metabolism by mercuric chloride in the rat kidney: Reversal by selenium. *Biochem. Pharmacol.* 31, 3093–3100, 1982.
159. Schuliga, M., Chouchane, S., and Snow, E.T., Upregulation of glutathione-related genes and enzyme activities in cultured human cells by sublethal concentrations of inorganic arsenic. *Toxicol. Sci.* 70, 183–192, 2002.

160. Barchowsky, A., Roussel, R.R., Klei, L.R., James, P.E., Ganju, N., Smith, K.R., and Dudek, E.J., Low levels of arsenic trioxide stimulate proliferative signals in primary vascular cells without activating stress effector pathways. *Toxicol. Appl. Pharmacol.* 159, 65–75, 1999.
161. Lu, J. and Holmgren, A., Selenoproteins. *J. Biol. Chem.* 284, 723–727, 2009.
162. Dalton, T.P., Chen, Y., Schneider, S.N., Nebert, D.W., and Shertzer, H.G., Genetically altered mice to evaluate glutathione homeostasis in health and disease. *Free Radic. Biol. Med.* 37, 1511–1526, 2004.
163. Cai, J., Huang, Z.-Z., and Lu, S.C., Differential regulation of γ-glutamylcysteine synthetase heavy and light subunit gene expression. *Biochem. J.* 326, 167–172, 1997.
164. Dickinson, D.A., Levonen, A.-L., Moellering, D.R., Arnold, E.K., Zhang, H., Darley-Usmar, V.M., and Forman, H.J., Human glutamate cysteine ligase gene regulation through the electrophile response element. *Free Radic. Biol. Med.* 37, 1152–1159, 2004.
165. Moinova, H.R. and Mulcahy, R.T., An electrophile responsive element (EpRE) regulates β-naphthoflavone induction of the human γ-glutamylcysteine synthetase regulatory subunit gene: Constitutive expression is mediated by an adjacent AP-1 site. *J. Biol. Chem.* 273, 14683–14689, 1998.
166. Mulcahy, R.T., Wartman, M.A., Bailey, H.H., and Gipp, J.J., Constitutive and β-naphthoflavone-induced expression of the human γ-glutamylcysteine synthetase heavy subunit gene is regulated by a distal antioxidant response element/TRE sequence. *J. Biol. Chem.* 272, 7445–7454, 1997.
167. Zipper, L.M. and Mulcahy, R.T., Erk activation is required for Nrf2 nuclear localization during pyrrolidine dithiocarbamate induction of glutamate cysteine ligase modulatory gene expression in HepG2 cells. *Toxicol. Sci.* 73, 124–134, 2003.
168. Gopalakrishnan, A. and Kong, A.-N.T., Anticarcinogenesis by dietary phytochemicals: Cytoprotection by Nrf2 in normal cells and cytotoxicity by modulation of transcription factors NF-κB and AP-1 in abnormal cancer cells. *Food Chem. Toxicol.* 46, 1257–1270, 2008.
169. Kensler, T.W., Wakabayashi, N., and Biswal, S., Cell survival responses to environmental stresses via the Keap1-Nrf2-ARE pathway. *Annu. Rev. Pharmacol. Toxicol.* 47, 89–116, 2007.
170. Nerland, D.E., The antioxidant/electrophile response element motif. *Drug Metab. Rev.* 39, 235–48, 2007.

10 Ionic and Molecular Mimicry and the Transport of Metals

Christy C. Bridges and Rudolfs K. Zalups

CONTENTS

10.1 Introduction 242
10.2 Arsenic 243
10.2.1 Endocrine Disruption and As 243
10.2.2 Intestinal Transport of As 243
10.2.3 Renal Transport of As 244
10.2.4 Hepatic Transport of As 245
10.3 Cadmium 246
10.3.1 Endocrine Disruption and Cd^{2+} 246
10.3.2 Intestinal Transport of Cd^{2+} 247
10.3.3 Renal Transport of Cd^{2+} 249
10.3.4 Hepatic Transport of Cd^{2+} 252
10.3.5 Testicular Handling of Cd^{2+} 254
10.3.6 Intercellular Mimicry with Cd^{2+} 254
10.4 Lead 254
10.5 Mercury 257
10.5.1 Inorganic Mercury 258
10.5.1.1 Intestinal Transport of Hg^{2+} 258
10.5.1.2 Renal Transport of Hg^{2+} 259
10.5.1.3 Hepatic Transport of Hg^{2+} 260
10.5.1.4 Transport of Hg^{2+} in the Placenta 261
10.5.1.5 Intracellular Mimicry with Hg^{2+} 261
10.5.2 Methylmercury 262
10.5.2.1 Transport of CH_3Hg^+ in the Brain 262
10.5.2.2 Transport of CH_3Hg^+ in Erythrocytes 262
10.5.2.3 Intestinal Transport of CH_3Hg^+ 263
10.5.2.4 Renal Transport of CH_3Hg^+ 264
10.5.2.5 Hepatic Transport of CH_3Hg^+ 265
10.5.2.6 Transport of CH_3Hg^+ in the Placenta 265
10.6 Selenium 266
10.7 Oxyanions of Toxic Metals 268
10.7.1 Arsenate 268
10.7.2 Chromate 268
10.7.3 Molybdate 269
10.7.4 Vanadate 269
10.8 Summary 270
Acknowledgments 270
References 270

10.1 INTRODUCTION

Toxic, nonessential metals, such as mercury, cadmium, and lead, have no known nutritive value. Accordingly, in most animal species, there appear to be no dedicated mechanisms for the uptake of these metals into target cells. Yet, many studies have proven that these toxic metals do indeed gain entry into various target cells in affected tissues and organs [1–5]. In recent years, the concepts of molecular mimicry and ionic mimicry have been postulated as mechanisms by which certain toxic metal species can gain entry into target cells. Molecular mimicry has been defined previously as "the phenomenon whereby the bonding of metal ions to nucleophilic groups on certain biomolecules results in the formation of organo-metal complexes that can behave or serve as structural and/or functional homologs of other endogenous biomolecules or the molecule to which the metal ion has bonded" (Figure 10.1) [1–4]. Molecular mimicry appears to play a significant role in the entry of certain metals into cells that become intoxicated by one or more of these metals.

Ionic mimicry has been defined as "the ability of an unbound, cationic species of a metal to behave or serve as a structural and/or functional mimic of another (usually an essential) element at the site of a carrier protein, ion channel, enzyme, transcription factor, and/or metal-binding protein" [1,3,5,6]. Molecular and ionic mimics may be classified as structural and/or functional mimics. A structural mimic refers to an elemental or molecular species that is similar in size and shape to another element or molecule. A functional mimic is one that can elicit the same effect (i.e., physiological response) as the native element or molecule. Structural and functional mimicry will be discussed in relation to molecular and/or ionic mimicry and the specific metal involved in these phenomena.

A considerable amount of scientific data on molecular and ionic mimicry has been published in recent years. Even so, there are many unanswered questions. This chapter will focus on known and putative mechanisms by which several toxic metals gain access to the intracellular compartments of target cells affected adversely by these metals. Evidence supporting the phenomena of molecular and/or ionic mimicry for selected toxic metals will be outlined individually by species of metal and the organs, tissues, and cells involved in the process.

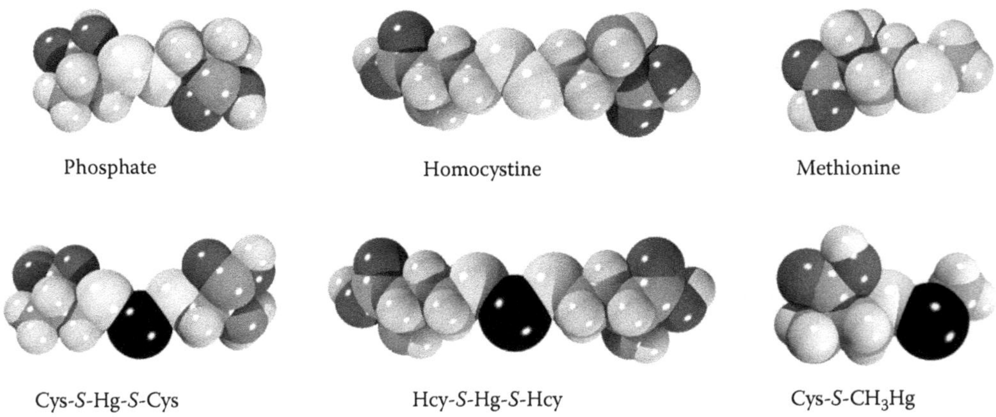

FIGURE 10.1 Space-filled models of selected mercuric conjugates implicated in molecular mimicry. There are significant similarities in chemical structure and molecular shape between the cysteine (Cys) *S*-conjugate of methylmercury (CH_3Hg-*S*-Cys) and the amino acid methionine. Also, note the similarities between the Cys *S*-conjugate of inorganic mercury (Cys-*S*-Hg-*S*-Cys) and the amino acid cystine, and the homocysteine (Hcy) *S*-conjugate of inorganic mercury (Hcy-*S*-Hg-*S*-Hcy) and homocystine. A significant body of evidence supports the hypothesis that mercuric conjugates of certain amino acids (such as Cys, Hcy, or Met) may act as molecular mimics of naturally occurring amino acids that are similar in structure and shape to the mercuric complexes.

10.2 ARSENIC

Arsenic (As) is a highly toxic element found naturally in the earth's crust. Organic As is formed when As ions combine with carbon and hydrogen; inorganic As is formed when As combines with the anionic species of oxygen, chlorine, or sulfur [7]. Data from animal studies have shown that organic forms of As are less toxic than the same dose of inorganic As [7]. There are few data on the effects of organic forms of As in humans. Inorganic As has long been recognized as a human poison and carcinogen [7,8] and the toxicological effects of inhalation and oral exposure to this metal are well documented. Numerous studies have shown that As has deleterious effects in the respiratory [9–14], neurological [9,15–18], and cardiovascular [19–22] systems. The most common and well-characterized symptom of As poisoning is the development of skin lesions and warts [10,13,21,23–26]. Interestingly, As, as arsenic trioxide (As_2O_3) has been shown to be an effective chemotherapeutic agent for the treatment of acute promyelocytic leukemia [27,28] and certain autoimmune diseases [29].

Human exposure to As occurs through environmental, dietary, and occupational sources [30]. In areas with rocky terrains, contamination of drinking water with inorganic As is a major environmental issue. The significance of this issue was emphasized after millions of Bangladesh residents were poisoned by consumption of well water contaminated with As [31]. In the United States, contamination of soil with inorganic As is also a concern, especially in mountainous areas. Childhood exposure to inorganic As may result from playing in, and/or accidentally ingesting, contaminated soil. Many people in the United States, however, are exposed to As via ingestion of certain fish and various types of seafood containing high levels of inorganic and/or organic As [30]. With regard to occupational sources of As in the United States, approximately 90% of all As used in industry is utilized in the "pressure treatment" of wood. Workers involved in the treatment process or exposed to dust produced when the wood is cut are at higher risk for As poisoning [7].

10.2.1 Endocrine Disruption and As

Interestingly, As in the form of sodium arsenite ($NaAsO_2$) appears able to act as a functional mimic of estrogen (estradiol) at the site of the estrogen receptor. Experiments in human breast cancer (MCF-7) cells have demonstrated that following exposure to $NaAsO_2$, As interacts with the hormone-binding domain of the estrogen receptor and results in activation of this receptor [32]. In addition, As has been shown to increase the transcription of the progesterone receptor gene and increase the rate of cell growth. Furthermore, As blocked the binding of estradiol to the estrogen receptor, suggesting that As and estradiol are binding to the same site. These data suggest that exposure to As may be a significant risk factor for the development of breast cancer and other hormone-related diseases. In contrast, more recent studies utilizing MCF-7 cells exposed to $NaAsO_2$ were unable to show that As activates the estrogen receptor [33,34]. In these studies, exposure of cells to $NaAsO_2$ resulted in a downregulation of the estrogen receptor, and subsequently a reduction in the growth-stimulatory effect of estrogen. Clearly, additional studies are necessary to elucidate fully the effect of As on the estrogen receptor.

10.2.2 Intestinal Transport of As

The molecular mechanisms responsible for the absorption of inorganic As following ingestion of contaminated food and water remain unclear. Studies in human intestinal (Caco-2) cells exposed to As_2O_3 indicate that absorption of As is a carrier-mediated process [35]. Possible carriers involved in this process include organic anion transporting polypeptides (OATP), phosphate transporters, and the divalent metal transporter (DMT1). Intestinal isoforms of OATP have not been implicated in the transport of As; however, OATP-C, which is present on the sinusoidal membrane of hepatocytes, has been shown to play a role in the hepatic uptake of As [36,37]. Given the similarity among members of the OATP family, it is plausible that one or more intestinal isoforms of OATP are capable of mediating the transport of As in the intestine in a manner similar to that of OATP-C in the liver.

Another possible mechanism for the intestinal uptake of inorganic As is the divalent metal transporter (DMT1/DCT1/Nramp2). This carrier is localized in the luminal plasma membrane of enterocytes [38] and has been implicated in the uptake of other toxic metals, such as cadmium (Cd^{2+}) [39,40]. Recent studies in mice showed that the decreased expression of DMT1 in the intestine corresponded to a decrease in the content of As in the intestine, serum, and liver [41]. Although these studies did not test directly the ability of DMT1 to transport As, they provide the strongest support, to date, for the hypothesis that As is a transportable substrate of DMT1.

10.2.3 Renal Transport of As

Studies of As transport within the kidney suggest that As is transported actively across the apical and basolateral plasma membranes of certain renal tubular epithelial cells. It has been suggested that the multidrug resistance-associated protein 1 (MRP1) is involved in this process. *In vivo* experiments using BALB/c and C57BL/6 mice showed that the renal expression of MRP1 increased after exposure of mice to sodium arsenite (NaAs) [42]. MRP1 is a multispecific transporter that is localized in the basolateral membrane of proximal tubular cells [43], and thus may play a role in the transport of As from within the intracellular compartment of cells into the blood. Administration of MK-571 (an inhibitor of MRP1) to mice resulted in a significant increase in NaAs-induced renal injury. Kimura and colleagues [42] concluded that MRP1 is critical for both the renal elimination of As and the prevention of renal injury induced by acute exposure to As. The molecular mechanisms by which MRP1 expression was enhanced following exposure of mice to As were examined later using BALB/c mice [42]. It was shown that exposure to NaAs activates the expression of interferon-γ [44]. Kimura and colleagues [44] hypothesized that the induction of interferon-γ protects renal epithelial cells against the toxic effects of As by inducing the expression of MRP1, which can then transport As out of tubular epithelial cells into the peritubular blood.

The actual species of As that is transported by MRP1 has been examined in several studies. It has been shown that As can form bonds with cysteine (Cys) [45], including the Cys moiety of glutathione (GSH). When three molecules of GSH bond to one As atom, arsenic triglutathione ($As(GS)_3$) is formed [46,47]. Since GSH is present in blood at micromolar concentrations, it is likely that $As(GS)_3$ is formed and present in blood. Given this, it is not surprising that $As(GS)_3$ and methylarsenic diglutathione ($CH_3As(GS)_2$) are present in the urine of mice exposed to As [48]. These data correlate with a recent study suggesting that $As(GS)_3$, $CH_3As(SG)_2$, and dimethylarsinous glutathione ($(CH_3)_2As(SG)$) are key species of As delivered to the kidney for excretion in urine [49]. Interestingly, it has been suggested that the speciation of As is dependent on the route of exposure, dose, and form of As to which one is exposed [50]. Intracellular concentrations of GSH are also in the micromolar range; thus $As(GS)_3$ may also be formed within cells and subsequently transported across the basolateral membrane by MRP1 into the peritubular blood. Indeed, Leslie and colleagues [51] demonstrated that $As(GS)_3$ is a transportable substrate of MRP1.

In addition to MRP1, another isoform of the MRP family, MRP2, may play a role in the export of inorganic As from within proximal tubular cells. MRP2 is localized in the luminal plasma membrane of proximal tubular cells [52] and has been shown to mediate the export of a broad range of substrates [53]. Based on *in vivo* experiments in killifish showing that upregulation of MRP2 expression increased the tolerance of proximal tubules to As, it was suggested that MRP2 was capable of removing As from proximal tubular cells [54]. Experiments in cultured human embryonic kidney cells (HEK-293) demonstrated that enhanced expression of MRP2 reduced the cellular accumulation of As [55]. These data suggest that MRP2 is capable of removing As from the intracellular compartment of proximal tubular cells. P-glycoprotein, which is encoded by the multidrug resistance 1 (*mdr1*) gene, may also play a role in the export of As complexes [56]. Indeed, *mdr1* knockout mice are more susceptible to arsenic toxicity [57]. Findings from studies of HEK-293 cells suggested that aquaporin 3 (AQP3) plays a role in the uptake of As into renal tubular cells [55]. AQP3 is localized in the basolateral plasma membrane of cortical, outer medullary, and inner medullary collecting ducts [58].

Less is known about the mechanisms by which As is taken up by renal tubular cells. Recent studies in cultured porcine proximal tubular cells (LLC-PK1) suggest that OATP-C and MRP2 work in tandem to transport organic anions from the basolateral to the apical membrane of proximal tubular cells [59]. These carriers have been shown individually to transport As, and thus may work together to mediate the transepithelial transport and eventual urinary excretion of this metal. Other members of the OATP family are also expressed in proximal tubular cells [60] and may participate in As transport, although this theory has yet to be tested.

10.2.4 Hepatic Transport of As

The ability of hepatocytes to export As has been the focus of numerous studies. *In vivo* studies in rats indicate that the urinary and biliary excretion of As is dependent on the intracellular level of GSH [61–63]. When GSH levels were depleted, the amount of As excreted in urine and bile was reduced significantly [61]. Furthermore, when the activity of γ-glutamyltranspeptidase was inhibited, thereby preventing the breakdown of GSH and its metabolites, the excretion of As remained unchanged. This lack of effect suggests that As is not transported as a conjugate of one of the products formed by the catalysis of GSH (e.g., Cys or cysteinylglycine), but rather it is transported as an As–GSH complex [61]. Based on these data, we can postulate that one or more transporters capable of mediating the transport of GSH and GSH complexes are involved in the transport of As.

Numerous studies have provided strong evidence for a role of the MRPs in the transport of As–GSH complexes [64–68]. Hepatocellular carcinoma (HeLa) cells overexpressing MRP1 were found to be more resistant to the effects of As than corresponding control cells [64]. Since MRP1 is an export protein and is localized in the sinusoidal membrane of hepatocytes [69], it is therefore possible that MRP1 may mediate the transport of As out of hepatocytes into blood. In support of this theory, studies in the soil nematode *Caenorhabditis elegans* showed that inactivation of the *mrp1* gene increases the sensitivity of these organisms to As [70]. Additional studies in a line of lung cancer (SW-1573/S1) cells transfected stably with MRP1 provided strong evidence that arsenic and GSH are cotransported by this carrier, possibly as an As–GSH complex [68]. Indeed, additional studies in HeLa cells overexpressing MRP1 suggest that As, as $As(SG)_3$, is a transportable substrate of this carrier [51]. Other studies showed that rat hepatocytes cultured in arsenic-containing media for 18 weeks became tolerant to As and this tolerance was due, in part, to an increase in the expression of MRP1 [67].

The elimination of As–GSH compounds at the canalicular membrane of hepatocytes appears to involve MRP2. *In vivo* studies in rats showed that hepatocellular necrosis following exposure to As was worse when MRP2 expression was lowered, while overexpression of MRP2 resulted in increased excretion of As [66]. In addition, long-term exposure of cultured hepatocytes to As resulted in an increase in the expression of MRP2 [71]. Indeed, experiments using Wistar and MRP2-deficient rats (TR^-) with cannulated bile ducts demonstrated that MRP2 is involved in the transport of As–GSH complexes across the biliary canaliculus of hepatocytes [65].

Since MRP1 and MRP2 have been shown to transport a growing number of various substrates, the role of molecular mimicry in the MRP-mediated transport of As is unclear. There are two possible explanations for the transport of As by MRPs. First, As complexes may act as mimics of one or more substrates that are transported normally by the MRPs. Alternatively, as As contamination in the environment has become more prevalent, MRPs may have adapted to the changing needs of the cells in which they are localized, consequently broadening their substrate specificities.

The mechanisms by which As is taken up by hepatocytes remain unclear. One possible candidate for this transport is OATP-C. This carrier is localized in the sinusoidal membrane of hepatocytes [72], and has been implicated in the uptake of As by renal tubular cells [37]. A role for OATP in the hepatic uptake of As has not yet been examined.

10.3 CADMIUM

Cadmium (Cd) is a group IIB element that is found naturally in the environment. The ionic form of Cd (Cd^{2+}) usually combines with oxygen (cadmium oxide, CdO), chlorine (cadmium chloride, $CdCl_2$), or sulfur (cadmium sulfate, $CdSO_4$). An estimated 30,000 tons of Cd are released into the atmosphere each year; approximately 4000–13,000 tons come from human activities. Cd^{2+} in the environment does not break down; therefore, there is increasing risk of human exposure to this metal [73].

Human exposure to Cd^{2+} occurs primarily through the ingestion of contaminated food or water and/or the inhalation of cigarette smoke. Major sources of dietary Cd^{2+} include fish, liver, grains, leafy vegetables, potatoes, and other root vegetables. On average, a person in the United States will consume approximately 30 μg of Cd^{2+} per day, with 1–3 μg of that absorbed by the gastrointestinal tract. Of a more serious nature is the Cd^{2+} inhaled from cigarette smoke, most of which is in the form of CdO [73,74]. Each cigarette contains approximately 1–2 μg of Cd^{2+}. Forty to sixty percent of inhaled Cd^{2+} is absorbed within the lungs directly into the systemic circulation [73,75,76]. Exposure to Cd^{2+} on a chronic basis can cause adverse effects in numerous tissues and organs, including the kidneys, liver, lung, pancreas, testis, placenta, and bone [5,72,77–94].

Cd may act as a functional mimic of certain biomolecules such as estrogen or may act as a structural mimic in order to gain access to target cells. One possible route of Cd^{2+} entry into cells is via membrane transporters involved normally in the uptake of nutritive metals, such as calcium, iron, and zinc. This uptake has been proposed recently to occur through a mechanism of ionic mimicry [5], whereby Cd^{2+} mimics the divalent cationic species of one or more of these nutritive metals at the binding site of one or more carrier proteins and/or channels that transport these metals.

There is also experimental evidence supporting the hypothesis that Cd^{2+} can form linear II coordinate covalent complexes with certain sulfhydryl-containing biomolecules, such as GSH, Cys, or Hcy [95,96]. These complexes may serve or behave as molecular mimics of endogenous amino acids, oligopeptides, organic anions, or organic cations at the site of membrane transporters of these types of substrates.

Cd^{2+} may also gain access to some epithelial cells via receptor-mediated endocytosis of a Cd^{2+}–protein complex (e.g., Cd^{2+}–metallothionein (MT), Cd^{2+}–albumin). Each of these potential mechanisms of uptake will be discussed in relation to individual organs in the following sections.

10.3.1 Endocrine Disruption and Cd^{2+}

In the breast, Cd^{2+} has been characterized as an endocrine-disrupting chemical (EDC). An EDC is defined as a natural or synthetic agent that can mimic the action of an endogenous compound [97]. Indeed, several studies have shown that Cd^{2+} acts as a mimic of estrogen at the site of the estrogen receptor [98–100]. Studies utilizing a breast cancer cell line (MCF-7) showed that exposure of cells to either $CdCl_2$ or estradiol elicited the same physiological effect. These findings suggest that Cd^{2+} and estradiol can independently activate the estrogen receptor [98]. Interestingly, exposure of cells to $CdCl_2$ also increased the transcription and translation of DNA and mRNA, respectively, encoding the progesterone receptor and increased the growth rate of the exposed cells [98]. Cd^{2+} appears to activate the estrogen receptor by interacting with cysteine residues in the hormone-binding domain of the receptor [99–101]. It is important to note that since Cd^{2+} does not normally exist in an unbound state in physiological solutions [5], studies in which cells were treated directly with $CdCl_2$ may not be accurate representations of the physiological environment found *in vivo*. However, subsequent *in vivo* studies in rats provided strong support for the previous *in vitro* studies by demonstrating that Cd^{2+} behaves as a functional mimic of estrogen in the uterus and mammary gland [102]. Intraperitoneal injection of rats with $CdCl_2$ resulted in a 1.9-fold increase in uterine weight and an increase in milk protein synthesis in the mammary gland. Both these effects are consistent with normal estrogen receptor activation. The addition of an antiestrogen (estrogen antagonist) blocked the effects of Cd^{2+} on these tissues, suggesting that Cd^{2+} and estradiol utilize the same pathway [102]. Collectively, these data indicate that Cd^{2+} may act as a functional mimic of estrogen at the site of the estrogen receptor. The ability of Cd^{2+} to act as a mimic of estrogen is important clinically in that overstimulation of the estrogen receptor may lead to estrogen-related diseases such as breast cancer [103–105].

Interestingly, Cd^{2+} has been shown to activate the androgen receptor [106], possibly via a mechanism similar to that of the estrogen receptor [107].

10.3.2 Intestinal Transport of Cd^{2+}

The majority of ingested Cd^{2+} is absorbed by the duodenum and the proximal jejunum [108]. Interestingly, Cd^{2+} (1.71 Å) and Fe^{2+} (1.72 Å) ions have nearly identical ionic radii. Considering this similarity and the fact that the duodenum is a major site of Fe^{2+} [109–112] and Cd^{2+} absorption, it is logical to postulate that these two cations may be taken up by enterocytes, and perhaps other target cells, via one or more of the same transport mechanisms. Support for this theory comes from *in vivo* experiments in rats showing that Cd^{2+} interferes with the intestinal absorption of Fe^{2+} [113–117].

In recent years, DMT1 has been identified in the luminal plasma membrane of enterocytes as a major mechanism for the intestinal uptake of iron (Fe^{2+}) [38,118–123]. It is clear that dietary Fe^{2+} status has a significant effect on the expression of DMT1 [124–129] and the uptake of Cd^{2+} in the intestine [130–134]. Given this finding, it was hypothesized that DMT1 participates in the uptake of Cd^{2+} [5,40,135,136]. However, DMT1 has been characterized as an ion transporter [120]. Cd^{2+} is not found normally as an unbound cation anywhere in the body. Yet, it is important to note that Cd^{2+} can dissociate from its bound ligands under certain conditions [137]. Considering this, we can hypothesize that when Cd^{2+} is presented to a stronger electrophilic ligand, it is released from the carrier ligand to bind to DMT1 or other transporters at the plasma membrane of the enterocyte. This release could theoretically provide an explanation for Cd^{2+} being transported across the luminal plasma membrane as a cationic species. Indeed, numerous studies have provided evidence for the transport of Cd^{2+} by DMT1 [39,40,136,138–141].

Tallkvist and colleagues [141] carried out experiments in an immortalized line of colonic (Caco-2) cells in which exposure to Fe^{2+} reduced the expression of DMT1. Interestingly, the reduction in DMT1 expression corresponded to a reduction in the uptake of Cd^{2+}. These data suggest that the expression of DMT1 is a determining factor in the uptake of Cd^{2+}. These data were confirmed by *in vivo* studies in rats [40,142]. Rats were fed a diet deficient or supplemented with Fe^{2+} for four or eight weeks, following which they were given an oral dose of $CdCl_2$. The mRNA levels of DMT1 and the content of Cd^{2+} in the small intestine of the animals fed the Fe^{2+}-deficient diet were significantly greater than those in animals that received the Fe^{2+}-supplemented diet. These data provide strong evidence supporting the hypothesis that DMT1 mediates the intestinal uptake of Cd^{2+}. *In vitro* studies in Caco-2 cells provided additional support for the hypothesis that Cd^{2+} is a substrate of DMT1 [143]. Experiments utilizing Caco-2 cells in which the expression of DMT1 was "knocked down" showed a decrease in the uptake of Fe^{2+}, lead (Pb^{2+}), and Cd^{2+}. The most direct evidence for a role of DMT1 in the transport of Cd^{2+} comes from experiments using oocytes from *Xenopus laevis* oocytes [39,120]. In oocytes expressing DMT1, the presence of Cd^{2+} in the extracellular medium was shown to invoke a large inward current, indicating that Cd^{2+} is a substrate of DMT1 [120]. Additional studies also utilizing oocytes expressing DMT1 showed that the inward flux of Cd^{2+} is a saturable process and was significantly greater than that in control oocytes [39]. These data indicate that DMT1 plays an important role in the uptake of Cd^{2+} in enterocytes. Although DMT1 is likely not the sole mechanism for the uptake of Cd^{2+} by the intestine [144], it appears that it plays a major role in the intestinal uptake of this metal.

Interestingly, studies carried out in human epithelial cells from intestinal crypts (HIEC) suggest that DMT1 does not participate in the uptake of Cd^{2+} in crypt cells [145]. Rather, the uptake of Cd^{2+} by these cells appears to occur through a zinc (Zn^{2+})-activated pathway [146]. This pathway, however, is not thought to contribute significantly to the overall intestinal absorption of Cd^{2+} [146].

A Zn^{2+} transporter has also been implicated in the uptake of Cd^{2+} in surface enterocytes. Studies using apical membrane vesicles from pig small intestine indicate that Zn^{2+} and Cd^{2+} share a binding site on a transporter that does not take up Ca^{2+} [147,148]. It has also been shown that Cd^{2+} competes with Zn^{2+} for a transporter that is independent of DMT1 [136]. It has been hypothesized that Cd^{2+} may enter surface enterocytes by acting as a mimic of Zn^{2+} at the site of a Zn^{2+} transporter, such as the human, zinc-regulated zinc transporter 1 (hZTL1). This transporter, which is present in the luminal plasma membrane of enterocytes, is responsible for the inward transport of Zn^{2+} in these cells [149]. In addition, three other Zn^{2+} transporters, ZIP8, rZIP10, and ZIP14, have been localized recently in the intestine and have been implicated in the uptake of Cd^{2+} in other tissues [150–155]. The role of these carriers in the intestinal uptake of Cd^{2+} has not yet been examined.

Calcium (Ca^{2+}) channels present on the luminal plasma membrane may also play a role in the uptake of Cd^{2+} by enterocytes. Since the atomic radius of Cd^{2+} (1.71 Å) is significantly smaller than that of Ca^{2+} (2.23 Å) [156], it seems possible that Cd^{2+} can act as a mimic of Ca^{2+} at the site of Ca^{2+} channels in order to gain entry into hepatocytes. We postulate that Cd^{2+} may serve or behave as a mimic of Ca^{2+} at the site of one or more Ca^{2+} transporters as a means to gain entry into enterocytes. The ability of Cd^{2+} to compete with Ca^{2+} at the site of membrane Ca^{2+} channels has been well established in numerous *in vitro* models [157–161]. In addition, *in vivo* studies in rats demonstrated that in conditions where dietary Ca^{2+} is restricted chronically, the intestinal uptake of orally administered Cd^{2+} is increased greatly [162,163]. Min and colleagues [164] proposed that Cd^{2+} uptake in the intestine may be mediated by the intestinal calcium transporter 1 (CaT1). This carrier is present on the apical plasma membrane of enterocytes [165] and may play a role in the absorption of ingested Cd^{2+}.

The actual process by which ionic Cd^{2+} is transported into intestinal epithelial cells has been postulated to be a two-step process [166–169]. The first step is thought to involve binding of Cd^{2+} to the plasma membrane. This binding was susceptible to chelators, such as EDTA, but was insensitive to changes in temperature. The second step was shown to be sensitive to temperature, but not to chelators, and likely represented the actual entry of Cd^{2+} across the luminal plasma membrane into the intracellular compartment of the enterocytes. This second step is thought to be a carrier-mediated process.

In addition to transport of Cd^{2+} as an ion, this metal may be taken up by enterocytes as a conjugate of a thiol-containing molecule. Cd^{2+} readily forms linear II coordinate covalent complexes with thiol-containing biomolecules such as Cys (Cys-*S*-Cd-*S*-Cys) or GSH (G-*S*-Cd-*S*-G) [95]. Since ingested food contains an abundance of such molecules, it is likely that some of the ingested Cd^{2+} is presented to the plasma membrane of enterocytes as a complex of one or more of these molecules. Although a certain amount of ligand exchange occurs with Cd^{2+} ions (allowing the transport of the cationic species of Cd^{2+}), it is highly probable that a substantial fraction of the Cd^{2+} taken up at the luminal plasma membrane of enterocytes is, at least initially, bound to thiol-containing ligand. The transport of *S*-conjugates of Cd^{2+} has not been studied directly in the intestine; however, we hypothesize that since the molecules Cys-*S*-Cd-*S*-Cys and G-*S*-Cd-*S*-G are structurally similar to cystine and glutathione disulfide (G-*S*-*S*-G), which are found endogenously, these complexes may serve as mimics of cystine and G-*S*-*S*-G at the site of membrane transporters for these molecules. It should be noted that numerous studies have shown that mercuric conjugates of GSH are transported out of the liver into the biliary tree, where they are broken down to form mercuric conjugates of Cys before entering the duodenum [170–172]. Experimental evidence suggests that Cd conjugates of GSH are likely handled in the same manner. Indeed, the findings of Cherian and Vostal [173] indicate that the hepatobiliary transport of Cd^{2+} involves a Cd^{2+} conjugate of GSH. Based on these findings and the fact that duodenal contents are rich in amino acids and peptides, it is reasonable to suggest that a significant fraction of absorbed Cd^{2+} is taken up as a conjugate of these ligands by one or more amino acid and/or peptide transporters. To our knowledge, the involvement of these amino acid carriers in the transport of Cd^{2+}–thiol complexes has not been investigated.

Endocytosis of Cd–protein complexes may also be a route of intestinal uptake of Cd^{2+}. Cd^{2+} can bind to the metal-binding protein MT in the intestinal lumen [137,174,175]. These Cd–MT complexes are taken up as intact complexes by enterocytes [176], most likely by an endocytic pathway. It appears that the accumulation of Cd^{2+} is lower when animals are administered Cd–MT than when animals are exposed to $CdCl_2$ [176].

Owing to the similarities in the uptake of Fe^{2+} and Cd^{2+} at the luminal plasma membrane of enterocytes, we postulate that the basolateral transport of these two metals may involve a common mechanism. One candidate is the metal protein transporter 1 (MTP1, Ferroportin 1, Ireg1). This carrier has been identified in the basolateral membrane of enterocytes, and is thought to mediate the export of Fe^{2+} across the basolateral plasma membrane into portal circulation [177–180]. Recent studies in rats and mice indicate that an increase in the expression of MTP1 is associated with an increase in the absorption of Cd^{2+} [142,181]. These studies suggest that the transport of Cd^{2+} out of cells at the basolateral membrane occurs via MTP1.

An additional mechanism for the exit of Cd^{2+} across the basolateral membrane of enterocytes is a Zn^{2+} transporter. Recent studies have provided strong evidence supporting the theory that Zn^{2+} and Cd^{2+} can utilize a common transporter [136,182–184]. One possible mechanism for the basolateral efflux of Cd^{2+} is the Zn^{2+} transporter, ZnT [185,186]. This carrier is thought to mediate the basolateral export of Zn^{2+} from within enterocytes into the circulation [187]. Owing to the basolateral localization of this transporter and the association between Zn^{2+} and Cd^{2+} transport, we hypothesize that this carrier may play a role in the outward flux of Cd^{2+} across the basolateral membrane of enterocytes. Multiple isoforms of ZnT (1, 4, 5, 6, and 7) have been localized in the intestinal tract and may play a role in the absorption of Cd^{2+}; however, the ability of the individual carriers to transport Cd^{2+} remains untested [185,187].

Cd^{2+} may also utilize Ca-ATPase as a mechanism for basolateral export from enterocytes. *In vitro* evidence suggests that Cd^{2+} is a competitive inhibitor of Ca^{2+} at the site of the intestinal Ca-ATPase present on the basolateral plasma membrane of enterocytes [188,189]. It has been suggested that Cd^{2+} utilizes this carrier for its transport out of the enterocyte into portal and systemic circulation. Since Cd^{2+} has been shown to act as an ionic mimic of Ca^{2+} in other organs and cellular models [157–161], it is reasonable to suggest that it may also do so at the site of the intestinal Ca-ATPase.

10.3.3 Renal Transport of Cd^{2+}

The kidney is one of the primary organs affected adversely following chronic exposure to Cd^{2+} [73,190]. The majority of Cd^{2+} entering the nephron is absorbed by the epithelial cells lining the proximal tubule. Studies in rats have shown that as much as 95% of the administered dose of Cd^{2+} is taken up by proximal tubular cells [191,192]. The mechanisms by which Cd^{2+} is taken up by these cells, however, remain unclear. Cd^{2+} has a strong affinity for thiol-containing biomolecules, such as GSH and Cys; therefore, it is likely that Cd^{2+} is presented to the luminal membrane of proximal tubular cells as G-*S*-Cd-*S*-G or Cys-*S*-Cd-*S*-Cys. G-*S*-Cd-*S*-G conjugates are likely not taken up as intake complexes. The brush-border enzymes γ-glutamyltransferase and cysteinylglycinase, which are responsible for the catabolism of intraluminal GSH, are abundant in the luminal plasma membrane of proximal tubular cells. Cys-*S*-Cd-*S*-Cys, which is the primary product formed by the actions of these enzymes on G-*S*-Cd-*S*-G, is the most likely species of Cd^{2+} that is taken up at the luminal plasma membrane of proximal tubular epithelial cells. Indeed, *in vivo* microperfusion of rat proximal tubules demonstrated that tubular absorption of Cd^{2+} increased by 82% when Cd^{2+} was coperfused with Cys [193]. Additional *in vivo* studies in rats have also shown that when $CdCl_2$ is coadministered subcutaneously with excess Cys, there is greater accumulation of Cd^{2+} in the epithelial cells lining the proximal tubule [194,195]. Similarly, other *in vivo* studies in rats showed that when Cd^{2+} and GSH or Cys were injected simultaneously, the renal uptake of Cd^{2+} increased significantly [196]. These studies also demonstrated that there is at least one luminal and one basolateral mechanism involved in the renal uptake of Cd^{2+} [196]. Specific mechanisms for the uptake of Cys-*S*-Cd-*S*-Cys by proximal tubular cells or any other type of epithelial cell have not been characterized directly. However, since both Cd^{2+} and Hg^{2+} (which are both group IIB metals) have similar properties, similar mechanisms may be involved in their uptake by target cells. Cys-*S*-Hg-*S*-Cys is a structural mimic of cystine, and is transported across the luminal plasma membrane of proximal tubular epithelial cells by the cystine transporter, system $b^{0,+}$ [197]. Given that Cd^{2+} appears to form complexes with Cys, which are similar structurally to Cys-*S*-Hg-*S*-Cys, we can hypothesize that Cys-*S*-Cd-*S*-Cys and Cys-*S*-Hg-*S*-Cys may be taken up by the same or similar mechanism. Indeed, studies utilizing isolated perfused proximal tubules indicate that Cd^{2+}, as a conjugate of Cys, is taken up readily at the apical membrane of proximal tubular cells [198]. This uptake was inhibited by the addition of cystine or lysine, suggesting that at least part of this uptake occurs via an amino acid transporter, such as system $b^{0,+}$ [198].

Another possible route of Cd^{2+} entry into proximal tubular cells is receptor-mediated endocytosis [5,199,200]. Within the body, Cd^{2+} may exist as a Cd–MT complex, which may originate from the liver [201] or intestinal tract [202,203]. Cd–MT is filtered freely at the glomerulus and may be taken up by the epithelial cells of the proximal tubule via an endocytotic mechanism [202,204–208]. Indeed, the cells of the proximal convoluted tubule are the primary sites affected adversely by Cd–MT [94,192,193,200,204,205,209,210]. This pathway is likely the major route of entry for Cd^{2+} into proximal tubular cells [211]. Megalin, a receptor involved in the absorption of filtered proteins [212], is thought to be involved in the proximal tubular absorption of Cd–MT complexes. Indeed, MT has been shown to bind to megalin [213]. Recent studies using cultured cells from S1 proximal tubules suggest that megalin is involved in the endocytosis of Cd–MT [214]. In addition, the GTPase ADP-ribosylation factor, ARF6, which regulates endocytosis and vesicular trafficking, appears to be involved in the uptake of Cd–MT [215]. Once internalized, it is thought that Cd–MT is delivered to endosomes or lysosomes, where Cd^{2+} dissociates from MT and is subsequently transported into the cytosol via the lysosomal isoform of DMT1 [216].

Cd^{2+} may also be taken up into proximal tubular cells in a cationic form, as evidenced by studies in isolated perfused proximal tubules from rabbits [217] and cultured proximal tubular cells [218–223]. As these experiments utilized $CdCl_2$, the data obtained may not accurately represent physiological processes that occur *in vivo*. This pathway probably represents a small fraction of the total uptake of Cd^{2+}.

A Zn^{2+} transporter has also been implicated in the proximal tubular uptake of Cd^{2+}. *In vivo* studies in rats demonstrated that administration of Zn^{2+} protects the kidney from the nephrotoxic effects that occur following exposure to Cd^{2+} [224,225]. Data from micropuncture studies in rats showed that the addition of Zn^{2+} reduced the accumulation of Cd^{2+} in proximal tubules. It was hypothesized that the Zn^{2+} transporters ZIP1, 2, and 3 may play a role in the uptake of Cd^{2+} into proximal tubules [191]. These transporters are localized in the apical plasma membrane of proximal tubular cells and their activity can be blocked by the presence of Cd^{2+} [226]. Recently, another Zn^{2+} transporter, ZIP8 (SLC39A8), has been identified as a carrier of Cd^{2+} [155]. Overexpression of this carrier, which is localized in the apical plasma membrane of proximal tubular cells, was found to enhance Cd^{2+}-induced damage to proximal tubules [155]. A second Zn^{2+} transporter has also been implicated recently in the uptake of Cd^{2+} at the luminal membrane of proximal tubular cells. Girijashanker and colleagues [151] have identified ZIP14 (SLC39A14) (splice variants ZIP14A and ZIP14B) in the apical membrane of Madin-Darby canine kidney (MDCK) cells and have implicated it in the transport of Zn^{2+}, Cd^{2+}, and manganese (Mn^{2+}). In addition to ZIP8 and ZIP14, rZIP10 (SLC39a10) has been identified as a Zn^{2+} transporter in the luminal plasma membrane of proximal tubular cells. When overexpressed in LLC-PK_1 cells, the ability of rZIP10 to take up Zn^{2+} was inhibited by Cd^{2+} [153]. Although the uptake of Cd^{2+} by these carriers has not been characterized directly, inhibition studies provide strong evidence for a role of these transporters in the uptake of Cd^{2+} by proximal tubular cells.

Another possible mechanism for the uptake of Cd^{2+} by proximal tubular cells is a Mn^{2+} transporter. Studies in various lines of cultured cells, including osteoblasts, proximal and distal convoluted tubular cells, HeLa cells, PC12 cells, and Caco-2 cells, indicate that a high-affinity Mn^{2+} transporter, distinct from DMT1, is involved in the uptake of Cd^{2+} [227–229]. Interestingly, the ability of Mn^{2+} to protect cells against toxicological effects of Cd^{2+} was most pronounced in cells derived from the proximal convoluted tubule [228], which is a major site of Cd^{2+} accumulation.

In other organs, DMT1 has been identified as a mechanism for the uptake of Cd^{2+} [135]. Since the uptake of Cd^{2+} by the proximal tubule can be blocked by Mn^{2+} and Zn^{2+}, which are both substrates of DMT1, it is logical to suggest that DMT1 may play a role in the uptake of Cd^{2+} into proximal tubular cells. Within proximal tubular cells, DMT1 has been localized in the apical plasma membrane [230] and cytoplasm [231], specifically in late endosomes and lysosomes [216,232]. It has been shown recently that alternative splicing of DMT1 regulates the subcellular localization of this carrier [233]. Both isoforms of DMT1 appear to be present in proximal tubular cells, with one localized in the apical plasma membrane and the other present in late endosomes and lysosomes [216,233,234]. It has been hypothesized that this pattern of expression for DMT1 is important for the intracellular handling of Cd–MT complexes. Following uptake by receptor-mediated endocytosis and subsequent delivery to endosomes and lysosomes, it appears that Cd^{2+} is dissociated from MT and transported into the cytoplasm via DMT1 [199,216].

Not all of the Cd^{2+} present in the luminal compartment of the nephron is taken up by the proximal nephron; a fraction is absorbed by the distal nephron. However, the distal tubular uptake of Cd^{2+} probably represents a small percentage of the total amount of absorbed Cd^{2+}. Indeed, immunolocalization experiments demonstrate that DMT1 is present in the luminal plasma membrane of epithelial cells lining the ascending thick limb of the loop of Henle, the distal convoluted tubule, and the principal cells of the cortical collecting duct [231]. Additional support for this hypothesis comes in part from experiments demonstrating that Cd^{2+} is a potent inhibitor of Fe^{2+} uptake in cells derived from the distal nephron [159,235]. Using an immortalized line of mouse distal convoluted tubular cells, Friedman and Gesek [159] demonstrated that Fe^{2+} is able to significantly inhibit the uptake of Cd^{2+}. Olivi and colleagues [235] used MDCK cells to show that Cd^{2+} and Fe^{2+} are able to competitively inhibit the uptake of each other. In addition, these investigators showed that the uptake of Cd^{2+} was greater in renal fibroblasts (HEK-293) that overexpress DMT1 than in corresponding wild-type cells. Collectively, these findings suggest that Cd^{2+} may act as an ionic mimic of Fe^{2+} at the site of the luminal transporter, DMT1, in epithelial cells of the distal nephron and collecting duct.

Ca^{2+} channels may also play a role in the absorption of Cd^{2+} into epithelial cells lining the distal tubule. In other organs such as liver and intestine, Cd^{2+} appears capable of utilizing Ca^{2+} channels to gain entry into cells [157–161]. Therefore, it is reasonable to suggest that the Ca^{2+} channels present in the distal nephron may serve as a route of entry for Cd^{2+}. Indeed, *in vitro* studies in an immortalized line of mouse distal tubular cells have implicated Ca^{2+} channels in the transport of Cd^{2+} [159]. Exposure of these cells to parathyroid hormone, which promotes Ca^{2+} absorption in the distal nephron [236,237], markedly increased the cellular uptake of Cd^{2+}. Furthermore, the observed Cd^{2+} uptake was inhibited by the Ca^{2+} channel antagonist nifedipine and was enhanced by the Ca^{2+} channel agonist BAY K 8644 [159]. These data suggest that Cd^{2+} may enter distal tubular epithelial cells via mechanisms of ionic mimicry whereby Cd^{2+} mimics Ca^{2+} at the site of Ca^{2+} channels. In addition, studies using MT knockout cells suggested that the T-type calcium channel Cacnα_{1G} is responsible for the uptake of Cd^{2+} [238]. This subunit is localized in the apical plasma membrane of cells of epithelial cells along the distal nephron [239]. Additional *in vivo* studies from rats also found that the presence of Cd^{2+} inhibited the absorption of Ca^{2+} in the distal nephron [191]. The authors of this study suggested that the epithelial Ca^{2+} channel (ECaC) may be involved in the uptake of Cd^{2+} into epithelial cells of the distal nephron [191]. Even so, a previous study suggested that Cd^{2+} does not compete directly with Ca^{2+}, but rather binds to a high-affinity inhibitory site on the transporter [240]. Clearly, additional studies are necessary to elucidate fully this route of entry.

The basolateral membrane of proximal tubular cells also serves as a point of entry for Cd^{2+}. A plausible mechanism for this transport may be Ca^{2+} channels. When LLC-PK_1 cells were grown on permeable supports and their basolateral surfaces were exposed to Cd^{2+} in the presence of high concentrations of Ca^{2+}, the accumulation of Cd^{2+} was reduced significantly [241]. It is also thought that Cd–thiol conjugates present in the blood may be transported across the basolateral membrane of proximal tubular epithelial cells. Although specific mechanisms for this route of uptake have not yet been identified, potential candidates for this transport are members of the organic anion transporter (OAT) family. OAT1 and OAT3 are present in the basolateral membrane of proximal tubular cells [242–244] and have been shown to mediate the transport of Hg^{2+} as a conjugate of Cys, Hcy, or NAC [245–247]. It is believed that at least some of the Cd^{2+} in blood, especially after acute exposures, is conjugated to GSH, Cys, and/or NAC. These Cd^{2+} complexes are structurally similar to those of Hg^{2+}; therefore, we hypothesize that OAT1 and OAT3 play a role in the uptake of thiol conjugates of Cd^{2+}. Interestingly, Cd^{2+} absorbed by proximal tubular cells may be transported across the basolateral membrane from within the cells back into circulation. It has been proposed that a Zn^{2+} transporter, possibly Znt1 [248] or ZIP5 [249], on the basolateral plasma membrane plays a role in this process [250].

Cd^{2+} may also be secreted from within renal tubular cells into the tubular lumen for elimination in the urine. *In vivo* studies in mice exposed to Cd^{2+} demonstrated that treatment with various chelators resulted in extraction of Cd^{2+} (as measured by the appearance of Cd^{2+} in urine and the disappearance of Cd^{2+} from the kidney) [251]. Calcium disodium diethylenetriaminepentaacetic acid (DTPA), disodium ethylenediaminetetraacetic acid (EDTA), 2,3-dimercaptosuccinic acid (DMSA), pencillamine (PEN), and *N*-acetylcysteine (NAC) were all effective chelators of Cd^{2+} [251,252]. The specific mechanisms by which these chelators mediated the extraction of Cd^{2+} from renal tissue were not determined; however, possible candidates involved in this extraction are MRP2 and MRP4. In the kidneys, these export proteins are localized exclusively in the apical plasma membrane of proximal tubular cells [52,253]. MRP2 and MRP4 have broad substrate specificities and have been shown to transport large, bulky substrates [254]. Given these findings, it seems likely that these carriers may play a role in the renal excretion of Cd^{2+} complexes.

10.3.4 Hepatic Transport of Cd^{2+}

Upon absorption of Cd^{2+} by the intestine, it is delivered to the liver via portal blood, where it is then taken up avidly by hepatocytes [73]. As much as 60% of a nontoxic dose of Cd^{2+} (5 μmol/kg) has been shown to accumulate in the liver of rats within 1 h after intravenous administration [196]. Although the mechanisms by which Cd^{2+} is taken up across the sinusoidal membrane of hepatocytes remain unclear, several hypotheses to explain this transport have been proposed. One possibility is that Cd^{2+} may mimic other elements or metals at the site of membrane transporters or channels (via some form of ligand exchange reaction) in the sinusoidal membrane of hepatocytes. It is also thought that Cd^{2+} may bind to proteins that are taken up into hepatocytes via receptor-mediated endocytosis.

One membrane transporter that may be involved in the sinusoidal uptake of the cationic form of Cd^{2+} is DMT1, an ion transporter that is involved in the uptake of Fe^{2+} [119,120]. In hepatocytes, DMT1 is localized in the sinusoidal membrane [123]. Although DMT1 has not been implicated directly in the uptake of Cd^{2+} in hepatocytes, competition studies suggest that it is involved in the hepatic uptake of this metal [163]. *In vivo* studies in mice with iron-deficiency anemia showed that the hepatic uptake of Cd^{2+} was significantly greater in these mice than in control animals [163], which is most likely due to an increase in the expression of DMT1. Also, DMT1 has been shown to mediate the uptake of Cd^{2+} in enterocytes [40,136,141] and distal tubular cells [159,235]. Given the role of DMT1 in the uptake of Cd^{2+} in other organs, it is logical to hypothesize that this carrier may also mediate the uptake of Cd^{2+} in hepatocytes. The participation of DMT1 in the hepatic uptake of

Cd^{2+} suggests that this transport involves a mechanism of ionic mimicry, whereby Cd^{2+} mimics Fe^{2+} to gain access to the cytosolic compartment of hepatocytes.

The metal transport protein 1 (MTP1), which has been detected in liver and has a function similar to that of DMT1 [177], may also be involved in the hepatocellular uptake of Cd^{2+}. Recent *in vivo* studies have suggested a role for MTP1 in the uptake of Cd^{2+} [181]; however, no direct evidence supporting this notion was provided.

Up to 30% of the hepatocellular uptake of Cd^{2+} may involve Ca^{2+} channels [161]. *In vitro* studies utilizing primary cultures of rat hepatocytes [157,158,255] and cultured immortalized hepatocytes (WRL-68) [161] indicate that Cd^{2+} may be transported through Ca^{2+} channels. Not surprisingly, treatment of cells with the Ca^{2+} channel antagonists diltiazem or verapamil blocked the uptake of Cd^{2+} into the hepatocytes. Blazka and Shaikh [157] concluded that Cd^{2+} gained entry into these cells via voltage-gated L-type Ca^{2+} channels.

In addition to Ca^{2+} channels, Cd^{2+} may also be taken up into hepatocytes via a Zn^{2+} transporter. *In vitro* studies have shown that the uptake of Cd^{2+} into primary cultures of rat hepatocytes can be inhibited by Zn^{2+} [255]. ZIP8 has been identified recently in the liver and appears to play a role in the uptake of Cd^{2+} in the testes and kidney [150]. This carrier has not been localized to a specific membrane of hepatocytes, but it is likely responsible for some of the Cd^{2+} transport that occurs in these cells [256]. In addition, ZIP14, a Zn^{2+} and Mn^{2+} transporter present in the liver, may also mediate a fraction of Cd^{2+} uptake by hepatocytes [151].

Receptor-mediated and fluid-phase endocytosis account for a large amount of membrane turnover and fluid absorption in hepatocytes [257]. The endocytosis of transferrin and ferritin is one of the best-characterized forms of receptor-mediated endocytosis in hepatocytes and represents a major pathway in the hepatic uptake of Fe^{2+} [258–260]. It is thought that Cd^{2+} can substitute for Fe^{2+} at the site of DMT1; therefore, one would not be surprised to find that this substitution may also occur at the site of one or more Fe^{2+}-binding proteins, such as ferritin. Indeed, ferritin has been shown to bind Cd^{2+} [261,262]. Therefore, it has been hypothesized that Cd–ferritin complexes may be endocytosed by hepatocytes by a means similar to that characterized for Fe^{2+}–ferritin [5]. It is also important to note that albumin, which is the most abundant plasma protein, is a potential carrier of Cd^{2+} in blood, and thus endocytosis of albumin–Cd^{2+} complexes may serve as a route for the hepatic entry of Cd^{2+}. Furthermore, Cd–MT complexes formed in the intestine [137,174,175] may also be taken up by hepatocytes via receptor-mediated endocytosis.

It is important to note that within hepatocytes, much of the absorbed Cd^{2+} becomes bound to MT. When hepatocellular necrosis and/or apoptosis is/are induced, Cd^{2+}–MT complexes are thought to be released into systemic circulation [202]. Some of this Cd^{2+} is delivered to the kidneys, where it is filtered by glomeruli and is then reabsorbed by the epithelial cells of the proximal tubule [202,206,208]. Importantly, Cd^{2+}–MT has been implicated as the primary form of Cd^{2+} that causes the nephropathy induced by Cd^{2+} [94,192,193,200,204,205,209,210,263].

A number of other transporters may also be involved in the transport of Cd^{2+} across the sinusoidal membrane of hepatocytes. These include the organic anion transporting polypeptides OATP1, OATP2, and OATP4 [264–266], the organic cation transporter OCT1 [267], MRP1, and amino acid or peptide transporters that are localized to the sinusoidal plasma membrane. Although these carriers have not been implicated in the transport of metals into or out of hepatocytes, most of them are multispecific transporters and may be able to carry metal ions or conjugated forms of metals.

Various lines of evidence indicate that a fraction of the Cd^{2+} that enters into hepatocytes is secreted into the bile, which is subsequently delivered to the duodenum for excretion in the feces [173,268,269]. Thin-layer chromatographic analyses of bile from $CdCl_2$-treated animals showed that Cd^{2+}, as a conjugate of GSH, is transported from the hepatocyte into bile via a carrier-mediated process [173]. Although a specific carrier has not been identified for this transport, the most likely candidate is MRP2. G-*S*-Cd-*S*-G is structurally similar to G-*S*-*S*-G, which is a substrate of MRP2. Considering this and the fact that MRP2 is localized in the canalicular membrane of hepatocytes [268–271], we hypothesize that G-*S*-Cd-*S*-G may act as a mimic of G-*S*-*S*-G at the site of MRP2.

10.3.5 Testicular Handling of Cd^{2+}

Exposure to Cd^{2+} in males may also have detrimental effects in the testes [272–276]. Acute testicular necrosis and the destruction of seminiferous tubules have been shown to occur in rats and mice after a single injection of Cd^{2+}. Until recently, the mechanisms by which Cd^{2+} enters testicular cells were unclear. It appears that in order to gain access to these cells, Cd^{2+} acts as a mimic of Zn^{2+} at the site of one or more Zn^{2+} transporters [150,277–279]. Experiments using interstitial cells isolated from the testes of rats have demonstrated that Cd^{2+} uptake by cells could be inhibited by Zn^{2+} and *N*-ethylmaleimide, a sulfhydryl-alkylating agent [278]. These findings suggest that the testicular uptake of Cd^{2+} may not only be due to the activity of a Zn^{2+} transporter, but may also be mediated by a transporter that carries thiol conjugates of Cd^{2+} (possibly as G-*S*-Cd-*S*-G or Cys-*S*-Cd-*S*-Cs). The latter may be an amino acid transporter, of which many have been identified in the testes [280]. Additional studies examining the interaction between Zn^{2+} and Cd^{2+} have shown that there are passive and active mechanisms involved in the uptake of Cd^{2+}. Although the passive uptake of Cd^{2+} was not affected by Zn^{2+}, the addition of Zn^{2+} was able to inhibit significantly the active component of Cd^{2+} uptake [279]. Subsequent *in vivo* experiments in mice demonstrated that the uptake of Cd^{2+} in the testes was a saturable process that was competitively inhibited by Zn^{2+}, but not by Ca^{2+} [277]. More recently, the Zn^{2+} transporter ZIP8 (SLC39A8) has been identified in vascular endothelial cells of the testes and is thought to be one of the mechanisms by which Cd^{2+} gains entry into testicular cells [150,256]. Additional support for this theory comes from *in vivo* studies using a line of cells from transgenic mice overexpressing ZIP8. In these mice, testicular necrosis was enhanced (compared to normal mice) following exposure to Cd^{2+}, suggesting that the overexpression of ZIP8 may mediate increased accumulation of Cd^{2+} [155]. A second ZIP transporter, rZIP10, has been identified in the testes and, although uptake of Cd^{2+} by this carrier has not been shown directly, its activity can be blocked by Cd^{2+}, suggesting that this carrier may also be involved in the uptake of Cd^{2+} [153].

A number of other xenobiotic transporters, including DMT1, MRPs, OAT2, and a zinc transporter (Zntl), have been detected in the testes [181,281,282]. Although there are no data directly implicating any of these carriers in the testicular transport of Cd^{2+}, it is possible that one or more of these transport proteins may play a role in the entry of Cd^{2+} into cells of the testis.

10.3.6 Intercellular Mimicry with Cd^{2+}

In vitro exposure of epithelial cells to nontoxic doses of Cd^{2+} has been shown to disrupt junctional complexes between cells [283–287]. Studies using LLC-PK_1 cells demonstrated that following exposure to Cd^{2+}, cells separated from each other but did not lose their attachments to the surface upon which they were cultured [283–285]. In addition, the transepithelial resistance of the monolayer was lowered significantly, suggesting that junctional complexes between cells were disrupted [283–286]. Numerous studies have since identified E-cadherin, a Ca^{2+}-dependent adhesion molecule and component of adhering junctions, as a primary target of Cd^{2+} toxicity [285,288–291]. It appears that Cd^{2+} acts as a structural mimic of Ca^{2+} at a Ca^{2+}-binding site on the E-cadherin molecule [289,291], leading to an inability to form functional junctional complexes. The ability of Cd^{2+} to disrupt junctional complexes between cells may play a role in the ability of Cd^{2+} to induce carcinogenesis in various tissues [292].

10.4 LEAD

Lead (Pb) occurs naturally in the earth's crust and may exist as a lead salt (Pb^{2+}) or as metallic Pb. Occupational exposure to Pb^{2+} can occur as a result of welding, manufacturing Pb-containing batteries, Pb smelting and refining, and the production of pottery. It is estimated that between 0.5 and 1.5 million workers are exposed to Pb^{2+} each year in their workplaces [293]. Children are particularly sensitive to the effects of Pb^{2+}, and are usually exposed by playing with and ingesting soil contaminated with Pb^{2+}. Exposure to paints containing Pb^{2+} occurs either via ingestion or by dermal contact.

Exposure to Pb^{2+} can be detrimental to the nervous, circulatory, skeletal, renal, hematopoietic, and endocrine systems [293–296]. Pb^{2+} poisoning is more common in children than in adults and is characterized by neurological symptoms such as headache, convulsions, ataxia, learning disorders, and hyperactive behavior [295,297,298]. The skeletal system, specifically bone, is a major site for the accumulation of Pb^{2+}. This accumulation appears to alter the ability of bone cells to respond to hormones and may also result in alterations in the plasma levels of 1,25-dihydroxy-vitamin D_3 [295,299]. Nephropathy or renal adenocarcinoma may also result from long-term exposure to Pb^{2+} [295,300–306].

Even though the clinical consequences of Pb^{2+} exposure are serious, little is known about the mechanisms by which Pb^{2+} enters target cells. Based on the information available currently, we can identify several potential routes of Pb^{2+} entry into cells. Associations between the uptake of Pb^{2+} and Fe^{2+}, and Pb^{2+} and Ca^{2+} have been reported in the literature, suggesting that ionic mimicry plays a role in the cellular uptake of Pb^{2+}. In addition, anion exchangers and receptor-mediated endocytosis have also been implicated in the cellular uptake of Pb^{2+}.

An inverse relationship between plasma levels of Fe^{2+} and Pb^{2+} has been shown by several studies. Exposure of Fe^{2+}-deficient rats and humans to Pb^{2+} results in a higher accumulation of Pb^{2+} [307–309]. This accumulation may be due to DMT1. This carrier is upregulated in Fe^{2+}-deficient conditions and may be capable of mediating the uptake of Pb^{2+} into cells.

DMT1 is a likely candidate for the uptake of Pb^{2+} into tissues. Indeed, in cultured human fibroblasts and yeast, overexpression of DMT1 increased the uptake of Pb^{2+} into these cells [310]. This Pb^{2+} transport was inhibited by iron, which is a known substrate of DMT1 [119,120]. In contrast, studies carried out in Caco-2 cells in which the expression of DMT1 had been "knocked down" showed that the uptake of Pb^{2+} into these cells was not altered [143]. Interestingly, in one study in which DMT1 was knocked down, the uptake of Pb^{2+} was not affected [143]. The authors of these studies suggest that, in addition to DMT1, there is another transporter responsible for the uptake of Pb^{2+}. Indeed, studies in rat astroglial cells indicate that there are at least two mechanisms involved in the uptake of Pb^{2+} [311]. At an extracellular pH of 5.5, the predominant transport mechanism appeared to be DMT1, while at pH 7.4, uptake of Pb^{2+} appeared to involve an anion transporter. Additional support for DMT1 as a mechanism for entry of Pb^{2+} comes from studies of metal transport following neuronal injury [312]. After neuronal injury, an increased expression of DMT1 has been observed [313], which in turn leads to an increase in the accumulation of divalent metals, such as Pb^{2+}, in neurodegenerative areas of the brain [312]. Considering these collective findings, it appears that Pb^{2+} is a transportable substrate of DMT1, probably via a mechanism of molecular mimicry where Pb^{2+} mimics Fe^{2+} at the site of this transporter.

Pb^{2+} may cross the basolateral plasma membrane of various types of epithelial cells and enter the blood via MTP1 (Ireg1; Ferroportin) [135]. This carrier is present on the basolateral plasma membrane of epithelial cells in organs involved in iron homeostasis (kidney, intestine, liver, and placenta) [177,179]. Given the role of MTP1 in the transport of Fe^{2+} and the ability of Pb^{2+} to be carried by other Fe^{2+} transporters (e.g., DMT1), it seems likely that Pb^{2+} is also a substrate for MTP1 at the basolateral surface of cells.

There is a clear association between Pb^{2+} and Ca^{2+}, as has been shown in numerous *in vivo* and *in vitro* models [314–324]. Studies in rats [311,313,314,316], ligated intestinal loops of the chicken [325], and perfused intestinal loops of the mouse [326] have shown that Pb^{2+} absorption is inversely related to dietary Ca^{2+}. Interestingly, it has also been shown that high dietary intake of Ca^{2+} can lead to lower plasma levels of Pb^{2+} [314–316]. High doses of Ca^{2+} were associated with lowered blood levels of Pb^{2+} in rats [314,316] and humans [315]. The interaction between Pb^{2+} and Ca^{2+} leads to the hypothesis that Pb^{2+} may gain entry into cells through one or more of the different types of Ca^{2+} channels expressed in various cells in the body. Moreover, since the ionic radius of Pb^{2+} (1.19 Å) is similar to that of Ca^{2+} (1.00 Å), it is possible that this Pb^{2+} may mimic Ca^{2+} at the site of Ca^{2+} transporters. Indeed, it has been shown that Pb^{2+} can be carried into cells via Ca^{2+} channels [327–331]. In cells isolated from bovine adrenal medulla, it was shown that compounds that depolarize the cell and subsequently open Ca^{2+} channels promote the uptake of Pb^{2+} [331]. These studies also showed that the flux of Pb^{2+} through Ca^{2+} channels is at least 10-fold greater than that observed for Ca^{2+}. This difference in flux is not surprising since the ionic radius of Pb^{2+} is smaller than that of Ca^{2+}. Additional experiments in cultured rat pituitary GH_3 cells, rat C_6 glioma cells, 301 cells (a subclone of human embryonic kidney (HEK) 293 cells) [327], and bovine brain capillary endothelial cells [332] indicate that Pb^{2+} enters these cells via channels that are activated by the reduction of intracellular stores of Ca^{2+}. These experiments also showed that the uptake of Pb^{2+} was a time- and concentration-dependent process and was inhibited significantly by the addition of Ca^{2+}. The opposite is also true. Studies utilizing rainbow trout found that the accumulation of Ca^{2+} within the gills of trout was reduced significantly by the presence of Pb^{2+} [329]. Increasing the concentration of Ca^{2+} in the water had a protective effect and reduced the accumulation of Pb^{2+}, suggesting that Pb^{2+} and Ca^{2+} compete for the same mechanism(s). Other studies utilizing cultured HEK-293 cells transfected with either L-, N-, or R-type Ca^{2+} channels suggest that the ability of these channels to take up Ca^{2+} is inhibited by the presence of Pb^{2+} [333,334]. The most direct evidence supporting a role for Ca^{2+} channels in the uptake of Pb^{2+} comes from a recent study using PC12 (rat pheochromocytoma) and HeLa (human cervical adenocarcinoma) cells [335]. Using live cell imaging, the authors of this study showed that Pb^{2+} enters cells via store-operated Ca^{2+} channels (SOCs) that are associated with the structural/regulatory elements of these channels, Orai1 and STIM1. A recent study also showed that Pb^{2+} binds to the binding site of the Ca^{2+} transporter in an opportunistic manner, and can adapt to structurally diverse binding geometries [336].

The Ca^{2+}-ATPase may also play a role in the uptake of Pb^{2+}. Experiments in human erythrocytes indicate that Pb^{2+} may substitute for Ca^{2+} as a ligand for Ca^{2+}-ATPase [337]. Additional studies in human red blood cell ghosts provided direct evidence indicating that Pb^{2+} is actively transported by a Ca^{2+}-ATPase [338].

Anion exchangers have also been implicated in the transport of Pb^{2+} [311,330,339,340]. Using resealed erythrocyte ghosts, Simons [340] demonstrated that over 90% of the uptake of Pb^{2+} could be attributed to transport via an anion exchanger. The rate of Pb^{2+} uptake was shown to be proportional to the external concentrations of Pb^{2+} and HCO_3^-. This was not surprising considering recent evidence showing that the export of organic anions (via OATP) is accompanied by HCO_3^- [341]. Compounds known to block anion exchangers (DIDS, 4-acetamido-4′-isothiocyanostilbene-2,2′-disulfonic acid (SITS), phloretin, furosemide, and bumetanide) effectively inhibited the uptake of Pb^{2+}; therefore, it was concluded that the transport of Pb^{2+} into erythrocytes is most likely mediated by an anion exchanger [322,330]. This conclusion was supported by subsequent studies in human erythrocytes [339]. In contrast, the uptake of Pb^{2+} into MDCK cells appeared to be an active process that was not affected by the addition of DIDS. Therefore, it was suggested that the transport of Pb^{2+} into erythrocytes occurs via an anion exchange mechanism while the uptake of Pb^{2+} by MDCK cells is thought to utilize a different mechanism [339]. Interestingly, the uptake of Pb^{2+} into cultured rat astroglial cells was inhibited by DIDS, and thus appears to occur via anion exchange [311].

Endocytosis of Pb^{2+}–protein complexes may also serve as a route for the entry of this metal into cells. The cytosolic protein α-2-microglobulin has been shown to bind Pb^{2+} [342,343]. In addition, diazapine-binding inhibitor (DBI) and thymosin β-4 have been identified as Pb^{2+}-binding proteins (PbBPs) in the human kidney [344,345]. It has been proposed that Pb^{2+}–PbBP complexes enter cells, such as proximal tubular cells, by endocytosis. Once inside the cell, it is thought that these Pb^{2+}–PbBP complexes are delivered first to lysosomes and then to the nucleus. It is unclear if mimicry is involved in the internalization of Pb^{2+}–PbBP complexes.

Pb^{2+} also appears to be transported as a conjugate of GSH [346]. Experiments in rats with cannulated bile ducts indicate that Pb^{2+} can be transported as a conjugate of GSH [346]. Rats treated with compounds that reduce the intracellular levels of GSH (diethylmaleate (DEM), cyclohexene oxide, or methyliodide) secreted significantly lower levels of Pb^{2+} into the bile. Although the transport of a GSH *S*-conjugate of Pb^{2+} was not demonstrated directly, these data suggest that such a complex may be a transportable form of Pb^{2+}, possibly at the site of a G-*S*-*S*-G transporter such as MRP2.

Pb^{2+} may also act as a functional mimic of endogenous ions at intracellular binding sites for certain endogenous ions. Several studies have shown that Pb^{2+} can act as a functional mimic of Ca^{2+} at the site of calmodulin [347–349], a protein that plays a role in the regulation of intracellular Ca^{2+} [350]. Pb^{2+} may also act as a substitute of Ca^{2+} at the site of a calcium–calmodulin-dependent phosphodiesterase [351] and at the site of Ca^{+2}-binding proteins (CaBP I and II) [352]. The interaction between Ca^{2+} and Pb^{2+} at sites of cellular junctions is also of significant importance. Many junctional complexes require Ca^{2+} in order to maintain their integrity. At these sites, Pb^{2+} may act as a structural and/or a functional mimic of Ca^{2+}. If acting as a structural mimic only, one would expect the functionality of cellular junctional complexes to be compromised. Interestingly, exposure to high levels of Pb^{2+} has been shown to result in the disruption of the blood–brain barrier [353], leading to edema and possibly brain damage. Importantly, the endothelial cells in the brain accumulate Pb^{2+} preferentially [354], which can account, in part, for the severe neurological symptoms observed with Pb^{2+} poisoning.

10.5 MERCURY

Mercury (Hg) is a toxic metal pollutant that is found in numerous environmental and occupational settings. It can exist in elemental (metallic), inorganic, and/or organic forms. Elemental mercury (Hg^0) is unique among all metals in that it is a liquid at room temperature. Hg^0 can be released readily into the atmosphere as Hg vapor because of its high vapor pressure. Inorganic forms of Hg, as mercurous (Hg^{1+}) or mercuric (Hg^{2+}) ions, commonly combine with anionic species of chlorine, sulfur, or oxygen to form mercurous or mercuric salts. The primary cation of Hg found in environmental settings is Hg^{2+}. Organic mercuric compounds form when mercuric ions bind covalently with carbon atoms of certain small organic functional groups such as methyl, ethyl, or phenyl groups. The most frequently encountered form of organic Hg in the environment is methylmercury (CH_3Hg^+). It is formed primarily via methylation of inorganic mercuric ions by microorganisms present in soil and water [4,355–357].

All forms of Hg are toxic to humans. However, the extent of the adverse effects induced by Hg depends on the form of Hg at the time of exposure, the duration of exposure, and the route of exposure. Exposure to all forms of Hg has been shown to cause deleterious effects in the cardiovascular system [358–361], gastrointestinal system [362–365], liver [365–367], kidneys [365,367–369], and neurological system [366,370–372].

Humans are exposed to Hg in numerous occupational and environmental settings as well as from dental amalgams, and medicinal and dietary sources [4,355,356,373]. However, the majority of human exposure results from the ingestion of food and water contaminated with CH_3Hg^+. Major predatory freshwater and saltwater fish, such as northern pike, salmon, swordfish, and shark, can contain high levels of CH_3Hg^+, largely related to the longevity of fish in contaminated waters. On ingestion of contaminated tissue of fish, CH_3Hg^+ is released and absorbed readily by the gastrointestinal tract of humans and other mammals [355]. After entering systemic circulation, mercuric ions can then be delivered to target organs.

It is important to note that within biological systems, mercurous or mercuric ions do not exist as inorganic salts or in an unbound, "free" ionic state [374]. Mercuric ions have a very high affinity for thiol-containing biomolecules, such as GSH, Cys, Hcy, NAC, and albumin. Thus, in biological systems, mercuric ions are always bound to one or more of these compounds. Several studies have shown that in the presence of excess low molecular weight, thiol-containing molecules, mercuric ions will bind to these molecules in a linear II, coordinate covalent manner (Figure 10.1) [375–377]. Mercuric–thiol conjugates formed under these conditions appear to be stable in an aqueous environment from pH 1 to 14 [376].

Numerous studies have implicated a mechanism of molecular mimicry in the uptake of thiol conjugates of inorganic and organic mercuric ions in selective target cells. Interestingly, mercuric conjugates of some nonprotein thiols, such as Cys, Hcy, or GSH, are structurally similar to endogenous compounds such as cystine, homocystine, glutathione disulfide (G*SS*G), or methionine. Consequently, it is thought that some mercuric species access target cells by serving as molecular mimics of endogenous compounds at the sites of carrier proteins involved in the transport of these compounds.

10.5.1 Inorganic Mercury

10.5.1.1 Intestinal Transport of Hg^{2+}

Gastrointestinal absorption of Hg^{2+}, although inefficient, occurs following the consumption of contaminated food and/or liquids. Foulkes [168] suggested that the uptake of Hg^{2+} from the intestinal lumen is dependent on the composition of the contents of the intestinal lumen. In other words, the mechanisms by which Hg^{2+} is taken up are dependent on the ligands to which Hg^{2+} is bound. Ingested food contains a great number of thiol-containing molecules, such as amino acids and peptides, to which Hg^{2+} may bind. Hg^{2+}–thiol complexes formed in the gastrointestinal tract may serve as structural or functional mimics of some of the endogenous molecules (such as amino acids and/or polypeptides) that are absorbed along the small intestine. Given the prevalence of amino acid and peptide transporters in enterocytes along the entire length of the intestine [378,379], it is reasonable to hypothesize that Hg^{2+}–thiol complexes may be taken up by one or more of these carriers. Not surprisingly, the duodenum appears to be the initial site of Hg^{2+} absorption [380].

Studies of the intestinal handling of Hg^{2+} in blue crabs demonstrated that there are passive and active mechanisms involved in the uptake of Hg^{2+} across plasma membranes of enterocytes [381,382]. In addition, studies in isolated perfused intestine of rainbow trout showed that there are two distinct mechanisms involved in the absorption of Hg^{2+}: an active pathway and an amiloride-sensitive pathway [383]. Based on experiments in which $HgCl_2$ was added directly to everted sacs of rat jejunum, Foulkes and Bergman [384] suggested that the intestinal absorption of Hg^{2+} is a two-step process in which Hg^{2+} first binds to the plasma membrane in the form of an anion such as $HgCl_3^-$. Second, Hg^{2+} traverses the plasma membrane and is internalized. Possible mechanisms for this uptake include amino acid and peptide carriers, Fe^{2+} transporters, and Zn^{2+} transporters.

Within the lumen of the intestine, ingested Hg^{2+} likely becomes bound to amino acids or peptides found in digested food. The complexes formed by this binding may be taken up by active and/or facilitated mechanisms involving amino acid and/or peptide transporters. Given the prevalence of amino acid transporters in the luminal plasma membrane of enterocytes and recent evidence implicating amino acid transporters in the transport of Cys *S*-conjugates of Hg (Cys-*S*-Hg-*S*-Cys) in renal proximal tubular cells, it is reasonable to postulate that amino acid and/or peptide transporters play a role in the intestinal absorption of Hg^{2+}. Unfortunately, there is no direct evidence to support this theory.

A small fraction of Hg^{2+} uptake may occur following ligand exchange where the mercuric ion is removed from its thiol carrier and is taken up by ion transporters such as DMT1. This transporter has been identified on the apical membrane of enterocytes [38] and may be able to mediate the transport of mercuric ions. Studies in mice showed that a decrease in the expression of intestinal DMT1 corresponded to a decrease in the intestinal accumulation of Hg^{2+} [41]. The ability of DMT1 to transport Hg^{2+} has not been shown directly.

Another possible mechanism for the intestinal absorption of Hg^{2+} is ZIP8. This Zn^{2+} carrier, which is localized in the intestine, is inhibited by Cd^{2+} and Hg^{2+} [150,152,154]. The exact localization of ZIP8 in the intestine and its ability to transport Hg^{2+} have not been determined.

The intestine also plays an important role in the elimination of Hg^{2+}. Two mechanisms appear to be involved in the fecal elimination of Hg^{2+}: (1) transcellular and/or paracellular secretion of Hg^{2+}

across enterocytes [385] and (2) delivery of Hg^{2+} from bile [383,385,386]. Data from *in vivo* studies in rats with cannulated or ligated bile ducts indicate that a substantial fraction of the total pool of Hg^{2+} that is excreted in the feces is due to intestinal secretion of Hg^{2+} from the blood into the intestinal lumen [387]. Secretion of Hg^{2+} by enterocytes may involve the transport of a Hg^{2+}–thiol complex, which could act as a mimic or a structural homolog of an endogenous molecule that is secreted normally by enterocytes. Amino acid transporters are potential mechanisms for this secretion. Given that many of these transporters are actually counter-exchangers, they have the potential to transport substrates both into and out of cells. Consequently, Hg^{2+}–thiol complexes, such as Cys-*S*-Hg-*S*-Cys, may utilize these carriers to enter and exit enterocytes, possibly through a mechanism involving molecular mimicry. Additionally, members of the MRP family may be involved in this transport. MRP3 is localized in the basolateral membrane of enterocytes and may participate in the transport of Hg^{2+} from blood into the intracellular compartment of enterocytes [388–391]. On the luminal membrane, MRP2 may mediate the export of Hg^{2+} from within cells into the intestinal lumen. Both these carriers have been shown to be involved in the intestinal secretion of organic anions [392]. Owing to the multispecific nature of these carriers and their subcellular localization, it is possible that they are involved in the intestinal secretion of Hg–thiol complexes. Transcripts encoding other members of the MRP family, including MRP4, MRP5, MRP6, and MRP7, have also been identified in enterocytes [393]; however, the cellular localization of these proteins has not been determined. Furthermore, the ability of these isoforms to transport Hg^{2+} has not been examined.

10.5.1.2 Renal Transport of Hg^{2+}

The primary site of uptake and accumulation of Hg^{2+} is the kidney, specifically the proximal tubule [4]. The accumulation of this metal in the kidneys is very rapid, with as much as 50% of a nontoxic dose present there within a few hours after exposure [394]. Until recently, the mechanisms by which Hg^{2+} is taken into renal tubular epithelial cells were largely unknown.

Experimental evidence indicates that there are luminal and basolateral mechanisms involved in the uptake of Hg^{2+} by proximal tubular cells [197,245–247,395–409]. On the luminal membrane, the majority of uptake appears to be dependent on the actions of γ-glutamyltransferase and cysteinylglycinase. Numerous *in vivo* experiments have demonstrated that inhibition of γ-glutamyltransferase significantly reduces the renal uptake and accumulation of systemically administered Hg^{2+} [396,410–413]. Therefore, it was hypothesized that GSH *S*-conjugates of Hg^{2+} (G-*S*-Hg-*S*-G) entering the luminal compartment of the proximal tubule are degraded rapidly and efficiently by γ-glutamyltransferase and cysteinylglycinase to yield thiol *S*-conjugates of Hg^{2+}, primarily Cys-*S*-Hg-*S*-Cys. Indeed, studies in brush-border membrane vesicles (isolated from the renal cortex and outer stripe of the outer medulla of rats) indicate that mercuric ions are taken up more readily when they are in the form of Cys-*S*-Hg-*S*-Cys than when they are conjugated to GSH or 2,3-dimercaptopropane-1-sulfonate (DMPS) [406]. Moreover, studies in suspensions of rabbit proximal tubules [414,415] and isolated perfused proximal tubules from rabbits [416,417] provide additional evidence for the luminal uptake of Cys-*S*-Hg-*S*-Cys. Given the similarities in size and shape between Cys-*S*-Hg-*S*-Cys and the amino acid cystine, we have hypothesized that Cys-*S*-Hg-*S*-Cys can mimic cystine at the site of one or more cystine transporters located in the luminal plasma membrane of proximal tubular epithelial cells. Subsequent experiments in isolated, perfused proximal tubules from rabbits showed that there is at least one Na^+-dependent and one Na^+-independent amino acid carrier involved in the transport of Cys-*S*-Hg-*S*-Cys [417].

System $b^{0,+}$ is a heterodimeric transporter comprised of two subunits: $b^{0,+}AT$ and rBAT [418,419]. Owing to its high affinity for cystine and neutral and basic amino acids, it is an excellent target for the uptake of Cys-*S*-Hg-*S*-Cys. *In vitro* studies utilizing type II Madin-Darby canine kidney (MDCKII) cells stably transfected with system $b^{0,+}$ indicate that Cys-*S*-Hg-*S*-Cys is a transportable substrate of this carrier [197]. The uptakes of Cys-*S*-Hg-*S*-Cys and cystine were each inhibited by the same amino acids, indicating that these two molecular species are substrates for the same carrier, that is, system $b^{0,+}$. This study also demonstrated that mercuric conjugates of GSH (G-*S*-Hg-*S*-G), *N*-acetylcysteine

(NAC-*S*-Hg-*S*-NAC), and cysteinylglycine (CysGly; CysGly-*S*-Hg-*S*-CysGly) are not readily transported by this carrier [197]. A subsequent study utilizing the same experimental system indicated that Hcy *S*-conjugates of Hg^{2+} (Hcy-*S*-Hg-*S*-Hcy), which are structurally similar to Cys-*S*-Hg-*S*-Cys and homocystine, are also substrates for system $b^{0,+}$ [197]. Together, these data provide strong evidence supporting the hypothesis that Cys-*S*-Hg-*S*-Cys and Hcy-*S*-Hg-*S*-Hcy are molecular mimics of the amino acids cystine and homocystine, respectively, at the site of system $b^{0,+}$.

Uptake of Hg^{2+} at the basolateral plasma membrane accounts for approximately 40–60% of proximal tubular uptake [396–399,404,405,407,420]. In experiments where glomerular filtration of rats was reduced to negligible levels, a 40% decrease in the renal tubular uptake of Hg^{2+} was observed, indicating that the remaining 60% of total Hg^{2+} uptake occurs at the basolateral membrane [407]. Data from these experiments also demonstrated that the addition of *para*-aminohippurate (PAH), which specifically inhibits members of the OAT family [421–426], effectively inhibited the uptake of Hg^{2+}. These findings indicate that the majority of the basolateral uptake of Hg^{2+} is likely mediated by one or more OATs. There is a substantive body of experimental evidence indicating that mercuric conjugates of Cys, Hcy, and NAC are taken up by OAT1 and OAT3 via a mechanism involving molecular mimicry. Both these transporters have been localized to the basolateral plasma membrane of proximal tubular epithelial cells [243,244]. The majority of current evidence indicates that OAT1 is the major mechanism for the uptake of Hg^{2+} at the basolateral plasma membrane of proximal tubular cells [396–399,404,405,409,420,427]. Indeed, recent findings from MDCK cells (stably transfected with OAT1) showed that Cys-*S*-Hg-*S*-Cys [246], NAC-*S*-Hg-*S*-NAC [245], and Hcy-*S*-Hg-*S*-Hcy [247] are substrates of this transporter. In addition, experiments utilizing *X. laevis* oocytes expressing either OAT1 or OAT3 have implicated each of these carriers in the uptake of Cys-*S*-Hg-*S*-Cys [245,246]. Collectively, these data strongly support a role for OAT1 and OAT3 in the basolateral uptake of certain mercuric complexes.

It is well documented that mercuric ions can be extracted from renal tubular cells following treatment with a metal chelator, either DMPS or DMSA [94,428–431]. This extraction process appears to involve a direct secretory process whereby mercuric ions move directly from the blood into the tubular lumen [432]. Interestingly, the secretion of Hg^{2+} from proximal tubular cells was shown to be dependent on the cotransport of GSH [413]. MRP2 is a member of the ATP-binding superfamily, is localized in the luminal plasma membrane of proximal tubular cells [51,433], and has been shown to transport some substrates in a GSH-dependent manner [254]. Indeed, indirect evidence from MRP2-deficient rats (*Eisai hyperbilirubinemic*) suggests that MRP2 plays a role in the hepatobiliary secretion of Hg^{2+} [386]. It was therefore hypothesized that MRP2 plays a similar role in the epithelial cells of the proximal tubule. A recent study utilizing proximal tubules isolated from killifish suggests that exposure to $HgCl_2$ increases the expression of MRP2 [269]. This study, however, did not directly measure the ability of MRP2 to transport Hg^{2+}. Moreover, the tubules were exposed to $HgCl_2$, which is not an accurate representation of normal *in vivo* conditions, wherein Hg^{+2} is usually bound to thiol-containing molecules. Recent findings from studies in MRP2-deficient rats (TR^-) exposed to $HgCl_2$ and treated subsequently with DMPS or DMSA provide strong evidence for a role of MRP2 in the DMPS- and DMSA-mediated extraction of mercuric ions [434,435]. It appears that DMPS and DMSA are taken up at the basolateral membrane of proximal tubular cells via OAT1 and the sodium-dependent dicarboxylate transporter, respectively [436–438]. It is hypothesized that once internalized, DMPS and DMSA form complexes with intracellular Hg^{2+}; these complexes are thought to then utilize MRP2 for transport across the luminal plasma membrane of proximal tubular cells into the tubular lumen. Indeed, studies using membrane vesicles prepared from Sf9 cells transfected with human MRP2 provided direct evidence for the ability of MRP2 to transport both DMPS- and DMSA-*S*-conjugates of Hg^{2+} [434,435].

10.5.1.3 Hepatic Transport of Hg^{2+}

The transport of Hg^{2+} across the sinusoidal membrane into hepatocytes is not understood completely. Since hepatocytes possess some of the same transporters that have been implicated in the transport of Hg^{2+} in other organs, it can be postulated that a transporter of G-*S*-*S*-G or one or more amino acid transporters may be involved in the transport of Hg^{2+} across the sinusoidal membrane of hepatocytes. It is well established that Hg^{2+} forms complexes with GSH and/or amino acids, such as Cys and Hcy. Various amino acid carriers have been identified in the liver [280,439]; however, it is presently unclear which, if any, of these carriers are present on the sinusoidal membrane. In addition, several members of the MRP family, which transport conjugates of GSH, have been identified in the sinusoidal membrane of hepatocytes [440], but these isoforms have not been implicated in the transport of Hg^{2+}.

Much more information is available on the transport of Hg^{2+} across the canalicular plasma membrane of hepatocytes. Early studies of Hg^{2+} transport suggested that the hepatobiliary transport of Hg^{2+} is dependent on hepatocellular concentrations of GSH [171,385,441–445]. In *in vivo* studies in which rats were pretreated with buthionine sulfoximine (BSO) or DEM, each decreasing the cellular levels of GSH, the uptake and/or retention of intravenously administered Hg^{2+} increased in hepatocytes [445]. It appears that in the absence of adequate levels of cytosolic GSH, mercuric ions are unable to exit hepatocytes and consequently accumulate in the intracellular compartment. Based on these data, it appears that the cellular levels of GSH play an important role in the transport of Hg^{2+} out of hepatocytes. Although the specific mechanisms responsible for the transport of mercuric ions from within hepatocytes into the biliary canaliculus have not been demonstrated directly, experimental evidence indicates that Hg^{2+} forms conjugates with GSH, which appear to be transported into the biliary canalicular compartment. G-*S*-Hg-*S*-G is structurally similar to glutathione disulfide (G-*S*-*S*-G); therefore, this mercuric complex may serve as a molecular mimic of G-*S*-*S*-G at the site of a G-*S*-*S*-G transporter. Indeed, the multidrug resistance protein 2 (MRP2), which is a carrier of G-*S*-*S*-G, has been identified on the canalicular membranes of hepatocytes [270,271,446,447]. Recently, MRP2 has been implicated in the transport of DMPS- and DMSA-*S*-conjugates of Hg^{2+} [435]; therefore, it is likely that it acts as a carrier for other conjugates of Hg^{2+}.

10.5.1.4 Transport of Hg^{2+} in the Placenta

Mercuric ions have been shown to accumulate in the placentas of pregnant women exposed to this metal [448–450]. Surprisingly, the mechanisms that mediate the transport of Hg^{2+} across the placenta are poorly understood. Experiments in brush-border membrane vesicles from human placenta suggest that an amino acid transporter may be involved in the uptake of Hg^{2+} in this organ [451]. These experiments demonstrated that the Na^+-dependent uptake of alanine was inhibited significantly by $HgCl_2$; it was therefore suggested that one or more amino acid transporters may be involved in this uptake. As these studies used $HgCl_2$ rather than a Hg–thiol complex, such as Cys-*S*-Hg-*S*-Cys, the results may not accurately reflect physiological processes that occur *in vivo*. They do, however, provide valuable preliminary data. Based on these data and the prevalence of amino acid transporters in the placenta [452,453], we can postulate that Hg^{2+}, as a thiol conjugate, may mimic a structurally similar amino acid and may be utilized as a substrate by one or more amino acid transporters. This proposed mechanism of molecular mimicry may be similar to that demonstrated in other tissues (e.g., system $b^{0,+}$ in proximal tubular cells, system L in endothelial and glial cells of the blood–brain barrier).

10.5.1.5 Intracellular Mimicry with Hg^{2+}

Just as some Hg–thiol complexes can mimic endogenous molecules at the site of membrane proteins present in the plasma membrane, these complexes may also mimic molecules in the intracellular compartments of cells. Since Cys-*S*-Hg-*S*-Cys appears to mimic cystine at the site of an amino acid transporter on the plasma membrane [197], it is not surprising to find that this conjugate also acts as a mimic of this amino acid at binding sites of intracellular molecules that utilize cystine as a substrate. An example of this mimicry has been demonstrated by the preliminary findings of Cooper and colleagues [454]. These findings showed that Cys-*S*-Hg-*S*-Cys mimics cystine at the binding site of the intracellular enzyme γ-cystathionase, which is activated normally by the binding of cystine or cystathionine. Interestingly, Cys-*S*-Hg-*S*-Cys may interact with the binding site of γ-cystathionase and inactivate the enzyme. Because of this inactivation, it was concluded that Cys-*S*-Hg-*S*-Cys behaves as a structural, but not a functional, intracellular mimic of cystine and cystathionine. These findings suggest that the ability of molecular species of metals to mimic endogenous molecules may have serious, deleterious effects on intracellular processes.

10.5.2 Methylmercury

10.5.2.1 Transport of CH_3Hg^+ in the Brain

The brain and central nervous system (CNS) are the primary target sites where the adverse affects of CH_3Hg^+ are manifested [355,455]. Accordingly, a large number of studies have focused on mechanisms by which CH_3Hg^+ gains access to the CNS and, more specifically, how it crosses the blood–brain barrier. Like Hg^{2+}, CH_3Hg^+ does not exist as a free, unbound cation in biological systems [374], but rather is found conjugated to thiol-containing biomolecules such as GSH, Cys, Hcy, or NAC [1]. Initial studies utilizing homogenates of rat cerebrum have demonstrated that the primary nonprotein thiol bound to CH_3Hg^+ is GSH [456]. This may be one route by which CH_3Hg^+ gains access to cells. Subsequent studies in rats and primary cultures of bovine brain endothelial cells have revealed a possible role for Cys in the transport of CH_3Hg^+ across the blood–brain barrier [457–459]. Specifically, coadministration of Cys with CH_3Hg^+ has been shown to increase the uptake of CH_3Hg^+ into capillary endothelial cells of the blood–brain barrier. Interestingly, experiments in rats have demonstrated that the uptake of CH_3Hg^+ is inhibited significantly by the neutral amino acid phenylalanine [459–462]. The data from these experiments led to the hypothesis that Cys *S*-conjugates of CH_3Hg^+ (CH_3Hg-*S*-Cys) are transportable substrates of a neutral amino acid transporter in the capillary endothelium of the blood–brain barrier. *In vivo* studies in rat brain [457] and *in vitro* studies in bovine cerebral capillary endothelial cells [457] have demonstrated that the uptake of CH_3Hg-*S*-Cys is inhibitable by neutral amino acids, providing additional support for the theory that CH_3Hg-*S*-Cys is taken up by a neutral amino acid carrier. The investigators in these studies suggested that CH_3Hg-*S*-Cys behaves as a "mimic" of an amino acid in order to cross the blood–brain barrier. Indeed, CH_3Hg-*S*-Cys is structurally similar to the amino acid methionine [463,464], which is a substrate of the neutral amino acid carrier, system L [280].

System L is present in the basolateral plasma membrane of many types of transporting epithelia [280]. Interestingly, in endothelial cells lining the blood–brain barrier, this carrier is found in both apical and basolateral plasma membranes [465]. This transporter has a broad substrate specificity [466], which may allow it to utilize CH_3Hg-*S*-Cys as a substrate. Indeed, *in vivo* studies in rats [467] and *in vitro* studies utilizing primary cultures of rat astrocytes [468,469] provided evidence that CH_3Hg-*S*-Cys is a transportable substrate of system L. Additional studies have shown that Hcy *S*-conjugates of CH_3Hg^+ (CH_3Hg-*S*-Hcy) may also be substrates of system L [470].

Since these initial studies were carried out, investigators have cloned two isoforms of system L: LAT1 and LAT2 [471–473]. These transporters are heterodimeric proteins, comprising a heavy chain, 4F2hc, and a light chain, LAT1 or LAT2, bound together by a disulfide bond [280,474]. With this knowledge it has become possible to characterize further the role of system L in the uptake of CH_3Hg-*S*-Cys. *X. laevis* oocytes expressing LAT1 or LAT2 were utilized to study directly the involvement of each carrier in the transport of CH_3Hg-*S*-Cys [475]. These data provided the first line of direct molecular evidence showing that CH_3Hg-*S*-Cys is a transportable substrate of LAT1 and LAT2 [475]. These data also provide substantive evidence for the phenomenon of molecular mimicry in that CH_3Hg-*S*-Cys appears to mimic methionine at the site of system L.

10.5.2.2 Transport of CH_3Hg^+ in Erythrocytes

Additional data related to the cellular handling of GSH *S*-conjugates of CH_3Hg^+ (CH_3Hg-*S*-G) have been obtained from studies in erythrocytes. When experiments were performed at 5°C, it appeared that multiple systems were involved in the transport of CH_3Hg-*S*-G. The primary mechanism was thought to involve one or more OATs. In addition, a D-glucose diffusive transporter, a cysteine facilitated transporter, and a Cl^- transporter may also play a role in the transport of CH_3Hg-*S*-G [476]. Based on these findings, it appears that CH_3Hg-*S*-G mimics an endogenous substrate of each of these transporters in order to gain access to the erythrocyte. The authors of this study acknowledged that the systems involved in the uptake of CH_3Hg-*S*-G at 5°C may differ from those responsible for this uptake at physiological temperatures. In later studies, Wu [477] measured the uptake of CH_3Hg-*S*-G at 5°C and 20°C and concluded that OAT is the primary mechanism of CH_3Hg-*S*-G uptake at both temperatures. Additional studies in which probenecid inhibited the uptake of CH_3Hg-*S*-G confirmed previous data indicating that this conjugate is a transportable substrate of OAT [478]. It was also reported that system N, system y^+, and the oligopeptide H^+ transport system were not involved in this process [478]. These data support the hypothesis that molecular mimicry at the site of one or more transporters plays a role in the transport of CH_3Hg-*S*-G in erythrocytes.

10.5.2.3 Intestinal Transport of CH_3Hg^+

Oral consumption of food and water contaminated with CH_3Hg^+ is the primary route by which humans are exposed to this form of mercury. Thus, a thorough understanding of the mechanisms involved in the intestinal absorption of CH_3Hg^+ is important in order to understand fully the transport and toxicity of CH_3Hg^+ in the body. Studies in ligated rat intestinal segments have demonstrated that the uptake of CH_3Hg-*S*-Cys and CysGly *S*-conjugates of CH_3Hg^+ (CH_3Hg-*S*-CysGly) was 1.5 times greater than the uptake of CH_3Hg-*S*-G [479]. Interestingly, when γ-glutamyltransferase was inhibited, the uptake of CH_3Hg-*S*-G by enterocytes was reduced by 50%, suggesting that some CH_3Hg-*S*-G is not absorbed as an intact molecule. CH_3Hg-*S*-CysGly, which is produced by the action of γ-glutamyltransferase on GSH, is likely degraded further at the luminal plasma membrane of enterocytes to yield CH_3Hg-*S*-Cys. Any CH_3Hg-*S*-CysGly, however, that escapes degradation to CH_3Hg-*S*-Cys may be transported by a peptide transporter present in the luminal membrane. In the intestine, di- and tripeptide transporters are the primary means for the uptake of amino acids. Given this, and the similarity between CH_3Hg-*S*-CysGly and a small peptide, it is possible that this complex mimics an endogenous di- or tripeptide to gain access to enterocytes. Alternatively, CH_3Hg-*S*-Cys has been identified as the primary species of CH_3Hg^+ that is secreted into the intestine from bile and appears to be reabsorbed rapidly by enterocytes [480]. CH_3Hg-*S*-Cys is structurally similar to methionine and may mimic this amino acid at the site of one of the intestine's many amino acid carriers.

Additional data from experiments in ligated segments of rat intestine demonstrated that treatment with probenecid and acivicin significantly reduced the luminal uptake of CH_3Hg^+ [479]. It was concluded that there are two independent transport systems for the uptake of CH_3Hg^+, as CH_3Hg-*S*-G, across the luminal plasma membrane of enterocytes. One of these mechanisms is dependent on the activity of γ-glutamyltransferase, while the other appears to be inhibited by probenecid. These findings lead one to postulate that one or more OATs, which are inhibited by probenecid, may be involved in the intestinal uptake of CH_3Hg^+. It is important, however, to note that probenecid is not a specific inhibitor of OAT, and may act upon other transporters. To date, OAT2 and OAT10 are the only members of the OAT family that have been detected in the intestine [481,482]. In addition, several members of the OATP family [482], in addition to a novel organic anion transporter-like protein (OATLP1), have been identified in the intestine [483]. Currently, there are no published data regarding the ability of any of these carriers to transport Hg^{2+}.

The efflux of CH_3Hg^+ across the basolateral membrane of the enterocyte is less clear. Studies using isolated, perfused catfish intestines showed that CH_3Hg-*S*-Cys is taken up rapidly by enterocytes and transported across the wall of the intestine [484]. The investigators in this study suggest that there are one or more active-transport carrier proteins involved in this transport. Competitive inhibition experiments provided indirect evidence that a neutral amino acid transporter, possibly system L, plays a role in the flux of CH_3Hg-*S*-Cys across the basolateral membrane of intestinal enterocytes. This transport is probably the result of molecular mimicry wherein CH_3Hg-*S*-Cys mimics methionine at the site of system L.

Another possible mechanism for the basolateral export of Hg^{2+} may relate to GSH. Foulkes [485] suggested that the intracellular concentrations of GSH are responsible for regulating the efflux of CH_3Hg^+ from the enterocyte into the blood. There are few data available to characterize this step of CH_3Hg^+ uptake. The apparent similarities between the structures of CH_3Hg-*S*-G and GSH lead one to postulate that the basolateral transport of CH_3Hg-*S*-G may utilize a GSH transporter.

10.5.2.4 Renal Transport of CH_3Hg^+

Although the primary target of CH_3Hg^+ is the CNS, this form of mercury also induces significant detrimental effects in other organs, such as the kidney [94,486–493]. Until recently, it was unclear as to how this organo-metal complex is taken up by renal tubular epithelial cells. Richardson and Murphy [494] suggested that the renal tubular uptake of CH_3Hg^+ is dependent on the cellular concentration of GSH. Moreover, several studies have shown that the renal uptake and accumulation of CH_3Hg^+ increase after coadministration of CH_3Hg^+ and GSH [495,496]. It is thought that γ-glutamyltransferase and cysteinylglycinase, present in the luminal membrane of proximal tubular cells, act on CH_3Hg-*S*-G to yield CH_3Hg-*S*-Cys [4]. It should be noted that the methylmercuric ion remains bound to the sulfur atom of Cys during the catabolism of GSH [497]. Inhibition of γ-glutamyltransferase by acivicin decreased the uptake of CH_3Hg^+ into renal tubules and increased the urinary excretion of GSH and CH_3Hg^+ [410,412,496–503]. Furthermore, data from mice deficient in γ-glutamyltransferase also showed that less CH_3Hg^+ is absorbed by renal tubular cells [170]. These data indicate that the catabolism of the CH_3Hg-*S*-G complex is required for the luminal absorption of CH_3Hg^+ by proximal tubular cells.

Given the structural similarities between CH_3Hg-*S*-Cys and methionine, it is likely that this conjugate is taken up into proximal tubular cells by an amino acid transporter. Indeed, studies utilizing oocytes from *X. laevis* provided direct evidence for the ability of the amino acid transporter, system $B^{0,+}$, to transport CH_3Hg-*S*-Cys and Hcy *S*-conjugates of CH_3Hg^+ (CH_3Hg-*S*-Hcy) [504]. This carrier is present on the luminal plasma membrane of proximal tubular cells [505] and has been shown to mediate the Na^+-dependent transport of many neutral and cationic amino acids, including methionine (Met) [506,507]. Interestingly, the substrate specificity of system $B^{0,+}$ is similar to that of system $b^{0,+}$, which has been shown to transport conjugates of Hg^{2+} [197,395].

Tanaka and colleagues [496] showed that in addition to luminal mechanisms, there are basolateral mechanisms involved in the renal tubular uptake of CH_3Hg^+. Uptake of CH_3Hg^+ at the basolateral membrane appears to involve a multispecific carrier, such as OAT1. In the kidneys, this transporter is localized exclusively in the basolateral membrane of proximal tubular epithelial cells [243,245], and it has been shown to mediate the uptake of NAC and DMPS *S*-conjugates of CH_3Hg^+ (CH_3Hg-*S*-NAC and CH_3Hg-*S*-DMPS, respectively) [508]. Recently, the uptake of CH_3Hg^+ has been characterized in a line of MDCK cells stably transfected with human OAT1. These data provide direct evidence for the ability of OAT1 to mediate the uptake of CH_3Hg-*S*-Cys, CH_3Hg-*S*-Hcy, and CH_3Hg-*S*-NAC [400–402].

Interestingly, it appears that a fraction of CH_3Hg^+ present in systemic circulation is oxidized to Hg^{2+} before and/or after it enters the proximal tubular epithelial cells of the kidney [94,491,492,509–511]. These findings lead one to suggest that at least some of the mercuric ions taken up in the kidneys after exposure to CH_3Hg^+ may be due to the transport of some chemical form of Hg^{2+} rather than CH_3Hg^+.

As with Hg^{2+}, it is well documented that chelators are able to extract mercuric ions following exposure to CH_3Hg^+. The mechanisms by which CH_3Hg^+ is extracted from proximal tubular cells, however, were unclear until recently. *In vivo* studies using TR^- rats have implicated MRP2 in chelator-mediated extraction of mercuric ions. Studies in TR^- rats exposed to CH_3Hg^+ and treated subsequently with NAC provide support for the hypothesis that MRP2 mediates the transport of CH_3Hg-*S*-NAC from within proximal tubular cells into the tubular lumen [512]. Experiments using membrane vesicles from TR$^-$ rats provided direct evidence suggesting that CH_3Hg-*S*-NAC is a transportable substrate of MRP2 [512]. Additional studies in TR$^-$ rats indicate that MRP2 plays a role in the DMPS- and DMSA-mediated elimination of CH_3Hg^+ from proximal tubular cells [513]. Findings from experiments using inside-out membrane vesicles from Sf9 cells transfected with human MRP2 provide direct evidence indicating that DMPS- and DMSA-*S*-conjugates of CH_3Hg^+ are transportable substrates of MRP2 [513]. Collectively, these data provide strong support for the hypothesis that MRP2 plays an important role in the renal elimination of mercuric ions following exposure to CH_3Hg^+.

10.5.2.5 Hepatic Transport of CH_3Hg^+

CH_3Hg^+, absorbed by the intestine, is delivered to the liver via portal blood. Little is known about the mechanisms by which CH_3Hg^+ in the blood crosses the sinusoidal membrane of hepatocytes. *In vivo* studies in rats have shown that the uptake and accumulation of CH_3Hg^+ is enhanced when Cys or GSH is either coadministered or subsequently administered with CH_3Hg^+ [462]. A subsequent study in cultured human hepatocytes (HepG2 cells) demonstrated that hepatocellular uptake of CH_3Hg^+ is faster when it is presented to cells as CH_3Hg-Cys [172]. These studies suggest that CH_3Hg-Cys is the most likely species of CH_3Hg^+ taken up at the sinusoidal membrane of hepatocytes and that this uptake occurs via system L.

Current evidence regarding the transport of CH_3Hg^+ from hepatocytes into the biliary canaliculus indicates that this metal is transported in association with GSH [441,443,514–516]. Indeed, early studies in hepatic tissues indicate that the majority of CH_3Hg^+ within hepatocytes is bound to GSH [517]. Magos and colleagues [518] demonstrated that increasing hepatic levels of GSH increased the excretion of GSH and CH_3Hg^+ into bile. Refsvik [519] showed that when the hepatic and biliary levels of GSH are reduced, the accumulation of CH_3Hg^+ in the liver is also reduced. It appears that the intracellular concentration of GSH has a significant effect on the transport of CH_3Hg^+. It is plausible to suggest that CH_3Hg-*S*-G is formed within hepatocytes and transported subsequently into bile at the canalicular membrane. CH_3Hg-*S*-G is structurally similar to GSH; therefore, it is possible that this conjugate may mimic GSH at the site of a GSH transporter on the canalicular membrane. Indeed, it has been suggested that a GSH transport system on the canalicular membrane serves a primary role in the biliary secretion of CH_3Hg-*S*-G [520]. MRP2 has since been identified as a GSH transporter on the canalicular membrane of hepatocytes [521–525] and likely plays an important role in the export of CH_3Hg^+. Indeed, recent data from TR$^-$ rats suggest that MRP2 plays a role in the hepatobiliary elimination of mercuric ions following exposure to CH_3Hg^+ [434,435,512].

After being secreted into the bile, CH_3Hg^+ may be reabsorbed along the biliary tree as a conjugate of GSH or one of its metabolites, CysGly and/or Cys [526]. Indeed, CH_3Hg^+ appears to be absorbed by ductal epithelial cells more readily when it is administered as a complex of GSH or Cys [527]. Once in the biliary tree, it appears that CH_3Hg-*S*-G is catabolized sequentially by γ-glutamyltransferase and cysteinylglycinase to yield CH_3Hg-*S*-Cys, which can be reabsorbed, both by cells lining the bile ducts and by enterocytes in the intestine [171]. Although the actual mechanisms involved in the uptake of CH_3Hg^+ conjugates along the biliary tree have not been determined, it is reasonable to hypothesize that CH_3Hg-Cys is acting as a mimic of an amino acid at the site of an amino acid carrier such as system L. A number of various amino acid transporters, including system L (LAT3) [528], have been identified in the liver and biliary tree [280,439]; however, the exact localization of each transporter has not been determined.

10.5.2.6 Transport of CH_3Hg^+ in the Placenta

One of the most publicized and serious toxicological consequences of CH_3Hg^+ exposure is the deleterious neurological effect observed in fetuses whose mothers were exposed to this form of mercury during pregnancy [529–536]. CH_3Hg^+ crosses the placenta readily and accumulates in the fetus [450,533,537] and placenta [448] at levels higher than those in maternal tissues and blood. Yet, little is known about the mechanism(s) by which this metal is taken up and transported across this organ.

Kaijiwara and colleagues [538] have shown that CH_3Hg^+ is transported across the rat placenta by a neutral amino acid carrier in a time- and dose-dependent manner. They demonstrated that coinjection with methionine increased the uptake of CH_3Hg^+. It was proposed that this increase may be the result of the intracellular conversion of methionine to Cys, which may subsequently combine with CH_3Hg^+ to form the readily transportable conjugate CH_3Hg-*S*-Cys. This conjugate may then mimic methionine at the site of system L to gain access to the placenta. Accordingly, the authors concluded that the uptake of CH_3Hg^+ in placenta is mediated by the neutral amino acid carrier system L [538]. The exact species of CH_3Hg^+ that was transported was not determined in this study, nor was there direct evidence supporting the conclusion that system L was involved in this transport. However, since the two isoforms of system L, LAT1 and LAT2, have been shown to mediate the transport of CH_3Hg-*S*-Cys across the epithelial cells of and the astrocytes associated with the blood–brain barrier [467,468,470,475], it is logical to hypothesize that this same carrier is also involved in the uptake of CH_3Hg-*S*-Cys in placenta. In the placenta, LAT1 is localized in the apical (maternal) plasma membrane of the trophoblasts while LAT2 is found in the basolateral (fetal) membrane [453]. These transporters are important participants in the transfer of nutrients from the maternal to the fetal circulation. It should be noted that a number of other protein carriers, including OATs, have been identified in the placenta [539]. The role of these other transporters in the transport of CH_3Hg^+ has not been examined; however, they should be considered as possible mechanisms for this transport.

In pregnant rats exposed to CH_3Hg^+, Aremu and colleagues [540] demonstrated that treatment with NAC facilitated the removal of mercuric ions from fetal and placental tissues. Similarly, in a separate study in which rats were exposed to CH_3Hg^+ and subsequently treated with DMPS or DMSA, it was shown that these two chelators were capable of extracting mercuric ions from placental and fetal tissues [541]. MRP2, present in the maternal (apical) membrane of trophoblasts [542], may play a role in this extraction. The mechanism on the basolateral membrane of trophoblasts that mediates the uptake of mercuric ions from fetal circulation into the placenta has not been identified. One possible mechanism is OAT4, which has been localized in the basolateral membrane of placental trophoblasts [542,543]. The ability of OAT4 to transport mercuric ions has not yet been examined.

10.6 SELENIUM

Selenium (Se) is an essential element that is commonly found in rock formations and soil. It is rarely found in its elemental form in the environment; rather it is usually present as sodium selenite and sodium selenate. Although Se, with a recommended daily allowance for Se of 55 μg/day, is essential for the proper function of intracellular antioxidant enzymes, it can be harmful to humans when consumed in excess [544,545]. Se is often found in paint, certain types of glass, vitamin supplements, plastics, and fungicides [544]. In addition, the ash released from the burning of coal is a major source of Se contamination in the environment.

Exposure to Se compounds, such as hydrogen selenide and selenium dioxide, results in serious injury to the respiratory tract [546,547], cardiovascular system [548], gastrointestinal tract [548–550], liver [546,547], and nervous system [549,551]. In addition, ingestion of excess selenium has been shown to cause vision impairment, paralysis, and respiratory failure in livestock [544].

Little is known about the mechanisms by which Se enters mammalian cells. Data from everted sacs of rat ileum indicate that the uptake of selenate across the luminal plasma membrane of enterocytes may be driven by a Na^+ gradient [552]. Since selenate is structurally similar to sulfate, this uptake may be mediated by the sodium/sulfate symporter NaSi-1, which is found at high levels in the ileum [553]. Indeed, additional studies have demonstrated that Se is a transportable substrate of NaSi-1 [554,555]. Similarly, another sodium/sulfate cotransporter, NaS2, has been identified in other organs, including the kidney, liver [556,557], and placenta [558], and may be involved in the transport of Se.

Studies in isolated rat enterocytes have shown that the uptake of Se is stimulated by intracellular and extracellular GSH [559]. Interestingly, when the catabolism of GSH was inhibited, the uptake of Se was reduced. These studies suggest that the transport of Se in enterocytes is dependent on the products of GSH metabolism [559]. In the liver, it appears that Se is transported from hepatocytes into bile as a conjugate of GSH (G*S*-Se-*S*G) [560]. Additional *in vivo* studies have demonstrated that Cys, in addition to GSH, stimulates the uptake of Se into isolated perfused segments of distal jejunum [561,562] and ligated intestinal loops [563]. Interestingly, in *in vitro* studies using the mucosal surface of isolated jejunum, GSH inhibited the uptake of Se while Cys stimulated its uptake. The authors of this study suggested that the catabolism of GSH occurs more efficiently in *in vivo* systems. They concluded that the transportable species of Se are not GSH *S*-conjugates, but rather Cys *S*-conjugates of Se [562]. Indeed, subsequent studies in primary cultures of enterocytes isolated from sheep showed that Se–Cys complexes are transported readily across the luminal plasma membrane of these cells [564]. This transport was inhibited by the presence of various amino acids, indicating that an amino acid transporter is probably involved in the uptake of this metal, perhaps by a mechanism involving molecular mimicry, similar to that utilized in the proximal tubular uptake of Cys-*S*-Hg-*S*-Cys. Additional studies in cultured human intestinal cells (Caco-2) indicate that the addition of GSH increased the uptake of selenite (SeO_3^{-2}) [565]. These studies also showed that cystine and Met conjugates of Se were taken up by these cells. Interestingly, the uptake of Se–Met conjugates was inhibited by Met, which suggests that a common carrier is involved in the transport of both compounds. The uptake of Se–cystine complexes was not inhibited by Cys [565]. The effect of cystine on the uptake of Se–cystine complexes was not examined in this study.

Se, like Cd, has been shown to behave as a functional mimic of estrogen at the site of the estrogen receptor. Studies in MCF-7 cells have shown that Se, in the form of sodium selenite, binds to the hormone-binding domain of the estrogen receptor and activates this receptor [566]. Exposure to Se was also found to increase the transcription of the progesterone receptor [566]. These findings have many important implications for the treatment and prevention of hormone-related diseases.

Another important function of Se is its role as an insulin mimic. According to Stapleton [567], an insulin mimic is any agent that can effectively elicit the effect of insulin on cells. The ability of Se to act as a mimic of insulin was first examined in isolated rat hepatocytes [568]. Data from these experiments showed that selenate is capable of stimulating the transport of glucose in hepatocytes in a dose-dependent manner. This transport was thought to be associated with the insertion of the glucose transporters GLUT-1 and GLUT-2 into the plasma membrane of hepatocytes, a response similar to that induced by insulin. The effect of selenate on the uptake of glucose has also been studied in rat soleus muscle [569]. The results of these studies indicate that selenate stimulated the uptake of glucose and increased the rate of aerobic and anaerobic glycolysis. Subsequent *in vivo* studies in streptozotocin-induced diabetic rats [570–572] and mice [572] showed that treatment with selenate effectively lowers the plasma levels of glucose.

Insulin is also involved in the regulation of cellular metabolic processes, such as carbohydrate and fatty acid metabolism [573], glycogen synthesis [574], glycolysis [575], gluconeogenesis [576,577], fatty acid biosynthesis [578,579], and the pentose phosphate pathway [580]. Several studies have provided evidence for the ability of selenate to act as a mimic of insulin in the above metabolic processes [572,575,581,582]. Indeed, oral administration of selenate to diabetic rats restored the expression of enzymes involved in glycogen synthesis and gluconeogenesis to normal levels [582]. Similarly, treatment of diabetic rats with selenate also restored the expression of enzymes involved in fatty acid biosynthesis and the pentose phosphate pathway to normal levels [572,575]. One of the characteristic signs of diabetes is hyperlipidemia. Interestingly, selenate has been shown to reduce the plasma levels of glucose, lipids, triglycerides, cholesterol, and free fatty acids in diabetic rats [581,583]. Furthermore, treatment of diabetic rats with sodium selenite prevented metabolic defects, such as increases in lipid peroxidation and nitric oxide products, and myocardial alterations that are observed in diabetic animals [584–587]. Studies using chick-embryo hepatocytes indicate that Se is not an insulin mimetic under all conditions [588]. The clinical implications of these collective findings have yet to be determined. Indeed, some of the effects of Se may be due to the ability of this metal to increase the expression of GSH-peroxidase [589]. Data from a recent study showed that mice overexpressing GSH-peroxidase had higher levels of insulin than control mice [590].

10.7 OXYANIONS OF TOXIC METALS

It has been noted that endogenous oxyanions, such as monovalent phosphate and sulfate, are structurally similar to oxyanions of several toxic metals [1,6]. The molecular structures of arsenate and vanadate are very similar to that of monovalent phosphate, while the molecules chromate, molybdate, and selenate are similar in shape and size to sulfate (Figure 10.2). Therefore, it is not surprising that the oxyanionic forms of toxic metals have been found to mimic monovalent phosphate or sulfate at the site of some membrane carrier proteins.

10.7.1 Arsenate

Arsenate has been shown to interact with monovalent phosphate at the sites of several different carriers. Experiments in osteosarcoma cells (ROS 17/2.8 [591]), bovine and rabbit renal brush-border membrane vesicles [592,593], vesicles from the matrix of chicken cartilage [594], and the luminal plasma membrane of rat intestine [595] have shown that arsenate will compete with phosphate at the site of the sodium-dependent phosphate cotransporter. Additionally, arsenate has been shown to competitively inhibit phosphate at the site of an uncharacterized sodium-independent transporter [596,597]. The ubiquitously expressed phosphate transporters (PiT-1 and PiT-2) had a low affinity for As; however, the intestinal sodium-phosphate transporter (NaPi-IIb) appears to have a very high affinity for this metal. In addition, the renal isoforms of the sodium–phosphate carrier (NaPi-IIa and NaPi-IIc), which are localized in the luminal plasma membrane of proximal tubular cells [598,553], may also be involved in the uptake of arsenate from the lumen of proximal tubular cells into the intracellular compartment [599]. However, since NaPI-IIa and NaPI-IIc have a low affinity for As, they likely do not represent a major pathway of As uptake. In earthworms, As toxicity was reduced by adding phosphate to the soil in which earthworms were raised. These findings suggest that phosphate competes with As for uptake into cells [600]. Interestingly, arsenate does not inhibit the transport of phosphate at the sodium-dependent phosphate exchanger in human erythrocytes [601], indicating that As does not act as a mimic of phosphate under all conditions.

Arsenate has also been shown to act as a mimic of phosphate at intracellular sites. For example, a phosphate molecule is donated normally to 3-phosphoglyceraldehyde to form 1,3-diphosphoglycerate, which in turn donates a phosphate molecule to ADP to form ATP. Arsenate has been shown to substitute for the molecule of phosphate that is donated to 3-phosphoglyceraldehyde, which results in the disruption of ATP synthesis [602].

10.7.2 Chromate

Chromium has been shown to be a competitive inhibitor of sulfate at the site of sulfate transporters in human erythrocytes [603,604], placenta [558,605,606], and kidney [607,608]. In kidney, chromium strongly inhibited the uptake of sulfate via the sodium–sulfate cotransporter (NaSi-1) [608, 609] and via the kidney sulfate transporter (SAT-1) [607], which are both localized in the brush-border membrane of renal tubules. This inhibition was hypothesized to occur through a competitive mechanism. Chromium has also been shown to interfere with the transport of sulfate via the placenta-specific sulfate transporter (NaS2) [558]. Taken together, these data indicate that chromium can act as a mimic of sulfate at the site of transporters that mediate the uptake of sulfate.

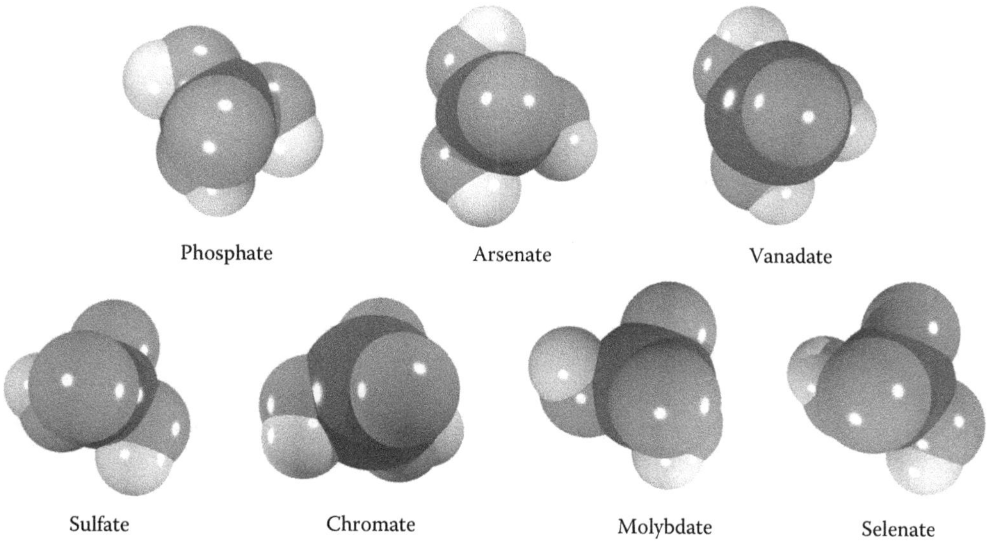

FIGURE 10.2 Space-filled models of selected oxyanions implicated in molecular mimicry. Note the similarities in structure between monovalent phosphate and the oxyanionic forms of the toxic metals arsenic (arsenate) or vanadium (vanadate). Both arsenate and vanadate have been shown to mimic phosphate at the site of phosphate transporters. In addition, the structure of sulfate is shown in comparison with the structures of selenate, molybdate, and chromate, which are homologous to sulfate. There is evidence indicating that these oxyanions can mimic sulfate at the site of transporters responsible for its uptake.

10.7.3 Molybdate

Molybdate has a tetrahedral structure that is similar to the structures of sulfate (Figure 10.2). Early studies of sulfate transport in rat ileum indicate that molybdate competitively inhibits the uptake of sulfate, suggesting that these two compounds are substrates for the same transporter [610,611]. Subsequent studies in membrane vesicles from human placenta have shown that molybdate is a competitive inhibitor of sulfate [606]. Furthermore, data from brush-border membrane vesicles from the proximal tubules of sheep demonstrate that molybdate and sulfate utilize the same transport system [611]. Additional support for these findings comes from *in vivo* studies in the renal proximal tubules of rats showing that molybdate is transported by luminal and basolateral sulfate transporters [612,613]. Collectively, these findings suggest that molybdate can act as a functional mimic of the oxyanion sulfate at the sites of several different extracellular transport proteins.

10.7.4 Vanadate

The oxyanion of vanadium, vanadate (VO_4^{3-}), appears to interfere with the transport of monobasic phosphate at the sites of several different carriers. According to Clarkson [1], studies of vanadate in erythrocytes indicate that the inward flux is inhibitable both by phosphate and by agents that inhibit the Cl^-/HCO_3^- exchanger. More recent findings from studies of vanadate in HEK-293 cells demonstrated that vanadate can inhibit the renal uptake of phosphate at the site of the NaPi-3 cotransporter [614]. In addition, vanadate may act as a mimic of a phosphate molecule on the cytoplasmic side of Na^+-K^+ ATPase [615–617]. The binding of a phosphate molecule, derived from ATP, is essential for the activity of this exchanger. Substitution of this phosphate with vanadate results in the inhibition of Na^+-K^+ ATPase [615–617]. Another possible carrier for vanadate is DMT1. *In vivo* studies in mouse pups indicate that higher brain levels of vanadium were associated with an increase in the expression of DMT1 [618].

Interestingly, vanadate has insulin-like effects, but the mechanisms by which it mediates these effects are unclear. Vanadate lowers serum glucose levels, improves insulin sensitivity, and increases glucose uptake into cells [619–621]. The effects of vanadate on insulin sensitivity may be due more to the ability of this metal to increase intracellular levels of Mg, which are determinants of insulin action, than to the ability of vanadate to mimic insulin [622]. Furthermore, vanadate also induces the recruitment of the glucose transporter GLUT-4 to the plasma membrane of adipocytes [623].

10.8 SUMMARY

As humans are becoming exposed to greater amounts of toxic metals in the workplace and environment, it is imperative that we understand how these metals affect essential cellular processes and, ultimately, human health. Molecules containing toxic metals have found ways of mimicking endogenous molecules in order to gain access to cells via essential transporters. Furthermore, some metals have been shown to mimic endogenous intracellular molecules, thereby interfering with essential cellular processes. Understanding how toxic metals interact with extracellular and intracellular proteins will prove to be important to the development of treatment regimens of diseases induced by exposure to these metals.

ACKNOWLEDGMENTS

The authors would like to thank Ms Emilie Bullard and Ms Lucy Joshee for editorial assistance. This work was supported by grants from the National Institutes of Health (National Institute of Environmental Health Sciences; NIEHS) awarded to C.C. Bridges (ES015511) and R.K. Zalups (ES05980).

REFERENCES

1. Clarkson, T.W., Molecular and ionic mimicry of toxic metals, *Annu. Rev. Pharmacol. Toxicol.*, 32, 545, 1993.
2. Ballatori, N., Transport of toxic metals by molecular mimicry, *Environ. Health Perspect.*, 110(Suppl. 5), 689, 2002.
3. Bridges, C.C. and Zalups, R.K., Molecular and ionic mimicry and the transport of toxic metals, *Toxicol. Appl. Pharmacol.*, 204, 274, 2005.
4. Zalups, R.K., Molecular interactions with mercury in the kidney, *Pharmacol. Rev.*, 52, 113, 2000.
5. Zalups, R.K. and Ahmad, S., Molecular handling of cadmium in transporting epithelia, *Toxicol. Appl. Pharmacol.*, 186, 163, 2003.
6. Wetterhahn-Jennette, K., The role of metals in carcinogenesis: Biochemistry and metabolism, *Environ. Health Perspect.*, 40, 233, 1981.
7. Agency for Toxic Substance and Disease Registry, Toxicological Profile for Arsenic, U.S. Department of Health and Humans Services, Public Health Service, Centers for Disease Control, Atlanta, GA, 2007.
8. Environmental Protection Agency, Integrated Risk Information System (IRIS) on Arsenic. National Center for Environmental Assessment, Office of Research and Development, Washington, DC, 1999.
9. Civantos, D.P., Rodriquez, A.L., Aguado-Borruey, J.M., and Narvaez, J.A., Fulminant malignant arrhythmia and multiorgan failure in acute arsenic poisoning, *Chest*, 108, 1774, 1995.
10. Dunlap, L.G., Perforations of the nasal septum due to inhalation of arsenous oxide, *J. Am. Med. Assoc.*, 76, 568, 1921.
11. Lundgren, K.D., Damage to respiratory organs in workers in a smelting plant, *Nord. Hyg. Tidskr.*, 3, 66, 1954.
12. Morton, W.E. and Caron, G.A., Encephalopathy: An uncommon manifestation of workplace arsenic poisoning? *Am. J. Ind. Med.*, 15, 1, 1989.
13. Pinto, S.S. and McGill, C.M., Arsenic trioxide exposure in industry, *Ind. Med. Surg.*, 22, 281, 1953.
14. Sandstrom, A.I.M., Wall, S.G.I., and Taube, A., Cancer incidence and mortality among Swedish smelter workers, *Brit. J. Ind. Med.*, 46, 82, 1989.
15. Armstrong, C.W., Stroube, R.B., Rubio, T., Siudyla, E.A., and Miller, G.B., Jr., Outbreak of fatal arsenic poisoning caused by contaminated drinking water, *Arch. Environ. Health*, 39, 276, 1984.

16. Blom, S., Lagerkvist, B., and Linderholm, H., Arsenic exposure to smelter workers: Clinical and neurophysiological studies, *Scand. J. Work. Environ. Health*, 11, 265, 1985.
17. Feldman, R.G., Niles, C.A., Kelly-Hayes, M., Sax, D.S., Dixon, W.J., Thompson, D.J., and Landau, E., Peripheral neuropathy in arsenic smelter workers, *Neurology*, 29, 939, 1979.
18. Lagerkvist, B.J. and Zetterlund, B., Assessment of exposure to arsenic among smelter workers: A five-year follow up, *Am. J. Ind. Med.*, 25, 477, 1994.
19. Cullen, N.M., Wolf, L.R., and St. Clair, D., Pediatric arsenic ingestion, *Am. J. Emerg. Med.*, 13, 432, 1995.
20. Jensen, G.E. and Hansen, M.L., Occupational arsenic exposure and glycosylated haemoglobin, *Analyst*, 123, 77, 1998.
21. Lagerkvist, B., Linderholm, H., and Nordberg, G.F., Vasospastic tendency and Raynaud's phenomenon in smelter workers exposed to arsenic, *Environ. Res.*, 39, 465, 1986.
22. Lagerkvist, B.E.A., Linderholm, H., and Nordberg, G.F., Arsenic and Raynaud's phenomenon: Vasospastic tendency and excretion of arsenic in smelter workers before and after the summer vacation, *Int. Arch. Occup. Environ. Health*, 60, 361, 1988.
23. Bickley, L.K. and Papa, C.M., Chronic arsenicism with vitiligo, hyperthyroidism, and cancer, *N. J. Med.*, 86, 377, 1989.
24. Holmqvist, I., Occupational arsenical dermatitis: A study among employees at a copper ore smelting work including investigations of skin reactions to contact with arsenic compounds, *Acta. Derm. Venerol.*, 31(Suppl. 26), 1, 1951.
25. Mohamed, K.B., Occupational contact dermatitis from arsenic in a tin-smelting factory, *Contact Dermatitis*, 38, 224, 1988.
26. Perry, K., Bowler, R.G., Buckell, H.M., Druett, H.A., and Schilling, R.S.F., Studies in the incidence of cancer in factory handling inorganic compounds of arsenic—II: Clinical and environmental investigations, *Brit. J. Ind. Med.*, 5, 6, 1948.
27. Niu, C., Yan, J., Yu, T., et al., Studies on treatment of acute promyelocytic leukemia with arsenic trioxide: Remission induction, follow-up, and molecular monitoring in 11 newly diagnosed and 47 relapsed acute promyelocytic leukemia patients, *Blood*, 94, 3315, 1999.
28. Soignet, S.L., Maslak, P., Wang, Z.G., et al., Complete remission after treatment of acute promyelocytic leukemia with arsenic trioxide, *New Engl. J. Med.*, 339, 1341, 1998.
29. Bobé, P. and Chelbi-Alix, M.K., New therapeutic perspectives for arsenic: From acute promyelocytic leukemia to autoimmune diseases, *Med. Sci (Paris)*, 24, 967, 2008.
30. Abernathy, C.O., Thomas, D.J., and Calderon, R.L., Health effects and risk assessment of arsenic, *J. Nutr.*, 133, 1536S, 2003.
31. Smith, A.H., Lingas, E.O., and Rahman, M., Contamination of drinking-water by arsenic in Bangladesh: A public health emergency, *Bull. World Health Organ.*, 78, 1093, 2000.
32. Stoica, A., Pentecost, E., and Martin, M.B., Effects of arsenite on estrogen receptor-α expression and activity in MCF-7 breast cancer cells, *Endocrinology*, 141, 3595, 2000.
33. Chow, S.K.Y., Chan, J.Y.W., and Fung, K.P., Suppression of cell proliferation and regulation of estrogen receptor α signaling pathway by arsenic trioxide on human breast cancer MCF-7 cells, *J. Endocrinol.*, 182, 325, 2004.
34. Davey, J.C., Bodwell, J.E., Gosse, J.A., and Hamilton, J.W., Arsenic as an endocrine disruptor: Effects of arsenic on estrogen receptor-mediated gene expression *in vivo* and in cell culture, *Toxicol. Sci.*, 98, 75, 2007.
35. Laparra, J.M., Velez, D., Barbera, R., Granero, L., Polache, A., Montoro, R., and Farre, R., Cytotoxic effect of As(III) in Caco-2 cells and evaluation of its human intestinal permeability, *Toxicol. In Vitro*, 20, 658, 2006.
36. Ito, K., Suzuki, H., Horie, T., and Sugiyama, Y., Apical/basolateral surface expression of drug transporters and its role in vectorial drug transport, *Pharm. Res.*, 22, 1559, 2005.
37. Lu, W.J., Tamai, I., Nezu, J., Lai, M.L., and Huang, J.D., Organic anion transporting polypeptide-C mediates arsenic uptake in HEK-293 cells, *J. Biomed. Sci.*, 13, 525, 2006.
38. Canonne-Hergaux, F., Gruenheid, P., Ponka, P., and Gros, P., Cellular and subcellular localization of the Nramp2 iron transporter in the intestinal brush border and regulation by dietary iron, *Blood*, 93, 4406, 1999.
39. Okubo, M., Yamada, K., Hosoyamada, M., Shibasaki, T., and Endou, H., Cadmium transport by human Nramp2 expressed in *Xenopus laevis* oocytes, *Toxicol. Appl. Pharmacol.*, 187, 162, 2003.
40. Park, J.D., Cherrington, N.J., and Klaassen, C.D., Intestinal absorption of cadmium is associated with divalent metal transporter 1 in rats, *Toxicol. Sci.*, 68, 288, 2002.
41. Ilbäck, N.G., Frisk, P., Tallkvist, J., Gadhasson, I.L., Blomberg, J., and Friman, G., Gastrointestinal uptake of trace elements are changed during the course of a common human viral (Coxsackievirus B3) infection in mice, *J. Trace Elem. Med. Biol.*, 22, 120, 2008.

42. Kimura, A., Ishida, Y., Wada, T., Yokoyama, H., Mukaida, N., and Kondo, T., MRP-1 expression levels determine strain-specific susceptibility to sodium arsenic-induced renal injury between C57BL/6 and BALB/c mice, *Toxicol. Appl. Pharmacol.*, 203, 53, 2005.
43. Evers, R., Zaman, G.J., van Deemter, L., Jansen, H., Calafat, J., Oomen, L.C., Oude Elferink, R.P., Borst, P., and Schinkel, A.H., Basolateral localization and export activity of the human multidrug resistance-associated protein in polarized pig kidney cells, *J. Clin. Invest.*, 97, 1211, 1996.
44. Kimura, A., Ishia, Y., Hayashi, T., Wada, T., Yokoyama, H., Sugaya, T., Mukaida, N., and Kondo, T., Interferon-γ plays protective roles in sodium arsenite-induced renal injury by up-regulating intrarenal multidrug resistance-associated protein 1 expression, *Am. J. Pathol.*, 169, 1118, 2006.
45. Kitchin, K.T. and Wallace, K., Arsenite binding to synthetic peptides based on the Zn finger region and the estrogen binding region of the human estrogen receptor-α, *Toxicol. Appl. Pharmacol.*, 206, 66, 2005.
46. Scott, N., Hatlelid, K.M., MacKenzie, N.E., and Carter, D.E., Reactions of arsenic(III) and arsenic(V) species with glutathione, *Chem. Res. Toxicol.*, 6, 102, 1993.
47. Delnomdedieu, M., Basti, M.M., Otvos, J.D., and Thomas, D.J., Reduction and binding of arsenate and dimethylarsinate by glutathione: A magnetic resonance study, *Chem. Biol. Interact.*, 90, 139, 1994.
48. Kala, S.V., Kala, G., Prater, C.I., Sartorelli, A.C., and Lieberman, M.W., Formation and urinary excretion of arsenic triglutathione and methylarsenic diglutathione, *Chem. Res. Toxicol.*, 17, 243, 2004.
49. Percy, A.J. and Gailer, J., Methylated trivalent arsenic-glutathione complexes are more stable than their arsenite analog, *Bioinorg. Chem. Appl.*, 2008. http://www.hindawi.com/journals/bca/2008/539082.html.
50. Cui, X., Kobayashi, Y., Hayakawa, T., and Hirano, S., Arsenic speciation in bile and urine following oral and intravenous exposure to inorganic and organic arsenics in rats, *Toxicol. Sci.*, 82, 478, 2004.
51. Leslie, E.M., Haimeur, A., and Waalkes, M.P., Arsenic transport by the human multidrug resistance protein (MRP1/ABCC1): Evidence that a tri-glutathione conjugate is required, *J. Biol. Chem.*, 279, 32700, 2004.
52. Schaub, T.P., Kartenbeck, J., Konig, J., Spring, H., Dorsam, J., Staehler, G., Storkel, S., Thon, W.F., and Keppler, D., Expression of the conjugate export pump encoded by the *mrp2* gene in the apical membrane of kidney proximal tubules, *J. Am. Soc. Nephrol.*, 8, 1213, 1997.
53. Zhou, S.F., Wang, L.L., Di, Y.M., Xue, C.C., Duan, W., Li, C.G., and Li, Y., Substrates and inhibitors of human multidrug resistance associated proteins and the implications in drug development, *Curr. Med. Chem.*, 15, 1981, 2008.
54. Miller, D.S., Shaw, J.R., Stanton, C.R., Barnaby, R., Karlson, K.H., Hamilton, J.W., and Stanton, B.A., MRP2 and acquired tolerance to inorganic arsenic in the kidney of killifish (*Fundulus heteroclitus*), *Toxicol. Sci.*, 97, 103, 2007.
55. Lee, T.C., Ho, I.C., Lu, W.J., and Huang, J.D., Enhanced expression of multidrug resistance-associated protein2 and reduced expression of aquaglyceroporin 3 in an arsenic-resistant human cell line, *J. Biol. Chem.*, 281, 18401, 2006.
56. Chin, K.V., Tanaka, S., Darlington, G., Pastan, I., and Gottesman, M.M., Heat shock and arsenite increase expression of the multidrug resistance (MDR1) gene in human renal carcinoma cells, *J. Biol. Chem.*, 265, 221, 1990.
57. Xie, Y., Liu, J., Liu, Y., Klaassen, C.D., and Waalkes, M.P., Toxicokinetic and genomic analysis of chronic arsenic exposure in multidrug resistance mdr1a/1b(−/ −) double knockout mice, *Mol. Cell Biochem.*, 255, 11, 2004.
58. Ecelbarger, C.A., MacNeil, G.G., and Greger, J.L., Aquaporin-3 water channel localization and regulation in rat kidney, *Am. J. Physiol.*, 269, F663, 1995.
59. Spears, K.J., Ross, J., Stenhouse, A., Ward, C.J., Goh, L.B., Wolf, C.R., Morgan, P., Ayrton, A., and Friedberg, T.H., Directional trans-epithelial transport of organic anions in porcine LLC-PK1 cells that co-express human OATP1B1 (OATP-C) and MRP2, *Biochem. Pharmacol.*, 69, 415, 2005.
60. Sekine, T., Miyazaki, H., and Endou, H., Molecular physiology of renal organic anion transporters, *Am. J. Physiol. Renal Physiol.*, 290, F251, 2006.
61. Csanaky, I. and Gregus, Z., Role of glutathione in reduction of arsenate and of gamma-glutamyltranspeptidase in disposition of arsenite in rats, *Toxicol.*, 207, 91, 2005.
62. Gao, Y., Pei, Q.L., Li, G.X., Han, G., Tian, F.J., Qin, X.J., Zhang, R., Hou, W.S., and Li, X.Y., Effects of MRP2-GSH cotransport system on hepatic arsenic metabolism in rats, *Zhonghua Lao Dong Wei Sheng Zhi Ye Bing Za Zhi*, 24, 278, 2006.
63. Gyurasics, A., Varga, F., and Gregus, Z., Glutathione-dependent biliary excretion of arsenic, *Biochem. Pharmacol.*, 42, 465, 1991.
64. Cole, S.P., Sparks, K.E., Fraser, K., Loe, D.W., Grant, C.E., Wilson, G.M., and Deeley, R.G., Pharmacological characterization of multidrug resistant MRP-transfected human tumor cells, *Cancer Res.*, 54, 5902, 1994.

65. Kala, S.V., Neely, M.W., Kala, G., Prater, C.I., Atwood, D.W., Rice, J.S., and Lieberman, M.W., The MRP2/cMOAT transporter and arsenic-glutathione complex formation are required for biliary excretion of arsenic, *J. Biol. Chem.*, 275, 33404, 2000.
66. Li, G.X., Pei, Q.L., Gao, Y., Liu, K.M., Nie, J.S., Han, G., Qiu, Y.L., and Zhang, W.P., Protective effects of hepatocellular canalicular conjugate export pump (Mrp2) on sodium arsenite-induce hepatic dysfunction in rats, *Exp. Toxicol. Pathol.*, 58, 447, 2007.
67. Liu, J., Chen, H., Miller, D.S., Saavedra, J.E., Keefer, L.K., Johnson, D.R., Klaassen, C.D., and Waalkes, M.P., Overexpression of glutathione *S*-transferase II and multidrug resistance transport proteins is associated with acquired tolerance to inorganic arsenic, *Mol. Pharmacol.*, 60, 302, 2001.
68. Zaman, G.J.R., Lankelma, J., van Tellingen, O., Beijnen, J., Dekker, H., Paulusma, C., Oude Elferink, R.P.J., Baas, F., and Borst, P., Role of glutathione in the export of compounds from cells by the multidrug-resistance associated protein, *Proc. Natl. Acad. Sci. USA*, 92, 7690, 1995.
69. Keppler, D., Leier, I., and Jedlitschky, G., Transport of glutathione conjugates and glucuronides by the multidrug resistance proteins MRP1 and MRP2, *Biol. Chem.*, 378, 787, 1997.
70. Broeks, A., Gerrard, B., Allikments, R., Dean, M., and Plasterk, R.H., Homologues of the human multidrug resistance genes MRP and MDR contribute to heavy metal resistance in the soil nematode *Caenorhabditis elegans*, *EMBO J.*, 15, 6132, 1996.
71. Vernhet, L., Seite, M.P., Allain, N., Guillouzo, A., and Fardel, O., Arsenic induces expression of the multidrug resistance-associated protein 2 (MRP2) gene in primary rat and human hepatocytes, *J. Pharmacol. Exp. Ther.*, 298, 234, 2001.
72. König, J., Cui, Y., Nies, A.T., and Keppler, D., A novel human organic anion transporting polypeptide localized to the basolateral hepatocyte membrane, *Am. J. Physiol. Gastrointest. Liver Physiol.*, 278, G156, 2000.
73. Agency for Toxic Substance and Disease Registry, Toxicological Profile for Cadmium, U.S. Department of Health and Humans Services, Public Health Service, Centers for Disease Control, Atlanta, GA, 2007.
74. Oberdorster, G., Pulmonary deposition, clearance and effects of inhaled soluble and insoluble cadmium compounds, *IARC Sci. Publ.*, 118, 189, 1992.
75. Elinder, C.G., Kjellstrom, T., Lind, B., Linnman, L., and Friberg, L., Cadmium in kidney cortex, liver and pancreas from Swedish autopsies, *Arch. Environ. Health*, 31, 292, 1976.
76. Lewis, G.F., Coughlin, L., Jusko, W., and Hartz, S., Contribution of cigarette smoking of cadmium accumulation in man, *Lancet*, 1, 291, 1972.
77. Bhattacharyya, M.H., Wilson, A.K., Tajan, S.S., and Jonah, M., Biochemical pathways in cadmium toxicity. In: R.K. Zalups and J. Koropatnick (Eds), *Molecular Biology and Toxicology of Metals*, p. 34, Taylor & Francis, London, 2000.
78. Diamond, G.L. and Zalups, R.K., Understanding renal toxicity of heavy metals, *Toxicol. Pathol.*, 26, 92, 1998.
79. Friberg, L., Elinder, C.G., Kjellstron, T., et al., General summary and conclusions and some aspects of diagnosis and treatment of chronic cadmium poisoning. In: L. Friberg, C.G. Elinder, T. Kjellstrom, et al. (Eds), *Cadmium and Health: A Toxicological and Epidemiological Appraisal*, Vol. 2, p. 247, CRC Press, Boca Raton, FL, 1986.
80. Goyer, R.A., Cherian, M.G., and Delaquerriere-Richardson, L., Correlation of parameters of cadmium exposure with onset of cadmium-induced nephropathy in rats, *J. Environ. Pathol. Toxicol. Oncol.*, 5, 89, 1984.
81. Habeebu, S.S., Liu, J., and Klaassen, C.D., Cadmium-induced apoptosis in mouse liver, *Toxicol. Appl. Pharmacol.*, 149, 203, 1998.
82. Jarup, L., Berglund, M., Elinder, C.G., Nordberg, G., and Vahter, M., Health effects of cadmium exposure: A review of the literature and a risk estimate, *Scand. J. Work Environ. Health*, 1, 1, 1998.
83. Kamiyama, T., Miyakawa, H., Li, J.P., Akiba, T., Liu, J.H., Liu, J., Marumo, F., and Sato, C., Effects of one-year cadmium exposure on livers and kidneys and their relation to glutathione levels, *Res. Commun. Mol. Pathol. Pharmacol.*, 88, 177, 1995.
84. Kazantizis, G., Some long-term effects of cadmium on the human kidney. In: *Cadmium 77. Proceedings of the 1st International Cadmium Conference in San Francisco, 1977*, p. 194, Metal Bulletin, Ltd., London, 1978.
85. Liu, J., Habeebu, S.S., Liu, Y., and Klaassen, C.D., Acute CdMT injection is not a good model to study chronic Cd nephropathy: Comparison of chronic $CdCl_2$ and CdMT exposure with acute CdMT injection in rats, *Toxicol. Appl. Pharmacol.*, 153, 48, 1998.
86. Liu, J., Liu, Y., Habeebu, S.S., and Klaassen, C.D., Susceptibility of MT-null mice to chronic $CdCl_2$-induced nephrotoxicity indicated that renal injury is not mediated by the CdMT complex, *Toxicol. Sci.*, 46, 197, 1998.

87. Liu, J., Liu, Y., Habeebu, S.M., Waalkes, M.P., and Klaassen, C.D., Chronic combined exposure to cadmium and arsenic exacerbates nephrotoxicity, particularly in metallothionein-I/II null mice, *Toxicology*, 147, 157, 2000.
88. Min, K.S., Kobayashi, K., Onosaka, S., Ohta, N., Okada, Y., and Tanaka, K., Tissue distribution of cadmium and nephropathy after administration of cadmium in several chemical forms, *Toxicol. Appl. Pharmacol.*, 86, 262, 1986.
89. Min, K.S., Onosaka, S., and Tanaka, K., Renal accumulation of cadmium and nephropathy following long-term administration of cadmium-metallothionein, *Toxicol. Appl. Pharmacol.*, 141, 102, 1996.
90. Nordberg, M. and Nordberg, G.F., Toxicological aspects of metallothionein, *Cell Mol. Biol.*, 46, 451, 2000.
91. Nordberg, G.F., Kjellstrom, T., and Nordberg, M., Kinetics and metabolism. In: L. Friberg, C.G. Elinder, and T. Kjellstrom (Eds), *Cadmium and Health: A Toxicological and Epidemiological Appraisal, Exposure, Dose, and Metabolism*, Vol. 1, p. 103, CRC Press, Boca Raton, FL, 1985.
92. Oteiza, P.I., Adonaylo, V.N., and Keen, C.L., Cadmium-induced testes oxidative damage in rats can be influenced by dietary zinc intake, *Toxicology*, 137, 13, 1999.
93. Sarkar, S., Yadav, P., and Bhatnagar, D., Lipid peroxidative damage on cadmium exposure and alterations in antioxidant system in rat erythrocytes: A study with relation to time, *Biometals*, 11, 153, 1998.
94. Zalups, R.K., Gelein, R.M., and Cherian, M.G., Shifts in the dose-effect relationship for the nephropathy induced by cadmium-metallothionein in rats after a reduction of renal mass, *J. Pharmacol. Exp. Ther.*, 262, 1256, 1992.
95. Rabenstein, D.L., Metal complexes of glutathione and their biological significance. In: Dolphin, O. Auramovibc, and R. Poulson (Eds), *Glutathione: Chemical, Biochemical, and Medical Aspects of Coenzymes and Cofactors*, Vol. 3, p. 147, Wiley, New York.
96. Rabenstein, D.L., Isab, A.A., Kadima, W., and Mohanakrishnan, P., A proton nuclear magnetic resonance study of the interaction of cadmium with human erythrocytes, *Biochim. Biophys. Acta.*, 762, 531, 1983.
97. Henson, M.C. and Chedrese, P.J., Endocrine disruption by cadmium, a common environmental toxicant with paradoxical effects on reproduction, *Exp. Biol. Med.*, 229, 383, 2004.
98. Garcia-Morales, P., Saceda, M., Kenney, N., Kim, N., Salomon, D.S., Gottardis, M.M., Soloman, H.B., Sholler, P.F., Jordan, V.C., and Martin, M.B., Effect of cadmium on estrogen receptor levels and estrogen-induced responses in human breast cancer cells, *J. Biol. Chem.*, 269, 16896, 1994.
99. Martin, M.B., Reiter, R., Pham, T., et al., Estrogen-like activity of metals in Mcf-7 breast cancer cells, *Endocrinology*, 144, 2425, 2003.
100. Stoica, A., Katzenellenbogen, B.S., and Martin, M.B., Activation of estrogen receptor-α by the heavy metal cadmium, *Mol. Endocrinol.*, 14, 545, 2000.
101. Nesatyy, V.J., Rutishauser, B.V., Eggen, R.I.L., and Suter, M.J.F., Identification of the estrogen receptor Cd-binding sites by chemical modification, *Analyst*, 130, 1087, 2005.
102. Johnson, M.D., Kenney, N., Stoica, A., et al., Cadmium mimics the *in vivo* effects of estrogen in the uterus and mammary gland, *Nat. Med.*, 9, 1081, 2003.
103. Brama, M., Gnessi, L., Basciani, S., et al., Cadmium induces mitogenic signaling in breast cancer cell by an ER alpha-dependent mechanism, *Mol. Cell Endocrinol.*, 264, 102, 2007.
104. Cantor, K.P., Stewart, P.A., Brinton, L.A., and Dosemeci, M., Occupational exposures and female breast cancer mortality in the United States, *J. Occup. Med.*, 37, 336, 1995.
105. McElroy, J.A., Shafer, M.M., Trentham-Dietz, A., Hampton, J.M., and Newcomb, P.A., Cadmium exposure and breast cancer risk, *J. Natl. Cancer Inst.*, 98, 869, 2006.
106. Martin, M.B., Voeller, H.J., Gelmann, E.P., et al., Role of cadmium in the regulation of AR gene expression and activity, *Endocrinol.*, 143, 263, 2002.
107. Takiguchi, M. and Yoshihara, S., New aspects of cadmium as endocrine disruptor, *Environ. Sci.*, 13, 107, 2006.
108. Andersen, O., Nielsen, J.B., Sorensen, J.A., and Scherrebeck, L., Experimental localization of intestinal uptake sites for metals (Cd, Hg, Zn, Se) *in vivo* in mice, *Environ. Health Perspect.*, 102 (Suppl. 3), 199, 1994.
109. Bothwell, T.H., Charlton, R.W., Cook, J.D., and Finch, C.A., *Iron Metabolism in Man*, p. 7, Blackwell Science, Oxford, 1979.
110. Crichton, R.R., *Inorganic Biochemistry of Iron Metabolism*, p. 29, Horwood, West Sussex, 1991.
111. Conrad, M.E. and Umbreit, J.N., Iron absorption and transport, *Am. J. Hematol.*, 64, 287, 2000.
112. Conrad, M.E. and Umbreit, J.N., Pathways of iron absorption, *Blood Cells, Mol. Dis.*, 29, 336, 2002.
113. Bunn, C.R. and Matrone, G., *In vivo* interactions of cadmium, copper, zinc and iron in the mouse and rat, *J. Nutr.*, 90, 395, 1966.

114. Hamilton, D.L. and Valberg, L.S., Relationship between cadmium and iron absorption, *Am. J. Physiol.*, 227, 1033, 1974.
115. Hill, C.H., Matrone, G., Payne, W.L., and Barber, C.W., *In vivo* interactions of cadmium with copper, zinc, and iron, *J. Nutr.*, 80, 227, 1963.
116. Leon, L. and Johnson, D.R., Role of iron in jejunal uptake of cadmium in the newborn rat, *J. Toxicol. Environ. Health*, 15, 687, 1985.
117. Schafer, S.G. and Forth, W., Effect of acute and subchronic exposure to cadmium on the retention of iron in rats, *J. Nutr.*, 114, 1989, 1984.
118. Griffiths, W.J.H., Kelly, A.L., Smith, S.J., and Cox, T.M., Localization of iron transport and regulatory proteins in human cells, *Q. J. Med.*, 93, 575, 2000.
119. Gruenheid, S., Cellier, M., Vidal, S., and Gros, P., Identification and characterization of a second mouse Nramp gene, *Genomics*, 25, 514, 1995.
120. Gunshin, H., Mackenzie, B., Berger, U.V., Gunshin, Y., Romero, M.F., Boron, W.F., Nussberger, S., Gollan, J.L., and Hediger, M.A., Cloning and characterization of a mammalian proton-coupled metal-ion transporter, *Nature*, 388, 482, 1997.
121. Ma, Y., Yeh, M., Yeh, K.Y., and Glass, J., Iron imports. V. Transport of iron through the intestinal epithelium, *Am. J. Physiol. Gastrointest. Liver Physiol.*, 290, G417, 2006.
122. Tandy, S., Williams, M., Leggett, A., Lopez-Jimenez, M., Dedes, M., Ramesh, B., Srai, S.K., and Sharp, P., Nramp2 expression is associated with pH-dependent iron uptake across the apical membrane of human intestinal Caco-2 cells, *J. Biol. Chem.*, 275, 1023, 2000.
123. Trinder, D., Oates, P.S., Thomas, C., Sadleir, J., and Morgan, E.H., Localization of divalent metal transporter 1 (DMT1) to the microvillus membrane of rat duodenal enterocytes in iron deficiency, but to hepatocytes in iron overload, *Gut*, 46, 270, 2000.
124. Anderson, G.J., Frazer, D.M., McKie, A.T., Wilkins, S.J., and Vulpe, C.D., The expression and regulation of the iron transport molecules hephaestin and IREG1: Implications for the control of iron export from the small intestine, *Cell Biochem. Biophys.*, 36, 137, 2002.
125. Martini, L.A., Tchack, L., and Wood, R.J., Iron treatment down regulates DMT1 and IREG1 mRNA expression in Caco-2 cells, *J. Nutr.*, 132, 693, 2002.
126. Rolfs, A., Bonkovsky, H.L., Kohlroser, J.G., McNeal, K., Sharma, A., Berger, U.V., and Hediger, M.A., Intestinal expression of genes involved in iron absorption in humans, *Am. J. Physiol. Gastrointest. Liver Physiol.*, 282, G598, 2002.
127. Wareing, M., Ferguson, C.J., Delannoy, M., Cox, A.G., McMahon, R.F., Green, R., Riccardi, D., and Smith, C.P., Altered dietary iron intake is a strong modulator of renal DMT1 expression, *Am. J. Physiol. Renal Physiol.*, 285, F1050, 2003.
128. Zoller, H., Koch, R.O., Theurl, I., Obrist, P., Pietrangelo, A., Montosi, G., Haile, D.J., Vogel, W., and Weiss, G., Expression of the duodenal iron transporters divalent-metal transporter 1 and ferroportin 1 in iron deficiency and iron overload, *Gastroenterology*, 120, 1412, 2001.
129. Zoller, H., Theurl, I., Koch, R., Kaser, A., and Weiss, G., Mechanisms of iron mediated regulation of the duodenal iron transporters divalent metal transporter 1 and ferroportin 1, *Blood Cells Mol. Dis.*, 29, 488, 2002.
130. Berglund, M., Akesson, A., Nermell, B., and Vahter, M., Intestinal absorption of dietary cadmium in women depends on body iron stores and fiber intake, *Environ. Health Perspect.*, 102, 1058, 1994.
131. Choudhury, H., Harvey, T., Thayer, W.C., Lockwood, T.F., Stiteler, W.M., Goodrum, P.E., Hassett, J.M., and Diamon, G.L., Urinary cadmium elimination as a biomarker of exposure for evaluating a cadmium dietary exposure-biokinetics model, *J. Toxicol. Environ. Health*, 63, 321, 2001.
132. Flanagan, P.R., McLellan, J.S., Haist, J., Cherian, G., Chamberlain, M.J., and Valberg, L.S., Increased dietary cadmium absorption in mice and human subjects with iron deficiency, *Gastroenterology*, 74, 841, 1978.
133. Fox, M.R., Jacobs, R.M., Jones, A.O., Fry, B.E., Jr., and Stone, C.L., Effects of vitamin C and iron on cadmium metabolism, *Ann. NY Acad. Sci.*, 355, 249, 1980.
134. Groten, J.P., Luten, J.B., and van Bladeren, P.J., Dietary iron lowers the intestinal uptake of cadmium-metallothionein in rats, *Eur. J. Pharmacol.*, 228, 23, 1992.
135. Bressler, J.P., Olivi, L., Cheong, J.H., Kim, Y., and Bannon, D., Divalent metal transporter 1 in lead and cadmium transport, *Ann. NY. Acad. Sci.*, 1012, 142, 2004.
136. Elisma, F. and Jumarie, C., Evidence for cadmium uptake through Nramp2: Metal speciation studies with Caco-2 cells, *Biochem. Biophys. Res. Commun.*, 20, 662, 2001.
137. Scheuhammer, A.M. and Cherian, M.G., Quantification of metallothioneins by a silver-saturation method, *Toxicol. Appl. Pharmacol.*, 82, 417, 1986.

138. Leazer, T.M., Liu, Y., and Klaassen, C.D., Cadmium absorption and its relationship to divalent metal transporter-1 in the pregnant rat, *Toxicol. Appl. Pharmacol.*, 185, 18, 2002.
139. Lecoeur, S., Huynh-Delerme, C., Blais, A., Duche, A., Tome, D., and Kolf-Clauw, M., Implication of distinct proteins in cadmium uptake and transport by intestinal cells HT-29, *Cell. Biol. Toxicol.*, 18, 409, 2002.
140. Picard, V., Govoni, G., Jabado, N., and Gros, P., Nramp 2 (DCT1/DMT1) expressed at the plasma membrane transports iron and other divalent cations into a calcein-accessible cytoplasmic pool, *J. Biol. Chem.*, 275, 35738, 2000.
141. Tallkvist, J., Bowlus, C.L., and Lonnerdal, B., DMT1 gene expression and cadmium absorption in human absorptive enterocytes, *Toxicol. Lett.*, 122, 171, 2001.
142. Ryu, D.Y., Lee, S.J., Park, D.W., Choi, B.S., Klaassen, C.D., and Park, J.D., Dietary iron regulates intestinal cadmium absorption through iron transporters in rats, *Toxicol. Lett.*, 152, 19, 2004.
143. Bannon, D.I., Aboundader, R., Lees, P.S.J., and Bressler, J.P., Effect of DMT1 knockdown in iron, cadmium, and lead uptake in Caco-2 cells, *Am. J. Physiol.*, 28, C44, 2003.
144. Suzuki, T., Momoi, K., Hosoyamada, M., Kimura, M., and Shibasaki, T., Normal cadmium uptake in microcytic anemia mk/mk mice suggests that DMT1 is not the only cadmium transporter *in vivo*, *Toxicol. Appl. Pharmacol.*, 227, 462, 2008.
145. Bergeron, P.M. and Jumarie, C., Characterization of cadmium uptake in human intestinal crypt cells HIEC in relation to inorganic metal speciation, *Toxicology*, 219, 156, 2006.
146. Bergeron, P.M. and Jumarie, C., Reciprocal inhibition of Cd(2 +) and Ca(2 +) uptake in human intestinal crypt cells for voltage-independent Zn-activated pathway, *Biochem. Biophys. Acta*, 1758, 702, 2006.
147. Tacnet, F., Watkins, D.W., and Ripoche, P., Studies of zinc transport into brush-border membrane vesicles isolated from pig small intestine, *Biochim. Biophys. Acta*, 1024, 323, 1990.
148. Tacnet, F., Watkins, D.W., and Ripoche, P., Zinc binding in intestinal brush-border membrane isolated from pig, *Biochim. Biophys. Acta*, 1063, 51, 1991.
149. Cragg, R.A., Christie, G.R., Phillips, S.R., Russi, R.M., Kury, S., Mathers, J.C., Taylor, P.M., and Ford, D., A novel zinc-regulated human zinc transporter, hZTL1, is localized to the enterocyte apical membrane, *J. Biol. Chem.*, 277, 22789, 2001.
150. Dalton, T.P., He, L., Wang, B., Miller, M.L., Jin, L., Stringer, K.F., Chang, X., Baxter, C.S., and Nebert, D.W., Identification of mouse SLC39A8 as the transporter responsible for cadmium-induced toxicity in the testis, *Proc. of Natl. Acad. Sci. USA*, 102, 3401, 2005.
151. Girijashanker, K., He, L., Soleimani, M., Reed, J.M., Li, H., Liu, Z., Wang, B., Dalton, T.P., and Nebert, D.W., Slc39a14, gene encodes ZIP14, a metal/bicarbonate symporter: Similarities to the ZIP8 transporter, *Mol. Pharmacol.*, 73, 1413, 2008.
152. He, L., Girijashanker, K., Dalton, T.P., Reed, J., Li, H., Soleimani, M., and Nebert, D.W., ZIP8, member of the solute-carrier-39 (SLC39) metal-transporter family: Characterization of transporter properties, *Mol. Pharmacol.*, 70, 171, 2006.
153. Kaler, P. and Prasad, R., Molecular cloning and functional characterization of novel zinc transporter rZIP10 (Slc39a10) involved in zinc uptake across rat renal brush-border membrane, *Am. J. Physiol. Renal Physiol.*, 292, F217, 2007.
154. Liu, A., Li, H., Soleimani, M., Girijashanker, K., Reed, J.M., He, L., Dalton, T.P., and Nebert, D.W., Cd^{2+} versus Zn^{2+} uptake by the ZIP8 HCO_3-dependent symporter: Kinetics, electrogenicity and trafficking, *Biochem. Biophys. Res. Commun.*, 365, 814, 2008.
155. Wang, B., Schneider, S.N., Dragin, N., et al., Enhanced cadmium-induced testicular necrosis and renal proximal tubule damage caused by gene-dose increase in a *Slc39a8*-transgenic mouse line, *Am. J. Physiol. Cell Physiol.*, 292, C1523, 2007.
156. Jacobson, K.B. and Turner, J.E., The interaction of cadmium and certain other metal ions with proteins and nucleic acids, *Toxicology*, 16, 1, 1980.
157. Blazka, M.E. and Shaikh, Z.A., Differences in cadmium and mercury uptakes by hepatocytes: Role of calcium channels, *Toxicol. Appl. Pharmacol.*, 110, 355, 1991.
158. Blazka, M.E. and Shaikh, Z.A., Cadmium and mercury accumulation in rat hepatocytes: Interactions with other metal ions, *Toxicol. Appl. Pharmacol.*, 113, 118, 1992.
159. Friedman, P.A. and Gesek, F.A., Cadmium uptake by kidney distal convoluted tubule cells, *Toxicol. Appl. Pharmacol.*, 128, 257, 1994.
160. Hinkle, P.M., Kinsella, P.A., and Osterhoudt, K.C., Cadmium uptake and toxicity via voltage-sensitive calcium channels, *J. Biol. Chem.*, 262, 16333, 1987.
161. Souza, V., Bucio, L., and Gutierrez-Ruiz, M.C., Cadmium uptake by a human hepatic cell line (WRL-68 cells), *Toxicology*, 120, 215, 1997.

162. Felley-Bosco, E. and Diezi, J., Dietary calcium restriction enhances cadmium-induced metallothionein synthesis in rats, *Toxicol. Lett.*, 60, 139, 1992.
163. Min, K.S., Iwata, N., Tetsutikawahara, N., Onosaka, S., and Tanaka, K., Effect of hemolytic and iron-deficiency anemia on intestinal absorption and tissue accumulation of cadmium, *Toxicol. Lett.*, 179, 48, 2008.
164. Minn, K.S., Ueda, H., and Tanaka, K., Involvement of intestinal calcium transporter 1 and metallothionein in cadmium accumulation in the liver and kidney of mice fed a low-calcium diet, *Toxicol. Lett.*, 176, 85, 2008.
165. Peng, J.B., Chen, X.Z., Berger, U.V., Vassilev, P.M., Tsukaguchi, H., Brown, E.M., and Hediger, M.A., Molecular cloning and characterization of a channel-like transporter mediating intestinal calcium absorption, *J. Biol. Chem.*, 274, 22739, 1999.
166. Foulkes, E.C., Interactions between metals in rat jejunum: Implications on the nature of cadmium uptake, *Toxicology*, 37, 117, 1985.
167. Foulkes, E.C., On the mechanism of transfer of heavy metals across cell membranes, *Toxicology*, 52, 263, 1988.
168. Foulkes, E.C., Transport of toxic heavy metals across cell membranes, *Proc. Soc. Exp. Biol. Med.*, 223, 234, 2000.
169. Foulkes, E.C. and McMullen, D.M., Kinetics of transepithelial movement of heavy metals in rat jejunum, *Am. J. Physiol.*, 253, G134, 1987.
170. Ballatori, N., Wang, W., and Lieberman, M.W., Accelerated methylmercury elimination in γ-glutamyltranspeptidase-deficient mice, *Am. J. Pathol.*, 152, 1049, 1998.
171. Dutczak, W.J. and Ballatori, N., γ-glutamyltransferase-dependent biliary-hepatic recycling of methylmercury in the guinea pig, *J. Pharmacol. Exp. Ther.*, 262, 619, 1992.
172. Wang, W., Clarkson, T.W., and Ballatori, N., γ-glutamyl transpeptidase and L-cysteine regulate methylmercury uptake by HepG2 cells, a human hepatoma cell line, *Toxicol. Appl. Pharmacol.*, 168, 72, 2000.
173. Cherian, M.G. and Vostal, J.J., Biliary excretion of cadmium in rat. I. Dose-dependent biliary excretion and the form of cadmium in the bile, *J. Toxicol. Environ. Health*, 2, 945, 1977.
174. Cherian, M.G. and Goyer, R.A., Metallothioneins and their role in the metabolism and toxicity of metals, *Life Sci.*, 23, 1, 1978.
175. Kägi, J.H.R. and Vallee, B.L., Metallothionein: A cadmium- and zinc-containing protein from equine renal cortex, *J. Biol. Chem.*, 235, 3460, 1960.
176. Sugawara, N., and Sugawara, C., Gastrointestinal absorption of Cd-metallothionein and cadmium chloride in mice, *Arch. Toxicol.*, 65, 689, 1991.
177. Abboud, S. and Haile, D.J., A novel mammalian iron-regulated protein involved in intracellular iron metabolism, *J. Biol. Chem.*, 275, 19906, 2000.
178. Donovan, A., Brownlie, A., Zhou, Y., et al., Positional cloning of zebrafish ferroportin 1 identifies a conserved vertebrate iron exporter, *Nature (London)*, 403, 776, 2000.
179. McKie, A.T., Marciani, P., Rolfs, A., et al., A novel duodenal iron-regulated transporter, IREG1, implicated in the basolateral transfer of iron to the circulation, *Mol. Cell.*, 5, 299, 2000.
180. Wessling-Resnick, M., Iron imports. III. Transfer of iron from the mucosa into circulation, *Am. J. Physiol. Gastrointest. Liver Physiol.*, 290, G1, 2006.
181. Kim, D.W., Kim, K.Y., Choi, B.S., Youn, P., Ryu, D.Y., Klaassen, C.D., and Park, J.D., Regulation of metal transporters by dietary iron, and the relationship between body iron levels and cadmium uptake, *Arch. Toxicol.*, 81, 327, 2007.
182. Cotzias, G.C., Borg, D.C., and Selleck, B., Specificity of zinc pathway in the rabbit: Zinc-cadmium exchange, *Am. J. Physiol.*, 201, 63, 1961.
183. Cotzias, G.C., Borg, D.C., and Selleck, B., Specificity of zinc pathway through the body: Turnover of Zn^{65} in the mouse, *Am. J. Physiol.*, 202, 359, 1962.
184. Jacobs, R.M., Jones, A.O., Fox, M.R., and Lener, J., Effects of dietary zinc, manganese, and copper on tissue accumulation of cadmium by Japanese quail, *Proc. Soc. Exp. Biol. Med.*, 172, 34, 1983.
185. McMahon, R.J. and Cousins, R.J., Regulation of the zinc transporter ZnT-1 by dietary zinc, *Proc. Natl. Acad. Sci. USA*, 95, 4841, 1998.
186. Palmiter, R.D. and Findley, S.D., Cloning and functional characterization of a mammalian zinc transporter that confers resistance to zinc, *EMBO J.*, 14, 639, 1995.
187. Yu, Y.Y., Kirschke, C.P., and Huang, L., Immunohistochemical analysis of Znt1, 4, 5, 6, and 7 in the mouse gastrointestinal tract, *J. Histochem. Cytochem.*, 55, 223, 2007.

188. Schoenmakers, T.J., Klaren, P.H., Flik, G., Lock, R.A., Pang, P.K., and Bonga, S.E., Actions of cadmium on basolateral plasma membrane proteins involved in calcium uptake by fish intestine, *J. Membrane Biol.*, 127, 161, 1992.
189. Verbost, P.M., Senden, M.H., and van Os, C.H., Nanomolar concentrations of Cd^{2+} inhibit Ca^{2+} transport systems in plasma membranes and intracellular Ca^{2+} stores in intestinal epithelium, *Biochim. Biophys. Acta*, 902, 247, 1987.
190. Friberg, L., Health hazards in the manufacture of alkaline accumulators with special reference to chronic cadmium poisoning, *Acta Med. Stand.*, 138(Suppl. 240), 1, 1950.
191. Barbier, O., Jacquillet, G., Tauc, M., Poujeol, P., and Cougnon, M., Acute study of interaction among cadmium, calcium, and zinc transport along the rat nephron *in vivo*, *Am. J. Physiol. Renal Physiol.*, 287, F1067, 2004.
192. Felley-Bosco, E. and Diezi, J., Fate of cadmium in rat renal tubules: A micropuncture study, *Toxicol. Appl. Pharmacol.*, 98, 243, 1989.
193. Felley-Bosco, E. and Diezi, J., Fate of cadmium in rat renal tubules: A microinjection study, *Toxicol. Appl. Pharmacol.*, 91, 201, 1987.
194. Murakami, M. and Webb, M., A morphological and biochemical study of the effects of L-cysteine on the renal uptake and nephrotoxicity of cadmium, *Brit. J. Exp. Pathol.*, 62, 115, 1981.
195. Murakami, M., Sano, K., and Webb, M., The effect of L-cysteine on the portion-selective uptake of cadmium in the renal proximal tubule, *Arch. Toxicol.*, 60, 365, 1987.
196. Zalups, R.K., Evidence for basolateral uptake of cadmium in the kidneys of rats, *Toxicol. Appl. Pharmacol.*, 164, 15, 2000.
197. Bridges, C.C., Bauch, C., Verrey, F., and Zalups, R.K., Mercuric conjugates of cysteine are transported by the amino acid transporter, system $b^{0,+}$: Implications of molecular mimicry, *J. Am. Soc. Nephrol.*, 15, 663, 2004.
198. Yang, Y., Barfuss, D.W., and Zalups, R.K., Potential mechanisms involved in the absorptive transport of cadmium in isolated perfused rabbit renal proximal tubules. *Tox Lett.* in press.
199. Erfurt, C., Roussa, E., and Thevenod, F., Apoptosis by Cd^{2+} or CdMT in proximal tubule cells: Different uptake routes and permissive role of endo/lysosomal CdMT uptake, *Am. J. Physiol. Cell Physiol.*, 285, C1367, 2003.
200. Murakami, M., Cain, K., and Webb, M., Cadmium-metallothionein-induced nephropathy: A morphological and autoradiographic study of cadmium distribution, the development of tubular damage and subsequent cell regeneration, *J. Appl. Toxicol.*, 3, 237, 1983.
201. Dorian, C., Gattone II, V.H., and Klaassen, C.D., Discrepancy between the nephrotoxic potencies of cadmium-metallothionein and cadmium chloride and the renal concentration of cadmium in the proximal convoluted tubules, *Toxicol. Appl. Pharmacol.*, 130, 161, 1995.
202. Dudley, R.E., Gammal, L.M., and Klaassen, C.D., Cadmium-induced hepatic and renal injury in chronically exposed rats: Likely role of hepatic cadmium-metallothionein in nephrotoxicity, *Toxicol. Appl. Pharmacol.*, 76, 150, 1985.
203. Jonah, M.M. and Bhattacharyya, M.H., Early changes in the tissue distribution of cadmium after oral but not intravenous cadmium exposure, *Toxicol.*, 58, 325, 1989.
204. Dorian, C., Gattone II, V.H., and Klaassen, C.D., Renal cadmium deposition and injury as a result of accumulation of cadmium-metallothionein (CdMT) by the proximal convoluted tubules: A light microscopy autoradiography study with ^{109}CdMT, *Toxicol. Appl. Pharmacol.*, 114, 173, 1992.
205. Dorian, C., Gattone II, V.H., and Klaassen, C.D., Accumulation and degradation of the protein moiety of cadmium-metallothionein (CdMT) in the mouse kidney, *Toxicol. Appl. Pharmacol.*, 117, 242, 1992.
206. Foulkes, E.C., Renal tubular transport of cadmium-metallothionein, *Toxicol. Appl. Pharmacol.*, 45, 505, 1978.
207. Nordberg, M., Jin, T., and Nordberg, G.F., Cadmium, metallothionein and renal tubular toxicity, *IARC Sci. Publ.*, 118, 293, 1992.
208. Webb, M., Role of metallothionein in cadmium metabolism. In: E.C. Foulkes (Ed.), *Cadmium*, p. 281, Springer, Berlin/New York, 1986.
209. Cherian, M.G. and Nordberg, M., Cellular adaptation in metal toxicology and metallothionein, *Toxicology*, 82, 70, 1983.
210. Nordberg, G.F., Goyer, R., and Nordberg, M., Comparative toxicity of cadmium-metallothionein and cadmium chloride on mouse kidney, *Arch. Pathol.*, 99, 192, 1975.
211. Thévenod, F., Nephrotoxicity and the proximal tubule: Insights from cadmium, *Nephron Physiol.*, 93, 87, 2003.

212. Christensen, E.I., and Birn, H., Megalin and cubilin: Synergistic endocytic receptors in renal proximal tubule, *Am. J. Physiol.*, 280, F562, 2001.
213. Klaassen, R.B., Crenshaw, K., Kozyraki, R., Verroust, P.J., Tio, L., Atrian, S., Allen, P.L., and Hammond, T.G., Megalin mediates renal uptake of heavy metal metallothionein complexes, *Am. J. Physiol.*, 287, F393, 2004.
214. Wolff, N.A., Abouhamed, M., Verroust, P.J., and Thévenod, F., Megalin-dependent internalization of cadmium-metallothionein and cytotoxicity in cultured renal proximal tubule cells, *J. Pharmacol. Exp. Ther.*, 318, 782, 2006.
215. Wolff, N.A., Lee, W.K., Abouhamed, M., and Thévenod, F., Role of ARF6 in internalization of metal-binding proteins, metallothionein and transferring, and cadmium-metallothionein toxicity in kidney proximal tubule cells, *Toxicol. Appl. Pharmacol.*, 230, 78, 2008.
216. Abouhamed, M., Wolff, N.A., Lee, W.-K., Smith, C.P., and Thévenod, F., Knockdown of endosomal/lysosomal divalent metal transporter 1 by RNA interference prevents cadmium-metallothionein-1 cytotoxicity in renal proximal tubule cells, *Am. J. Physiol. Renal Physiol.*, 293, F705, 2007.
217. Robinson, M.K., Barfuss, D.W., and Zalups, R.K., Cadmium transport and toxicity in isolated perfused segments of the renal proximal tubule, *Toxicol. Appl. Pharmacol.*, 121, 103, 1993.
218. Endo, T., Transport of cadmium across the apical membrane of epithelial cell lines, *Comp. Biochem. Physiol. C. Toxicol. Pharmacol.*, 131, 223, 2002.
219. Endo, T., Kimura, O., and Sakata, M., Bidirectional transport of cadmium across apical membrane of renal epithelial cell lines via H^+-antiporter and inorganic anion exchanger, *Toxicology*, 131, 183, 1998.
220. Endo, T., Kimura, O., and Sakata, M., pH-dependent transport of cadmium in rat renal brush border membrane vesicles: Cadmium efflux via H^+-antiport, *Toxicol. Lett.*, 99, 99, 1998.
221. Endo, T., Kimura, O., and Sakata, M., Cadmium uptake from apical membrane of LLC-PK1 cells via inorganic anion exchanger, *Pharmacol. Toxicol.*, 82, 230, 1998.
222. Endo, T., Kimura, O., Hatakeyama, M., Takada, M., and Sakata, M., Effects of zinc and copper on cadmium uptake by brush border membrane vesicles, *Toxicol. Lett.*, 91, 111, 1998.
223. Endo, T., Kimura, O., and Sakata, M., Further analysis of cadmium uptake from apical membrane of LLC-PK1 cells via inorganic anion exchanger, *Pharmacol. Toxicol.*, 84, 187, 1999.
224. Barbier, O., Dauby, A., Jacquillet, G., Tauc, M., Poujeol, P., and Cougnon, M., Zinc and cadmium interactions in a renal cell line derived from rabbit proximal tubule, *Nephron. Physiol.*, 99, 74, 2005.
225. Jacquillet, G., Barbier, O., Cougnon, M., Tauc, M., Namorado, M.C., Martin, D., Reyes, J.L., and Poujeol, P., Zinc protects renal function during cadmium intoxication in the rat, *Am. J. Physiol. Renal Physiol.*, 290, F127, 2006.
226. Dufner-Beattie, J., Langmade, S.J., Wang, F., Eide, D., and Andrews., G.K., Structure, function, and regulation of a subfamily of mouse zinc transporter genes, *J. Biol. Chem.*, 278, 50142, 2003.
227. Himeno, S., Yanagiya, T., Enomoto, S., Kondo, Y., and Imura, N., Cellular cadmium uptake mediated by the transport system for manganese, *Tohoku J. Exp. Med.*, 196, 43, 2002.
228. Martin, P., Fareh, M., Poggi, M.C., Boulukos, K.E., and Pognonec, P., Manganese is highly effective in protecting cells from cadmium intoxication, *Biochem. Biophys. Res. Commun.*, 351, 294, 2006.
229. Yanagiya, T., Imura, N., Enomoto, S., Kondo, Y., and Himeno, S., Suppression of a high-affinity transport system for manganese in cadmium-resistant metallothionein-null cells, *J. Pharmacol. Exp. Ther.*, 292, 1080, 2000.
230. Canonne-Hergaux, F. and Gros, P., Expression of the iron transporter DMT1 in kidney from normal and anemic mk mice, *Kidney Int.*, 62, 147, 2002.
231. Ferguson, C.J., Wareing, M., Ward, D.T., Green, R., Smith, C.P., and Riccardi, D., Cellular localization of divalent metal transporter DMT-1 in rat kidney, *Am. J. Physiol.*, 280, F803, 2001.
232. Tabuchi, M., Yoshimori, T., Yamaguchi, K., Yoshida, T., and Kishi, F., Human NRAMP2/DMT1, which mediates iron transport across endosomal membranes, is localized to late endosomes and lysosomes in HEp-2 cells, *J. Biol. Chem.*, 275, 22220, 2000.
233. Lam-Yuk-Tseung, S. and Gros, P., Distinct targeting and recycling properties of two isoforms of the iron transporter DMT1 (NRAMP2, Slc11A2), *Biochemistry*, 45, 2294, 2006.
234. Tabuchi, M., Tanaka, N., Nishida-Kitayama, J., Ohno, H., and Kishi, F., Alternative splicing regulates the subcellular localization of divalent metal transporter 1 isoforms, *Mol. Biol. Cell.*, 13, 4371, 2002.
235. Olivi, L., Sisk, J., and Bressler, J., Involvement of DMT1 in uptake of Cd in MDCK cells: Role of protein kinase C, *Am. J. Physiol.*, 281, C793, 2001.
236. Borke, J.L., Penniston, J.T., and Kumar, R., Recent advances in calcium transport by the kidney, *Semin. Nephrol.*, 10, 15, 1990.
237. Friedman, P.A., Renal calcium transport: Sites and insights, News, *Physiol. Sci.*, 3, 17, 1988.

238. Leslie, E.M., Liu, J., Klaassen, C.D., and Waalkes, M.P., Acquired cadmium resistance in metallothionein-I/II(−/ −) knockout cells: Role of the T-type calcium channel Cacnα_{1G} in cadmium uptake, *Mol. Pharmacol.*, 69, 629, 2006.
239. Andreasen, D., Hansen, P.B., Kwon, T.H., Nielsen, S., and Skøtt, O., The alpha(1G)-subunit of a voltage-dependent Ca(2 +) channel is localized in rat distal nephron and collecting duct, *Am. J. Physiol. Renal Physiol.*, 279, F997, 2000.
240. Peng, J.B., Chen, X.Z., Berger, U.V., Vassilev, P.M., Brown, E.M., and Hediger, M.A., A rat kidney-specific Ca transporter in the distal nephron, *J. Biol. Chem.*, 275, 28186, 2000.
241. Prozialeck, W.C. and Lamar, P.C., Surface binding and uptake of cadmium (Cd^{2+}) by LLC-PK_1 cells on permeable membrane supports, *Arch. Toxicol.*, 67, 113, 1993.
242. Anzai, N., Kanai, Y., and Endou, H., Organic anion transporter family: Current knowledge, *J. Pharmacol. Sci.*, 100, 411, 2006.
243. Kojima, R., Sekine, T., Kawachi, M., Cha, S.H., Suzuki, Y., and Endou, H., Immunolocalization of multispecific organic anion transporters, OAT1, OAT2, OAT3 in rat kidney, *J. Am. Soc. Nephrol.*, 13, 848, 2002.
244. Motohashi, H., Sakurai, Y., Saito, H., Masuda, S., Urakami, Y., Goto, M., Fukatsu, A., Ogawa, O., and Inui, K., Gene expression levels and immunolocalization of organic ion transporters in the human kidney, *J. Am. Soc. Nephrol.*, 13, 866, 2002.
245. Aslamkhan, A.G., Han, Y.H., Yang, X.P., Zalups, R.K., and Pritchard, J.B., Human renal organic anion transporter 1-dependent uptake and toxicity of mercuric thiol conjugates in Madin-Darby canine kidney cells, *Mol. Pharmacol.*, 63, 1, 2003.
246. Zalups, R.K., Aslamkhan, A.G., and Ahmad, S., Human organic anion transporter 1 mediated cellular uptake of cysteine-*S*-conjugates of inorganic mercury, *Kidney Int.*, 66, 251, 2004.
247. Zalups, R.K. and Ahmad, S., Homocysteine and the renal epithelial transport of inorganic mercury: Role of basolateral transporter organic anion transporter 1, *J. Am. Soc. Nephrol.*, 15, 2023, 2004.
248. Cousin, R.J. and McMohan, R.J., Integrative aspects of zinc transporters, *J. Nutr.*, 130, 13845, 2000.
249. Wang, F., Kim, B.E., Petris, M.J., and Eide, D.J. The mammalian ZIP5 protein is a zinc transporter that localizes to the basolateral surface of polarized cells, *J. Biol. Chem.*, 279, 51433–51441, 2004.
250. Kaur, J., Shamra, N., Attri, S., Gogia, L., and Prasad, R., Kinetic characterization of zinc transport process and its inhibition by cadmium in isolated rat renal basolateral membrane vesicles: *In vitro* and *in vivo* studies, *Mol. Cell. Biochem.*, 283, 169, 2006.
251. Klaassen, C.D., Waalkes, M.P., and Cantilena, L.R., Jr., Alteration of tissue disposition of cadmium by chelating agents, *Environ. Health Perspect.*, 54, 233, 1984.
252. Ottenwälder, H. and Simon, P., Differential effect of N-acetylcysteine on excretion of the metals Hg, Cd, Pb and Au, *Arch. Toxicol.*, 60, 401, 1987.
253. van Aubel, R.A., Smeets, P.H., Peters, J.G., Bindels, R.J., and Russel, F.G., The MRP4/ABCC4 gene encodes a novel apical organic anion transporter in human kidney proximal tubules: Putative efflux pump for urinary cAMP and cGMP, *J. Am. Soc. Nephrol.*, 13, 595, 2002.
254. Leslie, E.M., Deeley, R.G., and Cole, S.P.C., Multidrug resistance proteins: Role of P-glycoprotein, MRP1, MRP2, and BCRP (ABCG2) in tissue defense, *Toxicol. Appl. Pharmacol.*, 204, 216, 2005.
255. Shaikh, Z.A., Blazka, M.E., and Endo, T., Metal transport in cells: Cadmium uptake by rat hepatocytes and renal cortical epithelial cells, *Environ. Health Perspect.*, 103(Suppl. 1), 73, 1995.
256. Fujishiro, H., Okugaki, S., Kubota, K., Fujiyama, T., Miyataka, H., and Himeno, S., The role of ZIP8 down-regulation in cadmium-resistant metallothionein-null cells, *J. Appl. Toxicol.*, 2009 (Epub ahead of print).
257. Oka, J.A., Christensen, M.D., and Weigel, P.H., Hyperosmolarity inhibits galactosyl receptor-mediated but not fluid phase endocytosis in isolated rat hepatocytes, *J. Biol. Chem.*, 264, 12016, 1989.
258. Mack, U., Powell, L.W., and Halliday, J.W., Detection and isolation of a hepatic membrane receptor for ferritin, *J. Biol. Chem.*, 258, 4672, 1983.
259. Morgan, E.H. and Baker, E., Iron uptake and metabolisms by hepatocytes, *Fed. Proc.*, 45, 2810, 1986.
260. Osterloh, K. and Aisen, P., Pathways in the binding and uptake of ferritin by hepatocytes, *Biochim. Biophys. Acta*, 1011, 40, 1989.
261. Huebers, H.A., Huebers, E., Csiba, E., Rummel, W., and Finch, C.A., The cadmium effect on iron absorption, *Am. J. Clin. Nutr.*, 45, 1007, 1987.
262. Price, D.J. and Joshi, J.G., Ferritin binding of beryllium and other divalent metal ions, *J. Biol. Chem.*, 258, 10873, 1983.
263. Cherian, M.G., Goyer, R.A., and Delaquerriere-Richardson, L., Cadmium metallothionein induced nephropathy, *Toxicol. Appl. Pharmacol.*, 38, 399, 1976.

264. Cattori, V., van Montfoort, J.E., Stieger, B., Landmann, L., Meijer, D.K., Winterhalter, K.H., Meier, P.J., and Hagenbuch, B., Localization of organic anion transporting polypeptide 4 (Oatp4) in rat liver and comparison of its substrate specificity with Oatp1, Oatp2, and Oatp3, *Pflugers Arch.*, 443, 188, 2001.
265. Jacquemin, E., Hagenbuch, B., Steiger, C., Wolkoff, A.W., and Meier, P.J., Expression cloning of a rat liver (Na^+)-independent organic anion transporter, *Proc. Natl. Acad. Sci. USA*, 91, 133, 1994.
266. Kullak-Ublick, G.A., Hagenbuch, B., Steiger, B., Wolkoff, A.W., and Meier, P.J., Functional characterization of the basolateral rat liver organic anion transporting polypeptide, *Hepatology*, 20, 411, 1994.
267. Grundemann, D., Gorboulev, V., Gambaryan, S., Veyhl, M., and Koepsell, H., Drug excretion mediated by a new prototype of polyspecific transporter, *Nature*, 372, 549, 1994.
268. Leslie, E.M., Deeley, R.G., and Cole, S.P., Toxicological relevance of the multidrug resistance protein 1, MRP1 (ABCC1) and related transporters, *Toxicology*, 167, 3, 2001.
269. Terlouw, S.A., Graeff, C., Smeets, P.H.E., Fricker, G., Russel, R.G.M., Masereeuw, R., and Miller, D.S., Short- and long-term influences of heavy metals on anionic drug efflux from renal proximal tubule, *J. Pharmacol. Exp. Ther.*, 301, 578, 2002.
270. Akerboom, T.P., Inoue, M., Sies, H., Kinne, R., and Arias, I.M., Biliary transport of glutathione disulfide studied with isolated rat-liver canalicular-membrane vesicles, *Eur. J. Biochem.*, 141, 211, 1984.
271. Akerboom, T.P., Narayanaswami, V., Kunst, M., and Sies, H., ATP-dependent S-(2,4-dinitrophenyl)glutathione transport in canalicular plasma membrane vesicles from rat liver, *J. Biol. Chem.*, 266, 3147, 1991.
272. Gunn, S.A., Gould, T.C., and Anderson, W.A., The selective injurious response of testicular and epididymal blood vessels to cadmium and its prevention by zinc, *Am. J. Pathol.*, 42, 685, 1961.
273. Gunn, S.A., Gould, T.C., and Anderson, W.A., Cadmium-induced interstitial cell tumors in rats and mice and their prevention by zinc, *J. Natl. Cancer Inst.*, 31, 745, 1963.
274. Kar, A.B. and Das, R.P., Testicular changes in rats after treatment with cadmium chloride, *Acta Biol. Med. Ger.*, 5, 153, 1960.
275. Parizek, J., The destructive effect of cadmium ion on testicular tissue and its prevention by zinc, *J. Endocrinol.*, 15, 56, 1957.
276. Parizek, J. and Zahor, Z., Effect of cadmium salts on testicular tissue, *Nature*, 177, 1036, 1956.
277. King, L.M., Banks, W.A., and George, W.J., Differences in cadmium transport to the testis, epididymal, and brain in cadmium-sensitive and -resistant murine strains 129/J and A/J, *J. Pharmacol. Exp. Ther.*, 289, 825, 1999.
278. Waalkes, M.P. and Poirier, L.A., Interactions of cadmium with interstitial tissue of the rat testes. Uptake of cadmium by isolated interstitial cells, *Biochem. Pharmacol.*, 34, 2513, 1985.
279. Waalkes, M.P. and Perantoni, A., *In vitro* assessment of target cell specificity in cadmium carcinogenesis: Interactions of cadmium and zinc with isolated interstitial cells of the rat testes, *In vitro Cell Dev. Biol.*, 24, 558, 1988.
280. Wagner, C.A., Lang, F., and Bröer, S., Function and structure of heterodimeric amino acid transporters, *Am. J. Physiol. Cell Physiol.*, 281, C1077, 2001.
281. Augustine, L.M., Markelewicz, R.J., Boekelheide, K., and Cherrington, N.J., Xenobiotic and endobiotic transporter mRNA expression in the blood-testis barrier, *Drug Metab. Dispos.*, 33, 182, 2005.
282. Jabado, N., Canonne-Hergaux, F., Gruenheid, S., Picard, V., and Gros, P., Iron transporter Nramp2/DMT-1 is associated with the membrane of phagosomes in macrophages and Sertoli cells, *Blood*, 100, 2617, 2002.
283. Niewenhuis, R.J., Dimitru, C., and Prozialeck, W.C., Ultrastructural characterization of the early changes in intercellular junctions in response to cadmium (Cd^{2+}) exposure in LLC-PK1 cells. *Toxicol. Appl. Pharmacol.*, 142, 1, 1997.
284. Prozialeck, W.C. and Niewenhuis, R.J., Cadmium disrupts intercellular junctions and actin filaments in LLC-PK1 cells. *Toxicol. Appl. Pharmacol.*, 107, 81, 1991.
285. Prozialeck, W.C. and Niewenhuis, R.J., Cadmium disrupts Ca^{2+}-dependent cell-cell junctions and alters the pattern of E-cadherin immunofluorescence in LLC-PK1 cells. *Biochem. Biophys. Res. Commun.*, 181, 1118, 1991.
286. Prozialeck, W.C., Wellington, D.R., and Lamar, P.C., Comparison of the cytotoxic effects of cadmium chloride and cadmium-metallothionein in LLC-PK1 cells. *Life Sci.*, 53, PL337, 1993.
287. Prozialeck, W.C., Evidence that E-cadherin may be a target for cadmium toxicity in epithelial cells. *Toxicol. Appl. Pharmacol.*, 164, 231, 2000.
288. Duizer, E., Gilde, A.J., Versantvoort, C.H.M., and Groten, J.P., Effects of cadmium chloride on the paracellular barrier function of intestinal epithelial cell lines. *Toxicol. Appl. Pharmacol.*, 121, 160, 1999.
289. Prozialeck, W.C. and Lamar, P.C., Binding of cadmium to E-CAD1, a Ca^{2+} binding polypeptide analog of E-cadherin. *Life Sci.*, 58, PL325, 1996.

290. Prozialeck, W.C. and Lamar, P.C., Cadmium (Cd^{2+}) disrupts E-cadherin-dependent cell-cell junctions in MDCK cells. *In Vitro Cell Dev. Biol.*, 33, 516, 1997.
291. Prozialeck, W.C. and Lamar, P.C., Interaction of cadmium with a 13-residue polypeptide analog of a putative calcium-binding motif of E-cadherin. *Biochim. Biophys. Acta*, 1451, 93, 1999.
292. Pearson, C.A. and Prozialeck, W.C., E-cadherin, beta-catenin and cadmium carcinogenesis. *Med. Hypotheses*, 56, 573, 2001.
293. Agency for Toxic Substance and Disease Registry, Toxicological Profile for Lead, U.S. Department of Health and Humans Services, Public Health Service, Centers for Disease Control, Atlanta, GA, 2007.
294. Environmental Protection Agency, Supplement to the 2086 EPA air quality criteria for lead. Vol. 1, Addendum. EPA/600/8–89/049A, U.S. Office of Health and Environmental Assessment, Washington, DC, 1989.
295. Goyer, R.A., Lead toxicity: Current concerns, *Environ. Health Perspect.*, 100, 177, 1993.
296. National Research Council, Measuring lead exposure in infants, children and other sensitive populations, Washington, DC, *Natl. Acad.* 337, 1993.
297. Blackman, S.A., The lesions of lead encephalitis in children, *Bull. Johns Hopkins Hosp.*, 61, 1, 1937.
298. Needleman, H.L., Schell, A., Bellinger, D., Leviton, A., and Allred, E.N., The long-term effects of exposure to low doses of lead in childhood, *N. Engl. J. Med.*, 322, 83, 1990.
299. Pounds, J.G., Long, G.J., and Rosen, J.F., Cellular and molecular toxicity of lead in bone, *Environ. Health Perspect.*, 91, 57, 1991.
300. Baker, E.L., Goyer, R.A., Fowler, B.A., Khattr, U., Bernard, O.B., Adler, S., White, R., Babayan, R., and Feldman, R.G., Occupational lead exposure, nephropathy and renal cancer, *Am. J. Med.*, 1, 139, 1980.
301. Buchet, J.P., Roels, H., Bernard, A., and Lauwerys, R., Assessment of renal function of workers exposed to inorganic lead, cadmium or mercury vapor, *J. Occup. Med.*, 22, 741, 1980.
302. Cooper, W.C., Wong, O., and Kheifets, L., Mortality among employees of lead battery plants and lead-producing plants, 1947–1980, *Scand. J. Work Environ. Health*, 11, 331, 1985.
303. Fowler, B.A., Mechanisms of kidney cell injury from metals, *Environ. Health Perspect.*, 100, 57, 1992.
304. Goyer, R.A., The nephrotoxic effects of lead. In: P.H. Bach, F.W. Bonner, J.W. Bridges, and E.A. Lock, (Eds), *Nephrotoxicity: Assessment and Pathology,* p. 338. Wiley, Chichester, England, 1982.
305. Lilis, R., Long-term occupational lead exposure: Chronic nephropathy and renal cancer: A case report, *Am. J. Ind. Med.*, 2, 293, 1981.
306. Selevan, S.G., Landrigan, P.J., Stern, F.B., and Jones, J.H., Mortality of lead smelter workers, *Am. J. Epidemiol.*, 122, 673, 1985.
307. Brandman, A., Eskenazi, B., Sutton, P., Athanasoulis, M., and Goldman, L.R., Iron deficiency associated with higher blood lead in children living in contaminated environments, *Environ. Health Perspect.*, 109, 1079, 2001.
308. Schell, L.M., Denham, M., Stark, A.D., Gomez, M., Ravenscroft, J., Parsons, P.J., Aydermir, A., and Samelson, R., Maternal blood lead concentration, diet during pregnancy, and anthropometry predict neonatal blood lead in a socio-economically disadvantaged population, *Environ. Health Perspect.*, 111, 195, 2003.
309. Six, K.M. and Goyer, R.A., The influence of iron deficiency on tissue content and toxicity of ingested lead in the rat, *J. Lab. Clin. Med.*, 79, 128, 1972.
310. Bannon, D.I., Portnoy, M.E., Olivi, L., Lees, P.S., Culotta, V.C., and Bressler, J.P., Uptake of lead and iron by divalent metal transporter 1 in yeast and mammalian cells, *Biochem. Biophys. Res. Commun.*, 295, 978, 2002.
311. Cheong, J.H., Bannon, D., Olivi, L., Kim, Y., and Bressler, J., Different mechanisms mediate uptake of lead in a rat astroglial cell line, *Toxicol. Sci.*, 77, 334, 2004.
312. Ong, W.Y., He, X., Chua, L.H., and Ong, C.N., Increased uptake of divalent metals lead and cadmium into the brain after kainite-induced neuronal injury, *Exp. Brain Res.*, 173, 468, 2006.
313. Ong, W.Y., Ren, M.Q., Makjanic, J., Lim, T.M., and Watt, F., A nuclear microscopic study of elemental changes in the rat hippocampus after kainite-induced neuronal injury, *J. Neurochem.*, 72, 1574, 1999.
314. Barton, J.C., Active transport of lead-210 by everted segments of rat duodenum, *Am. J. Physiol.*, 247, G193, 1984.
315. Blake, K.C. and Mann, M., Effect of calcium and phosphorus on the gastrointestinal absorption of ^{203}Pb in man, *Environ. Res.*, 30, 188, 1983.
316. Bogden, J.D., Gertner, S.B., Christakos, S., Kemp, F.W., Yang, Z., Katz, S.R., and Chu, C., Dietary calcium modified concentrations of lead and other metals and renal calbindin in rats, *J. Nutr.*, 122, 1351,1992.

317. Goldstein, G.W., Lead encephalopathy: The significance of lead inhibition of calcium uptake by brain mitochondria, *Brain Res.*, 136, 185, 1977.
318. Goyer, R.A., Nutrition and metal toxicity, *Am. J. Clin. Nutr.*, 61(Suppl. 3), 646S, 1995.
319. Goyer, R.A., Toxic and essential metal interactions, *Annu. Rev. Nutr.*, 17, 37, 1997.
320. Kapoor, S.C. and van Rossum, G.D., Effects of Pb^{2+} added *in vitro* on Ca^{2+} movements in isolated mitochondria and slices of rat kidney cortex, *Biochem. Pharmacol.*, 33, 1771, 1984.
321. Mahaffey, K.R., Gartside, P.S., and Glueck, C.J., Blood lead levels and dietary calcium intake in 1–11 year-old children: The Second National Health and Nutrition Examination Survey, 1976–1980, *Pediatrics*, 78, 257, 1986.
322. Simons, T.J., Cellular interactions between lead and calcium, *Brit. Med. Bull.*, 42, 431, 1986.
323. Six, K.M. and Goyer, R.A., Experimental enhancement of lead toxicity by low dietary calcium, *J. Lab. Clin. Med.*, 83, 933, 1970.
324. Ziegler, E.E., Edwards, B.B., Jensen, R.L., and Mahaffey, K.R., Absorption and retention of lead by infants, *Pediatr. Res.*, 12, 29, 1978.
325. Mykkänen, H.M. and Wasserman, R.H., Gastro-intestinal absorption of lead (^{203}Pb) in chicks: Influence of lead, calcium and age, *J. Nutr.*, 111, 1757, 1981.
326. Flanagan, P.R., Hamilton, D.L., Haist, J., and Valberg, L.S., Inter-relationships between iron and lead absorption in iron-deficient mice, *Gastroenterology*, 77, 1074, 1979.
327. Kerper, L.E. and Hinkle, P.M., Cellular uptake of lead is activated by depletion of intracellular calcium stores, *J. Biol. Chem.*, 272, 8346, 1997.
328. Reuter, H., Calcium channel modulation by neurotransmitters, enzymes and drugs, *Nature*, 301, 569, 1983.
329. Rogers, J.T. and Wood, C.M., Characterization of bronchial lead-calcium interactions in the freshwater rainbow trout *Onchorhynchus mykiss*, *J. Exp. Biol.*, 207, 813, 2004.
330. Simons, T.J., Lead transport and binding by human erythrocytes *in vitro*, *Pflugers Arch.*, 423, 307, 1993.
331. Simons, T.J.B. and Pocock, G., Lead enters bovine adrenal medullary cells through calcium channels, *J. Neurochem.*, 48, 383, 1987.
332. Kerper, L.E. and Hinkle, P.M., Lead uptake in brain capillary endothelial cells: Activation by calcium store depletion, *Toxicol. Appl. Pharmacol.*, 146, 127, 1997.
333. Bernal, J., Lee, J.H., Cribbs, L.L., and Perez-Reyes, E., Full reversal of Pb^{++} block of L-type Ca^{++} channels requires treatment with heavy metal antidotes, *J. Pharmacol. Exp. Ther.*, 282, 172, 1997.
334. Peng, S., Hajela, R.K., and Atchison, W.D., Characteristics of block by Pb^{2+} of function of human neuronal L-, N-, and R-type Ca^{2+} channels transiently expressed in human embryonic kidney 293 cells, *Mol. Pharmacol.*, 62, 1418, 2002.
335. Chang, Y.F., Teng, H.C., Chen, S.Y., Wang, C.T., Chiou, S.H., Kao, L.S., Kao, F.J., Chiou, A., and Yang, D.M., Orai1-STIM1 formed store-operated Ca^{2+} channels (SOCs) as the molecular components needed for Pb^{2+} entry into living cells, *Toxicol. Appl. Pharmacol.*, 227, 430, 2008.
336. Kirberger, M. and Yang, J.J., Structural differences between Pb(2 +)- and Ca(2 +)-binding sites in proteins: Implications with respect to toxicity, *J. Inorg. Biochem.*, 2008 (Epub ahead of print).
337. Pfleger, H. and Wolf, H.U., Activation of membrane-bound high-affinity calcium ion-sensitive adenosine triphosphatase of human erythrocytes by bivalent metal ions, *Biochem J.*, 147, 359, 1975.
338. Simons, T.J., Active transport of lead by the calcium pump in human red cell ghosts, *J. Physiol.*, 405, 105, 1988.
339. Bannon, D.I., Olivi, L., and Bressler, J.P., The role of anion exchange in the uptake of Pb by human erythrocytes and Madin-Darby canine kidney cells, *Toxicology*, 147, 101, 2000.
340. Simons, T.J., The role of anion transport in the passive movement of lead across the human red cell membrane, *J. Physiol.*, 378, 287, 1986.
341. Leuthold, S., Hagenbuch, B., Mohebbi, N., Wagner, C.A., Meier, P.J., and Stieger, B., Mechanisms of pH-gradient driven transport mediated by organic anion polypeptide transporters, *Am. J. Physiol. Cell Physiol.*, 2009 (Epub ahead of print).
342. Fowler, B.A. and DuVal, G. Effects of lead on the kidney: Roles of high-affinity binding proteins, *Environ. Health Perspect.*, 91, 77, 1991.
343. Oskarsson, A., Squibb, K.S., and Fowler, B.A., Intracellular binding of lead in the kidney: The partial isolation and characterization of postmitochondrial lead binding components, *Biochem. Biophys. Res. Commun.*, 104, 290, 1982.
344. Smith, D.R., Khang, M.W., Quintanilla-Vega, B., and Fowler, B.A., Lead-binding polypeptides in human kidney cytosol: Diazepam binding inhibitor and thymosin β_4, *Toxicologist*, 14, 84, 1994.

345. Smith, D.R., Khang, M.W., Quintanilla-Vega, B., and Fowler, B.A., High affinity renal lead-binding proteins in environmentally exposed humans, *Chem. Biol. Interact.*, 115, 39, 1998.
346. Alexander, J., Aaseth, J., and Mikalsen, A., Excretion of lead in rat bile—the role of glutathione, *Acta Pharmacol. Toxicol.*, 59, 486, 1986.
347. Goldstein, G.W. and Ar, D., Lead activates calmodulin sensitive processes, *Life Sci.*, 33, 1001, 1983.
348. Haberman, E., Crowell, K., and Janicki, P., Lead and other metals can substitute for Ca^{2+} in calmodulin, *Arch. Toxicol.*, 54, 61, 1983.
349. Kern, M., Wisniewski, M., Cabell, L., and Audesirk, G., Inorganic lead and calcium interact positively in activation of calmodulin, *Neurotoxicol.*, 21, 353, 2000.
350. Stoclet, J.C., Derard, D., Kilhoffer, M.C., Lugnier, C., Miller, R., and Schaefer, P., Calmodulin and its role in intracellular calcium regulation, *Prog. Neurobiol.*, 23, 321,1987.
351. Richardt, G., Federolf, G., and Habermann, E., The interaction of aluminum and other metal ions with calcium-calmodulin-dependent phosphodiesterase, *Arch. Toxicol.*, 57, 257, 1985.
352. Richardt, G., Federolf, G., and Habermann, E., Affinity of heavy metal ions to intracellular Ca^{2+}-binding proteins, *Biochem. Pharmacol.*, 35, 1331, 1986.
353. Goldstein, G.W., Brain capillaries: A target for inorganic lead poisoning, *Neurotoxicology*, 5, 167, 1984.
354. Toews, A.D., Kolber, A., Hayward, J., Krigman, M.R., and Morell, P., Experimental lead encephalopathy in the suckling rat: Concentration of lead in cellular fractions enriched in brain capillaries, *Brain Res.*, 147, 131, 1978.
355. Agency for Toxic Substance and Disease Registry, Toxicological Profile for Mercury, U.S. Department of Health and Humans Services, Public Health Service, Centers for Disease Control, Atlanta, GA, 2007.
356. Clarkson, T.W. and Magos, L., The toxicology of mercury and its chemical compounds, *Crit. Rev. Toxicol.*, 36, 609, 2006.
357. Rooney, J.P.K., The role of thiols, dithiols, nutritional factors and interacting ligands in the toxicology of mercury, *Toxicology*, 234, 145, 2007.
358. Carmignani, M., Boscolo, P., Artese, L., Del Rosso, G., Porcelli, G., Felaco, M., Volpe, A.R., and Giuliano, G., Renal mechanisms in the cardiovascular effects of chronic exposure to inorganic mercury in rats, *Brit. J. Ind. Med.*, 49, 226, 1992.
359. Soni, J.P., Singhania, R.U., Bansal, A., and Rathi, G., Acute mercury vapor poisoning, *Indian Pediatr.*, 29, 365, 1992.
360. Wakita, Y., Hypertension induced by methylmercury in rats, *Toxicol. Appl. Pharmacol.*, 89, 144, 1987.
361. Warkany, J. and Hubbard, D.M., Acrodynia and mercury, *J. Pediatr.*, 42, 365, 1953.
362. Afonso, J. and deAlvarez, R., Effects of mercury on human gestation, *Am. J. Obstet. Gynecol.*, 80, 145, 1960.
363. Bluhm, R.E., Bobbitt, R.G., Welch, L.W., Wood, A.J. Bonfiglio, J.R., Sarzen, C., Heath, A.J., and Branch, R.A., Elemental mercury vapour toxicity, treatment, and prognosis after acute, intensive exposure in chloralkali plant workers: Part I., History, neuropsychological findings and chelator effects, *Hum. Exp. Toxicol.*, 11, 201, 1992.
364. Lundgren, K.D. and Swensson, A., Occupational poisoning by alkyl mercury compounds, *J. Industr. Hyg. Toxicol.*, 31, 190, 1949.
365. Murphy, M.J., Culliford, E.J., and Parsons, V., A case of poisoning with mercuric chloride, *Resuscitation*, 7, 35, 1979.
366. Jaffe, K.M., Shurtleff, D.B., and Robertson, W.O., Survival after acute mercury vapor poisoning—role of intensive supportive care, *Am. J. Dis. Child*, 137, 749, 1983.
367. Samuels, E.R., Heick, H.M.C., McLain, P.N., and Farant, J.P., A case of accidental inorganic mercury poisoning, *J. Anal. Toxicol.*, 6, 120, 1982.
368. Rowens, B., Geurrero-Betancourt, D., Gottlieb, C.A., Boyes, R.J., and Eichenhorn, M.S., Respiratory failure and death following acute inhalation of mercury vapor: A clinical and histologic perspective, *Chest*, 99, 185, 1991.
369. Yasutake, A., Hirayama, Y., and Inouye, M., Sex differences of nephrotoxicity by methylmercury in mice. In: P.H. Bach, N.J. Gregg, M.F. Wiks, and L. Delacruz (Eds), *Nephrotoxicity: Mechanisms, Early Diagnosis, and Therapeutic Management, Fourth International Symposium On Nephrotoxicity, 1989*, p. 389, Guilford, England; Marcel Dekker, Inc., New York., 1991.
370. Kutsuna, M. (Ed.), Minamata disease. In: *Study Group of Minamata Disease*, Kumamoto University, Japan, 1968.
371. Lin, J.L. and Lim, P.S., Massive oral ingestion of elemental mercury, *J. Toxicol. Clin. Toxicol.*, 31, 487, 1993.
372. Tsubaki, T. and Takahashi, H., *Recent Advances in Minamata Disease Studies*, Kodansha, Ltd., Tokyo, Japan, 1986.

373. Risher, J.F. and DeRosa, C.T., Inorganic: The other mercury, *J. Environ. Health*, 70, 9, 2007.
374. Hughes, W.L., A physiochemical rationale for the biological activity of mercury and its compounds, *Ann. NY Acad. Sci.*, 65, 454, 1957.
375. Canty, A.J., Colton, R., D'Agostino, A.P., and Traeger, J.C.A., Positive and negative ion electrospray mass spectrometric studies of some amino acids and glutathione, and their interactions with alkali metal ions and methylmercury(II), *Inorg. Chim. Acta*, 223, 103, 1994.
376. Fuhr, B. and Rabenstein, D.L., Nuclear magnetic resonance studies of the solution chemistry of metal complexes. IX. The binding of cadmium, zinc, lead, and mercury by glutathione, *J. Am. Chem. Soc.*, 95, 6944, 1973.
377. Rubino, F.M., Verduci, C., Giampiccolo, R., Pulvirenti, S., Brambilla, G., and Colombi, A., Molecular characterization of homo- and heterodimeric mercury(II)-*bis*-thiolates of some biologically relevant thiols by electrospray ionization and triple quadrupole tandem mass spectrometry, *J. Am Soc. Mass Spectrom.*, 15, 288, 2004.
378. Dave, M.H., Schulz, N., Zecevic, M., Wagner, C.A., and Verrey, F., Expression of heterodimeric amino acid transporters along the intestine, *J. Physiol*, 558, 597, 2004.
379. Ganapathy, V., Ganapathy, M.E., and Leibach, F.H, Intestinal transport of peptides and amino acids. In: K.E. Barrett and M. Donowitz (Eds), *Current Topics in Membranes*, p. 379, Academic Press, New York, 2001.
380. Endo, T., Nakaya, S., Kimura, R., and Murata, T., Gastrointestinal absorption of inorganic mercuric compounds *in vitro*, *Toxicol. Appl. Pharmacol.*, 83, 187, 1984.
381. Andres, S., Laporte, J.M., and Mason, R.P., Mercury accumulation and flux across the gills and the intestine of the blue crab (*Callinectes sapidus*), *Aquat. Toxicol.*, 56, 303, 2002.
382. Laporte, J.M., Andres, S., and Mason, R.P., Effect of ligands and other metals on the uptake of mercury and methylmercury across the gills and the intestines of the blue crab (*Callinectes sapidus*), *Comp. Biochem. Physiol. C*, 131, 185, 2002.
383. Hoyle, I. and Handy, R.D., Dose-dependent inorganic mercury absorption by isolated perfused intestine of rainbow trout, *Oncorhynchus mykiss*, involves both amiloride-sensitive and energy-dependent pathways, *Aquat. Toxicol.*, 25, 147, 2005.
384. Foulkes, E.C. and Bergman, D., Inorganic mercury absorption in mature and immature rat jejunum: Transcellular and intercellular pathways *in vivo* and in everted sacs, *Toxicol. Appl. Pharmacol.*, 120, 89, 1993.
385. Zalups, R.K., Intestinal handling of mercury in the rat: Implications of intestinal secretion of inorganic mercury following biliary ligation or cannulation, *J. Toxicol. Environ. Health A.*, 53, 615, 1998.
386. Sugawara, N., Lai, Y.R., Sugaware, C., and Arisono, K., Decreased hepatobiliary secretion of inorganic mercury, its deposition and toxicity in the Eisai hyperbilirubinemic rat with no hepatic canalicular organ anion transporter, *Toxicology*, 126, 23, 1998.
387. Zalups, R.K., Barfuss, D.W., and Lash, L.H., Relationships between alterations in glutathione metabolism and the disposition of inorganic mercury in rats: Effects of biliary ligation and chemically induced modulation of glutathione status, *Chem. Biol. Interact*, 123, 171, 1999.
388. Prime-Chapman, Fearn, R.A., Cooper, A.E., Moore, V., and Hirst, B.H., Differential multidrug resistance-associated protein 1 through 6 isoform expression and function in human intestinal epithelial Caco-2 cells, *J. Pharmacol. Exp. Ther.*, 311, 476, 2004.
389. Rost, D., Mahner, S., Sugiyama, Y., and Stremmel, W., Expression and localization of the multidrug resistance-associated protein 3 in rat small and large intestine, *Am. J. Physiol. Gastrointest. Liver Physiol.*, 282, G720, 2002.
390. Shoji, T., Suzuki, H., Kusuhara, H., Watanabe, Y., Sakamoto, S., and Sugiyama, Y., ATP-dependent transport of organic anions into isolated basolateral membrane vesicles from rat intestine, *Am. J. Physiol. Gastrointest. Liver Physiol.*, 287, G749, 2004.
391. Yokooji, T., Murakami, T., Yumoto, R., Nagai, J., and Takano, M., Site-specific bidirectional efflux of 2,4-dinitrophenyl-S-glutatione, a substrate of multidrug resistance-associated proteins, in rat intestine and Caco-2 cells, *J. Pharm. Pharmacol.*, 59, 513, 2007.
392. Itagaki, S., Chiba, M., Kobayashi, M., Hirano, T., and Iseki, K., Contribution of multidrug resistance-associated protein 2 to secretory intestinal transport of organic anions, *Biol. Pharm. Bull.*, 31, 146, 2008.
393. Maher, J.M., Slitt, A.L., Cherrington, N.J., Cheng, X., and Klaassen, C.D., Tissue distribution and hepatic and renal ontogeny of the multidrug resistance-associated protein (MRP) family in mice, *Drug Metab. Dispos.*, 33, 947, 2005.
394. Zalups, R.K., Early aspects of the intrarenal distribution of inorganic mercury in the rabbit: Effect of unilateral nephrectomy and dose of mercuric chloride, *J. Toxicol. Environ. Health*, 33, 213, 1993.

395. Bridges, C.C. and Zalups, R.K., Homocysteine, system $b^{0,+}$, and the renal epithelial transport and toxicity of inorganic mercury, *Am. J. Pathol.*, 165, 1385, 2004.
396. Zalups, R.K., Organic anion transport and action of γ-glutamyltranspeptidase in kidney linked mechanistically to renal tubular uptake of inorganic mercury, *Toxicol. Appl. Pharmacol.*, 132, 289, 1995.
397. Zalups, R.K., Enhanced renal outer medullary uptake of mercury associated with uninephrectomy: Implication of a luminal mechanism, *J. Toxicol. Environ. Health*, 50, 173, 1997.
398. Zalups, R.K., Basolateral uptake of inorganic mercury in the kidney, *Toxicol. Appl. Pharmacol.*, 150, 1, 1998.
399. Zalups, R.K., Basolateral uptake of mercuric conjugates of *N*-acetylcysteine and cysteine in the kidney involves the organic anion transport system, *J. Toxicol. Environ. Health*, 54, 101, 1998.
400. Zalups, R.K. and Ahmad, S., Handling of cysteine *S*-conjugates of methylmercury in MDCK cells expressing human OAT1, *Kidney Int.*, 68, 1684, 2005.
401. Zalups, R.K. and Ahmad, S., Handling of the homocysteine *S*-conjugate of methylmercury by renal epithelial cells: Role of organic anion transporter 1 and amino acid transporters, *J. Pharmacol. Exp. Ther.*, 315, 896, 2005.
402. Zalups, R.K. and Ahmad, S., Transport of *N*-acetylcysteine *S*-conjugates of methylmercury in Madin-Darby canine kidney cells stably transfected with human isoform of organic anion transporter 1, *J. Pharmacol. Exp. Ther.*, 314, 1158, 2005.
403. Zalups, R.K. and Barfuss, D.W., Intrarenal distribution of inorganic mercury and albumin after co-administration, *Toxicol. Appl. Pharmacol.*, 40, 77, 1993.
404. Zalups, R.K. and Barfuss, D.W., Pretreatment with *p*-aminohippurate inhibits the renal uptake and accumulation of injected inorganic mercury in the rat, *Toxicology*, 103, 23, 1995.
405. Zalups, R.K. and Barfuss, D.W., Participation of mercuric conjugates of cysteine, homocysteine, and *N*-acetylcysteine in mechanisms involved in the renal tubular uptake of inorganic mercury, *J. Am. Soc. Nephrol.*, 9, 551, 1998.
406. Zalups, R.K. and Lash, L.H., Binding of mercury in renal brush-border and basolateral membrane-vesicles: Implication of a cysteine conjugate of mercury involved in the luminal uptake of inorganic mercury, *Biochem. Pharmacol.*, 53, 1889, 1997.
407. Zalups, R.K. and Minor, K.H., Luminal and basolateral mechanisms involved in the renal tubular uptake of inorganic mercury, *J. Toxicol. Environ. Health*, 46, 73, 1995.
408. Zalups, R.K., Robinson, M.K., and Barfuss, D.W., Factors affecting inorganic mercury transport and toxicity in the isolated perfused proximal tubule, *J. Am. Soc. Nephrol.*, 2, 866, 1991.
409. Zalups, R.K., Parks, L., Cannon, V.T., and Barfuss, D.W., Mechanisms of action of 2,3-dimercaptopropane-1-sulfonate and the transport, disposition, and toxicity of inorganic mercury in isolated perfused segments of rabbit proximal tubules, *Mol. Pharmacol.*, 54, 353, 1998.
410. Berndt, W.O., Baggett, J.M., Blacker, A., and Houser, M., Renal glutathione and mercury uptake by kidney, *Fund. Appl. Toxicol.*, 5, 832, 1985.
411. de Ceaurriz, J., Payan, J.P., Morel, G., and Brondeau, M.T., Role of extracellular glutathione and γ-glutamyltranspeptidase in the disposition and kidney toxicity of inorganic mercury in rats, *J. Appl. Toxicol.*, 14, 201, 1994.
412. Tanaka, T., Naganuma, A., and Imura, N., Role of γ-glutamyltranspeptidase in renal uptake and toxicity of inorganic mercury in mice, *Toxicology*, 60, 187, 1990.
413. Tanaka-Kagawa, T., Naganuma, A., and Imura, N., Tubular secretion and reabsorption of mercury compounds in mouse kidney, *J. Pharmacol. Exp. Ther.*, 264, 776, 1993.
414. Wei, H., Qiu, L., Divine, K.K., Ashbaugh, M.D., McIntyre, L.C., Jr., Fernando, Q., and Gandolfi, A.J., Toxicity and transport of three synthesized mercury-thiol-complexes in isolated rabbit renal proximal tubule suspensions, *Drug Chem. Toxicol.*, 22, 323, 1999.
415. Zalups, R.K. and Barfuss, D.W., Transport and toxicity of methylmercury along the proximal tubule of the rabbit, *Toxicol. Appl. Pharmacol.*, 121, 176, 1993.
416. Cannon, V.T., Zalups, R.K., and Barfuss, D.W., Molecular homology and the luminal transport of Hg^{++} in the renal proximal tubule, *J. Am. Soc. Nephrol.*,11, 394, 2000.
417. Cannon, V.T., Zalups, R.K., and Barfuss, D.W., Amino acid transporters involved in luminal transport of mercuric conjugates of cysteine in rabbit proximal tubule, *J. Pharmacol. Exp. Ther.*, 298, 780, 2001.
418. Palacin, M., Estevez, R., Bertran, J., and Zorzano, A., Molecular biology of mammalian plasma membrane amino acid transporters, *Physiol. Rev.*, 78, 969, 1998.
419. Palacin, M. Fernandez, E., Chillaron, J., and Zorzano, A., The amino acid transport system $b^{o,+}$ and cystinuria, *Mol. Memb. Biol.*, 18, 21, 2001.

420. Zalups, R.K. and Barfuss, D.W., Small aliphatic dicarboxylic acids inhibit renal uptake of administered mercury, *Toxicol. Appl. Pharmacol.*, 148, 183, 1998.
421. Ferrier, B., Martin, M., and Roch-Ramel, F., Effects of *p*-aminohippurate and pyrazinoate on the renal excretion of salicylate in the rat: A micropuncture study, *J. Pharmacol. Exp. Ther.*, 224, 451, 1983.
422. Pritchard, J.B., Coupled transport of *p*-aminohippurate by rat kidney basolateral membrane vesicles, *Am. J. Physiol.*, 255, F596, 1988.
423. Roch-Ramel, F., Besseghir, K., and Murer, H., Renal excretion and tubular transport of organic anions and cations. In: E.E. Windhager (Ed.), *Handbook of Physiology, Section 8, Renal Physiology*, p. 2189, Oxford University Press, New York, 1992.
424. Shimomura, A., Chonko, A.M., and Grantham, J.J., Basis for heterogeneity for para-aminohippurate secretion in rabbit proximal tubules, *Am. J. Physiol.*, 241, F430, 1981.
425. Ullrich, K.J., Rumrich, G., Fasold, G., and Kloss, S., Contraluminal para-aminohippurate (PAH) transport in the proximal tubule of the rat kidney, I. Kinetics, influence of cations, anions, and capillary preperfusion, *Pfluegers Arch.*, 409, 229, 1987.
426. Ullrich, K.J., Rumrich, G., Fasold, G., and Kloss, S., Contraluminal para-aminohippurate (PAH) transport in the proximal tubule of the rat kidney, II. Specificity: Aliphatic dicarboxylic acids, *Pfluegers Arch.*, 408, 38, 1987.
427. Zalups, R.K. and Barfuss, D.W., Renal organic anion transport system: A mechanism for the basolateral uptake of mercury-thiol conjugates along the pars recta of the proximal tubule, *Toxicol. Appl. Pharmacol.*, 182, 234, 2000.
428. Aposhian, H.V., DMSA and DMPS: Water soluble antidotes for heavy metal poisoning, *Ann. Rev. Pharmacol. Toxicol.*, 23, 193, 1983.
429. Aposhian, H.V., Mairorino, R.M., Rivera, M., et al., Human studies with the chelating agents DMPS and DMSA, *Clin. Toxicol.*, 30, 505–528, 1992.
430. Planas-Bohne, F., The effect of 2,3-dimercaptopropane-1-sulfonate and dimercaptosuccinic acid on the distribution and excretion of mercuric chloride in rats, *Toxicology*, 19, 275, 1981.
431. Ruprecht, J., *Scientific Monograph for Dimaval (DMPS)*, 7th ed., Heyltex Corporation, Houston, TX, 2008.
432. Diamond, G.L., Klotzbach, J.M., and Stewart, J.R., Complexing activity of 2,3-dimercapto-1-propanesulfonate and its disulfide auto-oxidation product in rat kidney, *J. Pharmacol. Exp. Ther.*, 246, 270, 1988.
433. Schaub, T.P., Kartenbeck, J., Konig, J., Spring, H., Dorsam, J., Staehler, G., Storkel, S., Thon, W.F., and Keppler, D., Expression of the MRP2 gene-encoded conjugate export pump in human kidney proximal tubules and in renal cell carcinoma, *J. Am. Soc. Nephrol.*, 10, 1159, 1999.
434. Bridges, C.C., Joshee, L., and Zalups, R.K., Multidrug resistance proteins and the renal elimination of inorganic mercury mediated by 2,3-dimercaptopropane-1-sulfonic acid and meso-2,3-dimercaptosuccinic acid, *J. Pharmacol. Exp. Ther.*, 324, 383, 2008.
435. Bridges, C.C., Joshee, L., and Zalups, R.K., MRP2 and the DMPS- and DMSA-mediated elimination of mercury in TR^- and control rats exposed to thiol *S*-conjugates of inorganic mercury, *Toxicol. Sci.*, 105, 211, 2008.
436. Bahn, A., Knabe, M., Hagos, Y., Rodiger, M., Godehardt, S., Graber-Neufeld, D.S., Evans, K.K., Burckhardt, G., and Wright, S.H., Interaction of the metal chelator 2,3-dimercaptor-1-propanesulfonate with the rabbit multispecific organic anion transporter 1 (rbOAT1), *Mol. Pharmacol.*, 62, 1128, 2002.
437. Burckhardt, B.C., Drinkuth, B., Menzel, C., Konig, A., Steffgen, J., Wright, S.H., and Burckhardt, G., The renal Na^+-dependent dicarboxylate transporter, NaDC-3 translocates dimethyl- and disulfhydryl-compounds and contributes to renal heavy metal detoxification, *J. Am. Soc. Nephrol.*, 13, 2628, 2002.
438. Islinger, F., Gekle, M., and Wright, S.H., Interaction of 2,3-dimercapto-1-propane sulfonate with the human organic anion transporter hOAT1, *J. Pharmacol. Exp. Ther.*, 299, 741, 2001.
439. Bode, B.P., Recent molecular advances in mammalian glutamine transport, *J. Nutr.*, 131, 2475S, 2001.
440. Deeley, R.G., Westlake, C., and Cole, S.P., Transmembrane transport of endo- and xenobiotics by mammalian ATP-binding cassette multidrug resistance proteins, *Physiol. Rev.*, 86, 849, 2006.
441. Ballatori, N. and Clarkson, T.W., Biliary transport of glutathione and methylmercury, *Am. J. Physiol.*, 244, G435, 1983.
442. Ballatori, N. and Clarkson, T.W., Dependence of biliary secretion of inorganic mercury on the biliary transport of glutathione, *Biochem. Pharmacol.*, 33, 1093, 1984.
443. Ballatori, N. and Clarkson, T.W., Sulfobromophthalein inhibition of glutathione and methylmercury secretion into bile, *Am. J. Physiol.*, 248, G238, 1985.
444. Ballatori, N. and Clarkson, T.W., Biliary secretion of glutathione and of glutathione-metal complexes, *Fund. Appl. Toxicol.*, 5, 816, 1985.

445. Zalups, R.K. and Lash, L.H., Depletion of glutathione in the kidney and the renal disposition of administered inorganic mercury, *Drug Metab. Dispos.*, 25, 516, 1997.
446. Büchler, M., König, J., Brom, M., Kartenbeck, J., Spring, H., Horie, T., and Keppler, D., cDNA cloning of the hepatocyte canalicular isoform of the multidrug resistance protein, cMrp, reveals a novel conjugate export pump deficient in hyperbilirubinemic mutant rats, *J. Biol. Chem.*, 271, 15091, 1996.
447. Keppler, D. and König, J., Hepatic canalicular membrane 5: Expression and localization of the conjugate export pump encoded by the MRP2 (cMRP/cMOAT) gene in liver, *FASEB J.*, 11, 509, 1997.
448. Ask, K., Akesson, A., Berglund, M., and Vahter, M., Inorganic mercury and methylmercury in placentas of Swedish women, *Environ. Health Perspect.*, 110, 523, 2002.
449. Inouye, M. and Kajiwara, Y., Placental transfer of methylmercury and mercuric mercury in mice, *Environ. Med.*, 34, 169, 1990.
450. Suzuki, T., Matsumoto, N., Miyama, T., and Katsunuma, H., Placental transfer of mercuric chloride, phenyl mercury acetate and methylmercury acetate in mice, *Ind. Health*, 5, 149, 1967.
451. Iioka, H., Moriyama, I., Oku, M., Hino, K., Itani, Y., Okamura, Y., and Ichijo, M., The effect of inorganic mercury on placental amino acid transport using microvillous membrane vesicles, *Nippon Sanka Fujinka Gakkai Zasshi*, 39, 202, 1987.
452. Jansson, T., Amino acid transporters in the human placenta, *Pediatr. Res.*, 49, 141, 2001.
453. Kudo, Y. and Boyd, C.A.R., Human placental amino acid transporter genes: Expression and function, *Reproduction*, 124, 593, 2002.
454. Cooper, A.J.L., Krasnikov, B.F., and Zalups, R.K., Potent inactivation of rat liver γ-cystathionase by mercury cysteine *S*-conjugate. A nefarious case of molecular mimicry. Unpublished.
455. World Health Organization, *Mercury, in Air Quality Guidelines*, 2nd ed., 2000, Ch. 6.9.
456. Thomas, D.J. and Smith, J.C., Partial characterization of a low-molecular weight methylmercury complex in rat cerebrum, *Toxicol. Appl. Pharmacol.*, 47, 547, 1979.
457. Aschner, M. and Clarkson, T.W., Uptake of methylmercury in the rat brain: Effects of amino acids, *Brain Res.*, 462, 31, 1988.
458. Aschner, M. and Clarkson, T.W., Methylmercury uptake across bovine brain capillary endothelial cells *in vitro*: The role of amino acids, *Pharmacol. Toxicol.*, 64, 293, 1989.
459. Hirayama, K., Effect of amino acids on brain uptake of methylmercury, *Toxicol. Appl. Pharmacol.*, 55, 318, 1980.
460. Hirayama, K., Transport mechanism of methylmercury, intestinal absorption, biliary excretion and distribution of methylmercury, *Kumamoto Med. J.*, 28, 151, 1975.
461. Hirayama, K., Effects of combined administration of thiol compounds and methylmercury chloride on mercury distribution in rats, *Biochem. Pharmacol.*, 34, 2030, 1985.
462. Foulkes, E.C., Metallothionein and glutathione as determinants of cellular retention and extrusion of cadmium and mercury, *Life Sci.*, 52, 11617, 1993.
463. Thomas, D.J. and Smith, J.C., Effect of co-administered low-molecular-weight thiol compounds on short-term distribution of methylmercury in the rat, *Toxicol. Appl. Pharmacol.*, 62, 104, 1982.
464. Jernelov, A., A new biochemical pathway for the methylation of mercury and some ecological implications. In: M.W. Miller and T.W. Clarkson (Eds), *Mercury, Mercurials, and Mercaptans*, p. 315, Thomas, Springfield, IL, 1973.
465. Landner, L., Biochemical model for the biological methylation of mercury suggested from methylation studies *in vivo* with *Neurospora crassa*, *Nat. Lond.*, 230, 452, 1971.
466. Betz, A.L. and Goldstein, G.W., Polarity of the blood-brain barrier: Neutral amino acid transport into isolated brain capillaries, *Science Washington DC*, 202, 225, 1978.
467. Oldendorf, W.H., Stereospecificity of blood-brain barrier permeability to amino acids, *Am. J. Physiol.*, 224, 967, 1973.
468. Kerper, L.E., Ballatori, N., and Clarkson, T.W., Methylmercury transport across the blood-brain barrier by an amino acid carrier, *Am. J. Physiol. (Regulatory Integrative Comp. Physiol.)*, 262, R761, 1992.
469. Aschner, M., Eberle, N.B., Goderie, S., and Kimelberg, H.K., Methylmercury uptake in rat primary astrocyte cultures: The role of the neutral amino acid transport system, *Brain Res.*, 521, 221, 1990.
470. Aschner, M., Eberle, N.B., and Kimelberg, H.K., Interactions of methylmercury with rat primary astrocyte cultures: Methylmercury efflux, *Brain Res.*, 554, 10, 1991.
471. Mokrzan, E.M., Kerper, L.E., Ballatori, N., and Clarkson, T.W., Methylmercury-thiol uptake into cultured brain capillary endothelial cells on amino acid system L, *J. Pharmacol. Exp. Ther.*, 272, 1277, 1995.
472. Kanai, Y., Segawa, H., Miyamoto, K., Uchino, H., Takeda, E., and Endou, H., Expression cloning and characterization of a transporter for large neutral amino acids activated by the heavy chain of 4F2 antigen, *J. Biol. Chem.*, 273, 23629, 1998.

473. Pineda, M., Fernandez, E., Torrents, D., Estevez, R., Lopez, C., Camps, M., Lloberas, J., Zorzano, A., and Palacin, M., Identification of a membrane protein, LAT-2, that co-expresses with 4F2 heavy chain, an L-type amino acid transport activity with broad specificity for small and large zwitterionic amino acids, *J. Biol. Chem.*, 274, 19738, 1999.
474. Prasad, P.D., Wang, H., Huang, W., Kekuda, R., Rajan, D.P., Leibach, F.H., and Ganapathy, V., Human LAT1, a subunit of system L amino acid transporter: Molecular cloning and transport function, *Biochem. Biophys. Res. Commun.*, 255, 283, 1999.
475. Chillaron, J., Roca, R., Valencia, A., Zorzano, A., and Palacin, M., Heteromeric amino acid transporters: Biochemistry, genetics, and physiology, *Am. J. Physiol. Renal Physiol.*, 281, F995, 2001.
476. Simmons-Willis, T.A., Koh, A.S., Clarkson, T.W., and Ballatori, N., Transport of a neurotoxicant by molecular mimicry: The methylmercury-L-cysteine complex is a substrate for human L-type large neutral amino acid transporter (LAT) 1 and LAT 2, *Biochem. J.*, 367, 239, 2002.
477. Wu, G., Screening of potential transport systems for methylmercury uptake in rat erythrocytes at 5° by use of inhibitors and substrates, *Pharmacol. Toxicol.*, 77, 169, 1995.
478. Wu, G., Methylmercury-cysteine uptake by rat erythrocytes: Evidence for several transport systems, *J. Appl. Toxicol.*, 16, 77, 1996.
479. Wu, G., Effect of probenecid on the transport of methylmercury in erythrocytes by the organic anion transport system, *Arch. Toxicol.*, 71, 218, 1997.
480. Urano, T., Iwasaki, A., Himeno, S., Naganuma, A., and Imura, N., Absorption of methylmercury compounds from rat intestine, *Toxicol. Lett.*, 50, 159, 1990.
481. Norseth, T. and Clarkson, T.W., Intestinal transport of ^{203}Hg-labeled methylmercury chloride, *Arch. Environ. Health*, 22, 568, 1971.
482. Bahn, A., Hagos, Y., Reuter, S., Balen, D., Brzica, H., Krick, W., Burchardt, B.C., Sabolic,́ I., and Burckhardt, G., Identification of a new urate and high affinity nicotinate transporter, hOAT10 (SLC22A13), *J. Biol. Chem.* 283, 16332, 2008.
483. Hilgendorf, C., Ahlin, G., Seithel, A., Artursson, P., Ungell, A.L., and Karlsson, J., Expression of thirty-six drug transporter genes in human intestine, liver, kidney, and organotypic cell lines, *Drug Metab. Dispos.*, 35, 1333, 2007.
484. Jung, S.M., Lee, W.K., Kwak, J.O., Jung, S.Y., Park, J., Kim, W.Y., Kim, J., and Cha, S.H., Identification of a novel murine organic anion transporter like protein 1 (OATLP1) expressed in the kidney, *Exp. Mol. Med.*, 38, 485, 2006.
485. Leaner, J.J. and Mason, R.P., Methylmercury accumulation and fluxes across the intestine of channel catfish, Ictalurus punctatus, *Comp. Biochem. Physiol. C. Toxicol. Pharmacol.*, 132, 247, 2002.
486. Foulkes, E.C., Metallothionein and glutathione as determinants of cellular retention and extrusion of cadmium and mercury, *Life Sci.*, 51, 11617, 1993.
487. Friberg, L., Studies on the metabolism of mercuric chloride and methylmercury dicyandiamide: Experiments on rats given subcutaneous injections with radioactive mercury (^{203}Hg), *AMA Arch. Ind. Health*, 20, 42, 1959.
488. Magos, L. and Butler, W.H., The kinetics of methylmercury administered repeatedly to rats, *Arch. Toxicol.*, 35, 25, 1976.
489. Magos, L., Peristianis, G.G., Clarkson, T.W., Brown, A., Preston, S., and Snowden, R.T., Comparative study of the sensitivity of male and female rats to methylmercury, *Arch. Toxicol.*, 48, 11, 1981.
490. Magos, L., Brown, A.W., Sparrow, S., Bailey, E., Snowden, R.T., and Skipp, W.R., The comparative toxicology of ethyl- and methylmercury, *Arch. Toxicol.*, 57, 260, 1985.
491. McNeil, S.I., Bhatnagar, M.K., and Turner, C.J., Combined toxicity of ethanol and methylmercury in rat, *Toxicology*, 53, 345, 1988.
492. Norseth, T. and Clarkson, T.W., Studies on the biotransformation of ^{203}Hg-labeled methylmercury chloride in rats, *Arch. Environ. Health*, 21, 717, 1970.
493. Norseth, T. and Clarkson, T.W., Biotransformation of methylmercury salts in the rat studied by specific determination of inorganic mercury, *Biochem. Pharmacol.*, 19, 2775, 1970.
494. Prickett, C.S., Laug, E.P., and Kunze, F.M., Distribution of mercury in rats following oral and intravenous administration of mercuric acetate and phenylmercuric acetate, *Proc. Soc. Exp. Biol. Med.*, 73, 585, 1950.
495. Richardson, R.J. and Murphy, S.D., Effect of glutathione depletion on tissue deposition of methylmercury in rats, *Toxicol. Appl. Pharmacol.*, 31, 505, 1975.
496. Alexander, J. and Aaseth, J., Organ distribution and cellular uptake of methylmercury in the rat as influenced by the intra- and extracellular glutathione concentration, *Biochem. Pharmacol.*, 31, 685, 1982.
497. Tanaka, T., Naganuma, A., Miura, N., and Imura, N., Role of testosterone in γ-glutamyltranspeptidase-dependent renal methylmercury uptake in mice, *Toxicol. Appl. Pharmacol.*, 112, 58, 1992.

498. Naganuma, A., Oda-Urano, N., Tanaka, T., and Imura, N., Possible role of hepatic glutathione in transport of methylmercury into mouse kidney, *Biochem. Pharmacol.*, 21, 291, 1988.
499. de Ceaurriz, J. and Ban, M., Role of gamma-glutamyltranspeptidase and beta-lyase in the nephrotoxicity of hexachloro-1,3-butadiene and methylmercury in mice, *Toxicol. Lett.*, 50, 249, 1990.
500. DiSimplicio, P., Gorelli, M., Ciuffreda, P., and Leonzio, C., The relationship between γ-glutamyltranspeptidase and Hg levels in Se/Hg antagonism in mouse liver and kidney, *Pharmacol. Res.*, 22, 515, 1990.
501. Gregus, Z., Stein, A.F., and Klaassen, C.D., Effect of inhibition of gamma-glutamyl-transpeptidase on biliary and urinary excretion of glutathione-derived thiols and methylmercury, *J. Pharmacol. Exp. Ther.*, 242, 27, 1987.
502. Mulder, K.M. and Kostyniak, P.J., Effect of L-(αS, 5S)-α-amino-3-chloro-4,5-dihydro-5-isoxazoleacetic acid on urinary excretion of methylmercury in the mouse, *J. Pharmacol. Exp. Ther.*, 234, 156, 1985.
503. Tanaka, T., Naganuma, A., Kobayashi, K., and Imura, N., An explanation for strain and sex differences in renal uptake of methylmercury in mice, *Toxicology*, 69, 317, 1991.
504. Yasutake, A., Hirayama, K., and Inoue, M., Mechanism of urinary excretion of methylmercury in mice, *Arch. Toxicol.*, 63, 479, 1989.
505. Bridges, C.C. and Zalups, R.K., System $B^{0,+}$ and the transport of thiol-S-conjugates of methylmercury, *J. Pharmacol. Exp. Ther.*, 319, 948, 2006.
506. Gonska, T., Hirsch, J.R., and Schlatter, E., Amino acid transport in the renal proximal tubule, *Amino Acids*, 19, 395, 2000.
507. Nakanishi, T., Hatanaka, T., Huang, W., Prasad, P.D., Leibach, F.H., Ganapathy, M.E., and Ganapathy, V., Na^+ and Cl^- coupled active transport of carnitine by the amino acid transporter $ATB^{0,+}$ from mouse colon expressed in HRPE cells and *Xenopus* oocytes, *J. Physiol. (Lond.)*, 532, 297, 2001.
508. Sloan, J.L. and Mager, S., Cloning and functional expression of a human Na^+ and Cl^- dependent neutral and cationic amino acid transporter $B^{0,+}$, *J. Biol. Chem.*, 274, 23740, 1999.
509. Koh, A.S., Simmons-Willis, T.A., Pritchard, J.B., Grassl, S.M., and Ballatori, N., Identification of a mechanism by which the methylmercury antidotes *N*-acetylcysteine and dimercaptopropane sulfonate enhance urinary metal excretion: Transport by the renal organic transporter-1, *Mol. Pharmacol.*, 62, 921, 2002.
510. Dunn, J.D. and Clarkson, T.W., Does mercury exhalation signal demethylation of methylmercury? *Health Physl.*, 38, 411, 1980.
511. Gage, J.C., Distribution and excretion of methyl and phenylmercury salts, *Brit. J. Ind. Med.*, 21, 197, 1964.
512. Omata, S., Sato, M., Sakimura, K., and Sugano, H., Time-dependent accumulation of inorganic mercury in subcellular fractions of kidney, liver, and brain or rats exposed to methylmercury, *Arch. Toxicol.*, 44, 231, 1980.
513. Madejczyk, M.S., Aremu, D.A., Simmons-Willis, T.A., Clarkson, T.W., and Ballatori, N., Accelerated urinary excretion of methylmercury following administration of its antidote *N*-acetylcysteine requires Mrp2/Abcc2, the apical multidrug resistance-associated protein, *J. Pharmacol. Exp. Ther.*, 322, 378, 2007.
514. Zalups, R.K. and Bridges, C.C., MRP2 involvement in renal proximal tubular elimination of methylmercury mediated by DMPS or DMSA, *Toxicol. Appl. Pharmacol.*, 235, 10, 2008.
515. Ballatori, N. and Clarkson, T.W., Developmental changes in the biliary excretion of methylmercury and glutathione, *Science*, 216, 61, 1982.
516. Refsvik, T., Excretion of methylmercury in rat bile: The effect of thioctic acid, thionalide, hexadecyl- and octadecylmercaptoacetate, *Acta Pharmacol. Toxicol. (Copenh)*, 50, 196, 1982.
517. Refsvik, T. and Norseth, T., Methylmercuric compounds in rat bile, *Acta Pharmacol. Toxicol.*, 36, 67, 1975.
518. Omata, S., Sakimura, K., Ishii, T., and Sugano, H., Chemical nature of a methylmercury complex with a low molecular weight in the liver cytosol of rats exposed to methylmercury chloride, *Biochem Pharmacol.*, 27, 1700, 1978.
519. Magos, L., Clarkson, T.W., and Allen, J., The interrelationship between non-protein bound thiols and the biliary excretion of methylmercury, *Biochem. Pharmacol.*, 27, 2203, 1978.
520. Refsvik, T., Excretion of methylmercury in rat bile: The effect of diethylmaleate cyclohexene oxide and acrylamide, *Acta Pharmacol. Toxicol.*, 42, 153, 1978.
521. Dutczak, W.J., Truong, A.T., and Ballatori, N., Transport of glutathione-methylmercury complex in rat liver canalicular membrane vesicles, *Toxicologist*, 13, 191, 1993.
522. Ballatori, N. and Dutczak, W.J., Identification and characterization of high and low affinity transport systems for reduced glutathione in liver cell canalicular membranes, *J. Biol. Chem.*, 269, 19731, 1994.
523. Ballatori, N. and Truong, T., Multiple canalicular transport mechanisms for glutathione *S*-conjugates, *J. Biol. Chem.*, 270, 3594, 1995.

524. Fernandez-Checa, J.C., Takikawa, H., Horie, T., Ookhtens, M., and Kaplowitz, N., Canalicular transport of reduced glutathione in normal and mutant Eisai hyperbilirubinemic rats, *J. Biol. Chem.*, 267, 1667, 1992.
525. Fernandez-Checa, J.C., Ookhtens, M., and Kaplowitz, N., Selective induction by phenobarbital of the electrogenic transport of glutathione and organic anions in rat liver canalicular membrane vesicles, *J. Biol. Chem.*, 268, 10836, 1993.
526. Garcia-Ruiz, C., Fernandez-Checa, J., and Kaplowitz, N., Bidirectional mechanism of plasma membrane transport of reduced glutathione in intact rat hepatocytes and membrane vesicles, *J. Biol. Chem.*, 267, 22256, 1992.
527. Ballatori, N., Glutathione mercaptides as transport forms of metals, *Adv. Pharmacol.*, 27, 271, 1994.
528. Dutczak, W.J., Clarkson, T.W., and Ballatori, N., Biliary-hepatic recycling of a xenobiotic, Gallbladder absorption of methylmercury, *Am. J. Physiol.*, 260, G873, 1991.
529. Babu, E., Kanai, Y., Chairoungdua, A., Kim, D.K., Iribe, Y., Tangtrongsup, S., Jutabha, P., Yuewei, L., Ahmed, N., Sakamoto, S., Anzai, N., Nagomori, S., and Endou, H., Identification of a novel system L amino acid transporter structurally distinct from heterodimeric amino acid transporters, *J. Biol. Chem.*, 278, 43838, 2003.
530. Amin-Zaki, L., Elhassani, L.S., Majeed, M.A., Clarkson, T.W., Doherty, R.A., and Greenwood, M., Intrauterine methylmercury poisoning in Iraq, *Pediatrics*, 54, 587, 1974.
531. Davidson, P.W., Strain, J.J., Myers, G.J., Thurston, S.W., Bonham, M.P., Shamlaye, C.F., Stokes-Riner, A., Wallace, J.M., Robson, P.J., Duffy, E.M., Georger, L.A., Sloane-Reeves, J., Cernichiari, E., Canfield, R.L., Cox, C., Huang, L.S., Janciuras, J., and Clarkson, T.W., Neurodevelopmental effects of maternal nutritional status and exposure to methylmercury from eating fish during pregnancy, *Neurotoxicology*, 29, 767, 2008.
532. Harada, M., Congenital Minamata disease: Intrauterine methylmercury poisoning, *Teratology*, 18, 285, 1978.
533. Harada, M., Minamata disease: Methylmercury poisoning in Japan caused by environmental pollution, *Crit. Rev. Toxicol.*, 25, 1, 1995.
534. Inouye, M. and Kajiwara, Y., Developmental disturbances of the fetal brain in guinea pigs caused by methylmercury, *Arch. Toxicol.*, 62, 15, 1988.
535. Kajiwara, Y. and Inouye, M., Effects of methylmercury and mercuric chloride on preimplantation mouse embryos *in vivo*, *Teratology*, 33, 231, 1986.
536. Kajiwara, Y. and Inouye, M., Inhibition of implantation caused by methylmercury and mercuric chloride in mouse embryos *in vivo*, *Bull. Environ. Contam. Toxicol.*, 49, 541, 1992.
537. Matsumoto, H., Koya, G., and Tekeuchi, T., Fetal Minamata disease. A neuropathological study of two cases of intrauterine intoxication by a methylmercury compound, *J. Neuropathol. Exp. Neurol.*, 24, 563, 1965.
538. Inouye, M., Murao, K., and Kajiwara, Y., Behavioral and neuropathological effects of prenatal methylmercury exposure in mice, *Neurobehav. Toxicol. Teratol.*, 7, 227, 1985.
539. Kajiwara, Y., Yasutake, A., Adachi, T., and Hirayama, K., Methylmercury transport across the placenta via neutral amino acid carrier, *Arch. Toxicol.*, 70, 310, 1996.
540. Leazer, T.M. and Klaassen, C.D., The presence of xenobiotic transporters in rat placenta, *Drug Metab. Dispos.*, 31, 153, 2003.
541. Aremu, D.A., Madejczyk, M.S., and Ballatori, N., *N*-acetylcysteine as a potential antidote and biomonitoring agent of methylmercury exposure, *Environ. Health Perspect.*, 116, 26, 2008.
542. Bridges, C.C., Joshee, L., and Zalups, R.K., Effect of DMPS and DMSA on the fetal and placental disposition of mercuric ions following exposure to methylmercury, *Placenta*, 30, 800, 2009.
543. St. Pierre, M.V., Serrano, M.E., Macias, R.I.R., Dubs, U., Hoechli, M., Lauper, U., Meier, P.J., and Marin, J.J.G., Expression of members of the multidrug resistance protein family in human term placenta, *Am. J. Physiol.*, 279, R1495, 2000.
544. Cha, S.H., Sekine, T., Kushuhara, H., Yu, E., Kim, J.Y., Kim, D.K., Sugiyama, Y., Kanai, Y., and Endou, H., Molecular cloning and characterization of multispecific organic anon transporter 4 expressed in placenta, *J. Biol. Chem.*, 275, 4507, 2000.
545. Agency for Toxic Substance and Disease Registry, Toxicological Profile for Selenium, U.S. Department of Health and Humans Services, Public Health Service, Centers for Disease Control, Atlanta, GA, 2007.
546. Combs, G.F., Jr. and Gray, W.P., Chemopreventive agents: Selenium, *Pharmacol. Ther.*, 79, 179, 1998.
547. Dudley, H.C. and Miller, J.W., Toxicology of selenium. VI. Effects of subacute exposure to hydrogen selenide, *J. Ind. Hygiene Toxicol.*, 23, 470, 1941.

548. Hall, R.H., Laskin, S., Frank, P., Maynard, E.A., and Hodge, H.C., Preliminary observations on toxicity of elemental selenium, *AM A Arch. Ind. Hyg. Assoc.*, 4, 458, 1951.
549. Wilson, H.M., Selenium oxide poisoning, *N. C. Med. J.*, 23, 73, 1962.
550. Glover, J.R., Selenium in human urine: A tentative maximum allowable concentration for industrial and rural populations, *Ann. Occup. Hyg.*, 10, 3, 1967.
551. Holness, D.L., Taraschuk, I.G., and Nethercott, J.R., Health status of copper refinery workers with specific reference to selenium exposure, *Arch. Environ. Health*, 44, 291, 1989.
552. Clinton, M., Jr., Selenium fume exposure, *J. Ind. Hyg. Toxicol.*, 29, 225, 1947.
553. Arduser, F., Wolffram, S., and Scharrer, E., Active absorption of selenate by rat ileum, *J. Nutr.*, 115, 1203, 1985.
554. Murer, H., Markovich, D., and Biber, J., Renal and small intestinal sodium-dependent symports of phosphate and sulfate, *J. Exp. Biol.*, 196, 167, 1994.
555. Bush, A.E., Waldegger, S., Herzer, T., Biber, J., Markovich, D., Murer, H., and Lang, F., Electrogenic cotransport of Na^+ and sulfate in *Xenopus* oocytes expressing the cloned Na^+/SO_4^{2-} transport protein NaSi-1, *J. Biol. Chem.*, 269, 12407, 1994.
556. Lee, A., Beck, L., and Markovich, D., The human renal sodium sulfate cotransporter (SLC13A1; hNaSi-1) cDNA and gene: Organization, chromosomal localization, and functional characterization, *Genomics*, 70, 354, 2000.
557. Dawson, P.A., Pirlo, K.J., Steane, S.E., Nguyen, K.A., Kunzelmann, K., Chien, Y.J., and Markovich, D., The rat Na^+-sulfate cotransporter rNaS2: Functional characterization, tissue distribution, and gene (slc13a4) structure, *Pflugers Arch.*, 450, 263, 2005.
558. Dawson, P.A., Pirlo, K.J., Steane, S.E., Kunzelmann, K., Chien, Y.J., and Markovich, D., Molecular cloning and characterization of the mouse Na^+ sulfate cotransporter gene (*Slc13a4*): Structure and expression, *Genes Genet. Syst.*, 81, 265, 2006.
559. Miyauchi, S., Srinivas, S.R., Fei, Y.J., Gopal, E., Umapathy, N.S., Wang, H., Conway, S.J., Ganapathy, V., and Prasad, P.D., Functional characteristics of NaS2, a placental-specific Na^+-coupled transporter for sulfate and oxyanions of the micronutrients selenium and chromium, *Placenta*, 27, 550, 2006.
560. Anundi, I., Hogberg, J., and Stahl, A., Absorption of selenite in the rat small intestine: Interactions with glutathione, *Acta Pharmacol. Toxicol. (Copenh)*, 54, 273, 1984.
561. Gyurasics, A., Perjési, P., and Gregus, Z., Role of glutathione and methylation in the biliary excretion of selenium, the paradoxical effect of sulfobromophthalein, *Biochem. Pharmacol.*, 56, 1381, 1998.
562. Scharrer, E., Senn, E., and Wolffram, S., Stimulation of mucosal uptake of selenium from selenite by some thiols at various sites of rat intestine, *Biol. Trace Elem. Res.*, 33, 109, 1992.
563. Senn, E., Scharrer, E., and Wolffram, S., Effects of glutathione and of cysteine on intestinal absorption of selenium from selenite, *Biol. Trace Elem. Res.*, 33, 103, 1992.
564. Vendeland, S.C., Deagen, J.T., and Whanger, P.D., Uptake of selenotrisulfides of glutathione and cysteine by brush border membranes from rat intestines, *J. Inorg. Biochem.*, 47, 131, 1992.
565. Wurmli, R., Wolffram, S., Stingelin, T., and Scharrer, E., Stimulation of mucosal uptake of selenium from selenite by L-cysteine in sheep small intestine, *Biol. Trace Elem. Res.*, 20, 75, 1989.
566. Leblondel, G., Mauras, Y., Cailleux, A., and Allain, P., Transport measurements across Caco-2 monolayers of different organic and inorganic selenium: influence of sulfur compounds, *Biol. Trace Elem. Res.*, 83, 191, 2001.
567. Stoica, A., Pentecost, E., and Martin, M.B., Effects of selenite on estrogen receptor-α and activity in MCF-7 breast cancer cells, *J. Cell. Biol.*, 79, 282, 2000.
568. Stapleton, S.R., Selenium: An insulin-mimetic, *Cell Mol. Life Sci.*, 57, 1874, 2000.
569. Ezaki, O., The insulin-like effects of selenate in rat adipocytes, *J. Biol. Chem.*, 265, 1124, 1990.
570. Furnsinn, C., English, R., Ebner, K., Nowotny, P., Vogle, C., and Waldhausl, W., Insulin-like vs. non-insulin stimulation of glucose metabolism by vanadium, tungsten, and selenium compounds in rat muscle, *Life Sci.*, 59, 1989, 1996.
571. Berg, E.A., Wu, J.Y., Campbell, L., Kagey, M., and Stapleton, S.R., Insulin-like effects of vanadate and selenate on the expression of glucose-6-phosphate dehydrogenase and fatty acid synthase in diabetic rats, *Biochimie*, 12, 919, 1995.
572. McNeill, J.H., Delgatty, H.L.M., and Battell, M.L., Insulin-like effects of sodium selenate in streptozotocin-induced diabetic rats, *Diabetes*, 40, 1675, 1991.
573. Ghosh, R., Mukherjee, B., and Chatterjee, M., A novel effect of selenium on streptozotocin-induced diabetic mice, *Diabetes Res.*, 25, 165, 1994.
574. O'Brien, R.M. and Granner, D.K., Regulation of gene expression by insulin, *Physiol. Rev.*, 76, 1109, 1996.

575. Shepherd, P.R., Navé, B.T., and Siddle, K., Insulin stimulation of glycogen synthesis and glycogen synthase activity is blocked by wortmannin and rapamycin in 3T3-L1 adipocytes: Evidence for the involvement of phosphoinositide-3 kinase and p70 ribosomal protein-S6 kinase, *Biochem. J.*, 305, 25, 1995.
576. Magnuson, M.A., Andreone, T.L., Printz, R.L., Koch, S., and Granner, D.K., Rat glucokinase gene: Structure and regulation by insulin, *Proc. Natl. Acad. Sci. USA*, 86, 4838, 1989.
577. O'Brien, R.M., Lucas, P.C., Forest, C., Magnuson, M.A., and Granner, D.K., Identification of a sequence in the PEPCK gene that mediates a negative effect of insulin on transcription, *Science*, 249, 533, 1990.
578. Short, J.M., Wynshaw-Boris, A., Short, H.P., and Hanson, R.W., Characterization of the phosphoenolpyruvate carboxykinase (GTP) promoter-regulatory region, *J. Biol. Chem.*, 261, 9721, 1986.
579. Moustaid, N., Beyer, R.S., and Sul, H.S., Identification of an insulin response element in the fatty acid synthase promoter, *J. Biol. Chem.*, 269, 5629, 1994.
580. Stapleton, S.R., Mitchell, D.A., Salati, L.M., and Goodridge, A.G., Triiodothyronine stimulates transcription of the fatty acid synthase gene in chick embryo hepatocytes in culture, *J. Biol. Chem.*, 265, 18442, 1990.
581. Wagle, A., Jivraj, S., Garlock, G., and Stapleton, S.R., Insulin regulation of glucose-6-phosphate dehydrogenase gene expression is rapamycin sensitive and requires phosphatidylinositol 3-kinase expression, *J. Biol. Chem.*, 273, 14968, 1998.
582. Battell, M.L., Delgatty, H.L.M., and McNeill, J.H., Sodium selenate corrects glucose tolerance and heart function in STZ diabetic rats, *Mol. Cell Biochem.*, 179, 27, 1998.
583. Becker, D.J., Reul, B., Ozcelikay, A.T., Buchet, J.P., Henquin, J.C., and Brichard, S.M., Oral selenate improves glucose homeostasis and partly reverses abnormal expression of liver glycogenic and gluconeogenic enzymes in diabetic rats, *Diabetologia*, 39, 3, 1996.
584. Hwang, D., Seo, S., Kim, Y., Kim, C., Shim, S., Jee, S., Lee, S., Jang, M., Kim, M., Yim, S., Lee, S.K., Kang, B., Jang, I., and Cho, J., Selenium acts as an insulin-like molecule for the down-regulation of diabetic symptoms via endoplasmic reticulum stress and insulin signaling proteins in diabetes-induced non-obese diabetic mice, *J. Biosci.*, 32, 723, 2007.
585. Ayaz, M., Can, B., Ozdemir, S., and Turan, B., Protective effect of selenium treatment on diabetes-induced myocardial structural alterations, *Biol. Trace Elem. Res.*, 89, 215, 2002.
586. Ayaz, M., Ozdemir, S., Ugur, M., Vassort, G., and Turan, B., Effects of selenium on altered mechanical and electrical cardiac activities of diabetic rat, *Arch. Biochem. Biophys.*, 426, 83, 2004.
587. Ayza, M., Celik, H.A., Aydin, H.H., and Turan, B., Sodium selenite protects against diabetes-induced alterations in the antioxidant defense system of the liver, *Diabetes Metab. Res. Rev.*, 22, 295, 2006.
588. Ayaz, M. and Turan, B., Selenium prevents diabetes-induced alterations in $[Zn^{2+}]_I$ and metallothionein level of rat heart via restoration of redox cycle, *Am. J. Physiol. Heart Circ. Physiol.*, 290, H1071, 2006.
589. Zhu, Y., Goodridge, A.G., and Stapleton, S.R., Zinc, vanadate and selenate inhibit the tri-iodothyronine-induced expression of fatty acid synthase and malic enzyme in chick-embryo hepatocytes in culture, *Biochem. J.*, 303, 213, 1994.
590. Wang, X.D., Vatamaniuk, M.Z., Wang, S.K., Roneker, C.A., Simmons, R.A., and Lei, X.G., Molecular mechanisms for hyperinsulinaemia induced by overproduction of selenium-dependent glutathione peroxidase-1 in mice, *Diabetologia*, 51, 1515, 2008.
591. Campbell, S.C., Aldibbiat, A., Marriott, C.E., Landy, C., Ali, T., Ferris, W.F., Butler, C.S., Shaw, J.A., and Macfariane, W.M., Selenium stimulates pancreatic beta-cell gene expression and enhances islet function, *FEBS Lett.*, 582, 2333, 2008.
592. Caverzasio, J., Selz, T., and Bonjour, J.P., Characteristics of phosphate transport in osteoblast like cells, *Calcif. Tissue Int.*, 43, 83, 1988.
593. Schali, C., Vaughn, D.A., and Fanestil, D.D., Reconstitution of the partially purified renal phosphate (Pi) transporter, *Biochem. J.*, 235, 189, 1986.
594. Vachon, V., Delisle, M.C., Laprade, R., and Beliveau, R., Reconstruction of the renal brush-border membrane sodium/phosphate co-transporter, *Biochem J.*, 278, 543, 1991.
595. Montessuit, C., Caverzasio, J., and Bonjour, J.P., Characterization of a Pi transport system in cartilage matrix vesicles. Potential role in the calcification process, *J Biol. Chem.*, 266, 17791, 1991.
596. Ishizawa, T., Tsuji, A., Tamai, I., Teraski, T., Hosoi, K., and Fukatsu, S., Sodium and pH dependent carrier-mediated transport of antibiotic, fosfomycin, in the rat intestinal brush-border membrane, *J. Pharmacobiobyn.*, 13, 292, 1990.
597. Azzarolo, A.M., Ritchie, G., and Quamme, G.A., Some characteristics of sodium-independent phosphate transport across renal basolateral membranes, *Biochim. Biophys. Acta*, 1064, 229, 1991.

598. Quamme, G.A., Walker, J.J., and Yan, T.S., pH gradient stimulated phosphate transport in outer medullary brush-border membranes, *Am. J. Physiol.*, 257, F639, 1989.
599. Elger, M., Werner, A., Herter, P., Kohl, B., Kinne, R.K., and Hentschel, H., Na-P(i) cotransport sites in proximal tubule and collecting tubule of winter founder (Pleuronectes americanus), *Am. J. Physiol.*, 274, F374, 1998.
600. Villa-Bellosta, R. and Sorribas, V., Role of rat sodium/phosphate cotransporters in the cell membrane transport of arsenate, *Toxicol. Appl. Pharmacol.*, 2008 (Epub ahead of print).
601. Lee, B.T. and Kim, K.W., Arsenic accumulation and toxicity in the earthworm *Eisenia fetida*, affected by chloride and phosphate, *Environ. Toxicol. Chem.*, 27, 2488, 2008.
602. Shoemaker, D.G., Bender, C.A., and Gunn, R.B., Sodium-phosphate cotransport in human red blood cells. Kinetics and role in membrane metabolism, *J. Gen. Physiol.*, 92, 449, 1988.
603. De Master, E.G. and Mitchell, R.A.A., A comparison of arsenate and vanadate as inhibitors or uncouplers of mitochondrial and glycolytic energy metabolism, *Biochemistry*, 12, 3616, 1973.
604. Alexander, J. and Aaseth, J., Uptake of chromate in human red blood cells and isolated rat liver cells: The role of the anion carrier, *Analyst*, 120, 931, 1995.
605. Ormos, G. and Manyai, S., Anion transport of the red blood cell under nonequilibrium conditions, *J. Physiol.*, 276, 501, 1978.
606. Boyd, C.A.R. and Shennan, D.B., Sulphate transport into vesicles prepared from human placental brush border membranes: Inhibition by trace element oxides, *J. Physiol.*, 379, 367, 1986.
607. Shennan, D.B., Selenium (selenate) transport by human placental brush border membrane vesicles, *Brit. J. Nutr.*, 59, 13, 1988.
608. Markovich, D. and James, K.M., Heavy metals mercury, cadmium, and chromium inhibit the activity of the mammalian liver and kidney sulfate transporter sat-1, *Toxicol. Appl. Pharmacol.*, 154, 181, 1999.
609. Markovich, D. and Knight, D., Renal Na-Si cotransporter NaSi-1 is inhibited by heavy metals, *Am. J. Physiol. Renal Physiol.*, 274, F283, 1998.
610. Cardin, C.J. and Mason, J., Sulphate transport by rat ileum: Effect of molybdate and other anions, *Biochim. Biophys. Acta*, 394, 46, 1975.
611. Mason, J. and Cardin, C.J., The competition of molybdate and sulphate ions for a transport system in the ovine small intestine, *Res. Vet. Sci.*, 22, 313, 1977.
612. Ryan, J., McKillen, M., and Mason, J., Sulphate/molybdate interactions: *In vivo* and *in vitro* studies on the group VI oxyanion transport system in ovine renal tubule epithelial cells, *Ann. Rech. Vet.*, 18, 47, 1987.
613. David, C. and Ullrich, K.J., Substrate specificity of the luminal Na^+-dependent sulphate transport system in the proximal renal tubule as compared to the contraluminal sulphate exchange system, *Pflugers Arch.*, 421, 455, 1992.
614. Timmer, R.T. and Gunn, R.B., Phosphate transport by the human renal cotransporter NaPi-3 expressed in HEK-293 cells, *Am. J. Physiol.*, 174, C757, 1998.
615. Cantly, L.C., Jr., Cantley, L.G., and Josephson, L., A characterization of vanadate interactions with the (Na,K)-ATPase: Mechanistic and regulatory implications, *J. Biol. Chem.*, 253, 7361, 1978.
616. Cantley, L.C., Resh, M.D., and Guidotti, G., Vanadate inhibits the red cell (Na^+, K^+) ATPase from the cytoplasmic side, *Nature*, 272, 552, 1978.
617. Karlish, S.J.D., Beauge, L.A., and Glynn, I.M., Vanadate inhibits (Na^+, K^+) ATPase by blocking conformational change of the unphosphorylated form, *Nature*, 282, 333, 1979.
618. Garcia, S.J., Gellein, K., Syversen, T., and Aschner, M., Iron deficient and manganese supplemented diets alter metals and transporters in the developing rat brain, *Toxicol. Sci.*, 95, 205, 2007.
619. Dubyak, G.R. and Kleinzeller, A., The insulin-mimetic effects of vanadate in isolated rat adipocytes, *J. Biol. Chem.*, 255, 5306, 1980.
620. Goldfine, A.B., Simonson, D.C., Folli, F., Patti, M.E., and Kahn, C.R., Metabolic effects of sodium metavanadate in humans with insulin-dependent and noninsulin-dependent diabetes mellitus *in vivo* and *in vitro* studies, *J. Clin. Endocrinol. Metab.*, 80, 3311, 1995.
621. Orvig, C., Thompson, K.H., Battell, M., and McNeill, J.H., Vanadium compounds as insulin mimics, *Met. Ions Biol. Syst.*, 31, 575, 1995.
622. Barbagallo, M., Dominguez, L.J., and Resnick, L.M., Insulin-mimetic action of vanadate. Role of intracellular magnesium, *Hypertension.*, 38, 701, 2001.
623. Päquet, M.R., Romanek, R.J., and Sargeant, R.J., Vanadate induces the recruitment of GLUT-4 glucose transporter to the plasma membrane of rat adipocytes, *Mol. Cell Biochem.*, 109, 149, 1992.

11 Heavy Metal Transport and Detoxification in Crustacean Gastrointestinal and Renal Epithelial Cells

Gregory A. Ahearn

CONTENTS

11.1 Introduction 295
11.2 Gastrointestinal System 296
  11.2.1 Hepatopancreas 297
    11.2.1.1 Role of Metallothionein and Glutathione in Metal Detoxification 297
    11.2.1.2 Role of Precipitation Granules in Metal Detoxification 298
    11.2.1.3 Cation Transport Processes of Crustacean Digestive Tract and Renal Organs 299
    11.2.1.4 Metal Transport Mechanisms of Crustacean Hepatopancreatic Plasma Membranes and Intracellular Organelles 300
  11.2.2 Intestine 310
    11.2.2.1 Sodium and Chloride Transport and Intestinal Electrophysiology 310
    11.2.2.2 Heavy Metal Transport across Perfused Crustacean Intestine 311
11.3 Kidneys 314
  11.3.1 Transport of Water, Sodium, and Chloride, and Role in Osmoregulation, by the Kidney 315
  11.3.2 Antennal Gland Secretion of Magnesium and Sulfate 316
  11.3.3 Calcium Reabsorption by Crustacean Antennal Glands 316
  11.3.4 Secretion of Organic Anion and Cation Metabolites by Antennal Glands 317
  11.3.5 Heavy Metal Regulation by the Renal OAT1 319
Acknowledgments 320
References 320

11.1 INTRODUCTION

Potentially toxic environmental heavy metals may enter animals across the integument [1,2], by way of the gills [3–5], or through the gut in food [6,7]. Elimination of metals from the animal may occur across the same tissues and by way of the kidneys. Trace amounts of certain metals (e.g., zinc, copper, iron) are needed by most animals for a variety of metabolic processes, but high concentrations of all metals can be toxic and must be regulated. Aquatic invertebrates possess a variety of organ and cellular detoxification processes that reduce the concentrations of potentially toxic metals circulating in the blood, while maintaining proper concentrations of those metals needed for specific physiological and biochemical functions. These regulatory mechanisms include (1) physiological

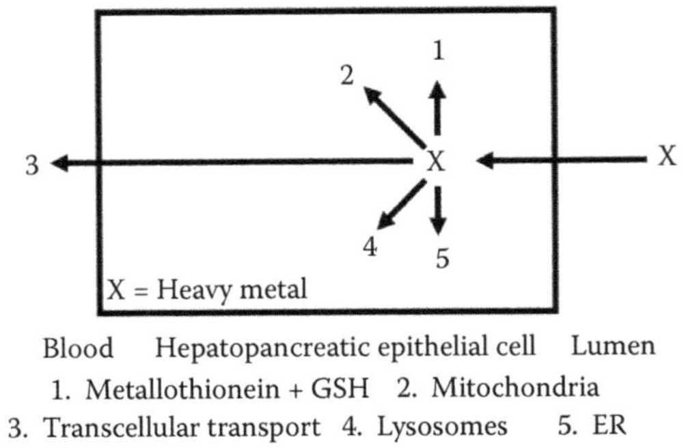

FIGURE 11.1 Possible fates of luminal heavy metals (X) in crustacean hepatopancreatic epithelial cells. Metal (X) crosses the apical membrane and may complex with metallothionein or glutathione (GSH), enter mitochondria, lysosomes, endoplasmic reticulum, or be transferred to the blood across the basolateral membrane. Other fates are possible. (From Ahearn, G.A., Mandal, P.K., and Mandal, A., *J. Comp. Physiol. B* 174, 439, 2004. With kind permission from Springer Science + Business Media.)

transport processes balancing metal excretion rates with uptake rates from the environment, (2) intracellular sequestration mechanisms for metals involving high-affinity binding sites on low-molecular-weight peptides known as glutathione and metallothionein and their subsequent excretion through the lysosomal endomembrane system, and (3) intracellular sequestration processes for metals involving specific vacuoles or lysosomes producing solid metallic phosphate or sulfate granules that may subsequently undergo exocytosis for elimination. This review will focus on metal transport, sequestration, and detoxification processes associated with three regulatory organs in selected crustaceans: (1) the hepatopancreas; (2) the intestine; and (3) the kidneys (antennal glands).

Figure 11.1 illustrates the possible fates of ingested monovalent, divalent, or trivalent cationic metals that pass across the brush-border or luminal membrane of a lobster hepatopancreatic epithelium. Similar fates and processes are likely to occur for metal uptake and handling by integumentary, gill, and renal epithelial cells. In these epithelial cells, cytoplasmic metallic ions (X) may (1) associate with glutathione or metallothionein [8], (2) be transported into mitochondria [9], (3) undergo efflux across the basolateral cell membrane to the blood [10], (4) be accumulated by lysosomes [11,12], or (5) be transferred into the ER [13]. Environmental heavy metals may be transported across crustacean epithelial cells either by using transport proteins normally accommodating ions such as sodium or calcium or by employing specific undefined metal transport proteins. This review will address the hypothesis that epithelial cells of crustacean hepatopancreas, intestine, and kidneys represent tissues of heavy metal homeostasis and act as an ensemble to help regulate appropriate organismic concentrations of metals for physiological and biochemical requirements.

11.2 GASTROINTESTINAL SYSTEM

The digestive system of crustaceans is a simple tube that is divided into functional regions with accessory organs arising from the tube at anterior or posterior locations. The arthropods exhibit a digestive tract divided into foregut, midgut, and hindgut. The first two areas are chitin lined and derived from embryonic ectoderm, while the midgut is without a chitinous lining and develops from endoderm. The crustacean midgut is further separated into the hepatopancreas and the intestine. The former is a complex organ composed of a multitude of epithelial-lined, blind-ended, tubules that perform a variety of functional roles in the living animal, including secretion of digestive enzymes, digestion of dietary elements, absorption of the resulting digestion products, emulsification of lipids, storage of carbohydrates, lipids, and heavy metals, and synthesis of the blood pigment, hemocyanin. The intestine, by contrast, is a simple epithelial-lined tube connecting the stomach with the rectum. This structure receives the digestion products of both the stomach and hepatopancreas prior to fecal formation and excretion from the organism.

11.2.1 Hepatopancreas

The arthropod hepatopancreas is composed of an epithelium with four major cell types, E-cells, F-cells, B-cells, and R-cells, and each is believed to contribute to the overall functions of digestion, absorption, secretion, emulsification, osmoregulation, and detoxification. Hepatopancreatic E-cells are mitotic and function much like mammalian intestinal crypt cells, giving rise to two distinct cell lines: an absorptive line generating new R-cells that act much like intestinal columnar cells, and a secretory line composed of F-cells manufacturing and secreting digestive enzymes like pancreatic acinar cells or stomach chief cells [14–16]. A second type of digestive cell, B-cells, is derived from F-cells and appears to play a significant role in the intracellular digestion and excretion of digestive breakdown products and xenobiotics [17].

11.2.1.1 Role of Metallothionein and Glutathione in Metal Detoxification

Metallothioneins and glutathione are proteins that are found in virtually all major invertebrate phyla as well as in all vertebrates. The former proteins are water soluble, heat stable, and have a molecular mass of approximately 6000–7000 Da. They generally possess about 60 amino acids, which are rich in cysteine and essentially lack aromatic amino acids, histidine, and methionine. Because of the high cysteine content, this group of proteins exhibits a high affinity for metal ions and can selectively bind them from very low intracellular concentrations. Glutathione is a tripeptide, composed of the amino acid residues glycine, glutamate, and cysteine. An increase in environmental metal concentrations leads to the induction of metallothionein and glutathione content through genomic stimulation so that the cytoplasmic level of these proteins keeps pace with the accelerated uptake of xenobiotics [18]. A major physiological role for metallothioneins is to serve as a reservoir of cations (such as copper and zinc), which are used in synthesizing apoenzymes [8,19–21]. In the lobster (*Homarus americanus*) and the blue crab (*Callinectes sapidus*), three metallothionein encoding genes have been identified as a result of their interaction with copper, zinc, and cadmium and have been designated as CuMT-I, CuMT-2, and CuMT-3 [22,23]. While all three isoforms may be involved in regulatory processes involving metal binding in the cytoplasm, only the CuMT-3 form appears to play a role in the activation of hemocyanin through the transfer of copper to apohemocyanin. These studies went on to show that the blue crab has four genes encoding different metallothioneins that individually become activated by the presence of copper, zinc, or cadmium [24]. Finally, there appears to be a tissue specificity to the expression of metallothionein isoforms in crustaceans. Crabs exposed to cadmium in the water accumulate the metal in the gills bound to MT-2, while MT-1 is absent [25]. In contrast, when crabs were fed diets containing cadmium, both hepatopancreatic isoforms MT-1 and MT-2 are present and both have a role in binding the metal.

In addition to their critical role in passing essential metals to enzymes involved in a variety of catalytic and respiratory functions, metallothioneins and glutathione, by their inductive capabilities, serve a key role in heavy metal detoxification when intracellular metal concentrations exceed those necessary for metabolic functions. In this detoxification role, metallothioneins and glutathione are intimately involved with the lysosomal system of cells. It has long been appreciated that lysosomes in mollusks [18,19,26–28], crustaceans [29], and annelids [30] contain high concentrations of heavy metals [8]. The mechanism by which metals bound to metallothioneins or glutathione are able to cross lysosomal membranes and become sequestered there from the rest of the cytoplasm is not currently understood. In lysosomes the metals are generally associated with an insoluble polymer of lipid peroxidation processes called lipofuscin. In one instance approximately 80% of copper accumulated in lysosomes of the mussel digestive gland (i.e., hepatopancreas) is bound to SH-rich proteins, which likely are remnants of cytoplasmic metallothioneins or glutathione [19]. Therefore, metallothioneins and the tripeptide glutathione bind metals in the cytoplasm by processes that show both metal and tissue specificity followed by transfer of these metal–protein associates into lysosomes where the amino acid sequences of proteins may be degraded and the freed metals are able to associate with a variety of trapped anionic elements simultaneously transferred there as part of a metal detoxification process.

11.2.1.2 Role of Precipitation Granules in Metal Detoxification

Epithelial cells of the gastrointestinal tract, kidneys, and gills of freshwater, marine, and terrestrial invertebrates, of virtually every phylum, ubiquitously contain heavy metal-containing solid concretions or granules within membrane-enclosed organelles such as lysosomes or vacuoles [8,31–33]. While these structures appear universal in distribution among the invertebrates, the physiological/biochemical events responsible for their formation are largely unknown [8]. The compositions of many of these concretions in representative invertebrates from several phyla have been examined using x-ray microprobe analysis. These studies have shown that the granules contain either calcium or heavy metal cations, such as zinc, copper, and iron, complexed with sulfur and phosphorus [27,33–37]. As a result of this complex formation between cationic metals and polyvalent inorganic anions, a potentially toxic heavy metal cation is removed from the cytoplasm and sequestered within the organelle membrane in an insoluble, detoxified form. Subsequent long-term storage or exocytotic events may extrude the concretions from the cell followed by organismic excretory mechanisms that deposit the metal back into the environment.

Figure 11.2 is a generic representation of four types of concretions that have been identified in invertebrate epithelial cells by microprobe analysis [38]. As indicated in this figure, metals appear to be sequestered into three different types of intracellular concretion vacuoles that may or may not include the chemical components of metallothionein (M). Type A vacuoles appear to accumulate zinc and phosphorus, Type B vacuoles contain cadmium, copper, mercury, and silver along with sulfur, and Type C vacuoles are restricted to iron and the respective anion associated with these iron-containing vacuoles is undetermined. A fourth type of concretion (Type D) is extracellular and is composed of calcium and carbonate. These apparent distinguishing properties between vacuoles and concretion compositions suggest that specific transport proteins may occur on membranes of the different vacuoles for both metals and complexing anions.

Viarengo [8] suggests that detoxification vacuoles may possess a similar calcium sequestering mechanism for storing this divalent cation as occurs in the ER. In this instance calcium, and related heavy metal divalent cations, could be transported from the cytoplasm to the vacuolar contents by an ATP-dependent, calcium-ATPase active transport protein. This calcium pump would be

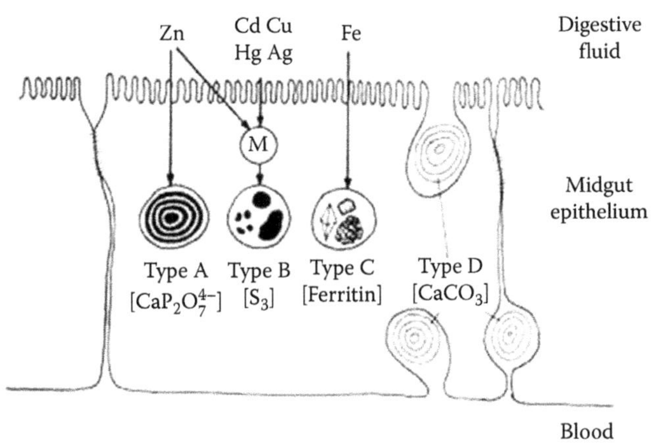

FIGURE 11.2 Types and composition of cellular vacuolar- or lysosomal-enclosed concretions in invertebrate cells. Four types of intracellular (Types A, B, C) and extracellular (Type D) concretions involving both cations and anions have been described with elemental analysis. Cationic metals associate with concretions that appear to illustrate specificity toward both cations and anions. M = metallothionein. (From Hopkin, S.P., *Ecophysiology of Metals in Terrestrial Invertebrates*, Elsevier Applied Science, London, 1989. With permission.)

accompanied by one or more anion transporters on the vacuolar membrane to accumulate sulfate, phosphate, or oxalate within the vacuole and thereby provide the complexing negatively charged ion needed for an insoluble precipitate to be formed. Other transport proteins have been described in the literature for copper, zinc, and iron transport across cellular plasma membranes and organellar membranes, and some of these specific metal carrier systems may also occur on concretion vacuolar membranes [39–42]. Until recently, there have been no experimental studies examining the nature of the metal transport systems present in vacuolar membranes that are responsible for the accumulation of metal cations within these membrane systems. Similarly, there are no published studies describing the physiological processes responsible for the transfer of sulfur, phosphorus, or any other multivalent anions, such as oxalate, across concretion vacuolar membranes. As a result, details of the processes responsible for sequestration, and therefore detoxification by concretion formation, within invertebrate epithelial cells have remained unclear.

11.2.1.3 Cation Transport Processes of Crustacean Digestive Tract and Renal Organs

A wide variety of epithelial and nonepithelial vertebrate cell types possess a Na^+/H^+ exchange protein in their plasma membranes that catalyzes the net uptake of extracellular sodium for the net extrusion of cytoplasmic protons [43,44]. Reported biological functions of this antiport mechanism include the regulation of intracellular pH, cell volume, and transcellular transport of sodium and bicarbonate [43–45]. In the cells of vertebrate animals this plasma membrane protein is highly conservative in its physiological properties across a range of cell types, universally displaying an electroneutral $1Na^+/1H^+$ exchange stoichiometry, which is energized by the cation illustrating the predominant transmembrane driving force. Sodium/proton antiporter processes in brush-border epithelial membrane vesicles of the gill, hepatopancreas, and antennal glands (kidneys) from freshwater and marine crustaceans, and from pyloric caeca of asteroid echinoderms significantly depart physiologically from the mammalian $1Na^+/1H^+$ transport paradigm, exhibiting an electrogenic cation exchange with a transport stoichiometry of $2Na^+/1H^+$ [46–51]. This invertebrate transport protein shares an amiloride sensitivity with antiporters of vertebrate cells, but the mechanism of amiloride inhibition of cation exchange is very dissimilar between the animal groups. In lobster (*H. americanus*) hepatopancreas and kidney brush-border membrane vesicles (BBMV), amiloride blocks $2Na^+/1H^+$ exchange by binding to two external cation sites, while only single monovalent cation binding sites are inhibited in vertebrate cells [48,50].

Exchanges of calcium in crustaceans take place between the hemolymph and environment across epithelial cell layers of the integument, gill, gut, and antennal gland. As shown in Figure 11.3, the

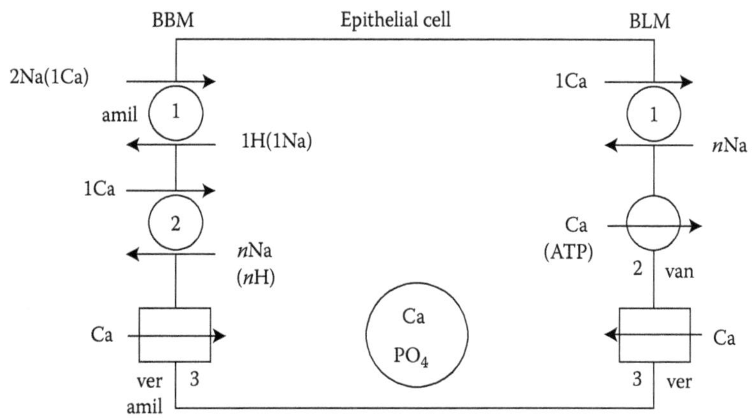

FIGURE 11.3 Apical and basolateral calcium transport processes of lobster (*H. americanus*) hepatopancreatic epithelium. Two carrier-mediated transport systems and a cation channel occur on both membranes. Calcium and metals may be stored within vacuoles or lysosomes complexed with sulfur or phosphorus. (Data from Ahearn, G.A. and Zhuang, Z., *Physiol. Zool.* 69, 383, 1996; Ahearn, G.A. et al., *J. Exp. Biol.* 196, 319, 1994.)

hepatopancreatic epithelium of the lobster (*H. americanus*) possesses three independent brush-border membrane transport proteins, which control the uptake of calcium from the gastrointestinal lumen [52–54]: (1) an amiloride-sensitive, electrogenic $2Na^+/1H^+$ antiporter (see earlier) exchanging either $2Na^+/1H^+$ or $1Ca^{2+}/1H^+$; (2) an amiloride-insensitive, electroneutral, cation antiporter exchanging $1Ca^{2+}/2Na^+$ or $1Ca^{2+}/2H^+$; and (3) a verapamil- and amiloride-inhibited calcium channel. The basolateral membrane of this digestive organ also possesses three calcium transfer processes for cation exchange between cytoplasm and blood [54]: (1) an electrogenic $1Ca^{2+}/3Na^+$ antiporter with relatively low apparent calcium binding affinity; (2) a high-affinity calcium ATPase for cellular efflux; and (3) a verapamil-inhibited calcium channel for possible calcium uptake from the blood during premolt. The contribution of each transport system to calcium uptake by epithelial cells of this organ and its transfer from one side of the epithelium to the other during different molt stages have been described [54]. Similar calcium transport processes for the brush-border membrane of kidney epithelial cells in the lobster were also described [48,55]. Calcium uptake by BBMV from starfish (*Pycnopodium helianthoides*) pyloric caeca or lobster (*H. americanus*) hepatopancreas was significantly reduced by the addition of zinc or cadmium to the external medium [56,57]. Both metals acted as competitive inhibitors of calcium uptake by the respective membrane preparations and suggested that both calcium and the metal cations shared at least one of the carrier processes previously described for ion uptake at this cell border. Since the apparent affinity constants (K_m) for calcium influx in lobster vesicles was approximately 300 μM, while that for cadmium inhibition of calcium influx (K_i) was 66 μM, the conclusion was reached that the divalent cation transport systems exhibit a higher apparent binding affinity for metals than for calcium.

11.2.1.4 Metal Transport Mechanisms of Crustacean Hepatopancreatic Plasma Membranes and Intracellular Organelles

11.2.1.4.1 Epithelial Brush-Border Membrane

Copper and iron uptakes across the apical membrane of lobster hepatopancreatic epithelial cells have been investigated using the fluorescent dye Phen Green and isotopic iron ($^{55}Fe^{2+}$), in conjunction with purified epithelial cell types and both BBMV and basolateral membrane vesicles (BLMV). Centrifugal elutriation was used to generate purified suspensions of the four hepatopancreatic epithelial cell types (E, F, B, and R cells) and copper influx into each cell type was measured after equilibrating the cells with Phen Green [11]. All four cell types absorbed copper from external concentrations ranging from 0.1 to 8 μM, but significant differences in transport rate occurred between the cell types. External calcium (0–10 mM) stimulated the uptake of external copper in a saturable fashion, suggesting the occurrence of carrier-mediated metal uptake. Addition of the Ca^{2+} channel blocker verapamil (30 μM) to the external medium reduced the uptake rate of copper by all four cell types. External zinc (0–1000 nM) was a competitive inhibitor of copper influx in E and R cells, suggesting that the two metals shared the same binding and transport mechanism. A model was proposed which suggested that copper may enter all hepatopancreatic epithelial cell types by a divalent cation antiport process that exchanges intracellular Ca^{2+} (or other cations) with either external copper or zinc (Figure 11.4). Verapamil-sensitive Ca^{2+} channels may allow access of external calcium to cytoplasmic exchange sites on the antiporter or to activator sites on the same transport protein.

Further evidence for a brush-border divalent cation exchanger, facilitating the uptake of metals from the gastrointestinal lumen, was described for iron uptake in lobster hepatopancreatic BBMV [58]. In one experiment, BBMV preloaded with calcium were found to display greater uptake rates of $^{55}Fe^{2+}$ than vesicles lacking internal calcium, suggesting antiport between the divalent cations. In another experiment, BBMV were preloaded with $^{59}Fe^{2+}$ and subsequently transferred to media containing mannitol, Fe^{2+}, or Cd^{2+}. In this instance, efflux media containing either iron or cadmium increased the efflux of $^{59}Fe^{2+}$ from the preloaded vesicles by trans-stimulation compared to the medium with mannitol, suggesting that metal exchange occurred via a common carrier protein.

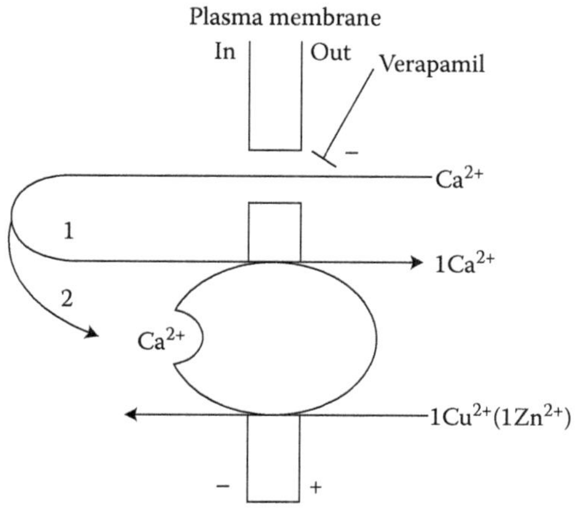

FIGURE 11.4 Model for the involvement of external calcium and zinc in copper influx into lobster hepatopancreatic epithelial cells. External calcium may enter hepatopancreatic cells across the plasma membrane through a verapamil-sensitive cation channel and then this cation may serve as an antiport substrate, exchanging (process #1) with external Cu^{2+} in an electroneutral fashion or allosterically activating (process #2) the antiport of external Cu^{2+} with Ca^{2+} or other intracellular cations (e.g., Mg^{2+}, Na^+, H^+, etc.). External Zn^{2+} acts as a competitive inhibitor of Cu^{2+} influx in the presence of external calcium and probably shares the external binding site on the transport protein with Cu^{2+}. (From Chavez-Crooker, P., Garrido, N., and Ahearn, G.A., *J. Exp. Biol.* 204, 1433, 2001.)

Additional support for a broad-spectrum, heavy-metal antiporter in apical membranes of invertebrate gastrointestinal epithelial cells was reported for pyloric caecal cells of the starfish *Pycnopodia helianthoides* [57]. In this study, $^{45}Ca^{2+}$ uptake by purified BBMV of the pyloric caeca was stimulated by an outwardly directed H^+ gradient and this stimulation was enhanced by the simultaneous presence of an induced membrane potential (inside negative; K^+/valinomycin). Zn^{2+} was a mixed inhibitor of $^{45}Ca^{2+}$ influx by carrier-mediated transport, and Mn^{2+}, Cu^{2+}, Fe^{2+}, and Mg^{2+} were also effective inhibitors of $^{45}Ca^{2+}$ uptake, but the mechanism(s) of inhibition by these other cations was(were) not disclosed. An equilibrium shift experiment showed that both Na^+ and Zn^{2+} were able to exchange with equilibrated $^{45}Ca^{2+}$ in these vesicles, suggesting that both monovalent and divalent cations were able to enter pyloric caecal cells through a common carrier-mediated transport system (Figure 11.5).

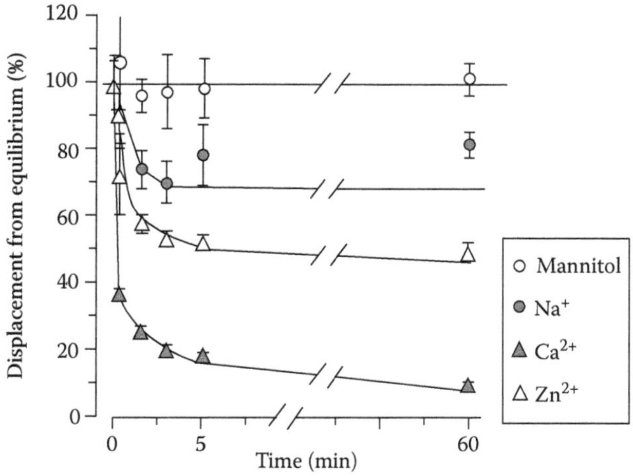

FIGURE 11.5 Equilibrium shift experiment demonstrating trans-stimulation of 0.1 mM $^{45}Ca^{2+}$ efflux from equilibrated pyloric caecal BBMV by external Na^+, Ca^{2+}, or Zn^{2+} (mannitol serving as a control). (Data from Zhuang, Z. and Ahearn, G.A., *J. Exp. Biol.* 198, 1207, 1995.)

These studies as a group suggest that calcium and metal uptake by digestive tract epithelial cells from dietary constituents in invertebrates likely include an exchange carrier protein as shown in Figure 11.4, which facilitates the uptake of a wide variety of luminal metals (e.g., Zn^{2+}, Fe^{2+}, Cd^{2+}, Cu^{2+}) in exchange for cytoplasmic H^+, Na^+, or Ca^{2+} ions. Once inside the epithelial cells, these metals may be sequestered and detoxified by a variety of physiological and biochemical processes as described below, or transferred across the basolateral epithelial cell membrane to the blood where they may influence biological events in many tissues.

11.2.1.4.2 Epithelial Basolateral Membrane

Heavy metals accumulated in crustacean epithelial cells and not bound to cytosolic metallothioneins and glutathione or sequestered and detoxified within organelles such as mitochondria, ER, or lysosomes may exit the cell across the basolateral cell membrane and enter the blood for distribution throughout the animal, where excessive concentrations may lead to pathological effects. As displayed in Figure 11.3, crustacean epithelial cells possess at least three potential basolateral proteins that may interact with cytosolic divalent heavy metal cations: a plasma membrane calcium ATPase (PMCA), a sodium–calcium exchanger (NCX), and a verapamil-sensitive calcium channel. In addition, metal-specific transport proteins may also be present on this membrane. A study was undertaken, using purified BLMV and $^{65}Zn^{2+}$, to assess whether the heavy metal zinc was capable of using any of the reported hepatopancreatic basolateral calcium transport systems [10]. In the absence of sodium, $^{65}Zn^{2+}$ influx was a hyperbolic function of external zinc concentration and followed carrier-mediated, Michaelis–Menten kinetics (Figure 11.6). This carrier transport was stimulated by the addition of 150 μM ATP and inhibited by the simultaneous presence of 150 μM ATP and 250 μM vanadate. Calcium ion was shown to be a competitive inhibitor of $^{65}Zn^{2+}$ transport in the presence of ATP but in the absence of sodium, with an inhibitory constant (e.g., K_i) of 205 nM calcium. These data suggest that zinc was transported across the basolateral membrane by a vanadate-inhibited ATPase carrier system that may be the same system transporting calcium across this part of the cell (e.g., PMCA). Alternatively, it is also possible that a Menkes-type copper-transporting ATPase [42,59–62] that accepts both copper and zinc may be responsible for the efflux of metals across this cell border in crustaceans. In mammals, this ATPase protein transports copper out of the intestine to the blood, and impairment of its function leads to a metal buildup within the gut and a deficiency in this metal for metal-requiring enzymes.

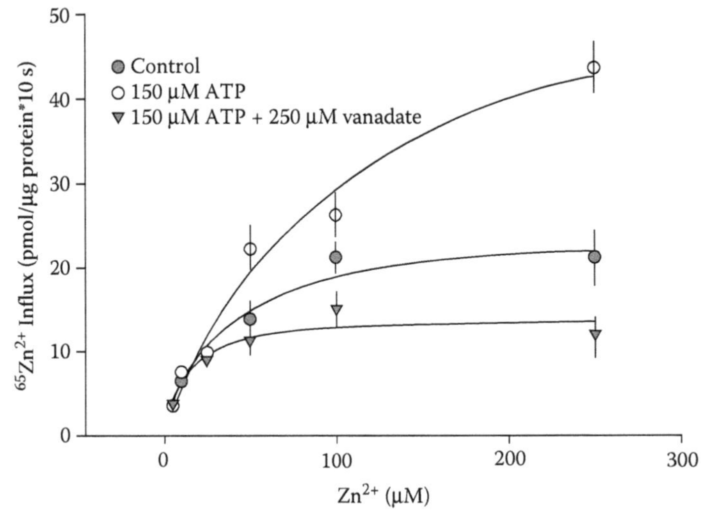

FIGURE 11.6 Kinetics of $^{65}Zn^{2+}$ influx (10 s uptakes) in a sodium buffer in the presence and absence of 150 μM ATP and 250 μM vanadate. Symbols are means ±1 SEM, $n = 3–5$ replicates per mean. (From Capo, J.A. et al., *J. Comp. Physiol. B* 175, 13, 2005. With kind permission from Springer Science + Business Media.)

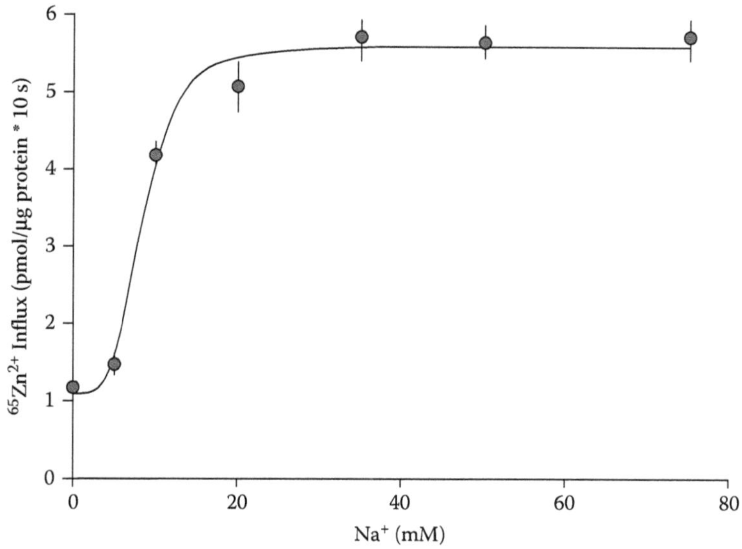

FIGURE 11.7 Effect of variable intravesicular sodium concentration on hepatopancreatic BLMV influx (10 s uptakes) of 25 μM $^{65}Zn^{2+}$. The sigmoidal curve was drawn through the data using Sigma Plot software assuming a fit to the Hill equation. Symbols are means ±1 SEM, $n = 3$–5 replicates per mean. (From Capo, J.A. et al., *J. Comp. Physiol. B* 175, 13, 2005. With kind permission from Springer Science + Business Media.)

In the absence of ATP, $^{65}Zn^{2+}$ influx into hepatopancreatic BLMV was a sigmoidal function of preloaded vesicular sodium concentration and exhibited a Hill coefficient of 4.03 ± 1.14, consistent with an exchange of 3 Na^+/1 Zn^{2+} (Figure 11.7). This transporter was also competitively inhibited by calcium in the absence of ATP, illustrating a K_i of 2.47 μM calcium. Both this experiment and the previous one suggest that $^{65}Zn^{2+}$ is transported by a basolateral calcium-ATPase (PMCA) and 3Na^+/1Ca^{2+} exchanger (NCX) and that calcium and zinc may compete for binding and transport from cytoplasm to hemolymph by these processes *in vivo.*

11.2.1.4.3 Metal Transport and Sequestration by Epithelial Mitochondria

As Figure 11.1 indicates, once an external metal is transported into a crustacean epithelial cell, it may be complexed with metallothionein or glutathione in the cytoplasm, or be transported into three organelles: mitochondria, ER, and lysosomes. Within each of the organelles the metal may associate with organic or inorganic compounds and be retained there for various periods of time. In many cases these complexes involving the cationic metal form insoluble precipitates that are not free to leave the organelle, and thereby effectively detoxify the metal. In some instances the buildup of these precipitates may interfere with the normal functioning of the organelle and present separate problems that are not associated with systemic metal toxicity.

The mitochondrion is an organelle that removes heavy metals from epithelial cytoplasm and detoxifies them by sequestering it as an insoluble precipitate. In the 1970s Lehninger and his colleagues showed that blue crab hepatopancreatic mitochondria could store large amounts of calcium phosphate in the organellar matrix (Figure 11.8), and at extreme concentrations these precipitates formed needle-like deposits that destroyed the organelles [63–66]. The calcium transport processes of lobster hepatopancreatic mitochondria were characterized in 1999 [9], and the effects of copper on these calcium transporters were examined in 2002 [67]. Pure hepatopancreatic mitochondria displayed $^{45}Ca^{2+}$ uptake by an apparent electrogenic, ruthenium red-inhibited transport process that was sensitive to cytoplasmic pH and heavy metals such as zinc [9]. $^{45}Ca^{2+}$ efflux from mitochondria took place by an apparent diltiazem-inhibited, electroneutral 2Na^+/1Ca^{2+} antiporter and an apparent diltiazem-insensitive 2H^+/1Ca^{2+} antiporter. Both antiporters were capable of exchanging preloaded calcium for external zinc.

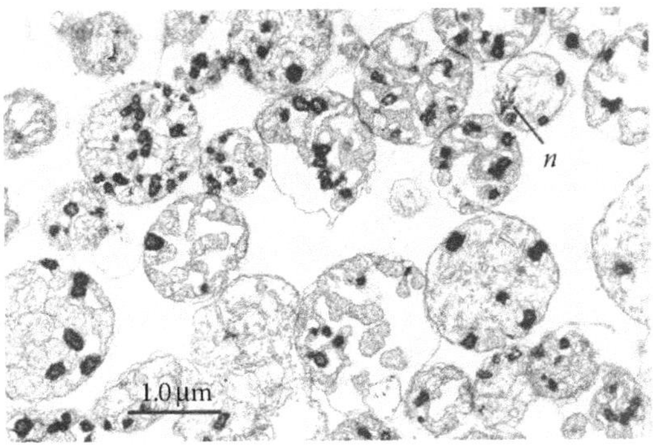

FIGURE 11.8 Crab hepatopancreas mitochondria massively loaded *in vitro* with calcium and phosphate, and fixed with glutaraldehyde and OsO_4. Almost every mitochondrion contains several large, discrete electron-dense deposits and a few exhibit needle-like deposits (*n*) that increase in frequency with an elevation in ion concentration. Sections were stained with uranyl acetate and lead. ×21,000. (© Chen, C.H. et al., originally published in *J. Cell Biol.* 61, 301–315, 1974. With permission.)

The copper-specific fluorescent dye Phen Green was used to investigate the mechanisms of copper uptake by dye-equilibrated hepatopancreatic mitochondria [67]. This study showed that all the mitochondrial calcium transport systems previously characterized were able to transport the heavy metal as well (Figure 11.9). Once inside hepatopancreatic epithelial cells, a mitochondrial ruthenium red-sensitive uniporter transport system removes either calcium or metal from the cytoplasm and accumulates the cations within the organelle. Metals, such as copper, may also enter or leave mitochondria by diltiazem-sensitive and diltiazem-insensitive antiporters that exchange the metal for either monovalent or divalent cations such as sodium or calcium. The conclusion of this study was that hepatopancreatic mitochondria are capable of serving as metal sequestration depots, with metals replacing calcium on uniporter and cation-exchange transport systems.

11.2.1.4.4 Metal Transport and Sequestration by Epithelial ER

In vertebrates [68] and invertebrates [69–71], the Sarco/ER Ca^{2+}-ATPase (SERCA) transporter transfers calcium ion from the cytoplasm to the respective organellar lumen against steep concentration gradients. It is not known to what extent this calcium ATPase transports other ions, such as heavy metals, into these organelles when they accumulate in the cytoplasm. In order to assess the

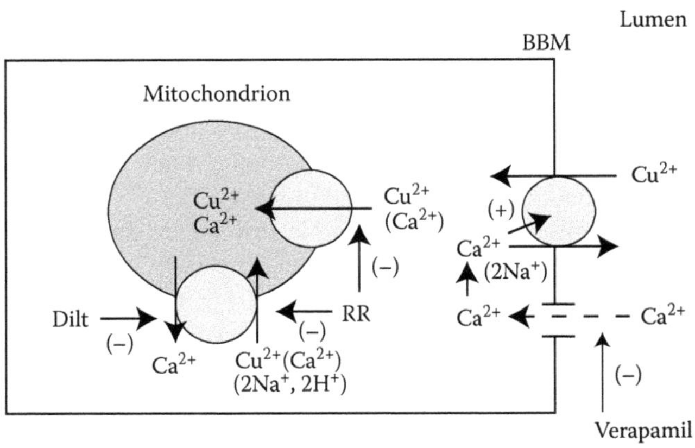

FIGURE 11.9 Calcium and copper transport processes in hepatopancreatic epithelial cells and their mitochondria. BBM, brush-border membrane; RR, ruthenium red; Dilt, diltiazem. (Data from Klein, M.J. and Ahearn, G.A., *J. Exp. Zool.* 283, 147, 1999; Chavez-Crooker, P., Garrido, N., and Ahearn, G.A., *J. Exp. Biol.* 205, 405, 2002.)

TABLE 11.1
Effect of 1 mM ATP on $^{45}Ca^{2+}$ and $^{65}Zn^{2+}$ Influx Kinetic Constants in Hepatopancreatic ER Membrane Vesicles

Ion	Treatment	K_m^n (μM)	J_{max} (pmol/mg prot. × s)	Hill Coefficient (n)
Ca^{2+}	Control	10.38 ± 1.01	14.75 ± 1.27	2.53 ± 0.46
Ca^{2+}	+ATP	12.76 ± 0.91	25.46 ± 1.45[a]	1.96 ± 0.15
Zn^{2+}	Control	38.63 ± 0.52	19.36 ± 0.17	1.81 ± 0.13
Zn^{2+}	+ATP	43.85 ± 4.68	29.28 ± 1.43[a]	1.58 ± 0.19

Values are means ±1 SEM. Kinetic constants were obtained using Sigma Plot software. Each constant was obtained using 3–5 replicates per point and the experiment was repeated once with similar qualitative results.

[a] Significantly different from the control condition ($P < 0.05$).

possibility that the SERCA was capable of transporting metals, a study was undertaken, using a centrifugation method to purify hepatopancreatic ER, and the uptake processes of this preparation for $^{45}Ca^{2+}$ and $^{65}Zn^{2+}$ were subsequently investigated [13]. The time course of $^{45}Ca^{2+}$ uptake by ER vesicles was stimulated by the addition of 1 mM ATP, and was subsequently sharply reduced in the presence of both 1 mM ATP and 100 μM vandate, suggesting that a P-type ATPase may be involved with calcium accumulation by the vesicles. $^{45}Ca^{2+}$ influx into ER vesicles was a sigmoidal function of external calcium activity and displayed a Hill coefficient of approximately 2. External ATP (1 mM) increased calcium influx J_{max}, whereas addition of either 250 μM vanadate or 10 μM thapsigargin significantly ($P < 0.01$) reduced calcium influx J_{max} while retaining a Hill coefficient near 2.0. The maximal transport velocity of sigmoidal $^{45}Ca^{2+}$ influx by ER vesicles was also significantly lowered by the addition of either 25 μM Zn^{2+} or Cu^{2+} to the external medium.

Influx of $^{65}Zn^{2+}$ into ER vesicles was also a sigmoidal function of external zinc concentration, exhibiting a Hill coefficient of approximately 2.0 [13]. The addition of 1 mM ATP significantly ($P < 0.01$) increased influx J_{max}, whereas the addition of 1 mM ATP plus 250 μM vanadate, or 1 mM ATP plus 10 μM thapsigargin, significantly ($P < 0.01$) reduced sigmoidal influx J_{max}. A summary of the kinetic constants of $^{45}Ca^{2+}$ and $^{65}Zn^{2+}$ influxes into lobster ER vesicles is shown in Table 11.1 and suggests that calcium and zinc appear to be handled by the ER in similar ways, but with the apparent transport binding affinity greater for calcium than for the metal. Thapsigargin is a SERCA inhibitor, while vanadate inhibits P-type ATPases. Since both inhibitors were effective in reducing the influxes of both $^{45}Ca^{2+}$ and $^{65}Zn^{2+}$, it is likely that the SERCA transporter was responsible for accumulating the two cations.

11.2.1.4.5 Transport and Sequestration by Epithelial Lysosomes

11.2.1.4.5.1 Transport of Ionic Metals While electron microscopic and x-ray microanalysis studies of invertebrate epithelia have shown the presence of a variety of heavy metals in epithelial vacuoles and lysosomes [13,33], little direct experimental evidence has been published describing the mechanisms by which metals may cross vacuolar or lysosomal membranes. Havelaar et al. [72] described a vanadate-insensitive heavy metal transporter on lysosomal membranes from rat liver that resembled a P-type cation ATPase and had a specificity for silver, copper, and cadmium. In mammals Wilson's copper ATPase [73–76] is involved with copper transport from liver to bile and is localized on organellar membranes. Patients with Wilson's disease have impaired excretion of copper from the liver due to impaired ATPase activity.

Three studies have recently been conducted with lobster hepatopancreatic lysosomal membrane vesicles (LMV) to assess the nature of heavy metal transport by membranes of these organelles and the characteristics of independent transport processes for complexing anions that form detoxifying concretion precipitates with the metals within the organelles [12,77,78]. In the first paper, the copper-specific fluorescent dye Phen Green was used to quantify transmembrane fluxes of copper [77], as described for the plasma membranes of isolated hepatopancreatic epithelial cells in suspension [58] and for purified mitochondrial suspensions from the same tissue [68]. Results of this study indicated the presence of a vanadate-sensitive, calcium-stimulated copper ATPase in membranes of these organelles that displayed high-affinity Michaelis–Menten, carrier-mediated transport kinetics, in contrast to the sigmoidal influx kinetics disclosed for zinc transport into hepatopancreatic ER vesicles discussed previously [13]. Together with a putative bafilomycin-sensitive V-ATPase in the membrane of the same organelles, importing hydrogen ion into the organellar interior, the copper ATPase may function as part of a physiological mechanism for precipitate formation between metallic cations and imported anions. These ionic precipitate complexes may then act as a sink for excess metals and thereby reduce the circulating concentrations of these elements.

The second, more recent paper [12] characterized the transport processes responsible for $^{65}Zn^{2+}$ uptake by lobster hepatopancreatic lysosomal vesicles and extended the functional information disclosed in the preceding paper using Phen Green to study lysosomal copper uptake. This study confirmed the Michaelis–Menten kinetics of metal influx into lysosomal vesicles, but in addition showed that the zinc transporter was stimulated by the addition of 1 mM ATP to the external medium and this stimulation was markedly reduced when either 250 μM vanadate or 10 μM thapsigargin was added to the incubation medium with 1 mM ATP. Cadmium and copper were shown to be competitive inhibitors of $^{65}Zn^{2+}$ influx with inhibitory constants (e.g., K_i) of 68 and 33 μM, respectively. Because of the basic kinetic differences for metal influxes between lobster hepatopancreatic lysosomal and ER vesicles, in conjunction with their shared sensitivities to both vanadate and thapsigargin, a paper was published that suggested that both lobster organelles may display organelle-specific isoforms of the same SERCA-like transporter capable of transporting a wide variety of divalent cations, including calcium, copper, zinc, and cadmium [79].

A working model was proposed in 2006 that described the nature of divalent cation influx into hepatopancreatic lysosomal vesicles as deduced from the results of the papers described above (Figure 11.10). This model suggests that zinc, copper, calcium, and cadmium can be transported into hepatopancreatic lysosomes by a SERCA-like, proton antiporter that is energized by both the

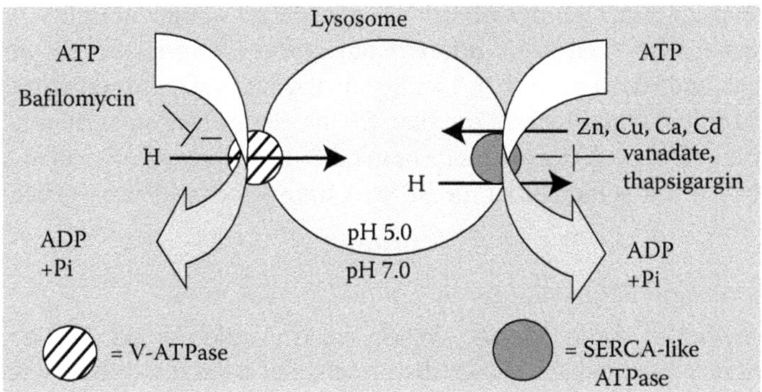

FIGURE 11.10 Working model of $^{65}Zn^{2+}$ transport by lobster hepatopancreatic LMV. Model suggests that zinc, copper, calcium, and cadmium are transported into lysosomes in epithelial cells by a proton antiporter protein that is energized by both the transmembrane proton gradient and cytoplasmic ATP. Lysosomal hydrogen ions are delivered to the interior of this organelle *in vivo* by a bafilomycin-sensitive V-ATPase, creating an acidic interior compared to that of the cytoplasm. A transmembrane proton gradient, directed outward, stimulates the uptake of metals in the absence of ATP and therefore appears to provide a driving force for cation exchange in the absence of a direct energy source applied to the metal transporter. In the presence of ATP, the P-type ATPase inhibitor vanadate all but abolishes metal accumulation by lysosomal vesicles, and thapsigargin significantly reduces the uptake of zinc under these conditions. (From Mandal, P.K., Mandal, A., and Ahearn, G.A. *J. Exp. Biol.* 305A, 203, 2006.)

transmembrane proton gradient and cytoplasmic ATP. Lysosomal hydrogen ion is delivered to the organelle interior by a bafilomycin-sensitive V-ATPase [77], creating an acidic interior compared to the cytoplasm. A transmembrane proton gradient, directed outward, stimulated the uptake of metals in the absence of ATP and could provide the sole driving force for cation exchange. In the presence of ATP, the P-type ATPase inhibitor vanadate significantly reduced metal accumulation in the absence of a pH gradient [12]. The ER SERCA-specific inhibitor thapsigargin also significantly reduced zinc uptake in the absence of other driving forces [12]. It was tentatively suggested that metal transport by hepatopancreatic lysosomal vesicles occurred by a lysosomal-specific isoform of the ER SERCA calcium ATPase. This identify can only be confirmed with molecular studies.

While the model in Figure 11.10 suggests a means by which divalent cationic metals can be transferred from the hepatopancreatic epithelial cytoplasm to the lysosomal interior, it does not clarify how complexing anions, which form insoluble precipitates with the transported metals, enter lysosomes and associate there with the cations. Recently, a study was reported that helped define the means by which anions move into lysosomes and complex with metallic cations, effectively detoxifying them [78]. In this study hepatopancreatic lysosomal vesicles loaded with SO_4^{2-} or PO_4^{3-} had a threefold greater steady-state accumulation of $^{65}Zn^{2+}$ than similar vesicles loaded with mannitol, Cl^-, or oxalate^{2-}. Zinc uptake in the presence of ATP was proton gradient enhanced and electrogenic, exhibiting an apparent exchange stoichiometry of $1Zn^{2+}/3H^+$. $^{35}SO_4^{2-}$ and ^{14}C-oxalate^{2-} uptakes were both enhanced in vesicles loaded with intravesicular Cl^- compared to vesicles containing mannitol, suggesting the presence of anion countertransport. $^{35}SO_4{}^{2-}$ influx was a sigmoidal function of external [SO_4^{2-}] with 25 mM internal [Cl^-] or with several intravesicular pH values (e.g., 7.0, 8.0, and 9.0). In all cases Hill coefficients of approximately 2.0 were obtained, suggesting that two sulfate ions exchange with single Cl^- or OH^- ions. $^{36}Cl^-$ influx was a sigmoidal function of external [Cl^-] with intravesicular pH values of 7.0 and 9.0. A Hill coefficient of 2.0 was also obtained, suggesting the exchange of two Cl^- for one OH^-. ^{14}C-oxalate influx was a hyperbolic function of external [oxalate^{2-}] with 25 mM internal [Cl^-], suggesting a 1:1 exchange of oxalate^{2-} for Cl^-. As a group, these experiments suggest the presence of an anion exchange mechanism in lysosomal membranes exchanging monovalent for polyvalent anions.

Polyvalent inorganic anions (SO_4^{2-} and PO_4^{3-}) are known to associate with metals inside vesicles from microprobe analysis; a detoxification model that summarizes the above functional results in this chapter, and those publications preceding it, was presented, which suggests how these anions may contribute to concretion formation through precipitation with metals at appropriate vesicular pH (Figure 11.11). As outlined in this model, at least three membrane proteins are associated with the transport of metals and complexing anions into hepatopancreatic epithelial lysosomes. Protein 1 is a V-ATPase that transfers protons into the vesicle interior, creating a decrease in pH, an accumulation of hydrogen ions, and an inside-positive membrane potential. The outwardly directed proton gradient and positive vesicular interior provide the driving force for the asymmetric exchange of cytosolic divalent metals for intravesicular hydrogen ions by an ATP-dependent Zn^{2+}-ATPase or a $3H^+/1Zn^{2+}$ exchanger (Protein 2). Polyvalent cytosolic anions such as sulfate or phosphate exchange with intravesicular monovalent anions such as Cl^- or OH^- by a second asymmetric antiporter (Protein 3), which uses the membrane potential as a driving force for exchange. Both divalent metals and polyvalent anions increase in concentration inside vesicles at acidic pH and are retained because they cannot be accommodated on the intravesicular binding sites of the exchangers. Divalent metals and polyvalent anions form precipitates (concretions) as V-ATPase decreases in activity and intravesicular pH rises.

11.2.1.4.5.2 Transport of Heavy Metal–Thiol Conjugates The regulation of blood metal concentrations in mammals is partially due to secretory activities of the renal proximal tubule. In this location a mechanism for the secretion of divalent metals as conjugates with thiol-containing molecules such as glutathione or the amino acid cysteine has been documented (Figure 11.12, [80]). As shown in Figure 11.12, metals such as mercury, bound to thiol agents in the blood, are transferred

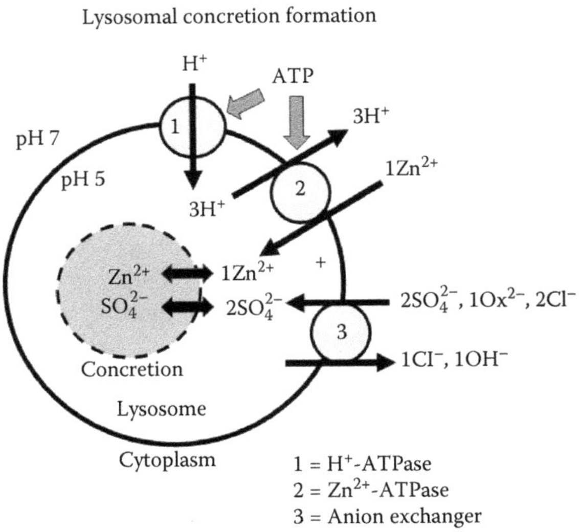

FIGURE 11.11 Working model of the role of polyvalent anions in hepatopancreatic lysosomal heavy metal sequestration and detoxification. Membrane-bound ATP-dependent, V-ATPase (Protein 1) transfers protons into the vesicle interior, creating a decrease in pH, an accumulation of hydrogen ions, and an inside-positive membrane potential. The outwardly directed proton gradient and positive vesicular interior provide the driving force for the asymmetric exchange of cytosolic divalent metals for intravesicular hydrogen ions by an ATP-dependent Zn^{2+}-ATPase or a $3H^+/1Zn^{2+}$ exchanger (Protein 2). Polyvalent cytosolic anions such as sulfate^{2-} or phosphate^{3-} exchange with intravesicular monovalent anions such as Cl^- or OH^- by a second asymmetric antiporter (Protein 3), which uses the membrane potential as a driving force for exchange. Both divalent metals and polyvalent anions increase in concentration inside vesicles at acidic pH and are retained because they cannot be accommodated on the intravesicular binding sites of the exchangers. Divalent metals and polyvalent anions form precipitates (concretions) as the V-ATPase decreases in activity and the intravesicular pH rises. (From Ahearn, G.A. et al., *J. Comp. Physiol. B* 174, 439, 2004.)

across the basolateral epithelial cell membrane of the proximal tubule epithelium by way of the organic anion transporter (OAT1) in exchange for intracellular dicarboxylates, such as α-ketoglutarate. The α-ketoglutarate can be generated by metabolism within the cell or can be brought into the cell from the blood by a colocalized Na^+-dependent dicarboxylate symporter that uses the transmembrane sodium gradient, established by the basolateral ATP-dependent, Na^+/K^+-ATPase, to accumulate the dicarboxylate against a concentration gradient. In this model, the α-ketoglutarate is recycled through the membrane as a means of transferring the metal across the cell layer for its eventual elimination in the urine. The OAT1 transport protein identified in mammal renal proximal tubule epithelial basolateral membranes is also localized in other vertebrate tissues where it serves a variety of transport functions related to a relatively broad substrate specificity [81].

Recent evidence suggests that an OAT-like transport system may occur on crustacean hepatopancreatic lysosomal membranes and serve as an alternative mechanism for sequestering and detoxifying heavy metals bound to cytosolic thiol agents. Hepatopancreatic LMV from the American lobster (*H. americanus*) were loaded with 1 mM α-ketoglutarate and used to study the effects of metal-binding compounds on zinc uptake (Roggenbeck and Ahearn, unpublished observations). Transport studies using $^{65}Zn^{2+}$ indicated that the addition of 1 mM L-cysteine to outside media caused a threefold increase in zinc uptake. Inhibition studies with 1 mM *p*-aminohippuric acid (PAH) in the presence of either 1 mM L-cysteine or 1 mM glutathione caused a twofold decrease in zinc uptake (Figure 11.13). These data suggested the possible presence of an OAT in lobster lysosomal membranes. Western blot results, using antibodies to mammalian OAT transporters, showed the presence of an orthologous OAT-like protein (approximate MW of 80 kDa) signal from the lobster (Roggenbeck and Ahearn, unpublished observations) (Figure 11.14). A cross-reacting signal from

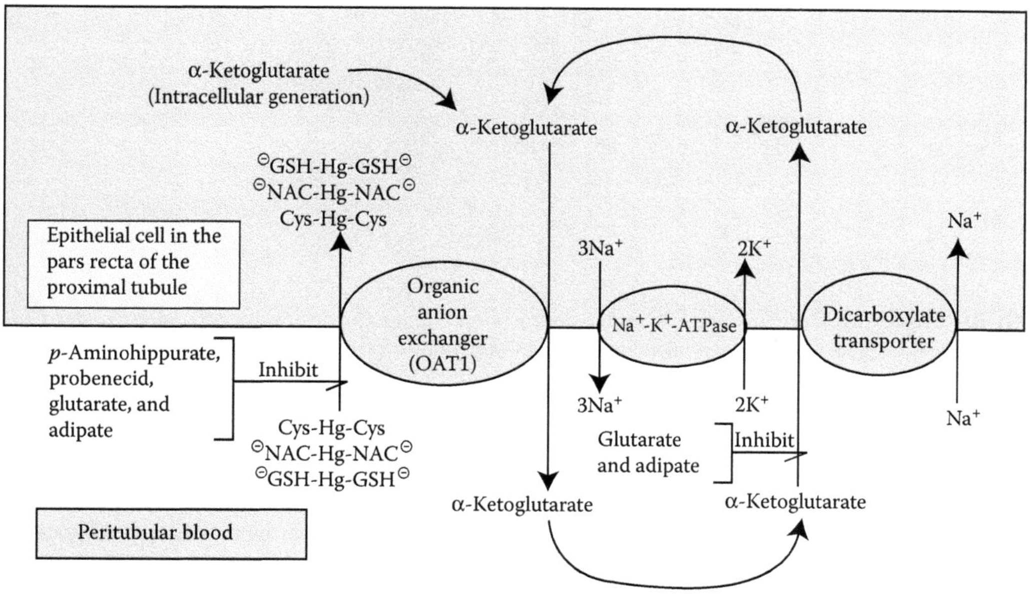

FIGURE 11.12 Diagrammatic representation of the putative roles of both the dicarboxylate/organic anions exchanger (OAT1) and the sodium-dependent dicarboxylate transporter in the basolateral uptake of inorganic mercury (Hg^{2+}) along the pars recta of the proximal tubule. Metabolically produced intracellular α-ketoglutarate (AKG) creates a chemical gradient facilitating the movement of this dicarboxylate out of the cell at the basolateral membrane by exchanging with organic anions in the peritubular blood at OAT1. Some of the exchanged AKG re-enters the epithelial cell once again across the basolateral membrane by symport with sodium and provides additional intracellular substrate for further exchange with organic anions in the blood. Inorganic mercury may complex with the amino acid L-cysteine or the tripeptide glutathione in the blood and employ the OAT1 exchanger to enter the proximal tubular epithelial cell across the basolateral membrane as part of a secretion process regulating the blood concentration of this metal. The exchange of these mercury conjugates with AKG can be blocked with other substrates of the OAT1 transporter such as *p*-aminohippurate (PAH), probenecid, glutarate, and adipate. (From Zalaps, R.K. and Barfuss, D.W., *Toxicol. Appl. Pharmacol.* 182, 234, 2002. With permission.)

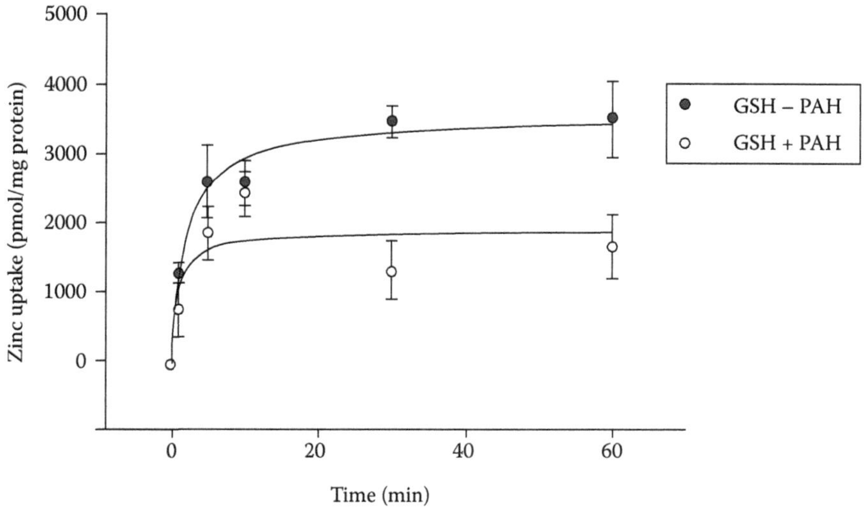

FIGURE 11.13 Effect of 1 mM PAH on 1 mM glutathione-stimulated (GSH) 25 μM $^{65}Zn^{2+}$ uptake by lobster (*H. americanus*) hepatopancreatic LMV. Vesicles were preloaded with 1 mM α-ketoglutarate (Roggenbeck and Ahearn, unpublished observations).

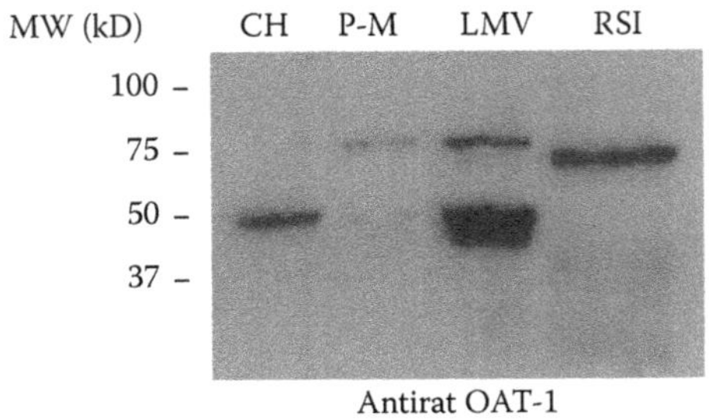

FIGURE 11.14 Western blot using a rabbit antirat OAT1 antibody on lobster (*H. americanus*) LMV. CH = crude homogenate; PM = total membranes; LMV = pure lysosomal membrane vesicles; RSI = rat small intestine control at 80 kDa (Roggenbeck and Ahearn, unpublished observations).

LMV of approximately 50 kDa was also observed, which is believed to correspond to the unglycosylated form of the lobster OAT-like protein. These preliminary observations suggest a possible role of an OAT-like transport system in the long-term sequestration of metal–thiol conjugates in invertebrates via lysosomes as a means of heavy metal detoxification. Localization of an OAT-like protein in lobster hepatopancreas lysosomal membranes is in agreement with a report of an approximately 48 kDa, presumably unglycosylated, OAT family protein (OAT3) residing and functioning in the lysosomal membrane of rat liver [82].

11.2.2 Intestine

The crustacean intestine is an epithelial-lined tube linking the pyloric region of the stomach with the hindgut. It consists of a single layer of columnar, mitochondrial-rich epithelial cells surrounded by circular and longitudinal muscles [83–85]. A large, dorsal artery extending from the heart supplies it with oxygen and nutrients from the blood side of the organ. It is derived from embryonic endoderm and is unchitinized, distinguishing it from either the foregut (stomach) or hindgut, which arise from embryonic ectoderm and contain a luminal chitin layer. Because of its simple tubular structure, this organ is easily manipulated in an experimental setting.

11.2.2.1 Sodium and Chloride Transport and Intestinal Electrophysiology

Perfused tubular preparations (Figure 11.15, [86–89]) and flattened tissue sheets mounted in Ussing-type chambers [90] of freshwater and marine crustacean intestine have been studied for their ion transport capabilities. These studies showed that the intestine of the freshwater prawn, *Macrobrachium rosenbergii*, exhibits net mucosal to serosal (MS) fluxes of both sodium and chloride [86] as a result of a strict coupling of an apical uptake process that transfers two sodium ions across the mucosal membrane for each chloride ion. This coupling process was shown to involve an allosteric cooperative cotransport of sodium, chloride, and calcium from intestinal lumen to epithelial cytosol [87,88]. It was proposed that this route of salt uptake in this freshwater crustracean likely significantly contributed to its hyperosmotic regulation.

Intestines from freshwater, brackish-water, and marine crabs, and from marine lobsters all exhibit a small transepithelial electrical potential, with the serosal side of the tissue being positive relative to the lumen [90–92]. Coincident with this transmural potential difference in the blue crab (*Callinectes sapidus*), the intestine also illustrates a significant net flux of sodium from lumen to blood [92]. However, the net flow of sodium across the intestine of this crab appears too small to contribute significantly to the ion balance of the animal [92]. On the other hand, in the lobster (*H. americanus*), the small transmural potential difference across its perfused intestine co-occurred with a net absorption of water, and the suggestion was made that this water transfer was likely coupled with the movement of a positively charged ion, likely sodium [91].

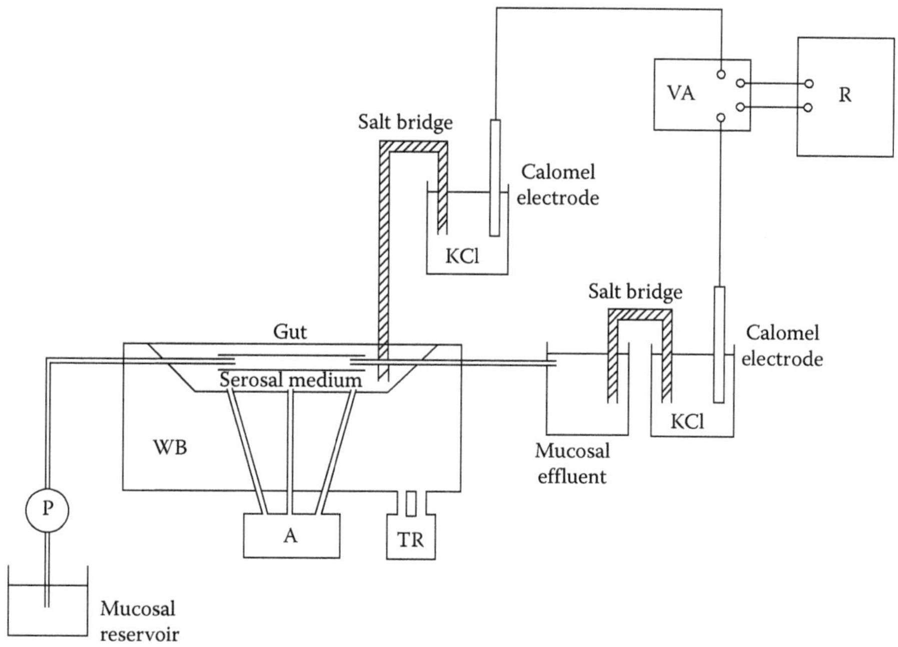

FIGURE 11.15 Experimental apparatus used to measure transmural electrical potential differences, ionic fluxes, and transmural transport of organic solutes across perfused intestines of shrimp and lobster. A peristaltic pump (P) was used to perfuse saline solution from a mucosal reservoir through glass capillary tubing or 16-gauge hypodermic needles to ligated intestine. Following passage through the gut, saline was collected by way of a second tube or needle emptying into a mucosal effluent container. Saturated KCl-agar salt bridges connected both mucosal effluent and serosal media via 3 M KCl solutions and calomel electrodes to a voltage amplifier (VA)-dual-pen recorder (R) complex for electrical estimations. Serosal medium was kept aerated (A) throughout incubation and its temperature was controlled (24°C) using a surrounding water bath (WB) and temperature regulator (TR). (From Ahearn, G.A., *Am. J. Physiol.* 239, C1, 1980. Used with permission from the American Physiological Society.)

11.2.2.2 Heavy Metal Transport across Perfused Crustacean Intestine

Perfused tubular intestines, using an apparatus similar to that shown in Figure 11.15, were used to characterize heavy metal transport into, and across, isolated organs from the crab (*Callinectes sapidus*) [93,94] and the lobster (*H. americanus*) [6,7,95]. Mercury and methylmercury were transported into, and across, perfused crab intestine by a variety of pathways that were both passive and active. A number of organic ligands added to the perfusate significantly affected the transfer of Hg across the organ from lumen to blood. L-Cysteine (1 μM) significantly increased 250 nM Hg flux to the blood, while thiourea (100 μM) and humic acid (9 μM) both decreased metal uptake [94]. Several other metals including arsenic, cadmium, zinc, lead, and selenium used independently or as a mixture were shown to either decrease (arsenic, zinc, selenium) or increase (cadmium and lead) the accumulation of Hg in intestinal tissues of this animal.

Perfusion of isolated lobster (*H. americanus*) intestine with a physiological saline containing 20 μM $^{65}Zn^{2+}$ led to a small, but significant, MS flux of the metal (Figure 11.16, [6]). Addition of 20 μM L-histidine to the perfusate, along with the labeled metal, doubled the transmural transport rate of $^{65}Zn^{2+}$ from lumen to bath. Lastly, when a potentially competing metal, Cu^{2+}, was added to the perfusate along with L-histidine, the $^{65}Zn^{2+}$ flux returned to control levels. Because the addition of Cu^{2+} reduced the stimulation of $^{65}Zn^{2+}$ transport caused by luminal L-histidine, the involvement of a carrier process in this activity was considered likely. Additional experiments showed that the transmural MS flux of $^{65}Zn^{2+}$ was a hyperbolic function of luminal zinc concentration and followed

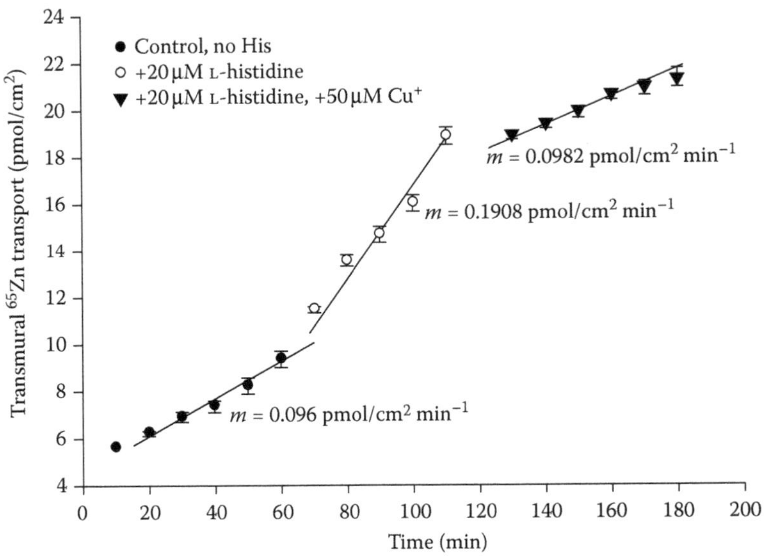

FIGURE 11.16 Effect of 20 μM L-histidine and 50 μM Cu^+ on the time course of transmural transport of 20 μM $^{65}Zn^{2+}$ across perfused intestine of American lobster (*H. americanus*). Values displayed are means ±1 SEM of three replicates per time point. Slopes of lines were calculated as regression lines using Sigma Plot software and the *m*-values on the figure as the slopes obtained from these analyses. (From Conrad, E.M. and Ahearn, G.A., *J. Exp. Biol.* 208, 287, 2005.)

the Michaelis–Menten equation for carrier-mediated transport [6], and luminal addition of the amino acid increased the maximal transport velocity of the metal by 30% over control values lacking the amino acid [7].

Detailed investigation of the effect of L-histidine on transmural transport of $^{65}Zn^{2+}$ involved the comparison of MS and SM (serosa to mucosa isotope movement) fluxes of the metal in the presence and absence of luminal stimulating amino acid to determine the direction and magnitude of net metal flux across the tissue [7]. The results of these experiments are displayed in Figure 11.17a and b and indicate the zinc fluxes in both directions across lobster intestine that occurred by carrier-mediated transport. Figure 11.17a suggests that the addition of L-histidine to the perfusate significantly increased the MS flux of the metal. Figure 11.17b indicates that there was a significant net absorptive flux of zinc across the intestine to the bath both in the presence and in the absence of L-histidine, but addition of the amino acid strongly stimulated the magnitude of this metal absorption.

A separate set of experiments in the studies with lobster intestine showed that the transmural MS flux of ^{3}H-L-histidine was a hyperbolic function of luminal L-histidine concentration and followed the Michaelis–Menten relationship [6]. The addition of luminal zinc tripled the maximal transport velocity of the transporter for the amino acid, whereas the addition of L-leucine reduced the maximal transport velocity to control levels.

The results displayed in Figures 11.16 and 11.17 strongly suggest a coupling in the absorption of zinc and L-histidine across lobster intestine. These data, and other experiments reported in Conrad and Ahearn [6,7], led to a working model for the synergistic interaction between the transports of these two substrates (Figure 11.18). This model proposed that the metal and amino acid were cotransported across the intestine as a functional unit on an apical membrane carrier protein that had binding sites for both substrates. An assumption for this model was that zinc and L-histidine formed a bis complex (Hist–Zn^{2+}–Hist) in solution and approached the binding sites on the transporter in this configuration. In the above publications it was shown that the dipeptide glycylsarcosine only inhibited the MS flux of ^{3}H-L-histidine when zinc was present and bis complexes were formed [6]. It was proposed that cotransport of the bis complex occurred on a dipeptide transport

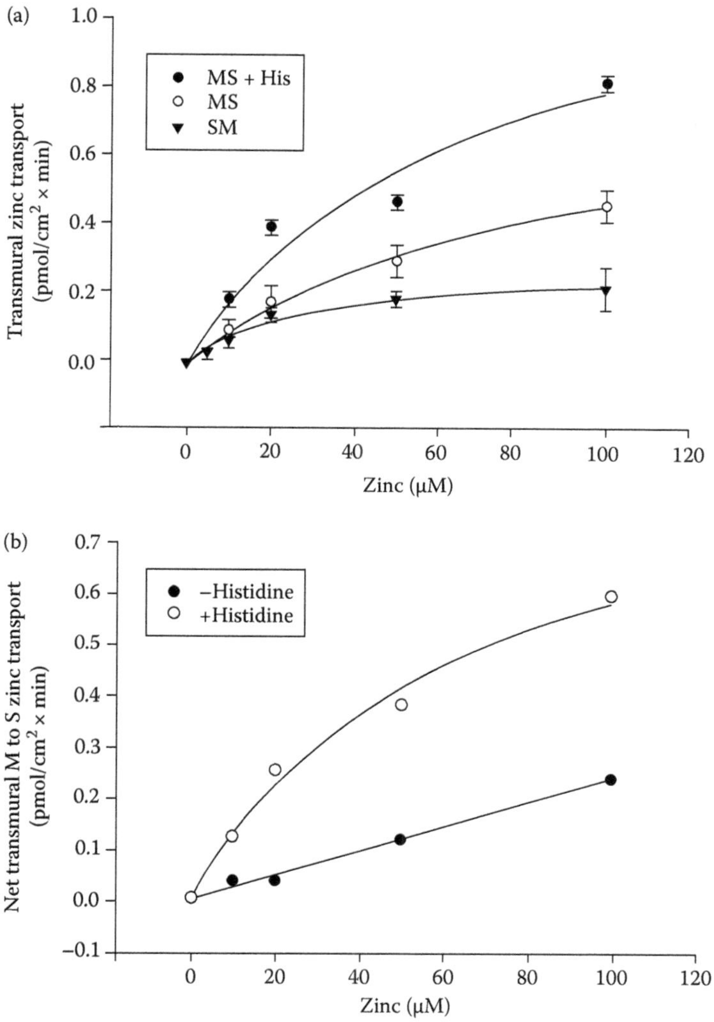

FIGURE 11.17 (a) Components of transmural $^{65}Zn^{2+}$ transport across perfused lobster (*H. americanus*) intestine in the presence and absence of luminal 20 μM L-histidine. Mucosal to serosal (M to S) and serosal to mucosal (S to M) fluxes of $^{65}Zn^{2+}$ were measured as functions of their respective *cis* substrate concentrations over 30-min incubations at each metal concentration. Readings were taken every 5 min during M to S fluxes and every 1 min for S to M fluxes. Slopes of the 30-min M to S uptake curve were used to determine fluxes toward the blood, while S to M fluxes toward the lumen at each concentration were assembled from the mean ± SEM of 1-min perfusate collections over the 30-min incubation. Vertical lines on the figure are SEM values of the respective means at each zinc concentration. Curves through the data were drawn using Sigma Plot software and the Michaelis–Menten equation for carrier transport (Data from Conrad, E.M. and Ahearn, G.A., *J. Comp. Physiol. B* 177, 297, 2007). (b) Effect of luminal 20 μM L-histidine on the magnitude of net M to S transport of $^{65}Zn^{2+}$ across perfused lobster (*H. americanus*) intestine. Net flux values at each concentration of zinc were calculated from the numbers shown in (a) by subtracting the S to M flux from the M to S flux of the metal in the presence and absence of luminal L-histidine. In both instances a significant ($P < 0.01$) net flux was toward the blood and was linear in the absence of luminal L-histidine and hyperbolic in the presence of the amino acid. (From Conrad, E.M. and Ahearn, G.A., *J. Comp. Physiol. B* 177, 297, 2007.)

carrier such as PEPT1 or PHT1 [96]. Under these conditions zinc in the bis complex acted as a mimic for a peptide bond and held the two amino acids in a configuration that resembled that of the normal substrates of the dipeptide carrier. Enhanced MS and net fluxes of both amino acid and metal occurred because when both substrates were present, the movement of each across the tissue

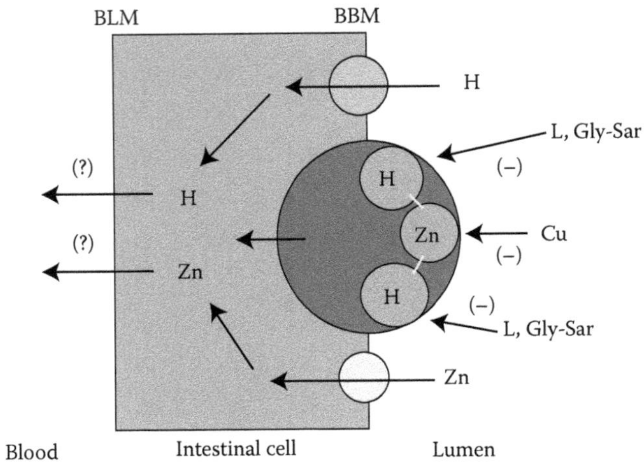

FIGURE 11.18 Working model of transmural MS transport of ^{3}H-L-histidine (His) and $^{65}Zn^{2+}$ across the perfused intestine of the American lobster (*H. americanus*). Three mucosal membrane carrier proteins are involved in the movement of these two solutes across the tissue: (1) a relatively specific L-histidine carrier that is not inhibited by L-leucine; (2) a relatively nonspecific dipeptide transporter that accepts two histidine molecules (His) linked to a zinc ion in an apparent bis-complex; and (3) a relatively specific zinc transporter that is inhibited by luminal cupric ion (Cu^{2+}). Luminal L-leucine (Leu) and glycyl-sarcosine (Gly-Sar) inhibit ^{3}H-L-histidine transport by interacting with the dipeptide carrier in a mixed type inhibition. Luminal copper (Cu^+ and Cu^{2+}) inhibits $^{65}Zn^{2+}$ transport by interacting with the dipeptide carrier in a mixed type inhibition. The model assumes that the three independent carriers are localized to the brush-border membrane (BBM) of intestinal epithelial cells. (From Conrad, E.M. and Ahearn, G.A., *J. Exp. Biol.* 208, 287, 2005.)

by cotransport added to the separate fluxes that took place by their respective independent transport systems. More recent experimental research has shown that the amino acid L-leucine may also use the intestinal dipeptide carrier to cross the tissue from lumen to blood when zinc is present in the perfusate as a result of bis complex formation as seen with L-histidine [95].

11.3 KIDNEYS

Crustacean kidneys are called antennal glands. They are paired structures that are located at the base of each antenna. They are composed of several regions that are anatomically and physiologically distinct and overall superficially resemble the microscopic mammalian nephron tubules except that the crustacean organs are far larger. The drawing in Figure 11.19 illustrates the parts of a crayfish antennal gland and the histological differences between the major segments [97]. As indicated in this figure, the antennal gland has a filtering apparatus called the coelomosac that functions much like the individual glomeruli of mammalian nephrons. The coelomosac removes soluble materials from the hemolymph and passes them to the labyrinth, proximal tubule, and distal tubule, where urine modification takes place. The distensible bladder is both a storage organ and a final site of urine modification. Each antennal gland terminates in a nephropore, through which passes the final urine product. As Figure 11.19 shows, the coelomosac possesses podocytes with few mitochondria, while cells of the labyrinth and tubule possess an extremely large number of these organelles to supply the power for active transport processes needed for essential nutrient return to the blood.

A number of historical and recent anatomical reviews of crustacean antennal glands at the histological, cytological, and ultrastructural levels have been published on a wide variety of species, and the interested reader is encouraged to review them for in-depth discussions of the structural variations that occur in these organs in animals from several different environments [97–111].

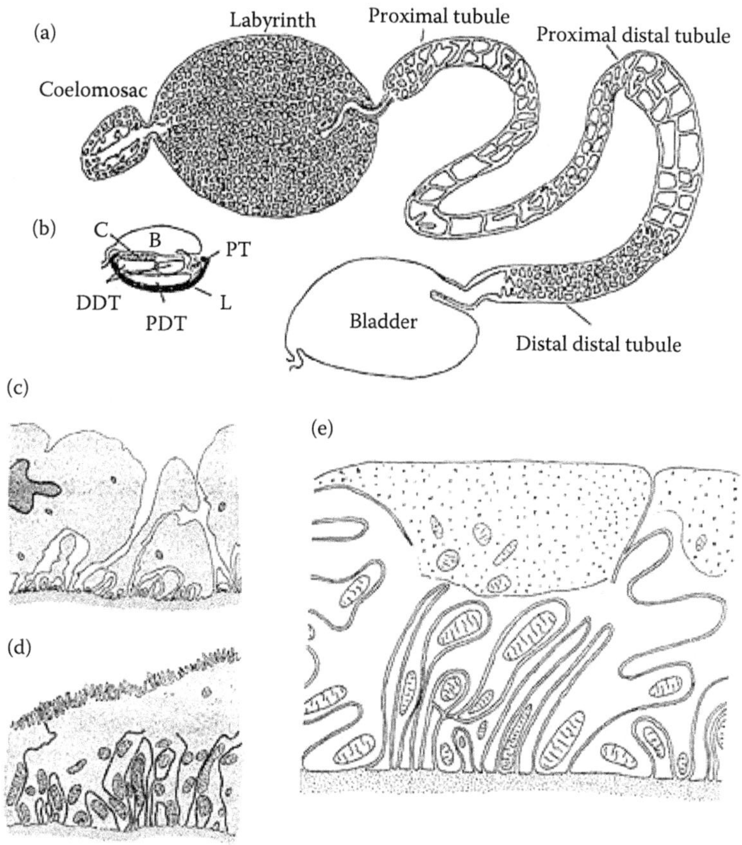

FIGURE 11.19 Structure of the antennal gland of the crayfish. (a) Different regions of the dissected tubule. (b) Spatial relationships *in situ*. Histology of the epithelial cells: (c) Podocytes of the labyrinth. (d) Cells of the labyrinth and proximal tubule. (e) Cells of the distal part of the distal tubule. B, bladder; C, coelomosac; DDT, distal region of the distal tubule; L, labyrinth; PDT, proximal part of the distal tubule; PT proximal tubule. (From Kirschner, L.B., *Annu. Rev. Physiol.* 29, 169, 1967. With permission.)

11.3.1 Transport of Water, Sodium, and Chloride, and Role in Osmoregulation, by the Kidney

For at least 50 years the different structural components of the crustacean kidney have been studied with respect to their excretory functions [111]. In addition, specific portions of the antennal glands have been compared to analogous regions of the vertebrate nephron in water and ion transport and in osmoregulatory activities. As with the vertebrate glomerulus, crustacean antennal glands are believed to form urine as a result of hemolymph filtration by the coelomosac [97,112]. In crayfish, the antennal gland is able to filter molecules with a molecular mass below 20 kDa, while the limiting molecular mass ranges between 50 and 100 kDa [113].

Like the vertebrate nephron, following filtration, the antennal glands appear to selectively absorb and secrete a wide variety of substances, tightly regulating their hemolymph concentrations in the process and producing a urine that may vary in composition and osmolarity. Early studies of the ion- and osmoregulatory role of different regions of the crustacean antennal glands were performed by micropuncture on the crayfish by Riegel [114–118]. In freshwater crayfish, the antennal gland proximal tubule and labyrinth brush-border membranes demonstrate isosmotic NaCl absorption [98,115], while the long distal segments and bladder absorb NaCl without water transfer

[114,115,119,120]. As a result of these activities, a dilute, hyposmotic urine is produced under these conditions that aids in organismic osmotic regulation in freshwater [105,121,122].

In intermolt crayfish, three enzymes or transport systems aid in reabsorbing more than 90% of filtered extracellular ions, carbonic anydrase, Na^+/K^+-ATPase, and Ca^{2+}-ATPase. The unidirectional ion influxes produced by these processes may be threefold greater than those displayed by the gills of the respective species even though the absorptive surface of the kidneys may only be one-fortieth that of the gills [120,123–126]. High Na^+/K^+-ATPase immunoreactivity has been found in the crayfish distal tubule, which contains cells with numerous mitochondria and extensive basal infoldings, while little or no enzyme labeling is associated with the coelomosac, suggesting a role of the former region in producing a dilute urine [101,107–109]. In contrast, marine crustacean kidneys exhibit a much lower Na^+/K^+-ATPase activity or lack the distal tubule part of the antennal gland, and therefore produce an isosmotic urine [123,127,128].

11.3.2 Antennal Gland Secretion of Magnesium and Sulfate

A major role of the antennal glands in marine species is the secretion of magnesium and sulfate, two divalent ions in high concentration in seawater [97,99,105,109c,110,129]. The secretion rates of these ions are regulated by the levels encountered in the environment, being lower in 50% seawater than in full-strength seawater [128]. The labyrinth of the fiddler crab *Uca mordax* [99] and the bladder in the crab *Cancer magister* [130] appear to be the respective locations of the secretory processes in the two species.

11.3.3 Calcium Reabsorption by Crustacean Antennal Glands

Crustacean antennal glands regulate hemolymph calcium concentrations through reabsorption of this cation from the filtrate throughout the molt cycle [125,131–133]. Membrane vesicle and immunolocalization studies of antennal glands from both marine (*H. americanus*) and freshwater (*Procambarus clarkii*) species have identified a number of distinct membrane proteins on both epithelial brush-border and basolateral membranes in each species that appear to be involved in moving calcium across the respective renal epithelia [48,55,120,133–135]. Figure 11.20 shows a working generic model of a crustacean epithelial cell from the antennal gland (or other ion regulatory structures of the gills and the gut), indicating the consensus view of calcium-regulating membrane proteins and intracellular structures that interact with calcium during its transcellular transfer or transient sequestration [135]. The apical membrane of these calcium-transporting epithelial cells exhibits the combination of a verapamil/nifedipine-sensitive calcium channel that allows the passive uptake of the cation from luminal contents. In addition, this cell border possesses amiloride-sensitive and amiloride-insensitive NCXs that may be electrogenic or electroneutral. The basolateral membrane also displays a calcium channel and NCXs, but in addition exhibits a high-affinity, ATP-dependent, calcium ATPase (PMCA) for extrusion of the cation from the cytosol. Within the cytoplasm, several structures may sequester divalent cations such as calcium or zinc, thereby helping to maintain very low intracellular ionic activities. These divalent cation-sequestering organelles include mitochondria [9], ER [13], and lysosomes [12].

Evidence has been presented showing that the epithelium of freshwater crustacean antennal glands exhibits greater calcium flux rates by the combination of the membrane transporters listed above than occur across gills or other epithelial structures in the same animal [132,134]. These large fluxes usually occur when the animal undergoes ecdysis, or molting, during which time they must rapidly recalcify a new exoskeleton and cannot afford to lose this cation in the urine [79,132]. To aid in this active reabsorption of filtered hemolymph calcium during the premolt phase of the molt cycle, there is evidence in crayfish (*Procambarus clarkii*) for a 10-fold increase in expression of the PMCA and a twofold increase in expression during postmolt, compared to its presence during intermolt [120].

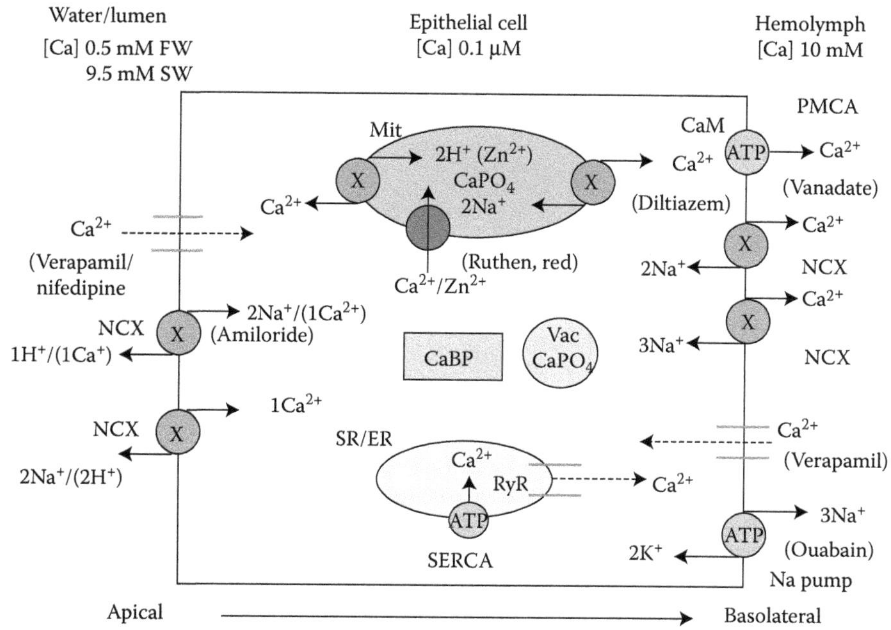

FIGURE 11.20 Crustacean epithelial model for apical-to-basolateral transcellular calcium transport and suggested subcellular calcium storage compartments. Pharmaceutical inhibitors are shown in parentheses. SR, sarcoplasmic reticulum; ER, endoplasmic reticulum; CaBP, calcium-binding protein; CaM, calmodulin; Mit, mitochondria, Vac, vacuole, FW, freshwater; SW, seawater; RyR, ryanodine receptor. (From Wheatly, M.G., Zanotto, F.P., and Hubbard, M.G., *Comp. Biochem. Physiol. B.* 132, 163, 2002. With permission.)

11.3.4 Secretion of Organic Anion and Cation Metabolites by Antennal Glands

The secretion of organic metabolites from the blood and their eventual excretion by crustacean antennal glands have been investigated for much of the last century [113,119,136–143]. In recent years much attention has been devoted to defining the cellular mechanisms present in crustacean antennal gland urinary bladder for the active secretion of organic anions and cations (reviewed in Refs. [142,143]). These studies have shown that the crustacean urinary bladder functions much as does the vertebrate renal proximal tubule in its handling of organic metabolites and xenobiotics.

Figure 11.21 displays a published model for the active secretion of the representative organic anion PAH by the urinary bladders of the crab *Cancer borealis* or the lobster *H. americanus* [139,141–143]. This model shows that several distinct carrier systems are linked in the bladder epithelial cells to bring about the net movement of PAH from blood to bladder lumen. In this model, the basolateral Na^+/K^+-ATPase creates a resting potential of approximately −65 mV, which is used as a driving force for the symport of sodium and α-ketoglutarate (aKG^{2-}) from the hemolymph to the bladder cytosol. Once inside the cell, aKG^{2-} acts as an antiport substrate that exchanges with PAH^- from the blood, again using the membrane potential for the electrogenic transport event. Cytoplasmic PAH^- exits the cell across the apical membrane by facilitated diffusion down the electrochemical potential into the bladder lumen. In this collection of transporters, the Na^+-di-carboxylate symporter, importing metabolic substrate from the blood, is a secondary active transport system, and the aKG^{2-}/PAH^- exchanger demonstrates tertiary coupling to the ATP-dependent $Na+/K^+$-ATPase. The overall result of this coupled process is the secretion of PAH^- from blood to bladder lumen.

The urinary bladders of crabs (*Cancer borealis*, *Cancer irroratus*) and lobsters (*H. americanus*) similarly possess an excretory process for organic cationic metabolites that are indirectly linked to the basolateral Na^+/K^+-ATPase in this organ. Figure 11.22 illustrates the basic components of organic cation secretion by crustacean antennal gland urinary bladder [140–143]. A common experimental substrate for this transport system is tetraethylammonium (TEA^+) [140]. TEA^+ transport by

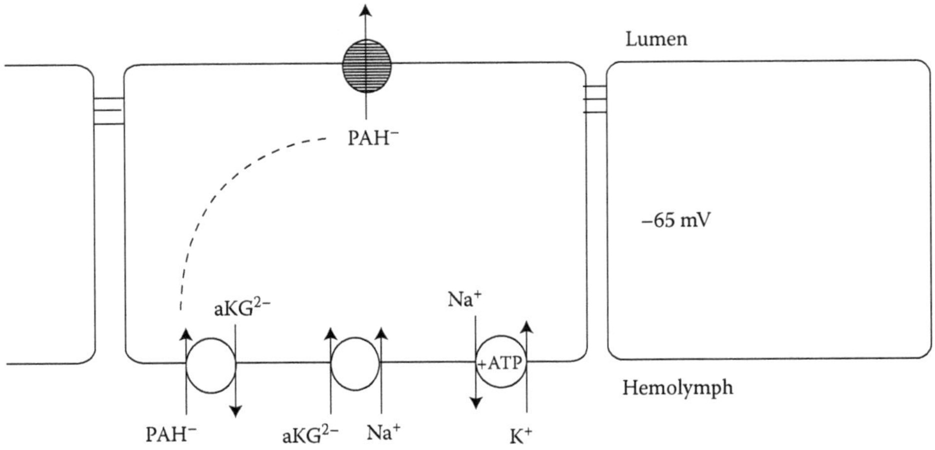

FIGURE 11.21 Proposed model for tertiary coupling of organic anion (*p*-aminohippuric acid, PAH^-) secretion to ATP-dependent ion transport. A sodium gradient established by Na^+/K^+-ATPase is used to drive the secondary uptake of α-ketoglutarate (aKG^{2-}). Efflux of aKG^{2-} is then coupled to uptake of PAH^-. Transapical movements of PAH^- are driven by the electrical gradient between cell and lumen. (From Pritchard, J.B. and Miller, D.S., *Am. J. Physiol.* 261, R1329, 1991. With permission.)

crustacean urinary bladder across the basolateral membrane occurs by a facilitated diffusion process using the transmembrane potential (e.g., −65 mV) as a driving force and is inhibited by quinine, which acts as a competitive inhibitor. A secondary active transport apical Na^+/H^+ exchanger is maintained by the basolateral Na^+K^+-ATPase and controls intracellular pH by extruding protons. The extruded hydrogen ions exchange with intracellular TEA^+ by tertiary active transport, resulting in the secretion of the organic cation into the bladder lumen. The overall result of this coupled transport suite of proteins is the secretion of TEA^+ from blood to urinary bladder lumen.

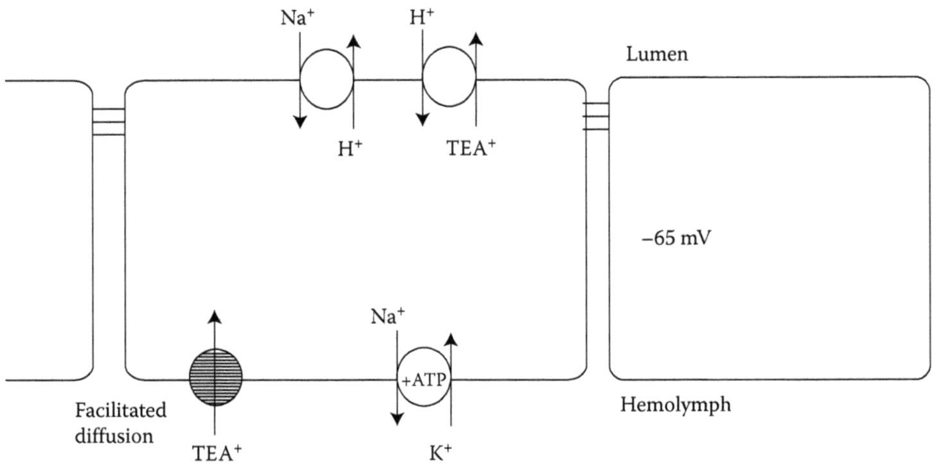

FIGURE 11.22 Suggested model for excretion of organic cations such as tetraethylammonium (TEA^+). The basolateral carrier is inhibitable by quinine and driven by the inside negative cell potential. TEA^+ is exchanged for H^+ across the apical membrane and protons are recycled through a Na^+/H^+ exchanger. (From O'Donnell, M.J. In: W.H. Dantzler (Ed.), *Comparative Physiology*, Vol. II, Section 13, Chapter 17, pp. 1207–1289, Am. Physiol. Soc., New York, 1997. With permission.)

11.3.5 Heavy Metal Regulation by the Renal OAT1

As discussed above, in vertebrates OAT1 has been localized to the renal proximal tubule basolateral cell membrane and, in that position, functions not only to secrete organic anions from the blood to the urine, but also to bind and transport heavy metals, such as mercury, from the blood to the epithelial cytosol as a thiol conjugate with L-cysteine and glutathione ([80], Figure 11.12). Evidence was presented above that this transporter may also be localized on crustacean hepatopancreatic lysosomal membranes and function to sequester and detoxify metals in this organelle as a result of conjugate formation between cytoplasmic zinc and either L-cysteine or glutathione (Figures 11.13 and 11.14).

Figures 11.23 and 11.24 suggest that an organic anion transport carrier protein may occur on the antennal gland urinary bladder basolateral cell membrane and transfers metabolites such as PAH from the blood to the renal epithelium cytosol via a transport process that appears identical to the vertebrate OAT1 suite of transport proteins (Figure 11.12). A series of experiments were conducted with lobster (*H. americanus*) antennal gland BLMV, using a centrifugation method previously applied to the hepatopancreas of the same animal [144,145], that examined the possible use of this crustacean organic anion transporter as a vehicle for heavy metal secretion as displayed by the OAT1 transport system (Roggenbeck and Ahearn, unpublished observations).

In the first experiment, these BLMV were preloaded with 1 mM α-ketoglutarate and were incubated for 40 min with 25 μM $^{65}Zn^{2+}$ in the presence and absence of 1 mM glutathione. As shown in Figure 11.23, those vesicles incubated with the tripeptide displayed a significantly greater uptake of $^{65}Zn^{2+}$ than did the control membranes, suggesting that conjugate formation between metal and glutathione may have occurred followed by enhanced zinc uptake. Addition of 1 mM PAH to glutathione-stimulated vesicles significantly decreased $^{65}Zn^{2+}$ uptake (data not shown). These results suggest the possible presence of an OAT1-like transport system on the basolateral antennal gland membrane, which may facilitate the secretion of metals from blood to urine.

Further evidence for the potential presence of an OAT1-like transporter in crustacean kidneys is presented as a Western blot in Figure 11.24, where an orthologous protein comigrating with the control rat small intestine (RSI) signal is observed using a mammalian anti-OAT1 antibody to probe

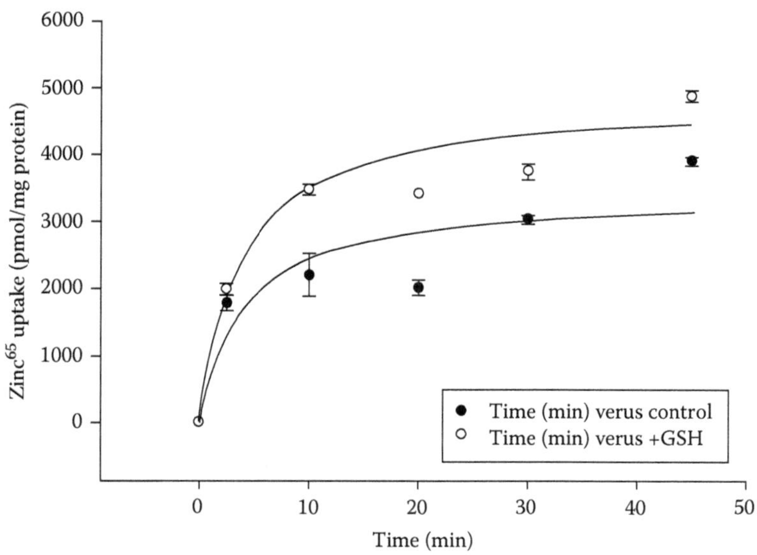

FIGURE 11.23 $^{65}Zn^{2+}$ uptake by American lobster (*H. americanus*) antennal gland BLMV preloaded with α-ketoglutarate and incubated with and without 1 mM glutathione (Roggenbeck and Ahearn, unpublished observations).

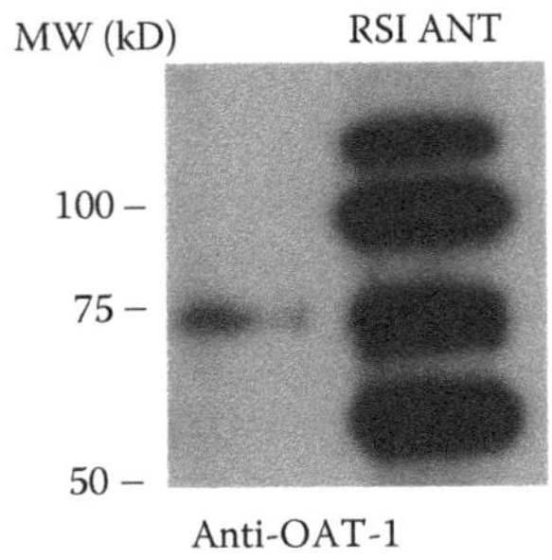

FIGURE 11.24 Immunoblot of American lobster (*H. americanus*) antennal gland homogenate showing reactivity with mammalian anti-OAT1 antibody. RSI is rat small intestine control. The 50 kDa signal likely corresponds to an unglycosylated form of the OAT-like protein, while the higher molecular weight signals correspond to glycosylated forms of the protein (Roggenbeck and Ahearn, unpublished observations).

antennal gland crude homogenate. In this blot the bands above and below that corresponding to the control band are likely unglycosylated and glycosylated forms of the protein.

These two sets of preliminary data extend the physiological studies previously performed on crustacean urinary bladder epithelium, as shown in Figure 11.21, suggesting the presence of an OAT1 in this organ performing a regulatory function for organic metabolites. The preliminary data also suggest that this transport system may function as a regulatory process for heavy metals as described for vertebrate renal tissues (Figure 11.12, [80]). Additional experimental information with this invertebrate renal organ will be needed to confirm these tentative results.

ACKNOWLEDGMENTS

Much of the published and unpublished work reported herein from the author's laboratory was supported by National Science Foundation grants IBN99-74569 and IBN04-21986.

REFERENCES

1. Ahearn, H.R.H., Ahearn, G.A., and Gomme, J., Effects of cadmium on integumentary uptake of ^{3}H-L-histidine in the marine polychaete worm, *Nereis succinea. J. Exp. Biol.* 203, 2877, 2000.
2. Peppler, J.E. and Ahearn, G.A., Effect of heavy metals on the uptake of ^{3}H-L-histidine by the polychaete *Nereis succinea. Comp. Biochem. Physiol. C.* 136, 181, 2003.
3. Bury, N.R., Grosell, M., Grover, A.K., and Wood, C.M., ATP-dependent silver transport across the basolateral membrane of rainbow trout gills. *Toxicol. Appl. Pharmacol.* 159, 1, 1999.
4. Grosell, M. and Wood, C.M., Copper uptake across rainbow trout gills: Mechanisms of apical entry. *J. Exp. Biol.* 205, 1179, 2002.
5. Verbost, P.M., Van Rooij, J., Flik, G., Lock, R.A.C., and Bonga, S.E.W., The movement of cadmium through freshwater trout branchial epithelium and its interference with calcium transport. *J. Exp. Biol.* 145, 185, 1989.
6. Conrad, E.M. and Ahearn, G.A., ^{3}H-L-histidine and $^{65}Zn^{2+}$ are co-transported by a dipeptide transport system in lobster (*Homarus americanus*) intestine. *J. Exp. Biol.* 208, 287, 2005.
7. Conrad, E.M. and Ahearn, G.A., Transepithelial transport of zinc and L-histidine across perfused intestine of American lobster, *Homarus americanus. J. Comp. Physiol. B* 177, 297, 2007.
8. Viarengo, A., Heavy metals in marine invertebrates: Mechanisms of regulation and toxicity at the cellular level. *Rev. Aquatic. Sci.* 1, 295, 1989.
9. Klein, M.J. and Ahearn, G.A., Calcium transport processes of lobster hepatopancreatic mitochondria. *J. Exp. Zool.* 283, 147, 1999.
10. Capo, J.A., Mandal, P.K., Eyyunni, S., and Ahearn, G.A., $^{65}Zn^{2+}$ transport by lobster hepatopancreatic basolateral membrane vesicles. *J. Comp. Physiol. B* 175, 13, 2005.
11. Chavez-Crooker, P., Garrido, N., and Ahearn, G.A., Copper transport by lobster hepatopancreatic epithelial cells separated by centrifugal elutriation: Measurements with the fluorescent dye, Phen Green. *J. Exp. Biol.* 204, 1433, 2001.

12. Mandal, P.K., Mandal, A., and Ahearn, G.A. $^{65}Zn^{2+}$ transport by lobster hepatopancreatic lysosomal membrane vesicles. *J. Exp. Biol.* 305A, 203, 2006.
13. Mandal, P.K., Mandal, A., and Ahearn, G.A., Physiological characterization of $^{45}Ca^{2+}$ and $^{65}Zn^{2+}$ transport by lobster hepatopancreatic endoplasmic reticulum. *J. Exp. Zool.* 303A, 515, 2005.
14. Al-Mohanna, S.Y. and Nott, J.A., R-cells and the digestive cycle in *Penaeus semisulcatus* de Haan (Crustacea: Decapoda). *Mar. Biol.* 95, 129, 1987.
15. Al-Mohanna, S.Y., Nott, J.A., and Lane, D.J.W., Mitotic E- and secretory F-cells in the hepatopancreas of the shrimp *Penaeus semisulcatus* (Crustacea: Decapoda). *J. Mar. Biol. Assoc. U.K.* 65, 901, 1985.
16. Paquet, F., Germain, P., and Fritsch, P., Etude ultrastructurale de la glande digestive du homard *Homarus gammarus*. Role des diverse types cellulaires dans les processus digestifs. *Cah. Biol. Mar.* 35, 15, 1993.
17. Al-Mohanna, S.Y. and Nott, J.A., B-cells and digestion in the hepatopancreas of *Penaeus semisulcatus* (Crustacea:Decapoda). *J. Mar. Biol. Assoc. U.K.* 66, 403, 1986.
18. Roesijadi, G., Metallothioneins in metal regulation and toxicity in aquatic animals. *Aquat. Toxicol.* 22, 81, 1992.
19. Viarengo, A., Moore, M.N., Pertica, M., Mancinelli, G., Zanicchi, G., and Pipe, R.K., Detoxification of copper in the cells of the digestive gland of mussel: The role of lysosomes and thioneins. *Sci. Toxicol. Environ.* 44, 135, 1985.
20. Viarengo, A., Moore, M.N., Mancinelli, G., Mazzucotelli, A., Pipe, R.K., and Farrar, S.V., Metallothioneins and lysosomes in metal toxicity and homeostasis in marine mussels: The effect of cadmium in the presence and absence of phenanthrene. *Mar. Biol.* 94, 251, 1987.
21. Virarengo, A. and Nott, J.A., Mechanisms of heavy metal cation homeostasis in marine invertebrates. *Comp. Biochem. Physiol.* 404C, 355, 1993.
22. Brouwer, M., Winge, D.R., and Gray, W.R., Structural and functional diversity of copper-metallothioneins from the American lobster *Homarus americanus*. *J. Inorg. Biochem.* 35, 289, 1989.
23. Brouwer, M., Syring, R., and Brouwer, T.H., Role of a copper-specific metallothionein of the blue crab, *Callinectes sapidus*, in copper metabolism associated with degradation and synthesis of hemocyanin. *J. Inorg. Biochem.* 88, 228, 2002.
24. Brouwer. M., Schlenk, D., Ringwood, A.H., and Brouwer-Hoexum, T., Metal-specific induction of metallothionein isoforms in the blue crab *Callinectes sapidus* in response to single- and mixed-metal exposure. *Arch. Biochem. Physiol.* 294, 461, 1992.
25. Brouwer, M., Enghild, J., Hoexum-Brouwer, T., Thogersen, I., and Truncali, A., Primary structure and tissue-specific expression of blue crab (*Callinectes sapidus*) metallothionein isoforms. *Biochem. J.* 311, 617, 1995.
26. George, S.G., Heavy metal detoxification in the mussel *Mytilis edulis*. Composition of Cd-containing kidney granules (tertiary lysosomes). *Comp. Biochem. Physiol.* 76C, 53, 1983.
27. George, S.G., Pirie, B.J.S., Cheyne, A.R., Coombs, T.L., and Grant, P.T., Detoxification of metals by marine bivalves: An ultrastructural study on the compartmentation of copper and zinc in the oyster *Ostrea edulis. Mar. Biol.* 45, 147, 1978.
28. Moore, M.N., Lysosomal responses to environmental chemicals in some marine invertebrates. In: C.S. Giam (Ed.), *Pollutant Effects on Marine Organisms*, pp. 143–165, D.C. Heath, Lexington, MA, 1977.
29. Chassard-Bouchaud, C., Role des lysosomes dans le phenomene de concentration du cadmium. Microanalyse par spectrographie des rayons X. C. R, *Acad. Sci. Paris* 293, 261, 1981.
30. Pirie, B., Liu, F., and George, S., Ultrastructural localization of copper and zinc in the polychaete *Nereis diversicolor*, from a highly contaminated estuary. *Mar. Environ. Res.* 17, 197, 1985.
31. Brown, B.E., The form and function of metal-containing "granules" in invertebrate tissues. *Biol. Rev.* 57, 621, 1982.
32. Mason, A.Z. and Simkiss, K., Sites of mineral deposition in metal-accumulating cells. *Exp. Cell Res.* 139, 383, 1982.
33. Al-Mohanna, S.Y. and Nott, J.A., The accumulation of metals in the hepatopancreas of the shrimp, *Penaeus semisulcatus* de Haan (Crustacea: Decapoda) during the moult cycle. *Proc. Confr. Mar. Environ. Pollution*, 1985, pp. 195–209.
34. Coombs, T. and George, S.G., Mechanisms of immobilization and detoxification of metals in marine organisms. In: D.S. McLusky and A.J. Berry (Eds), *Physiology and Behaviour of Marine Organisms*, pp. 179–187, Pergamon Press, New York, 1978.
35. Hopkin, S.P. and Nott, J.A., Some observations on concentrically structured, intracellular granules in the hepatopancreas of the shore crab *Carcinus maenas* (L.). *J. Mar. Biol. Ass. U.K.* 59, 867, 1979.
36. Mauri, M. and Orlando, E., Experimental study on renal concretions in the wedge shell *Donax trunculus* L. *J. Exp. Mar. Biol. U.K.* 63, 47, 1982.

37. Mason, A.Z., Simkiss, K., and Ryan, K.P., The ultrastructural localization of metals in specimens of *Littorina littorea* collected from clean and polluted sites. *J. Exp. Mar. Biol. U.K.* 64, 699, 1984.
38. Hopkin, S.P., *Ecophysiology of Metals in Terrestrial Invertebrates*, Elsevier Applied Science, London, 1989.
39. Bury, N.R., Walker, P.A., and Glover, C.N., Nutritive metal uptake in teleost fish. *J. Exp. Biol.* 206, 11–23, 2003.
40. Gunshin, H., Mackenzie, B., Berger, U.V., Gunshin, Y., Romero, M.F., Boron, W.F., Nussberger, S., Gollan, J.L., and Hediger, M.A., Cloning and characterization of a mammalian proton-coupled metal-ion transporter. *Nature* 388, 482, 1977.
41. McMahon, R.J. and Cousins, R.J., Mammalian zinc transporters. *J. Nutr.* 128, 667, 1998.
42. Vulpe, C.D. and Packman, S., Cellular copper transport. *Ann. Rev. Nutr.* 15, 293, 1995.
43. Grinstein, S., *Na/H Exchange*, CRC Press, Boca Raton, FL, 1988.
44. Rothstein, A., The Na/H exchange system in cell pH and volume control. *Rev. Physiol. Biochem. Pharmacol.* 112, 235, 1989.
45. Weinman, E.J. and Reuss, L., Na/H exchange at the apical membrane of *Necturus* gallbladder. *J. Gen. Physiol.* 80, 299, 1982.
46. Ahearn, G.A., The invertebrate electrogenic $2Na^+/1H^+$ exchanger: Polyfunctional epithelial workstation. *News Physiol. Sci.* 11, 31, 1996.
47. Ahearn, G.A. and Clay, L.P., Kinetic analysis of electrogenic 2Na/1H antiport in crustacean hepatopancreas. *Am. J. Physiol.* 257, R484, 1989.
48. Ahearn, G.A. and Franco, P., Na and Ca share the electrogenic 2Na/1H antiporter in crustacean antennal glands. *Am. J. Physiol.* 259, F758, 1990.
49. Ahearn, G.A. and Franco, P., Electrogenic 2Na/1H antiport in echinoderm gastrointestinal epithelium. *J. Exp. Biol.* 158, 495, 1991.
50. Ahearn, G.A., Franco, P., and Clay, L.P., Electrogenic 2Na/1H exchange in crustaceans. *J. Membr. Biol.* 116, 215, 1990.
51. Shetlar. R.E. and Towle, D.W., Electrogenic sodium-proton exchange in membrane vesicles from crab (*Carcinus maenas*) gill. *Am. J. Physiol.* 257, R924, 1989.
52. Ahearn, G.A. and Zhuang, Z., Cellular mechanisms of calcium transport in crustaceans. *Physiol. Zool.* 69, 383, 1996.
53. Zhuang, Z. and Ahearn, G.A., Calcium transport process of lobster hepatopancreatic brush border membrane vesicles. *J. Exp. Biol.* 119, 1195, 1996.
54. Zhuang, Z. and Ahearn, G.A. Energized Ca^{2+} transport by hepatopancreatic basolateral membranes of *Homarus americanus. J .Exp. Biol.* 201, 211, 1998.
55. Ahearn, G.A. and Franco, P., Ca transport pathways in brush border membrane vesicles of crustacean antennal glands. *Am. J. Physiol.* 264, R1206, 1993.
56. Ahearn, G.A., Zhuang, Z., Duerr, J., and Pennington, V., Role of the invertebrate electrogenic $2Na^+/1H^+$ antiporter in monovalent and divalent cation transport. *J. Exp. Biol.* 196, 319, 1994.
57. Zhuang, Z. and Ahearn, G.A., Ca^{2+} and Zn^{2+} are transported by the electrogenic 2 $Na^+/1H^+$ antiporter in echinoderm gastrointestinal epithelium. *J. Exp. Biol.* 198, 1207, 1995.
58. Aslamkhan, A.G. and Ahearn, G.A., Iron uptake by brush border membrane vesicles (BBMV) of the lobster (*Homarus americanus*) hepatopancreas. *J. Exp. Zool.* 295A, 145, 2003.
59. Menkes, J.H., Alter, M., Steigleder, G.K., Weakly, D.R., and Sung, J.H., A sex-linked recessive disorder with retardation of growth, peculiar hair, and focal cerebral and cerebellar degeneration. *Pediatrics* 29, 764, 1962.
60. Chelly, J., Turner, Z., Tonnesen, T., Patterson, A., and Ishikawa-Brush, Y., Isolation of a candidate gene for Menkes disease that encodes a potential heavy metal binding protein. *Nat. Genet.* 3, 14, 1993.
61. Mercer, J.F., Livingston, J., Hall, B., Paytner, J.A., and Begy, C., Isolation of a partial candidate gene for Menkes disease by positional cloning. *Nat. Genet.* 3, 20, 1993.
62. Vulpe, C., Levinson, B., Whitney, S., Packman, S., and Gilschier, J., Isolation of a candidate gene for Menkes disease and evidence that it encodes a copper-transporting ATPase. *Nat. Genet.* 3, 7, 1993.
63. Becker. G.L., Chen, C., Greenawalt, J.W., and Lehninger, A.L., Calcium phosphate granules in the hepatopancreas of the blue crab *Callinectes sapidus. J. Cell Biol.* 61, 316, 1974.
64. Becker, G.L., Termine, J.D., and Eanes, E.D., Comparative studies of intra- and extramitochondrial calcium phosphates from the hepatopancreas of the blue crab (*Callinectes sapidus*). *Calcif. Tissue Res.* 21, 105, 1976.
65. Chen, C.H. and Lehninger, A.L., Respiration and phosphorylation by mitochondria from the hepatopancreas of the blue crab (*Callinectes sapidus*). *Arch. Biochem. Biophys.* 154, 449, 1973.

66. Chen, C.H., Greenawalt, J.W., and Lehninger, A.L., Biochemical and ultrastructural aspects of Ca^{2+} transport by mitochondria of the hepatopancreas of the blue crab *Callincetes sapidus*. *J. Cell Biol.* 61, 301, 1974.
67. Chavez-Crooker, P., Garrido, N., and Ahearn, G.A., Copper transport by lobster (*Homarus americanus*) hepatopancreatic mitochondria. *J. Exp. Biol.* 205, 405, 2002.
68. MacLennan, D.H., Brandl, C.J., Korczak, B., and Green, N.M., Amino acid sequence of a $Ca^{2+}Mg^{2+}$-dependent ATPase from rabbit muscle sarcoplasmic reticulum, deduced from its complementary DNA sequence. *Nature* 316, 696, 1985.
69. Hagedorn, M., Weihrauch, D., Towle, D.W., and Ziegler, A., Molecular characterization of the smooth endoplasmic reticulum Ca^{2+}-ATPase of *Porcellio scaber* and its expression in sternal epithelia during the moult cycle. *J. Exp. Biol.* 206, 2167, 2003.
70. Wheatly, M., Gao, Y., Freire, N., and Zanotto, F.P., Subcellular calcium homeostasis during transepithelial calcium flux associated with crustacean molting: What role do SERCAs and PMCAs play? In: C.V. Gay, C.G. Dacke, and J.A. Danks (Eds), *Proceedings of the 3rd International Satellite Symposium on the Comparative Endocrinology of Calcium Regulation*, Cox, PA, Homepage http://www.sci.port.ac.uk/~cecr, 2001.
71. Zhang, Z., Chen, D., and Wheatly, M.G., Cloning and characterization of sarco/endoplasmic reticulum Ca2 + -ATPase (SERCA) from crayfish axial muscle. *J. Exp. Biol.* 203, 3411, 2000.
72. Havelaar, A.C., de Gast, I.L., Snijders, S., Beerens, C.E.M.T., Mancini, G.M.S., and Verheijen, F.W., Characterization of a heavy metal ion transporter in the lysosomal membrane. *FEBS Lett.* 436, 223, 1998.
73. Bull, P.C., Thomas, G.R., Rommens, J.M., Forbes, J.R., and Cox, D.W., The Wilson disease gene is a putative copper transporting P-type ATPase similar to the Menkes gene. *Nat. Genet.* 5, 327, 1993.
74. Petrukhin, K., Fischer, S.G., Pirastu, M., Tanzi, R.E., and Chernov, I., Mapping, cloning and genetic characterization of the region containing the Wilson disease gene. *Nat. Genet.* 5, 338, 1993.
75. Tanzi, R.E., Petrukhin, K., Chernov, I., Pellequer, J.L., and Ross, B., The Wilson disease gene is a copper transporting ATPase with homology to the Menkes disease gene. *Nat. Genet.* 5, 344, 1993.
76. Wilson, S.A.K., Progressive lenticular degeneration. A familial nervous disease associated with cirrhosis of the liver. *Brain* 34, 295, 1912.
77. Chavez-Crooker, P., Garrido, N., Pozo, P., and Ahearn, G.A., Copper transport by lobster (*Homarus americanus*) hepatopancreatic lysosomes. *Comp. Biochem. Physiol. Part C* 135: 107, 2003.
78. Sterling, K.M., Mandal, P.K., Roggenbeck, B.A., Ahearn, S.E., Gerencser, G.A., and Ahearn, G.A., Heavy metal detoxification in crustacean epithelial lysosomes: Role of anions in the sequestration process. *J. Exp. Biol.* 210, 3484, 2007.
79. Ahearn, G.A., Mandal, P.K., and Mandal, A., Mechanisms of heavy metal sequestration and detoxification in crustaceans: A review. *J. Comp. Physiol. B* 174, 439, 2004.
80. Zalaps, R.K. and Barfuss, D.W., Renal organic anion transport system: A mechanism for the basolateral uptake of mercury–thiol conjugates along the pars recta of the proximal tubule. *Toxicol. Appl. Pharmacol.* 182, 234, 2002.
81. Wright, S.H. and Dantzler, W.H., Molecular and cellular physiology of renal organic cation and anion transport. *Physiol. Rev.* 84, 987, 2004.
82. Breljak, D., Ljubojevic, M., Balen, D., Kusan, M., Anzai, N., Burckhardt, G., and Sabolic, I., Expression of organic anion transporter OAT3 in rat liver is androgen-dependent. Book of *Abstracts of the HDBMB 2006*/ Zrinka Kovarik (ed.), Zagreb, 206: p. 91, 2006.
83. Yonge, C.M., Studies on the comparative physiology of digestion. II. The mechanism of feeding digestion, and assimilation in *Nephrops norvegicus*. *Br. J. Exp. Biol.* 2, 343, 1924.
84. Dall, W., Hypo-osmoregulation in crustacean. *Comp. Biochem. Physiol.* 21, 653, 1967.
85. Mykles, D.L., The mechanism of fluid absorption at ecdysis in the American lobster, *Homarus americanus* and *H. gammarus*, and the Dungeness crab, *Cancer magister*. PhD thesis, University of California, Berkeley, 1979.
86. Ahearn, G.A., Maginniss, L.A., Song, Y.K., and Tornquist, A., Intestinal water and ion transport in freshwater malacostracan prawns (Crustacea). In: A.M. Jungreis, T. Hodges, A. Kleinzeller, and S. Schultz (Eds), *Water Relations in Membrane Transport in Plants and Animals*, pp. 129–142, Academic Press, New York. 1977.
87. Ahearn, G.A. and Tornquist, A., Allosteric cooperativity during intestinal cotransport of sodium and chloride in freshwater prawns. *Biochim. Biophys. Acta* 471, 273, 1977.
88. Ahearn, G.A., Allosteric cotransport of sodium, chloride, and calcium by the intestine of freshwater prawns. *J. Membr. Biol.* 42, 281, 1978.
89. Ahearn, G.A., Intestinal electrophysiology and transmural ion transport in freshwater prawns. *Am. J. Physiol.* 239, C1, 1980.

90. McNamara, J.C., Zanotto, F.P., and Onken, H., Adaptation to hypoosmotic challenge in brachyuran crabs: A microanatomical and electrophysiological characterization of the intestinal epithelia. *J. Exp. Zool.* 303A, 880, 2005.
91. Mykles, D.L., Ionic requirements of transepithelial potential difference and net water flux in the perfused midgut of the American lobster, *Homarus americanus. Comp. Biochem. Physiol.* 69A, 317, 1981.
92. Chu, H.K., Sodium transport across the perfused midgut and hindgut of the Blue Crab, *Callinectes sapidus*: The possible role of the gut in crustacean osmoregulation. *Comp. Biochem. Physiol.* 87A, 21, 1987.
93. Andres, S., Laporte, J.M., and Mason, R.P., Mercury accumulation and flux across the gills and the intestine of the blue crab (*Callinectes sapidus*). *Aquat. Toxicol.* 56: 303, 2002.
94. Laporte, J.M., Andres, S., and Mason, R.P., Effect of ligands and other metals on the uptake of mercury and methylmercury across the gills and intestine of the blue crab (*Callinectes sapidus*). *Comp. Biochem. Physiol. C* 131, 185, 2002.
95. Mullins, A.L. and Ahearn, G.A., Does L-leucine use a dipeptide transport protein for absorption by crustacean intestine? *FASEB J.* 22, 1202, 2, 2008.
96. Daniel, H. and Kottra, G., The proton oligopeptide cotransporter family SLC15 in physiology and pharmacology. *Pflugers Arch.* 447, 610, 2004.
97. Kirschner, L.B., Comparative physiology: Invertebrate excretory organs. *Annu. Rev. Physiol.* 29, 169, 1967.
98. Schmidt-Nielsen, B., Gertz, K.H., and Davis, L.E., Excretion and ultrastructure of the antennal gland of the fiddler crab *Uca mordax. J. Morphol.* 125, 473, 1968.
99. Schmidt-Nielsen, B. and Laws, D.F., Invertebrate mechanisms for diluting and concentrating the urine. *Ann. Rev. Physiol.* 25, 631, 1963.
100. Peterson, D.R. and Loizzi, R.F., Regional cytology and cytochemistry of the crayfish kidney tubule. *J. Morphol.* 141, 133, 1973.
101. Peterson, D.R. and Loizzi, R.F., Biochemical and cytochemical investigations of (Na^+-K^+)-ATPase in the crayfish kidney. *Comp. Biochem. Physiol. A* 49, 763, 1974.
102. Riegel, J.A. and Cook, M.A., Recent studies of excretion in crustaceans. *Fortschr. Zool.* 23, 48, 1975.
103. Icely, J.D. and Nott, J.A., The general morphology and fine structure of the antennary gland of *Corophium volutator* (Amphipoda: Crustacea). *J. Mar. Biol. Assoc. U.K.* 59, 745, 1979.
104. Johnson, P.T., *Histology of the Blue Crab, Callinectes sapidus. A Model for the Decapoda*, Praeger, New York, 1980.
105. Mantel, L.H. and Farmer, L.L., Osmotic and ionic regulation. In: L.H. Mantel (Ed.), *The Biology of Crustacea. Internal Anatomy and Physiological Regulation*, Vol. 5, p. 53, Academic Press, New York, 1983.
106. Fingerman, M., Glands and secretion. In: F.W. Harrison and A.G. Humes (Eds), *Microscopic Anatomy of Invertebrates. Decapod Crustacea*, Vol. 10, p. 345, Wiley-Liss, New York, 1992.
107. Khodabandeh, S., Charmantier, G., Blasco, C., Grousset, E., and Charmantier-Daures, M., Ontogeny of the antennal glands in the crayfish *Astacus leptodactylus* (Crustacea, Decapoda): Anatomical and cell differentiation. *Cell Tissue Res.* 319, 153, 2005a.
108. Khodabandeh, S., Kutnik, M., Aujoulat, F., Charmantier, G., and Chamantier-Daures, M., Ontogeny of the antennal glands in the crayfish *Astacus leptodactylus* (Crustacea, Decapoda): Immunolocalization of Na^+, K^+-ATPase. *Cell Tissue Res.* 319, 167, 2005b.
109. Khodabandeh, S., Charmantier, G., and Charmantier-Daures, M., Ultrastructural studies and Na^+,K^+-ATPase immunolocalization in the antennal urinary glands of the lobster *Homarus gammarus* (Crustacea, Decapoda). *J. Histochem. Cytochem.* 53, 1203, 2005c.
110. Freire, C.A., Cavassin, F., Rodrigues, E.N., Torres, A.H., and McNamara, J.C., Adaptive patterns of osmotic and ionic regulation, and the invasion of freshwater by the palaemonid shrimps. *Comp. Biochem. Physiol. A* 136, 771, 2003.
111. Freire, C., Onken, H., and McNamara, J.C., A structure–function analysis of ion transport in crustacean gills and excretory organs. *Comp. Biochem. Physiol. A* 151, 272, 2008.
112. Riegel, J.A., The influence of water-loading and low temperature on certain functional aspects of the crayfish antennal gland. *J. Exp. Biol.* 38, 291, 1961.
113. Kirschner, L.B. and Wagner, S., The site and permeability of the filtration locus in the crayfish antennal gland. *J. Exp. Biol.* 43, 385, 1965.
114. Riegel, J.A., Micropuncture studies of chloride concentration and osmotic pressure in the crayfish antennal gland. *J. Exp. Biol.* 40, 487, 1963.
115. Riegel, J.A., Micropuncture studies of the concentrations of sodium, potassium and inulin in the crayfish antennal gland. *J. Exp. Biol.* 42, 379, 1965.

116. Riegel, J.A., Micropuncture studies of formed-body secretion by the excretory organs of the crayfish, frog and stick insect. *J. Exp. Biol.* 44, 379, 1966a.
117. Riegel, J.A., Analysis of formed bodies in urine removed from the crayfish antennal gland by micropuncture. *J. Exp. Biol.* 44, 387, 1966b.
118. Riegel, J.A., Analysis of the distribution of sodium, potassium and osmotic pressure in the urine of crayfishes. *J. Exp. Biol.* 48, 587, 1968.
119. Holliday, C.W. and Miller, D.S., Cellular mechanisms of organic anion transport in crustacean renal tissue. *Am. Zool.* 24, 275, 1984.
120. Wheatly, M.G., Gao, Y., and Nade, M., Integrative aspects of renal epithelial calcium transport in crayfish: Temporal and spatial regulation of PMCA. *Int. Congr. Ser.* 1275, 96, 2004.
121. Parry, G., Excretion. In: T.H. Waterman (Ed.), *The Physiology of Crustacea. Metabolism and Growth*, Vol. 2, p. 341, Academic Press, New York, 1960.
122. Schoffeniels, E. and Dandrifosse, G., Osmoregulation: Aspects morphologiques et biochimiques. In: J. Forest, P. Pierre, G.-T. de Zoologie (Eds), *Anatomie, Systematique, Biologie, Tome VII: Crustaces, Frascicule I: Morphologie, Physiologie, Reproduction, Systematique*, p. 529, Masson, Paris, 1994.
123. Henry, R.P. and Wheatly, M.G., Interaction of respiration, ion regulation, and acid–base balance in the everyday life of aquatic crustaceans. *Am. Zool.* 32, 407, 1992.
124. Wheatly, M.G. and Henry, R.P., Branchial and antennal gland Na^+/K^+-dependent ATPase and carbonic anhydrase activity during salinity acclimation of the euryhaline crayfish *Pacifastacus leniusculus. J. Exp. Biol.* 133, 73, 1987.
125. Wheatly, M.G. and Toop, T., Physiological responses of the crayfish *Pacifastacus leniusculus* to environmental hyperoxia. II. Role of the antennal gland in acid–base and ion regulation. *J. Exp. Biol.* 143, 53, 1989.
126. Pavicic-Hamer, D. and Lucu, C., Activity of carbonic anhydrase in the antennal gland of lobster *Homarus gammarus* and crayfish *Astacus torrentium. Period. Biol.* 105, 367, 2003.
127. White, K.N. and Walker, G., The barnacle excretory organ. *J. Mar. Biol. Assoc. U.K.*, 61, 529, 1981.
128. Hessler, R.R. and Elofsson, R., Excretory system of *Hutchinsoniella macracantha* (Cephalocarida). *J. Crustacean Biol.* 11, 356, 1991.
129. Lockwood, A.P.M. and Riegel, J.A., The excretion of magnesium by *Carcinus maenas. J. Exp. Biol.* 51, 575, 1969.
130. Holliday, C.W., Magnesium transport by the urinary bladder of the crab, *Cancer magister. J. Exp. Biol.* 85, 187, 1980.
131. Wheatly, M.G. and Gannon, A.T., Ion regulation in crayfish—freshwater adaptations and the problem of molting. *Am. Zool.* 35, 49, 1995.
132. Rogers, J.V. and Wheatly, M.G., Accumulation of calcium in the antennal gland during the molting cycle of the freshwater crayfish *Procambarus clarkii. Invertebr. Biol.* 116, 248, 1997.
133. Wheatly, M.G., Pence, R.C., and Weil, J.R., ATP-dependent calcium uptake into basolateral vesicles from transporting epithelia of intermolt crayfish. *Am. J. Physiol.* 276, R566, 1999.
134. Gao, Y. and Wheatly, M.G., Characterization and expression of plasma membrane Ca^{2+} ATPase (PMCA3) in the crayfish *Procambarus clarkii* antennal gland during molting. *J. Exp. Biol.* 207, 2991, 2004.
135. Wheatly, M.G., Zanotto, F.P., and Hubbard, M.G., Calcium homeostasis in crustaceans: Subcellular Ca dynamics. *Comp. Biochem. Physiol. B.* 132, 163, 2002.
136. Binns, R., The physiology of the antennal gland of *Carcinus maenas* (L.). I. The mechanism of urine production. *J. Exp. Biol.* 51, 1, 1969.
137. Binns, R. and Peterson, A.J., Nitrogen excretion by the spiny lobster *Jasus edwardsi* (Hutton). The role of the antennal gland. *Biol. Bull.* 136, 147, 1969.
138. Cameron, J.N. and Batterton, C.V., Antennal gland function in the freshwater crab, *Callinectes sapidus:* Water, electrolyte, acid–base and ammonia excretion. *J. Comp. Physiol.* 123, 143, 1978.
139. Miller, D.S. and Holliday, C.W., PAH secretion in the urinary bladder of a crab *Cancer borealis. Am. J. Physiol.* 243, R147, 1982.
140. Miller, D.S. and Holliday, C.W., Organic cation secretion by *Cancer borealis* urinary bladder. *Am. J. Physiol.* 252, R153, 1987.
141. Miller, D.S., Smith, P.M., and Pritchard, J.B., Organic anion and cation transport in crab urinary bladder. *Am. J. Physiol.* 257, R501, 1989.
142. Pritchard, J.B. and Miller, D.S., Comparative insights into the mechanisms of renal organic anion and cation secretion. *Am. J. Physiol.* 261, R1329, 1991.

143. O'Donnell, M.J., Mechanisms of excretion and ion transport in invertebrates, In: W.H. Dantzler (Ed.), *Comparative Physiology*, Vol. II, Section 13, Chapter 17, pp. 1207–1289, Am. Physiol. Soc., New York, 1997.
144. Ahearn, G.A, Grover, M.L, Tsuji, R.T, and Clay, L.P., Proton-stimulated Cl/HCO_3 antiport by purified lobster hepatopancreatic basolateral membrane vesicles. *Am. J. Physiol.* 252, R859, 1987.
145. Gerencser, G.A., Ahearn, G.A., and Cattey, M.A., Sulfate/bicarbonate antiport by lobster hepatopancreatic basolateral membrane vesicles. *J. Exp. Zool.* 284, 158, 1999.

12 Metals and Cell Adhesion Molecules

Alan R. Parrish and Walter C. Prozialeck

CONTENTS

12.1 Introduction 327
12.2 Overview of Cell Adhesion Molecules 328
12.2.1 Integrins: Cell–Matrix Interactions 328
12.2.2 Homotypic Cell–Cell Adhesion 330
12.2.2.1 Tight Junction 330
12.2.2.2 Cadherin Complex 330
12.2.2.3 Gap Junctions 331
12.2.3 Heterotypic Cell–Cell Adhesion: Inflammation 332
12.3 Metal-Induced Disruption of Cell Adhesion 332
12.3.1 Integrins 332
12.3.2 Tight Junctions and Cadherin/Catenin Complexes 333
12.3.3 Gap Junctions 334
12.3.4 Inflammatory Adhesion Molecules 335
12.4 Potential Mechanisms 335
12.4.1 Gene Expression 335
12.4.2 Protein–Protein Interactions 336
12.4.3 Protein Phosphorylation 336
12.4.4 Protein Degradation 337
12.4.5 Disruption of the Actin Cytoskeleton 338
12.4.6 Oxidative Stress 338
12.5 Conclusions 338
References 339

12.1 INTRODUCTION

The importance of cell adhesion molecules as determinants of tissue architecture has long been recognized. These molecules mediate cell–cell and cell–substrate adhesion and thereby play critical roles in a variety of biological processes including development, tumorigenesis, and inflammation. Our understanding of cell–matrix and cell–cell adhesion has increased significantly in the last decade with the identification of the molecular components of adhesion complexes, as well as their structure and organization. Importantly, it has become apparent that adhesion molecules are not merely cellular "glue" but highly dynamic complexes that can play important roles in cell signaling. This chapter will highlight the growing volume of evidence showing that cell adhesion complexes may be important molecular targets for a variety of toxic metals such as manganese, cadmium, mercury, copper, lead, and arsenic. Specific examples of metal-induced disruption of adhesion complexes will be included, as well as the latest information on molecular mechanisms by which metals may influence cellular adhesion. While disruption of cell adhesion is a relatively new mechanistic consideration in metal toxicology and fewer studies have been conducted in this area than in other topics discussed in this book, it is clear that the actions of metals on cell adhesion molecules must be evaluated when considering the impact of metal exposure on human health.

12.2 OVERVIEW OF CELL ADHESION MOLECULES

Cell adhesion molecules are usually classified according to the types of cellular interactions that they mediate (cell–matrix, homotypic cell–cell, and heterotypic cell–cell). The following section provides a general overview of some of the more important molecules in each of the categories that have been implicated as possible targets of metal toxicity.

12.2.1 Integrins: Cell–Matrix Interactions

The integrins are cell-surface glycoproteins that were first identified as receptors for extracellular matrix proteins such as collagen, fibronectin, and laminin. Integrin receptors are heterodimers consisting of an α and β subunit (Figure 12.1). Each of these subunits has a large extracellular domain, a single transmembrane domain, and a short cytoplasmic domain that is ultimately linked to the actin cytoskeleton [1,2]. One notable exception is β4 integrin, which has an atypical cytoplasmic domain. Integrins are capable of regulating cell shape, behavior, and fate. Interestingly, integrin signaling is bidirectional, with both inside–out signaling, which regulates ligand affinity, as well as the classic outside–in signaling that regulates the cellular response to matrix adhesion [3–5]. At least 18 α and 8 β subunits have been identified in vertebrates, forming 24 unique integrin receptors with overlapping patterns of cellular expression and ligand affinity [1,2]. Our understanding integrin function has been facilitated, in part, by the generation of null mutations of each of the integrin subunits in mice [6]. While integrins are essential components of the focal adhesions that maintain cell–matrix contact, at least one integrin, α3β1, influences cell–cell adhesion as well [7]. In addition, a role for integrins in host cell–pathogen interaction is emerging [8,9]. These recent findings suggest that there are still much to learn about the diverse roles of integrins in normal and pathological conditions.

The cytoplasmic domain of integrin receptors is part of a large complex of signaling- and cytoskeletal-related proteins. Clustering of integrins and these proteins at sites of cell–extracellular matrix adhesion forms a specialized structure commonly referred to as a focal adhesion that attaches the cell to the extracellular matrix (Figure 12.1). Ligand binding stimulates the activity of several signaling pathways, including Rho GTPases, tyrosine phosphatases, cAMP-dependent protein kinase, protein kinase C (PKC), mitogen-activated protein kinase (MAPK), NFB, JNK, and Wnt [10–15]. Two of the more well-characterized integrin signaling molecules are focal adhesion kinase (FAK) and integrin-linked kinase (ILK). FAK is a nonreceptor tyrosine kinase that is tyrosine phosphorylated and activated by integrin-mediated adhesion, even though the protein does not directly associate with integrins [16,17]. Conversely, dephosphorylation occurs rapidly when cells are detached from the extracellular matrix. FAK is capable of binding to a number of other proteins, including c-src, PI-3-kinase, paxillin, talin, and p130cas. FAK has been demonstrated to play a critical role in the regulation of cell motility and apoptosis [1]. ILK was discovered based on its ability to interact with the β1 integrin subunit [18]. While it is a critical regulator of the protein:protein interactions that link integrins to the actin cytoskeleton [19], ILK can also regulate the activity of signaling pathways involved in cell cycle control, cell–cell adhesion, migration, and motility [20,21].

Integrin function is regulated by extracellular divalent cations, particularly calcium and magnesium; magnesium facilitates ligand binding, while calcium is inhibitory. Importantly, titration studies revealed that integrin activation influences metal ion affinity, demonstrating the bidirectional nature of integrin signaling [22]. Manganese and magnesium are functionally interchangeable in a number of systems [23]. Thus, it is not surprising that low concentrations of manganese induce adhesion and cell spreading [24,25]. Manganese-dependent neuronal differentiation of PC-3 cells is integrin dependent [26]. This process primarily involves αv integrin, as evidenced by the findings

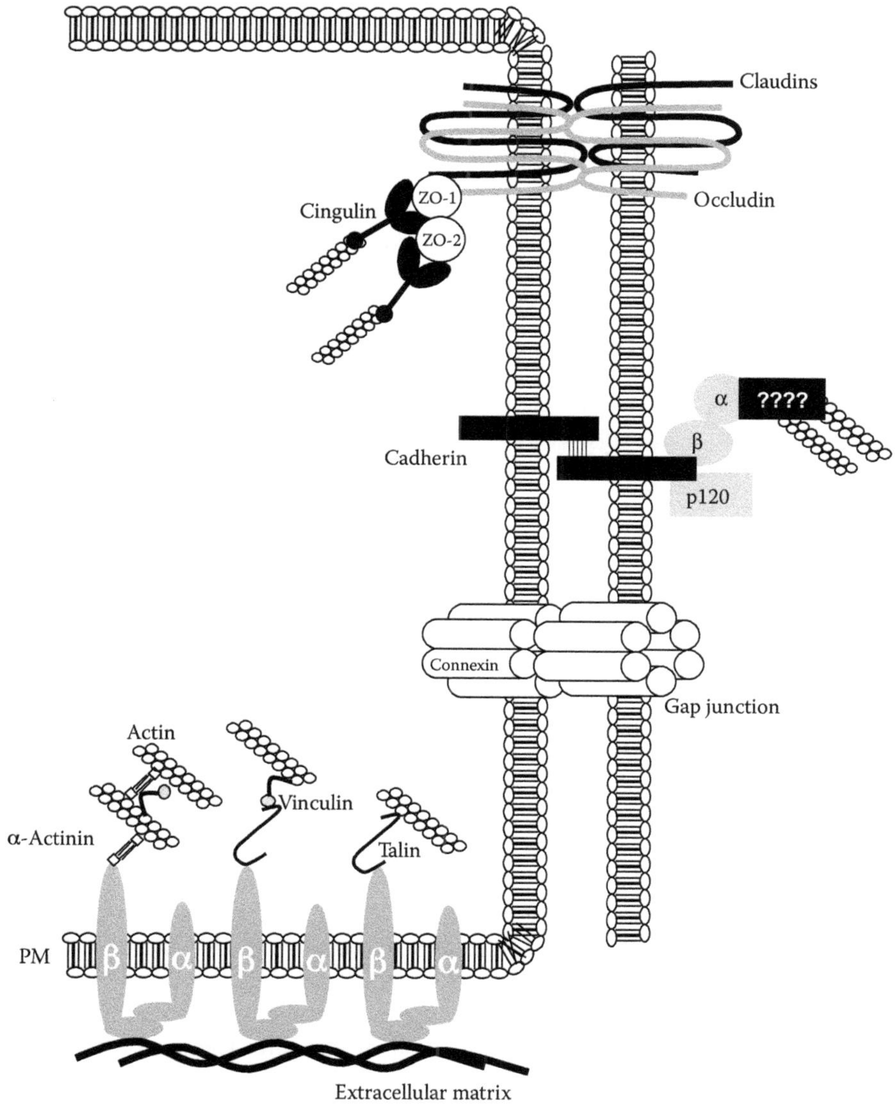

FIGURE 12.1 Representation of the cell adhesion complexes present in proximal tubular epithelial cells. Most proximal to the apical membrane is the tight junction that is composed of transmembrane (occludin and claudins) and cytosolic proteins (ZO-1 and ZO-2). ZO-1 binds directly to occludin, and is in turn bound by ZO-2; cingulin links the complex to the actin cytoskeleton. The adherens junction (cadherin/catenin complex) is located basal to the tight junction. Cadherins (E-, K-, and N-) are single transmembrane proteins whose cytoplasmic domain is bound by β- (or γ-catenin). α-Catenin does not directly bind cadherins; the complex requires the actin cytoskeleton via an unknown mechanism at this time. p120-Catenin regulates cadherin stability at the membrane. Gap junctions comprise connexins; six connexins form a connexon that is coupled to a connexon in an adjacent cell. Integrin receptors are heterodimers of α- and β-subunits that are linked to the actin cytoskeleton via a number of proteins including α-actinin, vinculin, or talin, and bind extracellular matrix proteins within the basement membrane.

that manganese failed to induce neurite outgrowth in αv-deficient PC12 cells and that neurite outgrowth induced by manganese was associated with αv upregulation [27]. By using GRGDS integrin-blocking peptides, Walowitz and Roth [28] showed that ERK1 and ERK2 phosphorylation is required for this manganese-induced effect. This positive effect on integrin activation by manganese is specific; cadmium, copper, nickel, and zinc do not induce neurite outgrowth in PC12 cells, while cobalt only partially mimics the effect of manganese [29]. This is a clear demonstration that metal-induced effects on adhesion molecules are not restricted to toxicological outcomes, but may also involve modulation of the normal homeostatic roles of the complexes.

12.2.2 Homotypic Cell–Cell Adhesion

12.2.2.1 Tight Junction

Tight junctions (zonula occludens) were first described as election dense plaques near the apical surface of polarized cells in which the plasma membranes of adjacent cells appeared to fuse. They are thought to be the main determinant of the epithelial barrier. The tight junction has three critical functions. The "fence" function restricts intradomain movement of membrane proteins and phospholipids and is essential for the development and maintenance of epithelial polarity. The "gate" function blocks the paracellular movement of solutes and is essential for the barrier properties of the epithelium. In addition, the tight junction is also a cell signaling nexus that regulates proliferation, differentiation, and dedifferentiation, cytoskeletal modulation, and vesicular trafficking within the epithelium [30–32].

Occludin, claudins, and junctional adhesion molecule (JAM) are all transmembrane components of tight junctions [33] (Figure 12.1). Occludin has four transmembrane domains, two extracellular domains, and a long cytoplasmic (carboxy-terminal) domain. Occludin is a phosphoprotein, and the hyperphosphorylated form is the main component within a functional tight junction [34]. Claudins are a multigene family of transmembrane proteins that are structurally similar to occludin (four transmembrane domains and two extracellular domains). At least 23 claudins have been identified in human tissues, although the potential for additional claudins exists via splice variants [35]. All carboxy-terminal tails of claudins contain a PDZ binding motif that may promote interaction with the cytoplasmic protein constituents of tight junctions. Importantly, claudin expression has been identified as the predominant regulator of tight junction permeability [35].

JAM is a member of the IgG superfamily that is structurally distinct from occludin and claudins. Based on homology with IgG family members, it is thought that JAM may play a role in the transmigration of inflammatory cells during inflammation [36]. CAR (coxsackievirus and adenovirus receptor) is a tight junction component that mediates homotypic cell adhesion, as well as infection by coxsackie β viruses and certain adenoviruses [37].

ZO-1, ZO-2, and ZO-3 are cytoplasmic proteins that are members of the membrane-associated guanylate kinase (MAGUK) gene superfamily. Common structural features of the ZO family of proteins include three PDZ domains, an SH3 domain, and a guanylate kinase homology region (GUK) [38]. ZO-1 can bind occludin, while both ZO-2 and ZO-3 bind to ZO-1, but not each other. It is thought that ZO-1 interacts with actin to link the tight junction complex to the cytoskeleton [39,40]; however, the ZO proteins also regulate gene transcription [33,41].

12.2.2.2 Cadherin Complex

The cadherin gene superfamily encodes transmembrane proteins that regulate calcium-dependent cell–cell adhesion [42–44]. Type-I (classical) cadherins comprise a large family of proteins that share significant structural conservation and include E-(epithelial) cadherin (liver cell adhesion molecule [L-CAM] or uvomorulin), N-(neuronal) cadherin (A-CAM), and P-(placental) cadherin [45]. The extracellular domain contains four highly conserved calcium-binding "cadherin repeats" (EC1–4) and one membrane proximal extracellular domain (EC5). Type-II (atypical) cadherins have a similar structure, but lack a conserved His-Ala-Val (HAV) sequence in the EC1 domain. The desmocollins and desmogleins are cadherin-like molecules that are components of desmosomes, a cell adhesion complex. The remaining cadherin molecules have low homology to type-I cadherins (less than 44% at the amino acid level) and have from 6 to 34 cadherin repeats. A functional cadherin adhesion complex requires interaction with cytoplasmic catenin proteins (Figure 12.1). α-Catenin does not directly bind to cadherins; rather, it interacts with the cytoplasmic domain of cadherins via β- or γ-catenin. p120-Catenin binds to the cadherin cytoplasmic domain and shares sequence homology with β- and γ-catenin, but does not bind α-catenin [46]. p120-Catenin has been demonstrated to regulate cadherin stability; the binding of p120 stabilizes cadherins, whereas the absence of p120 is associated with cadherin endocytosis and degradation [47,48]. It was long thought that α-catenin linked the cadherin/catenin complex to the cytoskeleton [49], a conclusion that was based on the observation that α-catenin was able to interact with a variety of actin-binding proteins, including α-actinin, vinculin, and actin [50–53]. However, this mechanism has recently been challenged based on recent studies suggesting that α-catenin does not stably interact with the actin cytoskeleton and cadherins simultaneously, but may regulate actin dynamics [54,55]. While the mechanism is still unclear, a connection between the cadherin/catenin complex and the actin cytoskeleton certainly exists [56]. Cadherin/catenin complexes play an important role in cytoskeletal organization and are critical for the establishment of epithelial cell polarity [57,58]. In addition, the actin cytoskeleton regulates cadherin function, as evidenced by the finding that cytochalasin D, an inhibitor of actin polymerization, weakens cell–cell contacts and reduces intercellular adhesion [59]. Therefore, the relationship between the cadherin/catenin complex and the actin cytoskeleton is bidirectional, with each participating in the regulation of the other.

β-Catenin also functions as a component of the wingless or Wnt nuclear signaling pathway and plays an important role in the regulation of gene expression [60–63]. Activation of Wnt signaling results in inhibition of GSK-3β activity via phosphorylation at ser9 [64,65] causing an accumulation of cytoplasmic β-catenin, which may then translocate to the nucleus and regulate transcription via the Tcf/Lef family of transcription factors [60]. Several genes regulated by β-catenin have been identified, including cyclin D1, c-myc, c-jun, and E-cadherin [66]. Conversely, phosphorylation of β-catenin by active GSK-3β targets the protein for ubiquitination and degradation [67]. In this context, GSK-3β activity is a key regulator of β-catenin transcriptional activity. In an elegant study, Gottardi and Gumbiner [68] showed that the balance between cadherin-mediated adhesion and Wnt signaling is not dictated by simple competition, but rather by the regulation of distinct forms of β-catenin. These data further suggest that the regulation of β-catenin transcriptional activity may involve much more than the simple dissociation from the cadherin/catenin complex and accumulation in the nucleus.

Even though tight junctions and cadherin/catenin complexes are distinct molecular structures, they are functionally interrelated. For example, the cadherins play important roles in the formation of tight junctions and the establishment of the epithelial barrier [58,69].

12.2.2.3 Gap Junctions

While technically not cell adhesion complexes, gap junctions are specialized structures comprising connexins (Cx) that allow for communication between cells. Connexins are a multigene family with 21 members in humans, ranging in size from 25 to 50 kDa [70]. The connexins have tissue- and/or cell-type specificity; however, most cells and organs express multiple connexins [70]. Each connexin has four transmembrane domains and two extracellular loops, each containing three critical cysteine residues that can form disulfide bonds. Six connexins assemble into a hemichannel (connexon) that interacts with a connexon on a neighboring cell to form a channel between the cells [71] (Figure 12.1). The channel (gap junction) allows for intercellular communication via the exchange of nutrients, metabolites, ions, and small molecules less than 1 kDa. Important molecules transported by gap junctions include ATP, glutamate, NAD^+, and prostaglandin E2 [72]. Recent work indicates that additional proteins are also present in gap junctions. The innexins were originally identified as structural components of gap junctions in *Drosophila*, but vertebrate homologues called pannexins have been identified [73,74]. The specific functions of these proteins are yet to be elucidated.

Gating is the mechanism by which the movement of ionic or nonionic molecules is restricted due to structural alterations of gap junctions. A number of mechanisms have been shown to regulate gap junction gating, including external (mechanical and shear stress, and voltage) and internal (cytoplasmic calcium) pathways [72,75,76]. Protein phosphorylation is also a critical regulator of gap junction function; connexins can be phosphorylated by a number of tyrosine (src) and serine/threonine kinases (protein kinases A, C, and G) [77,78].

12.2.3 Heterotypic Cell–Cell Adhesion: Inflammation

A variety of cell adhesion molecules including ICAMs, PECAM, VCAM, selectins, and integrins mediate the leukocyte-to-endothelial cell–cell interactions [79,80]. ICAMs are members of the Ig CAM superfamily and are ligands for the β2 integrin present on leukocytes [81]. ICAMs have two or more Ig-like repeats in the extracellular domain, a single transmembrane domain and a short cytoplasmic tail. Selectins play a major role in the adhesion of leukocyte-to-endothelial cells and platelets during inflammation. L- (leukocyte), E- (endothelial), and P- (platelet) selectin have conserved structures and mediate heterotypic cell adhesion via calcium-dependent mechanisms [1]. The recruitment of leukocytes and subsequent inflammatory processes involves intricate changes in the expression of a variety of leukocyte adhesion molecules. Given their role in inflammation it is not surprising that cytokines, including tumor necrosis factor α (TNFα) and interleukin-1β, are potent inducers of these molecules [82]. In the context of metals, it is noteworthy that desferrioxamine, an intracellular iron chelator, and neocuproine, an intracellular copper chelator, were found to inhibit TNFα-induced E-selectin, ICAM-1, and VCAM-1 expressions in endothelial cells, putatively via inhibiting SP-1 activation [83]. This suggests that endothelial activation during inflammation may be dependent on intracellular transition metal ions.

12.3 METAL-INDUCED DISRUPTION OF CELL ADHESION

12.3.1 Integrins

Several studies have demonstrated that integrins are molecular targets of several metals. In animal models of mercury-induced autoimmune disease, sublethal doses of inorganic mercury administered over a two week period resulted in the generation of antibodies to the glomerular basement membrane and proteinuria [84–86]. Antibodies targeting the α4 integrin subunit attenuated mercury-induced proteinuria [84]. However, antibodies to the β1 integrin subunit did not alter proteinuria, although they inhibited the autoimmune effects of mercury by reducing the circulating concentrations of antiglomerular basement membrane antibodies [85]. Interestingly, following subchronic cadmium intoxication in animals, antibodies to one integrin ligand, laminin, appeared in the blood [87]. However, unlike mercury, these cadmium-induced changes were not associated with the production of antibodies against the glomerular basement membrane.

Copper has also been show to have effects on integrin expression and localization. This has been demonstrated primarily through studies on uterine expression of integrins in women using intrauterine devices (IUDs) for the purpose of birth control. Initial studies demonstrated that the expression of αvβ3, but not α4, was decreased in biopsies of IUD users [88]. These results are supported by the finding that the expression of α3 and αv integrin subunits was decreased as assessed by immunofluorescence of biopsy samples of women using the T380 IUD [89]. In a set of *in vitro* experiments designed to examine the molecular basis of these findings, copper ions increased β3 integrin expression as assessed by flow cytometry [90]. There has been limited followup on these results, and the long-term impact of copper IUDs on integrin expression in the uterus—following cessation of birth control—has not yet been established. In addition, the mechanism underlying these clinical findings has not been explored.

In coculture experiments, zinc, nickel, and cobalt had a biphasic effect on polymorphonuclear leukocyte–endothelial cell adhesion, that is, enhancement at both low and high molar concentrations [91]. Keratinocyte migration, an integrin-dependent process, is enhanced by zinc, copper, and manganese [92]. Results of studies employing integrin-blocking antibodies indicate that the zinc and copper effect is mediated by α3, αv, and β1, while the manganese-induced migration requires α3 and β1 [93]. Inorganic lead inhibits FAK signaling in NRK cells as assessed by both cell adhesion and disassembly of focal adhesions [94]. Millimolar concentrations of manganese inhibit the spreading of human osteoblasts (MG-63 cell line) [95], suggesting that while low concentrations of manganese promote integrin function [24,25], inhibition of integrin function may contribute to manganese toxicity. Another recent study has shown that noncytotoxic concentrations of sodium arsenite disrupted focal adhesion in H9C cells and that this effect was associated with a decrease in the phosphorylation of FAK [96]. In addition, Lee et al. [97] have shown that arsenic alters the expression of β1 integrins in cultured human keratinocytes.

An interesting aspect of the metal–integrin connection is demonstrated by a series of studies by Conrad and Umbreit elucidating the role of integrins in iron absorption. In the initial studies, it was shown that radiolabeled iron was concentrated in integrins, and that mobilferrin coprecipitated with integrins [98]. Pulse chase studies suggest sequential passage of iron to integrin, mobilferrin, and ferritin. A monoclonal antibody against β3 integrin also immunoprecipitated the radiolabeled iron [99]. Interestingly, the authors also demonstrated that immunoprecipitates contained β1 integrin and radiolabeled zinc, suggesting that zinc absorption may also be integrin dependent. Using a blocking antibody strategy, Conrad et al. [100] established a definitive role of β3 integin in the absorption of ferric iron. However, the transport of ferrous iron did not appear to be integrin dependent [101]. Taken together, these results demonstrate a critical role of integrins in the absorption of metals, including zinc and iron, and suggest that the role of adhesion molecules in metal transport may be an important, yet overlooked, area of research.

12.3.2 Tight Junctions and Cadherin/Catenin Complexes

A number of studies have focused on the effects of metals on both tight junctions and the cadherin/catenin complex. Much of this work has come from investigations of the mechanisms of metal-induced nephrotoxicity. Previous publications have, in fact, reviewed the targeting of adhesion complexes in the kidney by metals and other nephrotoxicants [102–104].

Cadmium has been shown to disrupt E-cadherin-dependent cell adhesion in a number of cell lines, putatively via displacement of calcium from the extracellular binding domain. This hypothesis is supported by the finding that cadmium binds to synthetic and recombinant analogs of E-cadherin with a higher affinity than calcium [105,106]. Zimmerhackl et al. [107] have shown that cadmium is more toxic to LLC-PK1 cells (proximal tubule) than to MDCK cells (distal tubule/collecting duct), putatively via disruption of E-cadherin linkage to the actin cytoskeleton. *In vivo*, cadmium alters the pattern of N-cadherin localization in the kidneys of rats subchronically exposed to cadmium [108]. This effect occurs in the absence of changes in N-cadherin gene or protein expression [109]. In both *in vitro* and *in vivo* studies, the effects of cadmium on the cadherins occurred at doses and durations of exposure that did not produce overt evidence of necrotic or apoptotic cell death [103,108,110]. Jacquillet et al. [111] have also recently reported that endothelial cell–cell junctions may be disrupted during cadmium exposure. Expression of the endothelium-specific tight junction protein, claudin-5, was irregular and diminished in the glomeruli and small blood vessels of the kidneys from cadmium-treated rats. In an interesting *in vitro* study, cadmium caused translocation of β-catenin to the nucleus in a rat proximal tubular epithelial cell line [112]. In addition, several Wnt responsive genes, including c-myc, were upregulated; however, these studies used high doses (50 μm) that are associated with eventual cell death. Together these findings strongly suggest that the cadherins are relatively specific early targets of cadmium-induced renal injury and that cadmium interferes with the function of the cadherins without affecting the expression of the protein.

The ability of cadmium to disrupt cadherin/catenin complexes is not limited to the kidney. For example, cadmium caused alterations in the spatial distribution of VE-cadherin in endothelial cells, which was associated with inhibition of both migration and tubulogenesis [113]. Duizer et al. [114] demonstrated that the cellular localization of E-cadherin, as well as ZO-1 and occludin, was altered in an intestinal epithelial cell line (IEC-18) following cadmium challenge. The alterations in the distribution of these proteins correlated with loss of epithelial barrier function as assessed by trans-epithelial resistance and paracellular transport of mannitol and PEG-400. In the mouse lung, intra-tracheal administration of cadmium was associated with the rapid development of pulmonary edema and decreased E-cadherin in alveolar epithelial cells and decreased VE-cadherin in endothelial cells [115]. Importantly, these changes were not due to cell death as assessed by ethidium homodimer-1 staining in the lungs from the cadmium-treated animals. Cadherins may also be targeted by cadmium during development. Cadmium increased E-cadherin in yolk sacs, an effect that was not linked with alterations in the actin cytoskeleton, but was associated with embryotoxicity [116]. Thompson et al. [117] have recently shown that challenge of chick embryos with cadmium induced ventral body wall defects. These changes did not involve alterations in the expression or localization of E-cadherin, but were associated with the redistribution of β-catenin from the membrane to the cytoplasm, with some evidence of nuclear staining.

Other metals have also been shown to disrupt homotypic cell–cell adhesion. Potassium dichromate caused loss of E-cadherin, and disruption of actin cytoskeleton in cultured human keratinocytes [118]. In Caco-2 cells, copper caused a loss of epithelial permeability; interestingly the staining of occludin, ZO-1, or E-cadherin was not reduced, but there was a dramatic loss of F-actin staining [119]. In addition, bismuth, a nephrotoxic metal, was shown to specifically alter the cellular localization of N-, but not E-, cadherin in the mouse proximal tubular epithelium [120]. Interestingly, no changes in tight junction constituents were seen. Mercuric chloride also causes alterations in the cadherin/catenin complex in mice following *in vivo* challenge. These changes occurred prior to renal dysfunction and were associated with loss of proximal tubular epithelial cell polarity [121].

12.3.3 Gap Junctions

Results of several studies have demonstrated that certain metals can disrupt gap junctional intercellular communication (GJIC). Methylmercury (20 min, 30 μm) inhibited GJIC in primary cultures of rat proximal tubular cells as assessed by a dye coupling assay [122]. In primary cultures of rat proximal tubular cells exposed to 100 M cadmium chloride for 60 min, dye coupling revealed a decrease in GJIC that was associated with an increase in intracellular calcium [123]. It has been proposed that decreased GJIC represents an early event in mercuric chloride toxicity in MDCK cells as well [124]. In that study, 0.1 μM mercuric chloride (4 h) reduced GJIC by approximately 50% as measured by dye transfer between cells. Cadmium chloride also inhibited GJIC in the absence of cytotoxicity in liver epithelial cells (WB-344) [125]. These *in vitro* results are supported by the finding that *in vivo* challenge of mice with cadmium inhibited GJIC in the liver, concomitant with decreased Cx32 and Cx26 expressions [126].

While these studies suggest that gap junctions may be targets of metals such as mercury and cadmium, a number of studies suggest that gap junctions are not targets of lead toxicity. In primary cultures of osteoblast-like cells, it was shown that lead, either applied to the cells extracellularly or intracellularly, had no effect on GJIC [127]. The finding that lead does not affect GJIC is also supported by studies in astroglia, where lead did not affect intracellular communication [128], and REL liver cells, where lead acetate, chloride, monoxide, or sulfate did not inhibit GJIC [129]. In keratinocytes, lead acetate did not inhibit GJIC; however, both mercuric chloride (10 nM) and monomethylmercury (250 nM) inhibited gap junction function [130]. However, one study suggests that lead may inhibit GJIC. Using metabolic coupling, as assessed by metabolism of 6-thioguanine, of wild-type and mutant V79 cells, Trosko and coworkers demonstrated that arsenate, mercuric chloride, and lead acetate all inhibited GJIC [131]. While the weight of the evidence suggests that inhibition of GJIC is not a general mechanism of lead toxicity, there may be some cell- and/or organ-specificity of this response that needs to be explored further.

One of the earliest studies of metal-induced disruption of GJIC deserves mention. Salts of lead [$Pb(CH_3CO)_2$], cadmium [$Cd(CH_3COO)_2$], nickel [$NiSO_4$], and chromium [K_2CrO_4 and $CrCl_3$] only decreased GJIC at concentrations that elicited cell death [132]. This illustrates two important considerations for studies of metal-induced alterations in cell adhesion complexes. First, the speciation of the metal, especially *in vitro*, is an important, yet often overlooked, variable and, secondly, both concentration and time dependence must be considered in the entire context of effects (i.e., lethal versus nonlethal insult).

12.3.4 Inflammatory Adhesion Molecules

Several lines of evidence suggest that upregulation of inflammatory adhesion molecules may be an important mechanism of metal toxicity. Cadmium and mercury increased ICAM-1 expression in immortalized proximal tubule epithelial cells [133]. Importantly, both metals caused a decrease in secreted TGFβ-1 and addition of exogenous TGFβ-1 to the culture media attenuated the metal-induced increases of ICAM-1, suggesting that TGFβ-1 signaling may regulate inducible ICAM-1 levels in the kidney. The finding that cadmium induces ICAM-1 is not kidney specific; cadmium also enhanced ICAM-1 expression in brain endothelial cells [134,135] and in liver sinusoidal endothelial cells [136]. In the latter study, cadmium also induced E-selectin expression, suggesting that multiple inflammatory adhesion receptors may be targeted by this metal. However, in normal human keratinocytes, nickel, but not cobalt, chromium, or cadmium, increased ICAM-1 expression [137]. This suggests that cadmium-induced ICAM-1 exhibits organ specificity and that ICAM-1 is not induced by all metals. Zinc has also been shown to induce ICAM-1 gene expression in both fibroblasts and hematopoietic cells [138].

Much less is known about the impact of metals on other inflammatory adhesion molecules. In endothelial cells, nickel and cobalt, but not zinc and chromium, increased E-selectin expression, putatively via the NF-κB and AP-1 transcription factors [139]. In rats receiving a single injection of cadmium chloride (3 mg/kg), sinusoidal endothelial cells had a significant decrease in PECAM-1 expression 48 h postinjection followed by a marked increase in PECAM-1 expression 7 days postinjection [136]. In this study, increased PECAM-1 expression was thought to be linked to infiltration of inflammatory cells within the liver following cadmium exposure. While it is clear that a variety of metals increase the expression of ICAM-1, and E-selectin, the molecular pathways regulating these effects are poorly defined. Given the importance of inflammation in pathophysiology, future study of the impact of metals on these complexes will be an important component to our understanding of metal-induced toxicity.

12.4 POTENTIAL MECHANISMS

Studies on the impact of metals on cell adhesion molecules are a relatively new area of investigation. Consequently, the mechanisms underlying the effects of metals on these molecules are not well understood. Nevertheless, several potential mechanisms can be postulated (Figure 12.2).

12.4.1 Gene Expression

It is well established that metals can impact gene expression [140–146] raising the possibility that metals could regulate gene expression of cell adhesion molecules. In the context of metal-induced expression of inflammatory adhesion molecules, this appears to be a major mechanism, as demonstrated by the studies cited in the previous section. However, the cadmium-induced changes in N-cadherin localization in the rat kidney are not associated with significant changes in the level of N-cadherin gene expression [110]. Another study using gene microarray technology revealed no significant changes in cadherin expression following cadmium exposure in rats [147]. This suggests that alterations in gene expression may be a minor mechanism in acute metal-induced alterations in function and localization of homotypic cell adhesion molecules. However, it is still possible that regulation of adhesion molecule expression may be important in the metal-induced carcinogenesis and this is currently an area of active investigation.

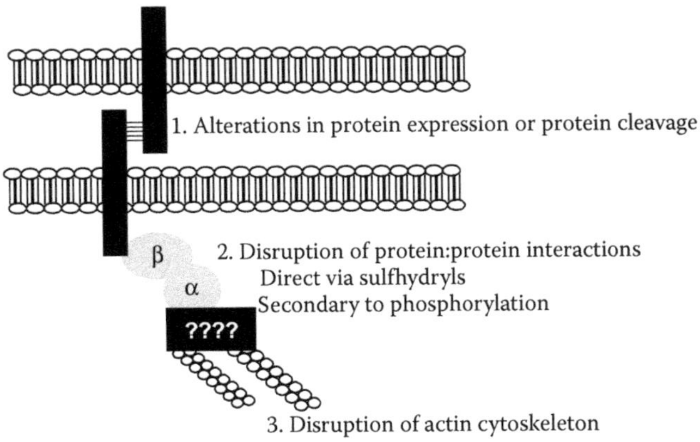

FIGURE 12.2 Potential mechanisms of metal-induced disruption of cell adhesion. Using a cadherin/catenin complex as a prototype, metals can disrupt adhesion complexes at several levels. As seen following cadmium (Data from Zimmerhackl, L.B., et al., *Am. J. Physiol.*, 275, F143, 1998) and mercury (Data from Jiang, J., et al., *Toxicol. Sci.*, 80, 170, 2004) challenge, metal exposure can result in the loss of adhesion proteins (1). In addition, metals could disrupt the critical protein–protein interactions required for cell adhesion (2); this could happen directly or indirectly via metal–induced kinase activation and subsequent protein phosphorylation. As such, both cadmium (Data from Zimmerhackl, L.B., et al., *Am. J. Physiol.*, 275, F143, 1998; Weidner, W.J., Waddell, D.S., and Sillman, A.J., *Arch. Toxicol.*, 74, 578, 2000) and mercury (Data from Jiang, J., et al., *Toxicol. Sci.*, 80, 170, 2004) have been demonstrated to disrupt cadherin:catenin protein interactions. Finally, disruption of the actin cytoskeleton could lead to loss of both cell–cell and cell–matrix adhesion (3); actin is disrupted by cadmium (Data from Lozsekova, A., et al., *Bratisl. Lek. Listy*, 103, 254, 2002; Mills, J.W. and Ferm, V.H., *Toxicol. Appl. Pharmacol.*, 101, 245, 1989; Wang, Z., Chin, T.A., and Templeton, D.M., *Cell Motil. Cytoskel.*, 33, 208, 1996; and Wang, Z. and Templeton, D.M., *Toxicol. Appl. Pharmacol.*, 139, 115, 1996), chromium (Data from Lozsekova, A., et al., *Bratisl. Lek. Listy*, 103, 254, 2002; Lee, K.W., et al., *Ann. N. Y. Acad. Sci.*, 1030, 258, 2004; and Gunaratnam, M. and Grant, M.H., *Toxicol. In Vitro*, 18, 245, 2004), and nickel. (Data from Lozsekova, A., et al., *Bratisl. Lek. Listy*, 103, 254, 2002; Gunaratnam, M. and Grant, M.H., *Toxicol. In Vitro*, 18, 245, 2004). It is important to note that each of these mechanisms may be a primary effect, or secondary to metal-induced oxidative stress.

12.4.2 Protein–Protein Interactions

Cell–matrix and cell–cell adhesion requires complexes consisting of a large number of proteins [148–152]. Therefore, disruption of protein–protein interactions is an attractive hypothesis for the metal-induced disruption of cellular adhesion. One relatively simple possibility is that metals have a higher affinity for sulfhydryl groups and could disrupt critical disulfide bonds that serve as scaffolds for protein–protein interactions and subsequent function [153]. For example, both integrins [153] and cadherins [154] have disulfide residues that must be preserved for function. Metals such as nickel [155] and mercury [156] can disrupt disulfides. While to date we know of no studies that have shown that metals can specifically target cell adhesion molecules, cadmium and mercury can disrupt the cadherin–catenin complex via alterations in cadherin–catenin interactions [107,121,157]; however, the cause-and-effect relationship in this process is not yet clear. Investigation of this mechanism will require careful attention to the speciation of metals, as well as the concentration of the metal that can access the adhesion complex.

12.4.3 Protein Phosphorylation

Phosphorylation is a key mechanism regulating the function of many cell adhesion molecules including integrins [158–160], the cadherin/catenin complex [161–163], and tight junctions [164–166]. Phosphorylation of proteins, at both tyrosine and serine/threonine, modulates the normal protein:protein interactions required for adhesive complexes and metals influence the activity of a number of kinases (Table 12.1).

TABLE 12.1
Metal-Induced Kinase Activities

Metals	Effects
Arsenic	Activate PKC *in vitro* [194]
Cadmium	Activate PKC *in vitro* and *in vivo* [185–189]
	Activate ERK, JNK, and p38 *in vitro* [219–221]
Copper	Activate JNK *in vitro* [226]
Manganese	Activate JNK and p38 *in vitro* [224,225]
Mercury	Activate c-src and Hck *in vitro* [209–211]
	Activate p38 *in vitro* [222,223]
Nickel	Activate JNK *in vitro* [227]

PKC represents a family of serine/threonine kinases that are well established as regulators of cadherin-mediated cell–cell adhesion, cytoskeletal organization, and cell permeability [167–170]. PKC also affects integrin adhesive and signaling activity [171–174] and PKC activation is associated with disruption of tight junctions [169,175]. Activation of PKC is often associated with a loss of cell–cell adhesion and an increase in permeability [176–181], as well as inhibition of GJIC [182–184]. Cadmium can activate PKC in a variety of cell types both *in vitro* [185,186] and *in vivo* [187–189]. A number of studies have also demonstrated that lead can stimulate PKC activity, particularly in neuronal cells [190–193]. Additional studies have shown that activation of PKC by sodium arsenite is associated with a loss of VE-cadherin from vascular endothelial cell–cell junctions [194].

The src family of tyrosine kinases has been well established as regulators of cell adhesion activity. Activation of src is associated with disruption of the cadherin/catenin complex [195–198], but may appear to have a lesser effect on tight junctions [199,200]. Activation of src is also associated with decreased endothelial permeability [201–204] and inhibition of GJIC [205–208]. Mercury has been shown to activate Hck, a src family member, in human mylomonocytic cells; this was postulated to be important in metal-induced immune dysfunction [209]. In addition, mercury can activate c-src via a novel redox-related mechanism [210,211].

The mitogen-activated protein kinase (MAPK) superfamily of serine/threonine kinases is composed of three major sets of kinases: the extracellular-receptor kinases (ERK1 and ERK2); the c-Jun N-terminal kinases/stress-activated protein kinases (JNK/SAPK), which include JNK1, JNK2, and JNK3; and the p38 MAPKs. Aberrant activity of MAPKs is linked to disruption of cadherin- and/or tight junction-regulated epithelial permeability [34,168,179,212–216], as well as inhibition of GJIC [217,218]. Cadmium is associated with activation of ERK [219], JNK [220], and p38 [221], while mercury is linked to activation of p38 [222,223]. Manganese has been demonstrated to activate JNK [224] and p38 [225] and both copper [226] and nickel [227] can activate JNK. Given the ubiquitous nature of the MAPK pathway, two important factors to consider in future studies of metal-induced activation of these kinases are the cell specificity of the response and the duration of activation; both of these variables have a tremendous impact on the overall response to kinase activation [228–234].

12.4.4 Protein Degradation

Studies of the metastatic phenotype of tumor cells have shown proteolytic degradation of adhesion molecules to be a mechanism of disruption of cell adhesion. Cadherins have been identified as substrates for several matrix metalloproteinases, including MMP-3 [235,236], MMP-7 [237,238], and MMP-14 (MT1-MMP) [239]. Disruption of tight junctions, including cleavage of occludin, is also mediated by MMPs [240–242]. While integrin subunits do not appear to be major substrates for MMPs, MMP-7 can cleave the β4 subunit [243]. It is clear, however, that integrins can regulate the expression and activity of these proteinases [244–246]. In the context of metals, nickel is associated with induction of MMP-2 expression and activity [247], cadmium inhalation *in vivo* increased MMP-2 and -9 activity in bronchoalveolar lavage fluid [248], and lead influenced MMP-9 expression in glial cells [249].

Caspases also are capable of cleaving cadherins and catenins [250–252]. As with the MMPs, relatively little is known about integrins as caspase substrates. However, integrins can regulate caspase localization and activity [253–255]. Several metals, including cadmium [256,257], lead [258], mercury [259], and chromium [260,261], have been shown to activate capases. However, in investigating metal-induced capsase activation, and subsequent cadherin or catenin cleavage, care must be taken to determine if alterations in cell adhesion are simply a manifestation of apoptotic cell death.

12.4.5 Disruption of the Actin Cytoskeleton

A fundamental characteristic of cell adhesion complexes is the intricate interaction between the complex and the cytoskeleton. In most cases this interaction involves predominantly the actin cytoskeleton. While the mechanism underlying this interaction may not be as simple as the linear protein–protein interactions originally hypothesized, it is undeniable that an intact actin cytoskeleton is essential for the normal adhesive function of integrins [14,262–264], tight junctions [39,40,265], and the cadherin/catenin complex [43,266,267]. A number of studies have demonstrated that metals can disrupt the actin cytoskeleton. Cadmium is associated with disruption of actin in MDCK cells [268], mesangial and vascular smooth muscle cells [269], skeletal muscle [270], and keratinocytes [118]. Hexavalent chromium has been shown to reorganize the actin cytoskeleton in fibroblasts [271], hepatocytes [272], and keratinocytes [118], while nickel affects the cytoskeleton in hepatocytes [272] and keratinocytes [118]. One of the challenges in establishing disruption of the actin cytoskeleton as a mechanism of underlying alterations in cell adhesion complexes is the fact that the relationship is bidirectional; disruption of both integrins and cadherins leads to actin reorganization. Therefore, a cause-and-effect relationship may become a chicken-or-the egg proposition.

12.4.6 Oxidative Stress

Oxidative stress is a significant and well-investigated mechanism involved in metal toxicity. As of the writing of this chapter, over 80 review articles alone on metals and oxidative stress are cited on PubMed. Importantly, each of the metals addressed in this book has the potential to interfere with normal redox status and induced oxidative stress. This is true for lead [273–275], manganese [276–278], copper [226,279,280], cadmium [281–283], zinc [284,285], iron [286–288], and mercury [289–291]. A number of studies have associated oxidative stress with disruption of integrins [292–294], tight junctions [152,294,295], and cadherin/catenin complexes [152,296,297]. It is important to note that oxidative stress may also be a component of each of the previous postulated mechanisms, oxidative stress can affect gene expression [298–300], protein–protein interactions [301], protein phosphorylation via ERKs, MAPKs, PKC, and src [302–307], MMP [308,309] and caspase activity [310–312], and actin cytoskeleton organization [306,313,314]. The appeal of this mechanism is that the literature clearly supports the hypothesis that metals can induce oxidative stress, which in turn can disrupt cell adhesion directly, or indirectly via the previously described mechanisms. However, it will require extensive studies to determine if this mechanism can account for the organ-, cellular- and complex-specificity of the disruption of cell adhesion by metals.

12.5 CONCLUSIONS

A great deal of evidence indicates that cell adhesion molecules and their associated proteins and signaling pathways may be primary targets in many types of metal-induced injury (Table 12.2). However, several important questions remain in this emerging field of research. One of the greatest gaps in our knowledge concerns the relevance of the many reported effects of metals on cell adhesion molecules *in vitro* to the toxic actions of metals *in vivo*. Resolving this issue will require additional studies utilizing *in vivo* models as well as *in vitro* models that more closely mimic the patterns of cell

TABLE 12.2
Metal-Induced Disruption of Cell Adhesion Complexes

Metals	Effects
Bismuth	*In vivo* disruption of N-cadherin localization [120]
Cadmium	*In vitro* disruption of E-cadherin [105–107,114]
	In vitro and *in vivo* decrease in GJIC [123,125,126]
	In vitro and *in vivo* induction of ICAM-1 [133–137]
	In vivo disruption of E-, N-, and VE-cadherin [108,109,115–117]
	In vivo induction of PECAM [136]
Chromium	*In vitro* disruption of E-cadherin [118]
Copper	Decreased integrin expression *in vitro* and *in vivo* [88–90]
	Increased epithelial permeability *in vitro* [119]
Manganese	Faciliate integrin function (μM) [24,25]
	Inhibit integrin function (mM) [95]
Mercury	*In vitro* disruption of GJIC [122,124]
	In vitro induction of ICAM-1 [133]
	In vivo disruption of N-cadherin [121]

adhesion molecule expression and the doses and durations of metal exposure *in vivo*. For example, many of the *in vitro* studies with toxic metals such as cadmium and mercury involved exposure of cells to micromolar concentrations of the free metal ions in physiological saline solutions. *In vivo*, target cells would usually be exposed to much lower concentrations of the metals in the physiological milieu of serum ultrafiltrates. Under these conditions, the metals would not exist in their free forms, but would exist in the form of complexes with molecules such as cysteine, glutathione, and even low molecular weight proteins [315,316]. It is also important to note that the specific patterns of cell adhesion molecule expression in many cell lines do not exactly mimic the patterns of expression in the corresponding tissue *in vivo*. For example, most of the proximal tubule-derived cell lines used for toxicity studies mainly express E-cadherin [317], whereas N-cadherin is the primary classical cadherin that is expressed in the proximal tubule *in vivo* [108,318–320]. Moreover, most of the proximal tubule-derived cell lines do not express the tight junction protein claudin-2 [317], which is the primary claudin that is present in the proximal tubule *in vivo* [321]. Finally, changes in cell adhesion molecule function occur in response to nonspecific cellular injury and inflammation. This makes it very difficult to determine if observed metal-induced changes in cell adhesion molecule function represent primary toxic effects or occur secondarily to the general metabolic derangement or even death of the injured cells. In light of this finding, it is clear that much work remains to be done in order to clarify the role of cell adhesion molecules in metal-induced injury.

REFERENCES

1. Juliano, R.L., Signal transduction by cell adhesion receptors and the cytoskeleton: Functions of integrins, cadherins, selectins, and immunoglobulin-superfamily members. *Annu. Rev. Pharmacol. Toxicol.*, 42, 283, 2002.
2. Luo, B.H., Carman, C.V., and Springer, T.A., Structural basis of integrin regulation and signaling. *Annu. Rev. Immunol.*, 25, 619, 2007.
3. Ginsberg, M.H., Partridge, A., and Shattil, S.J., Integrin regulation. *Curr. Opin. Cell Biol.*, 17, 509, 2005.
4. Arnaout, M.A., Goodman, S.L., and Xiong, J.P., Structure and mechanics of integrin-based cell adhesion. *Curr. Opin. Cell Biol.*, 19, 495, 2007.
5. Lad Y, Harburger D.S., and Calderwood D.A., Integrin cytoskeletal interactions. *Methods Enzymol.*, 426, 69–84, 2007.
6. Chen, C. and Sheppard, D., Identification and molecular characterization of multiple phenotypes in integrin knockout mice. *Methods Enzymol.*, 426, 291, 2007.

7. Zhang, F., et al., Distinct ligand binding sites in integrin alpha3beta1 regulate matrix adhesion and cell–cell contact. *J. Cell Biol.*, 163, 177, 2003.
8. Triantafilou, K., Takada, Y., and Triantafilou, M., Mechanisms of integrin-mediated virus attachment and internalization process. *Crit. Rev. Immunol.*, 21, 311, 2001.
9. Oldak, M., Smola-Hess, S., and Maksym, R., Integrin beta4, keratinocytes and papillomavirus infection. *Int. J. Mol. Med.*, 17, 195, 2006.
10. Novak, A., et al., Cell adhesion and the integrin-linked kinase regulate the LEF-1 and beta-catenin signaling pathways. *Proc. Natl. Acad. Sci. USA*, 95, 4374, 1998.
11. Troussard, A.A., et al., Cell-extracellular matrix interactions stimulate the AP-1 transcription factor in an integrin-linked kinase- and glycogen synthase kinase 3-dependent manner. *Mol. Cell Biol.*, 19, 7420, 1999.
12. Weaver, V.M., et al., Beta4 integrin-dependent formation of polarized three-dimensional architecture confers resistance to apoptosis in normal and malignant mammary epithelium. *Cancer Cell*, 2, 205, 2002.
13. Zhang, L., Bewick, M., and Lafrenie, R.M., Role of Raf-1 and FAK in cell density-dependent regulation of integrin-dependent activation of MAP kinase. *Carcinogenesis*, 23, 1251, 2002.
14. DeMali, K.A., Wennerberg, K., and Burridge, K., Integrin signaling to the actin cytoskeleton. *Curr. Opin. Cell Biol.*, 15, 572, 2003.
15. Homsy, J.G., et al., JNK signaling coordinates integrin and actin functions during Drosophila embryogenesis. *Dev. Dyn.*, 235, 427, 2006.
16. Hanks, S.K., et al., Focal adhesion protein tyrosine kinase phosphorylated in response to cell spreading on fibronectin. *Proc. Natl. Acad. Sci. USA*, 89, 8487, 1992.
17. Schaller, M.D., et al., pp125FAK, a structurally distinctive protein tyrosine-kinase associated with focal adhesions. *Proc. Natl. Acad. Sci. USA*, 89, 192, 1992.
18. Hannigan, G.E., et al., Regulation of cell adhesion and anchorage-dependent growth by a new beta 1-integrin linked protein kinase. *Nature*, 379, 91, 1996.
19. Wu, C., ILK interactions. *J. Cell Sci.*, 114, 2549, 2001.
20. Wu, C. and Dedhar, S., Integrin-linked kinase (ILK) and its interactors: A new paradigm for the coupling of extracellular matrix to actin cytoskeleton and signaling complexes. *J. Cell Biol.*, 155, 505, 2001.
21. Boulter, E. and Van Obberghen-Schilling, E., Integrin-linked kinase and its partners: A modular platform regulating cell-matrix adhesion dynamics and cytoskeletal organization. *Eur. J. Cell Biol.*, 85, 255, 2006.
22. Ajroud, K., et al., Binding affinity of metal ions to the CD11b A-domain is regulated by integrin activation and ligands. *J. Biol. Chem.*, 279, 25483, 2004.
23. Foradori A.C., et al., The discrimination between magnesium and manganese by serum proteins. *J. Gen. Physiol.*, 50, 2255, 1967.
24. Rabinovitch, M. and DeStefano, M.J., Manganese stimulates adhesion and spreading of mouse sarcoma I ascites cells. *J. Cell Biol.*, 59, 165, 1973.
25. Grinnell, F., Manganese-dependent cell-substratum adhesion. *J. Cell Sci.*, 65, 61, 1984.
26. Roth, J.A., et al., Mechanisms of manganese-induced rat pheochromocytoma (PC12) cell death and cell differentiation. *Neurotoxicology*, 23, 147, 2002.
27. Lein, P., et al., Manganese induces neurite outgrowth in PC12 cells via upregulation of alpha(v) integrins. *Brain. Res.*, 885, 220, 2000.
28. Walowitz, J.L. and Roth, J.A., Activation of ERK1 and ERK2 is required for manganese-induced neurite outgrowth in rat pheochromocytoma (PC12) cells. *J. Neurosci. Res.*, 57, 847, 1999.
29. Lin, W.H., et al., Manganese induces spreading and process outgrowth in rat pheochromocytoma (PC12) cells. *J. Neurosci. Res.*, 34, 546, 1993.
30. Lapierre, L.A., The molecular structure of the tight junction. *Adv. Drug Deliv. Rev.*, 41, 255, 2000.
31. Lee, D.B., Huang, E., and Ward, H.J., Tight junction biology and kidney dysfunction. *Am. J. Physiol. Renal Physiol.*, 290, F20, 2006.
32. Kohler, K. and Zahraoui, A., Tight junction: A co-ordinator of cell signalling and membrane trafficking. *Biol. Cell*, 97, 659, 2005.
33. Gonzalez-Mariscal, L., et al., Tight junction proteins. *Prog. Biophys. Mol. Biol.*, 81, 1, 2003.
34. Wong, V., Phosphorylation of occludin correlates with occludin localization and function at the tight junction. *Am. J. Physiol.*, 273, C1859, 1997.
35. Koval, M., Claudins—key pieces in the tight junction puzzle. *Cell Commun. Adhes.*, 13, 127, 2006.
36. Martin-Padura, I., et al., Junctional adhesion molecule, a novel member of the immunoglobulin superfamily that distributes at intercellular junctions and modulates monocyte transmigration. *J. Cell Biol.*, 142, 117, 1998.
37. Coyne, C.B. and Bergelson, J.M., CAR: A virus receptor within the tight junction. *Adv. Drug Deliv. Rev.*, 57, 869, 2005.

38. Fanning, A.S., et al., Protein interactions in the tight junction: The role of MAGUK proteins in regulating tight junction organization and function. In: W.J Nelson (Ed.), *Current Topics in Membranes*, Vol. 43, p. 211, San Diego, CA: Academic Press, 1996.
39. Fanning, A.S., et al., The tight junction protein ZO-1 establishes a link between the transmembrane protein occludin and the actin cytoskeleton. *J. Biol. Chem.*, 273, 29745, 1998.
40. Fanning, A.S., Ma, T.Y., and Anderson, J.M., Isolation and functional characterization of the actin binding region in the tight junction protein ZO-1. *FASEB J.*, 16, 1835, 2002.
41. Kavanagh, E., et al., Functional interaction between the ZO-1-interacting transcription factor ZONAB/DbpA and the RNA processing factor symplekin. *J. Cell Sci.*, 119, 5098, 2006.
42. Wu, Q. and Maniatis, T., A striking organization of a large family of human neural cadherin-like cell adhesion genes. *Cell*, 97, 779, 1999.
43. Goodwin, M. and Yap, A.S., Classical cadherin adhesion molecules: Coordinating cell adhesion, signaling and the cytoskeleton. *J. Mol. Histol.*, 35, 839, 2004.
44. Yap, A.S., Crampton, M.S., and Hardin, J., Making and breaking contacts: The cellular biology of cadherin regulation. *Curr. Opin. Cell Biol.*, 19, 508, 2007.
45. Kemler, R., Classical cadherins. *Semin. Cell Biol.*, 3, 149, 1992.
46. Reynolds, A.B., p120-catenin: Past and present. *Biochim. Biophys. Acta*, 1773, 2, 2007.
47. Ireton, R.C., et al., A novel role for p120 catenin in E-cadherin function. *J. Cell Biol.*, 159, 465, 2002.
48. Xiao, K., et al., Cellular levels of p120 catenin function as a set point for cadherin expression levels in microvascular endothelial cells. *J. Cell Biol.*, 163, 535, 2003.
49. Tsukita, S., et al., Molecular linkage between cadherins and actin filaments in cell-cell adherens junction. *Curr. Opin. Cell Biol.*, 4, 834, 1992.
50. Knudsen, K.A., et al., Interaction of α-actinin with the cadherin/catenin cell–cell adhesion complex via α-catenin. *J. Cell Biol.*, 130, 67, 1995.
51. Rimm, D.L., et al., α1(E)-catenin is an actin-binding and bundling protein mediating the attachment of F-actin to the membrane adhesion complex. *Proc. Natl. Acad. Sci. USA*, 92, 8813, 1995.
52. Watabe-Uchida, M., et al., α-catenin-vinculin interaction functions to organize the apical junctional complex in epithelial cells. *J. Cell Biol.*, 142, 847, 1998.
53. Imamura, Y., et al., Functional domains of α-catenin required for the strong state of cadherin-based cell adhesion. *J. Cell Biol.*, 144, 1311, 1999.
54. Drees, F., et al., Alpha-catenin is a molecular switch that binds E-cadherin-beta-catenin and regulates actin-filament assembly. *Cell*, 123, 903, 2005.
55. Yamada, S., et al., Deconstructing the cadherin-catenin-actin complex. *Cell*, 123, 889, 2005.
56. Gates, J. and Peifer, M., Can 1000 reviews be wrong? Actin, alpha-catenin, and adherens junctions. *Cell*, 123, 769, 2005.
57. Marrs, J.A., et al., Distinguishing roles of the membrane-cytoskeleton and cadherin mediated cell-cell adhesion in generating different Na^+, K^+-ATPase distributions in polarized epithelia. *J. Cell Biol.*, 123, 149, 1993.
58. Mays, R.W., Nelson, W.J., and Marrs, J.A., Generation of epithelial cell polarity: Roles for protein trafficking, membrane-cytoskeleton, and E-cadherin-mediated cell adhesion. *Cold Spring Harb. Symp. Quant. Biol.*, 60, 763, 1995.
59. Vasioukihn, V., et al., Directed actin polymerization is the driving force for epithelial cell–cell adhesion. *Cell*, 100, 209, 2000.
60. Cadigan, K.M. and Nusse, R., Wnt signaling: A common theme in animal development. *Genes Dev.*, 11, 3286, 1997.
61. Beavon, I.R., The E-cadherin–catenin complex in tumour metastasis: Structure, function and regulation. *Eur. J. Cancer*, 36, 1607, 2000.
62. Ben Ze'ev, A., Shtutman, M., and Zhurinsky, J., The integration of cell adhesion with gene expression: The role of beta-catenin. *Exp. Cell Res.*, 261, 75, 2000.
63. Nelson, W.J. and Nusse, R., Convergence of wnt, beta-catenin, and cadherin pathways. *Science*, 303, 1483, 2004.
64. Grimes, C.A. and Jope, R.S., The multifaceted roles of glycogen synthase kinase 3beta in cellular signaling. *Prog. Neurobiol.*, 65, 391, 2001.
65. Daugherty, R.L. and Gottardi, C.J., Phospho-regulation of beta-catenin adhesion and signaling functions. *Physiology*, 22, 303, 2007.
66. Willert, K. and Nusse, R., Beta-catenin: A key mediator of wnt signaling. *Curr. Opin. Genet. Dev.*, 8, 95–102, 2998.
67. Polakis, P., Casein kinase 1: A wnt'er of disconnect. *Curr. Biol.*, 12, R499, 2002.

68. Gottardi, C.J. and Gumbiner, B.M., Distinct molecular forms of beta-catenin are targeted to adhesive or transcriptional complexes. *J. Cell Biol.*, 167, 339, 2004.
69. Piepenhagen, P.A. and Nelson, W.J., Biogenesis of polarized epithelial cells during kidney development *in situ*: Roles of E-cadherin-mediated cell-cell adhesion and membrane cytoskeleton organization. *Mol. Biol. Cell*, 9, 3161, 1998.
70. Oyamada, M., Oyamada, Y., and Takamatsu, T., Regulation of connexin expression. *Biochim. Biophys. Acta*, 1719, 6, 2005.
71. Yeager, M. and Harris A.L., Gap junction channel structure in the early 21st century: Facts and fantasies. *Curr. Opin. Cell Biol.*, 19, 521, 2007.
72. Evans, W.H., De Vuyst, E., and Leybaert, L., The gap junction cellular internet: connexin hemichannels enter the signaling limelight. *Biochem. J.*, 397, 1, 2006.
73. Phelan, P., Innexins: Members of an evolutionarily conserved family of gap-junction proteins. *Biochim. Biophys. Acta*, 1711, 225, 2005.
74. Shestapalov, V.I. and Panchin, Y., Pannexins and gap junction protein diversity. *Cell. Mol. Life Sci.*, 65, 376, 2008.
75. Bukauskas, F.F. and Verselis, V.K., Gap junction channel gating. *Biochim. Biophys. Acta*, 1662, 42, 2004.
76. Saez, J.C. et al., Connexin-based gap junction hemichannels: Gating mechanisms. *Biochim. Biophys. Acta*, 1711, 215, 2005.
77. Moreno, A.P., Connexin phosphorylation as a regulatory event linked to channel gating. *Biochim. Biophys. Acta*, 1711, 164, 2005.
78. Moreno, A.P. and Lau, A.F., Gap junction channel gating modulated through protein phosphorylation. *Prog. Biophys. Mol. Biol.*, 94, 107, 2007.
79. Golias, C., et al., Leukocyte and endothelial cell adhesion molecules in inflammation focusing on inflammatory heart disease. *In Vivo*, 21, 757, 2007.
80. Weber, C., Fraemohs, L., and Dejana, E., The role of junctional adhesion molecules in vascular inflammation. *Nat. Rev. Immunol.*, 7, 467, 2007.
81. Hubbard, A.K. and Rothlein, R., Intercellular adhesion molecule-1 (ICAM-1) expression and cell signaling cascades. *Free Radic. Biol. Med.*, 28, 1379, 2000.
82. Meager, A., Cytokine regulation of cellular adhesion molecule expression in inflammation. *Cytokine Growth Factor Rev.*, 10, 27, 1999.
83. Zhang, W.J. and Frei, B., Intracellular metal ion chelators inhibit TNFalpha-induced SP-1 activation and adhesion molecule expression in human aortic endothelial cells. *Free Radic. Biol. Med.*, 34, 674, 2003.
84. Molina, A., et al., Prevention of mercuric chloride-induced nephritis in the brown Norway rat by treatment with antibodies against the alpha 4 integrin. *J. Immunol.*, 153, 2313, 1994.
85. Escudero, E., et al., Functional relevance of activated beta1 integrins in mercury-induced nephritis. *J. Am. Soc. Nephrol.*, 11, 1075, 2000.
86. Nieto, E., et al., Effects of mycophenolate mofetil in mercury-induced autoimmune nephritis. *J. Am. Soc. Nephrol.*, 13, 937, 2002.
87. Bernard, A., et al., Anti-laminin antibodies in Sprague-Dawley and brown Norway rats chronically exposed to cadmium. *Toxicology*, 31, 307, 1984.
88. Savaris, R., Zettler, C.G., and Ferrari, A.N., Expression of alpha4beta1 and alphavbeta3 integrins in the endometrium of women using the T200 copper intrauterine device. *Fertil. Steril.*, 74, 1102, 2000.
89. Oruc, S., et al., Changes in distribution patterns of integrins in endometrium in copper T380 intrauterine device users. *Acta Histochem.*, 107, 95, 2005.
90. Savaris, R. and Chies, J.A., Copper ions dynamically regulate beta3 integrin subunit expression in Ishikawa cells. *Contraception*, 67, 247, 2003.
91. Klein, C.L., Kohler, H., and Kirkpatrick, C.J., Increased adhesion and activation of polymorphonuclear neutrophil granulocytes to endothelial cells under heavy metal exposure *in vitro*. *Pathobiology*, 62, 90, 1994.
92. Tenaud, I., et al., *In vitro* modulation of keratinocyte wound healing integrins by zinc, copper and manganese. *Br. J. Dermatol.*, 140, 26, 1999.
93. Tenaud, I., et al., Zinc, copper and manganese enhanced keratinocyte migration through a functional modulation of keratinocyte integrins. *Exp. Dermatol.*, 9, 407, 2000.
94. Giuliani, R., et al., Focal adhesion molecules as potential target of lead toxicity in NRK-52E cell line. *FEBS Lett.*, 579, 6251, 2005.
95. Luthen, F., et al., Influence of manganese ions on cellular behavior of human osteoblasts *in vitro*. *Biomol. Eng.*, 24, 531, 2007.

96. Yancy, S.L., et al., Sodium arsenite exposure alters cell migration, focal adhesion localization and decreases tyrosine phosphorylation of focal adhesion kinase in H9C2 myoblasts. *Toxicol. Sci.*, 84, 278, 2005.
97. Lee, C.H., et al., Defective beta1-integrins expression in arsenical keratosis and arsenic-treated cultured human keratinocytes. *J. Cutan. Pathol.*, 33, 129, 2006.
98. Conrad, M.E. and Umbreit, J.N., A concise review: Iron absorption—the mucin-mobilferrin-integrin pathway. A competitive pathway for metal absorption. *Am. J. Hematol.*, 42, 67, 1993.
99. Conrad, M.E., et al., Alternate iron transport pathway: Mobilferrin and integrinin K562 cells. *J. Biol. Chem.*, 269, 7169, 1994.
100. Conrad, M.E., et al., Separate pathways for cellular uptake of ferric and ferrous iron. *Am. J. Physiol. Gastrointest. Liver Physiol.*, 279, G767, 2000.
101. Conrad, M.E. and Umbreit, J.N., Pathways of iron absorption. *Blood Cells Mol. Dis.*, 29, 336, 2002.
102. Parrish, A.R., Adhesion molecules in renal physiology and pathology. In: J.B. Tarloff and L.H. Lash (Eds), *Toxicology of the Kidney*. pp. 475–498. Boca Raton, FL: CRC Press, 2005.
103. Prozialeck, W.C., Evidence that E-cadherin may be a target for cadmium toxicity in epithelial cells. *Toxicol. Appl. Pharmacol.*, 164, 231, 2000.
104. Prozialeck, W.C. and Edwards, J.R., Cell adhesion molecules in chemically induced renal injury. *Pharmacol. Ther.*, 114, 74, 2007.
105. Prozialeck, W.C., Lamar, P.C., and Ikura, M., Binding of cadmium (Cd2+) to E-CAD1, a calcium-binding polypeptide analog of E-cadherin. *Life Sci.*, 58, L325, 1996.
106. Prozialeck, W.C. and Lamar, P.C., Interaction of cadmium (Cd(2+)) with a 13-residue polypeptide analog of a putative calcium-binding motif of E-cadherin. *Biochim. Biophys. Acta*, 1451, 93, 1999.
107. Zimmerhackl, L.B., et al., Cadmium is more toxic to LLC-PK1 cells than to MDCK cells acting on the cadherin-catenin complex. *Am. J. Physiol.*, 275, F143, 1998.
108. Prozialeck, W.C., Lamar, P.C., and Lynch, S.M., Cadmium alters the localization of N-cadherin, E-cadherin, and beta-catenin in the proximal tubule epithelium. *Toxicol. Appl. Pharmacol.*, 189, 180, 2003.
109. Prozialeck, W.C. and Lamar, P.C., Cadmium nephrotoxicity is associated with a loss of N-cadherin-mediated adhesion and alterations in epithelial polarity in the proximal tubule. *Toxicol. Sci.*, 84, S327, 2005.
110. Edwards, J.R., et al., Cadmium-induced disruption of proximal tubule cell adhesion is associated with redistribution of cell adhesion molecules and loss of epithelial polarity. *Toxicol. Sci.*, 90, S414, 2006.
111. Jacquillet, G., et al., Zinc protects renal function during cadmium intoxication in the rat. *Am. J. Physiol. Renal Physiol.*, 290, F127, 2006.
112. Thevenod, F., et al., Cadmium induces nuclear translocation of beta-catenin and increases expression of c-myc and Abcb1a in kidney proximal tubule cells. *Biometals*, 20, 807, 2007.
113. Woods, J.M., et al., Direct antangiogenic actions of cadmium on human vascular endothelial cells. *Toxicol. In Vitro*, epub ahead of print.
114. Duizer, E., et al., Effects of cadmium chloride on the paracellula barrier function of intestinal epithelial cell lines. *Toxicol. Appl. Pharmacol.*, 155, 117, 1999.
115. Pearson, C.A., Lamar, P.C., and Prozialeck, W.C., Effects of cadmium on E-cadherin and VE-cadherin in mouse lung. *Life Sci.*, 72, 1303, 2003.
116. Chen, B. and Hales, B.F., Cadmium-induced rat embryotoxicity *in vitro* is associated with an increased abundance of E-cadherin protein in the yolk sac. *Toxicol. Appl. Pharmacol.*, 128, 293, 1994.
117. Thompson, J., et al., Adherens junction breakdown in the periderm following cadmium administration in the chick embryo: Distribution of cadherins and associated molecules. *Reprod. Toxicol.*, 25, 39, 2008.
118. Lozsekova, A., et al., The effect of nickel sulphate, potassium dichromate, cobalt nitrate and cadmium sulphate on the proteins of cellular contacts and actin skeleton of cultivated human keratinocytes. *Bratisl. Lek. Listy*, 103, 254, 2002.
119. Ferruza, S., et al., Copper treatment alters the permeability of tight junctions in cultured human intestinal Caco-2 cells. *Am. J. Physiol.*, 277, G1138, 1999.
120. Leussink, B.T., et al., Loss of homotypic epithelial cell adhesion by selective N-cadherin displacement in bismuth nephrotoxicity. *Toxicol. Appl. Pharmacol.*, 175, 54, 2001.
121. Jiang, J., et al., Disruption of cadherin/catenin expression, localization, and interactions during HgCl2-induced nephrotoxicity. *Toxicol. Sci.*, 80, 170, 2004.
122. Yoshida, M., et al., Methylmercury inhibits gap junctional intercellular communication in primary cultures of rat proximal tubular cells. *Arch. Toxicol.*, 72, 192, 1998.
123. Fukumoto, M., et al., Effect of cadmium on gap junctional intercellular communication in primary cultures of rat renal proximal tubular cells. *Life Sci.*, 69, 247, 2001.

124. Aleo, M.F., et al., Antioxidant potential and gap junction-mediated intercellular communication as early biological markers of mercuric chloride toxicity in the MDCK cell line. *Toxicol. In Vitro*, 16, 457, 2002.
125. Jeon, S.H., Cho, M.H., and Cho, J.H., Effects of cadmium on gap junctional intercellular communication in WB-F344 rat liver epithelial cells. *Hum. Exp. Toxicol.*, 20, 577, 2001.
126. Jeong, S.H., Habeebu, S.S., and Klaassen C.D., Cadmium decreases gap junctional intercellular communication in mouse liver. *Toxicol. Sci.*, 57, 156, 2000.
127. Schirrmacher, K., et al., Effects of lead, mercury, and methylmercury on gap junctions and [Ca^{2+}] in bone cells. *Calcif. Tissue Int.*, 63, 134, 1998.
128. Legare, M.E., et al., Low-level lead exposure in cultured astroglia: Identification of cellular targets with vital fluorescent probes. *Neurotoxicology*, 14, 267, 1993.
129. Apostoli, P., et al., Effects of four inorganic lead compounds of the proliferation and junctional coupling of cultured REL liver cells. *Am. J. Ind. Med.*, 38, 340, 2000.
130. Zefferino, R., et al., The study of gap junctional intercellular communication in keratinocytes as screening of promoter effect induced by industrial and environmental toxic substances. *Med. Lav.*, 96, 222, 2005.
131. Loch-Caruso, R., Corcos, I.A., and Trosko, J.E., Inhibition of metabolic coupling by metals. *J. Toxicol. Environ. Health*, 32, 33, 1991.
132. Mikalsen, S.O., Effects of heavy metal ions on intercellular communication in Syrian hamster embryo cells. *Carcinogenesis*, 11, 1621, 1990.
133. Jiang, J., McCool, B.A., and Parrish, A.R., Cadmium- and mercury-induced intercellular adhesion molecule-1 expression in immortalized proximal tubule cells: Evidence for a role of decreased transforming growth factor-beta1. *Toxicol. Appl. Pharmacol.*, 179, 13, 2002.
134. Jeong, E.M., et al., Cadmium stimulates the expression of ICAM-1 via NF-kappaB activation in cerebrovascular endothelial cells. *Biochem. Biophys. Res. Commun.*, 320, 887, 2004.
135. Seok, S.M., et al., COX-2 is associated with cadmium-induced ICAM-1 expression in cerebrovascular endothelial cells. *Toxicol. Lett.*, 165, 212, 2006.
136. Mousa, S.A., Expression of adhesion molecules during cadmium hepatotoxicity. *Life Sci.*, 75, 93, 2004.
137. Gueniche, A., et al., Effect of various metals on intercellular adhesion molecule-1 expression and tumour necrosis factor alpha production by normal human keratinocytes. *Arch. Dermatol. Res.*, 286, 466, 1994.
138. Martinotti, S., et al., Heavy-metal modulation of the human intracellular adhesion molecule (ICAM-1) gene expression. *Biochim. Biophys. Acta*, 1261, 107, 1995.
139. Wagner, M., et al., Heavy metal ion induction of adhesion molecules and cytokines in human endothelial cells: The role of NF-kappaB, I kappaB-alpha and AP-1. *Pathobiology*, 65, 241, 1997.
140. Cousins, R.J., Metal elements and gene expression. *Annu. Rev. Nutr.*, 14, 449, 1994.
141. Bouton, C.M. and Pevsner, J., Effects of lead on gene expression. *Neurotoxicology*, 21, 1045, 2000.
142. Bouton, C.M., et al., Microarray analysis of differential gene expression in lead-exposed astrocytes. *Toxicol. Appl. Pharmacol.*, 176, 34, 2001.
143. Cheng, R.Y., et al., Microarray analysis of altered gene expression in the TM4 Sertoli-like cell line exposed to chromium(III) chloride. *Reprod. Toxicol.*, 16, 223, 2002.
144. Koizumi, S. and Yamada, H., DNA microarray analysis of altered gene expression in cadmium-exposed human cells. *J. Occup. Health*, 45, 331, 2003.
145. Kowara, R., et al., Microarray analysis of altered gene expression in murine fibroblasts transformed by nicke(II) to nickel(II)-resistant malignant phenotype. *Toxicol. Appl. Pharmacol.*, 205, 1, 2005.
146. Ayensu, W.K. and Tchounwou, P.B., Microarray analysis of mercury-induced changes in gene expression in human liver carcinoma (HepG2) cells: Importance in immune responses. *Int. J. Environ. Res. Public Health*, 3, 141, 2006.
147. Bartosiewicz, M.J., et al., Unique gene expression patterns in liver and kidney associated with exposure to chemical toxicants. *J. Pharmacol. Exp. Ther.*, 297, 895, 2001.
148. Howarth, A.G., Singer, K.L., and Stevenson, B.R., Analysis of the distribution and phosphorylation state of ZO-1 in MDCK and nonepithelial cells. *J. Membr. Biol.*, 137, 261, 1994.
149. Howarth, A.G. and Stevenson, B.R., Molecular environment of ZO-1 in epithelial and non-epithelial cells. *Cell Motil. Cytoskeleton*, 31, 323, 1995.
150. Andreeva, A.Y., et al., Protein kinase C regulates the phosphorylation and cellular localization of occludin. *J. Biol. Chem.*, 276, 38480, 2001.
151. Huber, A.H., et al., The cadherin cytoplasmic domain is unstructured in the absence of beta-catenin. A possible mechanism for regulating cadherin turnover. *J. Biol. Chem.*, 276, 12301, 2001.
152. Rao, R.K., et al., Tyrosine phosphorylation and dissociation of occludin-ZO-1 and E-cadherin-beta catenin complexes from the cytoskeleton by oxidative stress. *Biochem. J.*, 368, 471, 2002.

153. Jordan, P.A. and Gibbins, J.M., Extracellular disulfide exchange and the regulation of cellular function. *Antioxid. Redox. Signal.*, 8, 312, 2006.
154. Troyanovsky, R.B., Sokolov, E., and Troyanovsky, S.M., Adhesive and lateral E-cadherin dimers are mediated by the same interface. *Mol. Cell. Biol.*, 23, 7965, 2003.
155. Taniguchi, N., Alkyl- or arylthiolation of aryl iodide cleavage of the S-S bond of disulfide compound by nickel catalyst and zinc. *J. Org. Chem.*, 69, 6904, 2004.
156. Martson, A.W. and Wright, H.T., A method for covalent insertion of mercury into the cysteine disulfide bridges of proteins. *J. Biochem. Biophys. Meth.*, 9, 307, 1984.
157. Weidner, W.J., Waddell, D.S., and Sillman, A.J., Low levels of cadmium chloride alter the immunoprecipitation of corneal cadherin-complex proteins. *Arch. Toxicol.*, 74, 578, 2000.
158. Schlaepfer, D.D. and Hunter, T., Integrin signaling and tyrosine phosphorylation: just the FAKs? *Trends Cell Biol.*, 8, 151, 1998.
159. Vuori, K., Integrin signaling: Tyrosine phosphorylation events in focal adhesions. *J. Membr. Biol.*, 165, 191, 1998.
160. Fagerholm, S.C., Hilden, T.J., and Gahmberg, C.G., P marks the spot: Site-specific integrin phosphorylation regulates molecular interactions. *Trends Biochem. Sci.*, 29, 504, 2004.
161. Daniel, J.M. and Reynolds, A.B., Tyrosine phosphorylation and cadherin/catenin function. *Bioessays*, 19, 883, 1997.
162. Lilien, J. and Balsamo, J., The regulation of cadherin-mediated adhesion by tyrosine phosphorylation/dephosphorylation of beta-catenin. *Curr. Opin. Cell Biol.*, 17, 459, 2005.
163. Alema, S. and Salvatore, A.M., p120 Catenin and phosphorylation: Mechanisms and traits of an unresolved issue. *Biochim. Biophys. Acta*, 1773, 47, 2007.
164. Harhaj, N.S., et al., VEGF activation of protein kinase C stimulates occludin phosphorylation and contributes to endothelial permeability. *Invest. Opthalmol. Vis. Sci.*, 47, 5106, 2006.
165. Kago, T., et al., Cerebral ischemia enhances tyrosine phosphorylation of occuldin in brain capillaries. *Biochem. Biophys. Res. Commun.*, 339, 1197, 2006.
166. D'Souza, T., Indig, F.E., and Morin, P.J., Phosphorylation of claudin-4 by PKCepsilon regulates tight junction barrier function in ovarian cancer cells. *Exp. Cell Res.*, 313, 3364, 2007.
167. Vaaraniemi, J., et al., The effects of PMA and TFP and alterations in intracellular pH and calcium concentration on the membrane associations of phospholipid-binding proteins fodrin, protein kinase C and annexin II in cultured MDCK cells. *Biochim. Biophys. Acta*, 1189, 21, 1994.
168. Cereijido, M., Shoshani, L., and Contreras, R.G., Molecular physiology and pathophysiology of tight junctions. I. Biogenesis of tight junctions and epithelial polarity. *Am. J. Physiol. Gastrointest. Liver Physiol.*, 279, G477, 2000.
169. Clarke, H., Soler, A.P., and Mullin, J.M., Protein kinase C activation leads to dephosphorylation of occludin and tight junction permeability increase in LLC-PK1 epithelial cell sheets. *J. Cell Sci.*, 113, 3187, 2000.
170. Harhaj, N.S. and Antonetti, D.A., Regulation of tight junctions and loss of barrier function in pathophysiology. *Int. J. Biochem. Cell Biol.*, 36, 1206, 2004.
171. Defilippi, P., et al., Dissection of pathways implicated in integrin-mediated actin assembly. Involvement of protein kinase C, Rho GTPase, and tyrosine phosphorylation. *J. Biol. Chem.*, 272, 21726, 1997.
172. Rigot, V., et al., Integrin ligation and PKC activation are required for migration of colon carcinoma cells. *J. Cell Sci.*, 111, 3119, 1998.
173. Disatnik, M.H. and Rando, T.A., Integrin-mediated muscle cell spreading. The role of protein kinase C in outside-in and inside-out signaling and evidence of integrin cross-talk. *J. Biol. Chem.*, 274, 32486, 1999.
174. Ivaska, J., et al., Integin-protein kinase C relationships. *Biochem. Soc. Trans.*, 31, 90, 2003.
175. Mulllin, J.M., et al., Increased tight junction permeability can result from protein kinase C activation/translocation and act as a tumor promotional event in epithelial cancers. *Ann. N. Y. Acad. Sci.*, 915, 231, 2000.
176. Rosson, D., et al., Protein kinase C-alpha activity modulates transepithelial permeability and cell junctions in the LLC-PK1 epithelial cell line. *J. Biol. Chem.*, 272, 14950, 1997.
177. Defouw, L.M. and Defouw, D.O., Protein kinase C activity contributes to endothelial hyperpermeability during early angiogenesis in the chick chorioallantoic membrane. *Tissue Cell*, 33, 135, 2001.
178. Le, T.L., et al., Protein kinase C regulates endocytosis and recycling of E-cadherin. *Am. J. Physiol. Cell Physiol.*, 283, C489, 2002.
179. Yuan, S.Y., Protein kinase signaling in the modulation of microvascular permeability. *Vascul. Pharmacol.*, 39, 213, 2002.
180. Harrington, E.O., et al., Role of protein kinase C isoforms in rat epididymal microvascular endothelial barrier function. *Am. J. Respir. Cell Mol. Biol.*, 28, 626, 2003.

181. Aird, W.C., Endothelium as an organ system. *Crit. Care Med.*, 32, S271, 2004.
182. Enkvist, M.O. and McCarthy, K.D., Activation of protein kinase C blocks astroglial gap junction communication and inhibits the spread of calcium waves. *J. Neurochem.*, 59, 519, 1992.
183. Reynhout, J.K., Lampe, P.D., and Johnson, R.G., An activator of protein kinase C inhibits gap junction communication between cultured bovine lens cells. *Exp. Cell Res.*, 198, 337, 1992.
184. Leibold, E., Greim, H., and Schwarz, L.R., Inhibition of intercellular communication of rat hepatocytes by nafenopin: Involvement of protein kinase C. *Carcinogenesis*, 15, 1265, 1994.
185. Matsuoka, M. and Call, K.M., Cadmium-induced expression of immediate early genes in LLC-PK1 cells. *Kidney Int.*, 48, 383, 1995.
186. Long, G.J., The effect of cadmium on cytosolic free calcium, protein kinase C, and collagen synthesis in rat osteosarcoma (ROS 17/2.8) cells. *Toxicol. Appl. Pharmacol.*, 143, 189, 1997.
187. Beyersmann, D., Block, C., and Malviya, A.N., Effects of cadmium on nuclear protein kinase C. *Environ. Health Perspect.*, 102(S3), 177, 1994.
188. Bagchi, D. et al., Comparative *in vitro* and *in vivo* protein kinase C activation by selected pesticides and transition metal salts. *Toxicol. Lett.*, 91, 31, 1997.
189. Romare, A. and Lundholm, C.E., Cadmium-induced calcium release and prostaglandin E2 production in neonatal mouse calvaria are dependent on cox-2 induction and protein kinase C activation. *Arch. Toxicol.*, 73, 223, 1999.
190. Markovac, J. and Goldstein, G.W., Picomolar concentrations of lead stimulate brain protein kinase C. *Nature*, 334, 71, 1988.
191. Toscano, C.D. and Schanne, F.A., Lead-induced activation of protein kinase C in rat brain cortical synaptosomes. *Ann. N. Y. Acad. Sci.*, 919, 307, 2000.
192. Deng, W. and Poretz, R.D., Protein kinase C activation is required for the lead-induced inhibition of proliferation and differentiation of cultured oligodendroglial progenitor cells. *Brain Res.*, 929, 87, 2002.
193. Cheng, Y.J., Yang, B.C., and Liu, M.Y., Lead increases lipopolysaccharide-induced liver-injury through tumor necrosis factor-alpha overexpression by monocytes/macrophages: Role of protein kinase C and P42/44 mitogen-activated protein kinase. *Environ. Health Perspect.*, 114, 507, 2006.
194. Pereira, F.E., Coffin, J.D., and Beall, H.D., Activation of protein kinase C and disruption of endothelial monolayer integrity by sodium arsenite—potential mechanism in the development of atherosclerosis. *Toxicol. Appl. Pharmacol.*, 220, 164, 2007.
195. Hamaguchi, M., et al., p60v-src causes tyrosine phosphorylation and inactivation of the N-cadherin-catenin cell adhesion system. *EMBO J.*, 12, 307, 1993.
196. Irby, R.B. and Yeatman, T.J., Increased src activity disrupts cadherin/catenin-mediated homotypic adhesion in human colon cancer and transformed rodent cells. *Cancer Res.*, 62, 2669, 2002.
197. Nam, J.S., et al., Src family kinase inhibitor PP2 restores the E-cadherin/catenin cell adhesion system in human cancer cells and reduces cancer metastasis. *Clin. Cancer Res.*, 8, 2430, 2002.
198. Qi, J., et al., Involvement of src family kinases in N-cadherin phosphorylation and beta-catenin dissociation during transendothelial migration of melanoma cells. *Mol. Biol. Cell.*, 17, 1261, 2006.
199. Takeda, H. and Tsukita, S., Effects of tyrosine phosphorylation on tight junctions in temperature-sensitive v-src-transfected MDCK cells. *Cell Struct. Funct.*, 20, 387, 1995.
200. Gómez, S., et al., Independent regulation of adherens and tight junctions by tyrosine phosphorylation in Caco-2 cells. *Biochim. Biophys. Acta*, 1452, 121, 1999.
201. Eliceiri, B.P., et al., Selective requirement for src kinases during VEGF-induced angiogenesis and vascular permeability. *Mol. Cell.*, 4, 915, 1999.
202. Lee, Y.H., et al., Transforming growth factor-beta1 effects on endothelial monolayer permeability involve focal adhesion kinase/src. *Am. J. Respir. Cell Mol. Biol.*, 37, 485, 2007.
203. Gavard, J., Patel, V., and Gutkind, J.S., Angiopoietin-1 prevents VEGF-induced endothelial permeability by sequestering src through mDia. *Dev. Cell*, 14, 25, 2008.
204. Pyun, B.J., et al., Capsiate, a nonpungent capsaicin-like compound, inhibits angiogenesis and vascular permeability via a direct inhibition of src kinase activity. *Cancer Res.*, 68, 227, 2008.
205. Chang, C.C., et al., Potential role of the src gene product in inhibition of gap-junctional communication in NIH/3T3 cells. *Proc. Natl. Acad. Sci. USA*, 82, 5360, 1985.
206. Ito, S., et al., v-Src requires Ras signaling for the suppression of gap junctional intercellular communication. *Oncogene*, 25, 2420, 2006.
207. Lin, R., et al., v-Src tyrosine phosphorylation of connexin43: regulation of gap junction communication and effects on cell transformation. *Cell Commun. Adhes.*, 13, 199, 2006.
208. Shen, Y., et al., Src utilizes Cas to block gap junctional communication mediated by connexin43. *J. Biol. Chem.*, 282, 18914, 2007.

209. Robbins, S.M., Quintrell, N.A., and Bishop, J.M., Mercuric chloride activates the src-family protein tyrosine kinase, Hck in myelomonocytic cells. *Eur. J. Biochem.*, 267, 7201, 2000.
210. Pu, M., et al., Evidence of a novel redox-linked activation mechanism for the src kinase which is independent of tyrosine 527-mediated regulation. *Oncogene*, 13, 2615, 1996.
211. Senga, T., et al., The cysteine-cluster motif of c-src: Its role for the heavy metal-mediated activation of kinase. *Cancer Sci.*, 99, 571, 2008.
212. Bogatcheva, N.V., et al., Mitogen-activated protein kinases in endothelial pathophysiology. *J. Invest. Med.*, 51, 341, 2003.
213. Lui, W.Y., et al., TGF-beta3 regulates the blood-testis barrier dynamics via the p38 mitogen activated protein (MAP) kinase pathway: An *in vivo* study. *Endocrinology*, 144, 1139, 2003.
214. Borbiev, T., et al., p38 MAP kinase-dependent regulation of endothelial cell permeability. *Am. J. Physiol. Lung Cell Mol. Physiol.*, 287, L911, 2004.
215. Birukova, A.A., et al., MAP kinases in lung endothelial permeability induced by microtubule disassembly. *Am. J. Physiol. Lung Cell Mol. Physiol.*, 289, L75, 2005.
216. Basuroy, S., et al., MAPK interacts with occludin and mediates EGF-induced prevention of tight junction disruption by hydrogen peroxide. *Biochem. J.*, 393, 69, 2006.
217. Lee, K.W., et al., p38 is a key signaling molecule for H-ras-induced inhibition of gap junction intercellular communication in rat liver epithelial cells. *Ann. N. Y. Acad. Sci.*, 1030, 258, 2004.
218. Ruch, R.J., Trosko, J.E., and Madhukar, B.V., Inhibition of connexin43 gap junctional intercellular communication by TPA requires ERK activation. *J. Cell. Biochem.*, 83, 163, 2001.
219. Martin, P., et al., Low dose cadmium poisoning results in sustained ERK phosphorylation and caspase activation. *Biochem. Biophys. Res. Commun.*, 350, 803, 2006.
220. Matsuoka, M. and Igisu, H., Activation of c-Jun NH2-terminal kinase (JNK/SAPK) in LLC-PK1 cells by cadmium. *Biochem. Biophys. Res. Commun.*, 251, 527, 1998.
221. Galan, A., et al., Stimulation of p38 mitogen-activated protein kinase is an early regulatory event for the cadmium–induced apoptosis in human promonocytic cells. *J. Biol. Chem.*, 275, 11418, 2000.
222. Kim, S.H. and Sharma, R.P., Mercury-induced apoptosis and necrosis in murine macrophages: Role of calcium-induced reactive oxygen species and p38 mitogen-activated protein kinase signaling. *Toxicol. Appl. Pharmacol.*, 196, 47, 2004.
223. Kim, S.H., Bark, H., and Choi, C.H., Mercury induces multidrug resistance-associated protein gene through p38 mitogen-activated protein kinase. *Toxicol. Lett.*, 155, 143, 2005.
224. Hirata, Y., Adachi, K., and Kiuchi, K., Activation of JNK pathway and induction of apoptosis by manganese in PC12 cells. *J. Neurochem.*, 71, 1607, 1998.
225. Crittenden, P.L. and Filipov, N.M., Manganese-induced potentiation of *in vitro* proinflammatory cytokine production by activated microglial cells is associated with persistent activation of p38 MAPK. *Toxicol. In Vitro*, 22, 18, 2008.
226. Chen, S.H., et al., Apoptosis of cultured astrocytes induced by copper and neocuproine complex through oxidative stress and JNK activation. *Toxicol. Sci.*, 102, 138, 2008.
227. Ke, Q., Nickel compounds induce phosphorylation of histone H3 at serine 10 by activating JNK-MAPK pathway. *Carcinogenesis*, epub ahead of print.
228. Torii, S., et al., Regulatory mechanisms and function of ERK MAP kinases. *J. Biochem.*, 136, 557, 2004.
229. Karin, M. and Gallagher, E., From JNK to pay dirt: Jun kinases, their biochemistry, physiology and clinical importance. *IUBMB Life*, 57, 283, 2005.
230. Liu, J. and Lin, A., Role of JNK activation in apoptosis: A double-edged sword. *Cell Res.*, 15, 36, 2005.
231. Rubinfeld, H. and Seger, R., The ERK cascade: A prototype of MAPK signaling. *Mol. Biotechnol.*, 31, 151, 2005.
232. Zarubin, T. and Han, J., Activation and signaling of the p38 MAP kinase pathway. *Cell Res.*, 15, 11, 2005.
233. Bode, A.M. and Dong, Z., The functional contrariety of JNK. *Mol. Carcinogen.*, 46, 591, 2007.
234. Cuenda, A. and Rousseau, S., p38 MAP-kinases pathway regulation, function and role in human diseases. *Biochim. Biophys. Acta*, 1773, 1358, 2007.
235. Perl, A., et al., MMP-3 cleaves E-cadherin. *Nature*, 392, 190, 1998.
236. Noe, V., et al., Release of an invasion promoter E-cadherin fragment by matrilysin and stromelysin-1. *J. Cell Sci.*, 114, 111, 2001.
237. Ichikawa, Y., et al., Matrilysin (MMP-7) degrades VE-cadherin and accelerates accumulation of beta-catenin in the nucleus of human umbilical vein endothelial cells. *Oncol. Rep.*, 15, 311, 2006.
238. Lee, K.H., et al., Association of extracellular cleavage of E-cadherin mediated by MMP-7 with HGF-induced *in vitro* invasion in human stomach cancer cells. *Eur. Surg. Res.*, 39, 208, 2007.

239. Covington, M.D., Burghardt, R.C., and Parrish, A.R., Ischemia-induced cleavage of cadherins in NRK cells requires MT1-MMP (MMP-14). *Am. J. Physiol. Renal Physiol.*, 290, F43, 2006.
240. Reijerkerk, A., et al., Diapedesis of monocytes is associated with occludin disappearance in brain endothelial cells. *FASEB J.*, 20, 2550, 2006.
241. Gorodeski, G.I., Estrogen decrease in tight junctional resistance involves matrix-metalloproteinase-7-mediated remodeling of occludin. *Endocrinology*, 148, 218, 2007.
242. Yang, Y., et al., Matrix metalloproteinase-mediated disruption of tight junction proteins in cerebral vessels is reversed by synthetic matrix metalloproteinase inhibitor in focal ischemia in rat. *J. Cereber. Blood Flow Metabol.*, 27, 697, 2007.
243. von Bredow, D.C., et al., Cleavage of beta 4 integrin by matrilysin. *Exp. Cell Res.* 236, 341, 1997.
244. Seltzer, J.L., et al., Activation of 72-kDa type IV collagenase/gelatinase by normal fibroblasts in collagen lattices is mediated by integrin receptors but is not related to lattice contraction. *Exp. Cell Res.*, 213, 365, 1994.
245. Brooks, P.C., et al., Localization of matrix metalloproteinase MMP-2 to the surface of invasive cells by interaction with integrin alpha v beta 3. *Cell*, 85, 683, 1996.
246. Morini, M., et al., The alpha 3 beta 1 integrin is associated with mammary carcinoma cell metastasis, invasion, and gelatinase B (MMP-9) activity. *Int. J. Cancer*, 87, 336, 2000.
247. Perfetto, B., et al., Analysis of the signal transduction pathway of nickel-induced matrix metalloproteinase-2 expression in the human keratinocytes *in vitro*: Preliminary findings. *J. Cutan Pathol.*, 34, 441, 2007.
248. Kirschvink, N., et al., Repeated cadmium nebulizations induce pulmonary MMP-2 and MMP-9 production and emphysema in rats. *Toxicology*, 211, 36, 2005.
249. Lahat, N., et al., Inorganic lead enhances cytokine-induced elevation of matrix metalloproteinase MMP-9 expression in glial cells. *J. Neuroimmunol.*, 132, 123, 2002.
250. Brancolini, C., et al., Dismantling cell-cell contacts during apoptosis is coupled to a caspase-dependent proteolytic cleavage of beta-catenin. *J. Cell Biol.*, 139, 759, 1997.
251. Brancolini, C., Sgorbissa, A., and Schneider, C., Proteolytic processing of the adherens junctions components beta-catenin and gamma-catenin/plakoglobin during apoptosis. *Cell Death Differ.*, 5, 1042, 1998.
252. Steinhusen, U., et al., Cleavage and shedding of E-cadherin after induction of apoptosis. *J. Biol. Chem.*, 276, 4972, 2001.
253. Menon, B., et al., Expression of the cytoplasmic domain of beta1 integrin induces apoptosis in adult rat ventricular myocytes (ARVM) via the involvement of caspase-8 and mitochondrial death pathway. *Basic Res. Cardiol.*, 101, 485, 2006.
254. Rajeswari, J. and Pande, G., Direct association between capase 3 and alpha5beta1 integrin and its role during anoikis of rat fibroblasts. *Cell Biol. Int.*, 30, 963, 2006.
255. Hess, F., et al., Integrin-linked kinase interacts with caspase-9 and -8 in an adhesion-dependent manner for promoting radiation-induced apoptosis in human leukemia cells. *Oncogene*, 26, 1372, 2007.
256. Lee, W.K., Torchalski, B., and Thevenod, F., Cadmium-induced ceramide formation triggers calpain-dependent apoptosis in cultured kidney proximal tubule cells. *Am. J. Physiol. Cell Physiol.*, 293, C839, 2007.
257. Mao, W.P., et al., Cadmium induces apoptosis in human embryonic kidney (HEK) 293 cells by capase-dependent and -independent pathways acting on mitochondria. *Toxicol. In Vitro*, 21, 343, 2007.
258. Xu, J., Ji, L.D., and Xu, L.H., Lead-induced apoptosis in PC 12 cells: Involvement of p53, Bcl-2 family and caspase-3. *Toxicol. Lett.*, 166, 160, 2006.
259. Sutton, D.J. and Tchounwou, P.B., Mercury-induced externalization of phosphatidylserine and caspase 3 activation in human liver carcinoma (HepG2) cells. *Int. J. Environ Res. Public Health*, 3, 38, 2006.
260. Catelas, I. et al., Cytotoxic and apoptotic effects of cobalt and chromium ions on J774 macrophages—implication of caspase-3 in the apoptotic pathway. *J. Mater. Sci. Mater. Med.*, 12, 949, 2001.
261. Petit, A., et al., Effect of cobalt and chromium ions on bcl-2, bax, caspase-3, caspase-8 expression in human U937 macrophages. *Biomaterials*, 25, 2013, 2004.
262. Brakebusch, C. and Fassler, R., The integrin-actin connection, an eternal love affair. *EMBO J.*, 22, 2324, 2003.
263. Blystone, S.D., Integrating an integrin: A direct route to actin. *Biochim. Biophys. Acta*, 1692, 47, 2004.
264. Wiesner, S., Legate, K.R., and Fassler, R., Integin-actin interactions. *Cell Mol. Life Sci.*, 62, 1081, 2005.
265. Luton, F., The role of EFA6, exchange factor for Arf6, for tight junction assembly, functions, and interaction with the actin cytoskeleton. *Methods Enzymol.*, 404, 332, 2005.
266. Adams, C. L., et al., Mechanisms of epithelial cell-cell adhesion and cell compaction revealed by high-resolution tracking of E-cadherin-green fluorescent protein. *J. Cell Biol.*, 142, 1105, 1998.
267. Koch, A.W., Manzur, K.L., and Shan, W., Structure-based models of cadherin-mediated cell adhesion: The evolution continues. *Cell Mol. Life Sci.*, 61, 1884, 2004.

268. Mills, J.W. and Ferm, V.H., Effect of cadmium on F-actin and microtubules of Madin-Darby canine kidney cells. *Toxicol. Appl. Pharmacol.*, 101, 245, 1989.
269. Wang, Z., Chin, T.A., and Templeton, D.M., Calcium-independent effects of cadmium on actin assembly in mesangial and vascular smooth muscle cells. *Cell Motil. Cytoskel.*, 33, 208, 1996.
270. Wang, Z. and Templeton, D.M., Cellular factors mediate cadmium-dependent actin depolymerization. *Toxicol. Appl. Pharmacol.*, 139, 115, 1996.
271. Rudolf, E., et al., Hexavalent chromium disrupts the actin cytoskeleton and induces mitochondria-dependent apoptosis in human dermal fibroblasts. *Toxicol. In Vitro*, 19, 713, 2005.
272. Gunaratnam, M. and Grant, M.H., Damage to F-actin and cell death induced by chromium VI and nickel in monolayer cultures of rat hepatocytes. *Toxicol. In Vitro*, 18, 245, 2004.
273. Ahamed, M. and Siddiqui M.K., Low level lead exposure and oxidative stress: Current opinions. *Clin. Chim. Acta*, 383, 57, 2007.
274. Conterato, G.M., et al., Effect of lead acetate on cytosolic thioredoxin reductase activity and oxidative stress parameters in rat kidneys. *Basic Clin. Pharmacol. Toxicol.*, 101, 96, 2007.
275. Xu, J., et al., Lead induces oxidative stress, DNA damage and alteration of p53, Bax and Bcl-2 expressions in mice. *Food Chem. Toxicol.*, 46, 1488, 2008.
276. Erikson, K.M., et al., Manganese exposure and induced oxidative stress in the rat brain. *Sci. Total Environ.*, 334, 409, 2004.
277. Erikson, K.M., et al., Alterations of oxidative stress biomarkers due to in utero and neonatal exposures of airborne manganese. *Biol. Trace Elem. Res.*, 111, 199, 2006.
278. Milatovic, D., et al., Manganese induces oxidative impairment in cultured rat astrocytes. *Toxicol. Sci.*, 98, 198, 2007.
279. Galhardi, C.M., et al., Toxicity of copper intake: Lipid profile, oxidative stress and susceptibility to renal dysfunction. *Food. Chem. Toxicol.*, 42, 2053, 2004.
280. Uria-Adams, J.Y. and Keen, C.L., Copper, oxidative stress, and human health. *Mol. Aspects Med.*, 26, 268, 2005.
281. Nemmiche, S., Chabane-Sari, D., and Guiraud, P., Role of alpha-tocopherol in cadmium-induced oxidative stress in Wistar rat's blood, liver and brain. *Chem. Biol. Interact.*, 170, 221, 2007.
282. Thijssen, S., et al., Low cadmium exposure triggers a biphasic oxidative stress response in mice kidneys. *Toxicology*, 236, 29, 2007.
283. Nzengue, Y., et al., Oxidative stress and DNA damage induced by cadmium in the human keratinocyte HaCaT cell line: Role of glutathione in resistance to cadmium. *Toxicology*, 243, 193, 2008.
284. Rudolf, E., Rudolf, K., and Cervinka, M., Zinc induced apoptosis in HEP-2 cancer cells: The role of oxidative stress and mitochondria. *Biofactors*, 23, 207, 2005.
285. Rudolf, E., Depletion of ATP and oxidative stress underlie zinc-induced cell injury. *Acta Med.*, 50, 4, 2007.
286. Agarwal, R., Iron, oxidative stress, and clinical outcomes. *Pediatr. Nephrol.*, epub ahead of print.
287. Galaris, D. and Pantopoulos, K., Oxidative stress and iron homeostasis: Mechanistic and health aspects. *Crit. Rev. Clin. Lab Sci.*, 45, 1, 2008.
288. Reeder, B.J., Hider, R.C., and Wilson, M.T., Iron chelators can protect against oxidative stress through ferryl heme reduction. *Free Radic. Biol. Med.*, 44, 264, 2008.
289. Woods, J.S. and Ellis, M.E., Up-regulation of glutathione synthesis in rat kidney by methylmercury. Relationship to mercury-induced oxidative stress. *Biochem. Pharmacol.*, 50, 1719, 1995.
290. Nava, M., et al., Melatonin attenuates acute renal failure and oxidative stress induced by mercuric chloride in rats. *Am. J. Physiol. Renal Physiol.*, 279, F910, 2000.
291. Messer, R.L., et al., Mercury (II) alters mitochondrial activity of monocytes at sublethal doses via oxidative stress mechanisms. *J. Biomed. Mater. Res. B Appl. Biomater.*, 75, 257, 2005.
292. Smith, J.W., et al., Redox control of integrin adhesion receptors. *Methods Enzymol.*, 353, 156, 2002.
293. Lamari, F., et al., Cell adhesion and integrin expression are modulated by oxidative stress in EA.hy 926 cells. *Free Radic. Res.*, 41, 812, 2007.
294. Usatyuk, P.V., Parinandi, N.L., and Natarajan, V., Redox regulation of 4-hydroxy-2-nonenal-mediated endothelial barrier dysfunction by focal adhesion, adherens, and tight junction proteins. *J. Biol. Chem.*, 281, 35554, 2006.
295. Meyer, T.N., et al., Reassembly of the tight junction after oxidative stress depends on tyrosine kinase activity. *J. Biol. Chem.*, 276, 22048, 2001.
296. Parrish, A.R., et al., Chemically induced oxidative stress disrupts the E-cadherin/catenin cell adhesion complex. *Toxicol. Sci.*, 51, 80, 1999.
297. Schmelz, M., Schmid, V.J., and Parrish, A.R., Selective disruption of cadherin/catenin complexes by oxidative stress in precision-cut mouse liver slices. *Toxicol. Sci.*, 61, 389, 2001.

298. Morel, Y. and Barouki, R., Repression of gene expression by oxidative stress. *Biochem. J.*, 342, 481, 1999.
299. Karin, M., et al., Oxidative stress and gene expression: The AP-1 and NF-kappaB connections. *Biofactors*, 15, 87, 2001.
300. Eyries, M., Collins, T., and Khachigian, L.M., Modulation of growth factor gene expression in vascular cells by oxidative stress. *Endothelium*, 11, 133, 2004.
301. Smith, M.A. and Perry, G., Alzheimer disease: Protein:protein interaction and oxidative stress. *Bol. Estud. Med. Biol.*, 44, 5, 1996.
302. Gonzalez-Rubio, M., et al., Oxidative stress induces tyrosine phosphorylation of PDGF alpha- and beta-repectors and pp60c-src in mesangial cells. *Kidney Int.*, 50, 164, 1996.
303. Kurata, S., Selective activation of p38 MAPK cascade and mitotic arrest caused by low level oxidative stress. *J. Biol. Chem.*, 275, 23413, 2000.
304. Chen, D.B., et al., Tyrosine phosphorylation of caveolin 1 by oxidative stress is reversible and dependent on the c-src tyrosine kinase but not mitogen-activated protein kinase pathways in placental artery endothelial cells. *Biol. Reprod.*, 73, 761, 2005.
305. Matos, T.J., et al., Role of oxidative stress in ERK and p38 MAPK activation induced by the chemical sensitizer DNFB in a fetal skin dendritic cell line. *Immunol. Cell Biol.*, 83, 607, 2005.
306. Perez, L.M., et al., Oxidative stress induces actin-cytoskeletal and tight-junctional alterations in hepatocytes by a Ca^{2+}-dependent, PKC-mediated mechanism: Protective effect of PKA. *Free Radic. Biol. Med.*, 40, 2005, 2006.
307. Zhou, Y., et al., Oxidative stress-induced intestinal epithelial cell apoptosis is mediated by p38 MAPK. *Biochem. Biophys. Res. Commun.*, 350, 860, 2006.
308. Valentin, F., et al., Oxidative stress activates MMP-2 in cultured MMP-2 in cultured human coronary smooth muscle cells. *Fundam. Clin. Pharmacol.*, 19, 661, 2005.
309. Yu, F., et al., Induction of MMP-9 expression and endothelial injury by oxidative stress after spinal cord injury. *J. Neurotrauma*, 25, 184, 2008.
310. Carmody, R.J. and Cotter, T.G., Oxidative stress induces caspase-independent retinal apoptosis *in vitro*. *Cell Death Differ.*, 7, 282, 2000.
311. Yuan, J., et al., Involvement of cytochrome c release and caspase-3 activation in the oxidative stress-induced apoptosis in human tendon fibroblasts. *Biochim. Biophys. Acta*, 1641, 35, 2003.
312. Ryan, L., O'Callaghan, U.C., and O'Brien, N.M., Generation of an oxidative stress precedes caspase activation during 7beta-hydroxycholesterol-induced apoptosis in U937 cells. *J. Biochem. Mol. Toxicol.*, 18, 50, 2004.
313. Huot, J., et al., HPS27 phosphorylation-mediated resistance against actin fragmentation and cell death induced by oxidative stress. *Cancer Res.*, 56, 273, 1996.
314. Milzani, A., DalleDonne, I., and Colombo, R., Prolonged oxidative stress on actin. *Arch. Biochem. Biophys.*, 339, 267, 1997.
315. Diamond, G.L. and Zalups, R.K., Understanding renal toxicity of heavy metals. *Toxicol. Pathol.*, 26, 92, 1998.
316. Zalups, R.K. and Ahmad, S., Molecular handling of cadmium in transporting epithelia. *Toxicol. Appl. Pharmacol.*, 186, 163, 2003.
317. Prozialeck, W.C., et al., Epithelial barrier characteristics and expression of cell adhesion molecules in proximal tubule-derived cell lines commonly used for *in vitro* toxicity studies. *Toxicol. In Vitro*, 20, 942, 2006.
318. Nouwen, E.J., et al., Stage- and segment-specific expression of cell-adhesion molecules N-CAM, A-CAM, and L-CAM in the kidney. *Kidney Int.*, 44, 147. 1993.
319. Jung, K.Y., et al., Loss of N-cadherin and alpha-catenin in the proximal tubules of aging male Fischer 344 rats. *Mech. Ageing Dev.*, 125, 445, 2004.
320. Prozialeck, W.C., Lamar, P.C., and Appelt, D.M., Differential expression of E-cadherin, N-cadherin and beta-catenin in proximal and distal segments of the rat nephron. *BMC Physiol.*, 4, 10, 2004.
321. Reyes, J.L., et al., The renal segmental distribution of claudins changes with development. *Kidney Int.*, 62, 476, 2002.

13 Iron Metabolism and Disease

Giada Sebastiani and Kostas Pantopoulos

CONTENTS

13.1 Physiology of Iron Homeostasis 351
13.2 Hormonal Regulation of Iron Traffic 353
13.3 Malregulation of Iron Homeostasis 354
13.3.1 Iron Deficiency 354
13.3.2 Defective Iron Traffic 354
13.3.3 Iron Overload (Hemochromatosis) 354
13.4 Hereditary Hemochromatosis 355
13.5 Juvenile Hemochromatosis 356
13.6 A Unifying Concept for the Pathogenesis of HH 356
13.7 Ferroportin Disease 359
13.8 Other Rare Hereditary Causes of Systemic Iron Overload 359
13.9 Iron Overload in the CNS 360
13.10 Mitochondrial Iron Overload 360
13.11 Transfusional Siderosis and Chelation Therapy 361
13.12 Iron Overload States Secondary to Chronic Liver Diseases 362
13.13 Iron Overload in Chronic Hepatitis C 363
13.14 Iron Overload in Alcoholic Liver Disease 364
13.15 Iron Overload in Nonalcoholic Fatty Liver Disease 365
13.16 Hereditary Hyperferritinemia-Cataract Syndrome 365
13.17 Stepwise Decisional Tree for the Clinical Management of Iron Overload States 366
13.18 Concluding Remarks 368
Acknowledgment 368
References 368

13.1 PHYSIOLOGY OF IRON HOMEOSTASIS

Iron is an indispensable constituent of virtually all cells, but also a potential biohazard [1,2]. The adult human body contains 3–5 g of iron (45–55 mg/kg of body weight in females and males, respectively). The vast majority (~60–70%) is distributed in erythroid cells within the heme prosthetic group of hemoglobin and serves in oxygen binding. Likewise, significant amounts of iron are localized in muscle cells within myoglobin, which also binds oxygen. Other cell types contain lower quantities of heme or nonheme iron (such as iron–sulfur clusters) in cytochromes and various other metalloproteins. Excess of body iron is stored in liver hepatocytes and Kupffer cells, within ferritin and its degradation product hemosiderin.

Even though iron is the second most abundant transition metal on the Earth's crust (after aluminum), its bioavailability is limited, partly due to the low solubility of oxidized ferric (Fe^{3+}) ions under aerobic conditions. Thus, it is not surprising that humans (and mammals in general) lack any specific mechanisms for iron excretion and iron homeostasis is regulated at the level of acquisition [3–5]. During prenatal development, the fetus acquires iron from the mother. The transport of iron across the placental syncytiotrophoblasts is mediated by ferroportin, a ferrous (Fe^{2+}) ion exporter. Developing children and adults absorb heme and inorganic iron from the diet by duodenal enterocytes (Figure 13.1). The absorption of heme may involve the folate transporter HCP1 [6,7], while inorganic ferric iron is reduced in the intestinal lumen to ferrous by ferric reductases (such as Dcytb) and absorbed by the apical divalent metal transporter 1 (DMT1). Internalized heme is degraded by heme oxygenase 1 (HO-1) and absorbed iron is released at the basolateral site into the bloodstream by ferroportin. Ferrous iron is then oxidized to ferric by the ferroxidases hephaestin and ceruloplasmin, and captured by circulating transferrin for delivery to tissues upon binding to transferrin receptor 1 (TfR1). In adulthood, a healthy individual absorbs 1–2 mg/day of dietary iron to compensate either for nonspecific losses by cell desquamation in the skin and the intestine or for bleeding. Menstruating females physiologically lose iron from the blood and tend to have lower body iron stores than males and postmenopausal females.

Erythropoiesis requires ~30 mg/day of iron, primarily provided by recycling from effete red blood cells. These undergo phagocytosis by reticuloendothelial macrophages, which in turn metabolize hemoglobin and heme and release iron into circulation via ferroportin, to replenish the dynamic transferrin pool. Circulating transferrin is ~30% saturated and contains, at steady state, ~3 mg of iron, which is recycled >10 times/day.

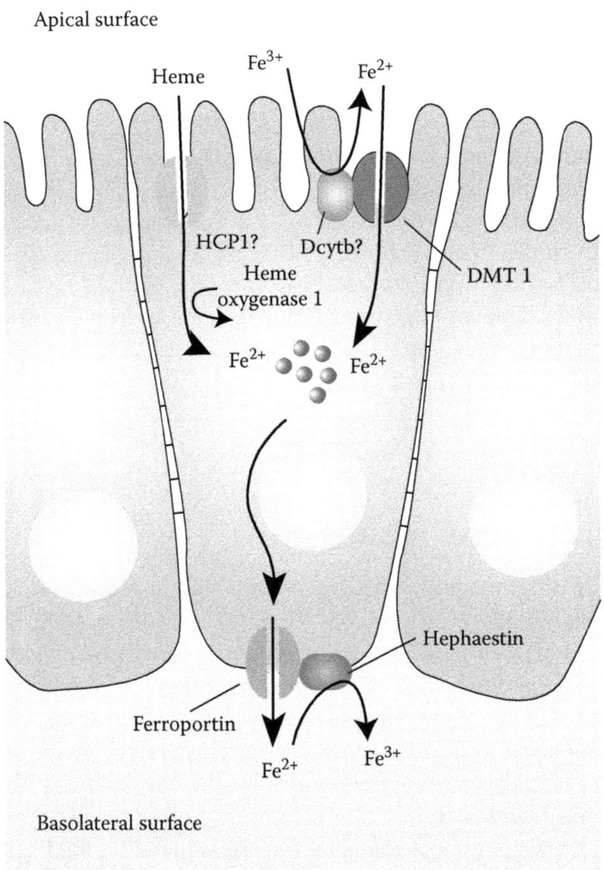

FIGURE 13.1 Dietary iron absorption by duodenal enterocytes. Ferric (Fe^{3+}) iron is reduced to ferrous (Fe^{2+}) in the intestinal lumen, possibly by the ferric reductase Dcytb and transported across the apical membrane of the enterocyte by DMT1. Heme is transported possibly via the folate transporter HCP1 and iron is liberated enzymatically within the enterocyte, following heme degradation by heme oxygenase 1 (HO1). Ferrous iron is exported from the basolateral membrane to the circulation by ferroportin and undergoes reoxidation by hephaestin.

13.2 HORMONAL REGULATION OF IRON TRAFFIC

The efflux of iron from enterocytes and macrophages to the bloodstream via ferroportin is critical for maintenance of iron homeostasis. This process is negatively regulated by hepcidin, a circulating hormonal peptide that binds to ferroportin and promotes its internalization and lysosomal degradation [8]. The hormone is synthesized in hepatocytes as a propeptide precursor, which is converted to a cysteine-rich bioactive molecule of 25 amino acids by proteolytic processing [9]. Mature hepcidin is secreted into plasma and controls iron traffic in response to alterations in iron stores, erythropoiesis, hypoxia, and inflammation. Hepcidin levels increase by iron intake and inflammation [10,11], inhibiting duodenal iron absorption, and promoting iron retention within macrophages (Figure 13.2). By contrast, hepcidin levels decrease in iron deficiency, hypoxia, or phlebotomy-induced anemia [12], facilitating duodenal iron absorption and iron release from macrophages (Figure 13.2). In iron deficiency or hypoxia, duodenal iron absorption is also stimulated independently of the hepcidin/ferroportin axis, by a mechanism involving the hypoxia-inducible factor HIF-2α-mediated transcriptional activation of DMT1 and Dcytb [13].

The expression of hepcidin is transcriptionally regulated by iron- and cytokine-dependent pathways. The inflammatory cytokines IL-6 and IL-1 induce hepcidin transcription via STAT3 [14–16]. The iron-dependent pathway involves proximal and distal promoter elements [17–19] and bone morphogenetic protein (BMP) signaling [20]. BMP-2, -4, -6, and -9 are potent inducers of hepcidin transcription, in conjunction with the BMP coreceptor hemojuvelin [21–24], which may assemble

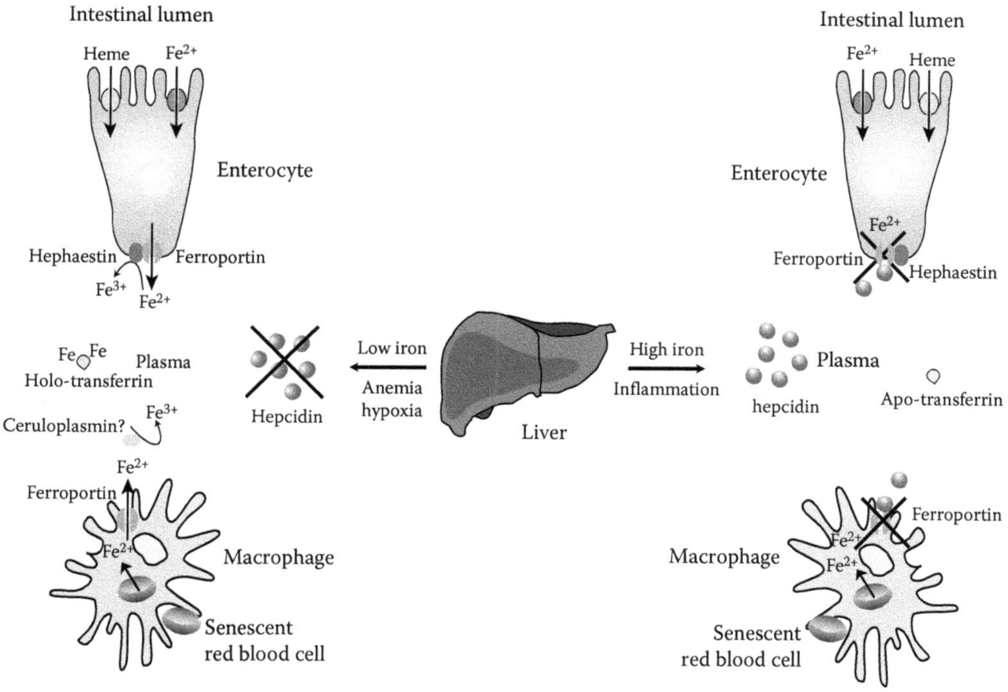

FIGURE 13.2 Regulation of iron efflux from duodenal enterocytes and reticuloendothelial macrophages by hepcidin. Enterocytes absorb heme or inorganic dietary iron and macrophages phagocytose iron-loaded senescent red blood cells. Both cell types release ferrous iron (Fe^{2+}) into plasma via ferroportin, which is incorporated into transferrin following oxidation into the ferric form (Fe^{3+}) via hephaestin or ceruloplasmin. The secretion of the iron-regulatory hormone hepcidin from the liver in response to high body iron stores or inflammatory signals results in internalization and degradation of ferroportin, and retention of iron within enterocytes and macrophages. A decrease in body iron stores, a requirement of iron for erythropoiesis, or hypoxia inhibits hepcidin expression, permitting dietary iron absorption by enterocytes and iron release from macrophages.

into a putative iron-sensing complex, possibly also consisting of BMP receptors, transferrin receptor 2 (TfR2), and the hemochromatosis protein HFE [25] (see below). Basal hepcidin transcription involves the CCAAT/enhancer-binding protein (C/EBPα) [26].

13.3 MALREGULATION OF IRON HOMEOSTASIS

In healthy subjects, the rate of dietary iron absorption and the levels of body iron are tightly controlled [3–5]. Disruption of iron homeostasis can lead to pathological conditions, including iron deficiency, defective iron traffic, or iron overload [27].

13.3.1 Iron Deficiency

A failure to meet physiological requirements of iron for erythropoiesis by dietary absorption results in depletion of iron stores and culminates in iron deficiency anemia (IDA) [28]. This is caused by diminished availability of dietary iron, malabsorption of iron, or chronic blood losses related to confounding disorders. IDA can lead to adverse effects, such as disability, impaired thermoregulation, immune dysfunction, and neurocognitive defects. Nutritional iron deficiency affects an estimated 2 billion people worldwide, mostly in developing countries, and poses a serious health care problem [29]. The management of IDA involves iron supplementation and fortification of foods.

13.3.2 Defective Iron Traffic

In prolonged inflammation, encountered in infectious or autoimmune disorders or cancer, iron may become limiting for erythropoiesis even in the presence of adequate stores. The anemia of chronic disease (ACD) is characterized by the retention of iron within reticuloendothelial macrophages and by decreased dietary iron absorption [30]. The ensuing diversion of iron from circulation into storage sites leads to hypoferremia. In the context of the inflammatory response, this may constitute part of the organism's strategy to deplete pathogens from an essential nutrient. The development of ACD is to a large extent based on a pathological increase in hepcidin levels via IL-6/STAT3 signaling [14–16], which blocks iron egress from macrophages and enterocytes via ferroportin. Other factors contributing to the disease progression include the altered expression of iron metabolism genes and the inhibition of proliferation of erythroid progenitor cells, triggered by inflammatory cytokines. ACD is the most frequent anemia among hospitalized patients in industrial countries. It is not life-threatening *per se*, but deteriorates the patient's quality of life. Treatment of the underlying disease poses the best therapeutic option. Pharmacological correction of ACD can be achieved by administration of recombinant erythropoietin and, in acute cases, by blood transfusion. Iron therapy is controversial because it may promote bacteremia; however, it may be useful if ACD is combined with IDA. The development of pharmacological inhibitors of the hepcidin pathway is promising for the treatment of ACD, especially in the subset of patients that poorly responds to erythropoietin.

A rare, hereditary form of IDA has been described that is refractory to oral iron supplementation and partially unresponsive to parenteral iron therapy. The molecular defect is based on mutations in the transmembrane serine protease matriptase (TMPRSS6), which is associated with aberrant iron traffic due to pathological increase in hepcidin levels [31,32]. A similar phenotype was observed in mice with targeted [33] or chemically induced [34] disruption of the TMPRSS6 gene.

13.3.3 Iron Overload (Hemochromatosis)

The accumulation of excess iron in tissues is toxic and promotes oxidative stress and cell damage. Excessive iron deposition in the liver predisposes for fibrosis, cirrhosis, and hepatocellular carcinoma (HCC) [35–37], and may exacerbate other types of liver disease [38,39]. Iron overload is also associated with cardiomyopathy, diabetes mellitus, hypogonadism, arthritis, and skin pigmentation [27]. Primary iron overload is linked to hereditary defects in the hepcidin pathway. On the other hand, secondary iron overload develops as a side effect of another disease. Transfusional siderosis results from repeated blood transfusions required for the treatment of various anemias, while mild to moderate secondary iron overload is common in chronic liver diseases. The phenotypes of iron overload disorders and the underlying pathogenic mechanisms are described below.

13.4 HEREDITARY HEMOCHROMATOSIS

The term "hereditary hemochromatosis" (HH) comprises a heterogeneous group of disorders characterized by chronic hyperabsorption of dietary iron at a rate that may reach 8–10 mg/day [40,41]. This results in gradual saturation of transferrin with iron (from physiological ~30% up to 100%) and the buildup of a pool of redox-active non-transferrin-bound iron (NTBI) in plasma. The toxic NTBI is eventually taken up by tissues and iron accumulates within parenchymal cells. Organ damage can be easily prevented by therapeutic phlebotomy, which reduces the iron burden. Early diagnosed patients undergoing phlebotomy have a normal life. In the absence of treatment, clinical symptoms are manifested in the fourth or fifth decade.

The most frequent type of HH (also known as HH type 1) is an autosomal recessive disease, related to mutations in the hemochromatosis gene HFE [40,41]. This was elucidated in 1996 by linkage disequilibrium and haplotype analysis from a large cohort of patients with hereditary iron overload [42]. Previous genetic studies provided evidence that the HFE gene is located close to the major histocompatibility complex (MHC) locus on the short arm of chromosome 6 [43,44]. In fact, HFE encodes an atypical MHC class I molecule, which interacts with β_2-microglobulin for processing via the Golgi network and expression on the cell surface. In contrast to typical MHC class I proteins, HFE forms a smaller groove between the $\alpha 1$ and $\alpha 2$ subunits, which is thought to be insufficient for peptide antigen presentation [45]. HFE is expressed in a wide range of cell types, including hepatocytes, macrophages, and enterocytes.

The majority of HH patients bear an HFE C282Y point mutation. This disrupts a disulfide bond that is necessary for the association of HFE with β_2-microglobulin. Consequently, HFE C282Y fails to reach the plasma membrane [46,47]. Unprocessed mutant HFE is trapped in the endoplasmic reticulum (ER) and eventually gets degraded by the proteasome [48]. Further, HFE mutations and polymorphisms (such as H63D and, to a smaller extent, S65C substitutions) have been linked to HH, especially in the compound heterozygous state with C282Y. Homozygosity for the HFE C282Y genotype is highly prevalent (1:200) in populations of Northern European ancestry; however, it is questionable whether the high prevalence is accompanied by comparatively high clinical penetrance [35,40,41,49,50]. It appears that apart from HFE mutations, the manifestation of hemochromatosis requires the contribution of additional, yet incompletely understood environmental, genetic, and/or epigenetic factors [51].

Nevertheless, the ablation of HFE [52,53] or β_2-microglobulin [54,55] is sufficient to promote iron overload in mice. A similar phenotype has been documented in orthologous HFE C282Y [56] and H63D [57] knock-in mice. Interestingly, the degree of iron overload depends on the genetic background of the animals [58–60]. Some of the differences have been attributed to variable expression of Mon1, a protein involved in the trafficking of ferroportin [61]. Taken together, the animal studies corroborate genetic and clinical data on the significance of HFE in the control of body iron homeostasis.

Clearly, HFE dysfunction accounts for the vast majority of HH cases in Caucasians. It should, however, be noted that some rare forms of hereditary iron overload are unrelated to pathogenic HFE mutations [62]. A clinical phenotype indistinguishable from that of HH type 1 is caused by mutations in the TfR2 gene, at the chromosome 7q22 [63,64]. Unlike its related homologue TfR1, TfR2 does not have a physiological role in the acquisition of transferrin-bound iron and its expression is restricted mainly to hepatocytes [65]. The first reported case of TfR2-related HH (classified as HH type 3) was due to a nonsense mutation leading to a premature termination codon (Y250X) [66]. A TfR2 knock-in mouse model with the orthologous Y250 substitution recapitulates the hemochromatosis phenotype [67], while mice with liver-specific disruption of TfR2 also develop iron overload [68]. A wide range of additional TfR2 mutations have been associated with HH type 3 [64].

13.5 JUVENILE HEMOCHROMATOSIS

Juvenile hemochromatosis (JH) is a rare form of hereditary iron overload with autosomal recessive transmission pattern (classified as HH type 2), characterized by early onset [69,70]. While the pattern of iron deposition is similar to that observed in HH of type 1 or 3, the rate of dietary iron absorption is higher. This results in early iron overload with clinical manifestations of hypogonadism, cardiomyopathy, and diabetes in the late teens and early twenties. The major locus of the JH gene was mapped to the 1q chromosome [71,72], but 1q-unlinked genotypes were also identified [73,74].

The molecular basis of the 1q-unlinked JH (HH subtype 2B) was elucidated first. The disease is caused by complete silencing of the iron-regulatory hormone hepcidin due to mutations in the hepcidin gene (called HAMP) on chromosome 19 [75]. The first reported cases were caused by either a nonsense mutation leading to a premature termination codon (R56X) or a frameshift mutation giving rise to aberrant prohepcidin [75]. Likewise, the disruption of hepcidin is sufficient to promote severe tissue iron overload in mice [76,77].

The 1q-linked subset of JH (HH subtype 2A) is due to pathogenic mutations in the HFE2 gene, encoding hemojuvelin (HJV) [78,79]. The most common is a G320V substitution. HJV is identical to repulsive guidance molecule c (RGMc). The RGMa and RGMb family members are expressed in neuronal cells and control the patterning of retinal axons during development [80]. By contrast, HJV is expressed in striated muscles and the liver [78], and apparently exerts an important regulatory function in systemic iron homeostasis. This view is also reinforced by the iron overload phenotype of $HJV^{-/-}$ mice [81,82], which represent another animal model for JH.

HJV-associated JH patients [78] and $HJV^{-/-}$ mice [81,82] fail to mount an iron-dependent increase in hepcidin expression despite pathological iron stores, indicating that HJV acts upstream of hepcidin and positively regulates its expression. Further experiments revealed that HJV is a BMP coreceptor and mediates signaling via the BMP pathway [21–24]. It appears that physiologically relevant iron-dependent signaling is primarily mediated by BMP6. Thus, $BMP6^{-/-}$ mice have low hepcidin levels and develop iron overload [83,84]. Moreover, a similar phenotype is observed in mice bearing liver-specific disruption of SMAD4, which acts downstream of BMPs [20]. HJV also interacts with neogenin [85], but this appears not to affect BMP signaling [24].

HJV is expressed on the cell surface and in perinuclear compartments and associates with membranes via a glycosylphosphatidylinositol (GPI) anchor [85,86]. It is glycosylated at Asn residues [86] and undergoes processing by complex mechanisms, possibly also including autocatalytic cleavage [85]. Differentiating muscle cells release a soluble isoform of HJV in extracellular media [86], which is also present in human serum and plasma [87]. It appears that the release of soluble HJV results from a cleavage at the C-terminus of the protein by furin [88,89]. A treatment of primary human hepatocyte cultures with preparations of recombinant soluble HJV decreased hepcidin mRNA levels [87]. Moreover, the administration of a soluble HJV.Fc fusion protein to mice decreased hepcidin expression *in vivo* [23], suggesting that cellular and soluble HJV have opposing functions. It appears that TMPRSS6, which is mutated in patients with iron refractory IDA [31,32], inhibits signaling to hepcidin by degrading cellular HJV [90].

13.6 A UNIFYING CONCEPT FOR THE PATHOGENESIS OF HH

The clinical, genetic, and biochemical studies with JH patients and mouse models demonstrated that JH is etiologically linked to severe hepcidin deficiency. Soon thereafter, it became clear that milder conditions of hereditary iron overload are also associated with misregulation of hepcidin. Thus, HH patients with mutations in the HFE [91,92] or TfR2 [93] genes exhibit inappropriately low hepcidin levels or blunted hepcidin responses to iron challenge [94], despite high transferrin saturation and elevated body iron stores. Similar results were obtained with $HFE^{-/-}$ [95–98] and $TfR2^{-/-}$ mice [99].

Taken together, the above findings suggest that the common denominator in the pathogenesis of all types of HH is defective hepcidin expression. Interestingly, the impairment of the hepcidin pathway quantitatively correlates with the degree of iron loading. Thus, complete genetic disruption of the HAMP gene or severe hepcidin deficiency, secondary to HJV mutations, led to early onset JH, the most aggressive form of hereditary iron overload. On the other hand, relatively milder hepcidin insufficiency, secondary to HFE or TfR2 mutations, leads to HH of types 1 or 3, respectively. Considering that pathological overexpression of hepcidin contributes to the development of ACD, the function of hepcidin is analogous to that of a rheostat that controls systemic iron traffic and homeostasis and needs to be maintained within a physiological window (Figure 13.3).

Evidently, the understanding of the pathogenesis of HH requires the elucidation of the mechanism for iron-dependent signaling to hepcidin, which involves HFE, TfR2, HJV, and BMPs. HFE, TfR2, and HJV may constitute part of a putative "iron-sensing complex" that responds to alterations in transferrin saturation and conveys BMP (primarily BMP6) signals to downstream targets, leading to transcriptional induction of hepcidin. Initial experimental evidence for this concept was provided by the capacity of HFE to physically interact with TfR2 [100], by analogy to the earlier established HFE–TfR1 interaction [101]. TfR2 is expressed in hepcidin-producing hepatocytes, and its stability increases in response to holo-transferrin [102,103]. Importantly, HFE controls hepcidin synthesis only when expressed in hepatocytes and not in other cell types [104,105]. Genetic experiments based on the modulation of the HFE–TfR1 interaction in transgenic mice suggested that HFE stimulates hepcidin expression when it is free of TfR1 [106]; importantly, the iron-mediated induction of hepcidin requires the physical interaction of HFE with TfR2 [107]. Further evidence supporting the hypothesis of a putative "iron-sensing complex" was provided by the interaction of HJV with BMP2 [21] and BMP6 [83].

These findings can be accommodated in the following working model: Under conditions where there is no need for hepcidin synthesis by the iron-dependent pathway (low serum iron), hepatocyte HFE is predominantly bound to TfR1 (Figure 13.4a). An increase in serum iron and transferrin saturation promotes the dissociation of HFE from TfR1 and concomitantly stabilizes TfR2 [102,103].

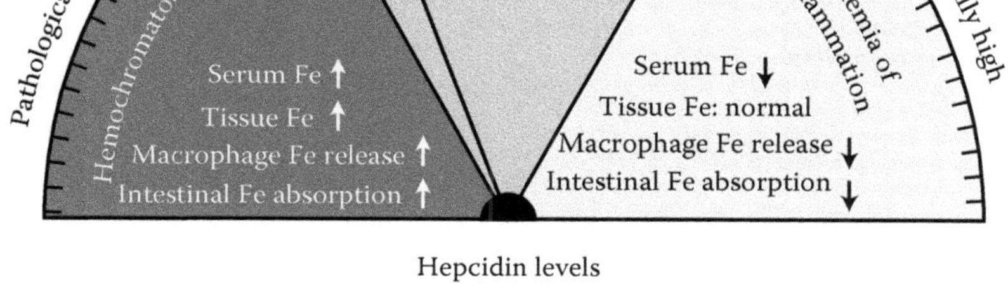

FIGURE 13.3 Hepcidin, a rheostat for systemic iron homeostasis. Physiological hepcidin levels indicate normal iron metabolism. Pathologically low hepcidin is associated with mild to severe hemochromatosis; the degree of iron loading is indirectly proportional to hepcidin levels. Pathologically high hepcidin is associated with the anemia of inflammation or ACD.

Liberated TfR1 becomes accessible for the binding of holo-transferrin, resulting in cellular iron acquisition. At the same time, HFE associates with stabilized TfR2 to form a putative "iron-sensing complex" that may also include HJV, BMP6 (possibly also other BMPs), and BMP receptors (BMPR). This results in recruitment of SMAD proteins and induction of hepcidin mRNA transcription (Figure 13.4b). According to this model, HFE senses alterations in transferrin saturation and controls iron uptake by hepatocytes, which indirectly triggers a systemic regulatory response via hepcidin. Excessive iron uptake by hepatocytes may be prevented by the degradation of TfR1 mRNA following iron-dependent inactivation of the iron-regulatory proteins IRP1 and IRP2 [108]; this would terminate signaling to hepcidin in a feedback loop. The assembly of the putative "iron-sensing complex" may be negatively regulated by matriptase-mediated degradation of HJV. Defects in this signaling pathway are predicted to result in systemic iron overload. The model illustrates a unifying concept for the pathogenesis of HH. Nevertheless, further biochemical experiments are required to dissect the putative "iron-sensing complex" and explore the validity of this model.

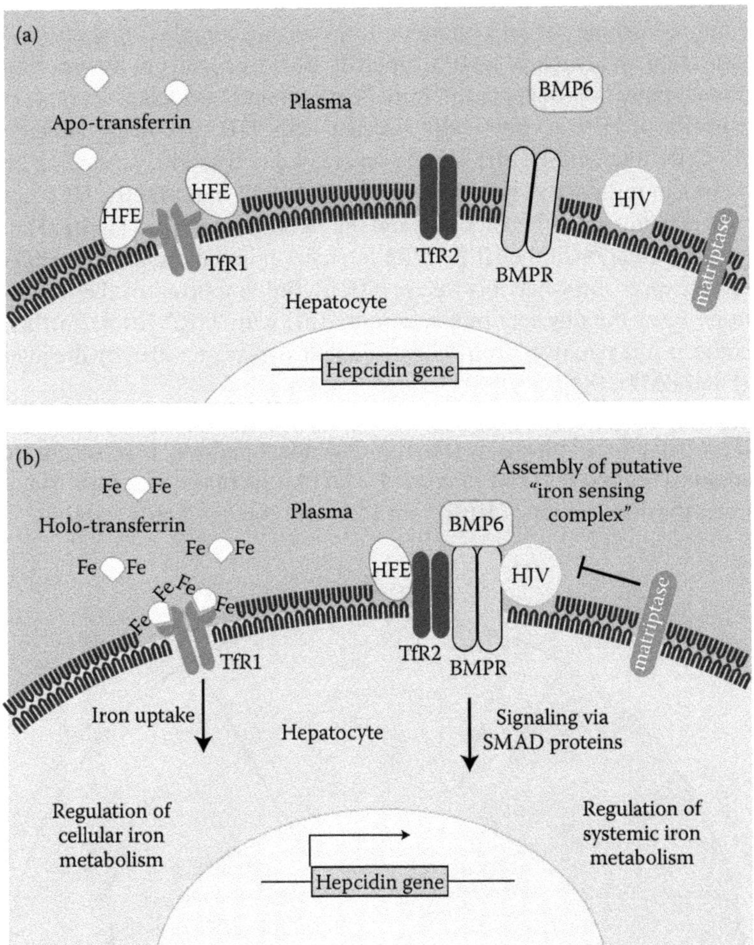

FIGURE 13.4 A model for HFE-mediated signaling to hepcidin in hepatocytes. (a) At low plasma iron concentration, HFE is bound to TfR1, and other proteins involved in signaling to hepcidin remain silent. (b) An increase in plasma iron levels results in displacement of HFE from TfR1, followed by iron uptake. This triggers the assembly of a putative "iron-sensing complex," comprising HFE, TfR2, BMP6, the BMP receptor (BMPR), and HJV, which mediates signaling to activate hepcidin transcription via SMAD proteins. Signaling to hepcidin is negatively regulated by the matriptase, which degrades HJV. According to this model, the hepatocyte integrates signals for regulation of iron metabolism at the cellular and systemic levels.

13.7 FERROPORTIN DISEASE

Considering the significance of the hepcidin pathway in the control of systemic iron homeostasis, it is also expected that mutations affecting the capacity of its target ferroportin to export iron are incompatible with health. The "ferroportin disease" defines a distinct form of hereditary iron overload (HH type 4) that is caused by mutations in the SLC40A1 gene on chromosome 2q32, encoding ferroportin [109]. It is more frequent than HH of types 2 or 3 and exhibits autosomal dominant transmission. The first reported cases were caused by missense mutations leading to N144H [110] and A77D [111] substitutions. Several additional missense mutations and deletions were subsequently documented. The ferroportin disease is heterogenous and its phenotypic hallmarks range from macrophage iron loading and low transferrin saturation (HH subtype 4A), to parenchymal iron loading and relatively high transferrin saturation (HH subtype 4B). Patients with HH subtype 4A show reduced tolerance to phlebotomy and become anemic on therapy, despite persistently elevated serum ferritin levels (marker of iron overload; see below). The phenotypic diversity reflects the effects of the mutation on ferroportin function. In subtype 4A, the mutations are associated with inappropriate ferroportin trafficking, which results in macrophage iron loading. In subtype 4B, the mutations impair the binding of hepcidin, promoting parenchymal iron loading. The autosomal dominant inheritance is probably linked to a dominant negative effect of mutations on the ferroportin dimer [112]. The *flatiron* mouse, carrying an H32R substitution in one allele of the SLC40A1 gene defines an animal model of the (subtype 4A) ferroportin disease [113].

13.8 OTHER RARE HEREDITARY CAUSES OF SYSTEMIC IRON OVERLOAD

Hereditary hypotransferrinemia or atransferrinemia due to mutations in the gene encoding the plasma iron carrier transferrin (on chromosome 3q21) is associated with microcytic anemia, increased dietary iron absorption, and tissue iron overload [114,115]. Similar symptoms have been observed in hypotransferrinemic (*hpx*) mice, which express inappropriately low levels of transferrin due to a spontaneous splicing defect [116]. As erythroid cells completely depend on the transferrin cycle for iron supply [117], hypotransferrinemia or atransferrinemia leads to ineffective erythropoiesis, which in turn triggers increased iron absorption. Excess iron accumulates in tissue parenchymal cells by mechanisms independent of the transferrin cycle.

Hereditary aceruloplasminemia is caused by mutations in the gene encoding ceruloplasmin, on chromosome 3q25 [118]. This plasma blue copper ferroxidase is necessary for iron efflux from macrophages and intestinal enterocytes via ferroportin, and the loading of transferrin. Aceruloplasminemia shows some phenotypic similarities with ferroportin disease (subtype 4A) such as iron overload in visceral organs (pancreas, spleen, and liver) with low serum iron and mild microcytic anemia. However, aceruloplasminemia is also characterized by neurological symptoms due to iron deposition in the brain [119]. It should be noted that in all other forms of hemochromatosis, the central nervous system (CNS) is spared from iron overload, as excess of iron does not cross the blood–brain barrier. In aceruloplasminemia, the brain iron accumulation may be related to the inactivation of an astrocyte-specific GPI-anchored ceruloplasmin isoform, which appears to control iron efflux from the CNS by stabilizing ferroportin [120,121]. The disease phenotype is recapitulated in mouse ceruloplasmin knockout models [122,123].

Deficiency in HO-1 may also lead to systemic iron overload. This enzyme catalyzes the catabolism of heme from senescent erythrocytes in macrophages (and other cell types) and therefore plays a key role in iron recycling [124]. In a single case report of HO-1 deficiency, a six-year-old Japanese boy presented with severe growth retardation, persistent hemolytic anemia, and inflammation and succumbed soon thereafter [125]. Iron deposits in the liver and kidney were documented at biopsy and autopsy. Mice with targeted disruption of HO-1 survive to adulthood, but display severe pathological features, including hepatic and renal iron overload [126].

In another single case report, an autosomal dominant form of systemic iron overload in a Japanese pedigree was associated with a point mutation (A49U) in the iron-responsive element (IRE) of H-ferritin mRNA [127]. The mutation is predicted to promote translational silencing of H-ferritin mRNA via increased binding of the iron-regulatory proteins IRP1 and IRP2 to the IRE (see Chapter 7). However, more work is required to establish whether there is any causative relationship between this effect and the disease phenotype.

13.9 IRON OVERLOAD IN THE CNS

Iron levels tend to increase in the aging brain, and iron imbalance correlates with the development of major neurodegenerative conditions, such as Parkinson's and Alzheimer's diseases [128]. The former is associated with iron deposition in the substantia nigra, and the latter, in neuritic plaques. The excess of redox-active iron promotes oxidative stress and renders the cells vulnerable to oxidative damage, contributing to neurodegeneration. It is possible that iron accumulation in these cell types is not an initial pathogenic event, but rather a secondary effect. Nevertheless, studies in animal models suggest that genetic or pharmacological reduction of the iron burden offers therapeutic benefits [129].

A hereditary etiology for brain iron overload has been established in aceruloplasminemia (discussed above), in neuroferritinopathy and, possibly, also in pantothenate kinase 2 (PANK2)-associated neurodegeneration. Neuroferritinopathy exhibits a dominant pattern of inheritance and is caused by a frameshift mutation in the gene encoding the L-ferritin chain [130]. The predicted C-terminal expansion of mutated L-ferritin may impair the assembly of holo-ferritin and iron storage; the accumulation of unshielded, redox-active iron eventually leads to neuronal cell death. A mouse model of neuroferritinopathy has been reported [131].

PANK2-associated neurodegeneration (formerly known as the Hallervorden–Spatz syndrome) is linked to deficiency in PANK2 [132]. This enzyme catalyzes the phosphorylation of pantothenate, which interacts with cysteine to form an intermediate metabolite in the coenzyme A biosynthesis pathway. The absence of phosphorylated pantothenate results in accumulation of cysteine, an amino acid with iron chelating capacity, and this may underlie the observed local iron overload in the globus pallidus of patients.

13.10 MITOCHONDRIAL IRON OVERLOAD

Several rare hereditary conditions are associated with local iron deposition in mitochondria. The diseases are caused by mutations in proteins involved in the biosynthesis of heme [133] or iron–sulfur clusters [134]. These pathways represent the major routes for cellular iron utilization and take place, at least partly, in mitochondria.

Sporadic and familial forms of X-linked sideroblastic anemia (XLSA) are linked to mutations in the erythroid-specific isoform of δ-aminolevulinic acid synthase (ALAS2) [135], the enzyme catalyzing the first step in the heme synthesis pathway. The disease is characterized by the presence of iron-loaded erythroblasts (sideroblasts), where excess iron is deposited in ring-shaped granules within mitochondria. Patients with mutations around the binding site of the ALAS2 cofactor 5′-pyridoxal phosphate usually respond to treatment with pyridoxine. By contrast, mutations in other domains of ALAS2 are refractory to pyridoxine treatment. Systemic iron overload may develop in severe cases requiring chronic blood transfusions (transfusional siderosis). Ineffective erythropoiesis, combined with increased iron absorption, may also lead to primary iron overload. The zebrafish *sauternes* (sau) mutant represents an animal model for XLSA [136].

Friedreich's ataxia (FRDA) is an autosomal recessive neurodegenerative disorder, caused by expansion of the GAA triplet repeat in the gene encoding frataxin; this results in diminished expression of the protein [137]. Clinical features include early onset gait and limb ataxia, muscle weakness, and cardiomyopathy, which are related to mitochondrial iron overload [138]. Frataxin is a nuclear encoded mitochondrial iron-binding protein that is involved in iron–sulfur cluster assembly, possibly as an iron donor [134]. The mitochondrial iron overload phenotype observed in FRDA patients is recapitulated in frataxin-deficient yeast [139] and mice [140]. Studies in a conditional frataxin knockout mouse model showed that the decrease in mitochondrial iron load by mitochondrial-specific chelators offers a potential therapeutic strategy [141].

X-linked sideroblastic anemia with ataxia (XLSA/A) is a rare condition, caused by missense mutations in the ATP-binding cassette of the mitochondrial transporter ABCB7 [142]. In addition to sideroblastic anemia, patients develop early onset cerebellar ataxia. Experiments in the yeast *Saccharomyces cerevisiae* suggest that the ABCB7 orthologue (Atm1) is involved in the transport of iron–sulfur cluster intermediates from the mitochondria to the cytosol for further processing [143]. The expression of human ABCB7 reverses mitochondrial iron overload in Atm1-deficient yeast [144]. The targeted disruption of mouse ABCB7 is embryonically lethal [145].

A splicing defect reducing the levels of glutaredoxin 5 (grx5) was recently linked to a sideroblastic-like form of microcytic anemia, and blood transfusion-related systemic iron overload [146]. Deficiency in grx5, which is required for iron–sulfur cluster assembly, was earlier associated with severe anemia and early embryonic lethality in the *shiraz* zebrafish mutant [147]. The *shiraz* phenotype is caused by a defect in the assembly of the iron–sulfur cluster of IRP1 in erythroid cells; the excess of apo-IRP1 represses expression of ALAS2 mRNA and thereby inhibits heme biosynthesis.

Finally, a hereditary myopathy with exercise intolerance and lactic acidosis is related to aberrant mitochondrial iron metabolism and mitochondrial iron overload. The disease is caused by low expression of the ISCU protein due to splicing defects [148,149]. ISCU functions as a scaffold in the complex pathway of iron–sulfur cluster assembly [134]. The levels of IRP1 in the muscles of patients were reduced, possibly as a result of iron-dependent destabilization of the apo-protein [150,151].

13.11 TRANSFUSIONAL SIDEROSIS AND CHELATION THERAPY

Repeated blood transfusions, applied for the treatment of various anemias (such as thalassemias, sideroblastic, aplastic, or hemolytic anemias, congenital dyserythropoietic anemias, or myelodysplastic syndromes) eventually lead to transfusional siderosis, a form of secondary iron overload [152]. Transfused red cells add substantial amounts of iron to the recipient's organism (~1 mg/mL of red cells). Ineffective erythropoiesis also triggers silencing of hepcidin and increased dietary iron absorption. In thalassemias, the transcriptional inhibition of hepcidin is mediated by upregulation of growth differentiation factor 15 (GDF15), a member of the transforming growth factor β (TGFβ) superfamily [153]. As a result of multiple red blood cell transfusions, iron deposits are primarily formed within reticuloendothelial macrophages. However, excess iron may also progressively accumulate within parenchymal cells, where it is thought to be more toxic. Iron deposition in myocardial fiber cells eventually leads to cardiomyopathy and heart failure, a common complication of secondary iron overload. Iron chelation therapy [154] can improve rapidly progressive heart failure and arrhythmias.

Iron chelating compounds consist of bidentate, tridentate, or hexadentate ligands, in which two, three, or six atoms, respectively, coordinate iron in an octahedral orientation [155,156]. Chelators neutralize the metal's redox capacity by occupying its coordination sphere and sterically inhibiting its interaction with free radicals. Hexadentate chelators tend to be more efficient than chelators with lower denticity, because they form more stable complexes and one chelator molecule fully occupies the coordination sites of one iron atom [157]. The full occupation of iron's coordination sites requires two or three molecules of tri- or bi-dentate chelators, respectively.

Desferrioxamine (DFO) is a clinically applied hexadentate chelator of the hydroxamate class [155]. It is naturally produced by *Streptomyces pilosus* as a siderophore, to capture iron for metabolic purposes. Over the past 40 years, DFO is being widely employed for the treatment of secondary iron overload. While the drug is generally safe and efficient, it has the disadvantage of not being orally active, and its delivery involves prolonged subcutaneous infusions. The poor oral bioavailability of DFO is due to its relatively large molecular size (mw 561) and hydrophilicity, which prevents it from crossing the plasma membrane of cells by passive diffusion. Cellular uptake of the drug is mediated by fluid phase endocytosis [158]. Iron-loaded DFO is excreted by the urine and stool.

Hundreds of compounds with iron chelating properties have been evaluated for their capacity to reduce iron burden in cell culture and animal models. However, only few of them have been approved for clinical use. Deferiprone (L1) is an orally administrable, lipophilic bidentate chelator that has been employed as an alternative to DFO therapy, but has also been associated with side effects, such as agranulocytosis, neutropenia, and liver dysfunction. Combined treatment with DFO and L1 poses a promising alternative to monotherapy [159]. Deferasirox is a recently developed lipophilic tridentate oral chelator, and long-term (3.5 years) clinical studies have validated its efficacy and safety for the treatment of transfusional siderosis [160].

13.12 IRON OVERLOAD STATES SECONDARY TO CHRONIC LIVER DISEASES

HH and transfusional siderosis are paradigms of heavy iron overload that can eventually lead to multiple organ failure, including liver disease. Interestingly, nonhemochromatotic chronic liver diseases are frequently associated with minimal to modest secondary iron overload, which may aggravate liver injury and predispose to hepatic fibrosis [161]. This has been documented in several conditions including chronic infection by hepatitis C (HCV) and B (HBV) viruses, alcoholic and nonalcoholic fatty liver diseases, the insulin-resistance hepatic iron overload syndrome, and *porphyria cutanea tarda*. Clinical studies suggest that secondary iron overload may have significant implications, particularly in the context of chronic HCV infection and alcoholic and nonalcoholic fatty liver diseases, which are highly prevalent in the general population. Progression toward end-stage liver disease is often unpredictable and depends on many cofactors. Iron overload has been suggested as such a cofactor, but the exact role of iron in disease pathogenesis remains unclear and awaits further investigation [39,162]. Table 13.1 summarizes the main features of primary and secondary iron overload states that lead to liver disease.

TABLE 13.1
Primary and Secondary Iron Overload States with Liver Disease: Etiopathogenetic and Clinical Differences

Primary
1. Pathogenetic role of genetics (HFE, hemojuvelin, ferroportin, transferrine receptor 2)
2. Primary regulatory role of hepcidin
3. Parenchymal distribution of hepatic iron in most cases (except HH subtype 4A)
4. Severe hepatic iron overload, systemic involvement of other target organs
5. Severe elevation of serum iron indices
6. Phlebotomy dramatically improves the course of disease
Secondary
1. Possible role of genetics (HFE mutations) as cofactor of the main liver disease
2. Role of hepcidin as secondary to HCV and alcohol-induced oxidative stress
3. Mixed distribution of hepatic iron (parenchymal and reticulo-endothelial)
4. Moderate hepatic iron overload in most cases, rarely severe iron overload; no systemic iron overload
5. Modest to severe elevation of serum iron indices
6. Phlebotomy possibly improves the course of disease and response to antiviral/metabolic therapy

13.13 IRON OVERLOAD IN CHRONIC HEPATITIS C

Chronic hepatitis C (CHC) constitutes a major health care issue with around 200 million individuals affected worldwide [163,164]. The natural course of CHC is characterized by progressive fibrosis in the inflamed liver, which may eventually lead to cirrhosis and end-stage liver complications including HCC. The progression of fibrosis in CHC is highly variable and several factors may be involved including alcohol, age, and gender (prognosis is worse for males infected at a young age). The role of iron in the progression of CHC is debatable. Several studies reported elevated iron indices in CHC patients [165–167]. Increased serum ferritin levels have been reported in 20–60%, while stainable hepatic iron deposits were detected in 3–38% of patients [166,168,169]. Furthermore, many cases of CHC, including those with elevated serum ferritin and/or transferrin saturation, showed no significant increase in hepatic iron concentration [166,169–171]. Generally, CHC-related iron overload is mild to moderate, with variable distribution (parenchymal [172–174], reticulo-endothelial [175–177], or mixed [168,178,179]), and often not sufficient to be hepatotoxic *per se* [39,170].

In clinical practice, the discrepancies between serum iron indices and hepatic iron concentration may be related to association of CHC with several conditions, in the absence of iron overload. Serum ferritin can be elevated as an acute phase protein because of the cytolytic necroinflammation that is common in CHC. Moreover, hepatic steatosis is frequent [180,181] and has been associated with raised serum ferritin [182,183]. Increased serum ferritin in CHC may be due to diabetes mellitus, which is commonly associated with HCV infection [184], but may also be multifactorial [185].

A large-scale study with 242 consecutive CHC patients showed hepatic iron deposition in 38.8% of cases, and significant iron deposition (≥ grade II on a four-grade scale) in only 10.7% of cases [168]. A cutoff of serum ferritin of 450 μg/L for males and 350 μg/L for females could exclude significant hepatic iron with high negative predictive value (93.4%). The role of HFE mutations in causing iron overload in HCV chronic infection has been studied in different studies, with somewhat discordant results (Table 13.2). A recent study deriving from the hepatitis C anti-viral long-term treatment to prevent cirrhosis (HALT-C) Trial investigated the relationship between HFE mutations, serum iron indices, hepatic iron concentration, and severity of liver disease [186]. To date, this is the largest study on this topic and it enrolled 1051 patients with CHC, mostly of white non-Hispanic ethnicity. The authors reported that HFE mutations were present in 34.5% of patients.

TABLE 13.2
Relationship between Liver Fibrosis, Iron Overload, and HFE Mutations in Chronic Hepatitis C

Reference	# Cases	Ethnicity	Association of Serum Iron Indices and HFE	Association of Hepatic Iron and HFE	Association of Liver Fibrosis and HFE
[172]	206	Italian	Yes	Yes	Yes
[173]	120	Mostly Swiss and Italian	No	No	No
[178]	242	Mostly Caucasian	No	No	No
[179]	164	Mostly Caucasian	No	No	No
[237]	137	Caucasian	Yes	Yes	Yes
[238]	135	Brazilian	Yes	Yes	Yes
[239]	119	Mostly Caucasian, non-Hispanic	Yes	Yes	Yes
[240]	316	Mostly White	Yes	Yes	Yes
[241]	401	Mostly German	Yes	Yes	Yes
[242]	184	White, non-Hispanic	Yes	No	No
[186]	1051	Mostly White, non-Hispanic	Yes	Yes	No

Patients with any HFE mutations had significantly higher hepatic iron concentration and serum iron indices than those without HFE mutations. However, fibrosis stage, necroinflammation, and steatosis grade were not significantly different between patients with and without HFE mutations.

Clinical data suggested that higher serum ferritin or hepatic iron may be associated with a diminished likelihood of response to antiviral therapy [174,175,186]. In an Italian multicenter study, iron removal by phlebotomy improved the rate of response to interferon [187]. On the other hand, a study of 242 patients found that the HFE C282Y mutation positively correlated with sustained response in a multivariate analysis [178]. Data deriving from the HALT-C Trial confirmed the increased susceptibility to antiviral therapy in patients carrying HFE mutations [186]. In this study, subjects harboring HFE mutations, particularly H63D, had significantly higher likelihood of both on-treatment virological responses (at 24 and 48 weeks) and sustained virological responses (24 weeks after the end of lead-in therapy) to retreatment with PEGylated interferon α-2a plus ribavirin. This could be due to effects of HFE mutations on protein structure and function, but other yet unidentified factors may also account for this finding. Interestingly, in a cell culture model, iron was shown to suppress subgenomic HCV replication by inactivating the viral polymerase NS5B [188,189]. Nevertheless, further work is required to examine whether this mechanism is relevant to clinical findings.

Considering the central role of hepcidin in the control of systemic iron homeostasis, it is reasonable to hypothesize that iron deposition in the liver of CHC patients may be related to effects of HCV on the hepcidin pathway. In clinical studies, hepcidin expression was found significantly lower in HCV- as compared to HBV-infected patients and HCV-negative controls [190]. Importantly, hepcidin expression was reversed upon HCV eradication with pegylated interferon plus ribavirin therapy, with a concomitant improvement of iron indices. A mouse model expressing an HCV polyprotein transgene exhibits iron accumulation in the liver with decreased hepcidin mRNA and serum prohepcidin levels, resulting in increased expression of ferroportin in the duodenum, spleen, and liver [191]. Nishina et al. proposed a model for HCV-induced iron overload where HCV, oxidative stress, and iron act in a "ménage á trois": HCV promotes oxidative stress, upregulating the C/EBPα homology protein (CHOP), an inhibitor of the transcription factor C/EBPα. Thus, the upregulation of CHOP prevents C/EBPα from binding to the hepcidin promoter and reduces hepcidin expression, leading to increased iron fluxes from the duodenum and macrophages and, eventually, hepatic iron deposition. Further experiments in an HCV replicon cell culture model showed that HCV-induced oxidative stress suppresses hepcidin expression through increased histone deacetylase activity [192].

13.14 IRON OVERLOAD IN ALCOHOLIC LIVER DISEASE

Alcoholic liver disease (ALD) is among the leading causes of end-stage chronic liver disease. Only a minority of heavy drinkers, estimated at 10–30%, develop advanced ALD [193,194]. Other factors are considered to act synergistically to enhance alcohol hepatotoxicity. Many patients with ALD present with elevated serum iron indices and hepatic iron deposition [195–197]. However, most ALD patients have normal or slightly elevated hepatic iron levels, frequently with a mixed parenchymal and reticuloendothelial pattern of distribution [198,199]. A mild degree of iron overload is sufficient to enhance alcohol-induced liver injury. Indeed, patients with HH and significant alcohol intake have a higher incidence of cirrhosis and HCC [200,201]. In a French trial that included 268 alcoholic patients, hepatic iron was positively correlated with fibrosis in a multivariate analysis of risk factors [202].

Clinical data suggest a genetic component to disease susceptibility. Significant associations have been reported between development of significant ALD and polymorphisms of the genes encoding cytochrome P450 and TNFα [203,204]. Studies assessing a possible role of HFE mutations to ALD susceptibility have yielded conflicting results and did not provide compelling evidence that the HFE genotype affects the severity of ALD [198,199,205]. A recent prospective study investigated the influence of HFE gene mutations and hepatic iron overload on HCC occurrence in patients with alcohol- and HCV-related cirrhosis [206]. Interestingly, hepatic iron overload and C282Y mutation were associated with a higher risk of HCC in patients with alcohol- but not HCV-related cirrhosis. Conceivably, in patients with alcoholic cirrhosis, liver iron overload is a major risk factor for the development of HCC.

Potential causes of hepatic iron overload in ALD include increased ingestion and intestinal absorption of dietary iron, upregulation of hepatic TfR1, secondary anemia due to hemolysis, hypersplenism, ineffective erythropoiesis, and hypoxemia due to intrapulmonary and portosystemic shunts [162,207,208]. Increased intestinal iron absorption may result from increased reduction of luminal ferric to ferrous iron, upregulation of DMT1 in duodenal enterocytes, and upregulation of ferroportin in duodenal enterocytes. Experiments in animal models suggest that the hepcidin/ferroportin pathway has a major role in the regulation of iron absorption in ALD. Thus, alcohol intake decreased hepcidin mRNA transcription in mice by an oxidative mechanism involving the downregulation of the transcription factor C/EBPα [209,210], by analogy to the effects of the HCV polyprotein [191].

13.15 IRON OVERLOAD IN NONALCOHOLIC FATTY LIVER DISEASE

Nonalcoholic fatty liver disease is increasingly recognized as the most prevalent liver disease in Western countries. It affects 31% of males and 16% of females in the US adult population [211]. The spectrum of the disease ranges from fatty liver alone to nonalcoholic steatohepatitis (NASH). Although fatty liver alone is considered nonprogressive, up to 20% of patients with NASH may develop cirrhosis and end-stage liver complications [212].

The pathogenic mechanisms leading to fatty liver or NASH are not completely understood. A prevalent theory implicates insulin resistance as a key contributor to nonalcoholic fatty liver disease, leading to hepatic steatosis, and perhaps also to NASH. Obesity, type 2 diabetes, hyperlipidemia, and other conditions associated with insulin resistance are generally present in patients with nonalcoholic fatty liver disease. A "two-hit" hypothesis was proposed, involving the accumulation of fat in the liver ("first hit"), together with a "second hit" that produces oxidative stress. Hepatic steatosis has been recognized as the first of two hits in the pathogenesis of NASH, since the presence of oxidizable fat within the liver is enough to trigger lipid peroxidation [213]. However, many patients with fatty liver do not progress to steatohepatitis. Potential second hits for the evolution toward NASH include all mechanisms contributing to the development of inflammation and fibrosis. The presumed factors initiating second hits are oxidative stress and subsequent lipid peroxidation, proinflammatory cytokines (particularly TNFα), and hormones derived from adipose tissue (adipokines) [214,215].

Elevation of serum iron indices has been found in several studies, while hepatic iron concentration was rarely elevated [203,204,216–219]. Iron has been proposed to contribute to the development or exacerbation of insulin resistance, which is the most important risk factor for development of fatty liver and NASH [182,183]. Clinical support for an important role of iron in NASH was provided by studies on iron-depletion therapy, which improved both transaminases and insulin sensitivity [220,221]. In a cohort of 153 patients with NASH-related cirrhosis, hepatic iron deposition was more frequent among cirrhotic patients with HCC than in HCC-free controls, suggesting a potential role of iron in the development of HCC [222]. A potential association of HFE mutations with the pathogenesis of NASH has also been investigated but with discordant results [183,217–219,223,224]. Explanations for the discrepancies include ascertainment bias, varied power of the studies, and possible ethnic differences in the study populations. HFE mutations may be part of a genetic pattern contributing to the progression of NASH in populations of Celtic origin.

13.16 HEREDITARY HYPERFERRITINEMIA-CATARACT SYNDROME

The hereditary hyperferritinemia-cataract syndrome (HHCS) is an autosomal dominant disorder, characterized by a marked (up to 20-fold) increase in serum ferritin levels in the absence of iron overload (or any abnormalities in body iron homeostasis) and early onset cataract. Ferritin is an intracellular iron storage protein, composed of H- and L-subunits [225]. Iron stimulates H- and L-ferritin mRNA translation by inactivating the repressors IRP1 and IRP2 from binding to their respective IRE [108]. A secreted form of L-ferritin circulates in the bloodstream but does not store iron and its function is unclear. The concentration of serum ferritin (normally <200 μg/L in females and <300 μg/L in males) serves as a marker for body iron. However, it is not always reliable because, apart from the translational derepression by iron, ferritin is also transcriptionally activated by inflammatory cytokines as an acute phase protein [226].

HHCS is etiologically linked to mutations in L-ferritin IRE that prevent the binding of IRP1 and IRP2 and results in unrestricted L-ferritin mRNA translation [227]. A series of HHCS-associated mutations in L-ferritin IRE, including deletions and point mutations, have been reported [228]. The severity of the HHCS phenotype correlates well with the degree of inhibition of IRP1 and IRP2 binding [229]. Nevertheless, clinical variability among individuals sharing the same mutation suggests the involvement of additional factors in disease progression [228]. Experiments in lymphoblastoid cell lines and in lens from HHCS patients (recovered from surgery) suggest that the overproduction of L-ferritin shifts the H-/L-equilibrium in holo-ferritin; this leads to the accumulation of L-homopolymers, which may underlie the pathogenesis of cataract [230]. The development of animal models may shed more light into this issue.

13.17 STEPWISE DECISIONAL TREE FOR THE CLINICAL MANAGEMENT OF IRON OVERLOAD STATES

The management of iron overload states should be a sequential process that initiates with the clinical suspicion for iron overload (Figure 13.5). Clinical features include asthenia, fatigue, arthralgias, skin pigmentation, impotence, diabetes, osteopenia, hepatomegaly, cardiac symptoms including rhythm disturbances and heart failure. The age at presentation may be indicative of non-HFE-related hemochromatosis that may be early or late, with symptoms mostly related to heart, liver, and endocrine glands. The most common clinical presentation of an iron overload state, which may result from a systematic screening or from a suspected iron overload, is an elevated serum ferritin value (>200 μg/L in females and >300 μg/L in males).

Before considering the possibility for primary iron overload, it is essential to exclude other frequent unrelated causes of hyperferritinemia, including metabolic syndrome (obesity or increased body mass index associated with one or more of the following: hypertension, non-insulin-dependent diabetes, hyperlipidemia, and hyperuricemia), and inflammatory or neoplastic conditions. The rare possibility of HHCS should also be excluded. The third step is to exclude causes of secondary iron overload, especially the most frequent ones that include iron-loading anemias and chronic liver diseases (see Table 13.1); the latter usually manifest with mild to moderate hyperferritinemia. Rare causes of iron overload may be suspected in the absence of any secondary iron overload state and when the classical presentation of HH is lacking.

The following step is to evaluate visceral iron excess using noninvasive techniques (magnetic resonance imaging) or, where not available, liver biopsy with Perls' staining. The absence of severe liver fibrosis can be predicted on the basis of clinical and biochemical variables; these are a combination of absence of hepatomegaly at clinical examination, normal aspartate aminotransferase, and serum ferritin <1000 μg/L [231]. In CHC, a cutoff of serum ferritin of 450 μg/L for males and of 350 μg/L for females could exclude significant hepatic iron overload (≥ grade II on a four-grade scale) with high certainty [168]. Liver biopsy serves not only for the diagnosis and quantification of hepatic iron deposition, but also for staging liver disease through semiquantification of liver fibrosis and necroinflammatory activity. Recently, many efforts have been dedicated to noninvasive diagnosis of liver fibrosis in chronic liver diseases using either instrumental devices or serum biochemical markers [232]. A widely employed instrumental approach involves the measurement of liver stiffness by transient elastography [233]. The Fibroscan device (Echosens, Paris) is composed of an

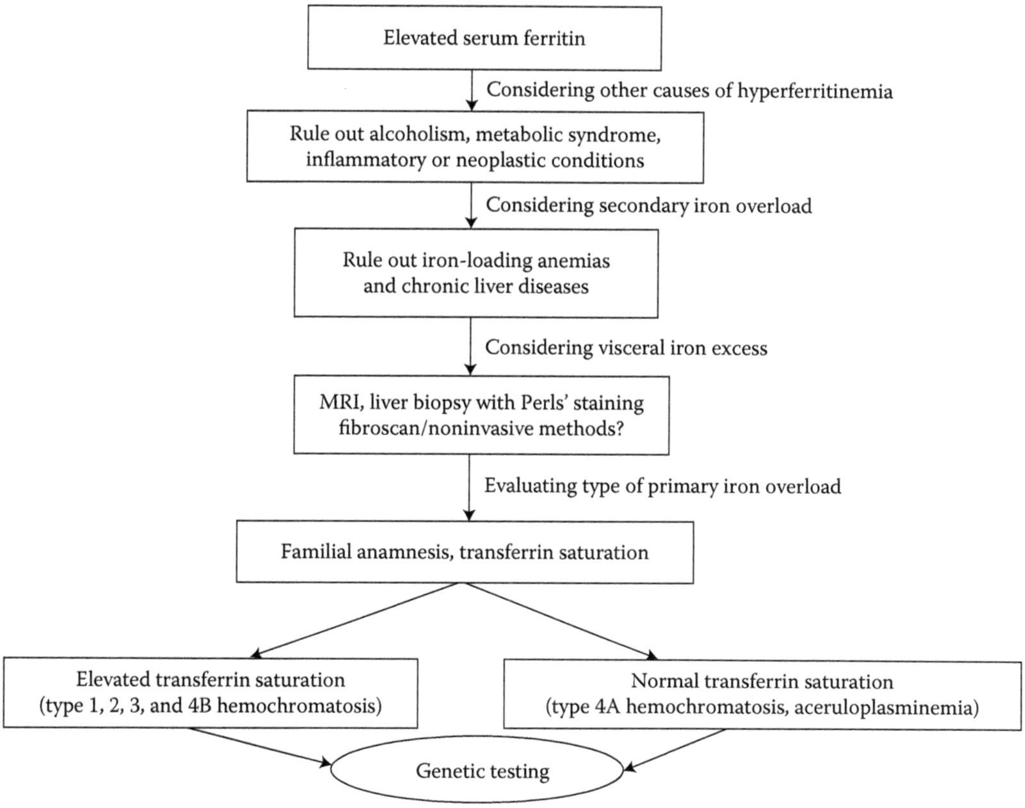

FIGURE 13.5 Stepwise decisional tree for the clinical management of iron overload states.

ultrasound transducer probe mounted on the axis of a vibrator. Vibrations are transmitted by the transducer, inducing an elastic shear wave that propagates through the underlying tissues. Pulse-echo ultrasound acquisition is used to follow the propagation of the shear wave and to measure its velocity, which is directly related to tissue stiffness: the stiffer the tissue, the faster the shear wave propagates. Liver stiffness values range from 2.5 to 75 kPa. Noninvasive staging of liver fibrosis is defined by cutoff values; according to various studies, the presence of cirrhosis is diagnosed by a cutoff value of 12.5–14.5 [234,235]. The overall accuracy ranges from 77% to 90%. The Fibrotest–Fibrosure test, which combines five serum parameters plus age and gender of the patient, has also been employed in CHC patients with an overall accuracy of 70–85%. To date, these new methods have mostly been used for CHC patients and only little data are available for HH. In a recent prospective study, HH patients had high Fibroscan values more often than controls, suggesting that noninvasive methods for liver fibrosis may also be applicable in the context of HH [236]. Nevertheless, longitudinal, prospective studies are necessary to validate these preliminary findings.

If primary iron overload is suspected, a careful familial anamnesis and the determination of all serum iron indices are necessary. Familial anamnesis, especially related to first degree relatives, may permit us to find other cases with late or early onset. In the case of elevated transferrin saturation (>60% in males and >50% in females) in Caucasians, the most likely diagnosis is HH of type 1, which can be confirmed by genotyping (C282Y homozygosity or other less frequent mutations, such as C282Y/H63D compound heterozygosity). If typical HFE-related genotypes are absent or the patient is not Caucasian, other types of HH must be considered: JH (HH of subtype 2A or 2B) in younger patients (<30 years) and HH type 3 or HH subtype 4B in older patients. When normal or low transferrin saturation is found (<45%), plasma ceruloplasmin should be determined to examine the possibility for hereditary aceruloplasminemia, especially in patients with anemia and/or neurological symptoms. If ceruloplasmin levels are normal, the most likely diagnosis is HH type 4A. Once the type of primary hemochromatosis has been defined, a family screening should be performed.

13.18 CONCLUDING REMARKS

The regulation of systemic and cellular iron metabolism is critical for health and its disruption leads to disease. We provided an overview of the various iron-related disorders from the perspective of the basic scientist and the clinician. Over the past few years, genetic and biochemical studies with hemochromatosis patients and animal models have uncovered the regulatory circuit that controls body iron homeostasis, which is dominated by the iron-regulatory hormone hepcidin. The overwhelming majority of clinically relevant cases of HH are characterized by inappropriately low hepcidin responses, while a pathological increase in hepcidin expression is associated with the ACD. Importantly, misregulation of the hepcidin pathway also appears to contribute, at least partially, to transfusional siderosis and to iron overload secondary to chronic liver disease. The development of drugs to pharmacologically control this pathway is expected to find applications in the management of iron-related disorders.

ACKNOWLEDGMENT

G.S. is funded by an unrestricted grant from Roche-Italia. K.P. is funded by the Canadian Institutes for Health Research (CIHR) and holds *Chercheur Nationaux,* a senior career award from the *Fonds de la recherche en santé du Quebéc* (FRSQ).

REFERENCES

1. Aisen, P., Enns, C., and Wessling-Resnick, M., 2001. Chemistry and biology of eukaryotic iron metabolism. *Int. J. Biochem. Cell Biol.* 33, 940–959.
2. Papanikolaou, G. and Pantopoulos, K., 2005. Iron metabolism and toxicity. *Toxicol. Appl. Pharmacol.* 202, 199–211.
3. Hentze, M.W., Muckenthaler, M.U., and Andrews, N.C., 2004. Balancing acts; molecular control of mammalian iron metabolism. *Cell* 117, 285–297.
4. Andrews, N.C. and Schmidt, P.J., 2007. Iron homeostasis. *Annu. Rev. Physiol.* 69, 69–85.
5. De Domenico, I., McVey Ward, D., and Kaplan, J., 2008. Regulation of iron acquisition and storage: Consequences for iron-linked disorders. *Nat. Rev. Mol. Cell Biol.* 9, 72–81.
6. Shayeghi, M., Latunde-Dada, G.O., Oakhill, J.S., Laftah, A.H., Takeuchi, K., Halliday, N., Khan, Y., Warley, A., McCann, F.E., Hider, R.C., Frazer, D.M., Anderson, G.J., Vulpe, C.D., Simpson, R.J., and McKie, A.T., 2005. Identification of an intestinal heme transporter. *Cell* 122, 789–801.
7. Qiu, A., Jansen, M., Sakaris, A., Min, S.H., Chattopadhyay, S., Tsai, E., Sandoval, C., Zhao, R., Akabas, M.H., and Goldman, I.D., 2006. Identification of an intestinal folate transporter and the molecular basis for hereditary folate malabsorption. *Cell* 127, 917–928.
8. Nemeth, E., Tuttle, M.S., Powelson, J., Vaughn, M.B., Donovan, A., Ward, D.M., Ganz, T., and Kaplan, J., 2004. Hepcidin regulates cellular iron efflux by binding to ferroportin and inducing its internalization. *Science* 306, 2090–2093.
9. Nemeth, E. and Ganz, T., 2006. Regulation of iron metabolism by hepcidin. *Annu. Rev. Nutr.* 26, 323–342.
10. Pigeon, C., Ilyin, G., Courselaud, B., Leroyer, P., Turlin, B., Brissot, P., and Loreal, O., 2001. A new mouse liver-specific gene, encoding a protein homologous to human antimicrobial peptide hepcidin, is overexpressed during iron overload. *J. Biol. Chem.* 276, 7811–7819.
11. Nemeth, E., Rivera, S., Gabayan, V., Keller, C., Taudorf, S., Pedersen, B.K., and Ganz, T., 2004. IL-6 mediates hypoferremia of inflammation by inducing the synthesis of the iron regulatory hormone hepcidin. *J. Clin. Invest.* 113, 1271–1276.
12. Nicolas, G., Chauvet, C., Viatte, L., Danan, J.L., Bigard, X., Devaux, I., Beaumont, C., Kahn, A., and Vaulont, S., 2002. The gene encoding the iron regulatory peptide hepcidin is regulated by anemia, hypoxia, and inflammation. *J. Clin. Invest.* 110, 1037–1044.

13. Shah, Y.M., Matsubara, T., Ito, S., Yim, S.H., and Gonzalez, F.J., 2009. Intestinal hypoxia-inducible transcription factors are essential for iron absorption following iron deficiency. *Cell Metab.* 9, 152–164; Mastrogiannaki, M., Matak, P., Keith, B., Simon, M.C., Vaulont, S., and Peyssonnaux, C., 2009. HIF-2alpha, but not HIF-1alpha, promotes iron absorption in mice. *J. Clin. Invest.* 119, 1159–1166.
14. Wrighting, D.M. and Andrews, N.C., 2006. Interleukin-6 induces hepcidin expression through STAT3. *Blood* 108, 3204–3209.
15. Pietrangelo, A., Dierssen, U., Valli, L., Garuti, C., Rump, A., Corradini, E., Ernst, M., Klein, C., and Trautwein, C., 2007. STAT3 is required for IL-6-gp130-dependent activation of hepcidin *in vivo*. *Gastroenterology* 132, 294–300.
16. Verga Falzacappa, M.V., Vujic Spasic, M., Kessler, R., Stolte, J., Hentze, M.W., and Muckenthaler, M.U., 2007. STAT3 mediates hepatic hepcidin expression and its inflammatory stimulation. *Blood* 109, 353–358.
17. Truksa, J., Peng, H., Lee, P., and Beutler, E., 2007. Different regulatory elements are required for response of hepcidin to interleukin-6 and bone morphogenetic proteins 4 and 9. *Brit. J. Haematol.* 139, 138–147.
18. Truksa, J., Lee, P., Peng, H., Flanagan, J., and Beutler, E., 2007. The distal location of the iron responsive region of the hepcidin promoter. *Blood* 110, 3436–3437.
19. Truksa, J., Lee, P., and Beutler, E., 2009. Two BMP responsive elements, STAT, and bZIP/HNF4/COUP motifs of the hepcidin promoter are critical for BMP, SMAD1, and HJV responsiveness. *Blood* 113, 688–695.
20. Wang, R.H., Li, C., Xu, X., Zheng, Y., Xiao, C., Zerfas, P., Cooperman, S., et al., 2005. A role of SMAD4 in iron metabolism through the positive regulation of hepcidin expression. *Cell Metab.* 2, 399–409.
21. Babitt, J.L., Huang, F.W., Wrighting, D.M., Xia, Y., Sidis, Y., Samad, T.A., Campagna, J.A., et al., 2006. Bone morphogenetic protein signaling by hemojuvelin regulates hepcidin expression. *Nat. Genet.* 38, 531–539.
22. Truksa, J., Peng, H., Lee, P., and Beutler, E., 2006. Bone morphogenetic proteins 2, 4, and 9 stimulate murine hepcidin 1 expression independently of Hfe, transferrin receptor 2 (Tfr2), and IL-6. *Proc. Natl. Acad. Sci. USA* 103, 10289–10293.
23. Babitt, J.L., Huang, F.W., Xia, Y., Sidis, Y., Andrews, N.C., and Lin, H.Y., 2007. Modulation of bone morphogenetic protein signaling *in vivo* regulates systemic iron balance. *J. Clin. Invest.* 117, 1933–1939.
24. Xia, Y., Babitt, J.L., Sidis, Y., Chung, R.T., and Lin, H.Y., 2008. Hemojuvelin regulates hepcidin expression via a selective subset of BMP ligands and receptors independently of neogenin. *Blood* 111, 5195–5204.
25. Pantopoulos, K., 2008. Function of the hemochromatosis protein HFE: Lessons from animal models. *World J. Gastroenterol.* 14, 6893–6901.
26. Courselaud, B., Pigeon, C., Inoue, Y., Inoue, J., Gonzalez, F.J., Leroyer, P., Gilot, D., et al., 2002. C/EBPalpha regulates hepatic transcription of hepcidin, an antimicrobial peptide and regulator of iron metabolism. Cross-talk between C/EBP pathway and iron metabolism. *J. Biol. Chem.* 277, 41163–41170.
27. Andrews, N.C., 1999. Disorders of iron metabolism. *N. Engl. J. Med.* 341, 1986–1995.
28. Clark, S.F., 2008. Iron deficiency anemia. *Nutr. Clin. Pract.* 23, 128–141.
29. Zimmermann, M.B. and Hurrell, R.F., 2007. Nutritional iron deficiency. *Lancet* 370, 511–520.
30. Weiss, G. and Goodnough, L.T., 2005. Anemia of chronic disease. *N. Engl. J. Med.* 352, 1011–1023.
31. Finberg, K.E., Heeney, M.M., Campagna, D.R., Aydinok, Y., Pearson, H.A., Hartman, K.R., Mayo, M.M., et al., 2008. Mutations in TMPRSS6 cause iron-refractory iron deficiency anemia (IRIDA). *Nat. Genet.* 40, 569–571.
32. Guillem, F., Lawson, S., Kannengiesser, C., Westerman, M., Beaumont, C., and Grandchamp, B., 2008. Two nonsense mutations in the TMPRSS6 gene in a patient with microcytic anemia and iron deficiency. *Blood* 112, 2089–2091.
33. Folgueras, A.R., Martin de Lara, F., Pendas, A.M., Garabaya, C., Rodriguez, F., Astudillo, A., Bernal, T., Cabanillas, R., Lopez-Otin, C., and Velasco, G., 2008. The membrane-bound serine protease matriptase-2 (Tmprss6) is an essential regulator of iron homeostasis. *Blood* 112, 2539–2545.
34. Du, X., She, E., Gelbart, T., Truksa, J., Lee, P., Xia, Y., Khovananth, K., Mudd, S., Mann, N., Moresco, E.M., Beutler, E., and Beutler, B., 2008. The serine protease TMPRSS6 is required to sense iron deficiency. *Science* 320, 1088–1092.
35. Adams, P.C. and Barton, J.C., 2007. Haemochromatosis. *Lancet* 370, 1855–1860.
36. Ramm, G.A. and Ruddell, R.G., 2005. Hepatotoxicity of iron overload: Mechanisms of iron-induced hepatic fibrogenesis. *Semin. Liver Dis.* 25, 433–449.
37. Kowdley, K.V., 2004. Iron, hemochromatosis, and hepatocellular carcinoma. *Gastroenterology* 127, S79–S86.

38. Pietrangelo, A., 2003. Iron-induced oxidant stress in alcoholic liver fibrogenesis. *Alcohol* 30, 121–129.
39. Sebastiani, G. and Walker, A.P., 2007. HFE gene in primary and secondary hepatic iron overload. *World J. Gastroenterol.* 13, 4673–4689.
40. Pietrangelo, A., 2004. Hereditary hemochromatosis—a new look at an old disease. *N. Engl. J. Med.* 350, 2383–2397.
41. Beutler, E., 2006. Hemochromatosis: Genetics and pathophysiology. *Annu. Rev. Med.* 57, 331–347.
42. Feder, J.N., Gnirke, A., Thomas, W., Tsuchihashi, Z., Ruddy, D.A., Basava, A., Dormishian, F., et al., 1996. A novel MHC class I-like gene is mutated in patients with hereditary haemochromatosis. *Nat. Genet.* 13, 399–408.
43. Simon, M., Bourel, M., Fauchet, R., and Genetet, B., 1976. Association of HLA-A3 and HLA-B14 antigens with idiopathic haemochromatosis. *Gut* 17, 332–334.
44. Jazwinska, E.C., Lee, S.C., Webb, S.I., Halliday, J.W., and Powell, L.W., 1993. Localization of the hemochromatosis gene close to D6S105. *Am. J. Hum. Genet.* 53, 347–352.
45. Lebrón, J.A., Bennet, M.J., Vaughn, D.E., Chirino, A.J., Snow, P.M., Mintier, G.A., Feder, J.N., and Bjorkman, P.J., 1998. Crystal structure of the hemochromatosis protein HFE and characterization of its interaction with transferrin receptor. *Cell* 93, 111–123.
46. Feder, J.N., Tsuchihashi, Z., Irrinki, A., Lee, V.K., Mapa, F.A., Morikang, E., Prass, C.E., et al., 1997. The hemochromatosis founder mutation in HLA-H disrupts β_2-microglobulin interaction and cell surface expression. *J. Biol. Chem.* 272, 14025–14028.
47. Waheed, A., Parkkila, S., Zhou, X.Y., Tomatsu, S., Tsuchihashi, Z., Feder, J.N., Schatzman, R.C., Britton, R.S., Bacon, B.R., and Sly, W.S., 1997. Hereditary hemochromatosis: Effects of C282Y and H63D mutations on association with β_2-microglobulin, intracellular processing, and cell surface expression of the HFE protein in COS-7 cells. *Proc. Natl. Acad. Sci. USA* 94, 12384–12389.
48. de Almeida, S.F., Fleming, J.V., Azevedo, J.E., Carmo-Fonseca, M., and de Sousa, M., 2007. Stimulation of an unfolded protein response impairs MHC class I expression. *J. Immunol.* 178, 3612–3619.
49. Allen, K.J., Gurrin, L.C., Constantine, C.C., Osborne, N.J., Delatycki, M.B., Nicoll, A.J., McLaren, C.E., et al., 2008. Iron-overload-related disease in HFE hereditary hemochromatosis. *N. Engl. J. Med.* 358, 221–230.
50. Waalen, J. and Beutler, E., 2008. Iron-overload-related disease in HFE hereditary hemochromatosis. *N. Engl. J. Med.* 358, 2293–2294 (author reply 2294–2295).
51. Beutler, E., 2007. Iron storage disease: Facts, fiction and progress. *Blood Cells Mol. Dis.* 39, 140–147.
52. Zhou, X.Y., Tomatsu, S., Fleming, R.E., Parkkila, S., Waheed, A., Jiang, J., Fei, Y., et al., 1998. HFE gene knockout produces mouse model of hereditary hemochromatosis. *Proc. Natl. Acad. Sci. USA* 95, 2492–2497.
53. Bahram, S., Gilfillan, S., Kühn, L.C., Moret, R., Schulze, J.B., Lebeau, A., and Schümann, K., 1999. Experimental hemochromatosis due to MHC class I HFE deficiency: Immune status and iron metabolism. *Proc. Natl. Acad. Sci. USA* 96, 13312–13317.
54. de Sousa, M., Reimao, R., Lacerda, R., Hugo, P., Kaufmann, S.H.E., and Porto, G., 1994. Iron overload in β_2-microglobulin-deficient mice. *Immunol. Lett.* 39, 105–111.
55. Rothenberg, B.E. and Voland, J.R., 1996. b_2 Knockout mice develop parenchymal iron overload: A putative role for class I genes of the major histocompatibility complex in iron metabolism. *Proc. Natl. Acad. Sci. USA* 93, 1529–1534.
56. Levy, J.E., Montross, L.K., Cohen, D.E., Fleming, M.D., and Andrews, N.C., 1999. The C282Y mutation causing hereditary hemochromatosis does not produce a null allele. *Blood* 94, 9–11.
57. Tomatsu, S., Orii, K.O., Fleming, R.E., Holden, C.C., Waheed, A., Britton, R.S., Gutierrez, M.A., et al., 2003. Contribution of the H63D mutation in HFE to murine hereditary hemochromatosis. *Proc. Natl. Acad. Sci. USA* 100, 15788–15793.
58. Fleming, R.E., Holden, C.C., Tomatsu, S., Waheed, A., Brunt, E.M., Britton, R.S., Bacon, B.R., Roopenian, D.C., and Sly, W.S., 2001. Mouse strain differences determine severity of iron accumulation in Hfe knockout model of hereditary hemochromatosis. *Proc. Natl. Acad. Sci. USA* 98, 2707–2711.
59. Levy, J.E., Montross, L.K., and Andrews, N.C., 2000. Genes that modify the hemochromatosis phenotype in mice. *J. Clin. Invest.* 105, 1209–1216.
60. Sproule, T.J., Jazwinska, E.C., Britton, R.S., Bacon, B.R., Fleming, R.E., Sly, W.S., and Roopenian, D.C., 2001. Naturally variant autosomal and sex-linked loci determine the severity of iron overload in beta 2-microglobulin-deficient mice. *Proc. Natl. Acad. Sci. USA* 98, 5170–5174.
61. Wang, F., Paradkar, P.N., Custodio, A.O., McVey Ward, D., Fleming, M.D., Campagna, D., Roberts, K.A., et al., 2007. Genetic variation in Mon1a affects protein trafficking and modifies macrophage iron loading in mice. *Nat. Genet.* 39, 1025–1032.

62. Pietrangelo, A., Montosi, G., Totaro, A., Garuti, C., Conte, D., Cassanelli, S., Fraquelli, M., Sardini, C., Vasta, F., and Gasparini, P., 1999. Hereditary hemochromatosis in adults without pathogenic mutations in the hemochromatosis gene. *N. Engl. J. Med.* 341, 725–732.
63. Pietrangelo, A., 2004. Non-HFE hemochromatosis. *Hepatology* 39, 21–29.
64. Wallace, D.F. and Subramaniam, V.N., 2007. Non-HFE haemochromatosis. *World J. Gastroenterol.* 13, 4690–4698.
65. Trinder, D. and Baker, E., 2003. Transferrin receptor 2: A new molecule in iron metabolism. *Int. J. Biochem. Cell Biol.* 35, 292–296.
66. Camaschella, C., Roetto, A., Cali, A., De Gobbi, M., Garozzo, G., Carella, M., Majorano, N., Totaro, A., and Gasparini, P., 2000. The gene TFR2 is mutated in a new type of haemochromatosis mapping to 7q22. *Nat. Genet.* 25, 14–15.
67. Fleming, R.E., Ahmann, J.R., Migas, M.C., Waheed, A., Koeffler, H.P., Kawabata, H., Britton, R.S., Bacon, B.R., and Sly, W.S., 2002. Targeted mutagenesis of the murine transferrin receptor-2 gene produces hemochromatosis. *Proc. Natl. Acad. Sci. USA* 99, 10653–10658.
68. Wallace, D.F., Summerville, L., and Subramaniam, V.N., 2007. Targeted disruption of the hepatic transferrin receptor 2 gene in mice leads to iron overload. *Gastroenterology* 132, 301–310.
69. Camaschella, C., Roetto, A., and De Gobbi, M., 2002. Juvenile hemochromatosis. *Semin. Hematol.* 39, 242–248.
70. Pietrangelo, A., 2006. Juvenile hemochromatosis. *J. Hepatol.* 45, 892–894.
71. Roetto, A., Totaro, A., Cazzola, M., Cicilano, M., Bosio, S., D'Ascola, G., Carella, M., et al., 1999. Juvenile hemochromatosis locus maps to chromosome 1q. *Am. J. Hum. Genet.* 64, 1388–1393.
72. Rivard, S.R., Lanzara, C., Grimard, D., Carella, M., Simard, H., Ficarella, R., Simard, R., et al., 2003. Juvenile hemochromatosis locus maps to chromosome 1q in a French Canadian population. *Eur. J. Hum. Genet.* 11, 585–589.
73. Papanikolaou, G., Politou, M., Roetto, A., Bosio, S., Sakelaropoulos, N., Camaschella, C., and Loukopoulos, D., 2001. Linkage to chromosome 1q in Greek families with juvenile hemochromatosis. *Blood Cells Mol. Dis.* 27, 744–749.
74. Papanikolaou, G., Papaioannou, M., Politou, M., Vavatsi, N., Kioumi, A., Tsiatsiou, P., Marinaki, P., Loukopoulos, D., and Christakis, J.I., 2002. Genetic heterogeneity underlies juvenile hemochromatosis phenotype: Analysis of three families of northern Greek origin. *Blood Cells Mol. Dis.* 29, 168–173.
75. Roetto, A., Papanikolaou, G., Politou, M., Alberti, F., Girelli, D., Christakis, J., Loukopoulos, D., and Camaschella, C., 2003. Mutant antimicrobial peptide hepcidin is associated with severe juvenile hemochromatosis. *Nat. Genet.* 33, 21–22.
76. Nicolas, G., Bennoun, M., Devaux, I., Beaumont, C., Grandchamp, B., Kahn, A., and Vaulont, S., 2001. Lack of hepcidin gene expression and severe tissue iron overload in upstream stimulatory factor 2 (USF2) knockout mice. *Proc. Natl. Acad. Sci. USA* 98, 8780–8785.
77. Lesbordes-Brion, J.C., Viatte, L., Bennoun, M., Lou, D.Q., Ramey, G., Houbron, C., Hamard, G., Kahn, A., and Vaulont, S., 2006. Targeted disruption of the hepcidin 1 gene results in severe hemochromatosis. *Blood* 108, 1402–1405.
78. Papanikolaou, G., Samuels, M.E., Ludwig, E.H., MacDonald, M.L., Franchini, P.L., Dube, M.P., Andres, L., et al., 2004. Mutations in HFE2 cause iron overload in chromosome 1q-linked juvenile hemochromatosis. *Nat. Genet.* 36, 77–82.
79. Lanzara, C., Roetto, A., Daraio, F., Rivard, S., Ficarella, R., Simard, H., Cox, T., et al., 2004. The spectrum of hemojuvelin gene mutations in 1q-linked juvenile hemochromatosis. *Blood* 11, 4317–4321.
80. Monnier, P.P., Sierra, A., Macchi, P., Deitinghoff, L., Andersen, J.S., Mann, M., Flad, M., et al., 2002. RGM is a repulsive guidance molecule for retinal axons. *Nature* 419, 392–395.
81. Niederkofler, V., Salie, R., and Arber, S., 2005. Hemojuvelin is essential for dietary iron sensing, and its mutation leads to severe iron overload. *J. Clin. Invest.* 115, 2180–2186.
82. Huang, F.W., Pinkus, J.L., Pinkus, G.S., Fleming, M.D., and Andrews, N.C., 2005. A mouse model of juvenile hemochromatosis. *J. Clin. Invest.* 115, 2187–2191.
83. Andriopoulos Jr, B., Corradini, E., Xia, Y., Faasse, S.A., Chen, S., Grgurevic, L., Knutson, M.D., et al., 2009. BMP6 is a key endogenous regulator of hepcidin expression and iron metabolism. *Nat. Genet.* 41, 482–487.
84. Meynard, D., Kautz, L., Darnaud, V., Canonne-Hergaux, F., Coppin, H., and Roth, M.P., 2009. Lack of the bone morphogenetic protein BMP6 induces massive iron overload. *Nat. Genet.* 41, 478–481.
85. Zhang, A.S., West, A.P., Jr., Wyman, A.E., Bjorkman, P.J., and Enns, C.A., 2005. Interaction of HJV with neogenin results in iron accumulation in HEK293 cells. *J. Biol. Chem.* 280, 33885–33894.
86. Kuninger, D., Kuns-Hashimoto, R., Kuzmickas, R., and Rotwein, P., 2006. Complex biosynthesis of the muscle-enriched iron regulator RGMc. *J. Cell Sci.* 119, 3273–3283.

87. Lin, L., Goldberg, Y.P., and Ganz, T., 2005. Competitive regulation of hepcidin mRNA by soluble and cell-associated hemojuvelin. *Blood* 106, 2884–2889.
88. Silvestri, L., Pagani, A., and Camaschella, C., 2008. Furin-mediated release of soluble hemojuvelin: A new link between hypoxia and iron homeostasis. *Blood* 111, 924–931.
89. Lin, L., Nemeth, E., Goodnough, J.B., Thapa, D.R., Gabayan, V., and Ganz, T., 2008. Soluble hemojuvelin is released by proprotein convertase-mediated cleavage at a conserved polybasic RNRR site. *Blood Cells Mol. Dis.* 40, 122–131.
90. Silvestri, L., Pagani, A., Nai, A., De Domenico, I., Kaplan, J., and Camaschella, C., 2008. The serine protease matriptase-2 (TMPRSS6) inhibits hepcidin activation by cleaving membrane hemojuvelin. *Cell Metab.* 8, 502–511.
91. Bridle, K.R., Frazer, D.M., Wilkins, S.J., Dixon, J.L., Purdie, D.M., Crawford, D. H., Subramaniam, V.N., Powell, L.W., Anderson, G.J., and Ram, G.A., 2003. Disrupted hepcidin regulation in HFE-associated haemochromatosis and the liver as a regulator of body iron homoeostasis. *Lancet* 361, 669–673.
92. Gehrke, S.G., Kulaksiz, H., Herrmann, T., Riedel, H.D., Bents, K., Veltkamp, C., and Stremmel, W., 2003. Expression of hepcidin in hereditary hemochromatosis: Evidence for a regulation in response to serum transferrin saturation and non-transferrin-bound iron. *Blood* 102, 371–376.
93. Nemeth, E., Roetto, A., Garozzo, G., Ganz, T., and Camaschella, C., 2005. Hepcidin is decreased in TFR2 hemochromatosis. *Blood* 105, 1803–1806.
94. Piperno, A., Girelli, D., Nemeth, E., Trombini, P., Bozzini, C., Poggiali, E., Phung, Y., Ganz, T., and Camaschella, C., 2007. Blunted hepcidin response to oral iron challenge in HFE-related hemochromatosis. *Blood* 110, 4096–4100.
95. Ahmad, K.A., Ahmann, J.R., Migas, M.C., Waheed, A., Britton, R.S., Bacon, B.R., Sly, W.S., and Fleming, R.E., 2002. Decreased liver hepcidin expression in the Hfe knockout mouse. *Blood Cells Mol. Dis.* 29, 361–366.
96. Muckenthaler, M., Roy, C.N., Custodio, A.O., Minana, B., DeGraaf, J., Montross, L.K., Andrews, N.C., and Hentze, M.W., 2003. Regulatory defects in liver and intestine implicate abnormal hepcidin and Cybrd1 expression in mouse hemochromatosis. *Nat. Genet.* 34, 102–107.
97. Constante, M., Jiang, W., Wang, D., Raymond, V.A., Bilodeau, M., and Santos, M.M., 2006. Distinct requirements for Hfe in basal and induced hepcidin levels in iron overload and inflammation. *Am. J. Physiol. Gastrointest. Liver Physiol.* 291, G229–G237.
98. Ludwiczek, S., Theurl, I., Bahram, S., Schumann, K., and Weiss, G., 2005. Regulatory networks for the control of body iron homeostasis and their dysregulation in HFE mediated hemochromatosis. *J. Cell. Physiol.* 204, 489–499.
99. Kawabata, H., Fleming, R.E., Gui, D., Moon, S.Y., Saitoh, T., O'Kelly, J., Umehara, Y., Wano, Y., Said, J.W., and Koeffler, H.P., 2005. Expression of hepcidin is down-regulated in TfR2 mutant mice manifesting a phenotype of hereditary hemochromatosis. *Blood* 105, 376–381.
100. Goswami, T. and Andrews, N.C., 2006. Hereditary hemochromatosis protein, HFE, interaction with transferrin receptor 2 suggests a molecular mechanism for mammalian iron sensing. *J. Biol. Chem.* 281, 28494–28498.
101. Feder, J.N., Penny, D.M., Irrinki, A., Lee, V.K., Lebron, J.A., Watson, N., Tsuchihashi, Z., Sigal, E., Bjorkman, P.J., and Schatzman, R.C., 1998. The hemochromatosis gene product complexes with the transferrin receptor and lowers its affinity for ligand binding. *Proc. Natl. Acad. Sci. USA* 95, 1472–1477.
102. Johnson, M.B. and Enns, C.A., 2004. Diferric transferrin regulates transferrin receptor 2 protein stability. *Blood* 104, 4287–4293.
103. Robb, A. and Wessling-Resnick, M., 2004. Regulation of transferrin receptor 2 protein levels by transferrin. *Blood* 104, 4294–4299.
104. Vujic Spasic, M., Kiss, J., Herrmann, T., Galy, B., Martinache, S., Stolte, J., Grone, H.J., Stremmel, W., Hentze, M.W., and Muckenthaler, M.U., 2008. Hfe acts in hepatocytes to prevent hemochromatosis. *Cell Metab.* 7, 173–178.
105. Vujic Spasic, M., Kiss, J., Herrmann, T., Kessler, R., Stolte, J., Galy, B., Rathkolb, B., et al., 2007. Physiologic systemic iron metabolism in mice deficient for duodenal Hfe. *Blood* 109, 4511–4517.
106. Schmidt, P.J., Toran, P.T., Giannetti, A.M., Bjorkman, P.J., and Andrews, N.C., 2008. The transferrin receptor modulates hfe-dependent regulation of hepcidin expression. *Cell Metab.* 7, 205–214.
107. Gao, J., Chen, J., Kramer, M., Tsukamoto, H., Zhang, A.S., and Enns, C.A., 2009. Interaction of the hereditary hemochromatosis protein HFE with transferrin receptor 2 is required for transferrin-induced hepcidin expression. *Cell Metab.* 9, 217–227.
108. Pantopoulos, K., 2004. Iron metabolism and the IRE/IRP regulatory system: An update. *Ann. N.Y. Acad. Sci.* 1012, 1–13.

109. Pietrangelo, A., 2004. The ferroportin disease. *Blood Cells Mol. Dis.* 32, 131–138.
110. Njajou, O.T., Vaessen, N., Joosse, M., Berghuis, B., van Dongen, J.W., Breuning, M.H., Snijders, P.J., et al., 2001. A mutation in SLC11A3 is associated with autosomal dominant hemochromatosis. *Nat. Genet.* 28, 213–214.
111. Montosi, G., Donovan, A., Totaro, A., Garuti, C., Pignatti, E., Cassanelli, S., Trenor, C.C., Gasparini, P., Andrews, N.C., and Pietrangelo, A., 2001. Autosomal-dominant hemochromatosis is associated with a mutation in the ferroportin (SLC11A3) gene. *J. Clin. Invest.* 108, 619–623.
112. De Domenico, I., Ward, D.M., Nemeth, E., Vaughn, M.B., Musci, G., Ganz, T., and Kaplan, J., 2005. The molecular basis of ferroportin-linked hemochromatosis. *Proc. Natl. Acad. Sci. USA* 102, 8955–8960.
113. Zohn, I.E., De Domenico, I., Pollock, A., Ward, D.M., Goodman, J.F., Liang, X., Sanchez, A.J., Niswander, L., and Kaplan, J., 2007. The flatiron mutation in mouse ferroportin acts as a dominant negative to cause ferroportin disease. *Blood* 109, 4174–4180.
114. Hayashi, A., Wada, Y., Suzuki, T., and Shimizu, A., 1993. Studies on familial hypotransferrinemia: Unique clinical course and molecular pathology. *Am. J. Hum. Genet.* 53, 201–213.
115. Beutler, E., Gelbart, T., Lee, P., Trevino, R., Fernandez, M.A., and Fairbanks, V.F., 2000. Molecular characterization of a case of atransferrinemia. *Blood* 96, 4071–4074.
116. Trenor, C.C., Campagna, D.R., Sellers, V.M., Andrews, N.C., and Fleming, M.D., 2000. The molecular defect in hypotransferrinemic mice. *Blood* 96, 1113–1118.
117. Levy, J.E., Jin, O., Fujiwara, Y., Kuo, F., and Andrews, N.C., 1999. Transferrin receptor is necessary for development of erythrocytes and the nervous system. *Nat. Genet.* 21, 396–399.
118. Yoshida, K., Furihata, K., Takeda, S., Nakamura, A., Yamamoto, K., Morita, H., Hiyamuta, S., Ikeda, S., Shimizu, N., and Yanagisawa, N., 1995. A mutation in the ceruloplasmin gene is associated with systemic hemosiderosis in humans. *Nat. Genet.* 9, 267–272.
119. Nittis, T. and Gitlin, J.D., 2002. The copper-iron connection: Hereditary aceruloplasminemia. *Semin. Hematol.* 39, 282–289.
120. Jeong, S.Y. and David, S., 2003. Glycosylphosphatidylinositol-anchored ceruloplasmin is required for iron efflux from cells in the central nervous system. *J. Biol. Chem.* 278, 27144–27148.
121. De Domenico, I., Ward, D.M., di Patti, M.C., Jeong, S.Y., David, S., Musci, G., and Kaplan, J., 2007. Ferroxidase activity is required for the stability of cell surface ferroportin in cells expressing GPI-ceruloplasmin. *EMBO J.* 26, 2823–2831.
122. Harris, Z.L., Durley, A.P., Man, T.K., and Gitlin, J.D., 1999. Targeted gene disruption reveals an essential role for ceruloplasmin in cellular iron efflux. *Proc. Natl. Acad. Sci. USA* 96, 10812–10817.
123. Patel, B.N., Dunn, R.J., Jeong, S.Y., Zhu, Q., Julien, J.P., and David, S., 2002. Ceruloplasmin regulates iron levels in the CNS and prevents free radical injury. *J. Neurosci.* 22, 6578–6586.
124. Ryter, S.W., Alam, J., and Choi, A.M., 2006. Heme oxygenase-1/carbon monoxide: From basic science to therapeutic applications. *Physiol. Rev.* 86, 583–650.
125. Yachie, A., Niida, Y., Wada, T., Igarashi, N., Kaneda, H., Toma, T., Ohta, K., Kasahara, Y., and Koizumi, S., 1999. Oxidative stress causes enhanced endothelial cell injury in human heme oxygenase-1 deficiency. *J. Clin. Invest.* 103, 129–135.
126. Poss, K.D. and Tonegawa, S., 1997. Heme oxygenase 1 is required for mammalian iron reutilization. *Proc. Natl. Acad. Sci. USA* 94, 10919–10924.
127. Kato, J., Fujikawa, K., Kanda, M., Fukuda, N., Sasaki, K., Takayama, T., Kobune, M., et al., 2001. A mutation, in the iron-responsive element of H ferritin mRNA, causing autosomal dominant iron overload. *Am. J. Hum. Genet.* 69, 191–197.
128. Zecca, L., Youdim, M.B., Riederer, P., Connor, J.R., and Crichton, R.R., 2004. Iron, brain ageing and neurodegenerative disorders. *Nat. Rev. Neurosci.* 5, 863–873.
129. Kaur, D., Yantiri, F., Rajagopalan, S., Kumar, J., Mo, J.Q., Boonplueang, R., Viswanath, V., et al., 2003. Genetic or pharmacological iron chelation prevents MPTP-induced neurotoxicity *in vivo*: A novel therapy for Parkinson's disease. *Neuron* 37, 899–909.
130. Curtis, A.R., Fey, C., Morris, C.M., Bindoff, L.A., Ince, P.G., Chinnery, P.F., Coulthard, A., et al., 2001. Mutation in the gene encoding ferritin light polypeptide causes dominant adult-onset basal ganglia disease. *Nat. Genet.* 28, 350–354.
131. Vidal, R., Miravalle, L., Gao, X., Barbeito, A.G., Baraibar, M.A., Hekmatyar, S.K., Widel, M., Bansal, N., Delisle, M.B., and Ghetti, B., 2008. Expression of a mutant form of the ferritin light chain gene induces neurodegeneration and iron overload in transgenic mice. *J. Neurosci.* 28, 60–67.
132. Zhou, B., Westaway, S.K., Levinson, B., Johnson, M.A., Gitschier, J., and Hayflick, S.J., 2001. A novel pantothenate kinase gene (PANK2) is defective in Hallervorden–Spatz syndrome. *Nat. Genet.* 28, 345–349.

133. Ponka, P., Schulman, H.M., and Cox, T.M., 1990. *Iron Metabolism in Relation to Heme Synthesis*, New York: McGraw-Hill.
134. Lill, R. and Muhlenhoff, U., 2008. Maturation of iron-sulfur proteins in eukaryotes: Mechanisms, connected processes, and diseases. *Annu. Rev. Biochem.* 77, 669–700.
135. Camaschella, C., 2008. Recent advances in the understanding of inherited sideroblastic anaemia. *Brit. J. Haematol.* 143, 27–38.
136. Brownlie, A., Donovan, A., Pratt, S.J., Paw, B.H., Oates, A.C., Brugnara, C., Witkowska, H.E., Sassa, S., and Zon, L.I., 1998. Positional cloning of the zebrafish sauternes gene: A model for congenital sideroblastic anaemia. *Nat. Genet.* 20, 244–250.
137. Campuzano, V., Montermini, L., Molto, M.D., Pianese, L., Cossee, M., Cavalcanti, F., Monros, E., et al., 1996. Friedreich's ataxia: Autosomal recessive disease caused by an intronic GAA triplet repeat expansion. *Science* 271, 1423–1427.
138. Pandolfo, M., 1999. Friedreich's ataxia: Clinical aspects and pathogenesis. *Semin. Neurol.* 19, 311–321.
139. Babcock, M., de Silva, D., Oaks, R., Davis-Kaplan, S., Jiralerspong, S., Montermini, L., Pandolfo, M., and Kaplan, J., 1997. Regulation of mitochondrial iron accumulation by Yfh1p, a putative homolog of frataxin. *Science* 276, 1709–1712.
140. Puccio, H., Simon, D., Cossee, M., Criqui-Filipe, P., Tiziano, F., Melki, J., Hindelang, C., Matyas, R., Rustin, P., and Koenig, M., 2001. Mouse models for Friedreich ataxia exhibit cardiomyopathy, sensory nerve defect and Fe-S enzyme deficiency followed by intramitochondrial iron deposits. *Nat. Genet.* 27, 181–186.
141. Whitnall, M., Rahmanto, Y.S., Sutak, R., Xu, X., Becker, E.M., Mikhael, M.R., Ponka, P., and Richardson, D.R., 2008. The MCK mouse heart model of Friedreich's ataxia: Alterations in iron-regulated proteins and cardiac hypertrophy are limited by iron chelation. *Proc. Natl. Acad. Sci. USA* 105, 9757–9762.
142. Allikmets, R., Raskind, W.H., Hutchinson, A., Schueck, N.D., Dean, M., and Koeller, D.M., 1999. Mutation of a putative mitochondrial iron transporter gene (ABC7) in X-linked sideroblastic anemia and ataxia (XLSA/A). *Hum. Mol. Genet.* 8, 743–749.
143. Kispal, G., Csere, P., Prohl, C., and Lill, R., 1999. The mitochondrial proteins Atm1p and Nfs1p are essential for biogenesis of cytosolic Fe/S proteins. *EMBO J.* 18, 3981–3989.
144. Csere, P., Lill, R., and Kispal, G., 1998. Identification of a human mitochondrial ABC transporter, the functional orthologue of yeast Atm1p. *FEBS Lett.* 441, 266–270.
145. Pondarre, C., Antiochos, B.B., Campagna, D.R., Clarke, S.L., Greer, E.L., Deck, K.M., McDonald, A., et al., 2006. The mitochondrial ATP-binding cassette transporter Abcb7 is essential in mice and participates in cytosolic iron-sulfur cluster biogenesis. *Hum. Mol. Genet.* 15, 953–964.
146. Camaschella, C., Campanella, A., De Falco, L., Boschetto, L., Merlini, R., Silvestri, L., Levi, S., and Iolascon, A., 2007. The human counterpart of zebrafish shiraz shows sideroblastic-like microcytic anemia and iron overload. *Blood* 110, 1353–1358.
147. Wingert, R.A., Galloway, J.L., Barut, B., Foott, H., Fraenkel, P., Axe, J.L., Weber, G.J., et al., 2005. Deficiency of glutaredoxin 5 reveals Fe-S clusters are required for vertebrate haem synthesis. *Nature* 436, 1035–1039.
148. Mochel, F., Knight, M.A., Tong, W.H., Hernandez, D., Ayyad, K., Taivassalo, T., Andersen, P.M., et al., 2008. Splice mutation in the iron-sulfur cluster scaffold protein ISCU causes myopathy with exercise intolerance. *Am. J. Hum. Genet.* 82, 652–660.
149. Olsson, A., Lind, L., Thornell, L.E., and Holmberg, M., 2008. Myopathy with lactic acidosis is linked to chromosome 12q23.3-24.11 and caused by an intron mutation in the ISCU gene resulting in a splicing defect. *Hum. Mol. Genet.* 17, 1666–1672.
150. Clarke, S.L., Vasanthakumar, A., Anderson, S.A., Pondarre, C., Koh, C.M., Deck, K.M., Pitula, J.S., Epstein, C.J., Fleming, M.D., and Eisenstein, R.S., 2006. Iron-responsive degradation of iron-regulatory protein 1 does not require the Fe-S cluster. *EMBO J.* 25, 544–553.
151. Wang, J., Fillebeen, C., Chen, G., Biederbick, A., Lill, R., and Pantopoulos, K., 2007. Iron-dependent degradation of apo-IRP1 by the ubiquitin-proteasome pathway. *Mol. Cell. Biol.* 27, 2423–2430.
152. Pippard, M.J., 1994. *Secondary Iron Overload*, London: W.B. Saunders Company Ltd.
153. Tanno, T., Bhanu, N.V., Oneal, P.A., Goh, S.H., Staker, P., Lee, Y.T., Moroney, J.W., et al., 2007. High levels of GDF15 in thalassemia suppress expression of the iron regulatory protein hepcidin. *Nat. Med.* 13, 1096–1101.
154. Chaston, T.B. and Richardson, D.R., 2003. Iron chelators for the treatment of iron overload disease: Relationship between structure, redox activity, and toxicity. *Am. J. Hematol.* 73, 200–210.
155. Tam, T.F., Leung-Toung, R., Li, W., Wang, Y., Karimian, K., and Spino, M., 2003. Iron chelator research: Past, present, and future. *Curr. Med. Chem.* 10, 983–995.

156. Kontoghiorghes, G.J., Pattichis, K., Neocleous, K., and Kolnagou, A., 2004. The design and development of deferiprone (L1) and other iron chelators for clinical use: Targeting methods and application prospects. *Curr. Med. Chem.* 11, 2161–2183.
157. Liu, Z.D. and Hider, R.C., 2002. Design of clinically useful iron(III)-selective chelators. *Med. Res. Rev.* 22, 26–64.
158. Doulias, P.T., Christoforidis, S., Brunk, U.T., and Galaris, D., 2003. Endosomal and lysosomal effects of desferrioxamine: Protection of HeLa cells from hydrogen peroxide-induced DNA damage and induction of cell-cycle arrest. *Free Radic. Biol. Med.* 35, 719–728.
159. Kattamis, A., 2005. Combined therapy with deferoxamine and deferiprone. *Ann. NY Acad. Sci.* 1054, 175–182.
160. Cappellini, M.D., 2008. Long-term efficacy and safety of deferasirox. *Blood Rev.* 22(Suppl. 2), S35–S41.
161. Adams, P., Brissot, P., and Powell, L.W., 2000. EASL international consensus conference on haemochromatosis. *J. Hepatol.* 33, 485–504.
162. Alla, V. and Bonkovsky, H.L., 2005. Iron in nonhemochromatotic liver disorders. *Semin. Liver Dis.* 25, 461–472.
163. Alberti, A., Chemello, L., and Benvegnu, L., 1999. Natural history of hepatitis C. *J. Hepatol.* 31(Suppl 1), 17–24.
164. Seeff, L.B., 2002. Natural history of chronic hepatitis C. *Hepatology* 36, S35–S46.
165. Di Bisceglie, A.M., Axiotis, C.A., Hoofnagle, J.H., and Bacon, B.R., 1992. Measurements of iron status in patients with chronic hepatitis. *Gastroenterology* 102, 2108–2113.
166. Arber, N., Konikoff, F.M., Moshkowitz, M., Baratz, M., Hallak, A., Santo, M., Halpern, Z., Weiss, H., and Gilat, T., 1994. Increased serum iron and iron saturation without liver iron accumulation distinguish chronic hepatitis C from other chronic liver diseases. *Dig. Dis. Sci.* 39, 2656–2659.
167. Bonkovsky, H.L., Banner, B.F., and Rothman, A.L., 1997. Iron and chronic viral hepatitis. *Hepatology* 25, 759–768.
168. Sebastiani, G., Vario, A., Ferrari, A., Pistis, R., Noventa, F., and Alberti, A., 2006. Hepatic iron, liver steatosis and viral genotypes in patients with chronic hepatitis C. *J. Viral Hepatitis* 13, 199–205.
169. Piperno, A., Vergani, A., Malosio, I., Parma, L., Fossati, L., Ricci, A., Bovo, G., Boari, G., and Mancia, G., 1998. Hepatic iron overload in patients with chronic viral hepatitis: Role of HFE gene mutations. *Hepatology* 28, 1105–1109.
170. Van Thiel, D.H., Friedlander, L., Fagiuoli, S., Wright, H.I., Irish, W., and Gavaler, J.S., 1994. Response to interferon alpha therapy is influenced by the iron content of the liver. *J. Hepatol.* 20, 410–415.
171. Sherrington, C.A. and Olynyk, J.K., 2002. Iron as a cofactor in chronic hepatitis C infection. *Liver* 22, 187–189.
172. Corengia, C., Galimberti, S., Bovo, G., Vergani, A., Arosio, C., Mariani, R., Redaelli, A., et al., 2005. Iron accumulation in chronic hepatitis C: Relation of hepatic iron distribution, HFE genotype, and disease course. *Am. J. Clin. Pathol.* 124, 846–853.
173. Negro, F., Samii, K., Rubbia-Brandt, L., Quadri, R., Male, P.J., Zarski, J.P., Baud, M., Giostra, E., Beris, P., and Hadengue, A., 2000. Hemochromatosis gene mutations in chronic hepatitis C patients with and without liver siderosis. *J. Med. Virol.* 60, 21–27.
174. Pirisi, M., Scott, C.A., Avellini, C., Toniutto, P., Fabris, C., Soardo, G., Beltrami, C.A., and Bartoli, E., 2000. Iron deposition and progression of disease in chronic hepatitis C. Role of interface hepatitis, portal inflammation, and HFE missense mutations. *Am. J. Clin. Pathol.* 113, 546–554.
175. Kaji, K., Nakanuma, Y., Sasaki, M., Unoura, M., Kobayashi, K., and Nonomura, A., 1995. Hemosiderin deposition in portal endothelial cells: A novel hepatic hemosiderosis frequent in chronic viral hepatitis B and C. *Hum. Pathol.* 26, 1080–1085.
176. Boucher, E., Bourienne, A., Adams, P., Turlin, B., Brissot, P., and Deugnier, Y., 1997. Liver iron concentration and distribution in chronic hepatitis C before and after interferon treatment. *Gut* 41, 115–120.
177. Brunt, E.M., Olynyk, J.K., Britton, R.S., Janney, C.G., Di Bisceglie, A.M., and Bacon, B.R., 2000. Histological evaluation of iron in liver biopsies: Relationship to HFE mutations. *Am. J. Gastroenterol.* 95, 1788–1793.
178. Distante, S., Bjoro, K., Hellum, K.B., Myrvang, B., Berg, J.P., Skaug, K., Raknerud, N., and Bell, H., 2002. Raised serum ferritin predicts non-response to interferon and ribavirin treatment in patients with chronic hepatitis C infection. *Liver* 22, 269–275.
179. Thorburn, D., Curry, G., Spooner, R., Spence, E., Oien, K., Halls, D., Fox, R., McCruden, E.A., MacSween, R.N., and Mills, P.R., 2002. The role of iron and haemochromatosis gene mutations in the progression of liver disease in chronic hepatitis C. *Gut* 50, 248–252.

180. Hourigan, L.F., Macdonald, G.A., Purdie, D., Whitehall, V.H., Shorthouse, C., Clouston, A., and Powell, E.E., 1999. Fibrosis in chronic hepatitis C correlates significantly with body mass index and steatosis. *Hepatology* 29, 1215–1219.
181. Hwang, S.J., Luo, J.C., Chu, C.W., Lai, C.R., Lu, C.L., Tsay, S. H., Wu, J.C., Chang, F.Y., and Lee, S.D., 2001. Hepatic steatosis in chronic hepatitis C virus infection: Prevalence and clinical correlation. *J. Gastroenterol. Hepatol.* 16, 190–195.
182. Fargion, S., Mattioli, M., Fracanzani, A.L., Sampietro, M., Tavazzi, D., Fociani, P., Taioli, E., Valenti, L., and Fiorelli, G., 2001. Hyperferritinemia, iron overload, and multiple metabolic alterations identify patients at risk for nonalcoholic steatohepatitis. *Am. J. Gastroenterol.* 96, 2448–2455.
183. Bugianesi, E., Manzini, P., D'Antico, S., Vanni, E., Longo, F., Leone, N., Massarenti, P., Piga, A., Marchesini, G., and Rizzetto, M., 2004. Relative contribution of iron burden, HFE mutations, and insulin resistance to fibrosis in nonalcoholic fatty liver. *Hepatology* 39, 179–187.
184. Lecube, A., Hernandez, C., Genesca, J., Esteban, J.I., Jardi, R., Garcia, L., and Simo, R., 2004. Diabetes is the main factor accounting for the high ferritin levels detected in chronic hepatitis C virus infection. *Diabetes Care* 27, 2669–2675.
185. Sebastiani, G., Vario, A., and Alberti, A., 2005. Diabetes is the main factor accounting for the high ferritin levels detected in chronic hepatitis C virus infection: Response to Lecube et al. *Diabetes Care* 28, 1838 (author reply 1838–1839).
186. Bonkovsky, H.L., Naishadham, D., Lambrecht, R.W., Chung, R.T., Hoefs, J.C., Nash, S.R., Rogers, T.E., et al., 2006. Roles of iron and HFE mutations on severity and response to therapy during retreatment of advanced chronic hepatitis C. *Gastroenterology* 131, 1440–1451.
187. Fargion, S., Fracanzani, A.L., Rossini, A., Borzio, M., Riggio, O., Belloni, G., Bissoli, F., et al., 2002. Iron reduction and sustained response to interferon-alpha therapy in patients with chronic hepatitis C: Results of an Italian multicenter randomized study. *Am. J. Gastroenterol.* 97, 1204–1210.
188. Fillebeen, C., Rivas-Estilla, A.M., Bisaillon, M., Ponka, P., Muckenthaler, M., Hentze, M.W., Koromilas, A.E., and Pantopoulos, K., 2005. Iron inactivates the RNA polymerase NS5B and suppresses subgenomic replication of hepatitis C virus. *J. Biol. Chem.* 280, 9049–9057.
189. Fillebeen, C., Muckenthaler, M., Andriopoulos, B., Bisaillon, M., Mounir, Z., Hentze, M.W., Koromilas, A.E., and Pantopoulos, K., 2007. Expression of the subgenomic hepatitis C virus replicon alters iron homeostasis in Huh7 cells. *J. Hepatol.* 47, 12–22.
190. Fujita, N., Sugimoto, R., Motonishi, S., Tomosugi, N., Tanaka, H., Takeo, M., Iwasa, M., et al., 2008. Patients with chronic hepatitis C achieving a sustained virological response to peginterferon and ribavirin therapy recover from impaired hepcidin secretion. *J. Hepatol.* 49, 702–710.
191. Nishina, S., Hino, K., Korenaga, M., Vecchi, C., Pietrangelo, A., Mizukami, Y., Furutani, T., et al., 2008. Hepatitis C virus-induced reactive oxygen species raise hepatic iron level in mice by reducing hepcidin transcription. *Gastroenterology* 134, 226–238.
192. Miura, K., Taura, K., Kodama, Y., Schnabl, B., and Brenner, D.A., 2008. Hepatitis C virus-induced oxidative stress suppresses hepcidin expression through increased histone deacetylase activity. *Hepatology* 48, 1420–1429.
193. Bonkovsky, H.L., Lambrecht, R.W., and Shan, Y., 2003. Iron as a co-morbid factor in nonhemochromatotic liver disease. *Alcohol* 30, 137–144.
194. Day, C.P. and Bassendine, M.F., 1992. Genetic predisposition to alcoholic liver disease. *Gut* 33, 1444–1447.
195. Ludwig, J., Hashimoto, E., Porayko, M.K., Moyer, T.P., and Baldus, W.P., 1997. Hemosiderosis in cirrhosis: A study of 447 native livers. *Gastroenterology* 112, 882–888.
196. Milman, N., Ovesen, L., Byg, K., and Graudal, N., 1999. Iron status in Danes updated 1994. I: Prevalence of iron deficiency and iron overload in 1332 men aged 40–70 years. Influence of blood donation, alcohol intake, and iron supplementation. *Ann. Hematol.* 78, 393–400.
197. Ioannou, G.N., Dominitz, J.A., Weiss, N.S., Heagerty, P.J., and Kowdley, K.V., 2004. The effect of alcohol consumption on the prevalence of iron overload, iron deficiency, and iron deficiency anemia. *Gastroenterology* 126, 1293–1301.
198. Grove, J., Daly, A.K., Burt, A.D., Guzail, M., James, O.F., Bassendine, M.F., and Day, C.P., 1998. Heterozygotes for HFE mutations have no increased risk of advanced alcoholic liver disease. *Gut* 43, 262–266.
199. Gleeson, D., Evans, S., Bradley, M., Jones, J., Peck, R.J., Dube, A., Rigby, E., and Dalton, A., 2006. HFE genotypes in decompensated alcoholic liver disease: Phenotypic expression and comparison with heavy drinking and with normal controls. *Am. J. Gastroenterol.* 101, 304–310.
200. Fargion, S., Fracanzani, A.L., Piperno, A., Braga, M., D'Alba, R., Ronchi, G., and Fiorelli, G., 1994. Prognostic factors for hepatocellular carcinoma in genetic hemochromatosis. *Hepatology* 20, 1426–1431.

201. Adams, P.C. and Agnew, S., 1996. Alcoholism in hereditary hemochromatosis revisited: Prevalence and clinical consequences among homozygous siblings. *Hepatology* 23, 724–727.
202. Raynard, B., Balian, A., Fallik, D., Capron, F., Bedossa, P., Chaput, J.C., and Naveau, S., 2002. Risk factors of fibrosis in alcohol-induced liver disease. *Hepatology* 35, 635–638.
203. Grove, J., Daly, A.K., Bassendine, M.F., and Day, C.P., 1997. Association of a tumor necrosis factor promoter polymorphism with susceptibility to alcoholic steatohepatitis. *Hepatology* 26, 143–146.
204. Grove, J., Brown, A.S., Daly, A.K., Bassendine, M.F., James, O.F., and Day, C. P., 1998. The RsaI polymorphism of CYP2E1 and susceptibility to alcoholic liver disease in Caucasians: Effect on age of presentation and dependence on alcohol dehydrogenase genotype. *Pharmacogenetics* 8, 335–342.
205. Lauret, E., Rodriguez, M., Gonzalez, S., Linares, A., Lopez-Vazquez, A., Martinez-Borra, J., Rodrigo, L., and Lopez-Larrea, C., 2002. HFE gene mutations in alcoholic and virus-related cirrhotic patients with hepatocellular carcinoma. *Am. J. Gastroenterol.* 97, 1016–1021.
206. Nahon, P., Sutton, A., Rufat, P., Ziol, M., Thabut, G., Schischmanoff, P.O., Vidaud, D., et al., 2008. Liver iron, HFE gene mutations, and hepatocellular carcinoma occurrence in patients with cirrhosis. *Gastroenterology* 134, 102–110.
207. Pascoe, A., Kerlin, P., Steadman, C., Clouston, A., Jones, D., Powell, L., Jazwinska, E., Lynch, S., and Strong, R., 1999. Spur cell anaemia and hepatic iron stores in patients with alcoholic liver disease undergoing orthotopic liver transplantation. *Gut* 45, 301–305.
208. Suzuki, Y., Saito, H., Suzuki, M., Hosoki, Y., Sakurai, S., Fujimoto, Y., and Kohgo, Y., 2002. Up-regulation of transferrin receptor expression in hepatocytes by habitual alcohol drinking is implicated in hepatic iron overload in alcoholic liver disease. *Alcohol. Clin. Exp. Res.* 26, 26S–31S.
209. Harrison-Findik, D.D., Schafer, D., Klein, E., Timchenko, N.A., Kulaksiz, H., Clemens, D., Fein, E., Andriopoulos, B., Pantopoulos, K., and Gollan, J., 2006. Alcohol metabolism-mediated oxidative stress down-regulates hepcidin transcription and leads to increased duodenal iron transporter expression. *J. Biol. Chem.* 281, 22974–22982.
210. Harrison-Findik, D.D., Klein, E., Crist, C., Evans, J., Timchenko, N., and Gollan, J., 2007. Iron-mediated regulation of liver hepcidin expression in rats and mice is abolished by alcohol. *Hepatology* 46, 1979–1985.
211. Clark, J.M. and Diehl, A.M., 2003. Nonalcoholic fatty liver disease: An underrecognized cause of cryptogenic cirrhosis. *JAMA* 289, 3000–3004.
212. Matteoni, C.A., Younossi, Z.M., Gramlich, T., Boparai, N., Liu, Y.C., and McCullough, A.J., 1999. Nonalcoholic fatty liver disease: A spectrum of clinical and pathological severity. *Gastroenterology* 116, 1413–1419.
213. Day, C.P. and James, O.F., 1998. Steatohepatitis: A tale of two "hits"? *Gastroenterology* 114, 842–845.
214. Tsukamoto, H., Horne, W., Kamimura, S., Niemela, O., Parkkila, S., Yla-Herttuala, S., and Brittenham, G.M., 1995. Experimental liver cirrhosis induced by alcohol and iron. *J. Clin. Invest.* 96, 620–630.
215. Duvnjak, M., Lerotic, I., Barsic, N., Tomasic, V., Virovic Jukic, L., and Velagic, V., 2007. Pathogenesis and management issues for non-alcoholic fatty liver disease. *World J. Gastroenterol.* 13, 4539–4550.
216. Bacon, B.R., Farahvash, M.J., Janney, C.G., and Neuschwander-Tetri, B.A., 1994. Nonalcoholic steatohepatitis: An expanded clinical entity. *Gastroenterology* 107, 1103–1109.
217. George, D.K., Goldwurm, S., MacDonald, G.A., Cowley, L.L., Walker, N.I., Ward, P.J., Jazwinska, E.C., and Powell, L.W., 1998. Increased hepatic iron concentration in nonalcoholic steatohepatitis is associated with increased fibrosis. *Gastroenterology* 114, 311–318.
218. Angulo, P., Keach, J.C., Batts, K.P., and Lindor, K.D., 1999. Independent predictors of liver fibrosis in patients with nonalcoholic steatohepatitis. *Hepatology* 30, 1356–1362.
219. Younossi, Z.M., Gramlich, T., Bacon, B.R., Matteoni, C.A., Boparai, N., O'Neill, R., and McCullough, A.J., 1999. Hepatic iron and nonalcoholic fatty liver disease. *Hepatology* 30, 847–850.
220. Facchini, F.S., Hua, N.W., and Stoohs, R.A., 2002. Effect of iron depletion in carbohydrate-intolerant patients with clinical evidence of nonalcoholic fatty liver disease. *Gastroenterology* 122, 931–939.
221. Valenti, L., Fracanzani, A.L., Dongiovanni, P., Bugianesi, E., Marchesini, G., Manzini, P., Vanni, E., and Fargion, S., 2007. Iron depletion by phlebotomy improves insulin resistance in patients with nonalcoholic fatty liver disease and hyperferritinemia: Evidence from a case-control study. *Am. J. Gastroenterol.* 102, 1251–1258.
222. Sorrentino, P., D'Angelo, S., Ferbo, U., Micheli, P., Bracigliano, A., and Vecchione, R., 2009. Liver iron excess in patients with hepatocellular carcinoma developed on non-alcoholic steato-hepatitis. *J. Hepatol.* 50, 351–357.

223. Bonkovsky, H.L., Jawaid, Q., Tortorelli, K., LeClair, P., Cobb, J., Lambrecht, R. W., and Banner, B.F., 1999. Non-alcoholic steatohepatitis and iron: Increased prevalence of mutations of the HFE gene in non-alcoholic steatohepatitis. *J. Hepatol.* 31, 421–429.
224. Chitturi, S., Weltman, M., Farrell, G.C., McDonald, D., Kench, J., Liddle, C., Samarasinghe, D., Lin, R., Abeygunasekera, S., and George, J., 2002. HFE mutations, hepatic iron, and fibrosis: Ethnic-specific association of NASH with C282Y but not with fibrotic severity. *Hepatology* 36, 142–149.
225. Harrison, P.M. and Arosio, P., 1996. The ferritins: Molecular properties, iron storage function and cellular regulation. *Biochim. Biophys. Acta* 1275, 161–203.
226. Torti, F.M. and Torti, S.V., 2002. Regulation of ferritin genes and protein. *Blood* 99, 3505–3516.
227. Beaumont, C., Leneuve, P., Devaux, I., Scoazec, J.-Y., Berthier, M., Loiseau, M.-N., Grandchamp, B., and Bonneau, D., 1995. Mutation in the iron responsive element of the L ferritin mRNA in a family with dominant hyperferritinaemia and cataract. *Nat. Genet.* 11, 444–446.
228. Roetto, A., Bosio, S., Gramaglia, E., Barilaro, M.R., Zecchina, G., and Camaschella, C., 2002. Pathogenesis of hyperferritinemia cataract syndrome. *Blood Cells Mol. Dis.* 29, 532–535.
229. Allerson, C.R., Cazzola, M., and Rouault, T.A., 1999. Clinical severity and thermodynamic effects of iron-responsive element mutations in hereditary hyperferritinemia-cataract syndrome. *J. Biol. Chem.* 274, 26439–26447.
230. Levi, S., Girelli, D., Perrone, F., Pasti, M., Beaumont, C., Corrocher, R., Albertini, A., and Arosio, P., 1998. Analysis of ferritins in lymphoblastoid cell lines and in the lens of subjects with hereditary hyperferritinemia-cataract syndrome. *Blood* 91, 4180–4187.
231. Guyader, D., Jacquelinet, C., Moirand, R., Turlin, B., Mendler, M.H., Chaperon, J., David, V., Brissot, P., Adams, P., and Deugnier, Y., 1998. Noninvasive prediction of fibrosis in C282Y homozygous hemochromatosis. *Gastroenterology* 115, 929–936.
232. Sebastiani, G. and Alberti, A., 2006. Non invasive fibrosis biomarkers reduce but not substitute the need for liver biopsy. *World J. Gastroenterol.* 12, 3682–3694.
233. Sandrin, L., Fourquet, B., Hasquenoph, J.M., Yon, S., Fournier, C., Mal, F., Christidis, C., et al., 2003. Transient elastography: A new noninvasive method for assessment of hepatic fibrosis. *Ultrasound Med. Biol.* 29, 1705–1713.
234. Castera, L., Vergniol, J., Foucher, J., Le Bail, B., Chanteloup, E., Haaser, M., Darriet, M., Couzigou, P., and De Ledinghen, V., 2005. Prospective comparison of transient elastography, Fibrotest, APRI, and liver biopsy for the assessment of fibrosis in chronic hepatitis C. *Gastroenterology* 128, 343–350.
235. Ziol, M., Handra-Luca, A., Kettaneh, A., Christidis, C., Mal, F., Kazemi, F., de Ledinghen, V., et al., 2005. Noninvasive assessment of liver fibrosis by measurement of stiffness in patients with chronic hepatitis C. *Hepatology* 41, 48–54.
236. Adhoute, X., Foucher, J., Laharie, D., Terrebonne, E., Vergniol, J., Castera, L., Lovato, B., et al., 2008. Diagnosis of liver fibrosis using FibroScan and other noninvasive methods in patients with hemochromatosis: A prospective study. *Gastroenterol. Clin. Biol.* 32, 180–187.
237. Smith, B.C., Gorve, J., Guzail, M.A., Day, C.P., Daly, A.K., Burt, A.D., and Bassendine, M.F., 1998. Heterozygosity for hereditary hemochromatosis is associated with more fibrosis in chronic hepatitis C. *Hepatology* 27, 1695–1699.
238. Martinelli, A.L., Franco, R.F., Villanova, M.G., Figueiredo, J.F., Secaf, M., Tavella, M.H., Ramalho, L.N., Zucoloto, S., and Zago, M.A., 2000. Are haemochromatosis mutations related to the severity of liver disease in hepatitis C virus infection? *Acta Haematol.* 102, 152–156.
239. Bonkovsky, H.L., Troy, N., McNeal, K., Banner, B.F., Sharma, A., Obando, J., Mehta, S., Koff, R.S., Liu, Q., and Hsieh, C.C., 2002. Iron and HFE or TfR1 mutations as comorbid factors for development and progression of chronic hepatitis C. *J. Hepatol.* 37, 848–854.
240. Tung, B.Y., Emond, M.J., Bronner, M.P., Raaka, S.D., Cotler, S.J., and Kowdley, K.V., 2003. Hepatitis C, iron status, and disease severity: Relationship with HFE mutations. *Gastroenterology* 124, 318–326.
241. Erhardt, A., Maschner-Olberg, A., Mellenthin, C., Kappert, G., Adams, O., Donner, A., Willers, R., Niederau, C., and Haussinger, D., 2003. HFE mutations and chronic hepatitis C: H63D and C282Y heterozygosity are independent risk factors for liver fibrosis and cirrhosis. *J. Hepatol.* 38, 335–342.
242. Kazemi-Shirazi, L., Datz, C., Maier-Dobersberger, T., Kaserer, K., Hackl, F., Polli, C., Steindl, P.E., Penner, E., and Ferenci, P., 1999. The relation of iron status and hemochromatosis gene mutations in patients with chronic hepatitis C. *Gastroenterology* 116, 127–134.

14 Metal Influences on Immune Function

Michael A. Lynes, Kathryn Pietrosimone, Gregory Marusov, Douglas V. Donaldson, Clare Melchiorre, Xiuyun Yin, David A. Lawrence, and Michael J. McCabe

CONTENTS

14.1 Introduction 379
14.2 Examples of Essential Metal Influences on Immune Mechanisms 380
14.2.1 Selenium 380
14.2.2 Iron as an Immune Regulator 380
14.2.3 The Role of Zinc in Immunological Function 381
14.2.4 Copper as an Immunomodulator 382
14.3 Immunotoxic Effects of Nonessential Metals 382
14.3.1 Metals, Cellular Thiols, and Oxidative Stress 383
14.3.2 Metal Influences on Apoptotic Signaling Pathways 386
14.3.3 Metallothionein: A Mediator of Metal-Dependent Immune Changes 389
14.3.4 Induction of MT by Metals 389
14.3.5 MT Synthesis and Distribution 390
14.3.6 Other Toxicant-Induced Homeostasis Mechanisms 392
14.3.7 Regulation of HSPs 392
14.3.8 HSPs and Immune Function 393
14.3.9 Metal-Induced HSPs in Immune Cells 394
14.3.10 Metal Exposure can Alter a Variety of Cellular Activities 395
14.3.11 Metal Effects on Protein Secretion 396
14.4 Metal-Mediated Immunosuppression 396
14.5 Toxic Metal-Associated Immune Activation 396
14.5.1 Metal-Induced Hypersensitivity 396
14.5.2 Heavy Metals and Autoimmunity 397
14.6 Detection and Assessment of Metal-Mediated Immunomodulation 398
Acknowledgment 400
References 400

14.1 INTRODUCTION

Essential metals can have beneficial influences on immune activities when present in appropriate concentrations, but can disturb the normal functions of the immune response when present in suboptimal amounts or in excess. Metal toxicants can also alter immune activities, in some cases depressing protective defenses against pathogens and biological toxins and in others producing excessive and damaging immune activities. The consequences of these alterations can be dramatic: conditions that diminish immune capacity can render an individual susceptible to significant infectious disease, while other conditions that produce inappropriate activation of immunity may result in chronic inflammation, hypersensitivities, debilitating autoimmune disease, and neoplasia. These changes arise from a variety of fundamental influences that metals can have on cellular biochemistry, gene regulation, and cellular homeostasis. This chapter is designed to explore the roles played by metals in both normal and dysfunctional immune activities.

14.2 EXAMPLES OF ESSENTIAL METAL INFLUENCES ON IMMUNE MECHANISMS

14.2.1 Selenium

Selenium (Se) was first discovered in 1817 by Berzelius [1] but was not classified as an essential metal until 1957 when selenium was found to prevent the onset of vitamin E-induced liver cirrhosis in rats [2]. The amount of selenium recommended for adult human consumption is 55 μg/day. Up to 300 μg/day may be beneficial in reducing the risk of certain cancers, but higher amounts are toxic [3]. Selenium is incorporated into selenoproteins as selenomethionine (SeMet) or selenocysteine (Sec) cotranslationally by special SeMet and Sec-specific tRNAs, and is found in at least 30 different selenium-dependent proteins [4]. Glutathione peroxidase (GPx) is an example of a selenium-containing protein that has an antioxidant role in the enzymatic management of hydrogen peroxide and the lipid hydroperoxides. There are five known forms of GPx in mammals, four of which are intracellular proteins, while the fifth, GPx3, is a secreted isoform [2,5]. GPx deficiencies can be associated with defective management of oxidants resulting in increased oxidative stress and functional defects in the immune response. Other important selenium-containing proteins include thioredoxin reductase, selenoprotein P, and selenoprotein W, which can also each contribute to the management of the cellular redox environment [6].

Immunological changes associated with selenium deficiency include increased susceptibility to viral and bacterial infections [7], decreased IgG and IgM titers [6], and other immunological abnormalities. Keshan's disease is associated with a cocksackievirus (CVB3/0) [7] and is endemic in certain parts of China where there are low levels of naturally occurring selenium. Selenium deficiency not only leads to immune deficits but is also linked to increased rates of mutation in some RNA viruses. This effect may be related to increased oxidative stress in the absence of sufficient GPx and can lead to decreased efficacy of the immune response to viral infection. A study of influenza infection in a selenium-deficient model showed that virus infection was associated with increased oxidative stress, decreased antioxidant enzyme activity, an increase in IL-6 production (indicative of inflammatory processes), and with reduced leukocyte counts [7,8]. In contrast, selenium supplementation can improve the immune response against viral infection. Supplementation studies have shown improved $CD4^+$ cell counts and decreased viral burdens in HIV-1 infections [7]. In the mid-1980s, recognition of selenium deficiency-associated disorders led the government of Finland (where dietary selenium was known to be among the lowest in the world) to mandate the addition of sodium selenate to agricultural fertilizers. Within a few years, dietary selenium levels were elevated across the country [9], with presumed benefits to the population [10].

14.2.2 Iron as an Immune Regulator

While iron is critical for many cellular and physiological functions, it can also be quite toxic due to its capacity to generate reactive oxygen species (ROS) via the Fenton reaction. Of the normal amount of dietary iron consumed per day (12–18 mg), only 1–2 mg of the iron is absorbed [11]. The immune system specifically regulates iron metabolism with two outcomes: to arm innate immune cells with ROS-producing capacity and to restrict iron access from pathogens. Immune cells can specifically contribute to this regulation of iron availability. As an example of this phenomenon, mice that carry a defect in the adaptive immune response (e.g., a mutation in the β_2-microglobulin gene that is essential for major histocompatibility complex [MHC] class I expression) are deficient in the ability to manage cellular iron levels (reviewed in Ref. [12]). Iron can be used by cells to activate NF-κB (a central transcription factor that acts in the inflammatory cascade of signal transduction) via the production of ROS and activation of IκB kinase (IKK) [13]. Iron levels also regulate the activity of inducible nitric oxide synthase (iNOS), an enzyme responsible for producing the oxidant nitric oxide. iNOS transcription is reduced when iron levels are high, which presumably serves to reduce oxidative stress. Excess iron has also been shown to impede the assembly of NADPH oxidase, another ROS-producing enzyme that is critical for innate immune functions [11].

The modulation of bioavailable iron is also thought to limit accessibility to iron for invading pathogens. Hepcidin, originally known as LEAP-1 (liver-expressed antimicrobial protein; [14]), is a peptide hormone regulator of iron [15] with important roles to play in inflammation [16]. Hepcidin can be upregulated by proinflammatory cytokines such as IL-6 and IL-1, and by stimulation through the Toll-like receptor 4 (TLR4) pathway.

Due to the increase in oxidative stress associated with iron overload, immune activities have the potential to become severely disrupted. For example, infection with hepatitis C virus (HCV) is associated with iron accumulation in the liver. This occurs because HCV can suppress the expression of hepcidin and prevent iron sequestration [13]. Excess iron in the liver has also been shown to increase HCV gene expression [13]. Because of the damaging affects associated with iron overload, iron chelators are currently being developed as antiviral therapies [13].

14.2.3 The Role of Zinc in Immunological Function

The relationship between zinc and the immune system is probably the most well documented of all the metals. It was first shown to be important in biological systems in 1869, when Raulin showed that it was required for the growth of *Aspergillus niger* [17]. There are over 300 enzymes that require zinc for enzymatic activity and at least 2000 transcription factors that incorporate zinc in their structure [17,18]. Zinc often interacts with proteins via interactions with cysteine and histidine resides. Zinc finger structural motifs are found in transcription factors that utilize zinc to form secondary structures that can interact with DNA.

Eukaryotic cells have been estimated to contain about 250 μM zinc. Uncomplexed or labile cytosolic zinc in the cell has been measured to be in the nanomolar to picomolar range [19,20] and is tightly regulated. While the biology of zinc proteins and enzymes is well established, the role of the labile zinc in the immune system is less well understood. Zinc is believed to participate in the regulation of cellular redox status. This is performed in concert with a number of zinc-binding proteins including metallothionein (MT) [21].

Zinc trafficking is complex. The transport of zinc in and out of cells and cellular compartments is mediated by two families of zinc transporter proteins called Zips (import) and ZnTs (export). As is the case with iron, the restriction of zinc availability to pathogens during infection can diminish the severity of infection. This is accomplished in the acute phase that follows initial infection and results from the production of IL-6. IL-6 upregulates Zip14 expression in the liver, causing plasma zinc levels to fall while liver zinc stores increase [19,22]. IL-6 also results in increased MT production, which may similarly alter metal distribution and bioavailability [23].

There are many immune disorders that are associated with zinc deficiency. Specific effects that have been observed include decreased IL-2 and interferon-γ (IFN-γ) production, changes in the levels of pro- and anti-inflammatory cytokines, decreased natural killer cell activity, reductions in phagocytic activity, and thymic atrophy (reviewed in Ref. [17]). It has been suggested that a nutritional deficiency of zinc can also skew the normal Th1 and Th2 balance [17,24]. Due to the array of problems associated with insufficient zinc as well as the apparent benefits of elevated zinc levels *in vivo* and *in vitro*, many studies have focused on the effects of zinc supplementation. Zn supplementation has been suggested to be beneficial in patients and in animal models of rheumatoid arthritis and has resulted in decreased circulating levels of proinflammatory cytokines [25,26]. It is worth noting that zinc supplementation does not appear to be beneficial in every context. In studies that examined zinc supplementation during bacterial sepsis it was found that the hydrocarbon chains of lipopolysaccharide (LPS) can become associated with labile zinc, resulting in a more rigid structure. Increased LPS rigidity promotes binding to TLR4 and results in an upregulation of proinflammatory cytokines, which can cause undesirable side effects [27].

14.2.4 Copper as an Immunomodulator

Humans require about 1.5–3 mg of copper per day, and absorption can be enhanced up to twofold by the addition of dietary methionine [28,29]. Normally, copper is sequestered by cells to prevent free copper ions from entering the blood stream. Like iron, copper can generate ROS. Copper can compete with zinc for zinc-binding catalytic sites and for zinc fingers in transcription factors, resulting in loss of function of these proteins [30]. There are a number of important copper-containing enzymes including Cu/Zn superoxide dismutase (Cu/Zn SOD; which is involved in the detoxification of ROS) and cytochrome *c* oxidase (which is required for mitochondrial respiration and electron transport). Copper transport is controlled by ATP7A and ATP7B, two P-type ATPases that transport copper across plasma membranes [30]. ATP7A normally functions to deliver Cu ions to copper-containing enzymes that are trafficking through the Golgi apparatus. During copper overload however, ATP7A can be localized to the cell surface so that copper can be shuttled out of the cell. Copper is carried to target proteins by copper-chaperone proteins. For example, ATOX1 is a Cu-chaperone that transports copper directly to ATP7A, while the copper chaperone for SOD1 (CCS) delivers copper directly to SOD1 [30]. Glutathione (GSH) can act as a copper-chaperone as well, transporting copper to bile or to MTs [30,31].

An important role that copper plays in the immune system is the regulation of ROS by antioxidant enzymes such as SOD1 [32]. Although information regarding the specific roles played by copper in the immune system is currently somewhat limited, there is substantial information on the impact of copper deficiencies on immunity. Insufficient copper intake (<1 mg/day) is associated with a wide variety of conditions ranging from decreased SOD activity [29] to an increased incidence of infection, decreased phagocytic capacity by innate immune cells, neutropenia, decreased natural killer cell activity, decreased respiratory burst, diminished antibody titers, and reduced $CD4^+$ T cell numbers [31]. Many mammalian species will also produce reduced levels of IL-2 if they are deficient in copper. This has made IL-2 a biomarker of Cu status [31], and it is thought that this IL-2 deficiency contributes to the observed reductions in T cell numbers [33].

14.3 IMMUNOTOXIC EFFECTS OF NONESSENTIAL METALS

In many ways, the influence of nonessential metals on biological systems is reminiscent of the effects of essential metals when they are present at levels outside the optimal range. This reflects the chemical similarities of metals and their propensities to interact with similar cellular partners. By displacing essential metals or by causing similar changes to cellular biochemistries, toxic metals will influence a broad range of fundamental cell behaviors. The immune system is a sentinel system of toxic metal exposures because disturbances in immunity can cause substantial and rapid changes to general health parameters. These changes can render an individual susceptible to infectious diseases or can activate damaging aspects of the immune response: where normal self-recognition (that is required for normal immune differentiation) progresses to autoimmune disease or where chronic inflammation produces bystander damage in nearby tissues.

It is important to emphasize that toxic effects of metals are not restricted or limited to a simple cytotoxic effect. Certainly, high doses of any metal, including the essential metals such as copper and zinc, can induce cell death that interferes with normal immune activities [34], but lower doses can be equally harmful by altering both regulatory and effector functions of the immune system [35]. There are at least four different mechanisms by which metals can modify normal cellular functions. Redox-active metals can enhance the production of ROS, causing lipid peroxidation and formation of reactive intermediates that can adduct nucleic acids and proteins. Additionally, metals such as mercury (Hg), cadmium (Cd), arsenic (As), and nickel (Ni) can directly bind to sulfhydryl groups of proteins and peptides. The major peptide bound by these metals is GSH. GSH is the major antioxidant in all living cells [36,37] and is known to decline with aging [38,39]. A third avenue of toxic metal influence is via replacement of an essential metal; numerous natural functions require a specific metal for normal function that are disturbed if another metal (perhaps with a higher binding affinity or if present in excess) displaces the essential metal in the native structure. A classic example of this phenomenon is GPx. When the selenium in its active site is displaced, enzymatic activity can be reduced. For example, GPx is inhibited by Cu^{2+} or Hg^{2+} (100%), as well as by Ni^{2+} (65%), Co^{2+} (39%), Zn^{2+} (31%), Ca^{2+} (25%), and Mg^{2+} (20%) [40]. A fourth mechanism is metal substitution for other cellular ions. This effect is exemplified by lead (Pb^{2+}) substituting for Ca^{2+}. Since most intracellular Ca^{2+} is bound, Pb can increase the level of free cytosolic Ca^{2+} [41], leading to oxidative stress in lymphocytes [42].

14.3.1 Metals, Cellular Thiols, and Oxidative Stress

Although metals can influence immune function by a number of different mechanisms, a common change in cellular biochemistry that has pervasive effects on immune activities is related to metal modification of cellular thiols (R–SH). The modification of thiols can be due to oxidation (R–S–S–R) or alkylation (R–S–R′) of cellular thiol targets. Depletion of cellular thiols can result in immunomodulation due to a restructuring of plasma membrane lipids and proteins (that contribute to receptor function and intracellular signal transduction circuits) of both lymphocytes and antigen-presenting cells (APCs). These changes can alter early signal transduction events associated with APC and T cell communication, subsequent activation of both cell types, proliferation or differentiation events, and/or cell trafficking within the body.

Oxidant/antioxidant parameters that change with exposure to many of the metals are similar in many ways to those that occur during the normal aging process, a process known to result in declining immune functions. Studies of the effects of aging or the effects of oxidants have been used to understand the mechanisms that underlie metal-mediated immunomodulation. The extracellular redox environment is intimately linked to cytosolic and nuclear redox conditions. Cellular thiol-containing proteins can influence lipid domains found in the membrane [43,44], and membranal lipids can influence the levels of cellular thiols [45]. Thus, oxidative stress can directly alter the lipid composition of the plasma membrane via lipid peroxidation [46]. Copper (Cu) has long been known to be able to induce lipid peroxidation [7] due to its redox cycling potential. Like Cu^{2+}, other metals, such as iron (Fe), chromium (Cr), vanadium (V), and cobalt (Co), can also generate reactive radicals and cause lipid peroxidation [47,48]. Many compelling theories that have been proposed to account for the aging process are related to oxidative stress [49–53] or loss of antioxidant capacity. Included among these theories for aging are the following: the cross-linkage theory [49], the radiation damage (or free radical) theory [50], and the autoimmune reactive theory [54]. All of these theories are connected to loss or modification of cellular thiols. Since metals are known to affect cross-linking [55], radical-induced damage [47,48], and autoimmune activity [56], it can be posited that the loss of immune function that accompanies aging is not an intrinsic process but could be affected by the accumulation of metals or modification of metalloproteins. A recent review of the current literature has concluded that the decline in immune functions with aging is due to both intrinsic and environmental influences [57].

Many signaling events required for T cell activation are thiol-sensitive, including signaling via CD4 through the cytosolic kinase $p56^{Lck}$ that acts very early in the signal transduction cascade, as well as phospholipase Cγ1 (PLCγ1) and a variety of other kinases [58–60]. The structural configuration of the transmembrane domains of the T cell receptor (TCR) and the associated CD3 and CD4 or CD8 molecules governs their association with intramembranal enzymes, cytoplasmic associates (e.g., p56^{Lck}), and the lipid domains of the plasma membrane. Both CD3 and CD4 undergo phosphorylation and internalization after T cell activation that is dependent upon their proximity to specific transmembranal and cytoplasmic kinases. Since the protein–protein, protein–lipid, and lipid–lipid interactions that occur near the transmembrane regions of membrane-associated proteins are sensitive to cellular thiols [61–67], they may influence the preferential activation of specific T cell subsets. There have been reports that Th1 and Th2 cells require the expression of distinct costimulatory signals (molecules in addition to Class II bound peptide) on APCs [68–70]. The CD4 molecule is of specific interest since it is known to be susceptible to critical intra- [59,69] and extracellular [71] thiol influences on activation; however, other surface molecules such as chemokine receptors [72], the CD45 family of molecules [73], and L-selectin (CD62L) [74] also participate in T cell activation in ways that are sensitive to thiol changes. Thus, metals with affinity for thiols could modulate the functions of these molecules. For example, vanadate, molybdate, and salts of zinc, copper, and mercury are all effective enzyme inhibitors of the protein tyrosine phosphatase (PTPase) activity of CD45 [75], a protein with a cysteine-rich extracellular domain.

It has been reported that not all immune activities decline with age at the same pace [76], in that $CD8^+$ cells decline more rapidly than $CD4^+$ cells. In fact, Th2 cells can become hyperactive in some older individuals. It is interesting to note that HIV seropositive individuals progress to ARC/AIDS more readily when IL-4 levels (produced by Th2 cells) predominate over IL-2 (produced by Th1 cells) [77]. Within the murine acquired immunodeficiency syndrome (MAIDS) model system, upregulation of Th2 activity also has been suggested [78]. When murine Th1 and Th2 clones are exposed to heavy metals (lead and mercury), which accumulate in the body, there is a preferential enhancement of Th2 activity [79]. The mechanisms involved in the differential activation of Th1 and Th2 cells with aging, HIV or heavy metals (all oxidative stressors) have not yet been delineated.

It has been reported that oxidative stress that affects APCs can allow a Th2-type immune response to predominate [80]. Surface expression of costimulatory molecules on APC or the ligands/receptors on the T cells after oxidative stress could be involved as could oxidative influences on IL-12 expression [80]. The ability of macrophages to regulate T cell activation was originally proposed in 1981 [81]. It has recently been reported that suppressive macrophages (which have a higher oxidative capacity than reducing capacity) can preferentially inhibit Th1 cells [82]. Some studies have suggested that Th1 cells require a higher density of antigen (MHC class II proteins folded around a peptide) than Th2 cells [83]. Increased expression of the MHC class II/peptide complex and the essential accessory molecules at sites of contact between APC and T cells [84] may be more of an activation requirement for Th1 than Th2 cells since Th1 cells appear to require a higher density of antigen to provoke activation of the T cell.

A change in membrane structure/fluidity could also influence the process of receptor/ligand redistribution that is required for activation. Since metal-induced ROS can alter membrane lipid structure and the fluidity of the membrane, it is reasonable to link metal effects to the activation of specific regulatory T cells. This function of APC does seem to be influenced by the cell membrane, in that ultraviolet light treatment (another ROS inducer) has been reported to cause a loss in APC clustering with T cells [85]. Cholesterol, a known modulator of membrane structure, also can modify antigen presentation [86]. Alterations in membrane structure/fluidity could provide an explanation for why oxidative stress exacerbates disease progression. Other reasons could involve intrinsic differences between the Th1 and Th2 cells themselves. Suppressive macrophages, which produce more oxidants than reducing equivalents, could be involved in apoptosis [87], and Th1 cells may be more sensitive. Oxidants [88,89] and thiol-binding metals [90] do seem to mutually promote Th2 development.

Further investigation of metal exposures and the age-related changes in the activation and functions of the Th1 and Th2 subsets is clearly needed, because there are conflicting reports in the literature on aging/oxidant immunomodulation. For example, production of IL-4 (from Th2 cells) has been reported to increase [91,92] or decrease [93,94] with aging. Most studies have employed murine cells and/or have assessed *in vitro* lymphocyte functions without consideration of the oxidative stress contributed by atmospheric oxygen dissolved in the culture media (this can represent about twice the oxygen concentration that a cell would naturally experience). A relatively consistent finding has been the declines in *in vitro* mitogen-stimulated T cell proliferation with use of lymphocytes from aged animals or aged humans; however, the mechanistic basis for this decline has been elusive. Even in this context, conflicting results have been obtained [95]. Some of the variations in results may reflect the fact that some studies use human cells and others use mouse cells. Lymphocytes from aged mice have less ability to increase their intracellular calcium after activation in comparison with young mice [96,97], and this age-related difference in calcium flux is not observed with human cells [98,99]. Since calcium flux can be modulated by the degree of oxidative stress on lymphocytes [42,60] and it is known that mouse and human cells differ in their thiol chemistries [100], these differences with regard to calcium flux and aging are not surprising and in fact support the possibility that changes in immune function with aging or after metal exposure may be associated with differential sensitivities to oxidative stress.

Loss of the proliferative ability of lymphocytes is most sensitive to loss of thiols found on the extracellular face of the plasma membrane [101]. These surface thiols increase [102] along with GSH levels [103] as lymphocytes become activated. Although *in vitro* activation of T cells from the elderly generally results in less IL-2 production in comparison with T cells from young individuals [104], this deficit can be overcome by preculture of the peripheral blood mononuclear cells (PBMCs) [105], by boosting with a vaccine [104] or by hormone treatment [106]. Nevertheless, overall activation as measured by proliferation (DNA synthesis) usually remains depressed in PBMCs from the elderly. It is interesting to note that as in depressed lymphoproliferative response of PBMCs from older donors, the lymphocytes from HIV-infected individuals show lower levels of GSH and other antioxidants [107–111]. The additive oxidative effects of aging and HIV infection could be an explanation for why the onset of AIDS in older individuals is more rapid [112]. It has been shown that non-HIV-infected "young" human T lymphocytes that have been depleted (90%) of reduced GSH with the specific glutamate cysteine ligase (GCL) inhibitor buthionine sulfoximine (BSO) were unable to synthesize DNA (progress through S-phase) after Con A stimulation; however, cellular activation events were not inhibited in these GSH-deficient cells, in that these cells could express IL-2, IL-2R, and transferrin receptor (TfR) [103]. Hence, T cell proliferation is intimately dependent on intracellular concentrations of GSH. Oxidative stress (or aging) may modify early signal transduction events as well as later events required for proliferation, and the differences in the sensitivities of these various events await better understanding.

It is interesting to note that enzymes and receptors that are associated with the plasma membrane that are intimately involved in lymphocyte activation (phospholipase A2 [PLA2] and protein kinase C [PKC]) have similar sulfhydryl regions that are believed to influence the interactions of these proteins with membrane phospholipids [66]. PKC has cysteine-rich domains (CRDs) within its regulatory and catalytic domains [113]. These domains are sensitive to exogenous modulation since heavy metals [114] and H_2O_2 [115] can each alter phorbol ester binding and kinase activity. Another thiol compound, β-mercaptoethanol, seems to be able to potentiate PKC translocation to the plasma membrane from the cytosol [116]. PLA2 modulates antigen presentation, suggesting that the antigen:MHC class II complex may be influenced by lipid structures [117], and the TfR has also been found to be dependent on its lipid environment for appropriate expression [118]. TfR [119] as well as the MHC class I and II molecules [120,121] are acylated (covalently associated with a fatty acid), which may facilitate specific membrane interactions. For currently unknown reasons, the expression of class I antigens is known to increase with aging [122] along with increased shedding into the plasma after activation [123].

Other regulatory plasma proteins, which derive their active conformation from intimate interaction of their hydrophobic domains with phospholipid fatty acid tails and are thiol-regulated, include adenylate cyclase [124], 5′-nucleotidase [125], insulin receptor [126], and Na^+/K^+ ATPase [127]. Changes in the lymphocytes' transport functions start early in the G1 phase of the cell cycle (Na^+, K^+, and Ca^{2+} changes occur within 1 min; glucose by 10 min; amino acids and nucleosides by 1 h). Since cell transport mechanisms directly utilize thiol-sensitive proteins [25,128–131] and lipid domains can influence these activities, thiol modulation is suggested to have a regulatory influence on multiple early membrane events involved in lymphocyte activation. As might be anticipated, many of the above activities change with aging. Na/K-ATPase activity decreases in lymphocytes from aged people [132]; membrane potential changes are less responsive with aged cells [133,134]; and nucleotide pools [135], kinase activity [136], protein phosphorylation patterns [137], and G protein involvements [138] have been reported to be modified by aging. Many of these processes are interconnected in the activation of T cells [139].

Altered structure/function of various lymphoid markers, enzymes, and signaling molecules have been implicated in the loss of immune reactivity with aging and metal modulation. In general, there is good agreement that oxidative damage of cellular constituents is associated with aging [140]. Protein oxidation occurs with aging [141]. Loss of reactive thiols with age can occur with decline in GSH [142], resulting in numerous physiological consequences, such as lower transcription factor activity [143]. The toxic effects of oxidants and metals are certainly not limited to immune cells [144]; however, as noted earlier, lymphocytes are especially sensitive to oxidative stress [145,146].

The initial signals delivered to cells are received by cell surface molecules. It is important to note that numerous surface receptors including growth factor receptors have sulfhydryl-rich domains or have numerous disulfides necessary for maintenance of their structural configuration as a receptor (e.g., adrenergic receptors [147], epidermal growth factor receptor [148], platelet-derived growth factor receptor [149], nerve growth factor receptor [150], insulin receptor [148], IL-2 receptor [151], and chemokine receptor CCR5, found on memory T cells, macrophages, and dendritic cells [DCs] [152]). The growth-promoting factor ADF (now known to be thioredoxin) is able to induce formation of the high-affinity IL-2 receptor (the associated holoreceptor of p55 or TAC and p75) [153,154]. Changing the sulfhydryl and disulfide locations of molecules could influence a molecule's overall conformation, its association with other molecules, and its topographical arrangement on the cell surface.

For cell-cycle progression to be initiated, lymphocytes must maintain the appropriate surface thiol composition [101,155]. Extensive disulfide cross-links prevent appropriate domain establishment or receptor signaling. Modulation of thiols on the surface and within the cell can alter lymphocyte activation [42,81,101,155], but the surface thiols appear to be particularly important. Pretreatment of lymphocytes with an impermeant oxidant (e.g., copper phenanthroline) causes loss of less than 5% of the total cellular thiols, but it completely prevents progression to G1b phase of the cell cycle [101].

The redox environment both inside and outside the cell is managed in large part by GSH. GSH is critical in maintaining a viable cellular redox state, which becomes modified during lymphocyte activation and proliferation [156,157]. GSH (and total thiol levels) have been shown to rise significantly following peripheral blood leukocytes activation by mitogens. Human lymphocytes significantly increase their GSH concentration within 4 h of activation [100]. By 8 h after activation there is a significant increase in surface thiols [102]. Interestingly, when GSH production is blocked, lymphocytes can progress to G1b but cannot synthesize DNA [100]. DNA synthesis was previously shown to be sensitive to thiol modulation [158]. At present, the major restriction at G1/S phase due to GSH depletion is not clear. Some possible mechanisms of this restriction are related to interference in the glutaredoxin/thioredoxin-mediated conversion of ribonucleotide reductase, inactivation of DNA polymerase I that is thiol-sensitive [159], or alteration of PCNA (proliferating cell nuclear antigen) activity or mitosis-promoting factor [160]. Cellular redox status is largely mediated by the ratio of GSH to oxidized glutathione (GSSG) [161] and may also contribute to the regulation of cellular transcription factors [162] and protein *S*-glutathionylation. The formation of the transcription factor AP-1 is sensitive to thiol modulation, and it has been reported that AP-1 DNA activity decreases with lymphocytes from aged mice [163].

14.3.2 Metal Influences on Apoptotic Signaling Pathways

In addition to the effects of metals and oxidants on the capacity of cells to maintain a functional cellular biochemistry, there is also an important cellular response to metals that results in apoptosis.

This process is a tightly regulated physiological process of selective cell deletion that is an essential aspect of normal immunoregulation, and is critical to the maintenance of self-tolerance as well as the homeostatic control of lymphoid and myeloid cell lineages [164,165]. Adverse consequences due to the loss of apoptotic control mechanisms within the immune system have been well documented in a number of contexts. An abnormal increase in the rate or the extent of apoptosis may contribute to immunodeficiencies, while a failure to undergo apoptosis may contribute to the development of autoimmune diseases or neoplasia [166]. Apoptotic cell death can proceed by an orchestrated sequence of events (i.e., programmed cell death), but it is also induced and regulated by a myriad of environmental stimuli including physiological effectors as well as toxic agents such as metals [167]. The molecular pathways by which apoptotic cell death proceeds are highly sensitive to metals [168,169], and accumulating data support that disruption of these events by metals has a significant impact on the immune system.

Cells that undergo apoptotic cell death are morphologically and biochemically distinct from healthy cells and from cells that undergo necrotic cell death as typified by the response to acute toxic insults [170,171]. The morphological features of apoptosis include chromatin condensation and margination to the nuclear periphery, decreased cell volume and cytoplasmic shrinkage, plasma membrane blebbing, and cellular and nuclear fragmentation into apoptotic bodies. In contrast to necrotic cell death, there is typically no loss of plasma membrane integrity associated with apoptosis. As such, the maintenance of plasma membrane integrity during apoptosis prevents the extensive inflammatory response resulting from the release of intracellular contents. The characteristic ingestion of apoptotic bodies by neighboring cells or phagocytes also limits the initiation of inflammation.

Apoptotic cell death is initiated, regulated, and executed through the activation of specific signaling pathways. There are positive and negative regulators of apoptotic signaling, and it is the balance between these events that determines cell fate. Metals have diverse effects on apoptotic signaling with examples of metals influencing proapoptotic signaling as well as examples of them interfering with apoptotic signaling. Apoptotic signaling pathways can be operationally divided into three main phases that are evolutionarily conserved [172]. In the first phase, apoptosis is initiated by external or internal stimuli with activation of the extrinsic and intrinsic apoptotic signaling pathways, respectively [173]. Signal initiation involves a complex orchestration of biochemical events involving positive and negative regulator proteins many of which, as mentioned above, are influenced by metals. Depending on the cell type as well as the initiating factor (i.e., internal or external signal), important regulatory elements include death receptors, proteases (e.g., caspases), mitochondria, Bcl-2 family proteins, and tumor suppressor gene products [174–176]. The second phase is characterized by the activation of the executioner proteases and the dismantling of the cell due to the degradation of death substrates, which includes vital structural proteins and chromatin [177]. The third phase, which may occur in concert with the cellular dismantling described in step two, involves the disposal of the cellular corpse via uptake and sequestration of the apoptotic bodies within phagocytic vesicles [178,179].

There are two distinct but ultimately overlapping signaling pathways by which apoptosis is initiated [173]. These are termed the intrinsic and the extrinsic pathways. Apoptosis triggered by internal signals (e.g., activation of the apoptosome characterized by Bcl-2/Apaf-1/caspase-9) occurs when a cell "loses touch" with its surroundings (e.g., growth factor withdrawal), sustains internal damage (e.g., oxidative stress and/or DNA damage), or simultaneously receives conflicting signals regarding cell cycle control. Apoptosis triggered by external signals (i.e., death receptor agonists) occurs via an instructive process mediated by members of the tumor necrosis factor (TNF) family including Fas (CD95) and its ligand (CD178) and TNF-related apoptosis inducing ligand (TRAIL) and its receptors TRAIL-R1 and -R2. CD95 is a 317-amino-acid type-I transmembrane glycoprotein that belongs to the TNF/nerve growth factor receptor family (also known as the death receptor superfamily). CD95 is the prototypic, and thus most studied, member of the death receptor superfamily [180]. A characteristic of death receptors is the presence of three extracellular cysteine-rich domains (CRDs); CRD2 and CRD3 contain the major contact surfaces for interaction with FasL (or surrogate anti-Fas antibody agonists), whereas CRD1 is a receptor self-association domain—termed the pre-ligand assembly domain (PLAD). In view of the fact that many metal ions bind free sulfhyryl groups with high affinity, it is conceivable that CRDs are targeted by metal ions or by the oxidants they may elicit. Indeed, there is precedent from the literature that certain metals coordinate with CRDs found on intracellular signaling proteins [181]. CD95 has no enzymatic function of its own; therefore, signaling is transduced via other proteins interacting with the cytoplasmic region of CD95 [182]. Signaling through CD95 is initiated upon interaction with CD95-L (CD178) or agonistic antibodies, which facilitate CD95 receptor clustering and the formation of a death-inducing signaling complex (DISC). DISC formation involves the recruitment of Fas-associated death domain protein (FADD) through homotypic DD—DD interactions between the carboxy-terminal regions of both CD95 and FADD. FADD, which bears another protein interaction module in its NH-terminus (the death effector domain—DED), serves as a signaling adapter by recruiting the initiator caspase procaspase-8 to the complex through homotypic interactions between DEDs on FADD and caspase-8. Recruitment of procaspase-8 to the DISC activates it and initiates the proteolytic cascade of effector caspases (e.g., caspase-3 and -7) that ultimately leads to the apoptotic dismantling of the cell. The ordered formation of the DISC is an essential feature of CD95 signaling; however, it has become apparent that the initial signaling events within the CD95 pathway are more complex than originally thought [180]. CD95 signal transduction is more than a linear-ordered sequence of events, as feedback within the pathway amplifies DISC formation. For example, both actin and active caspase-8 act in a positive feedback loop to amplify DISC formation. According to the latest model, CD95 signal initiation can be described as five distinct steps [183]. First, CD95 is present on the plasma membrane in an agonist-independent preassociated form, where the preassociation is mediated by interactions through the PLAD. The second step is the formation of SDS-stable CD95 microaggregates, which are agonist dependent and which can be detected with gel electrophoresis under nonreducing conditions. Formation of microaggregates is not dependent on caspases or actin filaments, and at this stage the aggregates cannot be seen as receptor clusters by immunofluorescence microscopy. The third step is formation of the DISC, which involves DD/DD interactions between CD95 and FADD. The fourth and fifth steps involve CD95 receptor clustering (as assessed by immunofluorescence) and internalization, respectively. These latter three events (including notably DISC formation) are dependent on the actin cytoskeleton, whereas formation of microaggregates is not.

The cytotoxicity of organic and inorganic mercury compounds is well documented. Several reports have indicated that mercurials induce either apoptotic or necrotic death in lymphoid as well as nonlymphoid cells. The concentration of mercury (high > low), its chemical form (organic > inorganic) and the cellular activation state (resting > activated) all influence its cytotoxic potential. While high concentrations of any form of mercury overtly induce necrosis, more moderate concentrations of mercury have been shown to induce apoptosis via activation of the intrinsic pathway. This likely occurs as a result of excessive generation of ROS, which disrupts the mitochondrial membrane leading to the release of cytochrome *c* and the formation of the apoptosome containing the initiator caspase-9. Understanding the cytotoxicity of mercury is important because the release of intracellular autoantigens upon mercury-induced cell death may trigger the activation of autoreactive T lymphocytes. On the other hand, lower exposure to noncytotoxic concentrations of inorganic mercury impairs activation of the extrinsic apoptotic program in human T cells [184]. The finding that low concentrations of inorganic mercury interfere with physiological cell death represents a paradigm shift with respect to implicating mercury toxicity in human disease, in that, cell survival instead of cell death is the outcome following exposure. Of key importance is the fact that CD95-mediated apoptosis is a critical regulator in maintaining immune system homeostasis and tolerance [185]. Defects within the CD95 signaling pathway can give rise to lupus-like autoimmune disorders. Genetic lesions of Fas in both mouse and humans have clearly identified this receptor as a key mediator in regulating lymphocyte homeostasis. Humans with Fas mutations present with lymphoproliferation, peripheral expansion of CD4-CD8-$\alpha\beta$ T cells and impaired lymphocyte apoptosis in a disorder known as autoimmune lymphoproliferative syndrome (ALPS). These findings are consistent with observations made in lupus-prone mouse models where CD95 is mutated spontaneously in the lpr mutation, or by gene-targeted mutations.

Treatment of T cell lines and human CD4+ lymphoblasts with noncytotoxic concentrations of inorganic mercury blocks CD95 signaling by preventing the proper formation of the DISC [186]. A CD95 receptor proximal signaling component, perhaps CD95, itself, appears to be targeted by mercury. Evidence supporting CD95 as a direct target comes from experiments demonstrating that the PLAD-dependent preassembly of CD95 oligomers is disrupted by mercury [187]. This same study showed selectivity for mercury disruption of CD95 signaling, in that, another Group IIb metal, zinc, had no effect on CD95 oligomerization. Others have shown that zinc inhibits downstream elements of the extrinsic apoptotic pathway, while targeting upstream elements of the intrinsic apoptotic pathway [188].

Using a mouse model of superantigen-induced activation of the extrinsic apoptotic selection program, following mercury intoxication *in vivo*, it was revealed that activation of caspase-8 and -3 was diminished in antigen-specific T cells [189]. This same study showed that the superantigen-induced induction of extrinsic apoptotic signaling was CD95 dependent, since caspase-8 and -3 activation was reduced by more than 60% in T cells derived from lpr CD95-deficient mice. This study illustrates the modulatory properties of mercury on T cell survival *in vivo*. Given the critical role for CD95-mediated apoptosis in controlling peripheral tolerance and regulating T cell homeostasis, these findings that Hg-intoxication affects death receptor signaling have implications for establishment and progression of autoimmunity.

14.3.3 Metallothionein: A Mediator of Metal-Dependent Immune Changes

Another contributor to the management of both essential and toxic heavy metals is the stress response protein metallothionein (MT). MTs represent a family of low-molecular-weight (6–7 kDa) proteins that are rich in cysteine residues (ca. 30 mol%). MT was originally recognized as a metal-binding protein in equine kidney [190], but MT-like proteins have since been identified in nearly every tissue in a wide variety of species across substantial evolutionary distances. There are four major MT isoforms in mammals (MT-I through -IV), of which MT-I and MT-II are the most widely studied and are expressed at basal levels in most organs and stages of development [191]. All four of the murine MT isoforms are encoded by genes on chromosome 8, while human MTs are encoded by a syntenic region composed of structural genes and pseudogenes on chromosome 16. While mouse MT-I and MT-II are usually expressed under similar conditions, MT-III is preferentially expressed in adult brain, and MT-IV is expressed in differentiating stratified squamous epithelium [192,193]. MT synthesis is induced by a wide range of inflammatory mediators and stress-related agents, including heavy metals, ROS, glucocorticoids, acute phase cytokines and IFN, and endotoxin.

14.3.4 Induction of MT by Metals

MT gene expression is controlled mainly at the level of transcription [194], although in cells selected for resistance to Cd, increased expression of MT protein can also be accomplished via MT gene amplification [195]. It is well established that both essential and toxic heavy metals can induce MT synthesis. The most potent inducers of MT are Zn and Cd, which can induce up to a 100-fold increase in hepatic MT levels [196]. Other heavy metal inducers of MT include essential metals such as Cu, Fe, and Mn, and nonessential metals such as Au, Bi, Co, Hg, In, Ni, and Pt [194,196,197].

While generally regarded as coordinately regulated, there are both species- and inducer-specific differences in MT-I and MT-II gene induction patterns. In rats, basal MT-II levels are higher than MT-I, while in mice both basal and metal-induced levels of MT-I mRNA are twice as high as MT-II mRNA [198,199]. Interestingly, treatment of rats with Cd induces MT-I and MT-II to similar levels, while Zn induces significantly higher levels of MT-II than MT-I. In humans, Cd increases both

MT-I and MT-II mRNAs, but Zn preferentially increases MT-II mRNA [200]. These differences may in part be attributed to distinct upstream elements in gene promoter regions. The 5′ upstream regions of MT-I and MT-II genes contain a TATA box and several different *cis*-acting response elements. In addition to upstream metal-responsive elements (MREs), glucocorticoid response elements (GREs), antioxidant response elements (AREs), and other response elements have also been identified in MT genes [194,195,201,202]. While some MT promoter elements may be unique to a particular MT isoform or species, MREs are present in multiple copies in the MT gene promoters of various isoforms in both invertebrates and vertebrates, suggesting that metal responsiveness is a critical feature of MT [203].

MT is normally saturated by Zn within cells, even though it has a higher affinity for scarcer heavy metals such as Cu and Cd; hence, it is possible that these heavy metal inducers bind to Zn-MT or other cellular Zn-storage proteins, displacing Zn and leading to subsequent activation of MTF-1 by the released Zn [204]. In accordance with this model, while induction of MT transcription by activated MTF-1 can be achieved by elevated Zn alone, induction by Cd, Cu, or H_2O_2 additionally requires the presence of Zn-saturated MT [205]. Like displacement of Zn by Cd or Cu, oxidation of Zn-MT can lead to the release of Zn and subsequent MTF-1 activation. While over 18 different metals can associate with MT, only Cu, Cd, Pb, Ag, Hg, and Bi can displace Zn [206,207].

Induction of MT genes by nonessential heavy metals can also occur via non-MRE-dependent mechanisms. There is an upstream stimulatory factor (USF) binding site in the mouse MT-I gene promoter that overlaps with an ARE consensus sequence that has been shown to mediate the response to oxidative stress [201]. The USF/ARE composite element increases basal expression of the mouse MT-I in *in vitro* transcription reactions [208] and *in vivo* in transiently transfected cells [209]. Stable and transient transfection assays in mouse hepatoma cells showed that the USF/ARE element contributes to Cd, but not Zn, induction of MT-I expression, as well as to basal promoter activity of the mouse MT-I gene [210]. In light of the vast literature supporting oxidative stress as an important mechanism of heavy-metal-induced immunotoxicity and carcinogenicity, these observations suggest that heavy metals may induce MT expression not only through MRE activation, but also via oxidative stress activation of AREs in MT gene promoters.

14.3.5 MT Synthesis and Distribution

Once translated, MT is found in almost every subcellular compartment, including mitochondria, endoplasmic reticulum (ER), nucleus, and cytoplasm [211,212]. MT is also found in extracellular spaces, including human serum and urine, and extracellular MT is present in elevated levels after toxic metal stress [213–215]. These observations suggest that MT may be secreted or otherwise released from stressed cells, despite the observation that most MT mRNA is associated with free polysomes, which has been interpreted to suggest the intracellular localization of the protein. It is possible that MT is specifically released from cells via an alternative pathway distinct from the common secretory pathway, as has been suggested for other stress proteins [215,216]. It is also possible that MT is released by passing through a damaged plasma membrane or from apoptotic or necrotic cells that have succumbed to overwhelming cellular stress [215].

The highest concentrations of MT in the body are found in the liver, kidney, intestine, and pancreas [217]. The main site of MT synthesis is the liver, although many nonhepatic cells, including those of the immune system, are also capable of producing MT under specific stimuli. Basal MT expression has been identified in both murine and human splenocytes and thymocytes, and specifically in human monocytes, T lymphocytes, B lymphocytes, and neutrophils [218,219]. Of these specific blood leukocyte populations, all but neutrophils appear to upregulate both MT mRNA and protein expression in response to Cd exposure [218]. Interestingly, polymorphonuclear cells (PMNs) are more resistant to Cd than mononuclear cells and they incorporate more cadmium, but the Cd is not bound to MT in PMNs [220].

There is good evidence that Zn is an important modulator of MT expression in mammalian immune cells. Chronic Zn administration to male Wistar rats leads to increases in total MT levels in both the thymus and spleen [221]. Monocytes exhibit the highest Zn content and the highest levels of MT mRNA and protein when compared with other leukocytes [218]. MTs are induced by zinc in cultured monocytes and in both T and B lymphocytes *in vitro* [219]. In both THP-1, a human monocyte cell line and in human PBMCs, increases of intracellular Zn levels lead to induction of increased MT mRNA and protein levels. Decreasing free Zn with the Zn chelator TPEN (*N*,*N*,*N*′,*N*′-tetrakis(2-pyridylmethyl)-ethylenedidiamine) leads to decreases in MT expression [222]. In addition to several human MT-I isoforms, a human MT isoform previously considered to be expressed only in fetal liver tissue, designated as "MT-0," is induced in PBMCs by zinc *in vitro* [223]. Human peripheral monocytes produced 19-fold and B lymphocytes sixfold more MTs than T lymphocytes in response to 125 μM Zn [224], suggesting that cells in which oxidants are produced as an effector mechanism for killing infectious pathogens may use MT to maintain redox status. Zinc induction of MT in human immune cells has also been documented *in vivo*. Zinc supplementation in humans leads to increased MT mRNA levels in both monocytes and PBMCs. When zinc supplementation was withdrawn, MT mRNA levels decreased [225]. *Ex vivo* activation of human $CD4^+$ T cells has been shown to increase free cytoplasmic Zn, and increased expression of several MT-I isoforms as well as MT-IIA was shown to be due to sustained Zn signaling in these cells [226].

Cadmium has been shown to induce MT expression in a number of mammalian immune cell types. Exposure of human peripheral blood lymphocytes and monocytes to Cd led to MT induction in both cell types, and similar results were found with a variety of human lymphoid cell lines derived from both T and B cells, although MT was induced in varying amounts depending on the particular cell line [227]. Similarly, T and B lymphocytes isolated from human peripheral blood were shown to be capable of Cd-induced MT production [228]. The kinetics of MT induction in human peripheral monocytes and in T and B lymphocytes by Cd or Zn are very rapid: detectable MT induction can be observed after only 2 h in some cases [224]. While Cd induced lower levels of MT than Zn in these studies, maximum MT levels were reached faster than with Zn. In addition to the immunotoxic effects of cadmium, the liver is a particularly sensitive target for Cd toxicity. It has been shown that liver Kupffer cells (a macrophage lineage) contain higher endogenous levels of MT than parenchymal cells, and $CdCl_2$ at 1 mg Cd/kg leads to intracellular accumulation of Cd and a concomitant increase in MT mRNA (sevenfold) and MT protein (twofold). MT is strongly expressed in several cell types in the lung, including alveolar macrophages, lymphocytes, alveolar type II epithelial cells, and fibroblasts [229].

There are a number of different mechanisms by which MT appears to modulate immune activities. MT may influence immune function by the redistribution of heavy metals both inside immune cells and between tissues as has been shown in instances where MT induction was shown to cause the redistribution of metals in embryonic and fetal rats [230]. Since MT plays an important role in zinc homeostasis, it can modulate immune function by changing Zn availability to the host of apoenzymes and transcription factors that are important to immune functioning, such as Spl and NF-κβ [231,232]. As a thiol-rich protein, MT can act as an antioxidant in either intracellular or extracellular environments to counteract the adverse effects of heavy metals. When induced by nonmetal agents (e.g., LPS, ROS, IFN-γ, etc.), MT can enlarge the metal reservoir. These changes could alter the availability of essential metals for proliferation and differentiation of cells in the immune system. In addition to the effects on the biochemistry of immune cells, MT can also act as an extracellular agent to stimulate lymphoproliferation [233], influence T cell [234] and B cell adaptive immunity [235], initiate chemotactic responses [236–238], promote wound healing after burn and brain injuries [239–241], and modify phagocytic cell activities [242]. Manipulations of MT can be used to modify the course of autoimmune disease [26,243,244], inflammatory processes [245,246], and infection [247].

14.3.6 Other Toxicant-Induced Homeostasis Mechanisms

Heat-shock proteins (HSPs) represent another family of highly conserved and ubiquitous proteins that are expressed in response to stress (including metal stress) in prokaryotes, eukaryotes, and even in archaebacteria [248]. The heat-shock response has also been observed in virtually every cell and tissue type of multicellular organisms, in explanted tissues, and in cultured cells [249]. Thus, while many studies of HSP induction and function were performed using mammalian cell lines, many of these findings have been extended to immune cells and other primary cell types. The term "stress proteins" has been applied to recognize the more general nature of this response, as glucose-regulated proteins (GRPs) are also considered members of the HSP family [250]. While GRPs are specifically located in the ER, other HSPs can be found in nearly every cellular compartment, including the cytosol, mitochondria, nucleus, nucleolus, ER, and outer and inner membranes [251].

14.3.7 Regulation of HSPs

Like the MT gene family, a number of stress-related response elements have been identified in the promoters of HSP genes. Sequence comparisons of *Drosophila* heat-shock promoters reveal a conserved sequence called the heat-shock element (HSE). In addition to ubiquitous HSEs, metal response elements (MREs) have also been identified in HSP gene promoters. The human HSP70 gene promoter contains a sequence homologous to the human MT-IIA MRE and this sequence is required for Cd induction of HSPs [252]. The expression of HSPs is under complex control, but many common pathways involve the activity of heat-shock factors (HSFs) and a group of zinc-dependent transcription factors [253]. This suggests another mechanism by which toxic metals can induce HSP expression. Like with Cd–MT- and Zn–MT-dependent MTF-1 activation, Cd or other heavy metals can bind to Zn–MT, thus displacing Zn and increasing the pool of free Zn available for activating HSFs.

In one study using primary rat hepatocytes, several metals were compared to traditional heat-shock treatment for their ability to induce mammalian HSP70 and HSP90. At the concentration of each metal that yielded maximal induction of each HSP, the relative efficacy of induction for both HSPs tested was heat (43.5°C) > As(III) > Cd > Hg > Zn > Ni [254]. In the same study, heat shock did not induce MT and the relative kinetics of MT induction were completely different from HSP induction for these metals, suggesting that mechanisms of induction for these classes of metal-responsive stress proteins differ. While Cd and Zn induced both MT and HSPs, As(III) induced more HSP than MT, and Ni was a good inducer of MT but a poor inducer of HSP synthesis. The same study found that some specific heavy metals, such as Fe, Pb, and Vn, are unable to induce HSP synthesis [254], although others have reported that lead can induce MT [255].

Different HSPs have different capacities to protect against the toxicity of specific metals. For example, HSP104, a member of the highly conserved HSP100 family, is required for tolerance to ethanol and sodium arsenite, but has no effect on tolerance to Cd or Cu [256]. HSPs are synthesized in response to low heavy metal concentrations that do not cause overt cellular toxicity, an observation that has prompted several researchers to propose HSP synthesis as a good biomarker for early exposure to metals [216,257]. In the rat hepatoma cell line, Reuber H35, sublethal doses of $CdCl_2$ lead to slight inhibition of general protein synthesis and induction of HSP70, HSP84, and HSP100 [258]. At higher doses that lead to severe inhibition of general protein synthesis and impaired cell survival, the induction of these three proteins does not further increase, but expression of the fully inducible HSP70 family member, HSP68, can be detected. Interestingly, induction of HSP68 mRNA can be detected at sublethal doses of $CdCl_2$ that are unable to induce detectable HSP68 protein levels, indicating translational regulation of HSP68 induction [258]. Thus, a comparative biomarker profile of mRNA and protein expression levels of various HSPs may have utility for discriminating between low and high levels of metal exposure.

14.3.8 HSPs and Immune Function

In recent years, a number of studies have shown that HSPs can modulate a variety of immune responses, including cellular and humoral immunity, inflammation and clearance of infection, anti-tumor immunity, and autoimmunity. The induction of HSPs has been reported for a number of immune cell types, including neutrophils [259], B and T lymphocytes [260,261], monocytes [262,263], and macrophages [264,265]. Increased HSP synthesis has also been observed after T lymphocyte activation with mitogens or cytokines [249,266,267]. As in other cell types, heat shock also induces HSPs in immune cells. In chickens, heat shock induces HSP expression in lymphocytes [268]. In mice, HSF-1 DNA binding activity and HSP70 synthesis are induced at significantly lower temperatures in the spleen relative to other tissues [269]. This lower temperature threshold (39°C) is within the physiological range for fevers and is not exhibited by B lymphocytes but is observed in T lymphocytes. This differential heat-shock temperature threshold is observed in different mouse strains and is also observed in T lymphocytes isolated from lymph nodes, suggesting that it is a general property of T lymphocytes that likely protects these cells from stress-induced apoptosis during the immune response to infection [269]. Similarly, in human macrophage-like cell lines, HSF-1 activation leads to HSP70 mRNA induction and protects cells from toxic injury [270].

In monocytes, HSP induction appears to protect against oxidative stress-induced cell death. Exogenous H_2O_2 induces HSP synthesis in human monocytes and alveolar macrophages [271]. Both HSP27 and HSP32 (a 32-kDa oxidation-specific stress protein that is also known as heme oxygenase) exhibit protective activity against oxidative stress [272,273]. Oxygen-free radicals generated during phagocytosis of opsonized sheep red blood cells induce a strong stress response in human monocytes–macrophages, inducing the classic HSPs, HSP70, as well as HSP32, HSP47, HSP65, HSP83-90, and HSP110 [274]. Heat shock and Cd both induce HSP synthesis and inhibit NADPH production in human neutrophils [275]. It is likely that HSP inhibition of superoxide production in neutrophils also reduces oxidative stress-induced cytotoxity in phagocytes [274]. In addition to connections with phagocytosis, there is significant evidence suggesting an important role played by HSPs in APC functioning within the immune system. HSP expression has been correlated with antigen processing and presentation by enhancing protein degradation [276], promoting proper protein folding and assembly processes [277], and by participating in transmembrane transport of proteins [216,248]. HSP70 family members appear to play a particularly important role in antigen processing and presentation. Processing of antigens is a complex process that involves the internalization of antigen into acidic compartments, proteolysis, and binding of peptides to the antigen-presenting molecule and all of these steps require the participation of molecular chaperones of the HSP70 family [278].

Like MTs, HSPs can be found in extracellular locations and on cell membranes following exposures to various stressors. The HSP70 family member, PBP72-74, is constitutively expressed on the cell membrane in both B cells and monocytes [279]. Other HSPs have been found on the surface of a variety of mammalian immune cell types, including monocytes, lymphoma cells, T cell lines, myeloid leukemia cell lines, either constitutively or after heat shock or viral infection [280,281]. Various immune cell types, including monocytes, macrophages, and B cells, have been shown to release HSPs extracellularly [262]. In B lymphoblastoid cell lines, HSP27, HSC70, HSP70, and HSP90 can all be detected in exosomes of untreated cells, and exposure of these cells to heat shock results in a marked increase in all of these HSPs in exosomes [282]. The ability to bind stress proteins is also shared by many types of immune cells, including DC and other APCs, and HSP binding to membrane receptors has been associated with altered immune cell function [283,284]. Interactions of HSP complexes with APCs and internalization of HSP by macrophages and DC induce cellular maturation, activation and migration, cytokine secretion, and upregulation of costimulatory molecules and MHC class II proteins [285,286]. HSP gp96 binds to a receptor on DC, and binding mediates maturation of DC, and receptor-mediated uptake is required for cross-presentation of gp96-associated peptides by DC [287]. Studies of HSP uptake by APCs reveal that both HSP70 and gp96 uptake is localized to clathrin-coated pits on macrophages. Hsp70 and gp96 are internalized in clathrin-coated vesicles in a receptor-dependent manner, and exogenous gp96 colocalizes with MHC class I molecules in endosomes following uptake [288]. These observations support a model for specific uptake of HSP–peptide complexes and subsequent transfer of the peptide antigen to MHC class I molecules for display on the cell surface [289]. Several receptors on APCs have been suggested to bind HSPs, including the scavenger receptors CD91 and CD36, TLR family members TLR2 and TLR4, and the costimulatory molecule CD40 [290].

14.3.9 Metal-Induced HSPs in Immune Cells

While many studies have utilized classic heat-shock treatment or phagocytosis to stimulate HSP induction in immune cells, heavy metal treatment has also been shown to induce HSP synthesis in some immune cell types. Exposure to several heavy metals leads to the generation of reactive free radicals and subsequent cellular oxidative stress, and it is likely that this is one mechanism of metal-induced HSP synthesis. Agents that induce oxidative injury, including both H_2O_2 and $CdCl_2$, induce HSP synthesis in murine peritoneal macrophages [291]. Similarly, exposure of human premonocytic U937 cells to cadmium induces the synthesis of HSPs, including heme oxygenase, which may protect against oxidative damage to DNA or proteins [292]. In this particular cell line, it has been suggested that oxidation-induced DNA damage and HSP synthesis proceed independently, as cadmium led to HSP induction but not oxidative DNA damage, while H_2O_2 induced DNA damage but no HSP synthesis at the exposure levels tested. Cell-specific differences in HSP induction have been observed in different types of human phagocytes. While cadmium induces HSP synthesis in both monocytes and neutrophils, these two cell types differ in the stress response to activation of the respiratory burst by NADPH oxidase [280,281]. Activation of the respiratory burst by phagocytosis of particulates leads to HSP synthesis in human monocytes only, while activation of NADPH oxidase by nonparticulate f-Met-Leu-Phe (FMLP) induces HSPs in PMNs only. In human lymphocytes, Cd activates transcription of both MT-IIA and HSP70 genes in a time- and dose-dependent manner, although MT-IIA is induced at lower concentrations than those required to induce HSP70 [293]. The immature T cell line CCRF-CEM was treated with two different concentrations of Cd (10 and 20 μM) and cDNA microarray analysis was used to identify cadmium-induced transcriptome alterations in lymphocytes [294]. The stress proteins HSP32, HSP60, and HSP70 were all among the 98 genes whose transcription was significantly upregulated in response to Cd. Heme oxygenase (HSP32) was also upregulated after only 6 h of Cd exposure, while HSP60 and HSP70 induction was not observed until 24 h after exposure. Heme oxygenase is also induced in murine peritoneal macrophages after exposure to sodium arsenite, in human cell lines after exposure to arsenite and cadmium ions, and in both rat and human DCs in response to cobalt protoporphyrin (CoPP) [295,296]. In RAW 264.7 murine macrophages, exposure to either soluble $CdCl_2$ or particulate cadmium oxide (CdO) led to increased *de novo* synthesis of HSP70 and HSP90 prior to cytotoxicity, and this HSP synthesis was accompanied by a decrease in phagocytic capacity and an increase in TNF-α levels [297].

Like cadmium, arsenic is known to generate many species of reactive free radicals, including hydrogen peroxide, hydroxyl radical, nitric oxide, and superoxide anion, and both *in vitro* and *in vivo* exposure to arsenicals has been shown to induce a number of stress response proteins, including MTs, ubiquitin, HSP27, HSP32, HSP60, HSP70, HSP90, and HSP110 [298]. The specific HSPs induced by arsenicals vary depending on the cell type and the oxidation state and chemical composition of the arsenical. Arsenite has been shown to induce MT in murine spleen [299] and in guinea pig alveolar macrophages where a 2-hour exposure to arsenite (5–150 mM) induces HSP27, HSP70, HSP90, and HSP110 [300]. The importance of arsenical-generated oxidative stress in toxicity of the metal is highlighted by the observation that various arsenicals induce heme oxygenase (HSP32) in many cell types, including immune cells. Arsenite induces HSP32 in murine peritoneal macrophages [296], in human lymphocytes [301], and in the human leukemic HL60 cells [295].

14.3.10 Metal Exposure Can Alter a Variety of Cellular Activities

Because of the array of critical cellular processes that are sensitive to metal effects, there is a broad range of consequences arising from immune cell activities. The effect of metals on naïve $CD4^+$ T cells exemplifies the ability of toxic metals to alter proliferation. In some experiments, mercury has an inhibitory effect on $CD4^+$ T cell proliferation, but under other conditions, mercury can stimulate lymphoproliferation. Both cadmium [302] and lead [303,304] can alter lymphoproliferation, and lead has been found to increase the number of cells in S + G2/M part of the cell cycle [303]. The increase of $CD4^+$ T cell proliferation that occurs in mixed lymphocyte cultures that are exposed to Pb has been termed "allo-enhancement" [303]. In this assay, the $CD4^+$ T cells recognize peptides that are presented in the context of the MHC proteins that are present on APCs. When APCs are activated in the presence of Pb and the Pb is then removed before direct contact with the $CD4^+$ T cells, increased proliferation still results. This suggests that the Pb needs to only interact with the APC to stimulate a $CD4^+$ T cell response [305]. The effects of Pb on APC to produce this allo-enhancement make the study of APC's antigen-processing function in the presence of Pb essential for understanding the proliferative response and the implications of *in vivo* Pb exposures for immune functioning.

Cadmium has been found to alter the process of T cell differentiation as it accumulates in the thymus [306]. Cd exposure results in changes to both negative and positive selection processes within the thymus, resulting in changes in the T cell subsets. While Cd alters T cell differentiation events in the thymus, cells in the spleen appear more resistant to this effect.

Heavy metals also affect the differentiation of B cells, independent of T cell differentiation. When B cells are stimulated with LPS while exposed to lead, they secrete more IgM than cells stimulated with LPS alone. B cells exposed to lead in the absence of LPS stimulation do not lead to IgM secretion [307], suggesting that while Pb itself cannot induce B cell differentiation to a plasma cell, it may be able to serve as a substitute for T cell factors that enhance B cell differentiation [307].

Trafficking of immune cells to and from an infection site is an early event in both the innate and adaptive phases of an immune response. This directional movement of cells under the influence of diffusible gradients (chemotaxis) is regulated in part by small chemotactic cytokines called chemokines. In order for immune cells to enter infected tissues, they must be able to pass through the endothelial cells that line the walls of the blood vessels. Adhesion molecules on the surface of immune cells and the endothelial cells have been found to play key roles in chemotaxis. When toxic metals disrupt these regulated processes, the immune response can be compromised.

Cadherins are calcium-dependent metalloproteins that are involved in some forms of cell adhesion. These metalloproteins span the cell membrane and contain an extracellular domain with calcium-binding sites and adhesive regions. The intracellular domain of cadherin interacts with actin, a cytoskeleton component, through catenins [308]. Classical cadherins are known as N, P, and E types [309]. Studies have shown that the cadherins can be altered by metal exposure [308,310] since some metals can displace calcium in these proteins [311]. This binding causes changes in the structural conformation of the cadherins and presumably their interactions with binding partners [310]. When porcine renal epithelial cells are exposed to Cd at 10–20 μm, the cells separate from one another due to a decrease in E-cadherin-mediated cell contact.

The disruption of adhesion molecules can severely cripple an immune cell's function. Peritoneal macrophages exposed to lead in the presence of LPS display a reduction in phagocytosis when compared to the same cells exposed to LPS alone. The *in vitro* substrate attachment of the macrophages also decreases in comparison with the control [312]. Mercury can also disrupt the adhesion of leukocytes to the substrate and can disrupt their ability to move [313]. Splenic macrophages attachment and the chemotactic response are inhibited by lead or arsenic, but the effect is even greater when the cells are exposed to both metals simultaneously [314].

14.3.11 Metal Effects on Protein Secretion

One of the essential mechanisms by which immune cells regulate immune activities is via the production of soluble cytokines that influence the function of other cells. When these regulatory processes are disturbed, a wide range of immune changes can result. These aspects of cellular differentiation are sensitive to exogenous metals. For example, lead reduces the secretion of IFN-γ and IL-2 [315], while it can increase IL-4 synthesis [316]. The exposure of PBMCs to low levels of Cd can also alter cytokine secretion profiles. In response to antibody stimulation in the presence of Cd, secretion of IL-1β, TNF-α, and IFN-γ decreased, while IL-4 secretion increased. Similar to Pb exposure, Cd exposure causes a shift toward a Th2 response in PBMC and the secretion patterns of these cytokines are reflected in the cytokine gene expression patterns [317].

14.4 METAL-MEDIATED IMMUNOSUPPRESSION

Ultimately, one common consequence of exposure to toxic heavy metals is the onset of immunodeficiency. The mechanisms by which this occurs can be myriad. Metals that interfere with signaling within critical cell populations or with the synthesis and secretion of cytokines by differentiating immune cells will diminish immune protection. Metal exposures that are associated with necrotic or apoptotic cell death of critical leukocyte populations will be immunosuppressive, but excessive cell death can also disturb the normal antigen processing of phagocytes by overwhelming the antigen-processing mechanism with dead cells. On the other hand, metals that stimulate irrelevant responses can have the effect of diluting other essential activities and thereby diminishing immune function by reducing the effective interactions between cells that are necessary for the normal differentiation of lymphocyte and monocyte populations.

14.5 TOXIC METAL-ASSOCIATED IMMUNE ACTIVATION

14.5.1 Metal-Induced Hypersensitivity

Hypersensitivity reactions are nonprotective immune responses that occur in response to environmental stimuli. These reactions with environmental stimuli induce an inflammatory response, which can lead to allergy and disease. Nickel and beryllium are the two metals most often linked with the onset of type IV hypersensitivity reactions, but other metals such as chromium, cobalt, copper, gold, iridium, mercury, palladium, platinum, rhodium, and titanium are increasingly recognized as causes of skin hypersensitivity. These hypersensitivity reactions with metals can be mediated by either CD4 or CD8 T cells [318,319], which recognize neoantigens that are created when metals complex with self-proteins. Reactions with these neoantigens can develop into allergic contact dermatitis or, following beryllium exposure, chronic beryllium disease (CBD), also known as beryliosis.

These hypersensitive reactions can be a consequence of exposure to beryllium metal alloys in dental amalgams [320]. Although amalgams have fallen out of favor due to their mercury content, many people continue to be exposed to this metal from existing tooth fillings. People are also exposed to beryllium through recycling of electronics, fossil fuel combustion, and other industrial processes. Beryllium exposure is considered an occupational hazard and leads to hypersensitivity in large populations of workers [321].

In a type IV hypersensitivity reaction, metal comes into contact with the skin and diffuses into the epidermal layer. This begins the elicitation phase in which a metal at a high concentration will provoke an immune response. In the elicitation phase, T cells mediate the immune response to the metal that has complexed with self-proteins. Immune cells bind this antigen and produce chemokines and cytokines, which serve to recruit additional effector T cells and macrophages to the site of the stimulus. Secondary exposure to the metal will again produce this immune response and subsequent inflammation [321]. Workers who are exposed to as little as 0.2 $\mu g/m^3$ of beryllium become sensitized to the metal 95% of the time [322]. When bronchoalveolar lymphocytes (BAL) are harvested from beryllium-exposed individuals, *in vitro* re-exposure stimulates the release of the cytokines interleukin (IL)-2 and IFN-γ [323]. The secretion of IL-4 and IL-7 was not observed after this re-exposure, which indicates that the Th1 subpopulation predominates cells in CBD [323,324]. A large portion of BAL cells in CBD patients are beryllium-specific $CD4^+$ T cells. However, the same proportion of beryllium-specific T cells cannot be found in the blood of CBD patients, which suggests that beryllium-specific $CD4^+$ T cells preferentially accumulate in the target organ [325,326].

Sensitivity to beryllium is tested through the beryllium-stimulated blood lymphocyte proliferation test (BeLPT). In this test, lymphocytes are taken from a patient and exposed to beryllium *in vitro*. The proliferation of T cells in response to beryllium indicates that the individual has been previously exposed to the metal and has developed an immune response specific for the metal-associated antigens [327]. Although a positive result from the BeLPT test indicates that a person is sensitized to the metal, it does not necessarily indicate a disease state [328].

Like other immune diseases with a genetic component for disease risk, the risk for CBD when exposed to beryllium is thought to also depend on genetic markers. CBD risk is strongly linked to the human leukocyte antigen HLA-DPB1Glu69. It has been found in 84–97% of disease cases [329–335]. Another gene associated with a risk of beryllium sensitization is HLA-DRAArg74 [328]. APCs that express HLA-DP with a glutamic acid at position 69 can present inhaled beryllium to beryllium-specific $CD4^+$ T cells [326].

14.5.2 Heavy Metals and Autoimmunity

Both experimental and clinical data in animals and humans show that some metals can stimulate an immune response to self-proteins and that this response can progress to autoimmune disease. Experiments with experimental animals have shown that mercury can induce autoimmune disease in susceptible animals. Brown Norway rats treated with $HgCl_2$ display a widespread tissue injury including thymic atrophy [336], splenomegaly, lymphoadenopathies, and a hyperimmunoglobulinemia (mainly IgG1 and IgE), ulceration of the skin, hepatic periportal mononuclear cell infiltration, necrotizing vasculitis in the gastrointestinal tract, and membranous glomerulonephropathy [337,338]. Rats also produced various autoantibodies, including anti-DNA, antiphospholipid, antiglomerular basement membrane, antilaminin 1, and antithyroglobin in response to $HgCl_2$ treatment [339]. However, the disease is self-limiting and experimental animals recovered spontaneously after 4 weeks, at which point the rats become resistant to subsequent challenges. $HgCl_2$ also induces autoimmunity in certain strains of mice. Unlike rats, mice produce highly specific autoantibodies directly toward nuclear antigens [340–344]. The main target of autoantibodies is fibrillarin, a 34-kDa ribonucleoprotein. In mice the autoimmunity can be re-induced by mercury treatment and the antinuclear autoantibodies persist for months even after discontinuing the treatment [345]. Mercury-induced autoimmunity in both rats and mice is genetically restricted. The induction of autoimmune disorders is restricted to rats with the RT-1n MHC haplotype, such as the Brown-Norway (BN) strain. On the other hand, rats with RT-11 MHC haplotype, such as Lewis strain rats (LEW), are resistant to mercury-induced autoimmunity [346]. Interestingly, mercury can prevent LEW rats from developing autoimmune diseases, such as experimental autoimmune encephalomyelitis (EAE) [347]. It was considered that TGF-β played an important role in this protection through inhibition of autoreactive anti-MHC class II T cells [348]. In mice, the induction of antifibrillarin autoantibodies is most pronounced in strains with I-As MHC alleles, such as SJL or A. SW [341,342]. However, several studies have demonstrated that mercury, in combination with other environmental factors or certain genetic predispositions, can accelerate the onset of systemic autoimmune disease in resistant strains with an autoimmune-prone genetic background [349,350]. While the mechanisms of mercury-induced autoimmunity are not yet understood, many studies have provided valuable insights. As a thiol-binding metal, Hg can tightly bind with thiols on GSH or MT to interfere with the ability of cells to control the intracellular redox state. There is evidence that $CD4^+$ T cells in BN rats exhibit lower levels of total thiols [351] than LEW rats. Intracellular thiol levels are important in determining Th1/Th2 cytokine expression. It was found that the resistant LEW rats have higher levels of IFN-γ and IL-12 [352,353] compared to those in BN rats, which may contribute to the resistance of LEW rats. As already noted, inorganic mercury protects against anti-CD95-induced apoptosis and suggests that cells unable to undergo apoptosis will eventually die by necrotic processes. These necrotic cells may potentiate the inflammatory processes that ultimately result in mercury-induced autoimmunity.

14.6 DETECTION AND ASSESSMENT OF METAL-MEDIATED IMMUNOMODULATION

There are a variety of approaches to the assessment of immune function in the presence of metals that have been employed over the years. Most commonplace are studies that evaluate the outcomes of metal exposure by measurements of specific immune activities such as antibody production as measured by enzyme-linked immunosorbent assay (ELISA) or other assays of antibody binding to cognate antigen. It is also commonplace to assess the effects of toxicants on lymphoid tissue cellularity using immunohistochemical analysis of the tissue or by dissociating the tissue and then evaluating the different leukocyte subpopulations using flow cytometry. In both of these approaches, the effects of metals on the numbers of cells, their viability, and the specific kinds of cells can be readily determined. It is also possible to use a modified ELISA assay to do functional assessments of the cells that are exposed to metals. In the ELISPOT assay [354], an antigen or anti-cytokine antibody can be immobilized on a semipermeable filter. Following a blocking stop in which further nonspecific protein adsorption is prevented, a specific cell population to be interrogated is incubated on the membrane. Activated cells can secrete an antibody that binds to the immobilized antigen or cytokines that can be captured by immobilized anti-cytokine antibody. This event can be detected with a secondary antibody that is covalently attached to an enzyme that converts a chromogenic substrate to a precipitating product. After incubation with the substrate, the spots of precipitated product that are indicative of specific secretory cells can be counted and compared to the number of input cells. Changes in cellular programming that occur as a result of metal exposures will be identified by changes in the numbers of cells with a particular functional phenotype.

Recent advances have resulted in new ways of evaluating the impact of metals (or other toxicants) on the immune response. For example, Luminex™-type microbead assays enable the assessment of multiple cytokines that have been produced after *in vivo* or *in vitro* metal exposure by their capture on a set of different microscopic beads that are each uniquely identifiable by their intrinsic spectral characteristics. On each of the different bead types, a different capture antibody is bound, and is capable of capturing a specific soluble analyte in amounts that are proportional to the concentrations of the analyte in the solution. Subsequently, the mixture of different bead types is mixed with a cocktail of fluorescently labeled matched-pair antibodies for each of the captured analyte types. These labeled antibodies will fluorescenate the beads in amounts proportional to the amount of captured analyte. Using flow cytometry, the beads can be categorized by bead type for the specific identification of what analyte is being measured and by fluorescence intensity to determine the analyte concentration. At present, the systems have a maximum capacity of about 50 simultaneous measurements in a sample preparation. Another system that is often used to characterize the molecular changes that are induced in the immune response by metal exposure is the gene expression DNA microarray. In this system, a set of DNA oligonucleotides that have been selected for their sequence complimentarity to the mRNA for a specific gene are coupled at particular regions of interest (ROIs) on a sensor chip. mRNA harvested from sample cells is enzymatically converted to fluorescently labeled cDNA and then incubated with the immobilized oligonucleotide ROIs. Where there is sequence homology, the complementary strand of fluorescent cDNA will hybridize to the oligonucleotide, and the amount of fluorescence that accumulates at a specific ROI will be indicative of the amount of activity for that specific gene. Of particular value in the assessment of metal effects, one can do a two-channel analysis, where the two different mRNA pools harvested from control and metal-exposed conditions are each individually labeled with a different fluorescent tag. The two labeled cDNA pools are mixed together before hybridizing to the oligonucleotide microarray and the relative fluorophore concentrations of each cDNA can be simultaneously quantified under identical conditions.

There are limitations to each of these approaches for assessing the impacts of metals on the immune response. One way of addressing these limitations can be accomplished with a new instrumentation platform that is based on the physical phenomenon of surface plasmon resonance (SPR). SPR occurs at a metal:dielectric interface, and is the result of coupling incident light to resonant electrons in the gold. In order to establish this resonance, specific conditions must be met. These conditions include an appropriate wavelength of illuminating light, a specific incident angle for that light, and by using light with the correct phase velocity. These conditions can be met in a microarray format if the surface of the gold sensor chip is corrugated (termed grating-coupled SPR imaging or GCSPRI) [355–357]. Because the SPR measurement is sensitive to the incident angle of illuminating light, changes in the local refractive index of the fluid above the sensor chip surface will shift the angle at which SPR occurs. Accumulating protein at the sensor chip surface (e.g., by antibody-mediated capture of cells or proteins) will have the effect of changing the SPR angle in a manner proportional to the refractive index change and the amount of analyte immobilized at the surface. Under these circumstances, different regions of the corrugated sensor chip can be assessed independently, enabling simultaneous measurement of hundreds to thousands of analytes in a single biological sample. While this approach produces data similar to that produced with other assay systems, it is also amenable to other applications that should enable a more refined evaluation of metal-mediated immune changes in the future. For example, it should be possible to characterize the patterns of cellular function that result from metal exposures, by evaluating the kinetics of the entire set of cytokines that are synthesized and secreted by normal and metal-intoxicated cells.

There is good evidence that metals not only change the array of cytokines synthesized by the body, but also alter the kinetics of secretion from the cells. For example, it has been shown that cadmium will change the patterns of methylation of genes [358] and that some cytokines (e.g., IL-4) [359] are sensitive to changes in DNA methylation. DNA methylation is an important mechanism of epigenetic control, a process by which differential gene expression within immune cells is regulated [360]. Changes in DNA methylation occur following arsenic and nickel exposure as well as after Cd exposure [358]. In the absence of Dnmtl (a DNA methyltransferase essential for normal T cell homeostasis and development), proliferation, survival, and differentiation of T cells were all impaired [361,362]. Naïve T cells go through stages of cytokine activation, commitment, silencing, and stabilization as they differentiate into Thl or Th2 $CD4^+$ subsets [363], and this gene regulation is controlled in part by mechanisms such as DNA methylation. DNA methylation regulates activation and silencing of the IFN-γ and IL-4 cytokine gene loci, resulting in Thl or Th2 cells accordingly [364].

Abberrant DNA methylation that follows some forms of toxicant exposure is especially important to the immune system because the genes that are dysregulated include both the IL-4 and IL-13 genes that are usually regulated by DNA methylation. These genes are found in a predominantly methylated form in naïve $CD4^+$ cells but in an unmethylated form in differentiated Th2 cells [359]. This shows the role DNA methylation has in enabling cytokine expression. The methylation of IL-4 and demethylation of IFN-γ result in $CD4^+$ differentiation to Thl, whereas the methylation of IFN-γ and demethylation of IL-4 result in $CD4^+$ differentiation to Th2. This expression of one gene and silencing of the other via methylation is the epigenetic component of the $CD4^+$ differentiation process [360]. Another gene controlled by methylation is the IL-2 gene, which is demethylated within 20 min of T cell activation [360]. MT is another gene whose expression levels are known to be inversely correlated with the level of promoter methylation. MT production in cells undergoing hypermethylation was significantly depressed [358].

It should be possible to combine the ability to capture and enumerate mammalian leukocytes with the ability to subsequently measure cytokines and other soluble macromolecules secreted or released from the captured cells [355]. Because the GCSPRI signals are collected in real time, the kinetics of leukocyte secretion of these macromolecules upon exposure to polyclonal activators may represent a robust biomarker signature of the various effects of metal toxicants. As a consequence of making such measurements, it may be possible to identify individuals who have been exposed to these toxicants or to identify those most susceptible to immune consequences of exposure, well in advance of global changes to the immune system. Moreover, it may be possible to manage those changes before they become severe.

ACKNOWLEDGMENT

The authors would like to acknowledge the NIEHS for support (ES007408 and ES016014).

REFERENCES

1. Papp, L. V., J. Lu, A. Holmgren, and K. K. Khanna, 2007. From selenium to selenoproteins: Synthesis, identity, and their role in human health. *Antioxid Redox Signal* 9: 775–806.
2. Behne, D. and A. Kyriakopoulos, 2001. Mammalian selenium-containing proteins. *Annu Rev Nutr* 21: 453–473.
3. Navarro-Alarcon, M. and C. Cabrera-Vique, 2008. Selenium in food and the human body: A review. *Sci Total Environ* 400: 115–141.
4. Arslan, M., P. Csermely, and C. Sőti, 2006. Protein homeostasis and molecular chaperones in aging. *Biogerontology* 7: 383.
5. Steinbrenner, H. and H. Sies, 2009. Protection against reactive oxygen species by selenoproteins. *Biochim Biophys Acta* 1790: 1478–1485.
6. Arthur, J. R., R. C. McKenzie, and G. J. Beckett, 2003. Selenium in the immune system. *J Nutr* 133: 1457S–1459S.
7. Hoffmann, P. R. and M. J. Berry, 2008. The influence of selenium on immune responses. *Mol Nutr Food Res* 52: 1273–1280.
8. Nelson, H. K., Q. Shi, P. Van Dael, E. J. Schiffrin, S. Blum, D. Barclay, O. A. Levander, and M. A. Beck, 2001. Host nutritional selenium status as a driving force for influenza virus mutations. *FASEB J* 15: 1846–1848.
9. Finley, J. W. 2006. Bioavailability of selenium from foods. *Nutr Rev* 64: 146–151.
10. Finley, J. W., C. D. Davis, and Y. Feng, 2000. Selenium from high selenium broccoli protects rats from colon cancer. *J Nutr* 130: 2384–2389.
11. Crichton, R. R., S. Wilmet, R. Legssyer, and R. J. Ward, 2002. Molecular and cellular mechanisms of iron homeostasis and toxicity in mammalian cells. *J Inorg Biochem* 91: 9–18.
12. Porto, G. and M. De Sousa, 2007. Iron overload and immunity. *World J Gastroenterol* 13: 4707–4715.
13. Drakesmith, H. and A. Prentice, 2008. Viral infection and iron metabolism. *Nat Rev Microbiol* 6: 541–552.
14. Krause, A., S. Neitz, H. J. Magert, A. Schulz, W. G. Forssmann, P. Schulz-Knappe, and K. Adermann, 2000. LEAP-1, a novel highly disulfide-bonded human peptide, exhibits antimicrobial activity. *FEBS Lett* 480: 147–150.
15. Ganz, T. 2003. Hepcidin, a key regulator of iron metabolism and mediator of anemia of inflammation. *Blood* 102: 783–788.
16. Park, C. H., E. V. Valore, A. J. Waring, and T. Ganz, 2001. Hepcidin, a urinary antimicrobial peptide synthesized in the liver. *J Biol Chem* 276: 7806–7810.
17. Prasad, A. S. 2008. Zinc in human health: Effect of zinc on immune cells. *Mol Med* 14: 353–357.
18. Laity, J. H. and G. K. Andrews, 2007. Understanding the mechanisms of zinc-sensing by metal-response element binding transcription factor-1 (MTF-1). *Arch Biochem Biophys* 463: 201–210.
19. Krezel, A., Q. Hao, and W. Maret, 2007. The zinc/thiolate redox biochemistry of metallothionein and the control of zinc ion fluctuations in cell signaling. *Arch Biochem Biophys* 463: 188–200.
20. Maret, W. 2004. Zinc and sulfur: A critical biological partnership. *Biochemistry* 43: 3301–3309.
21. Powell, S. R. 2000. The antioxidant properties of zinc. *J Nutr* 130: 1447S–1454S.
22. Eide, D. J. 2006. Zinc transporters and the cellular trafficking of zinc. *Biochim Biophys Acta* 1763: 711–722.

23. Kobayashi, K., J. Kuroda, N. Shibata, T. Hasegawa, Y. Seko, M. Satoh, C. Tohyama, et al. 2007. Induction of metallothionein by manganese is completely dependent on interleukin-6 production. *J Pharmacol Exp Ther* 320: 721–727.
24. Haase, H. and L. Rink, 2007. Signal transduction in monocytes: The role of zinc ions. *Biometals* 20: 579–585.
25. Haase, H., D. J. Mazzatti, A. White, K. H. Ibs, G. Engelhardt, S. Hebel, J. R. Powell, and L. Rink, 2007. Differential gene expression after zinc supplementation and deprivation in human leukocyte subsets. *Mol Med* 13: 362–370.
26. Youn, J., S. H. Hwang, Z. Y. Ryoo, M. A. Lynes, D. J. Paik, H. S. Chung, and H. Y. Kim, 2002. Metallothionein suppresses collagen-induced arthritis via induction of TGF-beta and down-regulation of proinflammatory mediators. *Clin Exp Immunol* 129: 232–239.
27. Wellinghausen, N., A. B. Schromm, U. Seydel, K. Brandenburg, J. Luhm, H. Kirchner, and L. Rink, 1996. Zinc enhances lipopolysaccharide-induced monokine secretion by alteration of fluidity state of lipopolysaccharide. *J Immunol* 157: 3139–3145.
28. Krupanidhi, S., A. Sreekumar, and C. B. Sanjeevi, 2008. Copper & biological health. *Indian J Med Res* 128: 448–461.
29. Milne, D. B. 1998. Copper intake and assessment of copper status. *Am J Clin Nutr* 67: 1041S–1045S.
30. Bertinato, J. and M. R. L'Abbe, 2004. Maintaining copper homeostasis: Regulation of copper-trafficking proteins in response to copper deficiency or overload. *J Nutr Biochem* 15: 316–322.
31. Bonham, M., J. M. O'Connor, B. M. Hannigan, and J. J. Strain, 2002. The immune system as a physiological indicator of marginal copper status? *Br J Nutr* 87: 393–403.
32. Dimayuga, F. O., C. Wang, J. M. Clark, E. R. Dimayuga, V. M. Dimayuga, and A. J. Bruce-Keller, 2007. SOD1 overexpression alters ROS production and reduces neurotoxic inflammatory signaling in microglial cells. *J Neuroimmunol* 182: 89–99.
33. Percival, S. S. 1998. Copper and immunity. *Am J Clin Nutr* 67: 1064S–1068S.
34. Oster, G. and M. P. Salgo, 1977. Copper in mammalian reproduction. *Adv Pharmacol Chemother* 14: 327–409.
35. Lawrence, D. A., M. J. McCabe, and M. Kowolenko, 1990. Metal influences on autoimmunity and infectious diseases. In *Metal Ions in Biology and Medicine*, P. Ollery, L.A. Poirier, M. Manfait, J. C. Etienne (eds). J. Libbey and Co., London, pp. 237–242.
36. Meister, A. 1992. On the antioxidant effects of ascorbic acid and glutathione. *Biochem Pharmacol* 44: 1905–1915.
37. Meister, A., O. W. Griffith, A. Novogrodsky, and S. S. Tate, 1979. New aspects of glutathione metabolism and translocation in mammals. *Ciba Found Symp* 72: 135–161.
38. Liu, H., H. Wang, S. Shenvi, T. M. Hagen, and R. M. Liu, 2004. Glutathione metabolism during aging and in Alzheimer disease. *Ann N Y Acad Sci* 1019: 346–349.
39. Maher, P. 2005. The effects of stress and aging on glutathione metabolism. *Ageing Res Rev* 4: 288–314.
40. Awasthi, Y. C., E. Beutler, and S. K. Srivastava, 1975. Purification and properties of human erythrocyte glutathione peroxidase. *J Biol Chem* 250: 5144–5149.
41. Schanne, F. A., T. L. Dowd, R. K. Gupta, and J. F. Rosen, 1989. Lead increases free Ca^{2+} concentration in cultured osteoblastic bone cells: Simultaneous detection of intracellular free Pb^{2+} by 19F NMR. *Proc Natl Acad Sci USA* 86: 5133–5135.
42. Duncan, D. D. and D. A. Lawrence, 1989. Differential lymphocyte growth-modifying effects of oxidants: Changes in cytosolic Ca^{2+}. *Toxicol Appl Pharmacol* 100: 485–497.
43. Bergmann, W. L., V. Dressler, C. W. Haest, and B. Deuticke, 1984. Cross-linking of SH-groups in the erythrocyte membrane enhances transbilayer reorientation of phospholipids. Evidence for a limited access of phospholipids to the reorientation sites. *Biochim Biophys Acta* 769: 390–398.
44. Chesney, R. W., N. Gusowski, and I. Zelikovic, 1986. Membrane fluidity and phospholipid composition in relation to sulfur amino acid intake in brush border membranes of rat kidney. *Pediatr Res* 20: 1305–1309.
45. Schachter, D., R. E. Abbott, U. Cogan, and M. Flamm, 1983. Lipid fluidity of the individual hemileaflets of human erythrocyte membranes. *Ann N Y Acad Sci* 414: 19–28.
46. Hu, M. L. and A. L. Tappel, 1992. Glutathione and antioxidants protect microsomes against lipid peroxidation and enzyme inactivation. *Lipids* 27: 42–45.
47. Stohs, S. J. and D. Bagchi, 1995. Oxidative mechanisms in the toxicity of metal ions. *Free Radic Biol Med* 18: 321–336.
48. Valko, M., H. Morris, and M. T. Cronin, 2005. Metals, toxicity and oxidative stress. *Curr Med Chem* 12: 1161–1208.

49. Bjorksten, J. 1958. A common molecular basis for the aging syndrome. *J Am Geriatr Soc* 8: 740–748.
50. Harman, D. 1956. Aging: A theory based on free radical and radiation chemistry. *J Gerontol* 11: 298–300.
51. Harman, D. 1993. Free radical involvement in aging. Pathophysiology and therapeutic implications. *Drugs Aging* 3: 60–80.
52. Masoro, E. 1990. The role of free radicals in senescence. *Annu Rev Gerontol Geriatr* 10: 183–197.
53. Sohal, R. S. and W. C. Orr, 1992. Relationship between antioxidants, prooxidants, and the aging process. *Ann N Y Acad Sci* 663: 74–84.
54. Walford, R. 1969. *The Immunological Theory of Aging*. Munksgaard, Copenhagen.
55. Nakashima, I., M. Y. Pu, A. Nishizaki, I. Rosila, L. Ma, Y. Katano, K. Ohkusu, et al. 1994. Redox mechanism as alternative to ligand binding for receptor activation delivering disregulated cellular signals. *J Immunol* 152: 1064–1071.
56. Lawrence, D. A. 1995. Metal ions in autoimmune and infectious diseases. In *Handbook of Metal–Ligand Interactions in Biological Fluids*, G. Berthon (ed). Marcel Dekker, Inc., New York, pp. 904–913.
57. Linton, P. J., S. P. Li, Y. Zhang, B. Bautista, Q. Huynh, and T. Trinh, 2005. Intrinsic versus environmental influences on T-cell responses in aging. *Immunol Rev* 205: 207–219.
58. Cemerski, S., A. Cantagrel, J. P. Van Meerwijk, and P. Romagnoli, 2002. Reactive oxygen species differentially affect T cell receptor-signaling pathways. *J Biol Chem* 277: 19585–19593.
59. Huse, M., M. J. Eck, and S. C. Harrison, 1998. A Zn^{2+} ion links the cytoplasmic tail of CD4 and the N-terminal region of Lck. *J Biol Chem* 273: 18729–18733.
60. Kanner, S. B., T. J. Kavanagh, A. Grossmann, S. L. Hu, J. B. Bolen, P. S. Rabinovitch, and J. A. Ledbetter, 1992. Sulfhydryl oxidation down-regulates T-cell signaling and inhibits tyrosine phosphorylation of phospholipase C gamma 1. *Proc Natl Acad Sci USA* 89: 300–304.
61. Cossins, A. R., M. Behan, G. Jones, and K. Bowler, 1987. Lipid–protein interactions in the adaptive regulation of membrane function. *Biochem Soc Trans* 15: 77–81.
62. Crain, R. C., G. V. Marinetti, and D. F. O'Brien, 1978. Topology of amino phospholipids in bovine retinal rod outer segment disk membranes. *Biochemistry* 17: 4186–4192.
63. Flaherty, L., D. Zimmerman, and S. S. Wachtel, 1979. H-Y antigen. Cell surface mapping and testosterone-induced supramolecular repatterning. *J Exp Med* 150: 1020–1027.
64. Jardieu, P., M. Akasaki, and K. Ishizaka, 1986. Association of I-J determinants with lipomodulin/macrocortin. *Proc Natl Acad Sci USA* 83: 160–164.
65. Klausner, R. D., A. M. Kleinfeld, R. L. Hoover, and M. J. Karnovsky, 1980. Lipid domains in membranes. Evidence derived from structural perturbations induced by free fatty acids and lifetime heterogeneity analysis. *J Biol Chem* 255: 1286–1295.
66. Maraganore, J. 1987. Structural elements for protein–phospholipid interactions may be shared in protein kinase C and phospholipid A2. *Trends Biochem Sci* 12: 176–177.
67. Meyer, H. W. 1983. Lipid domain structures in biological membranes. *Exp Pathol* 23: 3–26.
68. Magilavy, D. B., F. W. Fitch, and T. F. Gajewski, 1989. Murine hepatic accessory cells support the proliferation of Th1 but not Th2 helper T lymphocyte clones. *J Exp Med* 170: 985–990.
69. Matthias, L. J., P. T. Yam, X. M. Jiang, N. Vandegraaff, P. Li, P. Poumbourios, N. Donoghue, and P. J. Hogg, 2002. Disulfide exchange in domain 2 of CD4 is required for entry of HIV-1. *Nat Immunol* 3: 727–732.
70. Simon, J. C., P. D. Cruz, Jr., P. R. Bergstresser, and R. E. Tigelaar, 1990. Low dose ultraviolet B-irradiated Langerhans cells preferentially activate CD4+ cells of the T helper 2 subset. *J Immunol* 145: 2087–2091.
71. Barbouche, R., R. Miquelis, I. M. Jones, and E. Fenouillet, 2003. Protein-disulfide isomerase-mediated reduction of two disulfide bonds of HIV envelope glycoprotein 120 occurs post-CXCR4 binding and is required for fusion. *J Biol Chem* 278: 3131–3136.
72. Limatola, C., S. Di Bartolomeo, M. Catalano, F. Trettel, S. Fucile, L. Castellani, and F. Eusebi, 2005. Cysteine residues are critical for chemokine receptor CXCR2 functional properties. *Exp Cell Res* 307: 65–75.
73. Stahls, A. 1992. The sulfhydryl reagent *N*-ethylmaleimide induces hyperphosphorylation on tyrosine residues in the Jurkat T-cell line. *Biochem Biophys Res Commun* 187: 73–78.
74. Bennett, T. A., B. S. Edwards, L. A. Sklar, and S. Rogelj, 2000. Sulfhydryl regulation of l-selectin shedding: Phenylarsine oxide promotes activation-independent l-selectin shedding from leukocytes. *J Immunol* 164: 4120–4129.
75. George, R. J. and C. W. Parker, 1990. Preliminary characterization of phosphotyrosine phosphatase activities in human peripheral blood lymphocytes: Identification of CD45 as a phosphotyrosine phosphatase. *J Cell Biochem* 42: 71–81.

76. Tada, T. 1992. Nutrition and the immune system in aging: An overview. *Nutr Rev* 50: 360.
77. Ezzell, C. 1993. AIDS' unlucky strike: When T cells would rather switch than fight. *J. NIH Res.* 5: 59–64.
78. Gazzinelli, R. T., M. Makino, S. K. Chattopadhyay, C. M. Snapper, A. Sher, A. W. Hugin, and H. C. Morse, 3rd, 1992. CD4+ subset regulation in viral infection. Preferential activation of Th2 cells during progression of retrovirus-induced immunodeficiency in mice. *J Immunol* 148: 182–188.
79. Krieger, J. A., Y. Heo, and Lawrence, D. A. 1998. Oxidative stress and heavy metal modification of T lymphocyte subsets. In *T Lymphocyte Subpopulations in Immunotoxicology*, I. Kimber and Selgrade, M. J. K. (ed.). Wiley, Chichester, England, pp. 103–119.
80. Peterson, J. D., L. A. Herzenberg, K. Vasquez, and C. Waltenbaugh, 1998. Glutathione levels in antigen-presenting cells modulate Th1 versus Th2 response patterns. *Proc Natl Acad Sci USA* 95: 3071–3076.
81. Lawrence, D. A. 1980. Antigen activation of T cells. In *Handbook of Cancer Immunology*, 2nd ed. H. Waters (ed.). Garland STPM Press, New York, pp. 257–320.
82. Murata, Y., M. Amao, J. Yoneda, and J. Hamuro, 2002. Intracellular thiol redox status of macrophages directs the Th1 skewing in thioredoxin transgenic mice during aging. *Mol Immunol* 38: 747–757.
83. Janeway, C. A., Jr., S. Carding, B. Jones, J. Murray, P. Portoles, R. Rasmussen, J. Rojo, K. Saizawa, J. West, and K. Bottomly, 1988. CD4+ T cells: Specificity and function. *Immunol Rev* 101: 39–80.
84. Kupfer, A., S. J. Singer, C. A. Janeway, Jr., and S. L. Swain, 1987. Coclustering of CD4 (L3T4) molecule with the T-cell receptor is induced by specific direct interaction of helper T cells and antigen-presenting cells. *Proc Natl Acad Sci USA* 84: 5888–5892.
85. Krutmann, J., I. U. Khan, R. S. Wallis, F. Zhang, E. A. Rich, J. J. Ellner, and C. A. Elmets, 1990. Cell membrane is a major locus for ultraviolet B-induced alterations in accessory cells. *J Clin Invest* 85: 1529–1536.
86. Hughes, D. A., P. J. Townsend, and P. L. Haslam, 1992. Enhancement of the antigen-presenting function of monocytes by cholesterol: Possible relevance to inflammatory mechanisms in extrinsic allergic alveolitis and atherosclerosis. *Clin Exp Immunol* 87: 279–286.
87. Heigold, S. and G. Bauer, 2002. RAW 264.7 macrophages induce apoptosis selectively in transformed fibroblasts: Intercellular signaling based on reactive oxygen and nitrogen species. *J Leukoc Biol* 72: 554–563.
88. Frossi, B., M. De Carli, M. Piemonte, and C. Pucillo, 2008. Oxidative microenvironment exerts an opposite regulatory effect on cytokine production by Th1 and Th2 cells. *Mol Immunol* 45: 58–64.
89. King, M. R., A. S. Ismail, L. S. Davis, and D. R. Karp, 2006. Oxidative stress promotes polarization of human T cell differentiation toward a T helper 2 phenotype. *J Immunol* 176: 2765–2772.
90. de Vos, G., S. Abotaga, Z. Liao, E. Jerschow, and D. Rosenstreich, 2007. Selective effect of mercury on Th2-type cytokine production in humans. *Immunopharmacol Immunotoxicol* 29: 537–548.
91. Hobbs, M. V., W. O. Weigle, D. J. Noonan, B. E. Torbett, R. J. McEvilly, R. J. Koch, G. J. Cardenas, and D. N. Ernst, 1993. Patterns of cytokine gene expression by CD4+ T cells from young and old mice. *J Immunol* 150: 3602–3614.
92. Nagelkerken, L., A. Hertogh-Huijbregts, R. Dobber, and A. Drager, 1991. Age-related changes in lymphokine production related to a decreased number of CD45RBhi CD4+ T cells. *Eur J Immunol* 21: 273–281.
93. al-Rayes, H., W. Pachas, N. Mirza, D. J. Ahern, R. S. Geha, and D. Vercelli, 1992. IgE regulation and lymphokine patterns in aging humans. *J Allergy Clin Immunol* 90: 630–636.
94. Li, S. P. and R. A. Miller, 1993. Age-associated decline in IL-4 production by murine T lymphocytes in extended culture. *Cell Immunol* 151: 187–195.
95. Murasko, D. M. and I. M. Goonewardene, 1990. T-cell function in aging: Mechanisms of decline. *Annu Rev Gerontol Geriatr* 10: 71–96.
96. Grossmann, A., L. Maggio-Price, J. C. Jinneman, and P. S. Rabinovitch, 1991. Influence of aging on intracellular free calcium and proliferation of mouse T-cell subsets from various lymphoid organs. *Cell Immunol* 135: 118–131.
97. Miller, R. A., B. Philosophe, I. Ginis, G. Weil, and B. Jacobson, 1989. Defective control of cytoplasmic calcium concentration in T lymphocytes from old mice. *J Cell Physiol* 138: 175–182.
98. Grossmann, A., J. A. Ledbetter, and P. S. Rabinovitch, 1989. Reduced proliferation in T lymphocytes in aged humans is predominantly in the CD8+ subset, and is unrelated to defects in transmembrane signaling which are predominantly in the CD4+ subset. *Exp Cell Res* 180: 367–382.
99. Lustyik, G. and J. J. O'Leary, 1989. Aging and the mobilization of intracellular calcium by phytohemagglutinin in human T cells. *J Gerontol* 44: B30–B36.
100. Messina, J. P. and D. A. Lawrence, 1992. Effects of 2-mercaptoethanol and buthionine sulfoximine on cystine metabolism by and proliferation of mitogen-stimulated human and mouse lymphocytes. *Int J Immunopharmacol* 14: 1221–1234.

101. Duncan, D. D. and D. A. Lawrence, 1988. Four sulfhydryl-modifying compounds cause different structural damage but similar functional damage in murine lymphocytes. *Chem Biol Interact* 68: 137–152.
102. Lawrence, D. A., R. Song, and P. Weber, 1996. Surface thiols of human lymphocytes and their changes after *in vitro* and *in vivo* activation. *J Leukoc Biol* 60: 611–618.
103. Messina, J. P. and D. A. Lawrence, 1989. Cell cycle progression of glutathione-depleted human peripheral blood mononuclear cells is inhibited at S phase. *J Immunol* 143: 1974–1981.
104. McElhaney, J. E., B. L. Beattie, R. Devine, R. Grynoch, E. L. Toth, and R. C. Bleackley, 1990. Age-related decline in interleukin 2 production in response to influenza vaccine. *J Am Geriatr Soc* 38: 652–658.
105. Huang, Y. P., J. C. Pechere, M. Michel, L. Gauthey, M. Loreto, J. A. Curran, and J. P. Michel, 1992. *In vivo* T cell activation, *in vitro* defective IL-2 secretion, and response to influenza vaccination in elderly women. *J Immunol* 148: 715–722.
106. Suzuki, T., N. Suzuki, R. A. Daynes, and E. G. Engleman, 1991. Dehydroepiandrosterone enhances IL2 production and cytotoxic effector function of human T cells. *Clin Immunol Immunopathol* 61: 202–211.
107. Buhl, R., H. A. Jaffe, K. J. Holroyd, F. B. Wells, A. Mastrangeli, C. Saltini, A. M. Cantin, and R. G. Crystal, 1989. Systemic glutathione deficiency in symptom-free HIV-seropositive individuals. *Lancet* 2: 1294–1298.
108. Droge, W., H. P. Eck, M. Betzler, P. Schlag, P. Drings, and W. Ebert, 1988. Plasma glutamate concentration and lymphocyte activity. *J Cancer Res Clin Oncol* 114: 124–128.
109. Dworkin, B. M., W. S. Rosenthal, G. P. Wormser, L. Weiss, M. Nunez, C. Joline, and A. Herp, 1988. Abnormalities of blood selenium and glutathione peroxidase activity in patients with acquired immunodeficiency syndrome and aids-related complex. *Biol Trace Elem Res* 15: 167–177.
110. Eck, H. P., H. Gmunder, M. Hartmann, D. Petzoldt, V. Daniel, and W. Droge, 1989. Low concentrations of acid-soluble thiol (cysteine) in the blood plasma of HIV-1-infected patients. *Biol Chem Hoppe Seyler* 370: 101–108.
111. Passi, S., C. De Luca, M. Picardo, A. Morrone, and F. Ippolito, 1990. Blood deficiency values of polyunsaturated fatty acids of phospholipids, vitamin E and glutathione peroxidase as possible risk factors in the onset and development of acquired immunodeficiency syndrome. *G Ital Dermatol Venereol* 125: 125–130.
112. Phillips, A. N., C. A. Lee, J. Elford, A. Webster, G. Janossy, A. Timms, M. Bofill, and P. B. Kernoff, 1991. More rapid progression to AIDS in older HIV-infected people: The role of CD4+ T-cell counts. *J Acquir Immune Defic Syndr* 4: 970–975.
113. Parker, P. J., L. Coussens, N. Totty, L. Rhee, S. Young, E. Chen, S. Stabel, M. D. Waterfield, and A. Ullrich, 1986. The complete primary structure of protein kinase C—the major phorbol ester receptor. *Science* 233: 853–859.
114. Speizer, L. A., M. J. Watson, J. R. Kanter, and L. L. Brunton, 1989. Inhibition of phorbol ester binding and protein kinase C activity by heavy metals. *J Biol Chem* 264: 5581–5585.
115. Gopalakrishna, R. and W. B. Anderson, 1989. Ca^{2+}- and phospholipid-independent activation of protein kinase C by selective oxidative modification of the regulatory domain. *Proc Natl Acad Sci USA* 86: 6758–6762.
116. Fong, T. C. and T. Makinodan, 1989. Preferential enhancement by 2-mercaptoethanol of IL-2 responsiveness of T blast cells from old over young mice is associated with potentiated protein kinase C translocation. *Immunol Lett* 20: 149–154.
117. Falo, L. D., Jr., B. Benacerraf, and K. L. Rock, 1986. Phospholipase treatment of accessory cells that have been exposed to antigen selectively inhibits antigen-specific Ia-restricted, but not allospecific, stimulation of T lymphocytes. *Proc Natl Acad Sci USA* 83: 6994–6997.
118. Testa, U., P. G. Pelicci, P. Thomopoulos, M. Titeux, and H. Rochant, 1983. The number of the Trf receptors on human hematopoietic cell lines is influenced by membrane phospholipids. *Biochem Int* 7: 169–178.
119. May, W. S., E. G. Lapetina, and P. Cuatrecasas, 1986. Intracellular activation of protein kinase C and regulation of the surface transferrin receptor by diacylglycerol is a spontaneously reversible process that is associated with rapid formation of phosphatidic acid. *Proc Natl Acad Sci USA* 83: 1281–1284.
120. Kaufman, J. F., M. S. Krangel, and J. L. Strominger, 1984. Cysteines in the transmembrane region of major histocompatibility complex antigens are fatty acylated via thioester bonds. *J Biol Chem* 259: 7230–7238.
121. Simonis, S. and S. E. Cullen, 1986. Fatty acylation of murine Ia alpha, beta, and invariant chains. *J Immunol* 136: 2962–2967.
122. Janick-Buckner, D., C. J. Briggs, T. E. Meyer, N. Harvey, and C. M. Warner, 1991. Major histocompatibility complex antigen expression on lymphocytes from aging strain A mice. *Growth Dev Aging* 55: 53–62.

123. Saririan, K., A. Wali, R. P. Almeida, and C. Russo, 1993. Increased serum HLA class I molecule levels in elderly human responders to influenza vaccination. *Tissue Antigens* 42: 9–13.
124. Houslay, M. D. 1985. Regulation of adenylate cyclase (EC 4.6.I.I) activity by its lipid environment. *Proc Nutr Soc* 44: 157–165.
125. Dipple, I., L. M. Gordon, and M. D. Houslay, 1982. The activity of 5′-nucleotidase in liver plasma membranes is affected by the increase in bilayer fluidity achieved by anionic drugs but not by cationic drugs. *J Biol Chem* 257: 1811–1815.
126. Ginsberg, B. H., J. Jabour, and A. A. Spector, 1982. Effect of alterations in membrane lipid unsaturation on the properties of the insulin receptor of Ehrlich ascites cells. *Biochim Biophys Acta* 690: 157–164.
127. Sandermann, H., Jr. and B. A. Gottwald, 1983. Cooperative lipid activation of (Na + +K +)-ATPASE as a consequence of non-cooperative lipid–protein interactions. *Biochim Biophys Acta* 732: 332–335.
128. Boumendil-Podevin, E. F. and R. A. Podevin, 1977. Stimulatory and inhibitory effects of sulfhydryl reagent on *p*-aminohippuric acid transport by isolated renal tubules. *Biochim Biophys Acta* 467: 364–375.
129. Hare, J. D. 1975. Distinctive alterations of nucleoside, sugar and amino acid uptake of sulfhydryl reagents in cultured mouse cells. *Arch Biochem Biophys* 170: 347–352.
130. Kwock, L., D. F. Wallach, and K. Hefter, 1976. Involvement of sulfhydryl groups in the action of the insulin and radiation on thymocyte Na^+-dependent amino acid transport. *Biochim Biophys Acta* 419: 93–103.
131. Schoot, B. M., S. E. van Emst-de Vries, P. M. van Haard, J. J. de Pont, and S. L. Bonting, 1980. Studies on (Na++K+)-activated ATPase. XLVI. Effect of cation-induced conformational changes on sulfhydryl group modification. *Biochim Biophys Acta* 602: 144–154.
132. Witkowski, J. M., A. Mysliwski, and J. Mysliwska, 1985. Decrease of lymphocyte (Na+,K+)ATP-ase activity in aged people. *Mech Ageing Dev* 33: 11–17.
133. Damjanovich, S. and C. Pieri, 1991. Electroimmunology: Membrane potential, ion-channel activities, and stimulatory signal transduction in human T lymphocytes from young and elderly. *Ann N Y Acad Sci* 621: 29–39.
134. Varga, Z., N. Bressani, A. M. Zaia, L. Bene, T. Fulop, A. Leovey, N. Fabris, and S. Damjanovich, 1990. Cell surface markers, inositol phosphate levels and membrane potential of lymphocytes from young and old human patients. *Immunol Lett* 23: 275–280.
135. Borzi, R. M., P. Dal Monte, M. Uguccioni, R. Meliconi, and A. Facchini, 1992. Intracellular nucleotides of lymphocytes and granulocytes from normal ageing subjects. *Mech Ageing Dev* 64: 1–11.
136. Shi, J. and R. A. Miller, 1992. Tyrosine-specific protein phosphorylation in response to anti-CD3 antibody is diminished in old mice. *J Gerontol* 47: B147–B153.
137. Patel, H. R. and R. A. Miller, 1992. Age-associated changes in mitogen-induced protein phosphorylation in murine T lymphocytes. *Eur J Immunol* 22: 253–260.
138. Fulop, T., Jr., G. Barabas, Z. Varga, J. Csongor, M. Hauck, S. Szucs, I. Seres, A. Mohacsi, D. Kekessy, J. P. Despont, et al. 1992. Transmembrane signaling changes with aging. *Ann N Y Acad Sci* 673: 165–171.
139. Rudd, C. E., O. Janssen, Y. C. Cai, A. J. da Silva, M. Raab, and K. V. Prasad, 1994. Two-step TCR zeta/CD3-CD4 and CD28 signaling in T cells: SH2/SH3 domains, protein-tyrosine and lipid kinases. *Immunol Today* 15: 225–234.
140. Droge, W. 2002. The plasma redox state and ageing. *Ageing Res Rev* 1: 257–278.
141. Stadtman, E. R. 2001. Protein oxidation in aging and age-related diseases. *Ann N Y Acad Sci* 928: 22–38.
142. Thomas, J. A. and R. J. Mallis, 2001. Aging and oxidation of reactive protein sulfhydryls. *Exp Gerontol* 36: 1519–1526.
143. Webster, K. A., H. Prentice, and N. H. Bishopric, 2001. Oxidation of zinc finger transcription factors: Physiological consequences. *Antioxid Redox Signal* 3: 535–548.
144. Trump, B. F., I. K. Berezesky, M. W. Smith, P. C. Phelps, and K. A. Elliget, 1989. The relationship between cellular ion deregulation and acute and chronic toxicity. *Toxicol Appl Pharmacol* 97: 6–22.
145. Gerber, G. B. and Altman, K. I., 1970. *Radiation Biochemistry: Tissues and Body Fluids*. Academic Press, New York.
146. Schraufstatter, I., P. A. Hyslop, J. H. Jackson, and C. G. Cochrane, 1988. Oxidant-induced DNA damage of target cells. *J Clin Invest* 82: 1040–1050.
147. Malbon, C. C., George, S.T., and Moxham, C.P. 1987. Intramolecular disulfide bridges: Avenues to receptor activation. *Trends Biochem Sci* 12: 172–175.
148. Ullrich, A., J. R. Bell, E. Y. Chen, R. Herrera, L. M. Petruzzelli, T. J. Dull, A. Gray, L. Coussens, Y. C. Liao, M. Tsubokawa, et al. 1985. Human insulin receptor and its relationship to the tyrosine kinase family of oncogenes. *Nature* 313: 756–761.

149. Yarden, Y., J. A. Escobedo, W. J. Kuang, T. L. Yang-Feng, T. O. Daniel, P. M. Tremble, E. Y. Chen, M. E. Ando, R. N. Harkins, U. Francke, et al. 1986. Structure of the receptor for platelet-derived growth factor helps define a family of closely related growth factor receptors. *Nature* 323: 226–232.
150. Johnson, D., A. Lanahan, C. R. Buck, A. Sehgal, C. Morgan, E. Mercer, M. Bothwell, and M. Chao, 1986. Expression and structure of the human NGF receptor. *Cell* 47: 545–554.
151. Leonard, W. J., J. M. Depper, M. Kronke, R. J. Robb, T. A. Waldmann, and W. C. Greene, 1985. The human receptor for T-cell growth factor. Evidence for variable post-translational processing, phosphorylation, sulfation, and the ability of precursor forms of the receptor to bind T-cell growth factor. *J Biol Chem* 260: 1872–1880.
152. Blanpain, C., V. Wittamer, J. M. Vanderwinden, A. Boom, B. Renneboog, B. Lee, E. Le Poul, et al. 2001. Palmitoylation of CCR5 is critical for receptor trafficking and efficient activation of intracellular signaling pathways. *J Biol Chem* 276: 23795–23804.
153. Tagaya, Y., Y. Maeda, A. Mitsui, N. Kondo, H. Matsui, J. Hamuro, N. Brown, K. Arai, T. Yokota, H. Wakasugi, et al. 1989. ATL-derived factor (ADF), an IL-2 receptor/Tac inducer homologous to thioredoxin; possible involvement of dithiol-reduction in the IL-2 receptor induction. *EMBO J* 8: 757–764.
154. Tagaya, Y., M. Okada, K. Sugie, T. Kasahara, N. Kondo, J. Hamuro, K. Matsushima, C. A. Dinarello, and J. Yodoi, 1988. IL-2 receptor(p55)/Tac-inducing factor. Purification and characterization of adult T cell leukemia-derived factor. *J Immunol* 140: 2614–2620.
155. Noelle, R. J. and D. A. Lawrence, 1981. Modulation of T-cell function. II. Chemical basis for the involvement of cell surface thiol-reactive sites in control of T-cell proliferation. *Cell Immunol* 60: 453–469.
156. Fidelus, R. K. 1988. The generation of oxygen radicals: A positive signal for lymphocyte activation. *Cell Immunol* 113: 175–182.
157. Suthanthiran, M., M. E. Anderson, V. K. Sharma, and A. Meister, 1990. Glutathione regulates activation-dependent DNA synthesis in highly purified normal human T lymphocytes stimulated via the CD2 and CD3 antigens. *Proc Natl Acad Sci USA* 87: 3343–3347.
158. Holmgren, A. 1977. The function of thioredoxin and glutathione in deoxyribonucleic acid synthesis. *Biochem Soc Trans* 5: 611–612.
159. Lewin, B. 1980. *Gene Expression 2*. Wiley, New York.
160. Draetta, G. and D. Beach, 1988. Activation of cdc2 protein kinase during mitosis in human cells: Cell cycle-dependent phosphorylation and subunit rearrangement. *Cell* 54: 17–26.
161. Meister, A. 1983. Selective modification of glutathione metabolism. *Science* 220: 472–477.
162. Abate, C., L. Patel, F. J. Rauscher, 3rd, and T. Curran, 1990. Redox regulation of fos and jun DNA-binding activity in vitro. *Science* 249: 1157–1161.
163. Sikora, E., B. Kaminska, E. Radziszewska, and L. Kaczmarek, 1992. Loss of transcription factor AP-1 DNA binding activity during lymphocyte aging in vivo. *FEBS Lett* 312: 179–182.
164. Feig, C. and Peter, M. E., 2007. How apoptosis got the mmune system in shape. *Eur J Immunol* 37 S61–S70.
165. Giovannetti, A., M. Pierdominici, A. Di Iorio, R. Cianci, G. Murdaca, F. Puppo, F. Pandolfi, and R. Paganelli, 2008. Apoptosis in the homeostasis of the immune system and in human immune mediated diseases. *Curr Pharm Des* 14: 253–268.
166. Thompson, C. B. 1995. Apoptosis in the pathogenesis and treatment of disease. *Science* 267: 1456–1462.
167. Vaux, D. L. 2002. Apoptosis and toxicology—what relevance? *Toxicology* 181–182: 3–7.
168. Pulido, M. D. and A. R. Parrish, 2003. Metal-induced apoptosis: Mechanisms. *Mutat Res* 533: 227–241.
169. Rana, S. V. 2008. Metals and apoptosis: Recent developments. *J Trace Elem Med Biol* 22: 262–284.
170. Nelson, D. A. and E. White, 2004. Exploiting different ways to die. *Genes Dev* 18: 1223–1226.
171. Perl, M., C. S. Chung, and A. Ayala, 2005. Apoptosis. *Crit Care Med* 33: S526–S529.
172. Hengartner, M. O. 2000. The biochemistry of apoptosis. *Nature* 407: 770–776.
173. Danial, N. N. and S. J. Korsmeyer, 2004. Cell death: Critical control points. *Cell* 116: 205–219.
174. Jeong, S. Y. and D. W. Seol, 2008. The role of mitochondria in apoptosis. *BMB Rep* 41: 11–22.
175. Levine, B., S. Sinha, and G. Kroemer, 2008. Bcl-2 family members: Dual regulators of apoptosis and autophagy. *Autophagy* 4: 600–606.
176. Wilson, N. S., V. Dixit, and A. Ashkenazi, 2009. Death receptor signal transducers: Nodes of coordination in immune signaling networks. *Nat Immunol* 10: 348–355.
177. Li, J. and J. Yuan, 2008. Caspases in apoptosis and beyond. *Oncogene* 27: 6194–6206.
178. Henson, P. M. and D. A. Hume, 2006. Apoptotic cell removal in development and tissue homeostasis. *Trends Immunol* 27: 244–250.
179. Lauber, K., S. G. Blumenthal, M. Waibel, and S. Wesselborg, 2004. Clearance of apoptotic cells: Getting rid of the corpses. *Mol Cell* 14: 277–287.

180. Peter, M. E., R. C. Budd, J. Desbarats, S. M. Hedrick, A. O. Hueber, M. K. Newell, L. B. Owen, et al., 2007. The CD95 receptor: Apoptosis revisited. *Cell* 129: 447–450.
181. Irie, K., Y. Yanai, K. Oie, J. Ishizawa, Y. Nakagawa, H. Ohigashi, P. A. Wender, and U. Kikkawa, 1997. Comparison of chemical characteristics of the first and the second cysteine-rich domains of protein kinase C gamma. *Bioorg Med Chem* 5: 1725–1737.
182. Ashkenazi, A. and V. M. Dixit, 1999. Apoptosis control by death and decoy receptors. *Curr Opin Cell Biol* 11: 255–260.
183. Algeciras-Schimnich, A., L. Shen, B. C. Barnhart, A. E. Murmann, J. K. Burkhardt, and M. E. Peter, 2002. Molecular ordering of the initial signaling events of CD95. *Mol Cell Biol* 22: 207–220.
184. Whitekus, M. J., R. P. Santini, A. J. Rosenspire, and M. J. McCabe, Jr. 1999. Protection against CD95-mediated apoptosis by inorganic mercury in Jurkat T cells. *J Immunol* 162: 7162–7170.
185. Maniati, E., P. Potter, N. J. Rogers, and B. J. Morley, 2008. Control of apoptosis in autoimmunity. *J Pathol* 214: 190–198.
186. McCabe, M. J., Jr., M. J. Whitekus, J. Hyun, K. G. Eckles, G. McCollum, and A. J. Rosenspire, 2003. Inorganic mercury attenuates CD95-mediated apoptosis by interfering with formation of the death inducing signaling complex. *Toxicol Appl Pharmacol* 190: 146–156.
187. Ziemba, S. E., M. J. McCabe, Jr., and A. J. Rosenspire, 2005. Inorganic mercury dissociates preassembled Fas/CD95 receptor oligomers in T lymphocytes. *Toxicol Appl Pharmacol* 206: 334–342.
188. Ganju, N. and A. Eastman, 2003. Zinc inhibits Bax and Bak activation and cytochrome *c* release induced by chemical inducers of apoptosis but not by death-receptor-initiated pathways. *Cell Death Differ* 10: 652–661.
189. Laiosa, M. D., K. G. Eckles, M. Langdon, A. J. Rosenspire, and M. J. McCabe, Jr. 2007. Exposure to inorganic mercury *in vivo* attenuates extrinsic apoptotic signaling in Staphylococcal aureus enterotoxin B stimulated T-cells. *Toxicol Appl Pharmacol* 225: 238–250.
190. Margoshes, M. and Vallee, B. L. 1957. A cadmium binding protein from equine kidney cortex. *J. Am. Chem. Soc.* 79: 4813–4814.
191. Palmiter, R. D. 1998. The elusive function of metallothioneins. *Proc Natl Acad Sci USA* 95: 8428–8430.
192. Masters, B. A., C. J. Quaife, J. C. Erickson, E. J. Kelly, G. J. Froelick, B. P. Zambrowicz, R. L. Brinster, and R. D. Palmiter, 1994. Metallothionein III is expressed in neurons that sequester zinc in synaptic vesicles. *J Neurosci* 14: 5844–5857.
193. Quaife, C. J., S. D. Findley, J. C. Erickson, G. J. Froelick, E. J. Kelly, B. P. Zambrowicz, and R. D. Palmiter, 1994. Induction of a new metallothionein isoform (MT-IV) occurs during differentiation of stratified squamous epithelia. *Biochemistry* 33: 7250–7259.
194. Palmiter, R. D. 1987. Molecular biology of metallothionein gene expression. *Exp Suppl* 52: 63–80.
195. Andrews, G. K. 1990. Regulation of metallothionein gene expression. *Prog Food Nutr Sci* 14: 193–258.
196. Klaassen, C. D. and L. D. Lehman-McKeeman, 1989. Induction of metallothionein. *Int J Toxicol* 8: 1315–1321.
197. Durnam, D. M. and R. D. Palmiter, 1981. Transcriptional regulation of the mouse metallothionein-I gene by heavy metals. *J Biol Chem* 256: 5712–5716.
198. Lehman-McKeeman, L. D. and C. D. Klaassen, 1987. Induction of metallothionein-I and metallothionein-II in rats by cadmium and zinc. *Toxicol Appl Pharmacol* 88: 195–202.
199. Yagle, M. K. and R. D. Palmiter, 1985. Coordinate regulation of mouse metallothionein I and II genes by heavy metals and glucocorticoids. *Mol Cell Biol* 5: 291–294.
200. Richards, R. I., A. Heguy, and M. Karin, 1984. Structural and functional analysis of the human metallothionein-IA gene: Differential induction by metal ions and glucocorticoids. *Cell* 37: 263–272.
201. Dalton, T., R. D. Palmiter, and G. K. Andrews, 1994. Transcriptional induction of the mouse metallothionein-I gene in hydrogen peroxide-treated Hepa cells involves a composite major late transcription factor/antioxidant response element and metal response promoter elements. *Nucleic Acids Res* 22: 5016–5023.
202. Kelly, E. J., E. P. Sandgren, R. L. Brinster, and R. D. Palmiter, 1997. A pair of adjacent glucocorticoid response elements regulate expression of two mouse metallothionein genes. *Proc Natl Acad Sci USA* 94: 10045–10050.
203. Samson, S. L. and L. Gedamu, 1998. Molecular analyses of metallothionein gene regulation. *Prog Nucleic Acid Res Mol Biol* 59: 257–288.
204. Lichtlen, P. and W. Schaffner, 2001. Putting its fingers on stressful situations: The heavy metal-regulatory transcription factor MTF-1. *Bioessays* 23: 1010–1017.
205. Zhang, B., O. Georgiev, M. Hagmann, C. Gunes, M. Cramer, P. Faller, M. Vasak, and W. Schaffner, 2003. Activity of metal-responsive transcription factor 1 by toxic heavy metals and H2O2 *in vitro* is modulated by metallothionein. *Mol Cell Biol* 23: 8471–8485.

206. Kagi, J. H. and Y. Kojima, 1987. Chemistry and biochemistry of metallothionein. *Exp Suppl* 52: 25–61.
207. Nath, R., R. Kambadur, S. Gulati, V. K. Paliwal, and M. Sharma, 1988. Molecular aspects, physiological function, and clinical significance of metallothioneins. *Crit Rev Food Sci Nutr* 27: 41–85.
208. Stuart, G. W., P. F. Searle, H. Y. Chen, R. L. Brinster, and R. D. Palmiter, 1984. A 12-base-pair DNA motif that is repeated several times in metallothionein gene promoters confers metal regulation to a heterologous gene. *Proc Natl Acad Sci USA* 81: 7318–7322.
209. Carthew, R. W., L. A. Chodosh, and P. A. Sharp, 1987. The major late transcription factor binds to and activates the mouse metallothionein I promoter. *Genes Dev* 1: 973–980.
210. Li, Q., Hu, N., Daggett, M. A., Chu, W. A., Bittel, D., Johnson, J. A., and Andrews, G. K., 1998. Participation of upstream stimulator factor (USF) in cadmium-induction of the mouse metallothionein-I gene. *Nucleic Acids Res* 26: 5182–5189.
211. Banerjee, D., S. Onosaka, and M. G. Cherian, 1982. Immunohistochemical localization of metallothionein in cell nucleus and cytoplasm of rat liver and kidney. *Toxicology* 24: 95–105.
212. Goering, P. L. and C. D. Klaassen, 1983. Altered subcellular distribution of cadmium following cadmium pretreatment: Possible mechanism of tolerance to cadmium-induced lethality. *Toxicol Appl Pharmacol* 70: 195–203.
213. Falck, F. Y., Jr., L. J. Fine, R. G. Smith, J. Garvey, A. Schork, B. England, K. D. McClatchey, and J. Linton, 1983. Metallothionein and occupational exposure to cadmium. *Br J Ind Med* 40: 305–313.
214. Koyama, H., H. Satoh, S. Suzuki, and C. Tohyama, 1992. Increased urinary cadmium excretion and its relationship to urinary *N*-acetyl-beta-d-glucosaminidase activity in smokers. *Arch Toxicol* 66: 598–601.
215. Lynes, M. A., K. Zaffuto, D. W. Unfricht, G. Marusov, J. S. Samson, and X. Yin, 2006. The physiological roles of extracellular metallothionein. *Exp Biol Med (Maywood)* 231: 1548–1554.
216. Hightower, L. E. and P. T. Guidon, Jr. 1989. Selective release from cultured mammalian cells of heat-shock (stress) proteins that resemble glia-axon transfer proteins. *J Cell Physiol* 138: 257–266.
217. Coyle, P., J. C. Philcox, L. C. Carey, and A. M. Rofe, 2002. Metallothionein: The multipurpose protein. *Cell Mol Life Sci (CMLS)* 59: 627.
218. Harley, C. B., C. R. Menon, R. A. Rachubinski, and E. Nieboer, 1989. Metallothionein mRNA and protein induction by cadmium in peripheral-blood leucocytes. *Biochem J* 262: 873–879.
219. Mesna, O. J., I. L. Steffensen, A. Melhuus, H. Hjertholm, H. E. Heier, and R. A. Andersen, 1990. Induction of metallothionein production by zinc in human mononuclear cells. *Gen Pharmacol: The Vascular Syst* 21: 909.
220. Enger, M. D., C. E. Hildebrand, and C. C. Stewart, 1983. Cd^{2+} responses of cultured human blood cells. *Toxicol Appl Pharmacol* 69: 214–224.
221. Huerta, P., P. Galan, J. M. Teijon, and B. Ribas, 1989. Increase of metallothionein and other proteins in thymus and spleen during chronic zinc administration. *Toxicol Environ Chem* 25: 63–68.
222. Cao, J., J. A. Bobo, J. P. Liuzzi, and R. J. Cousins, 2001. Effects of intracellular zinc depletion on metallothionein and ZIP2 transporter expression and apoptosis. *J Leukoc Biol* 70: 559–566.
223. Soumillion, A., J. Van Damme, and M. De Ley, 1992. Cloning and specific polymerised-chain-reaction amplification of a third charge-separable human metallothionein isoform. *Eur J Biochem* 209: 999–1004.
224. Mesna, O. J., I. L. Steffensen, H. Hjertholm, and R. A. Andersen, 1995. Accumulation of metallothionein and its multiple forms by zinc, cadmium and dexamethasone in human peripheral T and B lymphocytes and monocytes. *Chem Biol Interact* 94: 225.
225. Cao, J. and R. J. Cousins, 2000. Metallothionein mRNA in monocytes and peripheral blood mononuclear cells and in cells from dried blood spots increases after zinc supplementation of men. *J. Nutr.* 130: 2180–2187.
226. Lee, W. W., D. Cui, M. Czesnikiewicz-Guzik, R. Z. Vencio, I. Shmulevich, A. Aderem, C. M. Weyand, and J. J. Goronzy, 2008. Age-dependent signature of metallothionein expression in primary CD4 T cell responses is due to sustained zinc signaling. *Rejuvenation Res* 11: 1001–1011.
227. Sone, T., S. Koizumi, and M. Kimura, 1988. Cadmium-induced synthesis of metallothioneins in human lymphocytes and monocytes. *Chem Biol Interact* 66: 61.
228. Yamada, H., S. Minoshima, S. Koizumi, M. Kimura, and N. Shimizu, 1989. Cadmium-induced synthesis of metallothioneins in human T and B cell purified by a fluorescence activated cell sorter. *Chem Biol Interact* 70: 117.
229. Hart, B. A., M. G. Cherian, and A. Angel, 1985. Cellular localization of metallothionein in the lung following repeated cadmium inhalation. *Toxicology* 37: 171–179.
230. Daston, G. P., G. J. Overmann, D. Baines, M. W. Taubeneck, L. D. Lehman-McKeeman, J. M. Rogers, and C. L. Keen, 1994. Altered Zn status by alpha-hederin in the pregnant rat and its relationship to adverse developmental outcome. *Reprod Toxicol* 8: 15–24.

231. Butcher, H. L., W. A. Kennette, O. Collins, R. K. Zalups, and J. Koropatnick, 2004. Metallothionein mediates the level and activity of nuclear factor kappa B in murine fibroblasts. *J Pharmacol Exp Ther* 310: 589–598.
232. Zeng, J., R. Heuchel, W. Schaffner, and J. H. Kagi, 1991. Thionein (apometallothionein) can modulate DNA binding and transcription activation by zinc finger containing factor Sp1. *FEBS Lett* 279: 310–312.
233. Lynes, M. A., J. S. Garvey, and D. A. Lawrence, 1990. Extracellular metallothionein effects on lymphocyte activities. *Mol Immunol* 27: 211–219.
234. Youn, J. and M. A. Lynes, 1999. Metallothionein-induced suppression of cytotoxic T lymphocyte function: An important immunoregulatory control. *Toxicol Sci* 52: 199–208.
235. Lynes, M. A., L. A. Borghesi, J. Youn, and E. A. Olson, 1993. Immunomodulatory activities of extracellular metallothionein. I. Metallothionein effects on antibody production. *Toxicology* 85: 161–177.
236. Gavett, S. H. and G. Oberdorster, 1994. Cadmium chloride and cadmium metallothionein-induced pulmonary injury and recruitment of polymorphonuclear leukocytes. *Exp Lung Res* 20: 517–537.
237. Hanada, K., H. Ishikawa, T. Imaizumi, I. Hashimoto, and H. Kimura, 1989. Effect of metallothionein on granulocyte chemotaxis and superoxide generation. *Dermatologica* 179(Suppl. 1): 143.
238. Yin, X., D. A. Knecht, and M. A. Lynes, 2005. Metallothionein mediates leukocyte chemotaxis. *BMC Immunol* 6: 21.
239. Chung, R. S., M. Penkowa, J. Dittmann, C. E. King, C. Bartlett, J. W. Asmussen, J. Hidalgo, et al., 2008. Redefining the role of metallothionein within the injured brain: Extracellular metallothioneins play an important role in the astrocyte-neuron response to injury. *J Biol Chem* 283: 15349–15358.
240. Chung, R. S., J. C. Vickers, M. I. Chuah, and A. K. West, 2003. Metallothionein-IIA promotes initial neurite elongation and postinjury reactive neurite growth and facilitates healing after focal cortical brain injury. *J Neurosci* 23: 3336–3342.
241. Morellini, N. M., N. L. Giles, S. Rea, K. F. Adcroft, S. Falder, C. E. King, S. A. Dunlop, et al., 2008. Exogenous metallothionein-IIA promotes accelerated healing after a burn wound. *Wound Repair Regen* 16: 682–690.
242. Youn, J., L. A. Borghesi, E. A. Olson, and M. A. Lynes, 1995. Immunomodulatory activities of extracellular metallothionein. II. Effects on macrophage functions. *J Toxicol Environ Health* 45: 397–413.
243. Penkowa, M. and J. Hidalgo, 2000. Metallothionein I + II expression and their role in experimental autoimmune encephalomyelitis. *Glia* 32: 247–263.
244. Penkowa, M. and J. Hidalgo, 2001. Metallothionein treatment reduces proinflammatory cytokines IL-6 and TNF-alpha and apoptotic cell death during experimental autoimmune encephalomyelitis (EAE). *Exp Neurol* 170: 1–14.
245. Kang, Y. J., G. Li, and J. T. Saari, 1999. Metallothionein inhibits ischemia-reperfusion injury in mouse heart. *Am J Physiol* 276: H993–H997.
246. Lynes, M. A., C. A. Richardson, R. McCabe, K. C. Crowthers, J. C. Lee, J. Youn, I. B. Schweitzer, and L. D. Shultz,1998. Metallothionein modulates the course of autoimmune disease in viable motheaten mice. *Toxicol Lett* 95: 179.
247. Mita, M., M. Satoh, A. Shimada, M. Okajima, S. Azuma, J. S. Suzuki, K. Sakabe, S. Hara, and S. Himeno, 2008. Metallothionein is a crucial protective factor against *Helicobacter pylori*-induced gastric erosive lesions in a mouse model. *Am J Physiol Gastrointest Liver Physiol* 294: G877–G884.
248. Lindquist, S. and E. A. Craig, 1988. The heat-shock proteins. *Annu Rev Genet* 22: 631–677.
249. Lindquist, S. 1986. The heat-shock response. *Annu Rev Biochem* 55: 1151–1191.
250. Hightower, L. E. 1991. Heat shock, stress proteins, chaperones, and proteotoxicity. *Cell* 66: 191.
251. Multhoff, G. and C. Botzler, 1998. Heat-shock proteins and the immune response. *Ann N Y Acad Sci* 851: 86–93.
252. Wu, B. J., R. E. Kingston, and R. I. Morimoto, 1986. Human HSP70 promoter contains at least two distinct regulatory domains. *Proc Natl Acad Sci USA* 83: 629–633.
253. Hatayama, T. and M. Hayakawa, 1999. Differential temperature dependency of chemical stressors in HSF1-mediated stress response in mammalian cells. *Biochem Biophys Res Commun* 265: 763–769.
254. Bauman, J. W., J. Liu, and C. D. Klaassen, 1993. Production of metallothionein and heat-shock proteins in response to metals. *Fundam Appl Toxicol* 21: 15–22.
255. Maitani, T., A. Watahiki, and K. T. Suzuki, 1986. Induction of metallothionein after lead administration by three injection routes in mice. *Toxicol Appl Pharmacol* 83: 211–217.
256. Sanchez, Y., J. Taulien,, K. A. Borkovich, and S. Lindquist, 1992. Hsp104 is required for tolerance to many forms of stress. *EMBO J.* 11: 2357–2364.

257. Anderson, N. G. F., S. L. Nance, M. A. Gemmell, S. L. Tollacsen, and N. G. Anderson, 1987. Effects of toxic agents at the protein level: Quantitative measurment of 213 mouse liver proteins following xenobiotic treatment. *Fundam Appl Toxicol* 8: 39–50.
258. Ovelgonne, J. H., J. E. M. Souren, F. A. C. Wiegant, and R. Van Wijk, 1995. Relationship between cadmium-induced expression of heatshock genes, inhibition of protein synthesis and cell death. *Toxicology* 99: 19.
259. Eid, N. S., R. E. Kravath, and K. W. Lanks, 1987. Heat-shock protein synthesis by human polymorphonuclear cells. *J Exp Med* 165: 1448–1452.
260. Ferris, D. K., A. Harel-Bellan, R. I. Morimoto, W. J. Welch, and W. L. Farrar, 1988. Mitogen and lymphokine stimulation of heat shock proteins in T lymphocytes. *Proc Natl Acad Sci USA* 85: 3850–3854.
261. Spector, N. L., A. S. Freedman, G. Freeman, J. Segil, J. F. Whitman, W. J. Welch, and L. M. Nadler, 1989. Activation primes human B lymphocytes to respond to heat shock. *J Exp Med* 170: 1763–1768.
262. Calderwood, S. K., S. S. Mambula, P. J. Gray, Jr., and J. R. Theriault, 2007. Extracellular heat shock proteins in cell signaling. *FEBS Lett* 581: 3689.
263. Davies, E. L., M. M. Bacelar, M. J. Marshall, E. Johnson, T. D. Wardle, S. M. Andrew, and J. H. Williams, 2006. Heat shock proteins form part of a danger signal cascade in response to lipopolysaccharide and GroEL. *Clin Exp Immunol* 145: 183–189.
264. Koga, T., A. Wand-Wurttenberger, J. DeBruyn, M. E. Munk, B. Schoel, and S. H. Kaufmann, 1989. T cells against a bacterial heat shock protein recognize stressed macrophages. *Science* 245: 1112–1115.
265. Polla, B. S., J. V. Bonventre, and S. M. Krane, 1988. 1,25-Dihydroxyvitamin D3 increases the toxicity of hydrogen peroxide in the human monocytic line U937: The role of calcium and heat shock. *J Cell Biol* 107: 373–380.
266. Craig, E. A. 1993. Chaperones: Helpers along the pathways to protein folding. *Science* 260: 1902–1903.
267. Lanks, K. W. 1986. Modulators of the eukaryotic heat shock response. *Exp Cell Res* 165: 1–10.
268. Morimoto, R. and E. Fodor, 1984. Cell-specific expression of heat shock proteins in chicken reticulocytes and lymphocytes. *J Cell Biol* 99: 1316–1323.
269. Gothard, L. Q., M. E. Ruffner, J. G. Woodward, O. K. Park-Sarge, and K. D. Sarge, 2003. Lowered temperature set point for activation of the cellular stress response in T-lymphocytes. *J Biol Chem* 278: 9322–9326.
270. Zhou, X., A. R. Patel, F. Perez, and D. A. Jurivich, 2009. Acetylcholinesterase inhibitor rivastigmine enhances cellular defenses in neuronal and macrophage-like cell lines. *Transl Res* 153: 132–141.
271. Healy, A. M., E. Mariethoz, L. Pizurki, and B. S. Polla, 1992. Heat shock proteins in cellular defense mechanisms and immunity. *Ann N Y Acad Sci* 663: 319–330.
272. Polla, B. S. 1991. The heat shock response in human phagocytes. *Immunol Lett* 30: 159.
273. Sirchia, R., A. Longo, and C. Luparello, 2008. Cadmium regulation of apoptotic and stress response genes in tumoral and immortalized epithelial cells of the human breast. *Biochimie* 90: 1578.
274. Clerget, M. and B. S. Polla, 1990. Erythrophagocytosis induces heat shock protein synthesis by human monocytes-macrophages. *Proc Natl Acad Sci USA* 87: 1081–1085.
275. Maridonneau-Parini, I., J. Clerc, and B. S. Polla, 1988. Heat shock inhibits NADPH oxidase in human neutrophils. *Biochem Biophys Res Commun* 154: 179–186.
276. Chiang, H. L., S. R. Terlecky, C. P. Plant, and J. F. Dice, 1989. A role for a 70-kilodalton heat shock protein in lysosomal degradation of intracellular proteins. *Science* 246: 382–385.
277. Flynn, G. C., T. G. Chappell, and J. E. Rothman, 1989. Peptide binding and release by proteins implicated as catalysts of protein assembly. *Science* 245: 385–390.
278. DeNagel, D. C. and S. K. Pierce, 1992. A case for chaperones in antigen processing. *Immunol Today* 13: 86–89.
279. Manara, G. C., P. Sansoni, L. Badiali-De Giorgi, G. Gallinella, C. Ferrari, V. Brianti, F. F. Fagnoni, C. L. Ruegg, G. De Panfilis, and G. Pasquinelli, 1993. New insights suggesting a possible role of a heat shock protein 70-kD family-related protein in antigen processing/presentation phenomenon in humans. *Blood* 82: 2865–2871.
280. Polla, B. S., E. MariÈthoz, D. Hubert, and C. Barazzone, 1995. Heat-shock proteins in host–pathogen interactions: Implications for cystic fibrosis. *Trends Microbiol* 3: 392.
281. Polla, B. S., H. Stubbe, S. Kantengwa, I. Maridonneau-Parini, and M. R. Jacquier-Sarlin, 1995. Differential induction of stress proteins and functional effects of heat shock in human phagocytes. *Inflammation* 19: 363–378.
282. Clayton, A., A. Turkes, H. Navabi, M. D. Mason, and Z. Tabi, 2005. Induction of heat shock proteins in B-cell exosomes. *J Cell Sci* 118: 3631–3638.

283. Calderwood, S. K., J. R. Theriault, and J. Gong, 2005. How is the immune response affected by hyperthermia and heat shock proteins? *Int J Hyperthermia* 21: 713–716.
284. Theriault, J. R., S. S. Mambula, T. Sawamura, M. A. Stevenson, and S. K. Calderwood, 2005. Extracellular HSP70 binding to surface receptors present on antigen presenting cells and endothelial/epithelial cells. *FEBS Lett* 579: 1951–1960.
285. Binder, R. J., K. M. Anderson, S. Basu, and P. K. Srivastava, 2000. Cutting edge: Heat shock protein gp96 induces maturation and migration of CD11c+ cells *in vivo*. *J Immunol* 165: 6029–6035.
286. Wallin, R. P., A. Lundqvist, S. H. More, A. von Bonin, R. Kiessling, and H. G. Ljunggren, 2002. Heat-shock proteins as activators of the innate immune system. *Trends Immunol* 23: 130–135.
287. Singh-Jasuja, H., H. U. Scherer, N. Hilf, D. Arnold-Schild, H. G. Rammensee, R. E. Toes, and H. Schild, 2000. The heat shock protein gp96 induces maturation of dendritic cells and down-regulation of its receptor. *Eur J Immunol* 30: 2211–2215.
288. Arnold-Schild, D., D. Hanau, D. Spehner, C. Schmid, H. G. Rammensee, H. de la Salle, and H. Schild, 1999. Cutting edge: Receptor-mediated endocytosis of heat shock proteins by professional antigen-presenting cells. *J Immunol* 162: 3757–3760.
289. Blachere, N. E., Z. Li, R. Y. Chandawarkar, R. Suto, N. S. Jaikaria, S. Basu, H. Udono, and P. K. Srivastava, 1997. Heat shock protein–peptide complexes, reconstituted *in vitro*, elicit peptide-specific cytotoxic T lymphocyte response and tumor immunity. *J Exp Med* 186: 1315–1322.
290. Rapp, U. K. and S. H. E. Kaufmann, 2003. Glucose-regulated stress proteins and antibacterial immunity. *Trends Microbiol* 11: 519.
291. Sato, H., T. Ishii, Y. Sugita, N. Tateishi, and S. Bannai, 1993. Induction of a 23 kDa stress protein by oxidative and sulfhydryl-reactive agents in mouse peritoneal macrophages. *Biochim Biophys Acta* 1148: 127–132.
292. Minisini, M. P., S. Kantengwa, and B. S. Polla, 1994. DNA damage and stress protein synthesis induced by oxidative stress proceed independently in the human premonocytic line U937. *Mutat Res* 315: 169–179.
293. Pellegrini, O., E. Davenas, L. Morin, G. T. Tsangaris, J. Benveniste, Y. Manuel, and Y. Thomas, 1994. Modulation of stress proteins by Cd^{2+} in a human T cell line. *Eur J Pharmacol* 270: 221–228.
294. Tsangaris, G. T., A. Botsonis, I. Politis, and F. Tzortzatou-Stathopoulou, 2002. Evaluation of cadmium-induced transcriptome alterations by three color cDNA labeling microarray analysis on a T-cell line. *Toxicology* 178: 135–160.
295. Taketani, S., H. Kohno, T. Yoshinaga, and R. Tokunaga, 1989. The human 32-kDa stress protein induced by exposure to arsenite and cadmium ions is heme oxygenase. *FEBS Lett* 245: 173–176.
296. Taketani, S., H. Sato, T. Yoshinaga, R. Tokunaga, T. Ishii, and S. Bannai, 1990. Induction in mouse peritoneal macrophages of 34 kDa stress protein and heme oxygenase by sulfhydryl-reactive agents. *J Biochem* 108: 28–32.
297. Goering, P. L., R. K. Kuester, A. R. Neale, M. S. Chapekar, T. G. Zaremba, E. A. Gordon, and V. M. Hitchins, 2000. Effects of particulate and soluble cadmium species on biochemical and functional parameters in cultured murine macrophages. *In Vitro Mol Toxicol* 13: 125–136.
298. Del Razo, L. M., B. Quintanilla-Vega, E. Brambila-Colombres, E. S. Calderón-Aranda, M. Manno, and A. Albores, 2001. Stress proteins induced by arsenic. *Toxicol Appl Pharmacol* 177: 132.
299. Kreppel, H., J. W. Bauman, J. I. E. Liu, J. M. J. McKim, and C. D. Klaassen, 1993. Induction of metallothionein by arsenicals in mice. *Toxicol. Sci.* 20: 184–189.
300. Cohen, D. S., E. Palmer, W. J. Welch, and D. Sheppard, 1991. The response of guinea pig airway epithelial cells and alveolar macrophages to environmental stress. *Am J Respir Cell Mol Biol* 5: 133–143.
301. Menzel, D. B., R. E. Rasmussen, E. Lee, D. M. Meacher, B. Said, H. Hamadeh, M. Vargas, H. Greene, and R. N. Roth, 1998. Human lymphocyte heme oxygenase 1 as a response biomarker to inorganic arsenic. *Biochem Biophys Res Commun* 250: 653–656.
302. Shen, H. M., S. Y. Dong, and C. N. Ong, 2001. Critical role of calcium overloading in cadmium-induced apoptosis in mouse thymocytes. *Toxicol Appl Pharmacol* 171: 12–19.
303. McCabe, M. J., Jr., K. P. Singh, and J. J. Reiners, Jr. 2001. Low level lead exposure *in vitro* stimulates the proliferation and expansion of alloantigen-reactive CD4(high) T cells. *Toxicol Appl Pharmacol* 177: 219–231.
304. Razani-Boroujerdi, S., B. Edwards, and M. L. Sopori, 1999. Lead stimulates lymphocyte proliferation through enhanced T cell–B cell interaction. *J Pharmacol Exp Ther* 288: 714–719.
305. Farrer, D. G., S. M. Hueber, and M. J. McCabe, Jr. 2005. Lead enhances CD4+ T cell proliferation indirectly by targeting antigen presenting cells and modulating antigen-specific interactions. *Toxicol Appl Pharmacol* 207: 125–137.

306. Viau, M., V. Collin-Faure, P. Richaud, J. L. Ravanat, and S. M. Candeias, 2007. Cadmium and T cell differentiation: Limited impact *in vivo* but significant toxicity in fetal thymus organ culture. *Toxicol Appl Pharmacol* 223: 257–266.
307. McCabe, M. J., Jr. and D. A. Lawrence, 1990. The heavy metal lead exhibits B cell-stimulatory factor activity by enhancing B cell Ia expression and differentiation. *J Immunol* 145: 671–677.
308. Prozialeck, W. C. 2000. Evidence that E-cadherin may be a target for cadmium toxicity in epithelial cells. *Toxicol Appl Pharmacol* 164: 231–249.
309. Angst, B. D., C. Marcozzi, and A. I. Magee, 2001. The cadherin superfamily: Diversity in form and function. *J Cell Sci* 114: 629–641.
310. Prozialeck, W. C., G. B. Grunwald, P. M. Dey, K. R. Reuhl, and A. R. Parrish, 2002. Cadherins and NCAM as potential targets in metal toxicity. *Toxicol Appl Pharmacol* 182: 255–265.
311. Prozialeck, W. C. and P. C. Lamar, 1999. Interaction of cadmium (Cd(2+)) with a 13-residue polypeptide analog of a putative calcium-binding motif of E-cadherin. *Biochim Biophys Acta* 1451: 93–100.
312. Bussolaro, D., F. Filipak Neto, R. Gargioni, L. C. Fernandes, M. A. Randi, E. Pelletier, and C. A. Oliveira Ribeiro, 2008. The immune response of peritoneal macrophages due to exposure to inorganic lead in the house mouse *Mus musculus. Toxicol In Vitro* 22: 254–260.
313. Contrino, J., P. Marucha, R. Ribaudo, R. Ference, P. E. Bigazzi, and D. L. Kreutzer, 1988. Effects of mercury on human polymorphonuclear leukocyte function *in vitro. Am J Pathol* 132: 110–118.
314. Bishayi, B. and M. Sengupta, 2006. Synergism in immunotoxicological effects due to repeated combined administration of arsenic and lead in mice. *Int Immunopharmacol* 6: 454–464.
315. Shen, X., K. Lee, and R. Konig, 2001. Effects of heavy metal ions on resting and antigen-activated CD4(+) T cells. *Toxicology* 169: 67–80.
316. Heo, Y., P. J. Parsons, and D. A. Lawrence, 1996. Lead differentially modifies cytokine production *in vitro* and *in vivo. Toxicol Appl Pharmacol* 138: 149–157.
317. Hemdan, N. Y., F. Emmrich, U. Sack, G. Wichmann, J. Lehmann, K. Adham, and I. Lehmann, 2006. The *in vitro* immune modulation by cadmium depends on the way of cell activation. *Toxicology* 222: 37–45.
318. Grabbe, S. and T. Schwarz, 1998. Immunoregulatory mechanisms involved in elicitation of allergic contact hypersensitivity. *Immunol Today* 19: 37–44.
319. Silvennoinen-Kassinen, S., I. Ikaheimo, J. Karvonen, M. Kauppinen, and M. Kallioinen, 1992. Mononuclear cell subsets in the nickel-allergic reaction *in vitro* and *in vivo. J Allergy Clin Immunol* 89: 794–800.
320. Haberman, A. L., M. Pratt, and F. J. Storrs, 1993. Contact dermatitis from beryllium in dental alloys. *Contact Dermat* 28: 157–162.
321. Forte, G., F. Petrucci, and B. Bocca, 2008. Metal allergens of growing significance: Epidemiology, immunotoxicology, strategies for testing and prevention. *Inflamm Allergy Drug Targets* 7: 145–162.
322. Madl, A. K., K. Unice, J. L. Brown, M. E. Kolanz, and M. S. Kent, 2007. Exposure-response analysis for beryllium sensitization and chronic beryllium disease among workers in a beryllium metal machining plant. *J Occup Environ Hyg* 4: 448–466.
323. Tinkle, S. S., L. A. Kittle, B. A. Schumacher, and L. S. Newman, 1997. Beryllium induces IL-2 and IFN-gamma in berylliosis. *J Immunol* 158: 518–526.
324. Fontenot, A. P., S. J. Canavera, L. Gharavi, L. S. Newman, and B. L. Kotzin, 2002. Target organ localization of memory CD4(+) T cells in patients with chronic beryllium disease. *J Clin Invest* 110: 1473–1482.
325. Amicosante, M. and A. P. Fontenot, 2006. T cell recognition in chronic beryllium disease. *Clin Immunol* 121: 134–143.
326. Fontenot, A. P., B. E. Palmer, A. K. Sullivan, F. G. Joslin, C. C. Wilson, L. A. Maier, L. S. Newman, and B. L. Kotzin, 2005. Frequency of beryllium-specific, central memory CD4+ T cells in blood determines proliferative response. *J Clin Invest* 115: 2886–2893.
327. Rossman, M. D., J. A. Kern, J. A. Elias, M. R. Cullen, P. E. Epstein, O. P. Preuss, T. N. Markham, and R. P. Daniele, 1988. Proliferative response of bronchoalveolar lymphocytes to beryllium. A test for chronic beryllium disease. *Ann Intern Med* 108: 687–693.
328. Saltini, C., L. Richeldi, M. Losi, M. Amicosante, C. Voorter, E. van den Berg-Loonen, R. A. Dweik, H. P. Wiedemann, D. C. Deubner, and C. Tinelli, 2001. Major histocompatibility locus genetic markers of beryllium sensitization and disease. *Eur Respir J* 18: 677–684.
329. Maier, L. A., D. S. McGrath, H. Sato, P. Lympany, K. Welsh, R. Du Bois, L. Silveira, et al., 2003. Influence of MHC class II in susceptibility to beryllium sensitization and chronic beryllium disease. *J Immunol* 171: 6910–6918.
330. McCanlies, E. C., J. S. Ensey, C. R. Schuler, K. Kreiss, and A. Weston, 2004. The association between HLA-DPB1Glu69 and chronic beryllium disease and beryllium sensitization. *Am J Ind Med* 46: 95–103.

331. Richeldi, L., K. Kreiss, M. M. Mroz, B. Zhen, P. Tartoni, and C. Saltini, 1997. Interaction of genetic and exposure factors in the prevalence of berylliosis. *Am J Ind Med* 32: 337–340.
332. Richeldi, L., R. Sorrentino, and C. Saltini, 1993. HLA-DPB1 glutamate 69: A genetic marker of beryllium disease. *Science* 262: 242–244.
333. Rossman, M. D., J. Stubbs, C. W. Lee, E. Argyris, E. Magira, and D. Monos, 2002. Human leukocyte antigen class II amino acid epitopes: Susceptibility and progression markers for beryllium hypersensitivity. *Am J Respir Crit Care Med* 165: 788–794.
334. Wang, Z., G. M. Farris, L. S. Newman, Y. Shou, L. A. Maier, H. N. Smith, and B. L. Marrone, 2001. Beryllium sensitivity is linked to HLA-DP genotype. *Toxicology* 165: 27–38.
335. Wang, Z., P. S. White, M. Petrovic, O. L. Tatum, L. S. Newman, L. A. Maier, and B. L. Marrone, 1999. Differential susceptibilities to chronic beryllium disease contributed by different Glu69 HLA-DPB1 and -DPA1 alleles. *J Immunol* 163: 1647–1653.
336. Kosuda, L. L., M. O. Hannigan, P. E. Bigazzi, J. H. Leif, and D. L. Greiner, 1996. Thymus atrophy and changes in thymocyte subpopulations of BN rats with mercury-induced renal autoimmune disease. *Autoimmunity* 23: 77–89.
337. Qasim, F. J., S. Thiru, P. W. Mathieson, and D. B. Oliveira, 1995. The time course and characterization of mercuric chloride-induced immunopathology in the brown Norway rat. *J Autoimmun* 8: 193–208.
338. Qasim, F. J., P. W. Mathieson, S. Thiru, and D. B. Oliveira, 1994. Use of methyl prednisolone and antioxidants in mercuric chloride-induced experimental vasculitis. *Clin Exp Immunol* 98: 66–70.
339. Pusey, C. D., C. Bowman, A. Morgan, A. P. Weetman, B. Hartley, and C. M. Lockwood, 1990. Kinetics and pathogenicity of autoantibodies induced by mercuric chloride in the brown Norway rat. *Clin Exp Immunol* 81: 76–82.
340. Hultman, P., L. J. Bell, S. Enestrom, and K. M. Pollard, 1992. Murine susceptibility to mercury. I. Autoantibody profiles and systemic immune deposits in inbred, congenic, and intra-H-2 recombinant strains. *Clin Immunol Immunopathol* 65: 98–109.
341. Hultman, P., L. J. Bell, S. Enestrom, and K. M. Pollard, 1993. Murine susceptibility to mercury. II. Autoantibody profiles and renal immune deposits in hybrid, backcross, and H-2d congenic mice. *Clin Immunol Immunopathol* 68: 9–20.
342. Hultman, P. and S. Enestrom, 1992. Dose–response studies in murine mercury-induced autoimmunity and immune-complex disease. *Toxicol Appl Pharmacol* 113: 199–208.
343. Monestier, M., M. J. Losman, K. E. Novick, and J. P. Aris, 1994. Molecular analysis of mercury-induced antinucleolar antibodies in H-2S mice. *J Immunol* 152: 667–675.
344. Pollard, K. M., D. K. Lee, C. A. Casiano, M. Bluthner, M. M. Johnston, and E. M. Tan, 1997. The autoimmunity-inducing xenobiotic mercury interacts with the autoantigen fibrillarin and modifies its molecular and antigenic properties. *J Immunol* 158: 3521–3528.
345. Hultman, P., S. J. Turley, S. Enestrom, U. Lindh, and K. M. Pollard, 1996. Murine genotype influences the specificity, magnitude and persistence of murine mercury-induced autoimmunity. *J Autoimmun* 9: 139–149.
346. Dubey, C., B. Bellon, and P. Druet, 1991. TH1 and TH2 dependent cytokines in experimental autoimmunity and immune reactions induced by chemicals. *Eur Cytokine Netw* 2: 147–152.
347. Pelletier, L., J. Rossert, R. Pasquier, H. Villarroya, R. Oriol, and P. Druet, 1991. $HgCl_2$-induced perturbation of the T cell network in experimental allergic encephalomyelitis. II. *In vivo* demonstration of the role of T suppressor and contrasuppressor cells. *Cell Immunol* 137: 379–388.
348. Bridoux, F., A. Badou, A. Saoudi, I. Bernard, E. Druet, R. Pasquier, P. Druet, and L. Pelletier, 1997. Transforming growth factor beta (TGF-beta)-dependent inhibition of T helper cell 2 (Th2)-induced autoimmunity by self-major histocompatibility complex (MHC) class II-specific, regulatory CD4(+) T cell lines. *J Exp Med* 185: 1769–1775.
349. Hansson, M. and M. Abedi-Valugerdi, 2003. Xenobiotic metal-induced autoimmunity: Mercury and silver differentially induce antinucleolar autoantibody production in susceptible H-2s, H-2q and H-2f mice. *Clin Exp Immunol* 131: 405–414.
350. Pollard, K. M. and P. Hultman, 1997. Effects of mercury on the immune system. *Met Ions Biol Syst* 34: 421–440.
351. Roos, A., N. Claessen, E. J. Schilder-Tol, M. A. Chand, J. J. Weening, and J. Aten, 1997. Thiol levels in CD134-defined subsets of rat T lymphocytes: Possible implications for HgC12-induced immune dysregulation. *Biochem Biophys Res Commun* 240: 452–457.
352. Gillespie, K. M., A. Saoudi, J. Kuhn, C. J. Whittle, P. Druet, B. Bellon, and P. W. Mathieson, 1996. Th1/Th2 cytokine gene expression after mercuric chloride in susceptible and resistant rat strains. *Eur J Immunol* 26: 2388–2392.

353. Mathieson, P. W. 1995. Mercury: God of Th2 cells? *Clin Exp Immunol* 102: 229–230.
354. Kalyuzhny, A. E. 2009. ELISPOT assay on membrane microplates. *Methods Mol Biol* 536: 355–365.
355. Jin, G. B., D. W. Unfricht, S. M. Fernandez, and M. A. Lynes, 2006. Cytometry on a chip: Cellular phenotypic and functional analysis using grating-coupled surface plasmon resonance. *Biosens Bioelectron* 22: 200–206.
356. Unfricht, D. W., S. L. Colpitts, S. M. Fernandez, and M. A. Lynes, 2005. Grating-coupled surface plasmon resonance: A cell and protein microarray platform. *Proteomics* 5: 4432–4442.
357. Zieziulewicz, T. J., D. W. Unfricht, N. Hadjout, M. A. Lynes, and D. A. Lawrence, 2003. Shrinking the biologic world—nanobiotechnologies for toxicology. *Toxicol Sci* 74: 235–244.
358. Takiguchi, M., W. E. Achanzar, W. Qu, G. Li, and M. P. Waalkes, 2003. Effects of cadmium on DNA-(Cytosine-5) methyltransferase activity and DNA methylation status during cadmium-induced cellular transformation. *Exp Cell Res* 286: 355–365.
359. Santangelo, S., D. J. Cousins, N. E. Winkelmann, and D. Z. Staynov, 2002. DNA methylation changes at human Th2 cytokine genes coincide with DNase I hypersensitive site formation during CD4(+) T cell differentiation. *J Immunol* 169: 1893–1903.
360. Sanders, V. M. 2006. Epigenetic regulation of Th1 and Th2 cell development. *Brain Behav Immun* 20: 317–324.
361. Lee, P. P., D. R. Fitzpatrick, C. Beard, H. K. Jessup, S. Lehar, K. W. Makar, M. Perez-Melgosa, et al., 2001. A critical role for Dnmt1 and DNA methylation in T cell development, function, and survival. *Immunity* 15: 763–774.
362. Makar, K. W. and C. B. Wilson, 2004. DNA methylation is a nonredundant repressor of the Th2 effector program. *J Immunol* 173: 4402–4406.
363. Grogan, J. L., M. Mohrs, B. Harmon, D. A. Lacy, J. W. Sedat, and R. M. Locksley, 2001. Early transcription and silencing of cytokine genes underlie polarization of T helper cell subsets. *Immunity* 14: 205–215.
364. Ansel, K. M., D. U. Lee, and A. Rao, 2003. An epigenetic view of helper T cell differentiation. *Nat Immunol* 4: 616–623.

Index

A

ABC. *See* ATP-binding cassette (ABC)
ABC-me. *See* ABC-mitochondrial erythroid (ABC-me)
ABC-mitochondrial erythroid (ABC-me), 172, 173
ACD. *See* Anemia of chronic disease (ACD)
ACE. *See* Angiotension converting enzyme (ACE)
Aceruloplasminemia, 359
Acrodermatitis enteropathica, 144, 183, 196
ACTH. *See* Adrenocorticotropic hormone (ACTH)
Adduct formation, 6, 20
  Cu (II)-PTSC reaction, 6
  fluorescent Zn^{2+} sensors, 20, 21
  internal oxidation reduction, 6
  ligand substitution reaction, 20
  mechanistic pathway, 20
  Zn^{2+} trafficking, 20
Adduct(s), 20
  DNA, 5
  Fe (III), 23
  formation, 6, 20
  redox-active metals, 383
  ternary, 22
Adeno*m*ajor *l*ate *t*ranscription *f*actor (MLTF), 146
Adrenocorticotropic hormone (ACTH), 202
AFP. *See* Alfafetoprotein (AFP)
ALA dehydratase (ALAD), 118
ALA synthetase (ALAS), 118
ALAD. *See* Amino-levulenic acid dehydratase (ALAD); ALA dehydratase (ALAD)
ALAS. *See* ALA synthetase (ALAS)
ALAS2. *See* δ-aminolevulinic acid synthase (ALAS2)
Alcoholic liver disease (ALD), 364
ALD. *See* Alcoholic liver disease (ALD)
Alfafetoprotein (AFP), 148
Allo-enhancement, 395
ALPS. *See* Autoimmune lympho proliferative syndrome (ALPS)
Amino-levulenic acid dehydratase (ALAD), 5, 114
Anemia of chronic disease (ACD), 354
Angiotension converting enzyme (ACE), 116
Antennal glands. *See* Kidneys
Antigen presenting cells (APCs), 383, 384, 395
*A*ntioxidant *r*esponse *e*lement (ARE), 146, 232, 390
APCs. *See* Antigen presenting cells (APCs)
Apical membrane. *See* Hepatic microanatomy—canalicular membrane
Apo-MT, 4, 8–9, 10, 25
  decomposing reaction, 11
  in vitro reactions, 10–11
  proteomic Zn^{2+} fraction, 10
  reactivity with Zn-proteome, 10
  substitution reactions, 16
  TPEN and EGTA extraction, 10
  Zn^{2+} deficiency, 11
  Zn-finger-DNA, 10–11
  Zn-proteome reactivity, 10
  Zn-Spl-DNA, 10, 11
Apoptosis, 7, 155, 387
  apoptotic cell death, 387
  apoptotic signaling, 387–388
  CD95 signal initiation, 388
  cytochrome *c*-activated caspase-(3), 155
  cytotoxicity of mercury, 388–389
  lead induced, 119
  morphological features, 387
  MTF-1, 148
  neoplasia, 387
  neuronal, 156
  spontaneous, 198
  T lymphocytes against, 393
  zinc deficiency, 2
  Zn^{2+} trafficking, 20
AQP. *See* Aquaporins (AQP)
Aquaporins (AQP), 60, 245
ARE. *See* *A*ntioxidant *r*esponse *e*lement (ARE)
Arsenic, 228, 243
  biliary excretion, 80
  binds to sulfhydryl groups, 383
  DNA methylation, 399
  endocrine disruption, 243
  GSH, 244
  hepatic transport, 245
  intestinal transport, 243–244
  kinase activities, 337
  mammalian MRP(2), 100
  metabolism, 228–229
  metal chelators, 64
  metal-binding ligands, 99–100
  mixed exposure, 120–121
  molecular mimicry, 269
  MRP, 244, 245
  reactive free radicals, 394
  renal transport, 244
ASGP-R. *See* Asialoglycoprotein receptors (ASGP-R)
Asialoglycoprotein receptors (ASGP-R), 88
ATPase, ($NA^+ + K^+$)-stimulated, 59, 226
  inactivation, 60
  mercury induced, 59–60
ATP-binding cassette (ABC), 42, 80, 133
Autoimmune lympho proliferative syndrome (ALPS), 388

B

Baby hamster kidney (BHK), 93
  proliferation, 191
  Znt(1), 93
BAL. *See* British Anti-Lewisite (BAL); Bronchoalveolar lymphocytes (BAL)
Basal surface. *See* Hepatic microanatomy—sinusoidal endothelium

Basolateral membrane vesicles (BLMV), 49, 300
Basolateral plasma membrane (BLM), 217
Basolateral uptake, 47, 50, 51
    dicarboxylate transporter, 48–49
    membrane transporters, 50
    mercuric conjugates, 49–50
    organic anion transport systems, 47–48
Bathophenanthrolinedisulfonate (BPS), 14
BBM. *See* Brush-border plasma membrane (BBM)
BBMV. *See* Brush-border membrane vesicles (BBMV)
BeLPT. *See* Beryllium-stimulated blood lymphocyte proliferation test (BeLPT)
Beryliosis. *See* Chronic beryllium disease (CBD)
Beryllium-stimulated blood lymphocyte proliferation test (BeLPT), 397
BHK. *See* Baby hamster kidney (BHK)
Bile
    acid, 94
    biliary excretion, 80, 99
    biliary-hepatic cycle, 85
    canalicular bile, 84, 85, 265
    hepatic bile, 84
    mercury accumulation, 63
    metal transport, 99–102
BLM. *See* Basolateral plasma membrane (BLM)
BLMV. *See* Basolateral membrane vesicles (BLMV)
BMP. *See* Bone morphogenetic protein (BMP)
BMP receptors (BMPR), 358
BMPR. *See* BMP receptors (BMPR)
BN strain. *See* Brown-Norway (BN) strain
Bone morphogenetic protein (BMP), 176, 353
Bovine serum albumin (BSA), 224
BPS. *See* Bathophenanthrolinedisulfonate (BPS)
British Anti-Lewisite (BAL), 64
Bronchoalveolar lymphocytes (BAL), 397
Brown-Norway (BN) strain, 397
Brush-border membrane vesicles (BBMV), 299
Brush-border plasma membrane (BBM), 216
BSA. *See* Bovine serum albumin (BSA)
BSO. *See* Buthionine sulfoximine (BSO); L-buthionine-*S*, *R*-sulfoximine (BSO)
Bulk flow. *See* Convection
Buthionine sulfoximine (BSO), 63, 215, 261, 385

C

CA. *See* Carbonic anhydrase (CA); Cornu ammonis (CA)
Cadherins, 329, 395
    adherens junction, 329
    cadherin/catenin complexes, 331
    Ca^{2+} displacement, 226
    cadmium, 333–334
    cell adhesion, 226
    classical, 395
    complex, 330–331, 333, 336, 338
    cytoplasmic catenin proteins, 330–331
    desmosomes components, 330
    E-cadherin, 226, 254
    metal induced disruption, 333
    N-cadherin, 333, 335, 339
    stabilization, 331
    tight junction, 229, 331, 333
    type-I, 330
    type-II, 330
    VE-cadherin, 333, 334
    Wnt signaling, 331
Cadmium, 246, 306, 333, 334, 336–338, 391, 395
    Ca^{2+} channels, 251
    Cd–ferritin complexes, 253
    Cd–MT complexes, 249, 250, 253
    DMT(1), 247
    endocrine disruption, 246–247
    hepatic transport, 252–253
    intercellular mimicry, 254
    intestinal transport, 247–249
    renal transport, 249–252
    testicular handling, 254
    Zn^{2+} transporter, 250
Calcium transporter 1 (CaT1), 248
Calprotectin, 199
cAMP. *See* Cyclic adenosine monophosphate (cAMP)
cAMP response element binding protein (CREB), 157
Canalicular membrane transporters, 99
    MRP2-mediated GSH complexes, 99–102
    sulfate exchanger, 102
Canalicular MRP (cMRP), 100
Canalicular multispecific organic anion transporter (cMOAT), 100
Canalicular plasma membrane (CPM), 217. *See also* Sinusoidal plasma membrane (SPM)
    hepatic transport, 99
    MRP2-mediated excretion, 99
    organic solute carriers, 102
    sulfate exchanger, 102
CAR. *See* Coxsackievirus and adenovirus receptor (CAR)
Carbonic anhydrase (CA), 6
CAT. *See* Catalase (CAT)
CaT1. *See* Calcium transporter 1 (CaT1)
Catalase (CAT), 116
Cation diffusion facilitator (CDF), 80, 185
CBD. *See* Chronic beryllium disease (CBD)
CCS. *See* Copper chaperone for superoxide dismutase (CCS)
CDDP. *See* *cis*-diamminedichloroplatin (II) (CDDP)
CDF. *See* Cation diffusion facilitator (CDF)
Cd-metallothionein (CdMT), 226
CdMT. *See* Cd-metallothionein (CdMT)
C/EBPα homology protein (CHOP), 364
Cell adhesion disruption, 332, 339
    cadherin/catenin complexes, 333
    gap junctions, 334
    inflammatory adhesion molecules, 335
    integrins, 332
    metal-induced kinase activities, 337
    tight junctions, 333
Cell adhesion molecules, 327, 328, 332
    actin cytoskeleton disruption, 338
    cadherin complex, 330–331, 333, 336, 338
    cell adhesion complexes, 329
    cell–matrix interactions, 328
    gap junctions, 331
    gene expression, 335
    heterotypic cell–cell adhesion, 332
    homotypic cell–cell adhesion, 330
    oxidative stress, 338
    potential mechanisms, 335, 336

protein degradation, 337–338
protein phosphorylation, 336–337
protein–protein interactions, 336
tight junction, 229, 331, 333
Cells, 2, 11, 14, 16, 97, 188, 287
Ehrlich, 12
LLC-PK_1, 21, 22
metabolome, 2
U(373), 13
Central nervous system (CNS), 115, 130, 156, 262, 359
Ceruloplasmin, 87, 353
aceruloplasminemia, 359
copper, 86, 97
hephaestin, 94
HIF-1, 92
iron oxidation, 88, 352
iron uptake, 91
Wilson's disease, 92
CHC. *See* Chronic hepatitis C (CHC)
Chelation therapy, 361
desferrioxamine (DFO), 361–362
iron chelating compounds, 361
Chelator dimercaprol. *See* British Anti-Lewisite (BAL)
Chelators, 361
Chemokines, 395
CHOP. *See* C/EBPα homology protein (CHOP)
Chronic beryllium disease (CBD), 396, 397
Chronic hepatitis C (CHC), 363
cis-diamminedichloroplatin (II) (CDDP), 154
Cisplatin. *See cis*-diamminedichloroplatin (II) (CDDP)
c-Jun N-terminal kinase (JNK), 192, 337
Claudins, 330
cMOAT. *See* Canalicular multispecific organic anion transporter (cMOAT)
cMRP. *See* Canalicular MRP (cMRP)
CNS. *See* Central nervous system (CNS)
CO. *See* Coproporphyrin oxidase (CO)
Cobalt protoporphyrin (CoPP), 394
Connexins (Cx), 331
Convection, 95
CoPP. *See* Cobalt protoporphyrin (CoPP)
Copper chaperone for superoxide dismutase (CCS), 183, 382
Copper uptake, 89–90
ATPase, 90
Ccc2 functions, 90
ceruloplasmin, 91–92
CTR1 and CTR(2), 81, 82, 89–91
hepatocellular model, 90
intracellular distribution, 91
mechanisms, 90, 91, 304, 306
yeast copper uptake mechanisms, 91
Coproporphyrin oxidase (CO), 118
Corepressor histone deacetylase complexes (HDACS), 150
Cornu ammonis (CA), 197
Coxsackievirus and adenovirus receptor (CAR), 330
CpG islands, 150
CPM. *See* Canalicular plasma membrane (CPM)
CRDs. *See* Cysteine-rich domains (CRDs)
CREB. *See* cAMP response element binding protein (CREB)
Cu (II)-PTSC. *See* Pyridoxalthiosemicarbazonato-Cu (II) (Cu (II)-PTSC)
Cx. *See* Connexins (Cx)
Cyclic adenosine monophosphate (cAMP), 157
Cys. *See* Cysteine (Cys)
Cysteine (Cys), 19, 37
Cysteine-rich domains (CRDs), 385, 387

D

DAT. *See* Dopamine transporter (DAT)
DBI. *See* Diazapine-binding inhibitor (DBI)
DCT-1. *See* Divalent cation transporterl (DCT-1)
Dcytb. *See* Duodenal cytochrome b (Dcytb)
DEA-NO. *See* Diethylaminenonoate (DEA-NO)
Death effector domain (DED), 388
Death receptors, 387
Death-inducing signaling complex (DISC), 388
DED. *See* Death effector domain (DED)
Delayed-type hypersensitivity (DTH), 118
δ-aminolevulinic acid synthase (ALAS2), 360
Delta-aminolevulinic acid dehydratase (ALD), 364
DEM. *See* Diethyl maleate (DEM)
Desferrioxamine (DFO), 14, 361–362
DFO. *See* Desferrioxamine (DFO)
Diazapine-binding inhibitor (DBI), 256
Dicarboxylate transporter, 48
α-ketoglutarate, 48
cotransport system, 48–49
Cys–mercury complex, 48
OAT, 48
pretreatment, 48
Diethyl maleate (DEM), 62, 215, 256
Diethylaminenonoate (DEA-NO), 25
Diffusion, 95
CDF, 185
facilitated, 318
simple, 95, 96
zinc transporters, 185
Dipeptidase (DP), 216
DISC. *See* Death-inducing signaling complex (DISC)
Disodium diethylenetriaminepentaacetic acid (DTPA), 252
Dithiocarbamates (DTC), 15
Divalent cation transporterl (DCT-1), 131. *See* Divalent metal transporter-1 (DMT1)
Divalent metal transporter-1 (DMT1), 128, 168, 243, 352
DMPS. *See* 2, 3-dimercapto-1-propanesulfonic acid (DMPS)
DMSA. *See* 2, 3-dimercaptosuccinic acid (DMSA)
DMT1. *See* Divalent metal transporter-1 (DMT1)
DNA methylation, 150, 399
Dopamine transporter (DAT), 133–134
DP. *See* Dipeptidase (DP)
DTC. *See* Dithiocarbamates (DTC)
DTH. *See* Delayed-type hypersensitivity (DTH)
DTNB. *See* 5, 5*N*-dithiobis (2-nitrobenzoate) (DTNB)
DTPA. *See* Disodium diethylenetriaminepentaacetic acid (DTPA)
Duodenal cytochrome b (Dcytb), 168

E

EAE. *See* Experimental autoimmune encephalomyelitis (EAE)
EAR. *See* Estimated average requirement (EAR)
ECaC. *See* Epithelial Ca^{2+} channel (ECaC)

EDC. *See* Endocrine-disrupting chemical (EDC)
EDTA. *See* Ethylenediaminetetraacetic acid (EDTA)
EGFR. *See* Epidermal growth factor receptor (EGFR)
Electron spin resonance (ESR), 6
Electrophile response element (EpRE). *See* *A*ntioxidant *r*esponse *e*lement (ARE)
ELISA. *See* Enzyme-linked immunosorbent assay (ELISA)
Emtin peptides. *See* Megalin
Endocrine-disrupting chemical (EDC), 246
Endocytosis, 249
  albumin–Cd^{2+} complexes, 253
  Cd–protein complexes, 249
  hepatic metal uptake, 80
  Pb^{2+}–protein complexes, 256
  proximal tubular absorption, 39
  receptor-mediated, 87–88, 168, 170, 250, 253
  transferrin and ferritin, 253
  transferrin-receptor-mediated, 80, 97
Endoplasmic reticulum (ER), 355, 390
Endosomal/lysosomal vesicles, 93, 96
  ATOX(1), 96
  ATP7A, 96, 97
  ATP7B, 96
  DMT(1), 97
  intravesicular copper, 97
  Menkes gene, 97
  metal sequestration mechanisms, 97
  STEAP(3), 97–98
  transferrin receptor-mediated endocytosis, 97
  Znt(2), 98
Enzyme-linked immunosorbent assay (ELISA), 398
Epidermal growth factor receptor (EGFR), 191
Epithelial Ca^{2+} channel (ECaC), 251
EpRE. *See* Electrophile response element (EpRE)
ER. *See* Endoplasmic reticulum (ER)
ERK. *See* Extracellular signal-regulated kinase (ERK); Extracellular-receptor kinases (ERK)
ESR. *See* Electron spin resonance (ESR)
Estimated average requirement (EAR), 201
Ethylenediaminetetraacetic acid (EDTA), 252
Experimental autoimmune encephalomyelitis (EAE), 397
Extracellular signal-regulated kinase (ERK), 191
Extracellular-receptor kinases (ERK), 337

F

FA. *See* Friedreich's ataxia [(FA); (FRDA)]
FADD. *See* Fas-associated death domain protein (FADD)
FAK. *See* Focal adhesion kinase (FAK)
Fas-associated death domain protein (FADD), 388
FC. *See* Ferrochelatase (FC)
Feline leukemia virus subgroup C receptor (FLVCR), 173
Ferritin, 171, 366
  endocytosis, 253
  hepatic iron uptake, 88
  HHCS, 365–366
  in iron homeostasis, 175
  iron mobilization, 171
  iron trafficking, 174
  mitochondrial, 172
  neuroferritinopathy, 360
  receptors, 88
  serum concentration, 366
  subunits, 171
Ferrochelatase (FC), 118–119
Ferroportin, 93
  disease, 359
  ferroportin 1 (FPN1), 169
  hepcidin, 94
  iron transport, 352, 353
Ferroportin disease, 359. *See also* Iron overload
Ferroportin 1 (FPN1), 169
Fet(3), 90. *See also* Ceruloplasmin
FHHNC. *See* Hypomagnesemia with hypercalciuria and nephrocalcinosis (FHHNC)
5,5*N*-dithiobis (2-nitrobenzoate) (DTNB), 24
Fluorescent resonance energy transfer (FRET), 25
Fluorescent Zn^{2+} sensor TSQ, 21
  LLC-PK_1 cells, 21–22
  LMW fractions, 22
  ternary adducts formation, 22
  Zinquin, 21
FLVCR. *See* Feline leukemia virus subgroup C receptor (FLVCR)
f-Met-Leu-Phe (FMLP), 394
FMLP. *See* f-Met-Leu-Phe (FMLP)
Focal adhesion kinase (FAK), 328
4,7-ρ-sulfonyl-phenyl-1,10-phenanthroline. *See* Bathophenanthrolinedisulfonate (BPS)
FPN1. *See* Ferroportin 1 (FPN1)
Frataxin, 173, 360–361
FRDA. *See* Friedreich's ataxia [(FA); (FRDA)]
FRET. *See* Fluorescent resonance energy transfer (FRET)
Friedreich's ataxia [(FA); (FRDA)], 167, 360
Functional polarity, 83–84

G

γ-aminobutyric acid (GABA), 134–135, 136
γ-glutamylcysteine synthetase (GCS), 56
γ-glutamyltranspeptidase (GGT), 214, 216, 220
γGSChc. *See* Gamma glutamyl-cysteine synthetase heavy chain (γGSChc)
GABA. *See* γ-aminobutyric acid (GABA)
GAD. *See* Glutamate decarboxylase (GAD)
Gamma glutamyl-cysteine synthetase heavy chain (γGSChc), 148
Gap junctional intercellular communication (GJIC), 334
Gap junctions, 331, 334
  connexon, 331
  gating, 331
  metal induced disruption, 334
  methylmercury, 334
Gastrointestinal system, 296
  arthropods, 296
  crustaceans, 296
  hepatopancreas, 297
  intestine, 310
Gating, 331
GCL. *See* Glutamate cysteine ligase (GCL)
GCs. *See* Glucocorticoid—hormones (GCs)
GCS. *See* Glutamate–cysteine ligase (GCS); γ-glutamylcysteine synthetase (GCS)
GDF15. *See* Growth differentiation factor 15 (GDF15)
GFPs. *See* Green fluorescent proteins (GFPs)

GFR. *See* Glomerular filtration rate (GFR)
GGT. *See* γ-glutamyltranspeptidase (GGT)
GIF. *See* Growth inhibitory factor (GIF)
GJIC. *See* Gap junctional intercellular communication (GJIC)
Globus pallidus (GP), 128
Glomerular filtration rate (GFR), 224
GLRX. *See* Glutaredoxin (GLRX)
Glucocorticoid
  hormones (GCs), 149
  receptor (GR), 149
  response elements (GREs), 149, 390
Glucose-regulated proteins (GRPs), 392
Glutamate cysteine ligase (GCL), 215, 385
Glutamate–cysteine ligase (GCS), 215, 216
Glutamate decarboxylase (GAD), 134
Glutaredoxin (GLRX), 216
Glutathione (GSH), 6, 10, 37, 80, 129, 213, 297, 383
  arsenic metabolism, 228–229
  biliary, 221
  cadmium toxicity, 226–228
  dependent functions, 215
  detoxification role, 297
  exogenous, 222
  gene expression regulation, 232
  hepatic, 218, 221
  hepatic transport, 218
  Hg interaction, 223
  inorganic mercury effects, 230
  interorgan metabolism, 214–212
  interorgan transport, 216–217
  mercury toxicity, 222
  metabolism, 215
  metal disposition, 221
  metal transport, 214
  metal-induced changes, 229–231
  metals as GSH *S*-conjugates, 220
  metals excretion, 221
  plasma thiols, 221
  redox cycle, 232
  renal, 218–220, 222
  renal growth (NPX) model, 225
  synthesis and degradation, 214
  tissue accumulation modulation, 221
Glutathione disulfide (GSSG), 24, 221, 258, 261
Glutathione peroxidase (GPX), 116, 380
Glycosylphosphatidylinositol (GPI) anchor, 356
GP. *See* Globus pallidus (GP)
GPI anchor. *See* Glycosylphosphatidylinositol (GPI) anchor
GPX. *See* Glutathione peroxidase (GPX)
GR. *See* Glucocorticoid—receptor (GR)
GRD. *See* GSSG reductase (GRD)
Green fluorescent proteins (GFPs), 25
GREs. *See* Glucocorticoid—response elements (GREs)
Growth differentiation factor 15 (GDF15), 361
Growth inhibitory factor (GIF), 188
GRPs. *See* Glucose-regulated proteins (GRPs)
GSH complexes, 99–100, 218, 224, 245
  cells overexpressing MRP(1), 100
  glutathione–platinum complex transport, 100
  intrabiliary degradation, 101
  lack Mrp2 activity, 100–101
  methylmercury reabsorption, 101–102
  MRP (1), 100, 101
  MRP (2), 100
  MRP2-mediated excretion, 99
GSH peroxidase (GPX), 56, 216, 268
GSH *S*-conjugates of mercury, 40, 41, 43
  ABC transport proteins, 42
  mercuric conjugates formation, 40, 42
  MRP localization, 42
  pool fraction, 42
GSH *S*-transferases (GSTs), 215
GSH, exogenous, 66, 222, 233
  extracellular ligand for mercury, 224
  intracellular GSH levels, 222–224
  mercury–GSH complex, 224
  protective effects, 66
GSH. *See* Glutathione (GSH)
GSSG reductase (GRD), 226
GSSG. *See* Glutathione disulfide (GSSG)
GSTs. *See* GSH *S*-transferases (GSTs)
Guanylate kinase (GUK), 330
GUK. *See* Guanylate kinase (GUK)

H

Haber–Weiss reaction, 167
Hard Soft Acid Base (HSAB) principle, 186
HAV sequence. *See* His-Ala-Val (HAV) sequence
HBV. *See* Hepatitis B viruses (HBV)
HCC. *See* Hepatocellular carcinoma (HCC)
HCP1. *See* Heme carrier protein-1 (HCP1)
HCV. *See* Hepatitis C viruses (HCV)
Hcy. *See* Homocysteine (Hcy)
HDACS. *See* Corepressor histone deacetylase complexes (HDACS)
HDL. *See* High-density lipid (HDL)
Heat-shock element (HSE), 392
Heat-shock factors (HSFs), 392
Heat-shock proteins (HSPs), 149, 392
  arsenic, 394–395
  and GR activity, 149
  immune function, 393–394
  metal induced, 394
  regulation, 392
  in renal tubular necrosis, 61
HEK. *See* Human embryonic kidney (HEK)
Heme carrier protein-1 (HCP1), 169
Heme metabolism, 60
  free radical-mediated oxidation, 60–61
  influence of mercury, 60
  porphyrin excretion, 61
  porphyrinuria, 60
Heme oxygenase 1 (HO-1), 169, 352
Hemochromatosis. *See* Iron overload
Hemojuvelin (HJV), 176, 356
Hepatic iron export, 93
  ferroportin, 93–94
  hepcidin, 94
  hephaestin, 93, 94
Hepatic microanatomy
  blood spaces separation, 82–83
  canalicular membrane, 82
  lateral domain, 82
  mammalian hepatocytes, 82
  paracellular pathway, 83
  sinusoidal endothelium, 82

Hepatic transport, 63, 81
arsenic, 245
bile secretion, 84
biliary-hepatic cycle, 84
cadmium, 252
canalicular membrane transporters, 99
determinants, 82
functional polarity, 83
GSH *S*-conjugates, 218
hepatic microanatomy, 82
hepatocytes, 81
in vitro systems, 81
inorganic mercury, 63
intracellular ligands, 96
membrane recycling, 87, 99
mercury, 260–261, 265
metal uptake, 96, 98
nuclear uptake, 98
paracellular pathway, 85–86
permeation, 95
plasma and intracellular protein binding, 86
plasma membrane vesicles separation, 81–82
sinusoidal membrane channels, 94
sinusoidal metal transporters, 89
sinusoidal metal uptake, 86
3D culture, 81
Xenopus oocytes, 82
Hepatic transport permeation, 95
convection, 95
diffusion, 95
high-affinity component, 95
metal binding, 96
metal substrate concentration, 95
protease contaminants, 95
transport mechanisms alteration, 96
uptake, 96
Hepatic transport sinusoidal metal transporters, 89
copper uptake transporters, 89
DMT1/SLC11A(2), 92
hepatic iron export, 93
organic solute carriers, 94
zinc uptake and efflux, 92
Hepatitis B viruses (HBV), 362
Hepatitis C viruses (HCV), 362, 381
Hepatocellular carcinoma (HCC), 87, 245, 354
Hepatopancreas, 297
arthropods, 297
calcium transport processes, 299, 300
cation transport processes, 299
concretions types, 298–299
glutathione, 297
luminal heavy metals, 296
metal transport mechanisms, 300
metallothionein, 297
mitochondria, 304
precipitation granules, 298
Hepcidin, 358, 364, 381
ACD treatment, 354
defective expression, 357
ferrportin regulation, 94
gene, 356
in hepatic iron export, 93–94
HFE-mediated signaling, 358
iron efflux regulation, 353
iron-dependent signaling, 357
iron metabolism, 175–176, 353
juvenile hemochromatosis, 356
levels, 169, 353, 357
Hephaestin, 93, 94, 169. *See also* Ceruloplasmin
iron oxidation, 352
Hereditary hemochromatosis (HH), 355. *See also* Hepcidin
cirrhosis, 364
hemochromatosis gene HFE, 355–356
HH type (1), 355
pathogenesis, 356–358
Hereditary Hyperferritinemia-Cataract Syndrome (HHCS), 365–366
HH. *See* Hereditary hemochromatosis (HH)
HHCS. *See* Hereditary Hyperferritinemia-Cataract Syndrome (HHCS)
HIEC. *See* Human epithelial cells from intestinal crypts (HIEC)
HIF. *See* Hypoxia-inducible factor (HIF)
High-density lipid (HDL), 128
His-Ala-Val (HAV) sequence, 330
HJV. *See* Hemojuvelin (HJV)
hK family. *See* Human kallikrein (hK) family
Homocysteine (Hcy), 37
HO-1. *See* Heme oxygenase 1 (HO-1)
HREs. *See* Hypoxia-responsive elements (HREs)
HSAB principle. *See* Hard Soft Acid Base (HSAB) principle
HSE. *See* Heat-shock element (HSE)
HSFs. *See* Heat-shock factors (HSFs)
HSPs. *See* Heat-shock proteins (HSPs)
Human embryonic kidney (HEK), 244, 256
Human epithelial cells from intestinal crypts (HIEC), 248
Human kallikrein (hK) family, 195
Hypersensitivity reactions, 396
BeLPT, 397
beryllium exposure, 396
type IV, 396
Hypomagnesemia with hypercalciuria and nephrocalcinosis (FHHNC), 86
Hypoxia-inducible factor (HIF), 92
Hypoxia-responsive elements (HREs), 92

I

IARC. *See* International Agency for Research against Cancer (IARC)
IDA. *See* Iron—deficiency anemia (IDA)
IEC-18. *See* Intestinal epithelial cell line (IEC-18)
IFN-γ. *See* Interferon-γ (IFN-γ)
IGF-1. *See* Insulin-like growth factor-1 (IGF-1)
IKK. *See* IκB kinase (IKK)
ILK. *See* Integrins—linked kinase (ILK)
Immune mechanisms
autoimmunity, 397–398
copper, 382
hypersensitivity, 396
immune activation, 396
immunomodulation, 398
immunosuppression, 396
iron, 380

metal influences, 380
selenium, 380
zinc, 381
Immune system, 199, 380, 382, 390
adaptive, 199–200
adverse consequences, 387
copper's role, 382
cytokine production, 200
DNA methylation, 399
HSP expression, 393
innate, 200
lead, 117
relationship with zinc, 381
zinc signaling, 199
Immunomodulation assessment, 398
cellular thiols depletion, 383
DNA methylation, 399
ELISA, 398
ELISPOT, 398
metal-mediated, 383, 398
microbead assays, 398
SPR, 399
Immunotoxic effects, 382
apoptotic signaling pathways, 386–389
cellular activities, 395
HSPs, 392–395
induction of MT, 389
metal-dependent immune changes, 389
MT synthesis, 390
oxidative stress, 383–386
protein secretion, 396
thiols modification, 383–386
Inducible nitric oxide synthase (iNOS), 381
Inhibitory factor-κB (IκB), 231
Initiator regions (InR), 146
Inorganic mercury, 36
dose-dependent toxic effects, 37
GSH, 38
luminal uptake, 40
proximal tubular uptake, 39
renal cellular death, 37–38
iNOS. *See* Inducible nitric oxide synthase (iNOS)
InR. *See* Initiator regions (InR)
Insulin, 267
Insulin-like growth factor-1 (IGF-1), 193
Insulin receptor (IR), 191
Integrins, 328, 329–330, 332
copper, 332
linked kinase (ILK), 328
metal induced disruption, 332–333
receptors, 328
signaling, 328
Interferon-γ (IFN-γ), 200, 381
International Agency for Research against Cancer (IARC), 119
Intestinal epithelial cell line (IEC-18), 333
Intestine, 198–199, 310
heavy metal transport, 311
L-histidine effect on zinc transport, 312–314
intestinal electrophysiology, 310
sodium and chloride transport, 310
Intracellular distribution, 53, 59, 91, 189
Intracellular GSH metabolism, 56–57
Intracellular MT metabolism, 55
Intracellular sequestration, 96, 296
COT(1), 98
endosomal/lysosomal vesicles, 96–98
metal-binding ligands, 96
mitochondrial metal uptake, 98
Na-independent efflux mechanism, 98
nuclear metal homeostasis, 98
Intracellular thiols, 62, 398
acute biliary ligation, 63
concentrations of GSH, 62
degradation of GSH, 63
intracellular levels of MT, 64
renal accumulation, 62–63
urinary excretion, 63
Intrauterine devices (IUDs), 332
Ionic mimicry, 220, 255
IR. *See* Insulin receptor (IR)
IRE. *See* Iron—response element (IRE)
Iron, 167, 365, 366, 381
deficiency anemia (IDA), 354
endosomal ferrireductase, 170–171
endosome on mitochondria, 171
homeostasis, 173, 175
LIP, 169
metabolism, 168
mitochondrial, 172–173
regulation, 174–175
regulatory proteins (IRPs), 168
response element (IRE), 92, 131, 168, 360
trafficking, 168, 169, 170, 174
transferrin (Tf), 169, 171
Iron homeostasis, 172, 173–175
defective iron traffic, 354
dietary iron absorption, 352
hepcidin, 357
hormonal regulation, 353
IRP1&2 role, 175
malregulation, 354
physiology, 351
Iron overload, 354–355, 362
alcoholic liver disease, 364–365
chronic hepatitis C, 363–364
clinical management, 366–368
CNS, 360
hereditary, 355
hereditary causes, 359
juvenile, 356
mitochondrial, 360
nonalcoholic fatty liver disease, 365
states, 362
systemic, 359
IRPs. *See* Iron—regulatory proteins (IRPs)
IUDs. *See* Intrauterine devices (IUDs)
IκB. *See* Inhibitory factor-κB (IκB)
IκB kinase (IKK), 381

J

JAK. *See* Janus kinases (JAK)
JAM. *See* Junctional adhesion molecule (JAM)
Janus kinases (JAK), 149
JH. *See* Juvenile hemochromatosis (JH)
JNK. *See* c-Jun N-terminal kinase (JNK)
Junctional adhesion molecule (JAM), 330

Juvenile hemochromatosis (JH), 356. *See also* Hepcidin
1q-unlinked JH, 356
HJV, 356

K

Kidney, 36, 37, 314
antennal gland, 315, 316, 317–318
basolateral mercury uptake, 47–51
calcium transport, 316, 317
dicarboxylate transporter, 48–49
dithiol chelators use, 52
DMPS, 52
DMSA, 52
endocytosis, 39
heavy metal regulation, 318–320
luminal plasma membrane, 43–46
mercuric deposits, 38–39
mercury complexes, 39, 40–42, 49–50
mercury effects, 36, 37
mercury in renal epithelial cells, 53
mercury transport, 38, 40–46
molecular “mimicry”, 38, 46, 47
MRP proteins, 51–52
organic anion transport systems, 47
osmoregulation, 315–316
role of liver, 52
thiols, 39
tubular uptake, 39, 52
zinc uptake, 319
Killer cell Ig-like receptors (KIR), 200
Kinase suppressor of Ras (KSR), 192
KIR. *See* Killer cell Ig-like receptors (KIR)
KSR. *See* Kinase suppressor of Ras (KSR)

L

Labile Iron Pool (LIP), 169, 171
L-buthionine-*S, R*-sulfoximine (BSO), 221
L-CAM. *See* Liver—cell adhesion molecule (L-CAM)
LDL. *See* Lowdensity lipoprotein (LDL)
Lead, 114, 254
ALAD, 115
calcium, 255–256
cancer, 119
cardiovascular system, 116–117
cellular signaling pathways, 117
cellular uptake, 114
clinical consequences, 255
DMT(1), 255
endocytosis, 256
exposure, 255
immune system, 117
intercellular mimicry, 257
intracellular binding, 114–115
mitigating factors, 119
nanomaterials, 121
nervous system, 115
reproductive system, 119
target organs, 115
transport, 114
Lead and immune system, 117
ALA dehydratase, 118
ALA synthetase, 118
cellular immune effects, 117–118
coproporphyrin oxidase, 118
ferrochelatase, 118–119
hematopoietic system, 118
humoral immune effects, 117
uroporphyrinogen decarboxylase, 118
Lead pharmacological considerations, 120
age, 120
diet and nutrition, 120
exposure level, 120
gender, 120
genetics, 120
mixtures exposures, 120–121
Pb × As interaction studies, 121
Pb × Cd, 121
Pb × Cd × As, 121
Lead transport in kidneys, 115
cytoplasm, 116
glomerular capillaries, 116
mitochondria, 116
nuclei, 116
proximal tubule cells, 116
LEAP-1. *See* Liver—expressed antimicrobial protein (LEAP-1)
LEW. *See* Lewis strain rats (LEW)
Lewis strain rats (LEW), 397
Ligands, 11, 13, 14, 19, 214
binding, 328
biometal complexes, 24
diethyldithiocarbamate, 15
dithiol ligands, 15
DTC, 15
intracellular, 96
metal binding, 13, 20
pharmacologic, 15
substitution, 4
thiolate, 18, 221
use, 14
zinc, 195
LIP. *See* Labile Iron Pool (LIP)
Lipopolysaccharide (LPS), 146, 382
B cell stimulation, 395
MT induction, 146–147, 153
Liver, 80
cell adhesion molecule (L-CAM), 330
expressed antimicrobial protein (LEAP-1), 381. *See* Hepcidin
hepatic function, 80
Wilson and Menkes disease genes, 80
LIV-1 subfamily of ZIP transporters (LZT), 133
LMV. *See* Lysosomal membrane vesicles (LMV)
LMW. *See* Low molecular weight (LMW)
Low molecular weight (LMW), 22
Lowdensity lipoprotein (LDL), 157
LPS. *See* Lipopolysaccharide (LPS)
Luminal plasma membrane, 43, 46
cleavage products, 43
Cys conjugates, 43, 44
Hcy formation, 45
Hcy mercuric conjugates, 44–45, 46
in vivo data, 43, 45
perfused S2 segments, 44
perfused tubular data, 43

proximal tubular cells, 45
remethylation pathway, 45
transsulfuration pathway, 45
Lysosomal membrane vesicles (LMV), 305
LZT. *See* LIV-1 subfamily of ZIP transporters (LZT)

M

Madin-Darby canine kidney (MDCK) cells, 45, 250, 259
MAG. *See* Myelin-associated glycoprotein (MAG)
MAGUK. *See* Membrane-associated guanylate kinase (MAGUK)
MAIDS. *See* Murine acquired immunodeficiency syndrome (MAIDS)
Major histocompatibility complex (MHC), 117, 193, 355, 381
Manganese, 127
abnormal levels, 134
astrocytes, 134
DMT(1), 131
dopamine transporter, 133–134
essentiality, 127
integrin activation, 329, 333
neurochemical alterations, 134–135
olfactory bulb, 131–133
oxidative stress, 135
Tf/Tf receptor, 130–131
toxicity, 128
transport, 129, 130
ZIP transporter proteins, 133
Manganism, 127, 129, 136. *See also* Parkinson's disease (PD)
GABA, 135
manganism-like PD, 134
symptoms, 128
MAO. *See* Monoamine oxidase (MAO)
MAP. *See* Mitogen-Activated Protein (MAP)
MAPK. *See* Mitogen-activated protein kinase (MAPK)
Matrix metalloproteinases (MMP), 337
MCK. *See* Muscle creatine kinase (MCK)
MCT. *See* Monocarboxylate transporter (MCT)
MDCK. *See* Madin-Darby canine kidney (MDCK)
mdr1 gene. *See* Multidrug resistance 1 (*mdr1*) gene
Megalin, 250
Membrane recycling, 87, 99
ASGP-R, 88
ATP7B-mediated transport, 99
chromatographic separation of vanadium, 88–89
direct transport via ATP7B, 99
efflux pathways, 84
endocytotic/exocytotic mechanisms, 87–88
ferritin receptors, 88
hepatic sinusoidal metal uptake, 84
hepatocyte bile transport pathways, 85
internalized endocytotic vesicles, 87
iron-binding proteins, 88
Menkes' syndrome, 99
metal transport, 87, 99
receptor-mediated endocytosis, 87, 88
transferrin endocytotic process, 88
transport as GSH complex, 99
tubulovesicular transport process, 87
Wilson's disease, 99
Membrane transporters, 50
activities, 225
canalicular, 99
function, 246
MDCK cells transfection, 50
OAT, 50
Membrane-associated guanylate kinase (MAGUK), 330
Mercuric conjugates of Cys, 43, 49
basolateral uptake of mercury, 50
bilateral ureteral ligation, 49–50
of Hcy and NAC, 45, 50
inorganic mercury, 50
organic *S*-conjugates, 49
renal tubular uptake, 50
Mercuric ions, 36. *See also* Mercury
affinity, 257
affinity constants, 37
biological impact, 36–37
bonding, 38
extraction, 260
luminal export, 51
molecular "mimicry", 38
molecular interactions, 60
organic mercurials, 37
predilection, 37
proximal tubule, 38
thiol-containing molecules, 37
thiol–mercury complexes formation, 37
transport, 47
Mercury, 36, 257
alterations in ($Na^+ + K^+$)-stimulated ATPase, 59
autoimmunity, 62
basolateral uptake, 260
calcium ions distribution, 59
cytoskeleton interactions, 61
effects in kidney, 36
elemental, 36
expression of stress proteins, 61
fecal elimination, 258
glomerular nephropathy, 62
GSH metabolism, 56
heme metabolism, 60
hepatic transport, 260, 265
inorganic forms, 36
inorganic mercury, 258
intestinal transport, 258–259, 263
intracellular mimicry, 261–262
lipid peroxidation, 57
methylmercury, 262
molecular interactions, 55, 60
molecular transport, 38
monomethyl, 36
MT metabolism, 55
organic forms, 36
oxidative stress, 57
renal mitochondrial function, 58
renal transport, 259–260, 264–265
system L, 262
transport in brain, 262
transport in erythrocyte, 262–264
transport in placenta, 261, 265–266
urinary excretion, 53
Mercury and cytoskeleton interactions, 61–62
methylmercury disruption, 61
microtubules disruption, 61

Mercury basolateral uptake, 47
conjugates of Cys, 49
conjugates of GSH, 49
conjugates of Hcy and NAC, 50
dicarboxylate transporter, 48
organic anion transport systems, 47
thiol *S*-conjugates, 50
Mercury luminal uptake, 40, 41, 44
axial heterogeneitym, 40
cysteinylglycinase, 42
cysteinylglycine, 42
dehydropeptidases inhibition, 42–43
γ-glutamyltransferase, 40
GSH degradation, 40
GSH *S*-conjugates of mercury, 40, 43
mercuric conjugates, 43, 44
molecular "mimicry", 46
pretreatment with acivicin, 40
Mercury, renal handling, 62
compensatory tubular hypertrophy, 67–68
functioning nephrons, 67
mercury toxicity, 67
thiols and renal accumulation, 62, 64
Mercury transport in kidneys, 38
basolateral uptake, 47
luminal uptake, 40
mercury intracellular distribution, 53
MRP proteins, 51
proximal tubular uptake, 39
renal disposition, 38
renal tubular uptake, 52
Metabolism, 2
Metal detoxification
glutathione, 297
metallothionein, 297
precipitation granules, 298
Metal ion deficiency, 13
cells incubation, 14
liberating Fe (III), 14
metal binding ligands utilization, 13, 14
metal chelating agent, 13, 14
PHEN, 15
TPEN, 14–15
Metal ion exchange, 5, 6, 16
bond breaking, 18
Cd-MT formation, 16–17
imidazole interaction, 19
interprotein, 18
kinetic analysis, 17
metalloenzymes, 16
metal-thiolate clusters, 17–18
mF(3), 19
NMR spectroscopy, 19
reaction with Zn-MT, 5–6
sequestration of Cd^{2+}, 17
transcription factor, 18
Zn9-TFIIIA reacts with Cd^{2+}, 19
Zn-finger proteins, 18–19
Zn-MT, 20
Zn-proteins with Cd^{2+}, 16
Metal ion trafficking, 2–3
adduct formation, 6
apo-MT, 4
conceptual problem, 3
copper trafficking, 4
cis-dichlorodiammine Pt (II) reaction, 5
ligand substitution, 4
mercurial localization, 5
metal ion exchange, 5–6
metal-ligand binding, 3, 4
redox reactions, 6–7
sulfhydryl ligands, 4–5
Metal-ligand binding, 3, 7
apo-MT, 8, 9
apo-MT reactivity, 10
compartments, 9
efflux process, 9
forward and reverse reactions, 3
heavy metal ions, 3
ligand substitution deactivation, 10
mammalian metallothionein, 8, 9
metal ion deficiency, 13
metal-thiolate cluster structure, 8
nutrient Zn^{2+} depletion, 7
pharmacologic ligands, 15
transition metal ions, 3, 4
x-ray crystallographic structure, 8
Zn^{2+} deficiency, 11
Zn^{2+} saturation, 9–10
Zn^{2+} trafficking, 7, 9
Zn^{2+} trafficking requirement, 7
Zn-proteome formation, 7–8
Metal regulatory elements (MREs), 146
Metal response elements (MREs), 55, 192, 390, 392
Metal-response element-binding transcription factor-1 (MTF-1), 18, 147, 183. *See also* Metal response elements (MREs)
activation, 390
MT gene transcription, 146
zinc sensors, 183, 192
Metal transport, hepatopancreas, 300
basolateral membrane, 302
brush-border membrane, 300
calcium and copper transport, 304
calcium uptake, 301
copper influx, 301
ER, 304–305
heavy metal–thiol conjugates transport, 307
intravesicular sodium concentration, 303
ionic metals transport, 305
lysosomes, 305
mitochondria, 303
OAT-like transport system, 308–310
polyvalent anions role, 308
zinc influx kinetics, 302
zinc transport working model, 306
Metal transport protein 1 (MTP1), 253. *See also* Divalent metal transporter-1 (DMT1); Ferroportin
Metal transporters, 80
CTR(1), 80
DMT(1), 80, 89, 128, 168, 243
IRE, 92
multispecific, 92
Nramp family, 89, 92
sinusoidal, 89
vanadate-insensitive heavy, 305
ZIP transporter proteins, 133
Znt(2), 98

Metal-unsaturated protein. *See* Apo-MT
Metallothionein (MT), 15, 22, 37, 86, 183, 186, 297, 381, 389
apo-MT, 15, 16
cadmium, 391
cancer, 153–155
cardiac toxicity, 155
C-terminal region, 16
detoxification role, 297
distribution, 390
epigenetic events, 150
functions, 151
gene structure, 145–146
gene transcription, 146–147
glucocorticoid induction, 149
immune function, 153
induction, 149–150, 389–390
isoforms, 145, 389
ligand (L) substitution reactions, 16
metal-thiolate connectivities, 16
MTF-1, 147–148
neurological function, 156–157
peptide, 16, 49–61
physiological role, 297
post-transcriptional regulation, 150–151
proteins, 144
redox cycle, 187
regulation, 188
SGLT, 15
synthesis, 390
thionylation, 187
tissue-specific MT expression, 151
trafficking, 188
zinc buffers, 151–153
zinc with sulfur ligands, 186
Methylcyclopentadienyl manganese tricarbonyl (MMT), 128
MHC. *See* Major histocompatibility complex (MHC)
Mimic, functional, 242, 246
endogenous ions, 257
estrogen, 243, 247, 267
oxyanion sulfate, 269
Mitogen kinase phosphatase (MKP), 193
Mitogen-Activated Protein (MAP), 193
Mitogen-activated protein kinase (MAPK), 157, 191, 227, 328, 337
MKP. *See* Mitogen Kinase Phosphatase (MKP)
MLTF. *See* Adenomajor late transcription factor (MLTF)
MMP. *See* Matrix metalloproteinases (MMP)
MMT. *See* Methylcyclopentadienyl manganese tricarbonyl (MMT)
Molecular "mimicry", 38, 44, 46, 47, 220, 233, 242
amino acid transporters, 46–47
CH_3Hg-*S*-Cys, 264
Cys *S*-conjugates, 46
DMT(1), 255
Hg–thiol complexes, 261
mercuric conjugates, 260
system $b^{0,+}$, 47
Molecular homology. *See* Molecular "mimicry"
Molecular interactions, 60
mercury and water channel, 60
in renal epithelial cells, 55
Molecular targets, 327, 332
cadmium, 226
MAP MKP activity, 193
PDE activity, 193
PTP activity, 192–193
signaling complexes, 193–194
transient binding of zinc, 192
Monoamine oxidase (MAO), 129
Monocarboxylate transporter (MCT), 133
MPP^+. *See* 1-methyl-4-phenylpyridium (MPP^+)
MPTP. *See* 1-methly-4-phenyl-1, 2, 3, 6-tetrahyrdopyridine (MPTP)
MREs. *See* Metal regulatory elements (MREs); Metal response elements (MREs)
MRP. *See* Multidrug resistance-associated proteins (MRP)
MS fluxes. *See* Mucosal to serosal (MS) fluxes
MT. *See* Metallothionein (MT)
MT KO. *See* MT-1/MT-2 knockout (MT KO)
MT-1/MT-2 knockout (MT KO), 153
MTF-1. *See* Metal-response element-binding transcription factor-1 (MTF-1)
Mucosal to serosal (MS) fluxes, 310
Multidrug resistance-associated proteins (MRP), 42, 51, 100, 261. *See also* ATP-binding cassette (ABC)
dithiol chelators use, 52
family, 218
GSH *S*-conjugates transport, 218
luminal export of mercuric ions, 51
MRP, 42, 52, 244
proximal tubular secretion, 42
Multidrug resistance 1 (*mdr1*) gene, 245
Murine acquired immunodeficiency syndrome (MAIDS),384
Muscle creatine kinase (MCK), 173
Myelin-associated glycoprotein (MAG), 157

N

N, N, N′, N′-tetrakis (2-pyridylmethyl)-ethylenedidiamine (TPEN), 10, 391
N-(6-methoxy-8-quinolyl)-*p*-toluensulfonamide (TSQ),21, 24
Na^+-taurocholate cotransporting polypeptide (NTCP), 94
NAC. *See N*-acetylcysteine (NAC)
N-acetylcysteine (NAC), 37, 218, 252
NADH. *See* Nicotine adenine dinucleotide (NADH)
NASH. *See* Nonalcoholic steatohepatitis (NASH)
National Center for Biotechnology Information (NCBI), 145
Natural killer (NK) cells, 200
Natural-resistance-associated macrophage protein (Nramp), 80
NCBI. *See* National Center for Biotechnology Information (NCBI)
NCX. *See* Sodium–calcium exchanger (NCX)
NEM. *See N*-ethylmaleimide (NEM)
Nephrectomized (NPX) rats, 68
Nephrons, functioning, 67
compensatory renal growth, 67–68
intracellular metabolisms, 68
intrarenal handling of mercury, 68
unilateral nephrectomy, 68
N-ethylmaleimide (NEM), 215

NF-κB. *See* Nuclear factor-kappa B (NF-κB)
Nicotine adenine dinucleotide (NADH), 23
Nitric oxide synthase (NOS), 117, 381
NK cells. *See* Natural killer (NK) cells
NMDA. *See* *N*-methyl D-aspartate (NMDA)
N-methyl D-aspartate (NMDA), 135
Nonalcoholic steatohepatitis (NASH), 365
  "two-hit" hypothesis, 365
Non-transferrin-bound iron (NTBI), 355
NOS. *See* Nitric oxide synthase (NOS)
NPX rats. *See* Nephrectomized (NPX) rats
Nramp. *See* Natural-resistance-associated macrophage protein (Nramp)
NRAMP-2. *See* Divalent metal transporter-1 (DMT1)
NTBI. *See* Non-transferrin-bound iron (NTBI)
NTCP. *See* Na^+-taurocholate cotransporting polypeptide (NTCP)
Nuclear factor-kappa B (NF-κB), 148, 200, 231

O

OAT. *See* Organic anion transporter (OAT)
OATLP1. *See* Organic anion transporter—like protein (OATLP1)
OATP. *See* Organic anion transporter—polypeptide (OATP)
Occludin, 330
OCT. *See* Organic cation transporter (OCT)
1, 10-Phenanthroline (PHEN), 15
1-methyl-4-phenyl-1, 2, 3, 6-tetrahyrdopyridine (MPTP), 129
1-methyl-4-phenylpyridium (MPP^+), 133
Open reading frames (ORFs), 145
ORFs. *See* Open reading frames (ORFs)
Organic anion transporter (OAT), 48, 252. *See also* Canalicular MRP (cMRP)
  like protein (OATLP1), 263
  OAT(1), 50, 308, 310
  OAT(3), 310
  polypeptide (OATP), 94, 133, 243
  pretreatment with PAH, 48
  renal burden of mercury, 47
  "stop-flow" conditions, 48
  substantial evidence, 47
  systems, 47
Organic cation transporter (OCT), 94, 253
Organic solute carriers, 80, 94, 102
  BSEP, 102
  MDR(1), 102
  MDR(3), 102
  NTCP, 94
  OATP, 94
  OCT, 94
Oxyanions, 80, 102
  arsenate, 268
  chromate, 268
  molybdate, 269
  space-filled models, 269
  toxic metals, 268
  vanadate, 269–270

P

PAH. *See* *p*-aminohippurate (PAH)
p-aminohippurate (PAH), 48, 219, 260, 308, 309. *See also* Organic anion transporter (OAT)
PANK2. *See* Pantothenate kinase 2 (PANK2)
Pantothenate kinase 2 (PANK2), 360
Paracellin (PCLN), 86
Paracellular pathway, 83, 85
  Ca^{2+} and Mg^{2+} concentrations, 85–86
  FHHNC, 86
  hepatic junctional complexes, 85, 86
  substances movement, 85
Parkinson's disease (PD), 127, 128
  mitochondrial damage, 135
  oxidative stress, 129
Pb^{2+}-binding proteins (PbBPs), 114, 119, 256
PbBPs. *See* Pb^{2+}-binding proteins (PbBPs)
PBMCs. *See* Peripheral blood mononuclear cells (PBMCs)
PC12 cells. *See* Pheochromocytoma (PC12) cells
PCLN. *See* Paracellin (PCLN)
PD. *See* Parkinson's disease (PD)
PDE. *See* Phosphodiesterase (PDE)
Peak expiratory flow rate (PEFR), 130
PEFR. *See* Peak expiratory flow rate (PEFR)
PEN. *See* Pencillamine (PEN)
Pencillamine (PEN), 252
Peripheral blood mononuclear cells (PBMCs), 385
Peripheral nervous system (PNS), 115. *See also* Central nervous system (CNS)
PHEN. *See* 1, 10-phenanthroline (PHEN)
Pheochromocytoma (PC12) cells, 184, 256
Phosphatase and tensin homolog (PTEN), 191
Phosphatidyl inositol-3-kinase (PI3K), 191
Phosphodiesterase (PDE), 193
Phospholipase A2 (PLA2), 385
Phospholipase Cγ1 (PLCγ1), 383
PI3K. *See* Phosphatidyl inositol-3-kinase (PI3K)
PKC. *See* Protein kinase C (PKC)
PLA2. *See* Phospholipase A2 (PLA2)
Placental growth factor (PlGF), 148
PLAD. *See* Preligand assembly domain (PLAD)
Plasma membrane calcium ATPase (PMCA), 302, 303
PLCγ1. *See* Phospholipase Cγ1 (PLCγ1)
PlGF. *See* Placental growth factor (PlGF)
PMCA. *See* Plasma membrane calcium ATPase (PMCA)
PMNs. *See* Polymorphonuclear cells (PMNs)
PNS. *See* Peripheral nervous system (PNS)
Polymorphonuclear cells (PMNs), 390
Preligand assembly domain (PLAD), 388
Pro-antioxidant, 187
Prostate-specific antigen (PSA), 195
Protein binding, 7, 86
  chelating agents, 86
  detoxification, 86
  kinetic lability, 86
  ligand substitution, 4
  metal-induced toxicity, 86
  plasma and intracellular, 86
  trace metals activity, 86
Protein kinase C (PKC), 192, 328, 385
  activation, 337
  CRDs, 385
  IκBα degradation, 153
  signaling, 190
Protein tyrosine phosphatase (PTPase), 384
Protein tyrosine phosphatase 1B (PTP1B), 188

Proximal tubular uptake, 39
albumin–mercury complexes, 39
cadmium, 250
endocytosis, 39
mercury, 39, 224, 260
PSA. *See* Prostate-specific antigen (PSA)
PTEN. *See* Phosphatase and tensin homolog (PTEN)
PTPase. *See* Protein tyrosine phosphatase (PTPase)
PTP1B. *See* Protein tyrosine phosphatase 1B (PTP1B)
Pyridoxalthiosemicarbazonato-Cu (II) (Cu (II)-PTSC), 6

R

Rat brain endothelial (RBE4) cells, 131
Rat glomerular mesangial cells (rGMCs), 116
RBE4 cells. *See* Rat brain endothelial (RBE4) cells
RBF. *See* Renal blood flow (RBF)
RDA. *See* Recommended dietary allowance (RDA)
Reactive oxygen species (ROS), 116, 149, 167, 227, 380
Recommended dietary allowance (RDA), 201
Redox reactions, 6, 22
with chromate, 7
with disulfide, 24
GSH redox cycle, 231–232
iron trafficking, 22–23, 24
MT redox cycle, 186, 187
with nitric oxide species, 24
O_2 activation, 7
oxidation–reduction reactions, 6–7
N-terminal lobe, 23
transferrin-bound Fe (III), 23–24
transferrin receptor, 23
Regions of interest (ROIs), 398
Renal blood flow (RBF), 225
Renal epithelial cells, 53, 55
arsenic transport, 244
de novo synthesis, 61
heat shock proteins, 61
heme metabolism, 60–61
intracellular distribution, 53, 59
intracellular metabolism, 55, 56–57
lipid peroxidation, 57, 58
mercury and cytoskeleton interactions, 61–62
mercury influence, 60
mercury-induced oxidative stress, 57, 58
molecular interactions, 60
($NA^+ + K^+$)-stimulated ATPase, 59–60
porphyrinuria, 60
renal lysosomal fraction, 53
stress proteins expression, 61
Renal mitochondrial function, 58
ATP depletion, 58
in vitro treatment, 58
in vivo treatment, 58
mercury effects, 58–59
Renal tubular uptake of mercury, 40, 50, 52, 260
hepatic content of GSH, 52
inorganic mercury, 47–48
liver role, 52
methylmercury, 264
Repulsive guidance molecule c (RGMc), 356
Retinal pigment epithelial (RPE) cells
RGMc. *See* Repulsive guidance molecule c (RGMc)
rGMCs. *See* Rat glomerular mesangial cells (rGMCs)
ROIs. *See* Regions of interest (ROIs)
ROS. *See* Reactive oxygen species (ROS)
RPE cells. *See* Retinal pigment epithelial (RPE) cells

S

SAPK. *See* Stress-activated protein kinases (SAPK)
Sarco/ER Ca^{2+} ATPase (SERCA), 304
like transporter, 306
thapsigargin, 305, 307
sc. *See* subcutaneous (sc)
Selenium, 266, 380
deficiency, 380
glutathione peroxidase (GPx), 380, 383
GSH, 267
GSH synthesis stimulation, 231
health impacts, 266
insulin mimic, 267
Se–cystine complexes, 267
SERCA. *See* Sarco/ER Ca^{2+} ATPase (SERCA)
SGLT. *See* Sodium–glucose cotransporters (SGLT)
SH. *See* Sulfhydryl (SH) groups
sHJV. *See* soluble HJV (sHJV)
Signal transducer and activator of transcription (STAT) kinases, 149
Single-nephron GFR (SNGFR), 225
Sinusoidal plasma membrane (SPM), 217
cadmium transport, 252
GSH *S*-conjugate transport, 218, 220
hepatic GSH transport, 218
mercury transport, 260–261, 265
metal transporters, 80
organic solute transporters, 94
receptors, 83
transport proteins, 83
Sinusoidal uptake, 86–87
Ca^{2+} influx, 95
calcium uptake, 95
cellular zinc uptake, 92, 93
copper uptake transporters, 89–92
DMT(1), 252
hepatic iron export, 93–94
hepatic phosphate uptake, 89
hepatic sulfate uptake, 89
metals interaction, 95
multispecific metal transporter, 92
organic solute carriers, 94, 102
phosphate transporters, 89
sinusoidal membrane channels, 94–95
sinusoidal metal transporters, 89
sulfate transporters, 89
transport during membrane recycling, 87–88
SIRS. *See* Systemic inflammatory response syndrome (SIRS)
6-hydroxydopamine (6-OHDA), 135
6-OHDA. *See* 6-hydroxydopamine (6-OHDA)
SLC39. *See* Solute-carrier-39 (SLC39)
SNAP. See *S*-nitrosylacetylpenicillamine (SNAP)
SNGFR. *See* Single-nephron GFR (SNGFR)
S-nitrosylacetylpenicillamine (SNAP), 2n, 25
SOCs. *See* Store-operated Ca^{2+} channels (SOCs)

SOD. *See* Superoxide dismutase (SOD)
Sodium–calcium exchanger (NCX), 302
crustacean antennal glands, 316
Sodium–glucose cotransporters (SGLT), 15
soluble HJV (sHJV), 176
recombinant, 356
Solute-carrier-39 (SLC39), 133, 184
acrodermatitis enteropathica, 144, 183
cadmium carrier, 227, 250, 254
cellular zinc uptake, 92
zinc transporters, 185
S100 protein psoriasin (S100A7), 199
SPM. *See* Sinusoidal plasma membrane (SPM)
SPR. *See* Surface plasmon resonance (SPR)
STAT kinases. *See* Signal transducer and activator of transcription (STAT) kinases
Store-operated Ca^{2+} channels (SOCs), 256
Stress-activated protein kinases (SAPK), 337
Stress proteins, 61, 392. *See also* Copper chaperone for superoxide dismutase (CCS); Heat-shock proteins (HSPs)
activation, 116, 392
de novo synthesis, 61
expression, 61
glucose-regulated proteins (GRPs), 392
heat shock proteins, 61
Heme oxygenase (HSP32), 394
HSPA(5), 116
immune function, 393
Structural mimic, 242
cadmium, 246
calcium, 254, 257
cystine, 250
subcutaneous (sc), 56
Sulfhydryl (SH) groups, 37
containing molecules, 5, 246
cysteine, 11, 24, 99, 221
glutathione pools, 25
GSH, 99
ligands, 8
mercury, 5, 86
metal-thiolate cluster, 17
PKC, 385
sulfhydryl-containing proteins, 226
Superoxide dismutase (SOD), 116
activity in heart, 155
copper chaperone, 183
Cu/Zn, 382
manganese, 127
Surface plasmon resonance (SPR), 399
System L, 262
isoforms, 262
methylmercury uptake, 265
molecular mimicry at, 264, 265, 266
role, 263
substrate, 262
Systemic inflammatory response syndrome (SIRS), 153

T

T. *See* Thionein (T)
T cell receptor (TCR), 193, 383
T helper (TH), 118, 199
tBHQ. *See* tertbutyl hydroquinone (tBHQ)
TCR. *See* T cell receptor (TCR)
TEA^+. *See* Tetraethylammonium (TEA^+)
tertbutyl hydroquinone (tBHQ), 149
Tetraethylammonium (TEA^+), 317–318
Tf. *See* Transferrin (Tf)
TFIID. *See* Transcription factor IID (TFIID)
TFIIIA. *See* Transcription factor IIIA (TFIIIA)
TfR. *See* Transferrin receptor (TfR)
TfR1. *See* Transferrin receptor 1 (TfR1)
TGFβ. *See* Transforming growth factor β (TGFβ)
TGF-β1. *See* Transforming growth factor β1 (TGF-β1)
TH. *See* T helper (TH)
Thiols
cellular, 222
endogenous, 60
exogenous, 68
extracellular, 58, 64–67
indirect antioxidant, 187
intracellular, 57, 62–64, 68
in mercury absorption, 39, 50, 57
modification, 383
nonprotein, 65, 221
plasma, 221
protein, 214
surface, 385, 386
Thiols modification, 383
aging, 383
calcium flux, 384
lymphocytes, 384–386
plasma membrane restructuring, 383
T cell activation, 383–384
Thiols, extracellular, 64
BAL, 64
captopril, 67
DMPS, 64, 65–66
DMSA, 64
dose-dependent protection, 64
GSH, 66
intrarenal accumulation, 67
low-molecular-weight thiols, 66, 67
mercury–GSH complex, 66–67
protective effects of DMPS, 66, 67
protective effects of DMSA, 67
protective effects of GSH, 66
ultrafiltrate of rabbit plasma, 66
Thionein (T), 185, 186. *See also* Metallothionein (MT)
Thionin, 186. *See also* Metallothionein (MT); Thionein (T)
Tight junctions, 83, 85, 330
Tissue transglutaminase (tTG), 148
TLR4. *See* Toll-like receptor 4 (TLR4)
TMDs. *See* Transmembrane domains (TMDs)
TNF. *See* Tumor necrosis factor (TNF)
TNF-α. *See* Tumor necrosis factor-α (TNF-α)
TNF-related apoptosis inducing ligand (TRAIL), 387
Toll-like receptor 4 (TLR4), 381
TPEN. *See N, N, N′*, N′-tetrakis (2-pyridylmethyl)-ethylenedidiamine (TPEN)
Trafficking, 381
iron, 22–24
zinc, 7, 11, 20

TRAIL. *See* TNF-related apoptosis inducing ligand (TRAIL)
Transcription factor IID (TFIID), 146
Transcription factor IIIA (TFIIIA), 18
Transferrin (Tf), 14, 87, 169–170
Transferrin receptor (TfR), 23, 80, 88, 131, 134, 385
Transferrin receptor 1 (TfR1), 169, 258, 352
Transforming growth factor β (TGFβ), 361
Transforming growth factor β1 (TGF-β1), 148
Transfusional siderosis, 355, 360, 361, 362
  chelation therapy, 361–362
"transhepatocellular" pathway, 85
Transient receptor potential (TRP), 129
Transmembrane domains (TMDs), 133
Transporter proteins, 381
  phosphate, 89
  zinc, 133
  Zips, 381
  ZnTs, 381
Transporters
  amino acid, 44, 46, 47, 94
  CDF proteins, 185
  copper uptake, 89–90
  dicarboxylate, 48
  14 Zip proteins, 185
  hepatic phosphate, 89
  hepatic sulfate, 89
  membrane, 39, 50
  multispecific metal, 92
  sodium-dependent, 44
  sodium-independent, 44, 45
  zinc, 8, 185
  ZnT proteins, 185
TRP. *See* Transient receptor potential (TRP)
TSQ. *See* *N*-(6-methoxy-8-quinolyl)-*p*-toluensulfonamide (TSQ)
tTG. *See* Tissue transglutaminase (tTG)
Tumor necrosis factor (TNF), 148, 387
Tumor necrosis factor-α (TNF-α), 201, 332
2-OG. *See* 2-oxoglutarate (2-OG)
2-oxoglutarate (2-OG), 219
2,3-dimercapto-1-propanesulfonic acid (DMPS), 37, 224, 259
2,3-dimercaptosuccinic acid (DMSA), 37, 252

U

Untranslated region (UTR), 168
Upstream stimulatory factor (USF), 390
Urinary excretion
  cumulative fecal excretion, 53
  DMPS, 54
  GSH, 40, 222
  mercury, 53, 54, 55, 222
  mercury with albumin, 53–54
  principal factors, 53
  sodium maleate, 54
UROD. *See* Uroporphyrinogen decarboxylase (UROD)
Uroporphyrinogen decarboxylase (UROD), 118
USF. *See* Upstream stimulatory factor (USF)
UTR. *See* Untranslated region (UTR)

V

Vascular endothelial growth factor (VEGF), 129–130, 148
VEGF. *See* Vascular endothelial growth factor (VEGF)

X

XLSA. *See* X-linked sideroblastic anemia (XLSA)
X-linked sideroblastic anemia (XLSA), 360

Z

Zinc, 182, 201, 335, 391
  cell biology, 189, 194
  chaperones, 183
  cytokine production, 200
  cytosolic zinc enzymes, 184
  deficiency, 199, 200
  essential for vision, 198
  eye, 198
  homeostasis, 199, 201
  immunological function, 381
  intestine, 198
  mammary gland epithelial cells, 198
  mutation, 183
  neurons, 197–198
  osteoblasts, 198
  pancreas, 196–197
  paneth cells, 198
  pro-antioxidant, 187
  prostate, 195
  proteome, 182
  secreting cells, 194
  sensor, 185
  skeletal tissue formation, 198
  skin, 199
  trafficking, 381
Zinc deficiency, 2, 144, 153, 199–200
  acrodermatitis enteropathica, 144, 183, 196
  apoptosis, 200
  effects, 2
  lymphocyte, 199
  mutated human Zip(4), 144, 183, 196
Zinc homeostasis, 184
  cellular, 183, 184
  metallothioneins, 186
  proteins, 183
  systemic, 201–202
  zinc sensors, 185
  zinc transporters, 185
Zinc sensors, 183, 185. *See also* Metal response element-binding transcription factor-1 (MTF-1)
Zinc signal, 194, 197
  action, 189
  biological process, 189, 191
  generation and occurrence, 189
  immune system, 199
  intracellular mechanisms, 190
  molecular targets, 189, 192
  pancreas zinc signaling, 196
  specialized cells, 194
  transduction pathways, 191, 194

Zinc signal transduction pathways, 191
EGF signaling, 191
insulin signaling, 191
MAPK, 191
PKC, 192
Zinc uptake, 92, 93, 144, 153, 300
Arabidopsis, 92–93
conjugate formation, 319
dietary zinc supplementation, 93
hypozincemia, 202
thapsigargin, 307
zinc efflux, 93
ZIP transporters, 92, 198
ZNT transporters, 93
Zn^{2+} trafficking, 11
cytosolic distribution, 11, 12
Ehrlich ascites carcinoma, 11, 12
kinetic lability, 12
ligand substitution reaction, 12
U373 cells, 13
Zn-MT detection, 13
Zincosomes, 195
ZIP. *See* Zrt/Irt-like protein (ZIP)
Zn-finger (Zn-FN), 19
Zn-FN. *See* Zn-finger (Zn-FN)
Zn-metallothionein (Zn-MT), 5, 24
DEA-NO, 24–25
reactions with DTNB, 24
reactions with nitric oxide, 24
SNAP, 25
thiol-disulfide interchange reaction, 24
Zn-MT. *See* Zn-metallothionein (Zn-MT)
Zn-protoporphyrin (ZPP), 119
Zonula occludens. *See* Tight junctions
ZPP. *See* Zn-protoporphyrin (ZPP)
Zrt/Irt-like protein (ZIP), 92, 133

For Product Safety Concerns and Information please contact our EU
representative GPSR@taylorandfrancis.com
Taylor & Francis Verlag GmbH, Kaufingerstraße 24, 80331 München, Germany

www.ingramcontent.com/pod-product-compliance
Ingram Content Group UK Ltd.
Pitfield, Milton Keynes, MK11 3LW, UK
UKHW050057200926
475702UK00019B/2883
* 9 7 8 1 1 3 8 3 7 2 6 5 8 *